Cancer Biomarkers in Body Fluids

Cancer Biomarkers in Body Fluids

Editor
Fabrizio Bianchi

MDPI • Basel • Beijing • Wuhan • Barcelona • Belgrade • Manchester • Tokyo • Cluj • Tianjin

Editor
Fabrizio Bianchi
Unit of Cancer Biomarkers
Fondazione IRCCS
Casa Sollievo della Sofferenza
San Giovanni Rotondo
Italy

Editorial Office
MDPI
St. Alban-Anlage 66
4052 Basel, Switzerland

This is a reprint of articles from the Special Issue published online in the open access journal *Cancers* (ISSN 2072-6694) (available at: www.mdpi.com/journal/cancers/special_issues/Biomarkers_Body-Fluids).

For citation purposes, cite each article independently as indicated on the article page online and as indicated below:

LastName, A.A.; LastName, B.B.; LastName, C.C. Article Title. *Journal Name* **Year**, *Volume Number*, Page Range.

ISBN 978-3-0365-6649-8 (Hbk)
ISBN 978-3-0365-6648-1 (PDF)

© 2023 by the authors. Articles in this book are Open Access and distributed under the Creative Commons Attribution (CC BY) license, which allows users to download, copy and build upon published articles, as long as the author and publisher are properly credited, which ensures maximum dissemination and a wider impact of our publications.

The book as a whole is distributed by MDPI under the terms and conditions of the Creative Commons license CC BY-NC-ND.

Contents

About the Editor ix

Preface to "Cancer Biomarkers in Body Fluids" xi

Elisa Dama, Tommaso Colangelo, Emanuela Fina, Marco Cremonesi, Marinos Kallikourdis and Giulia Veronesi et al.
Biomarkers and Lung Cancer Early Detection: State of the Art
Reprinted from: *Cancers* **2021**, *13*, 3919, doi:10.3390/cancers13153919 1

Karmele Valencia and Luis M. Montuenga
Exosomes in Liquid Biopsy: The Nanometric World in the Pursuit of Precision Oncology
Reprinted from: *Cancers* **2021**, *13*, 2147, doi:10.3390/cancers13092147 23

Mateusz Smolarz and Piotr Widlak
Serum Exosomes and Their miRNA Load—A Potential Biomarker of Lung Cancer
Reprinted from: *Cancers* **2021**, *13*, 1373, doi:10.3390/cancers13061373 39

Houssam Aheget, Loubna Mazini, Francisco Martin, Boutaïna Belqat, Juan Antonio Marchal and Karim Benabdellah
Exosomes: Their Role in Pathogenesis, Diagnosis and Treatment of Diseases
Reprinted from: *Cancers* **2020**, *13*, 84, doi:10.3390/cancers13010084 59

Masanori Oshi, Vijayashree Murthy, Hideo Takahashi, Michelle Huyser, Maiko Okano and Yoshihisa Tokumaru et al.
Urine as a Source of Liquid Biopsy for Cancer
Reprinted from: *Cancers* **2021**, *13*, 2652, doi:10.3390/cancers13112652 105

Cora Palanca-Ballester, Aitor Rodriguez-Casanova, Susana Torres, Silvia Calabuig-Fariñas, Francisco Exposito and Diego Serrano et al.
Cancer Epigenetic Biomarkers in Liquid Biopsy for High Incidence Malignancies
Reprinted from: *Cancers* **2021**, *13*, 3016, doi:10.3390/cancers13123016 121

Alessio Ugolini and Marianna Nuti
Rheumatoid Factor: A Novel Determiner in Cancer History
Reprinted from: *Cancers* **2021**, *13*, 591, doi:10.3390/cancers13040591 145

Milena Matuszczak, Jack A. Schalken and Maciej Salagierski
Prostate Cancer Liquid Biopsy Biomarkers' Clinical Utility in Diagnosis and Prognosis
Reprinted from: *Cancers* **2021**, *13*, 3373, doi:10.3390/cancers13133373 155

Enea Ferlizza, Rossella Solmi, Michela Sgarzi, Luigi Ricciardiello and Mattia Lauriola
The Roadmap of Colorectal Cancer Screening
Reprinted from: *Cancers* **2021**, *13*, 1101, doi:10.3390/cancers13051101 185

Gerit Theil, Paolo Fornara and Joanna Bialek
Position of Circulating Tumor Cells in the Clinical Routine in Prostate Cancer and Breast Cancer Patients
Reprinted from: *Cancers* **2020**, *12*, 3782, doi:10.3390/cancers12123782 205

Xavier Ruiz-Plazas, Antonio Altuna-Coy, Marta Alves-Santiago, José Vila-Barja, Joan Francesc García-Fontgivell and Salomé Martínez-González et al.
Liquid Biopsy-Based Exo-oncomiRNAs Can Predict Prostate Cancer Aggressiveness
Reprinted from: *Cancers* **2021**, *13*, 250, doi:10.3390/cancers13020250 227

Kyue-Yim Lee, Yoona Seo, Ji Hye Im, Jiho Rhim, Woosun Baek and Sewon Kim et al.
Molecular Signature of Extracellular Vesicular Small Non-Coding RNAs Derived from Cerebrospinal Fluid of Leptomeningeal Metastasis Patients: Functional Implication of miR-21 and Other Small RNAs in Cancer Malignancy
Reprinted from: *Cancers* **2021**, *13*, 209, doi:10.3390/cancers13020209 . . . . **247**

Emanuela Fina, Davide Federico, Pierluigi Novellis, Elisa Dieci, Simona Monterisi and Federica Cioffi et al.
Subpopulations of Circulating Cells with Morphological Features of Malignancy Are Preoperatively Detected and Have Differential Prognostic Significance in Non-Small Cell Lung Cancer
Reprinted from: *Cancers* **2021**, *13*, 4488, doi:10.3390/cancers13174488 . . . . **269**

John J. Park, Russell J. Diefenbach, Natalie Byrne, Georgina V. Long, Richard A. Scolyer and Elin S. Gray et al.
Circulating Tumor DNA Reflects Uveal Melanoma Responses to Protein Kinase C Inhibition
Reprinted from: *Cancers* **2021**, *13*, 1740, doi:10.3390/cancers13071740 . . . . **285**

Julie Earl, Emma Barreto, María E. Castillo, Raquel Fuentes, Mercedes Rodríguez-Garrote and Reyes Ferreiro et al.
Somatic Mutation Profiling in the Liquid Biopsy and Clinical Analysis of Hereditary and Familial Pancreatic Cancer Cases Reveals *KRAS* Negativity and a Longer Overall Survival
Reprinted from: *Cancers* **2021**, *13*, 1612, doi:10.3390/cancers13071612 . . . . **299**

Julie Earl, Emma Barreto, María E. Castillo, Raquel Fuentes, Mercedes Rodríguez-Garrote and Reyes Ferreiro et al.
Correction: Earl et al. Somatic Mutation Profiling in the Liquid Biopsy and Clinical Analysis of Hereditary and Familial Pancreatic Cancer Cases Reveals KRAS Negativity and a Longer Overall Survival. *Cancers* 2021, *13*, 1612
Reprinted from: *Cancers* **2021**, *13*, 3687, doi:10.3390/cancers13153687 . . . . **315**

Lucia Oton-Gonzalez, John Charles Rotondo, Carmen Lanzillotti, Elisa Mazzoni, Ilaria Bononi and Maria Rosa Iaquinta et al.
Serum HPV16 E7 Oncoprotein Is a Recurrence Marker of Oropharyngeal Squamous Cell Carcinomas
Reprinted from: *Cancers* **2021**, *13*, 3370, doi:10.3390/cancers13133370 . . . . **317**

Alba Loras, Cristina Segovia and José Luis Ruiz-Cerdá
Epigenomic and Metabolomic Integration Reveals Dynamic Metabolic Regulation in Bladder Cancer
Reprinted from: *Cancers* **2021**, *13*, 2719, doi:10.3390/cancers13112719 . . . . **335**

Beatriz Soldevilla, Angeles López-López, Alberto Lens-Pardo, Carlos Carretero-Puche, Angeles Lopez-Gonzalvez and Anna La Salvia et al.
Comprehensive Plasma Metabolomic Profile of Patients with Advanced Neuroendocrine Tumors (NETs). Diagnostic and Biological Relevance
Reprinted from: *Cancers* **2021**, *13*, 2634, doi:10.3390/cancers13112634 . . . . **373**

Liina Salminen, Elena Ioana Braicu, Mitja Lääperi, Antti Jylhä, Sinikka Oksa and Sakari Hietanen et al.
A Novel Two-Lipid Signature Is a Strong and Independent Prognostic Factor in Ovarian Cancer
Reprinted from: *Cancers* **2021**, *13*, 1764, doi:10.3390/cancers13081764 . . . . **397**

Annabel Meireson, Simon J. Tavernier, Sofie Van Gassen, Nora Sundahl, Annelies Demeyer and Mathieu Spaas et al.
Immune Monitoring in Melanoma and Urothelial Cancer Patients Treated with Anti-PD-1 Immunotherapy and SBRT Discloses Tumor Specific Immune Signatures
Reprinted from: *Cancers* **2021**, *13*, 2630, doi:10.3390/cancers13112630 . . . . **409**

Celina L. Szanto, Annelisa M. Cornel, Sara M. Tamminga, Eveline M. Delemarre, Coco C. H. de Koning and Denise A. M. H. van den Beemt et al.
Immune Monitoring during Therapy Reveals Activitory and Regulatory Immune Responses in High-Risk Neuroblastoma
Reprinted from: *Cancers* **2021**, *13*, 2096, doi:10.3390/cancers13092096 . . . . **431**

Michal Mego, Katarina Kalavska, Marian Karaba, Gabriel Minarik, Juraj Benca and Tatiana Sedlackova et al.
Plasma Nucleosomes in Primary Breast Cancer
Reprinted from: *Cancers* **2020**, *12*, 2587, doi:10.3390/cancers12092587 . . . . **449**

Mark Woollam, Luqi Wang, Paul Grocki, Shengzhi Liu, Amanda P. Siegel and Maitri Kalra et al.
Tracking the Progression of Triple Negative Mammary Tumors over Time by Chemometric Analysis of Urinary Volatile Organic Compounds
Reprinted from: *Cancers* **2021**, *13*, 1462, doi:10.3390/cancers13061462 . . . . **463**

About the Editor

Fabrizio Bianchi

Fabrizio Bianchi, M.Sc. in Biological Sciences, Ph.D. in Molecular and Biological Sciences, is Head of the Cancer Biomarkers Unit at the Research Hospital (IRCCS) Fondazione Casa Sollievo della Sofferenza. He is expert in the field of cancer transcriptomics and cancer biomarkers. Dr. Bianchi pioneered the identification of serum circulating microRNAs for lung cancer early detection. He worked as research fellow at the FIRC Institute for Molecular Oncology (Milan, Italy), and at the Program of Molecular Medicine of the European Institute of Oncology (Milan, Italy). He has been a visiting scientist at the Nuffield Department of Clinical and Laboratory Sciences at University of Oxford (England, UK). Dr. Bianchi is member of the American Association for Cancer Research (AACR), the European Association for Cancer Research (EACR), and member of the steering committee of the Italian Cancer Society (SIC).

Preface to "Cancer Biomarkers in Body Fluids"

Over the next 20 years, a sharp rise in cancer cases is expected, increasing from 18.1 million people diagnosed in 2018 to an expected 29.5 million people in 2040 (Global Cancer Observatory; WHO). Cancer burden can be reduced by promoting prevention campaigns, increasing early detection, and implementing personalized cancer therapies. In such a scenario, the identification of circulating biomarkers in body fluids is emerging as a breakthrough in cancer diagnostics for the relative ease of obtaining biological samples using minimally invasive procedures before, during, and after cancer treatment; additionally, the availability of groundbreaking technologies which perform high-throughput and informative biomolecular analyses on limiting sample amounts is boosting biomarkers screening studies. Recently, several scientific publications have provided proof of principle studies which show the great advantage of using circulating biomarkers to monitor exposure to cancer risk factors, increase the accuracy of cancer screening protocols, and detect actionable therapeutic targets. Liquid biopsy shows promise for cancer screening and diagnostics, though some technical challenges still remain.

Fabrizio Bianchi
Editor

Review

Biomarkers and Lung Cancer Early Detection: State of the Art

Elisa Dama [1,†], Tommaso Colangelo [1,†], Emanuela Fina [2], Marco Cremonesi [3], Marinos Kallikourdis [3,4], Giulia Veronesi [5,‡] and Fabrizio Bianchi [1,*,‡]

1. Cancer Biomarkers Unit, Fondazione IRCCS Casa Sollievo della Sofferenza, 71013 San Giovanni Rotondo, Italy; e.dama@operapadrepio.it (E.D.); t.colangelo@operapadrepio.it (T.C.)
2. Humanitas Research Center, IRCCS Humanitas Research Hospital, 20089 Rozzano, Milan, Italy; emanuela.fina@humanitasresearch.it
3. Adaptive Immunity Laboratory, IRCCS Humanitas Research Hospital, 20089 Rozzano, Milan, Italy; marco.cremonesi@humanitasresearch.it (M.C.); Marinos.Kallikourdis@humanitasresearch.it (M.K.)
4. Department of Biomedical Sciences, Humanitas University, 20072 Pieve Emanuele, Italy
5. Division of Thoracic Surgery, IRCCS San Raffaele Scientific Institute, 20132 Milan, Italy; veronesi.giulia@hsr.it
* Correspondence: f.bianchi@operapadrepio.it; Tel.: +39-08-8241-0954; Fax: +39-08-8220-4004
† Authors note: co-first authors.
‡ Authors note: co-last authors.

Simple Summary: Lung cancer is the leading cause of cancer death worldwide. Detecting lung malignacies promptly is essential for any anticancer treatment to reduce mortality and morbidity, especially in high-risk individuals. The use of liquid biopsy to detect circulating biomarkers such as RNA, microRNA, DNA, proteins, autoantibodies in the blood, as well as circulating tumor cells (CTCs), can substantially change the way we manage lung cancer patients by improving disease stratification using intrinsic molecular characteristics, identification of therapeutic targets and monitoring molecular residual disease. Here, we made an update on recent developments in liquid biopsy-based biomarkers for lung cancer early diagnosis, and we propose guidelines for an accurate study design, execution, and data interpretation for biomarker development.

Abstract: Lung cancer burden is increasing, with 2 million deaths/year worldwide. Current limitations in early detection impede lung cancer diagnosis when the disease is still localized and thus more curable by surgery or multimodality treatment. Liquid biopsy is emerging as an important tool for lung cancer early detection and for monitoring therapy response. Here, we reviewed recent advances in liquid biopsy for early diagnosis of lung cancer. We summarized DNA- or RNA-based biomarkers, proteins, autoantibodies circulating in the blood, as well as circulating tumor cells (CTCs), and compared the most promising studies in terms of biomarkers prediction performance. While we observed an overall good performance for the proposed biomarkers, we noticed some critical aspects which may complicate the successful translation of these biomarkers into the clinical setting. We, therefore, proposed a roadmap for successful development of lung cancer biomarkers during the discovery, prioritization, and clinical validation phase. The integration of innovative minimally invasive biomarkers in screening programs is highly demanded to augment lung cancer early detection.

Keywords: lung cancer; early diagnosis; biomarkers; liquid biopsy

1. Introduction

Lung cancer is an aggressive disease accounting for ~380,000 deaths/year only in Europe (WHO; http://gco.iarc.fr; accessed on 21 April 2021) and ~2 million deaths/year worldwide. With the COVID-19 pandemic, these rates are unfortunately expected to rise, mainly due to delays in screening, hospitalizations and therapies, which will cause a stage-shift for newly diagnosed lung tumors [1–3].

Detecting lung malignancies promptly is essential for any anticancer treatment to reduce mortality and morbidity, especially in high-risk individuals [4]. The US National Lung Screening Trial (NLST) and other non-randomized trials [5] demonstrated that Low-Dose Computed Tomography (LDCT) screening can reduce mortality (~20%). Recently, the European NELSON trial has observed a lung cancer mortality reduction of ~25% at 10 years and up to ~30% at 10 years [6]. The drawback of LDCT screening is the presence of uncertainties about high costs, risk of radiation exposure, and false positives observed in the screening population [7], which may obstacle a fully safe large scale implementation of the LDCT screening for lung cancer in Europe [8]. The false-positive rate is particularly problematic, as suspicious nodules may require invasive investigations, causing unnecessary morbidity and reduced acceptance of screening among at-risk individuals. Therefore, the integration of LDCT screening with innovative cancer biomarkers analyzable through minimally invasive approaches aimed to increase screening accuracy is highly demanded. Several pre-clinical studies have suggested that circulating molecules such as microRNA, DNA, proteins, autoantibodies in the blood, as well as circulating tumor cells (CTCs), could be potentially useful to diagnose lung cancer and increase screening accuracy [9–12]. In addition, some studies in actual lung cancer screening cohorts confirmed the diagnostic validity of measuring blood biomarkers for lung cancer early detection [13–15]. Yet, pitfalls and caveats emerged during validation of some proposed biomarkers for lung cancer early detection once applied to independent cohorts/multicenter studies and/or actual lung cancer screening cohorts, which highlight the need to establish a roadmap to develop effective biomarkers.

We reviewed the literature for the most promising biomarkers and relevant technical issues, of which here we present a summary with the aim to propose guidelines for an accurate study design and execution, and data interpretation for biomarker development. We hope that these guidelines will aid further research and facilitate the translation of circulating biomarkers into clinical setting.

2. Lung Cancer Biomarkers

In the last 10 years, there has been a sharp rise in published studies on lung cancer diagnostic biomarkers, with over 544 papers published only in the last 5 years (Figure 1A). However, a sizable fraction of these works relies on a relatively small cohort of samples analyzed, without validation of biomarkers in independent cohorts and, more importantly, in lung cancer screening trials. Ideally, robust biomarker(s) should facilitate the selection of at-risk individuals independently of risk factors such as age and smoking habits, and/or provide pathological information about indeterminate pulmonary nodules (IPNs) to aid clinical decision making, and/or provide predictive/prognostic information. Here, we focused on the most promising minimally invasive, reproducible and extensively validated biomarkers assessed in prospective studies, including lung cancer screening trials.

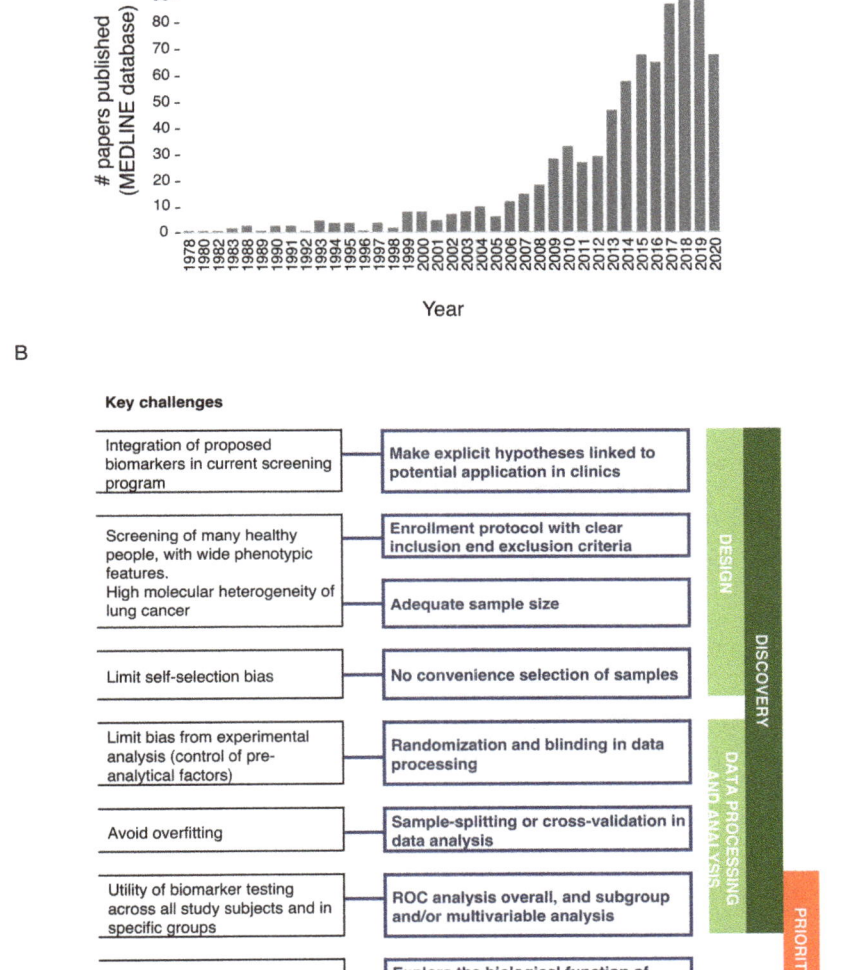

Figure 1. (**A**) Papers on lung cancer diagnostic biomarkers. PubMed free search engine which primarily accesses the MEDLINE database was interrogated (April 2021) by using 'advanced search' tool and with the following MESH terms: Lung neoplasms; Biomarkers; Diagnosis. (**B**) Schematic representation of best practice in biomarker development for early detection of lung cancer.

2.1. DNA-Based Biomarkers

Circulating tumor DNA (ctDNA) was extensively investigated in the latest years due to recent technological advances in the field of next-generation sequencing (NGS). Indeed, the NGS technologies allow the analysis of custom panels of genes (i.e., targeted gene panels, TGP) at an affordable cost (~330€ per sample; [16]) and the detection of mutant alleles presenting with low frequency (<1%; [11,17,18]), which is mandatory when dealing with ctDNA, i.e., underrepresented among the more abundant cell-free DNA (cfDNA) of hematopoietic origin. Although ctDNA was shown to be effective in the diagnosis of advanced lung cancer, the use of ctDNA for detection of early stage lung tumors is suboptimal (with sensitivity ranging from ~50%, [11,19] to 15% in the case of stage I NSCLC; [20]), which can be ascribable to the rare amount of ctDNA present in blood samples of stage I disease patients; indeed, the low proliferation/metabolic rate, and/or dismal tumor angiogenesis, and/or lack of necrotic areas of these localized and tiny tumor lesions all contribute to a reduced ctDNA shedding, as recent observations have suggested [21].

Furthermore, commercial TGPs are usually designed to track druggable cancer driver mutations in more advanced cancer which, therefore, can be underrepresented in early stage disease, i.e., characterized by lower intra-tumor genetic heterogeneity [22–24]. Consequently, the chance to capture nucleotide variants in ctDNA of stage I is dismal. As an alternative, some groups applied the CAncer Personalized Profiling by deep Sequencing (CAPP-Seq) [11] in liquid biopsies to overcome the limited sensitivity of more standard approaches. CAPP-seq introduced a preliminary bioinformatics approach to select target genes containing regions recurrently mutated in the cancer of interest [11]. Despite significant results reached by applying such technology to track molecular residual disease (MRD) during lung cancer therapy [25], the application of CAPP-seq for diagnosis of early stage lung cancer still resulted in a suboptimal sensitivity (~50% [21]). Whole-exome (WES) or whole-genome (WGS) sequencing [26] of ctDNA, covering the entire set of known human genes in order to overcome limitations of TGP, have been also attempted [27]. However, it should be kept in mind that the larger the gene panels, the more difficult it is to obtain high sensitivity for mutation calling and to maintain affordable costs. The high level of ctDNA fragmentation (~100–150 bp in size; [27,28]) should also be considered when designing libraries for NGS. Other caveats are related to the detection of ctDNA are related to clonal hematopoiesis (CH), i.e., an age-dependent process determining the accumulation of somatic mutations in hematopoietic stem and progenitor cells ultimately leading to the clonal expansion of mutated hematopoietic cells; CH accounts for the non-tumor derived mutations detected from plasma [29]. Therefore, it is worth considering to sequence matched white blood cell (WBC) DNA and cfDNA to determine the tumor specific fraction of cfDNA mutations.

Beyond detecting ctDNA mutations, other groups described methylation profiling of cfDNA as a source of innovative minimally invasive cancer biomarkers. A global hypomethylation of DNA is usually observed in cancer cells, yet hypermethylated regions overlapping with CpG islands promoters of tumor suppressor genes were also discovered and exploited to detect ctDNA [30]. The analysis cfDNA using specific methylation signatures to estimate the ctDNA fraction was indeed showed to be a valuable approach for diagnostic and prognostic purposes in lung cancer [31,32]. In a recent large trial with a multi-cancer cohort of over 6000 participants, the methylation profile of ctDNA was found to be highly specific (~99.3%) and to reach an acceptable sensitivity of 67.3% in a set of 12 cancer types and including lung cancer. However, sensitivity dropped down when analyses were limited to early-stage disease (39%; <25% in lung cancer) [33], thus suggesting the need for further investigation of cfDNA methylation signatures in actual lung cancer screening trials for refinement and validation.

2.2. RNA-Based Biomarkers

Different circulating RNA species (microRNA, miRNA; piwi-interacting RNAs, piRNA; transfer RNAs, tRNA; small nucleolar RNAs, snoRNA; small nuclear RNAs, snRNA) were identified in the human serum [34]. Circulating microRNAs (c-miRNAs) are predominant in the literature, and their remarkable stability in harsh conditions and resistance to circulating RNAses [35] make them ideal candidates for developing lung cancer biomarkers. C-miRNAs are released by virtually all human cells by passive (e.g., in apoptotic bodies, complexed with AGO proteins) and active (e.g., in exosomes [36]/microvescicles) mechanisms [37], and can influence tissue homeostasis by a sort of paracrine signaling [37] or by triggering pathogenic mechanisms including neoplastic transformation and tumor progression [38,39]. Indeed, tumor cells, cancer-associated fibroblasts (CAFs) and blood cells were found to release miRNAs in the microenvironment which then enter into the bloodstream [37,40].

Therefore, monitoring miRNA species and relative quantities in the blood represents a valid strategy for early diagnosis of lung cancer. Few studies underwent an extensive validation of c-miRNA as minimally invasive biomarkers for lung cancer early detection (Table 1). Montani et al. validated [13] a serum 13 c-miRNAs signature (miR-Test) by using the qRT-PRC in high-risk individuals ($n = 1115$; >20 pack-year smoking history, aged >50 years) enrolled in an Italian LDCT screening trial (the COSMOS study), which showed a sensitivity of 0.78, a specificity of 0.75, and an AUC of 0.85. Likewise, Sozzi et al. [14] validated a 24 c-miRNA signature (the MSC classifier) by using the qRT-PCR in plasma samples of high-risk subjects enrolled in another Italian LDCT screening study (the bioMILD study; $n = 939$ participants), with a sensitivity of 0.87 and a specificity of 0.81. Wozniac et al. [41] analyzed plasma samples of 100 non-small cell lung cancer (NSCLC) patients (stage I–IIIA) and 100 healthy subjects, using the same qRT-PCR technology as the one used by the Italian studies, and identified another set of 24 miRNAs showing a predicted AUC of 0.78 when accounting for overfitting [41]. In Table S1, we reported overlapping c-miRNAs in the various signatures identified by qRT-PCR. Notably, authors meta-analyzed the MSC classifier as well as another 34 c-miRNA signature identified by Bianchi et al. [9] (from which the miR-Test was derived) and reported an AUC of 0.70 and 0.78, respectively [41].

In multiethnic and multicentric studies on NSCLC patients and matched controls (lung cancer-free or with benign lung nodule individuals), Wang et al. [42] and Ying et al. [43], using the qRT-PCR, have identified two serum c-miRNA diagnostic signatures composed by 5 miRNAs each (miR-214 was commonly found; Table S1). Other studies using different screening platforms, such as microarray analysis of serum samples [44] or whole-blood samples [45], have identified lung cancer diagnostic c-miRNA using large cohorts of clinically detected lung cancer patients (Table 1).

Table 1. List of studies reporting the development of c-miRNA-based biomarkers diagnostic for lung cancer.

Authors	PubMed ID	miRNA (n)	AUC	Sample Type	LDCT
Boeri et al. [46]	21300873	13	0.88	Plasma	Yes
Sozzi et al. [14]	24419137	24	-[a]	Plasma	Yes
Bianchi et al. [9]	21744498	34	0.89	Serum	Yes
Montani et al. [13]	25794889	13	0.85	Serum	Yes
Wozniak et al. [41]	25965386	24	0.78[b]	Plasma	No
Shen et al. [47]	21864403	3	0.86	Plasma	No
Lin et al. [48]	28580707	3	0.87	Plasma	No
Chen et al. [49]	21557218	10	0.97	Serum	No
Wang et al. [42]	26629532	5	0.82	Serum	No
Ying et al. [43]	32943537	5	0.91–0.97	Serum	No
Zhu et al. [50]	27093275	4	0.97[c]	Serum	No
Nadal et al. [51]	26202143	4	0.99	Serum	No
Asakura et al. [44]	32193503	2	0.99	Serum	No
Fehlmann et al. [45]	32134442	15	-[d]	Blood	No

The number of miRNA (n) in each diagnostic signature is reported together with the performance (AUC, i.e., area under curve) and the type of biospecimen where biomarkers were derived (Serum or Plasma). LDCT, studies which performed validation of biomarkers on actual LD-CT screening trials (Yes). [a] Sensitivity, 88% and a specificity of 80%; [b] Predicted performance when applied to independent samples. [c] miRNAs combined with carcinoembryonic antigen (CEA). [d] Sensitivity, 82.8%, and a specificity of 93.5%. PubMed identifiers (PubMed ID) are reported to allow retrieving cited publications.

Despite the proven validity of most of these c-miRNA signatures for early diagnosis of lung cancer, there are still limitations in their application in medical laboratories. Challenging issues related to sample processing and miRNA profiling, pre-analytical and analytical standardization as well as the considerable cost of sophisticated technologies, make the translation of such biomarkers from the bench to the bedside very complicated.

Later, we will further discuss some of these limitations with the aim to provide guidelines for biomarker profiling and translation to the clinic.

2.3. Protein-Based Biomarkers

The ability of tumor antigens [12] and tumor-associated autoantibodies (TAABs) [52] in body fluids to serve as potential biomarkers for lung cancer early detection has been investigated for years. In 2015, Doseeva et al. [53] showed that the combined use of tumor antigens (CEA, CA-125, and CYFRA 21-1) and autoantibodies (NY-ESO-1) was accurate enough (sensitivity, 77%; and specificity, 80%) for the early detection of NSCLC among high-risk individuals. Analysis of CEA and CA-125 among others protein biomarkers (i.e., CA19-9, PRL, HGF, OPN, MPO and TIMP-1) were also included in a multi-analyte blood test (CancerSEEK; [19]), which increased the sensitivity in tumor detection when combined with ctDNA mutation profiling [19].

A large number of studies, systematically reviewed by Yang and colleagues [54], showed that lung cancer patients produce antibodies recognizing self-antigens (i.e., TAAbs). These TAAbs were tested as potential biomarkers for lung cancer detection at different stages of tumor progression. Among TAAbs, the New York esophageal squamous cell carcinoma-1 (NY-ESO-1) autoantibodies appeared to be most promising for NSCLC detection alone or in combination with other TAAbs [54]. However, the diagnostic utility would be more evident if patients affected by *bona fide* autoimmune disease could also be included in the analysis, in order to test whether TAAbs are actually specific for lung cancer.

Recently, the detection and quantification of complement activation fragments in plasma samples from high-risk individuals who underwent LDCT screening were found to be a valid strategy to identify lung cancer biomarkers [15]. A simple diagnostic model based on the quantification of complement-derived fragment C4c and cancer antigens, i.e., 21.1 (CYFRA 21-1) and C-reactive protein (CRP), was able to discriminate between benign and malignant pulmonary nodules (AUC, 0.86), with a high specificity (92%) in a cohort of individuals enrolled in a CT-screening program. This was an important finding due to the considerable fraction (~24%; [5]) of false positive findings by LDCT at the baseline. Authors also showed that the model combined with clinical factors can be valuable in patients with indeterminate pulmonary nodules (IPNs) to decide for more effective therapeutic strategies [55].

2.4. Immune Serum Conversion as Biomarker for Lung Precancerous Lesions

Quantification of inflammation, via measurement of systemic levels of pro-inflammatory cytokines released by activated immune cells, showed a correlation between inflammation and a higher risk for lung cancer incidence in smokers [56,57]. On the other hand, extensive independent analysis of cohorts of non-smokers confirmed the association between sustained inflammation and a higher risk of developing lung cancer [58–68]. In this sense, pro-inflammatory immune activity, which is reflected in the level of circulating cytokines, may be a contributing factor to tumorigenesis in the lung.

The immune system affects not only the tumorigenesis, but also the progression of the disease [69–71]. Thus, whilst research efforts have focused on inflammatory mediators for their potential roles as risk factors for lung cancer in healthy individuals, in parallel, inflammatory mediators have also been assessed for their role in tumor progression in patients with established tumors. Even early stage premalignant lesions are highly infiltrated by immune cells, suggesting that the immune system may affect the transition to malignant lesions [72]. Thus, inflammatory cytokines could drive the progression to malignancy. To date, a detailed and systematic characterization of circulating inflammatory cytokines in patients bearing premalignant lesions in the lungs is still largely missing.

Interestingly, in line with the hypothesis that chronic inflammation is detrimental during carcinogenesis and cancer progression, Ridker and colleagues have recently demonstrated that atherosclerotic patients treated systemically with canakinumab, an antibody inhibiting pro-inflammatory cytokine IL-1β, are protected from lung cancer development, most likely due to the reduction of pro-tumoral inflammation [73]. This seminal clinical finding further highlights how circulating immune mediators may be pivotal for lung cancer progression.

2.5. Circulating Tumor Cells (CTCs) for Lung Cancer Screening

In 2014, a ground-breaking paper showed that, by using a size-based enrichment technology (ISET®, Isolation by Size of Epithelial/Tumor cells), it was possible to detect cells with morphological features of malignancy (i.e., circulating tumor cells, CTCs) in blood samples of patients suffering of chronic obstructive pulmonary disease (COPD) [74]. The presence of CTCs was shown to anticipate the radiological diagnosis of stage I NSCLC [74], thus leading to an increasing interest around the diagnostic role of CTCs and their implementation as a possible biomarker in lung cancer screening programs.

CTCs can be defined as tumor cells in transit in the circulatory system. They originate from primary and secondary tumor sites and are endowed with the molecular features needed to overcome some of the numerous and challenging steps of the metastatic cascade, including intravasation, survival in the blood microenvironment and dissemination to distant organs [75,76]. CTCs are rare events, mixed with a huge number of other cell types, mainly erythrocytes (3.5–7 billion/mL) and leukocytes (4–11 million/mL), and occurring at a variable frequency, even less than 1 cell per milliliter of peripheral venous blood depending on the tumor type and stage [77,78].

CTC detection for lung cancer diagnosis was found to be promising in initial and explorative studies by Hofman and colleagues [10,74]. The same research group then launched a large multicenter prospective French trial (AIR study, NCT02500693), which enrolled a cohort of 614 high-risk subjects according to the NLST-UPSTF criteria (aged 55–74 years, 30 or more pack-year smoking history; current smokers or heavy smokers having quit in the last 15 years) in order to assess the diagnostic accuracy of CTCs detected by the ISET® technology. However, the sensitivity of CTC analysis in detecting 19 lung cancers found at first low-dose computed tomography (LDCT) scan was low, i.e., ~26% [79].

Encouraging results in terms of detection rate were recently obtained using a 4-color FISH test (Table 2) performed on the peripheral blood mononuclear cell (PBMC) fraction isolated by density gradient centrifugation. Through this technique, it was possible to detect cells with at least 2 polysomies or gains in 4 loci involved in the NSCLC tumorigenesis or prognosis (i.e., at 10q22.3, 3p22.1, 3q29 loci, or at chromosome 10 centromere) in 89% of 107 patients with ≤30 mm diameter pulmonary nodules. Contrariwise, none of the 100 lung cancer-free control cases were scored positive when the cut-off value was ≥3 cells with genome abnormalities. Overall, sensitivity was 88.8%, specificity 100%, and accuracy 94.2% [80]. Although the frequency and number of PBMCs with aneuploidy was higher in patients compared to controls, both the validity of a cut-off value of at least 3 cells with aneuploidy to call as CTC-positive a lung cancer patient and the significance of the presence of a maximum of 2 cells with aneuploidy in individuals at high risk for lung cancer should be confirmed in further case series. However, this paper suggests that looking at the entire PBMC population, rather than selecting specific subsets of cells, and using DNA-based detection techniques could considerably augment test sensitivity and specificity. In another work the introduction of alternative protein markers besides cytokeratins (CKs), such as the glycolysis enzyme hexokinase 2 (HK2), increased the detection of CTCs in a cohort of 18 stage III lung adenocarcinoma patients without clinical evidence of distant metastases from 39% when considering $CK^{pos}CD45^{neg}$ to 61% when considering $HK2^{high}CD45^{neg}$ cell subsets [81]. This suggests that using epithelial markers alone may not be sufficient to detect CTCs in non-metastatic setting, and that by adding other markers such as metabolic gene expression analysis can improve lung cancer diagnostic accuracy.

Table 2. Technical performance and clinical significance of circulating tumor cell (CTC) detection in early stage non-small cell lung cancer (NSCLC) patients and in screening programs.

Clinical Setting	n NSCLC Patients *	n Control Subjects *	CTC Enrichment and Detection Method	CTC Identification Criteria	Peripheral Blood Volume	n Target Cell-Positive Cases (Percentage)	Clinical Significance	Reference
preoperative	210 (191 stage I-III)	40 control subjects	EpCAM-based capture and expression of CK8-18-19 and CD45 (CellSearch) Size-based isolation by filtration through porous membranes (ISET, Rarecells) and staining with colorants for cytological samples	Round to oval morphology and CK+ CD45- for CellSearch Morphological features of malignancy for ISET	7 mL for CellSearch 10 mL for ISET	82 (39.3% stage I-III) by CellSearch 104 (49.5%; 49.7% stage I-III)0 control subjects by both technologies	EpCAM-positive selection is less sensitive than size-based isolation	Hofman V et al., Int J Cancer 2011 [82]
screening	0	245 cancer-free (168 COPD, 42 smokers, 35 non-smoking) subjects	Size-based isolation by filtration through porous membranes (ISET, Rarecells) Staining with colorants for cytological samples	Morphological features of malignancy	10 mL	5 COPD (3.0%) at first CT-scan→5 out of 5 confirmed diagnosis of lung cancer at subsequent scans	CTC detection anticipates lung cancer diagnosis by CT-scan screening (1 to 4 years)	Ilie M et al., Plos One 2014 [74]
screening	15 (advanced lung ca.)	32 GGO 19 no GGO	Antibody-based capture of EpCAM+ cells (GILUPI CellCollector, GILUPI) a) EpCAM/CK and CD45 expression by immunofluorescence, and morphological features by imaging analysis(b) Cancer-related gene panel mutations by NGS	EpCAM+/CK+ CD45- and mutated cancer genes	Estimated 1.5-3 liters	11 patients (73.3%) 5 GGO (15.63%) 0 no GGO	CTC can be detected in subjects with preneoplastic nodules and can differentiate GGO from no GGO	He Y et al., Sci Rep 2017 [83]
screening	0	High-risk individuals (smoking habits, age, chronic infections, PSA level) 3888	Microfluidics for flow rate-, surface interaction-, plasticity-, and elasticity-based cell separation (IsoPic, iCellate) Pan-CK and CD45 expression by immunofluorescence	CK+ CD45-	7.5 mL	107 (3.2%) patients	Detection frequency compatible with screening-detected lung cancer rate; follow-up needed to validate results	Castro J et al., Dis Markers 2018 [84]

Table 2. Cont.

Clinical Setting	n NSCLC Patients *	n Control Subjects *	CTC Enrichment and Detection Method	CTC Identification Criteria	Peripheral Blood Volume	n Target Cell-Positive Cases (Percentage)	Clinical Significance	Reference
screening	29 treatment-naive (stage I–IV)	31 high-risk w/ or w/o benign nodules 20 control subjects	Size-based isolation by filtration through filters with a syringe pump (CellSieve, Creatv MicroTech) CK8/18/19, EpCAM and CD45 expression assessed by Immunofluorescence	CK+/EpCAM+ CD45− (single cells or cluster of ≥2 cells)	7.5 mL	Single CTC: 29 patients (100%) 18 high-risk (58.1%) 0 control subjects CTC cluster: 12 patients (41.4%) 0 high-risk or control subjects	High detection rate of single target cells, good specificity of clusters	Manjunath Y et al., Lung Cancer 2019 [85]
screening	115 (97 stage I–III)	87 long-term smokers 20 healthy controls	Size-based isolation by filtration through filters with a syringe pump (CellSieve, Creatv MicroTech) CK8/18/19, EpCAM, CD14 and CD45 expression assessed by immunofluorescence	Cell diameter ≥30 μm, CK+/EpCAM+ CD14+/ CD45+	7.5 mL	88 patients (86.5%): 38 (65.5%) stage I, 13 (72.2%) stage II, 19 (90.5%) stage III 6 long-term smokers (6.9%) 0 healthy controls	High specificity and sensitivity of tumor-macrophage-hybrid cells	Manjunath Y et al., JTO 2020 [86]
screening	19 (Stage I–IV screening-detected)	592 LDCT-screened lung cancer-free heavy smokers	Size-based isolation by filtration through porous membranes (ISET, Rarecells) Staining with colorants for cytological samples	Morphological features of malignancy	10 mL	22 control cases (3.7%) 5 patients (26.3%)	CTC detection rate not sufficient for application in screening programs	Marquette CH et al., Lancet Respir Med 2020 [79]
preoperative	34 (non-metastatic)	20 lung cancer-free 10 benign lung nodules	Antibody-based capture of EpCAM+ cells (GILUPI CellCollector, GILUPI) (a) Cytokeratin CK7/19/panCK, PD-L1 and CD45 expression by immunofluorescence (b) DNA CNV by NGS	CK+ CD45− and DNA CNV	Estimated 1.5–3 liters [83]	18 patients (52.9%) 1 control case (3.3%)	Technical approach able to validate CTC authenticity	Duan G-C et al. OncoTargets and Therapy 2020 [87]
screening	107 (67% stage I–II)	100 lung cancer-free individuals	Ficoll density gradient collection of PBMC 4-color FISH with probes at 10q22.3/CEP10 and 3p22.1/3q29	Polysomy in at least two fluorescence channels	10 mL	95 patients (88.8%) 0 control subjects	Genetically abnormal circulating cells can be detected with high accuracy	Katz DL et al., Cancer Cytopathol 2020 [80]

* Evaluable for CTC analysis. Abbreviations: CK, cytokeratin; ISET, Isolation by Size of Epithelial/Tumor cells; COPD, chronic obstructive pulmonary disease; GGO, ground-glass opacity nodule; PSA, prostate-specific antigen; LDCT, low-dose computed tomography; NGS, next generation sequencing; CNV, copy-number variation; FISH, fluorescence in situ hybridization; PBMC, peripheral blood mononuclear cells.

Compared to cell-free circulating biomarkers, circulating cells represent an ideal and promising systemic 'surrogate' of a tissue as they offer the opportunity to investigate the entire cell at morphological, protein, RNA and DNA level, and to develop experimental models for functional studies. However, the analysis of CTCs in blood samples requires the enrollment of trained personnel and the acquisition of dedicated technologies to enrich blood samples and detect target cells unambiguously. Results of studies in the diagnostic and preoperative setting demonstrate that the accuracy and clinical validity of each kind of technical approach for CTC analysis is still variable and has to be carefully assessed and confirmed in large multicenter and validation trials.

3. A Roadmap to the Successful Development of Blood-Based Biomarkers for Lung Cancer Early Detection

The bottleneck for the successful translation of biomarkers to the clinical use generally lies in the suboptimal standardization in each step of the biomarker pipeline, including discovery, prioritization, and clinical validation. We prepared a summary of the main issues and the best practices in biomarker development (Figure 1B). The first fundamental step in biomarker discovery is establishing a high-quality design which includes making explicit hypotheses on the potential application/integration into current recommended screening programs as well as adopting enrollment protocols with clear inclusion and exclusion criteria for patients and controls. Moreover, heterogeneity (epidemiological, biological and molecular) needs to be considered as the driver for adequate sample size to fulfill the best design. Indeed, published studies often lack acceptable sample size with respect to the numerous phenotypic features that should be considered to widely represent the screening population [88], and the number of variables that should be analyzed to deconvolute the high level of genetic heterogeneity of lung cancer. To limit self-selection bias, instead of convenience selection of subjects (based on easy availability of the sample) [89], control populations should be identified based on matching criteria with the patients' cohort, and extensively represent the actual incidence and prevalence of lung cancer in the screening population.

In the absence of standards for handling specimens (collection, storage and processing) and controls for pre-analytical factors, randomization and blinding should be applied to reduce bias from the experimental analysis. Indeed, quality and reproducibility of biomarkers can be influenced by uncontrolled pre-analytical conditions (i.e., fasting, lipemia, partial hemolysis [90]) and by sample collection bias, especially when the biomarker is labile or sensitive to temperature fluctuation or handling conditions (i.e., type of collection tubes, centrifugation steps, long-term or short-term storage, freeze/thaw cycles; [91,92]). We therefore suggest performing initial pilot experiments to measure the stability of circulating biomarkers, i.e.: (i) by testing different samples collection strategies, using different collection tubes for serum or plasma collection [93–96]; (ii) quantifying how much hemolysis (partial or hidden) can influence biomarker concentration [97,98], (iii) checking if analyte concentration is influenced by fasting status [90], and (iv) testing if different storage conditions (short-term vs. long-term; +4 or $-20/-80$ °C or liquid nitrogen) can alter biomarker quantity and quality [90]. After such analyses, a standard operating procedure (SOP) for sample collection and handling should be defined and rigorously applied to the specific biomarkers screening study.

Nowadays, high-throughput data allow the identification of many biomarkers acting jointly on the risk of lung cancer; these markers can be easily combined in a single multivariable statistical model; moreover, to avoid the resulting possible overfitting (i.e., capturing noise instead of the true underlying data structure), machine learning approaches with sample-splitting or cross-validation should be considered [99]. The performance of a new biomarker for the early detection of cancer is easily measured by true-positive and false-positive rates, and summarized through receiver operating characteristic curves (ROC). However, the "average" performance is often presented in the literature, with ROC calculated across all study subjects, while subgroup and/or multivariable analysis should better reveal the utility of biomarker testing in specific groups (i.e., tumor stages, nodule density, histotypes).

Exploration of biomarkers' performance in subgroups could also help with ranking the selected candidates for clinical relevance. Moreover, when a new biomarker study is published, only limited discussion on the biological function of the candidates is reported, and assay/platform reproducibility and standardization are frequently lacking (see below). In our experience, an in-depth analysis of technical and biological variables which might have an impact on the detection and quantification of selected biomarkers should also be performed. For example, uncontrolled environmental conditions during sample processing could influence the quantification of biomarkers of interest. Marzi et al. [90] showed, by using an automated purification system based on spin columns for nucleic acid purification, that efficiency in miRNA extraction was inversely proportional to temperature increase during daily runs. Similar findings were also described by other research groups [100].

In the case of analysis of multiple biomarkers (e.g., DNA, RNA and protein), the collected samples (whole blood, plasma, serum) can be split in several aliquots which can be differently prioritized for processing based on stability of the biomarkers of interest; in case of RNA, which is more liable, the relevant sample aliquot can be processed immediately while other aliquots (for other biomarker types) can be processed subsequently. Likewise, the use of different extraction kits with or without additional centrifugation steps could affect quantities and species of the biomarkers of interest. Cheng et al. [101] showed that plasma samples can be contaminated by residual platelets, which impact most miRNA measurements (~70%), therefore authors suggested to add pre- or post-storage centrifugation steps in order to remove residual platelet contamination. Furthermore, miRNA quantities may vary depending on the kit used for extraction [102,103].

To keep track of the impact of these pre-analytical and analytical variables, we strongly recommend using endogenous and exogenous controls. In circulating miRNA, biomarker analysis measuring both endogenous controls (e.g., RNU6, RNU44, miR-16 [104]) and exogenous controls, e.g., synthetic miRNAs from other organisms (ath-miR-159a and/or cel-miR-39), allows monitoring sample degradation, extraction efficiency and performance of miRNA detection by using different screening platforms (e.g., qRT-PCR, ddPCR, microarray, NGS).

Lastly, the analytical translation in a clinically applicable platform and validation in a large prospective trial are both needed to complete validation of candidate biomarkers. Industrial and clinical partners could facilitate these phases, providing funding supports and know-how in large-scale test production, regulatory affairs and commercialization [88]. A major issue in the validation of biomarkers for lung cancer early detection is to prove its benefit in the context of screening programs, where lead- and length-time biases and overdiagnosis are peculiar. Therefore, the choice of the end-point is essential and, although biases could occur in interpreting causes of death, lung-cancer mortality reduction should represent the primary endpoint [99], then followed by the evaluation of overall mortality.

4. Overview of Platforms for Circulating Biomarkers Detection: A Focus on c-miRNA Detection

The performance of different screening platforms available in terms of sensitivity, specificity and reproducibility, as well as relative costs of analysis should also be considered in advance before starting biomarkers profiling. As previously described, c-miRNAs are the most discussed in the literature as promising biomarkers for lung cancer early diagnosis. Besides the several pre-analytical and analytical factors, which can impinge on the biomarker reliability as we previously discussed, some considerations should be made on the impact on the accuracy of c-miRNA biomarkers when using different experimental platforms and technologies for biomarkers detection.

To quantify c-miRNA expression, a variety of platforms have been developed so far, mainly based on quantitative PCR (qRT-PCR), microarray, or next-generation sequencing (NGS) technology. Recently, the efficiency and concordance of different miRNA profiling platforms were assessed [105–108]. In 2014, Mestdagh et al. [105] analyzed the expression level of 196 common miRNAs measured by 12 different application platforms to provide a sort of "miRNA quality control (miRQC)" analysis. They performed experiments with high and low RNA input amounts and organized output measurements into four groups to represent the various testing questions, i.e.: reproducibility, specificity, sensitivity, and accuracy [105]. Similar qRT-PCR platforms showed a different performance in terms of reproducibility and specificity [105]. Sensitivity, on the other hand, is very much technology-related, with qRT-PCR platforms (i.e., TaqMan Cards PreAmp; ThermoFisher) being superior to hybridization- (i.e., microarray) and sequencing-based platforms. Furthermore, the hybridization platforms displayed higher specificity, but lower detection rates compared to most of the qRT-PCR and sequencing platforms [105]. Overall, the authors reported that sensitivity and specificity have a deep and important inverse relationship [105].

Next-generation technologies are now also available for miRNA profiling. For example, Small RNA sequencing (RNA-Seq), in particular, was reported to be superior for discovery studies, but less useful for high-throughput or fast turnaround applications [105]. Furthermore, when various RNA isolation and library preparation protocols are used, the reproducibility of small RNA-seq is significantly and negatively affected [106,109]. Recently, Godoy et al. [107] evaluated a small RNA-seq method optimized for low-input samples [106,110,111] (i.e., liquid biopsy) to three relatively novel platforms, i.e., (i) the HTG Molecular's EdgeSeq miRNA Whole Tran-scriptome Assay (EdgeSeq), (ii) the Abcam's FirePlex (FirePlex), and (iii) NanoString's nCounter (nCounter). These three platforms were selected for their rapid turnaround time and ease of use, properties that are attractive for biomarker assays. The authors used pools of synthetic RNA oligonucleotides and standardized extracellular RNA human plasma samples to assess reproducibility, bias, specificity, sensitivity, and accuracy. Briefly, the authors concluded that: (i) small RNA-seq was the most accurate, sensitive and specific method with an AUC of 0.99 for miRNA detection, which was superior to EdgeSeq (AUC = 0.97), nCounter (AUC = 0.94) or FirePlex (AUC = 0.81); (ii) EdgeSeq was the most reproducible and had the least detection bias; and (iii) nCounter was less sensitive than small RNA-seq, EdgeSeq, and FirePlex. Recently, Hong LZ et al. [108] performed a systematic evaluation of multiple qPCR platforms (MiRXES ID3EAL, Qiagen miScript, TaqMan Cards preAMP, Exiqon LNA), nCounter technology (NanoString) and miRNA-Seq for microRNA biomarker discovery in human biofluids. Performance parameters such as reproducibility, detection rate, and inter-platform correlation were used to evaluate each technology. MiRXES qRT-PCR and miRNA-Seq platforms had an almost perfect reproducibility between runs, calculating the Concordance Correlation Coefficient (CCC = 0.99), while the other three qRT-PCR platforms had moderate inter-run concordance (CCC > 0.9), and the NanoString platform had poor inter-run concordance (CCC = 0.82). The MiRXES qRT-PCR and NanoString platforms detected the highest and the lowest number of miRNAs above the LLOQ (lower limit of quantification) in serum samples, respectively. The authors concluded that the miRNA-Seq technology is preferable for discovery, while targeted qRT-PCR for subsequent validation of candidate extracellular miRNA biomarkers is recommended.

Finally, the droplet digital PCR (ddPCR) technique is becoming the gold standard in the application of liquid biopsy due to a number of advantages: (i) it allows an absolute quantification by means of sample partitioning and Poisson statistics (an internal/external normalization is thus not required); (ii) it has a superior precision and sensitivity in detecting low-abundant targets; (iii) it is less affected by PCR inhibitors [112–115]. However, ddPCR is less frequently used for c-miRNA measurements due also to a restricted multiplexing capacity, longer turnaround time for sample processing, and higher costs. In Table 3, a summary of the pros and cons of c-miRNA screening technologies is provided.

Table 3. miRNA platforms comparison.

Method	Platform (Vendor)	Turnaround Time	Costs Per Sample	Panel Content (Human miRNA)	Reproducibility	SE	SP	ACC
qRT-PCR	miScript (Qiagen)	+++	$$	1066	Medium [105,108]	Medium [105]	Medium [105]	High [105]
	miRCURY Exiqon (Qiagen)	+++	$$	752	High [105]-Medium [108]	Medium [105,108]	High [105]-Medium [105]	High [105]
	TaqMan Cards preAMP (Life Technologies) [a]	+++	$$	754	Medium [108]-Low	High [94]-Medium [108]	High [105]-Medium [105]	High [105]
	TaqMan OpenArray (Life Technologies) [a]	+	$	754	Low [105]	Medium [105]	High [105]	High [105]
	SmartChip (WaferGen)	+++	$$	1036	High [105]	Low [105]	Medium [105]-Low [105]	Low [105]
	qScript (Quanta BioSciences)	+++	$$	489	High [105]	Medium [105]	High [105]-Medium [105]	High [105]
	miRXES ID3EAL (miRXES)	+++	$$	560	High [108]	High [108]	NA	NA
GeneChip miRNA arrays	microarray (Affymetrix)	++	$	Up-to-date content from miRBase 20	Medium [105]	NA	High [105]-Low [105]	Low [105]
	microarray (Agilent)	++	$	Up-to-date content from miRBase 21	High [105]	Low [105]	High [105]-Medium [105]	Low [105]
nCounter platform	nCounter (NanoString)	+	$	800	Medium [105]-Low [108]	Low [107,108]	High [105]-Medium [107]	Low [105]
sRNA-Seq (miRNA-seq)	TruSeq (Illumina)	++	$	Up to 2693 [b] (miRBase 22)	High [105,107,108]	High [107,108]-Medium [105]	High [105,107]	Medium [105]
	Ion Torrent (Life Technologies)	++	$	Up to 2693 [b] (miRBase 22)	Medium [105]	Medium [105]	Low [105]	Medium [105]
HTG EdgeSeq	HTG EdgeSeq (HTG Molecular Diagnostics) plus Illumina or Thermo Fisher Ion Torrent sequencers	+	$$	2083	High [107]	High [107]	High [107]	NA
Standard flow cytometer	FirePlex (Abcam)	+	$$	up to 65 miRNAs per well	Low [107]	Medium [107]	Low [107]	NA

Platform comparison. [a] Standard TaqMan MicroRNA Assays use a target-specific stem-loop reverse transcription primer; [b] Number of mature microRNAs in miRBase release 22; NA = not analyzed [105,107,108]. From $ to $$, qualitative scale of costs for sample processing. From + to +++, qualitative scale of turnaround time for sample processing.

5. Discussion

Cancer biomarkers substantially change the way we manage lung cancer patients by improving disease stratification using intrinsic molecular characteristics, identification of therapeutic targets and monitoring molecular residual disease. However, the application of biomarkers for lung cancer early diagnosis is still limited by a lack of substantial trial-like research studies where the accuracy of proposed biomarkers is analyzed in real-world datasets. Previous studies highlighted pros and cons of different circulating biomarkers proposed for lung cancer detection and possible integration in the clinical routine (reviewed in Seijo et al. [116]). Circulating biomarkers can be very effective to inform clinical decision making in the management of indeterminate pulmonary nodules (IPNs) and in the management of diagnosed and resected lung cancer patients. Current management of IPNs is largely based on watchful waiting and may imply a risk of disease dissemination. Nodules found on annual LDCT screening, which are frequently very small in size and hamper current biopsy techniques, may benefit from an integrated risk model, which includes the different sources of information: clinical, imaging and biomarkers. This type of integrated risk model might also inform decisions regarding screening intervals, personalized follow-up of lung cancer patients, and prognostication.

Here, we made an update on recent developments in liquid biopsy-based biomarkers for lung cancer early diagnosis and proposed a roadmap for optimal biomarkers identification and development. A limit of this study is that we opted for a focused analysis on extensively validated biomarkers in large cohorts of samples including lung cancer screening studies rather than describing all circulating biomarkers proposed in the literature.

We have also brought to light the current limitations in biomarker research, which can be briefly summarized in: (i) poorly designed studies for biomarker discovery and validation; (ii) uncontrolled pre-analytic and analytic variabilities lacking standard operating procedures; (iii) frequent lack of validation studies using independent cohorts of samples collected from lung cancer screening studies; and (iv) somewhat sophisticated technologies for biomarker profiling that are hard to transfer to the clinical setting.

Biomarker research clearly offers substantial help in the characterization of at-risk population subgroups for screening selection and—more importantly—in the identification of disease precursors, predictive and prognostic factors before signs and symptoms of the disease appear. In particular, the analysis of liquid biopsies (i.e., plasma/serum) is emerging as promising for the quantification of biomarkers through also the use of lab-on-chip technologies, which would allow a rapid disease detection/monitoring and a biological characterization at the bedside [117,118]. Furthermore, genomic and proteomic breath tests besides airway epithelium signatures, are being trialed for early and non-invasive diagnosis of cancer and pulmonary disease, in particular for lung cancer and COPD [119,120]. Likewise, new emerging RNA-based biomarkers such as long non-coding RNA (lncRNA), circular RNA (circRNA) and platelets mRNAs have been described circulating in the blood with a potential for lung cancer early detection (Tables S2–S4; Figure 2).

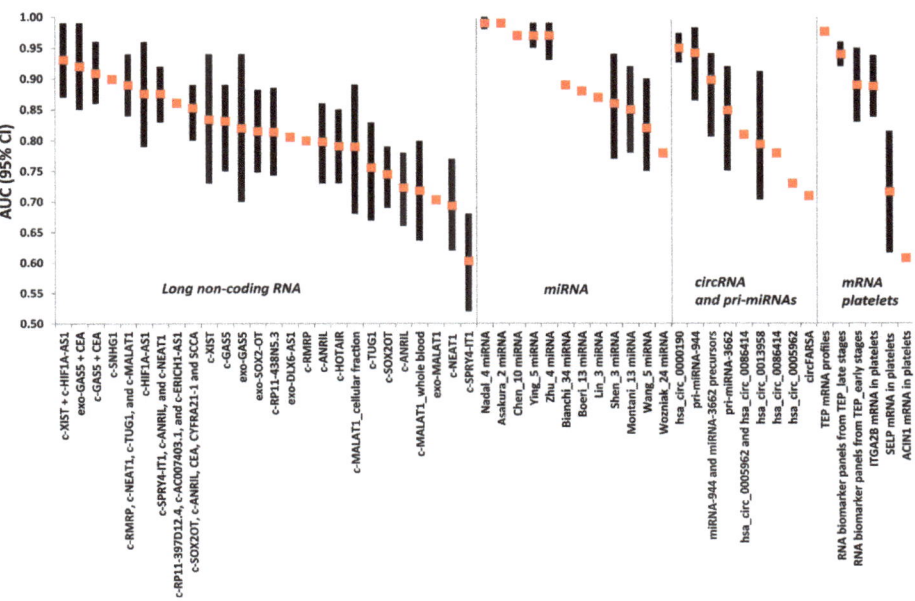

Figure 2. Forest plot showing the AUC and 95% confidence interval (when reported) for c-miRNA-based signatures (listed in Table 1) and other RNA-based biomarkers (listed in Tables S2–S4). Red squares represent the AUC for each marker and black vertical bars extend from the lower limit to the upper limit of the 95% confidence interval (95% CI).

6. Conclusions

Thus far, all these multi-source biomarkers have never been combined into a coordinated and comprehensive workup for screening, diagnosis and treatment decision. The main barrier consists of difficulties in organizing worldwide large-scale studies with centralized resources for data/sample collection and processing following standard operating procedures. In addition, it is urgent to develop innovative approaches using big data and artificial intelligence (AI) analytics, such as machine learning, to improve both lung cancer early detection, personalized prevention strategies, and early treatments. We therefore look forward for these next-generation biomarkers in lung cancer screening programs to ameliorate early diagnosis, prognosis, and therapeutic response.

Supplementary Materials: The following are available online at https://www.mdpi.com/article/10.3390/cancers13153919/s1, Table S1: List of circulating miRNA markers diagnostic for lung cancer analyzed by qRT-PCR, Table S2: List of circulating lncRNA markers diagnostic for lung cancer, Table S3: List of circulating circRNA and pri-miRNAs markers diagnostic for lung cancer, Table S4: List of mRNA platelets markers diagnostic for lung cancer.

Author Contributions: Conceptualization, G.V., and F.B.; methodology, E.D., T.C., E.F., M.C., M.K., G.V., F.B.; investigation, E.D., T.C., E.F., M.C., M.K., G.V., F.B.; writing—original draft preparation, E.D., T.C., E.F., M.C., M.K., G.V., F.B.; writing—review and editing, E.D., T.C., E.F., M.C., M.K., G.V., F.B.; visualization, E.D., T.C., E.F., G.V., F.B.; supervision, F.B. All authors have read and agreed to the published version of the manuscript.

Funding: This research was funded by by the Italian Ministry of Health [GR-2016-02363975 to F.B.; GR-2019-12370460 to T.C.; CLEARLY to G.V. and F.B.; Ricerca Corrente: "Analysis of promestastatic phenotypes in lung cancer cells for the identification of new therapeutic targets" to F.B.] and the Associazione Italiana Ricerca sul Cancro [IG-22827 to F.B]. T.C. and E.F. are supported by a fellowship from Umberto Veronesi Foundation.

Acknowledgments: We thank Chiara Di Giorgio for critically editing the manuscript.

Conflicts of Interest: The study funders had no role in the design of the study, the collection, analysis, and interpretation of the data, the writing of the manuscript, and the decision to submit the manuscript for publication. All of the authors declare no conflict of interest.

References

1. Degeling, K.; Baxter, N.N.; Emery, J.; Jenkins, M.A.; Franchini, F.; Gibbs, P.; Mann, G.B.; McArthur, G.; Solomon, B.J.; IJzerman, M.J. An Inverse Stage-Shift Model to Estimate the Excess Mortality and Health Economic Impact of Delayed Access to Cancer Services Due to the COVID-19 Pandemic. *Asia Pac. J. Clin. Oncol.* **2021**. [CrossRef]
2. Dinmohamed, A.G.; Visser, O.; Verhoeven, R.H.A.; Louwman, M.W.J.; van Nederveen, F.H.; Willems, S.M.; Merkx, M.A.W.; Lemmens, V.E.P.P.; Nagtegaal, I.D.; Siesling, S. Fewer Cancer Diagnoses during the COVID-19 Epidemic in The Netherlands. *Lancet Oncol.* **2020**, *21*, 750–751. [CrossRef]
3. Sud, A.; Jones, M.E.; Broggio, J.; Loveday, C.; Torr, B.; Garrett, A.; Nicol, D.L.; Jhanji, S.; Boyce, S.A.; Gronthoud, F.; et al. Collateral Damage: The impact on outcomes from cancer surgery of the Covid-19 pandemic. *Ann. Oncol.* **2020**, *31*, 1065–1074. [CrossRef]
4. McMahon, P.M.; Kong, C.Y.; Johnson, B.E.; Weinstein, M.C.; Weeks, J.C.; Kuntz, K.M.; Shepard, J.-A.O.; Swensen, S.J.; Gazelle, G.S. Estimating long-term effectiveness of lung cancer screening in the mayo ct screening study. *Radiology* **2008**, *248*, 278–287. [CrossRef]
5. Aberle, D.R.; Adams, A.M.; Berg, C.D.; Black, W.C.; Clapp, J.D.; Fagerstrom, R.M.; Gareen, I.F.; Gatsonis, C.; Marcus, P.M.; Sicks, J.D. Reduced lung-cancer mortality with low-dose computed tomographic screening. *New Engl. J. Med.* **2011**, *365*, 395–409. [CrossRef]
6. De Koning, H.J.; Van Der Aalst, C.M.; De Jong, P.A.; Scholten, E.T.; Nackaerts, K.; Heuvelmans, M.A.; Lammers, J.-W.J.; Weenink, C.; Yousaf-Khan, U.; Horeweg, N.; et al. Reduced lung-cancer mortality with volume ct screening in a randomized trial. *New Engl. J. Med.* **2020**, *382*, 503–513. [CrossRef] [PubMed]
7. Kinsinger, L.S.; Anderson, C.; Kim, J.; Larson, M.; Chan, S.H.; King, H.A.; Rice, K.L.; Slatore, C.G.; Tanner, N.T.; Pittman, K.; et al. Implementation of Lung Cancer Screening in the Veterans Health Administration. *JAMA Intern. Med.* **2017**, *177*, 399–406. [CrossRef] [PubMed]
8. Puggina, A.; Broumas, A.; Ricciardi, W.; Boccia, S. Cost-effectiveness of screening for lung cancer with low-dose computed tomography: A systematic literature review. *Eur. J. Public Health* **2016**, *26*, 168–175. [CrossRef] [PubMed]
9. Bianchi, F.; Nicassio, F.; Marzi, M.; Belloni, E.; Dall'olio, V.; Bernard, L.; Pelosi, G.; Maisonneuve, P.; Veronesi, G.; Di Fiore, P.P. A serum circulating mirna diagnostic test to identify asymptomatic high-risk individuals with early stage lung cancer. *EMBO Mol. Med.* **2011**, *3*, 495–503. [CrossRef]
10. Hofman, V.; Bonnetaud, C.; Ilie, M.I.; Vielh, P.; Vignaud, J.M.; Fléjou, J.F.; Lantuejoul, S.; Piaton, E.; Mourad, N.; Butori, C.; et al. Preoperative circulating tumor cell detection using the isolation by size of epithelial tumor cell method for patients with lung cancer is a new prognostic biomarker. *Clin. Cancer Res.* **2011**, *17*, 827–835. [CrossRef]
11. Newman, A.M.; Bratman, S.V.; To, J.; Wynne, J.F.; Eclov, N.C.; Modlin, L.A.; Liu, C.L.; Neal, J.W.; Wakelee, H.A.; Merritt, R.E.; et al. An ultrasensitive method for quantitating circulating tumor dna with broad patient coverage. *Nat. Med.* **2014**, *20*, 548–554. [CrossRef]
12. Tockman, M.S.; Gupta, P.K.; Myers, J.D.; Frost, J.K.; Baylin, S.B.; Gold, E.B.; Chase, A.M.; Wilkinson, P.H.; Mulshine, J.L. Sensitive and specific monoclonal antibody recognition of human lung cancer antigen on preserved sputum cells: A new approach to early lung cancer detection. *J. Clin. Oncol.* **1988**, *6*, 1685–1693. [CrossRef]
13. Montani, F.; Marzi, M.J.; Dezi, F.; Dama, E.; Carletti, R.M.; Bonizzi, G.; Bertolotti, R.; Bellomi, M.; Rampinelli, C.; Maisonneuve, P.; et al. MiR-Test: A blood test for lung cancer early detection. *J. Natl. Cancer Inst.* **2015**, *107*. [CrossRef]
14. Sozzi, G.; Boeri, M.; Rossi, M.; Verri, C.; Suatoni, P.; Bravi, F.; Roz, L.; Conte, D.; Grassi, M.; Sverzellati, N.; et al. Clinical utility of a plasma-based mirna signature classifier within computed tomography lung cancer screening: A correlative mild trial study. *J. Clin. Oncol.* **2014**, *32*, 768–773. [CrossRef]
15. Ajona, D.; Pajares, M.J.; Corrales, L.; Perez-Gracia, J.L.; Agorreta, J.; Lozano, M.D.; Torre, W.; Massion, P.P.; de-Torres, J.P.; Jantus-Lewintre, E.; et al. Investigation of complement activation product c4d as a diagnostic and prognostic biomarker for lung cancer. *JNCI J. Natl. Cancer Inst.* **2013**, *105*, 1385–1393. [CrossRef]
16. Van Nimwegen, K.J.M.; Van Soest, R.A.; Veltman, J.A.; Nelen, M.R.; Van Der Wilt, G.J.; Vissers, L.E.L.M.; Grutters, J.P.C. Is the $1000 genome as near as we think? A cost analysis of next-generation sequencing. *Clin. Chem.* **2016**, *62*, 1458–1464. [CrossRef]
17. Ståhlberg, A.; Krzyzanowski, P.M.; Jackson, J.B.; Egyud, M.; Stein, L.; Godfrey, T.E. Simple, multiplexed, PCR-based barcoding of dna enables sensitive mutation detection in liquid biopsies using sequencing. *Nucleic Acids Res.* **2016**, *44*, e105. [CrossRef]
18. Ståhlberg, A.; Krzyzanowski, P.M.; Egyud, M.; Filges, S.; Stein, L.; Godfrey, T.E. Simple multiplexed PCR-based barcoding of dna for ultrasensitive mutation detection by next-generation sequencing. *Nat. Protoc.* **2017**, *12*, 664–682. [CrossRef]
19. Cohen, J.D.; Li, L.; Wang, Y.; Thoburn, C.; Afsari, B.; Danilova, L.; Douville, C.; Javed, A.A.; Wong, F.; Mattox, A.; et al. Detection and localization of surgically resectable cancers with a multi-analyte blood test. *Science* **2018**, *359*, 926–930. [CrossRef]
20. Abbosh, C.; Birkbak, N.J.; Wilson, G.A.; Jamal-Hanjani, M.; Constantin, T.; Salari, R.; Le Quesne, J.; Moore, D.A.; Veeriah, S.; Rosenthal, R.; et al. Phylogenetic CtDNA Analysis Depicts Early-Stage Lung Cancer Evolution. *Nature* **2017**, *545*, 446–451. [CrossRef]

21. Chabon, J.J.; Hamilton, E.G.; Kurtz, D.M.; Esfahani, M.S.; Moding, E.J.; Stehr, H.; Schroers-Martin, J.; Nabet, B.Y.; Chen, B.; Chaudhuri, A.A.; et al. Integrating genomic features for non-invasive early lung cancer detection. *Nature* **2020**, *580*, 245–251. [CrossRef] [PubMed]
22. Jamal-Hanjani, M.; Wilson, G.A.; McGranahan, N.; Birkbak, N.J.; Watkins, T.B.K.; Veeriah, S.; Shafi, S.; Johnson, D.H.; Mitter, R.; Rosenthal, R.; et al. Tracking the evolution of non–small-cell lung cancer. *New Engl. J. Med.* **2017**, *376*, 2109–2121. [CrossRef]
23. Vitale, I.; Shema, E.; Loi, S.; Galluzzi, L. Intratumoral heterogeneity in cancer progression and response to immunotherapy. *Nat. Med.* **2021**, *27*, 212–224. [CrossRef]
24. Gerstung, M.; Jolly, C.; Leshchiner, I.; Dentro, S.C.; Gonzalez, S.; Rosebrock, D.; Mitchell, T.J.; Rubanova, Y.; Anur, P.; Yu, K.; et al. The evolutionary history of 2658 cancers. *Nature* **2020**, *578*, 122–128. [CrossRef]
25. Chaudhuri, A.A.; Chabon, J.J.; Lovejoy, A.F.; Newman, A.M.; Stehr, H.; Azad, T.D.; Khodadoust, M.S.; Esfahani, M.S.; Liu, C.L.; Zhou, L.; et al. Early detection of molecular residual disease in localized lung cancer by circulating tumor DNA profiling. *Cancer Discov.* **2017**, *7*, 1394–1403. [CrossRef]
26. Giroux Leprieur, E.; Hélias-Rodzewicz, Z.; Takam Kamga, P.; Costantini, A.; Julie, C.; Corjon, A.; Dumenil, C.; Dumoulin, J.; Giraud, V.; Labrune, S.; et al. Sequential CtDNA Whole-Exome Sequencing in Advanced Lung Adenocarcinoma with Initial Durable Tumor Response on Immune Checkpoint Inhibitor and Late Progression. *J. Immunother. Cancer* **2020**, *8*. [CrossRef]
27. Keller, L.; Belloum, Y.; Wikman, H.; Pantel, K. Clinical relevance of blood-based ctDNA snalysis: Mutation detection and beyond. *Br. J. Cancer* **2021**, *124*, 345–358. [CrossRef]
28. Zheng, Y.W.L.; Chan, K.C.A.; Sun, H.; Jiang, P.; Su, X.; Chen, E.Z.; Lun, F.M.F.; Hung, E.C.W.; Lee, V.; Wong, J.; et al. Non-hematopoietically Derived DNA is shorter than hematopoietically derived DNA in plasma: A tansplantation model. *Clin. Chem.* **2012**, *58*, 549–558. [CrossRef]
29. Chan, H.T.; Chin, Y.M.; Nakamura, Y.; Low, S.-K. Clonal hematopoiesis in liquid biopsy: From biological noise to valuable clinical implications. *Cancers* **2020**, *12*. [CrossRef]
30. Ehrlich, M. DNA Hypomethylation in cancer cells. *Epigenomics* **2009**, *1*, 239–259. [CrossRef]
31. Ooki, A.; Maleki, Z.; Tsay, J.-C.J.; Goparaju, C.; Brait, M.; Turaga, N.; Nam, H.-S.; Rom, W.N.; Pass, H.I.; Sidransky, D.; et al. A panel of novel detection and prognostic methylated DNA markers in primary non-small cell lung cancer and serum DNA. *Clin. Cancer Res.* **2017**, *23*, 7141–7152. [CrossRef]
32. Hulbert, A.; Jusue-Torres, I.; Stark, A.; Chen, R.; Rodgers, K.; Lee, B.; Griffin, C.; Yang, A.; Huang, P.; Wrangle, J.; et al. Early detection of lung cancer using DNA promoter hypermethylation in plasma and sputum. *Clin. Cancer Res.* **2017**, *23*, 1998–2005. [CrossRef]
33. Liu, M.C.; Oxnard, G.R.; Klein, E.A.; Swanton, C.; Seiden, M.V.; Liu, M.C.; Oxnard, G.R.; Klein, E.A.; Smith, D.; Richards, D.; et al. Sensitive and specific multi-cancer detection and localization using methylation signatures in cell-free DNA. *Ann. Oncol.* **2020**, *31*, 745–759. [CrossRef]
34. Umu, S.U.; Langseth, H.; Bucher-Johannessen, C.; Fromm, B.; Keller, A.; Meese, E.; Lauritzen, M.; Leithaug, M.; Lyle, R.; Rounge, T.B. A comprehensive profile of circulating RNAs in human serum. *RNA Biol.* **2018**, *15*, 242–250. [CrossRef] [PubMed]
35. Chen, X.; Ba, Y.; Ma, L.; Cai, X.; Yin, Y.; Wang, K.; Guo, J.; Zhang, Y.; Chen, J.; Guo, X.; et al. Characterization of microRNAs in serum: A novel class of biomarkers for diagnosis of cancer and other diseases. *Cell Res.* **2008**, *18*, 997–1006. [CrossRef] [PubMed]
36. Chevillet, J.R.; Kang, Q.; Ruf, I.K.; Briggs, H.A.; Vojtech, L.N.; Hughes, S.M.; Cheng, H.H.; Arroyo, J.D.; Meredith, E.K.; Gallichotte, E.N.; et al. Quantitative and stoichiometric analysis of the microRNA content of exosomes. *Proc. Natl. Acad. Sci. USA* **2014**, *111*, 14888–14893. [CrossRef] [PubMed]
37. Turchinovich, A.; Weiz, L.; Burwinkel, B. Extracellular MiRNAs: The mystery of their origin and function. *Trends Biochem. Sci.* **2012**, *37*, 460–465. [CrossRef]
38. Le, M.T.N.; Hamar, P.; Guo, C.; Basar, E.; Perdigão-Henriques, R.; Balaj, L.; Lieberman, J. MiR-200-containing extracellular vesicles promote breast cancer cell metastasis. *J. Clin. Invest.* **2014**, *124*, 5109–5128. [CrossRef] [PubMed]
39. Melo, S.A.; Sugimoto, H.; O'Connell, J.T.; Kato, N.; Villanueva, A.; Vidal, A.; Qiu, L.; Vitkin, E.; Perelman, L.T.; Melo, C.A.; et al. Cancer exosomes perform cell-independent microRNA biogenesis and promote tumorigenesis. *Cancer Cell* **2014**, *26*, 707–721. [CrossRef]
40. Pritchard, C.C.; Kroh, E.; Wood, B.; Arroyo, J.D.; Dougherty, K.J.; Miyaji, M.M.; Tait, J.F.; Tewari, M. Blood cell origin of circulating microRNAs: A cautionary note for cancer biomarker studies. *Cancer Prev. Res.* **2012**, *5*, 492–497. [CrossRef]
41. Wozniak, M.B.; Scelo, G.; Muller, D.C.; Mukeria, A.; Zaridze, D.; Brennan, P. Circulating microRNAs as non-invasive biomarkers for early detection of non-small-cell lung cancer. *PLoS ONE* **2015**, *10*, e0125026. [CrossRef]
42. Wang, C.; Ding, M.; Xia, M.; Chen, S.; Van Le, A.; Soto-Gil, R.; Shen, Y.; Wang, N.; Wang, J.; Gu, W.; et al. A five-MiRNA panel identified from a multicentric case–control study serves as a novel diagnostic tool for ethnically diverse non-small-cell lung cancer patients. *EBioMedicine* **2015**. [CrossRef]
43. Ying, L.; Du, L.; Zou, R.; Shi, L.; Zhang, N.; Jin, J.; Xu, C.; Zhang, F.; Zhu, C.; Wu, J.; et al. Development of a serum MiRNA panel for detection of early stage non-small cell lung cancer. *Proc. Natl. Acad. Sci. USA* **2020**, *117*, 25036–25042. [CrossRef] [PubMed]
44. Asakura, K.; Kadota, T.; Matsuzaki, J.; Yoshida, Y.; Yamamoto, Y.; Nakagawa, K.; Takizawa, S.; Aoki, Y.; Nakamura, E.; Miura, J.; et al. A miRNA-based diagnostic model predicts resectable lung cancer in humans with high accuracy. *Commun. Biol.* **2020**, *3*, 134. [CrossRef]

45. Fehlmann, T.; Kahraman, M.; Ludwig, N.; Backes, C.; Galata, V.; Keller, V.; Geffers, L.; Mercaldo, N.; Hornung, D.; Weis, T.; et al. Evaluating the use of circulating microRNA profiles for lung cancer detection in symptomatic patients. *JAMA Oncol.* **2020**, *6*, 714. [CrossRef]
46. Boeri, M.; Verri, C.; Conte, D.; Roz, L.; Modena, P.; Facchinetti, F.; Calabro, E.; Croce, C.M.; Pastorino, U.; Sozzi, G. MicroRNA signatures in tissues and plasma predict development and prognosis of computed tomography detected lung cancer. *Proc. Natl. Acad. Sci. USA* **2011**, *108*, 3713–3718. [CrossRef]
47. Shen, J.; Liu, Z.; Todd, N.W.; Zhang, H.; Liao, J.; Yu, L.; Guarnera, M.A.; Li, R.; Cai, L.; Zhan, M.; et al. Diagnosis of lung cancer in individuals with solitary pulmonary nodules by plasma microRNA biomarkers. *BMC Cancer* **2011**, *11*, 374. [CrossRef]
48. Lin, Y.; Leng, Q.; Jiang, Z.; Guarnera, M.A.; Zhou, Y.; Chen, X.; Wang, H.; Zhou, W.; Cai, L.; Fang, H.; et al. A classifier integrating plasma biomarkers and radiological characteristics for distinguishing malignant from benign pulmonary nodules. *Int. J. Cancer* **2017**, *141*, 1240–1248. [CrossRef] [PubMed]
49. Chen, X.; Hu, Z.; Wang, W.; Ba, Y.; Ma, L.; Zhang, C.; Wang, C.; Ren, Z.; Zhao, Y.; Wu, S.; et al. Identification of ten serum microRNAs from a genome-wide serum microRNA expression profile as novel noninvasive biomarkers for nonsmall cell lung cancer diagnosis. *Int. J. Cancer* **2012**, *130*, 1620–1628. [CrossRef]
50. Zhu, W.; Zhou, K.; Zha, Y.; Chen, D.; He, J.; Ma, H.; Liu, X.; Le, H.; Zhang, Y. Diagnostic value of serum miR-182, miR-183, miR-210, and miR-126 levels in patients with early-stage non-small cell lung cancer. *PLoS ONE* **2016**, *11*, e0153046. [CrossRef]
51. Nadal, E.; Truini, A.; Nakata, A.; Lin, J.; Reddy, R.M.; Chang, A.C.; Ramnath, N.; Gotoh, N.; Beer, D.G.; Chen, G. A novel serum 4-microRNA signature for lung cancer detection. *Sci. Rep.* **2015**, *5*, 12464. [CrossRef]
52. Zhong, L.; Peng, X.; Hidalgo, G.E.; Doherty, D.E.; Stromberg, A.J.; Hirschowitz, E.A. Identification of circulating antibodies to tumor-associated proteins for combined use as markers of non-small cell lung cancer. *Proteomics* **2004**, *4*, 1216–1225. [CrossRef] [PubMed]
53. Doseeva, V.; Colpitts, T.; Gao, G.; Woodcock, J.; Knezevic, V. Performance of a multiplexed dual analyte immunoassay for the early detection of non-small cell lung cancer. *J. Transl. Med.* **2015**, *13*, 55. [CrossRef] [PubMed]
54. Yang, B.; Li, X.; Ren, T.; Yin, Y. Autoantibodies as diagnostic biomarkers for lung cancer: A systematic review. *Cell Death Discov.* **2019**, *5*, 126. [CrossRef] [PubMed]
55. Ajona, D.; Remirez, A.; Sainz, C.; Bertolo, C.; Gonzalez, A.; Varo, N.; Lozano, M.D.; Zulueta, J.J.; Mesa-Guzman, M.; Martin, A.C.; et al. A model based on the quantification of complement C4c, CYFRA 21-1 and CRP exhibits high specificity for the early diagnosis of lung cancer. *Transl. Res.* **2021**. [CrossRef] [PubMed]
56. Allin, K.H.; Bojesen, S.E.; Nordestgaard, B.G. Baseline C-reactive protein is associated with incident cancer and survival in patients with cancer. *J. Clin. Oncol.* **2009**. [CrossRef] [PubMed]
57. Brenner, D.R.; Fanidi, A.; Grankvist, K.; Muller, D.C.; Brennan, P.; Manjer, J.; Byrnes, G.; Hodge, A.; Severi, G.; Giles, G.G.; et al. Inflammatory cytokines and lung cancer risk in 3 prospective studies. *Am. J. Epidemiol.* **2017**. [CrossRef]
58. Trichopoulos, D.; Psaltopoulou, T.; Orfanos, P.; Trichopoulou, A.; Boffetta, P. Plasma C-reactive protein and risk of cancer: A prospective study from Greece. *Cancer Epidemiol. Biomark. Prev.* **2006**. [CrossRef] [PubMed]
59. Il'yasova, D.; Colbert, L.H.; Harris, T.B.; Newman, A.B.; Bauer, D.C.; Satterfield, S.; Kritchevsky, S.B. Circulating levels of inflammatory markers and cancer risk in the health aging and body composition cohort. *Cancer Epidemiol. Biomark. Prev.* **2005**. [CrossRef]
60. Enewold, L.; Mechanic, L.E.; Bowman, E.D.; Zheng, Y.L.; Yu, Z.; Trivers, G.; Alberg, A.J.; Harris, C.C. Serum concentrations of cytokines and lung cancer survival in African Americans and Caucasians. *Cancer Epidemiol. Biomark. Prev.* **2009**. [CrossRef]
61. Silva, I.D.S.; De Stavola, B.L.; Pizzi, C.; Meade, T.W. Circulating levels of coagulation and inflammation markers and cancer risks: Individual participant analysis of data from three long-term cohorts. *Int. J. Epidemiol.* **2010**. [CrossRef]
62. Chaturvedi, A.K.; Caporaso, N.E.; Katki, H.A.; Wong, H.L.; Chatterjee, N.; Pine, S.R.; Chanock, S.J.; Goedert, J.J.; Engels, E.A. C-reactive protein and risk of lung cancer. *J. Clin. Oncol.* **2010**. [CrossRef] [PubMed]
63. Pine, S.R.; Mechanic, L.E.; Enewold, L.; Chaturvedi, A.K.; Katki, H.A.; Zheng, Y.L.; Bowman, E.D.; Engels, E.A.; Caporaso, N.E.; Harris, C.C. Increased levels of circulating interleukin 6, interleukin 8, C-reactive protein, and risk of lung cancer. *J. Natl. Cancer Inst.* **2011**. [CrossRef] [PubMed]
64. Pine, S.R.; Mechanic, L.E.; Enewold, L.; Bowman, E.D.; Ryan, B.M.; Cote, M.L.; Wenzlaff, A.S.; Loffredo, C.A.; Olivo-Marston, S.; Chaturvedi, A.; et al. Differential serum cytokine levels and risk of lung cancer between African and European Americans. *Cancer Epidemiol. Biomark. Prev.* **2016**. [CrossRef]
65. Shiels, M.S.; Pfeiffer, R.M.; Hildesheim, A.; Engels, E.A.; Kemp, T.J.; Park, J.H.; Katki, H.A.; Koshiol, J.; Shelton, G.; Caporaso, N.E.; et al. Circulating inflammation markers and prospective risk for lung cancer. *J. Natl. Cancer Inst.* **2013**. [CrossRef]
66. Shiels, M.S.; Shu, X.O.; Chaturvedi, A.K.; Gao, Y.T.; Xiang, Y.B.; Cai, Q.; Hu, W.; Shelton, G.; Ji, B.T.; Pinto, L.A.; et al. A prospective study of immune and inflammation markers and risk of lung cancer among female never smokers in Shanghai. *Carcinogenesis* **2017**. [CrossRef] [PubMed]
67. Siemes, C.; Visser, L.E.; Coebergh, J.W.W.; Splinter, T.A.W.; Witteman, J.C.M.; Uitterlinden, A.G.; Hofman, A.; Pols, H.A.P.; Stricker, B.H.C. C-Reactive protein levels, variation in the C-reactive protein gene, and cancer risk: The rotterdam study. *J. Clin. Oncol.* **2006**. [CrossRef] [PubMed]
68. Watson, J.; Salisbury, C.; Banks, J.; Whiting, P.; Hamilton, W. Predictive value of inflammatory markers for cancer diagnosis in primary care: A prospective cohort study using electronic health records. *Br. J. Cancer* **2019**. [CrossRef] [PubMed]

69. Bremnes, R.M.; Al-Shibli, K.; Donnem, T.; Sirera, R.; Al-Saad, S.; Andersen, S.; Stenvold, H.; Camps, C.; Busund, L.T. The role of tumor-infiltrating immune cells and chronic inflammation at the tumor site on cancer development, progression, and prognosis: Emphasis on non-small cell lung cancer. *J. Thorac. Oncol.* **2011**, *6*, 824–833. [CrossRef]
70. Bremnes, R.M.; Busund, L.T.; Kilver, T.L.; Andersen, S.; Richardsen, E.; Paulsen, E.E.; Hald, S.; Khanehkenari, M.R.; Cooper, W.A.; Kao, S.C.; et al. The role of tumor-infiltrating lymphocytes in development, progression, and prognosis of non-small cell lung cancer. *J. Thorac. Oncol.* **2016**, *11*, 789–800. [CrossRef]
71. Powell, H.A.; Iyen-Omofoman, B.; Baldwin, D.R.; Hubbard, R.B.; Tata, L.J. Chronic obstructive pulmonary disease and risk of lung cancer: The importance of smoking and timing of diagnosis. *J. Thorac. Oncol.* **2013**. [CrossRef]
72. Krysan, K.; Tran, L.M.; Grimes, B.S.; Fishbein, G.A.; Seki, A.; Gardner, B.K.; Walser, T.C.; Salehi-Rad, R.; Yanagawa, J.; Lee, J.M.; et al. The immune contexture associates with the genomic landscape in lung adenomatous premalignancy. *Cancer Res.* **2019**. [CrossRef] [PubMed]
73. Ridker, P.M.; MacFadyen, J.G.; Thuren, T.; Everett, B.; Libby, P.; Glynn, R.J.; Ridker, P.; Lorenzatti, A.; Krum, H.; Varigos, J.; et al. Effect of interleukin-1β inhibition with canakinumab on incident lung cancer in patients with atherosclerosis: Exploratory results from a randomised, double-blind, placebo-controlled trial. *Lancet* **2017**. [CrossRef]
74. Ilie, M.; Hofman, V.; Long-Mira, E.; Selva, E.; Vignaud, J.-M.; Padovani, B.; Mouroux, J.; Marquette, C.-H.; Hofman, P. "Sentinel" circulating tumor cells allow early diagnosis of lung cancer in patients with chronic obstructive pulmonary disease. *PLoS ONE* **2014**, *9*, e111597. [CrossRef] [PubMed]
75. Massagué, J.; Obenauf, A.C. Metastatic colonization by circulating tumour cells. *Nature* **2016**, *529*, 298–306. [CrossRef] [PubMed]
76. Castro-Giner, F.; Aceto, N. Tracking cancer progression: From circulating tumor cells to metastasis. *Genome Med.* **2020**, *12*, 31. [CrossRef]
77. Allard, W.J.; Matera, J.; Miller, M.C.; Repollet, M.; Connelly, M.C.; Rao, C.; Tibbe, A.G.J.; Uhr, J.W.; Terstappen, L.W.M.M. Tumor cells circulate in the peripheral blood of all major carcinomas but not in healthy subjects or patients with nonmalignant diseases. *Clin. Cancer Res.* **2004**, *10*, 6897–6904. [CrossRef]
78. Lianidou, E.S.; Strati, A.; Markou, A. Circulating tumor cells as promising novel biomarkers in solid cancers. *Crit. Rev. Clin. Lab. Sci.* **2014**, *51*, 160–171. [CrossRef]
79. Marquette, C.-H.; Boutros, J.; Benzaquen, J.; Ferreira, M.; Pastre, J.; Pison, C.; Padovani, B.; Bettayeb, F.; Fallet, V.; Guibert, N.; et al. Circulating tumour cells as a potential biomarker for lung cancer screening: A prospective cohort study. *Lancet Respir. Med.* **2020**, *8*, 709–716. [CrossRef]
80. Katz, R.L.; Zaidi, T.M.; Pujara, D.; Shanbhag, N.D.; Truong, D.; Patil, S.; Mehran, R.J.; El-Zein, R.A.; Shete, S.S.; Kuban, J.D. Identification of circulating tumor cells using 4-color fluorescence in situ hybridization: Validation of a noninvasive aid for ruling out lung cancer in patients with low-dose computed tomography–detected lung nodules. *Cancer Cytopathol.* **2020**, *128*, 553–562. [CrossRef]
81. Yang, L.; Yan, X.; Chen, J.; Zhan, Q.; Hua, Y.; Xu, S.; Li, Z.; Wang, Z.; Dong, Y.; Zuo, D.; et al. Hexokinase 2 discerns a novel circulating tumor cell population associated with poor prognosis in lung cancer patients. *Proc. Natl. Acad. Sci. USA* **2021**, *118*, e2012228118. [CrossRef] [PubMed]
82. Hofman, V.; Ilie, M.I.; Long, E.; Selva, E.; Bonnetaud, C.; Molina, T.; Vénissac, N.; Mouroux, J.; Vielh, P.; Hofman, P. Detection of circulating tumor cells as a prognostic factor in patients undergoing radical surgery for non-small-cell lung carcinoma: Comparison of the efficacy of the Cellsearch Assay^{TM} and the isolation by size of epithelial tumor cell method. *Int. J. Cancer* **2011**, *129*, 1651–1660. [CrossRef] [PubMed]
83. He, Y.; Shi, J.; Shi, G.; Xu, X.; Liu, Q.; Liu, C.; Gao, Z.; Bai, J.; Shan, B. Using the new CellCollector to capture circulating tumor cells from blood in different groups of pulmonary disease: A cohort study. *Sci. Rep.* **2017**, *7*, 9542. [CrossRef]
84. Castro, J.; Sanchez, L.; Nuñez, M.T.; Lu, M.; Castro, T.; Sharifi, H.R.; Ericsson, C. Screening circulating tumor cells as a noninvasive cancer test in 3388 individuals from high-risk groups (ICELLATE2). *Dis. Markers* **2018**, *2018*, 1–5. [CrossRef] [PubMed]
85. Manjunath, Y.; Upparahalli, S.V.; Suvilesh, K.N.; Avella, D.M.; Kimchi, E.T.; Staveley-O'Carroll, K.F.; Li, G.; Kaifi, J.T. Circulating tumor cell clusters are a potential biomarker for detection of non-small cell lung cancer. *Lung Cancer* **2019**, *134*, 147–150. [CrossRef]
86. Manjunath, Y.; Mitchem, J.B.; Suvilesh, K.N.; Avella, D.M.; Kimchi, E.T.; Staveley-O'Carroll, K.F.; Deroche, C.B.; Pantel, K.; Li, G.; Kaifi, J.T. Circulating giant tumor-macrophage fusion cells are independent prognosticators in patients with NSCLC. *J. Thorac. Oncol.* **2020**, *15*, 1460–1471. [CrossRef] [PubMed]
87. Duan, G.-C.; Zhang, X.-P.; Wang, H.-E.; Wang, Z.-K.; Zhang, H.; Yu, L.; Xue, W.-F.; Xin, Z.-F.; Hu, Z.-H.; Zhao, Q.-T. Circulating tumor cells a screening diagnostic marker for early-stage non-small cell lung cancer. *OncoTargets Ther.* **2020**, *13*, 1931–1939. [CrossRef]
88. Poste, G. Bring on the biomarkers. *Nature* **2011**, *469*, 156–157. [CrossRef]
89. Goossens, N.; Nakagawa, S.; Sun, X.; Hoshida, Y. Cancer biomarker discovery and validation. *Transl. Cancer Res.* **2015**, *4*, 256–269. [CrossRef]
90. Marzi, M.J.; Montani, F.; Carletti, R.M.; Dezi, F.; Dama, E.; Bonizzi, G.; Sandri, M.T.; Rampinelli, C.; Bellomi, M.; Maisonneuve, P.; et al. Optimization and standardization of circulating microRNA detection for clinical application: The miR-test case. *Clin. Chem.* **2016**, *62*, 743–754. [CrossRef]
91. Marton, M.J.; Weiner, R. Practical guidance for implementing predictive biomarkers into early phase clinical studies. *BioMed Res. Int.* **2013**, *2013*, 891391. [CrossRef] [PubMed]

92. Yin, P.; Lehmann, R.; Xu, G. Effects of pre-analytical processes on blood samples used in metabolomics studies. *Anal. Bioanal. Chem.* **2015**, *407*, 4879–4892. [CrossRef] [PubMed]
93. Zhang, S.; Zhao, Z.; Duan, W.; Li, Z.; Nan, Z.; Du, H.; Wang, M.; Yang, J.; Huang, C. The influence of blood collection tubes in biomarkers' screening by mass spectrometry. *Proteom. Clin. Appl.* **2020**, *14*, e1900113. [CrossRef] [PubMed]
94. Yang, S.; McGookey, M.; Wang, Y.; Cataland, S.R.; Wu, H.M. Effect of blood sampling, processing, and storage on the measurement of complement activation biomarkers. *Am. J. Clin. Pathol.* **2015**, *143*, 558–565. [CrossRef]
95. Risberg, B.; Tsui, D.W.Y.; Biggs, H.; De Almagro, A.R.-V.M.; Dawson, S.-J.; Hodgkin, C.; Jones, L.; Parkinson, C.; Piskorz, A.; Marass, F.; et al. Effects of collection and processing procedures on plasma circulating cell-free DNA from cancer patients. *J. Mol. Diagn.* **2018**, *20*, 883–892. [CrossRef] [PubMed]
96. Glinge, C.; Clauss, S.; Boddum, K.; Jabbari, R.; Jabbari, J.; Risgaard, B.; Tomsits, P.; Hildebrand, B.; Kääb, S.; Wakili, R.; et al. Stability of circulating blood-based microRNAs—Pre-analytic methodological considerations. *PLoS ONE* **2017**, *12*, e0167969. [CrossRef] [PubMed]
97. Poel, D.; Buffart, T.E.; Oosterling-Jansen, J.; Verheul, H.M.; Voortman, J. Evaluation of several methodological challenges in circulating miRNA qPCR studies in patients with head and neck cancer. *Exp. Mol. Med.* **2018**, *50*, e454. [CrossRef]
98. Kirschner, M.B.; Edelman, J.J.; Kao, S.C.; Vallely, M.P.; Van Zandwijk, N.; Reid, G. The impact of hemolysis on cell-free microRNA Biomarkers. *Front. Genet.* **2013**, *4*, 94. [CrossRef]
99. Maruvada, P. Joint national cancer institute-food and drug administration workshop on research strategies, study designs, and statistical approaches to biomarker validation for cancer diagnosis and detection. *Cancer Epidemiol. Biomark. Prev.* **2006**, *15*, 1078–1082. [CrossRef]
100. Sourvinou, I.S.; Markou, A.; Lianidou, E.S. Quantification of circulating miRNAs in plasma. *J. Mol. Diagn.* **2013**, *15*, 827–834. [CrossRef] [PubMed]
101. Cheng, H.H.; Yi, H.S.; Kim, Y.; Kroh, E.M.; Chien, J.W.; Eaton, K.D.; Goodman, M.T.; Tait, J.F.; Tewari, M.; Pritchard, C.C. Plasma processing conditions substantially influence circulating microRNA biomarker levels. *PLoS ONE* **2013**, *8*, e64795. [CrossRef] [PubMed]
102. El-Khoury, V.; Pierson, S.; Kaoma, T.; Bernardin, F.; Berchem, G. Assessing cellular and circulating miRNA recovery: The impact of the RNA isolation method and the quantity of input material. *Sci. Rep.* **2016**, *6*, 19529. [CrossRef] [PubMed]
103. Kloten, V.; Neumann, M.H.D.; Di Pasquale, F.; Sprenger-Haussels, M.; Shaffer, J.M.; Schlumpberger, M.; Herdean, A.; Betsou, F.; Ammerlaan, W.; Hällström, T.; et al. Multicenter evaluation of circulating plasma microRNA extraction technologies for the development of clinically feasible reverse transcription quantitative PCR and next-generation sequencing analytical work flows. *Clin. Chem.* **2019**, *65*, 1132–1140. [CrossRef] [PubMed]
104. Schwarzenbach, H.; Silva, A.M.; Calin, G.; Pantel, K. Data normalization strategies for microRNA quantification. *Clin. Chem.* **2015**, *61*, 1333–1342. [CrossRef]
105. Mestdagh, P.; Hartmann, N.; Baeriswyl, L.; Andreasen, D.; Bernard, N.; Chen, C.; Cheo, D.; D'Andrade, P.; DeMayo, M.; Dennis, L.; et al. Evaluation of quantitative miRNA expression platforms in the microRNA quality control (miRQC) study. *Nat. Methods* **2014**, *11*, 809–815. [CrossRef]
106. Giraldez, M.D.; Spengler, R.M.; Etheridge, A.; Godoy, P.M.; Barczak, A.J.; Srinivasan, S.; De Hoff, P.L.; Tanriverdi, K.; Courtright, A.; Lu, S.; et al. Comprehensive multi-center assessment of small RNA-seq methods for quantitative MiRNA profiling. *Nat. Biotechnol.* **2018**, *36*, 746–757. [CrossRef]
107. Godoy, P.M.; Barczak, A.J.; DeHoff, P.; Srinivasan, S.; Etheridge, A.; Galas, D.; Das, S.; Erle, D.J.; Laurent, L.C. Comparison of reproducibility, accuracy, sensitivity, and specificity of miRNA quantification platforms. *Cell Rep.* **2019**, *29*, 4212–4222.e5. [CrossRef]
108. Hong, L.Z.; Zhou, L.; Zou, R.; Khoo, C.M.; Chew, A.L.S.; Chin, C.-L.; Shih, S.-J. Systematic evaluation of multiple qPCR platforms, nanostring and miRNA-seq for microRNA biomarker discovery in human biofluids. *Sci. Rep.* **2021**, *11*, 4435. [CrossRef]
109. Srinivasan, S.; Yeri, A.; Cheah, P.S.; Chung, A.; Danielson, K.; De Hoff, P.; Filant, J.; Laurent, C.D.; Laurent, L.D.; Magee, R.; et al. Small RNA sequencing across diverse biofluids identifies optimal methods for exRNA isolation. *Cell* **2019**, *177*, 446–462.e16. [CrossRef]
110. Yeri, A.; Courtright, A.; Danielson, K.; Hutchins, E.; Alsop, E.; Carlson, E.; Hsieh, M.; Ziegler, O.; Das, A.; Shah, R.V.; et al. Evaluation of commercially available small RNAseq library preparation kits using low input RNA. *BMC Genom.* **2018**, *19*, 331. [CrossRef]
111. Rozowsky, J.; Kitchen, R.R.; Park, J.; Galeev, T.R.; Diao, J.; Warrell, J.; Thistlethwaite, W.; Subramanian, S.L.; Milosavljevic, A.; Gerstein, M. exceRpt: A comprehensive analytic platform for extracellular RNA profiling. *Cell Syst.* **2019**, *8*, 352–357.e3. [CrossRef] [PubMed]
112. Whale, A.S.; Jones, G.M.; Pavšič, J.; Dreo, T.; Redshaw, N.; Akyürek, S.; Akgöz, M.; Divieto, C.; Sassi, M.P.; He, H.-J.; et al. Assessment of digital PCR as a primary reference measurement procedure to support advances in precision medicine. *Clin. Chem.* **2018**, *64*, 1296–1307. [CrossRef] [PubMed]
113. Campomenosi, P.; Gini, E.; Noonan, D.M.; Poli, A.; D'Antona, P.; Rotolo, N.; Dominioni, L.; Imperatori, A.S. A comparison between quantitative PCR and droplet digital PCR technologies for circulating microRNA quantification in human lung cancer. *BMC Biotechnol.* **2016**, *16*, 60. [CrossRef]

114. Rački, N.; Dreo, T.; Gutierrez-Aguirre, I.; Blejec, A.; Ravnikar, M. Reverse transcriptase droplet digital PCR shows high resilience to PCR Inhibitors from Plant, Soil and Water Samples. *Plant Methods* **2014**, *10*, 42. [CrossRef]
115. Das Gupta, S.; Ndode-Ekane, X.E.; Puhakka, N.; Pitkänen, A. Droplet digital polymerase chain reaction-based quantification of circulating microRNAs using small RNA concentration normalization. *Sci. Rep.* **2020**, *10*, 9012. [CrossRef]
116. Seijo, L.M.; Peled, N.; Ajona, D.; Boeri, M.; Field, J.K.; Sozzi, G.; Pio, R.; Zulueta, J.J.; Spira, A.; Massion, P.P.; et al. Biomarkers in lung cancer screening: Achievements, promises, and challenges. *J. Thorac. Oncol.* **2019**, *10*, 343–357. [CrossRef] [PubMed]
117. Song, Y.; Huang, Y.-Y.; Liu, X.; Zhang, X.; Ferrari, M.; Qin, L. Point-of-Care technologies for molecular diagnostics using a drop of blood. *Trends Biotechnol.* **2014**, *32*, 132–139. [CrossRef]
118. Hayes, B.; Murphy, C.; Crawley, A.; O'Kennedy, R. Developments in Point-of-Care Diagnostic Technology for Cancer Detection. *Diagnostics* **2018**, *8*. [CrossRef]
119. Billatos, E.; Vick, J.L.; Lenburg, M.E.; Spira, A.E. The airway transcriptome as a biomarker for early lung cancer detection. *Clin. Cancer Res.* **2018**, *24*, 2984–2992. [CrossRef] [PubMed]
120. Konstantinidi, E.M.; Lappas, A.S.; Tzortzi, A.S.; Behrakis, P.K. Exhaled breath condensate: Technical and diagnostic aspects. *Sci. World J.* **2015**, *2015*, 435160. [CrossRef] [PubMed]

Review

Exosomes in Liquid Biopsy: The Nanometric World in the Pursuit of Precision Oncology

Karmele Valencia [1,2,3,4,*] and Luis M. Montuenga [1,2,3,5,*]

1. Program in Solid Tumors, Center for Applied Medical Research (CIMA), 31008 Pamplona, Spain
2. Consorcio de Investigación Biomédica en Red de Cáncer (CIBERONC), 28029 Madrid, Spain
3. Navarra Health Research Institute (IDISNA), 31008 Pamplona, Spain
4. Department of Biochemistry and Genetics, School of Sciences, University of Navarra, 31009 Pamplona, Spain
5. Department of Pathology, Anatomy and Physiology, School of Medicine, University of Navarra, 31009 Pamplona, Spain
* Correspondence: kvalencia@external.unav.es (K.V.); lmontuenga@unav.es (L.M.M.); Tel.: +34-948194700 (K.V. & L.M.M.); Fax: +34-948194714 (K.V. & L.M.M.)

Simple Summary: Exosomes are small vesicles of 100 nm in size that are released from every cell constantly. They contain different molecules (DNA, RNA, lipids, metabolites, etc.) that reflect the content of the cell they come from. Exosomes can be found in all biological fluids. In cancer, exosomes are involved in several events such as tumor growth, metastasis, and the immune response, by delivering their cargos to recipient cells. Due to their unique features, exosomes have become promising analytes in the field of liquid biopsy, which searches for biomarkers to manage different steps of the tumor process. We believe that exosomes will become an important tool in liquid biopsy in the near future. In this review we provide an updated literature compilation about exosomes as biomarkers in oncology and discuss their possibilities and limitations.

Abstract: Among the different components that can be analyzed in liquid biopsy, the utility of exosomes is particularly promising because of their presence in all biological fluids and their potential for multicomponent analyses. Exosomes are extracellular vesicles with an average size of ~100 nm in diameter with an endosomal origin. All eukaryotic cells release exosomes as part of their active physiology. In an oncologic patient, up to 10% of all the circulating exosomes are estimated to be tumor-derived exosomes. Exosome content mirrors the features of its cell of origin in terms of DNA, RNA, lipids, metabolites, and cytosolic/cell-surface proteins. Due to their multifactorial content, exosomes constitute a unique tool to capture the complexity and enormous heterogeneity of cancer in a longitudinal manner. Due to molecular features such as high nucleic acid concentrations and elevated coverage of genomic driver gene sequences, exosomes will probably become the "gold standard" liquid biopsy analyte in the near future.

Keywords: exosomes; cancer; liquid biopsy; biomarkers

1. Exosome Biogenesis and Composition—Reflecting Their Origin

Exosomes are extracellular vesicles (EVs) with a size range of ~40 to 160 nm (average ~100 nm) in diameter with an endosomal origin. All eukaryotic (and also prokaryotic) cells release exosomes as part of their active physiology [1].

Exosomes are generated in a process of sequential invagination of the plasma membrane that results in the formation of multivesicular bodies (MVBs), which can intersect with the trans-Golgi network, endoplasmic reticulum, or other intracellular vesicles, contributing to the content heterogeneity of exosomes. Within the cell, the MVB can either fuse with lysosomes or autophagosomes to be degraded or fuse with the plasma membrane to release the contained vesicles (exosomes). Exosome biogenesis is reflected in the presence of a variety of proteins either integrated in their membrane or as exosomal cargo: small

Rab family GTPases; annexins and flotillin; Alix, Tsg101, and ESCRT complex; tetraspanins CD9, CD63, and CD81; or heat shock proteins Hsp70 [2–5]. ExoCarta, an exosome database (http://exocarta.org/; accessed on 27 April 2021), has been developed to identify exosomal contents. Approximately 10,000 different proteins have been characterized in relation to the exosomal component [6]. Figure 1 shows the main exosome components.

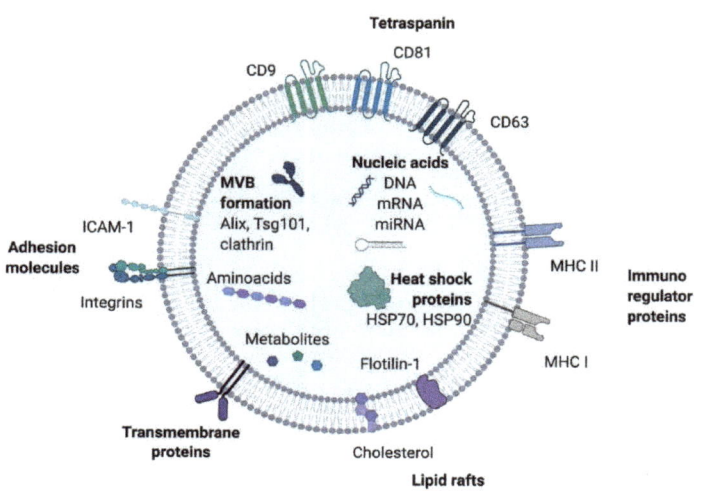

Figure 1. Components of an exosome. Exosomes contain a wide variety of molecules of different natures, such as nucleic acids, proteins, or lipids. All the content at both the membrane and soluble levels represents the cell of origin the exosome is release from.

How DNA is contained in exosomes is far from being resolved and is still controversial. It has been shown that DNA-containing micronuclei that originate from nuclear membrane collapse can interact with exosomal tetraspanins, leading to the shuttling of the DNA in MVBs [7]. Also, mitochondria produce vesicles containing mtDNA that reach the endolysosomal system to form MVBs (reviewed in [8,9]).

Exosome production varies depending on the cellular origin, metabolic status, and cellular microenvironment. One unresolved question about exosomes today is to distinguish tumoral-origin exosomes from non-tumoral counterparts. Moreover, it is still unclear how exactly the exosomal content is selected and loaded into vesicles and how exosomal trafficking is regulated. To solve these questions, it is crucial to fully understand the biology of exosomes. This better knowledge is an essential requirement for future clinical applications of exosomes as diagnostic (and even treatment) tools.

2. Exosomes: A Source of Biomarkers

The path towards more precise and personalized management of cancer patients is currently focused on the development of novel non-invasive biopsy technologies that are easy to obtain, may be repeated over time to follow longitudinally the progression of the disease, and may be able to reflect the phenotypic and genetic heterogeneity of the tumor. Liquid biopsy (LB) offers all of these potential benefits. LB is based on the search for biomarkers that may help clinical decision making. Those biomarkers may be applied to screening/early diagnosis, prognosis, prediction of response or resistance to treatments, detection of minimal residual disease, confirmation of relapse, disease monitoring, etc.

Among the different components that can be analyzed in liquid biopsy, the utility of EVs is particularly promising because of their presence in all biological fluids and their potential for multicomponent analyses. The concentration of analytes in membrane-surrounded vesicles may potentially allow for higher sensitivity and specificity over other types of liquid biopsy looking for single and even multiplexed free circulating biomarkers [10,11]. Exosomes are the most abundant analyte within the liquid biopsy, reaching 1×10^{11} particles per milliliter of blood. In an oncologic patient, up to 10% of all the circulating exosomes will be tumor-derived exosomes depending on tumor stage [12]. Figure 2 describes liquid biopsy analytes and their concentrations. Exosome content mirrors the features of its cell of origin in terms of DNA, RNA, lipids, metabolites, and cytosolic/cell-surface proteins. In addition, exosome content has a number of advantages in comparison to other liquid biopsy analytes. First, exosomes contain high-quality RNA that can be extracted from fresh or frozen fluids. Second, different types of RNA are contained in exosomes, including miRNA [13,14], piwi-interacting RNA, pseudo-genes, lncRNA, tRNA, and mRNA including different splice isoforms found in the cells of origin. Third, exosomes are released from viable tumor cells. Furthermore, their DNA recapitulates the entire genome and the mutational burden of the parental tumor, a great advantage compared to ctDNA, where DNA is fragmented. Evidently, it is significantly more difficult to obtain information about the specific DNA alterations pursued in a given analysis or, worse, to obtain the sequence of the entire genome from highly fragmented circulating DNA [15,16]. In addition, as exosomes contain both RNA and DNA (reflecting tumor mutations), the use of a single platform to study both molecular species is a clear advantage for finding rare or not-abundant mutations. Finally, the protein content of a single exosome reaches up to 400 unique proteins [17,18].

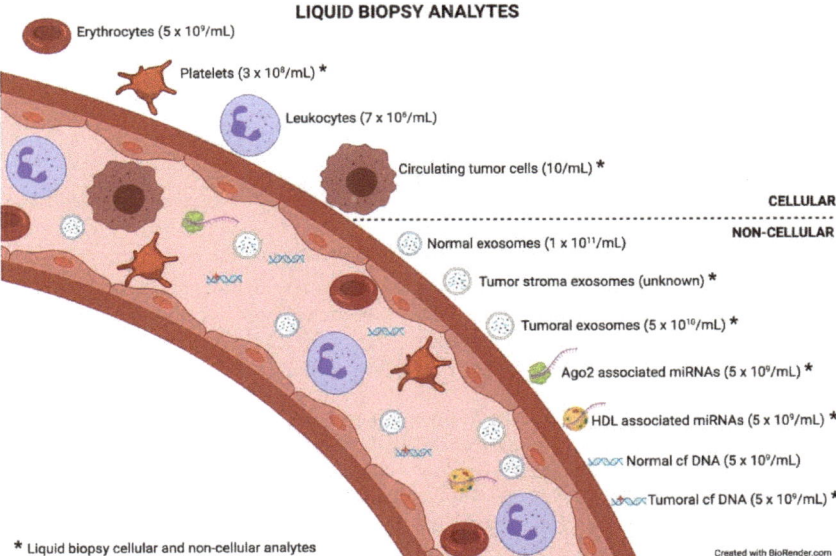

Figure 2. Liquid biopsy analytes. In the bloodstream, many components can be found, cellular or non-cellular in nature. Some of them constitute liquid biopsy analytes (marked with an asterisk). Higher concentrations of analytes (in parentheses) will facilitate isolation techniques and subsequent analysis. Data taken from [12].

Therefore, the fact that exosomes include several molecules that can be considered as potential biomarkers, alone or in combination, increases the possibility of success in the pursuit of a good LB biomarker, which is a clear advantage over the other LB analytes.

Moreover, the number of released exosomes could also be considered a clinical indicator itself (see Section 3).

Table 1 summarizes information rendered by different analytes of LB and shows the potential clinical applications of them as biomarkers.

Table 1. Liquid biopsy analytes: features, extractable information, and clinical applications as biomarkers. Table adapted from [19].

Traits	Liquid Biopsy Analyte				
	CTCs [1]	ctDNA [2]	Exosomes	ctRNA [3]	miRNA
Origin					
Viable cells	✔ [4]	✘ [5]	✔	? [6]	?
Apoptotic cells	✔	✔	?	?	?
Components					
DNA	✔	✔	✔	N.A. [7]	N.A.
RNA	✔	N.A.	✔	✔	✔
Proteins	✔	N.A.	✔	N.A.	N.A.
Metabolites	✔	N.A.	?	N.A.	N.A.
Extractable information					
Copy number variation	✔	✔	✔	✘	✘
Mutations	✔	✔	✔	✔	✘
Epigenetic information	✔	✔	✔	✘	✘
Fusion genes	✔	✔	✔	✔	✘
Splice variants	✔	✘	✔	✔	✘
Single-cell information	✔	✘	✘	✘	✘
Application in personalized medicine					
Diagnosis	✔	✔ [8]	✔	?	✔
Classification of molecular subtypes	✔	✔	?	?	✘
Clonal evolution tracking	✔	✔	?	✘	✘
Prognosis	✔	✔	✔	?	✔
Recurrence	✔	✔	✔	✔	✘
Predictive	✔	✔	✔	?	✘
Resistance prediction	✔	✔	✔	?	✘
Monitoring treatment	✔	✔	✔	?	?

[1] Circulating tumor cell; [2] circulating tumor DNA; [3] circulating tumor RNA; [4] yes; [5] no; [6] no data; [7] not applicable; [8] most probably.

3. Exosome Heterogeneity: An Unknown Wealth?

Exosomes constitute a heterogeneous population of vesicles. This heterogeneity arises from the combination of different parameters such as cellular origin, content, size, number, and functionality. These parameters interact directly with each other, making it very difficult to isolate one without entering the field of another. Within an organ, exosomes can be released from epithelial (tumoral or normal) cells, as well as from stromal cells, lymphocytes, etc. This discrimination could be possible due to the preservation of cell-type-specific membrane proteins on the exosome membrane. There have already been reports in the literature of some examples where, using well-known specific proteins found in exosomes, researchers were able to easily recognize and differentiate exosomes with breast or pancreatic origin [20,21].

The cellular origin of exosomes will determine their composition, at both the membrane and soluble levels. Therefore, the second factor that creates heterogeneity among exosomes is their content. The content of exosomes also varies in response to many factors. It responds to different cellular stages such us metabolic wellness [22]. Thus, an exosome's hallmarks will dynamically change as a result of the modifications that occur in their cell of origin. Moreover, tumor-derived exosomes (TEX) expressing different integrins or other molecules in their membrane have been related to different organotropisms similar to what is shown in tumor spreading cells [23,24] and, more interestingly, TEX are uptaken with greater affinity by certain cell types within an organ [25].

The content of an exosome is limited by its size. This brings us to the third parameter of heterogeneity. Exosomes are also a mixed population in terms of size. As previously mentioned, exosomes show a size range of ~40 to 160 nm. Therefore, a 150 nm diameter exosome will be able to contain a greater number of molecules than a smaller exosome. It is still unknown whether different exosome sizes respond to distinct cellular stages or cause diverse responses in target cells, but what have been reported in recent studies are significant differences in the number and size of exosomes in cancer patients depending on the studied biological fluid [26].

The number of exosomes released from a cell is another source of heterogeneity. Due to the constant influx of exosomes, the exosomal release–uptake dynamics of different cells, and the lack of fine characterization of exosome origin, it is difficult to ascertain whether the amount of TEX is different compared to that from normal cells. Historically, it has been demonstrated in vitro that tumor cells secrete more exosomes than their normal cell counterparts. Thus, different studies reported higher exosome protein amounts in cancer patients than in healthy controls [27] (reviewed in [10,28]). However, technological studies in breast cancer pointed to the opposite situation, where the capture of shed exosomes in a single-cell platform showed lower numbers of exosomes in tumor cells compared with tissue-matched, nontumorigenic cell-line-derived exosomes [29]. Such studies relied on different isolation methods, experimental designs, and quantification methods, facts that can easily disturb results. Therefore, further investigation is needed to clarify this important aspect of exosome biogenesis. The literature describes an increased number of total circulating exosomes in the peripheral blood of cancer patients and, surprisingly, their size and morphology are also altered compared to those of healthy donors [30]. More interestingly, recent studies showed significant differences in the number and size of exosomes in cancer patients depending on the studied biological fluid [26]. This fact highlights the importance of selecting an ideal bodily fluid as a tool for the search and study of exosome-based biomarkers in each given type of cancer. In summary, the underlying mechanism of these alterations during the tumor course is unclear.

The final source of heterogeneity we will refer to is exosome functionality. Exosomes show very diverse effects on the cells that uptake them. The consequences are so varied that we dedicate an epigraph below to exploring the most studied and characteristic outcomes (Figure 3).

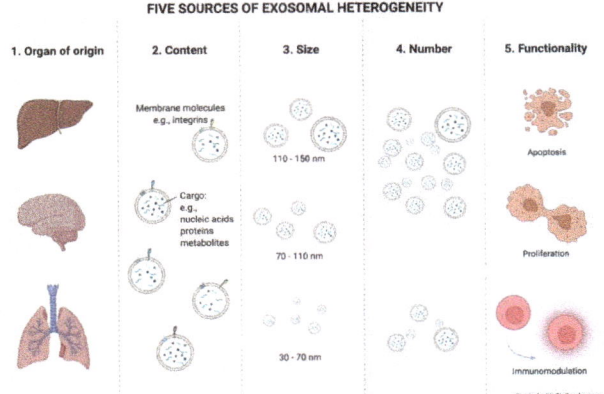

Figure 3. Five sources of exosomal heterogeneity. The heterogeneity of the exosomes results from the combination of five factors: the cell of origin from which they are released (organ and cell type of origin); their molecular composition; their size; their number; and the functionality triggered in recipient cells. Different combinations of these five factors make exosome heterogeneity highly complex.

Taken together, these data suggest that exosome heterogeneity might play a dual role in the characterization of a patient's tumor. On the one hand, the number and other above-mentioned hallmarks of exosomes could give us a clue about the tumoral stage and its possible progression, but on the other hand, this mix could dilute valuable information in their use as accurate biomarkers. Exosomal-related biomarkers are discussed below.

4. Sending a Message: The Role of Exosomes in Intercellular Communication

Exosomes have been shown to provide a natural mechanism for cell-to-cell communication, with a plethora of roles in physiology and pathology. In every communication process, a relationship between a sender and a receiver is established through the emission and reception of a "message" that will have an impact on the recipient. Exosomes are known to play a very important role in the communication process between tumor cells and their microenvironment. Recently, several groups visualized through elegant imaging techniques the process of exosome uptake in NSCLC [31] and breast cancer [32]. The content of tumor-released exosomes can be uptaken by other adjoining tumor cells, tumor-niche (stroma) cells, immune cells, or distal organ cells after travelling through the circulatory system.

There are still many questions about the role of exosomes in intercellular communication. For example, it is still unknown how different outcomes on receptor cells may be affected by uptake affinity differences between recipient cell types or by different modes of exosomal uptake (receptor-mediated endocytosis, direct binding, direct fusion, etc.) [33]. The regulation of the different potential cellular fates of the cargo transported by the uptaken exosomes is also not clearly known. The contents of the exosomes can be directly transferred to the degradation pathway or may be secreted into the endoplasmic reticulum and/or to the cytoplasm. Specific membrane transport mechanisms may be involved in these different inner cellular outcomes. Furthermore, it is plausible that depending on the nature of the exosomal cargo and the state of the recipient cell, the ability of the exosomal message to affect specific recipient cell functions may be variable, which makes the understanding and the study of the exosomal-based communication process even more complex.

As just mentioned above, exosomes have an impact on the recipient cells that will influence the development of the tumoral process. Exosomes have been described to be involved in different neoplastic stages such as tumor growth, metastasis, and resistance to therapy, contributing to different hallmark features of cancer (Figure 4) [34]. Many of these hallmarks will appear in the following section.

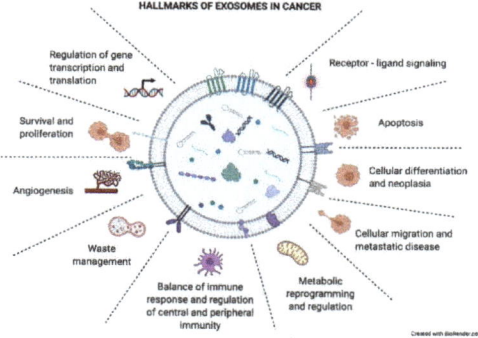

Figure 4. Hallmarks of exosomes in cancer. Tumor-derived exosomes have important functional roles in intercellular crosstalk, affecting the biology of their target cells in different manners. Through this crosstalk, exosomes drive the tumoral process and other pathological conditions. The picture summarizes responses that exosome uptake can trigger in the recipient cells (functional hallmarks).

In order to better explain the role of exosomes in intercellular communication and the effects that they trigger in recipient cells, we divided this epigraph into two sections, using the distance to which the receptor cell is located as the criterion. We focus on TEX examples as in the cancer field this is the central and most studied population of exosomes.

4.1. A Short-Range Shipment: The Role of Exosomes in the Tumor Microenvironment

Epithelial-to-mesenchymal transition (EMT) is a process through which cancer cells may become more proliferative and resistant and gain migratory and invasive properties [35]. TEX are thought to be partially responsible for this cellular plasticity, inducing EMT in adjoining tumor recipient cells [36] through the modulation of several well-known signaling pathways. Thus, regulation of Wnt/β–catenin or PI3K/AKT in human lung cancer cell lines [37,38] and modulation of the Hippo and ERK pathways [39,40] in hepatocellular carcinoma upon TEX uptake have been reported. Similarly, activation of AKT signaling triggered by TEX has been reported to induce EMT [41].

Classically, the field of exosomes has focused on understanding how TEX uptake by stromal cells modifies the tumor niche, modulating the microenvironment to favor tumor development. Most studies have focused on defining the functional changes in cancer-associated fibroblasts (CAFs) and immune cells [42].

In this sense, several systems of TEX-mediated immune suppression have been described. TEX carry ligands that bind to cognate receptors on immune cells, inducing tolerogenic signaling [43] and inhibiting tumor-specific T cells [44]. The response of activated T cells to TEX interaction triggers a reduction in both JAK expression and the response to IL-2 [45,46] which prevents them from proliferating. Furthermore, TEX carry CD39 and CD73, which activate the adenosine pathway, a well-known immunosuppressive factor that inhibits T-cell function [47,48]. More interestingly, TEX carrying FasL [49] or programmed death ligand 1 (PD-L1) induce the apoptosis of activated CD8+T cells by triggering both extrinsic and intrinsic apoptosis pathways [50]. Importantly, FAsL and PD-L1 exosomal expression levels correlate to spontaneous apoptosis of circulating T cells and to tumor prognosis [51]. Recently, it was reported that the suppression of exosomal PD-L1 induces systemic anti-tumor immunity and memory [52]. On the contrary, TEX lead the differentiation and expansion of Tregs [44,53]. TEX also modulate NK cytotoxicity by downregulating NKG2D expression, which suppresses NK cell activity [54]. Besides this, tumor-derived exosomes inhibit monocyte differentiation into DC cells [55], directly inhibiting DC bioactivity and inducing immune tolerance [56]. However, TEX skew monocyte differentiation into myeloid-derived suppressor cells (MDSCs) [57,58], which accumulate in murine tumor, spleen, peripheral blood, and lung in vivo [59]. This fact negatively affects antigen processing and presentation and produces several immunosuppressive inhibitory factors, including NO and ROS, causing TCR nitration or T-cell apoptosis [60]. Moreover, neutrophils that uptake TEX DNA increase IL-8 and tissue factor production, boosting tumor inflammation and paraneoplastic events (thrombosis) [61]. TEX also generate an immunosuppressive microenvironment by activating macrophages to a tumor-associated macrophage (TAM)-like phenotype [62,63]. Finally, the role of exosomes in the innate immune response has also been described in cancer. TEX were shown to harbor B cells and exert a decoy function limiting complement-mediated lysis and decreasing cytotoxicity against cancer cells [64].

TEX are also implicated in angiogenic remodeling, an essential step in tumor survival, growth, and dissemination, through favoring new vessel formation [25] or destroying the integrity of the endothelium and promoting vascular permeability and metastasis [65].

Tumor dissemination is also facilitated by TEX triggering matrix destruction by MMP1 activation [66].

The reciprocal exchange of exosomes between tumor cells and CAF has also been a focus of study. In this way, CAF-derived exosomes support the metabolic fitness of cancer cells growing as tumors through several known mechanisms, such as switching mitochondrial oxidative phosphorylation to glycolysis [67] or promoting motility via Wnt-planar cell polarity (PCP) signaling [68].

4.2. A Long-Range Shipment: The Role of Exosomes in Metastatic Organs

Primary tumor TEX can reach metastatic organs through the circulation (blood or lymphatic). It has been described how the pattern of integrins present in the exosome membrane determines TEX organotropism. Thus, breast cancer TEX bearing integrins α6β4 and α6β1 were associated with lung metastasis, while exosomal integrin αvβ5 was linked to liver metastasis [23]. Once exosomes have reached the metastatic organ, they are uptaken by specific cells in those organs, and the message they carry is translated by the receptor cells into microenvironment remodeling orders known as premetastatic niche preparation [69], an essential change for the nesting and engraftment of circulating tumor cells (CTCs) reaching the metastatic organ. The recruitment of bone marrow progenitor cells and macrophages to metastatic sites is one of the changes related to TEX that are involved in premetastatic niche formation and enhance metastatic potential [70,71]. Also, TEX prevent patrolling Ly6C low monocyte expansion, enabling immunosuppression and leading to metastasis [72]. The activation of cancer-associated fibroblasts (CAFs) is also involved in premetastatic niche formation. TEX can trigger TGF-b signaling pathways and thereafter initiate a program of differentiation of fibroblasts toward a myofibroblastic phenotype, altering the stroma which will be then responsible for supporting tumor growth, vascularization, and metastasis [73]. In turn, CAF-derived exosomes induce oxidative phosphorylation in metastatic breast cancer cells, contributing to their exit from the dormant state [74]. Figure 5 recapitulates the role of TEX locally and in distant organs.

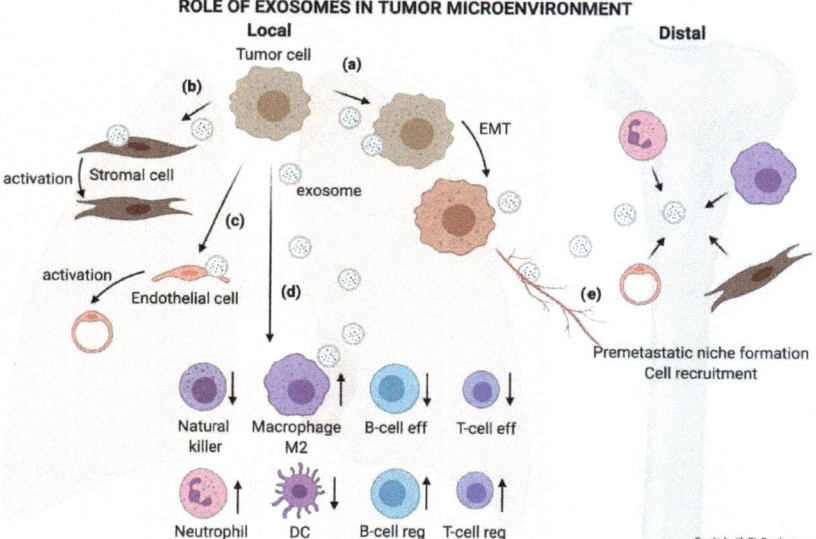

Figure 5. The role of exosomes in the tumor microenvironment. TEX communicate with their microenvironment. At a local level, exosomes can be uptaken by other adjoining tumor cells, favoring an EMT process (**a**). They may also modify the biology of stromal (**b**) and endothelial (**c**) cells by activating them to support the tumor. Exosomes received by immune cells favor an immunosuppressive microenvironment that helps tumor growth (**d**). Furthermore, exosomes can reach blood or lymphatic vessels and travel to distal organs, where they will educate native cells, preparing a premetastatic niche where subsequent circulating tumor cells (CTCs) will nest and grow (**e**).

5. TEX Biomarkers in Clinics: A List of Possibilities

One of the great proposals in the field is to study exosome contents as potential biomarkers. Despite biological fluids being composed of a complex mixture of molecules (RNA, DNA, and proteins), diagnostic approaches have traditionally focused on a single molecular species. In the case of exosomal cargo, the same trend has happened. RNA is the most abundant and studied exosomal component, being unusually stable thanks to its exosomal membrane confinement. In 2012, the National Institute of Health (NIH) dedicated a strategic Common Fund to the study of exosomal RNA (http://commonfund.nih.gov/Exrna/index; accessed on 27 April 2021). Since then, the interest in the field has continuously increased. The number of entries related to exosomes, RNA, and cancer in PubMed has increased more than 10-fold from 2012 to date.

Currently, the focus of translational studies is also turning to exosomal DNA assessment, with more than 200 publications being found in PubMed in 2020. Probably, one of the most specific hallmarks of cancer is DNA mutations, which can also be captured within exosomes [75]. Many recent works take advantage of existing technologies for circulating free DNA (ctDNA) detection in LB. The translational use of exosome DNA sequencing is an exciting approach that still needs to be fully explored and developed.

Proteins contained in exosomes also include altered proteins associated with cancer. In addition, exosomal surface proteins are related to the functional status of the cells comprising the tumor immune microenvironment, which may be important biomarkers for monitoring response to immunotherapies [52].

Due to their multifactorial content, exosomes constitute a unique tool to capture the complexity and enormous heterogeneity of cancer [10]. To bring exosome-based liquid biopsy diagnosis closer to the clinic, several high-throughput platforms have recently been developed. Among them, microfluidic devices based on antibody-capturing systems in microchips [76–80] seem to be the best option for clinical application [81]. These novel technical approaches aim to make exosome-based diagnostics cost and labor effective, by means of developing highly sensitive and reproducible detection devices to isolate and identify circulating cancer markers without using a large volume of sample and sparing the time-consuming ultracentrifuge-based isolation processes that are usually associated with exosome analysis.

Multicomponent diagnostic/prognostic applications based on exosomes are currently being considered. These high-throughput multiplexed analyses can also be combined with deep-learning-based interpretation methodologies, which will require pilot studies in large numbers of clinical samples [82]. These approaches may overcome the sensitivity and specificity of current biomarkers in a number of clinical situations. Moreover, the inclusion of exosome cargo analysis in the biomarker laboratory armamentarium may help to characterize not only the tumor but also its microenvironment, leading to a more accurate tumor description and understanding.

The table below (Table 2) summarizes a list of exosomal analytes proposed as biomarkers in the last three years. All of them have been studied in well-characterized cohorts of patients. Nevertheless, in a considerable proportion of these examples, especially in the case of miRNAs, validation in independent cohorts, together with robust statistical criteria and harmonized protocols, is still needed. The main disadvantage of working with exosomes is still the lack of technical consensus, which leads to poor inter-laboratory reproducibility of the results. Therefore, before exosome-based biomarkers become a clinical reality, major efforts have to be made to standardize every single procedure in exosomal-based biomarker studies: isolation, characterization, and analytical protocols [83].

Table 2. Examples of exosomal-derived potential biomarkers with clinical significance published in the last three years.

Exosomal miRNAs as Cancer Biomarkers				
miRNA	Cancer type	Clinical value	Biofluid	Reference
Let-7b-5p, -122-5p, -146b-5p, -210-3p, -215-5p	Breast cancer	Diagnosis	Plasma	[84]
miR-224	Hepatocellular carcinoma	Diagnosis/Prognosis	Serum	[85]
miR-106b, miR-1269a	Lung cancer	Diagnosis/Prognosis	Serum	[86,87]
miR-375, -1307	Ovarian cancer	Diagnosis	Serum	[88]
Exosomal lncRNAs as Cancer Biomarkers				
lncRNA	Cancer type	Clinical value	Biofluid	Reference
PCAT-1, UBC1 and SNHG16	Bladder cancer	Diagnosis/Prognosis	Urine	[89]
MALAT-1	Lung cancer	Diagnosis	Serum	[90]
Exosomal mRNA as Cancer Biomarkers				
mRNA	Cancer type	Clinical value	Biofluid	Reference
BRAF, KRAS (mutant)	Colorectal cancer	Diagnosis	Serum	[91]
Exosomal mutated DNA as Cancer Biomarkers				
DNA	Cancer type	Clinical value	Biofluid	Reference
IDH1	Glioblastoma	Diagnosis/Prognosis	Plasma	[92]
EGFR	Lung cancer	Diagnosis/Prognosis	Plasma/Bronchioalveolar lavage	[93–96]
BRAF	Melanoma	Therapeutic monitoring	Plasma	[97]
KRAS, P53	Pancreatic cancer	Diagnosis/Prognosis	Serum/Plasma	[98,99]
MYC, P53, MLH1, PTEN, AR	Prostate cancer	Diagnosis/Prognosis	Plasma	[100,101]
Exosomal proteins as Cancer Biomarkers				
Protein	Cancer type	Clinical value	Biofluid	Reference
PDL-1	Melanoma	Prognosis	Plasma	[102]

6. Future Perspectives and Challenges: The Dawn of a New Era

Liquid biopsy applications have been exponentially growing since 2010. According to RNCOS market research, the global liquid biopsy market is expected to reach 5 billion dollars by 2023 [103]. Among the different analytes in LB, circulating tumor DNA (ctDNA) seems to be the one with the most promising results in the field. The main bottleneck of ctDNA-based LB is to develop technologies sensitive enough to measure low amounts of ctDNA in circulation, particularly when early detection or minimal residual disease is pursued. Next-generation sequencing (NGS)-based technologies have reached a compromise between sensitivity and cost and they are already available in clinical laboratories. By August 2020, the FDA had approved the first two blood tests, Guardant360 CDx and FoundationOne Liquid CDx, as companion diagnostic tests that provide molecular information (mainly specific mutations or CNA) predictive for the effective use of associated drugs in NSCLC, prostate, breast, and ovary and for general tumor profiling in solid tumors [104,105]. Previous NGS-based tests approved for use in DNA extracted from FFPE or other tumor tissue samples have previously shown great efficacy as companion biomarkers.

Although many efforts have been made in detecting ctDNA in blood, it is worth mentioning that ctDNA seems to be mainly released passively from dying normal or tumor cells (necrosis or the different types of programmed cell death). It is also actively shed from neutrophils by the process called NETosis [106]. However, DNA can also be released within exosomes in an active and selective manner. In fact, it has been reported that more than 93% of amplifiable cfDNA in blood is in fact found as cargo of plasma exosomes [107]. Therefore, exosomes are potentially very valuable raw materials for more sensitive analysis of circulating DNA, as DNA is highly concentrated in exosomes released from tumor and other cells. To date, only a few studies have compared the clinical parameters of "gold standard" ctDNA and exosomal DNA (exoDNA). Only in pancreatic ductal adenocarcinoma, KRAS mutation detection in exoDNA was superior to ctDNA for prognosis [98,99]. It has also been shown that the combination of exoDNA/RNA and ctDNA has better sensitivity and specificity than ctDNA alone for EGFR T790M mutation detection in NSCLC [93,94] and BRAF V600E mutation detection in melanoma [97].

Table 3. Pros and cons of the main exosome isolation techniques.

Factors	Ultracentrifugation		Precipitation	Affinity		Microfluidic	Filtration	
	Differential	Gradient		Immune	Flow Cytometry		Ultrafiltration	Molecular Exclusion
Purity	low	high	low	high	high	high	low	high
Yield	medium	low	medium	medium	medium	low	medium	high
Specialized equipment	medium	medium	high	medium	low	low	high	high
Specialized user	medium	low	high	medium	medium	medium	high	high
RNA characterization	high	high	high	high	high	high	high	high
Protein characterization	medium	high	low	high	high	high	medium	high
Functional studies	medium	medium	low	medium	medium	medium	medium	high
Scalability	medium	low	high	medium	high	low	medium	medium
Time	medium	low	high	medium	low	medium	high	medium
Cost	high	medium	high	low	low	low	medium	medium

Despite the need for more studies in large patient cohorts to evaluate exoDNA as a circulating biomarker, preliminary data are very promising. High exosomal nucleic acid concentrations and elevated coverage of the genomic driver gene sequences will probably help to make the analysis of exosomes the "gold standard" LB DNA-based analyte in the near future.

In summary, although the field of exosomes in liquid biopsy is still immature, its potential for the very near future seems enormous, promising, and fascinating. The major hurdles for exosomal-based biomarkers to reach the clinic are the standardization and optimization of isolation and characterization methodologies and the validation of reported results in multiple independent cohorts.

In fact, there are many different techniques to isolate exosomes. They can be classified into five main groups according to the chemical or physical isolation system: centrifugation, precipitation, affinity binding, microfluidics, and molecular size-exclusion-based techniques. Each method has its pros and cons. In general, an exosome isolation technique with elevated yield numbers will render low exosome purity, and vice versa. Therefore, the isolation method may be adapted to respond to each specific need. It is important to take into consideration such factors as the type and amount of initial sample or the subsequent use of those isolated exosomes. Moreover, some other aspects will determine the final choice of the technique, e.g., the need for specialized equipment, cost, time, or scalability. Table 3 summarizes the pros and cons of the main exosome isolation technologies.

Although there is still no consensus on a standard isolation method, the International Society for Extracellular Vesicles (ISEV) is making an strong effort to achieve this aim [83].

Also, understanding of the regulatory mechanisms that control tumor-derived exosome heterogeneity that may influence the reproducibility of diagnostic outcomes is essential.

In addition, the development of liquid-biopsy-based multiparametric assays is expected to return large data sets of different nature (nucleic acids, proteins, etc.). For this reason, the implementation of artificial intelligence tools for data management and analysis, as well as the development of models that include all complex exosome-derived data, is starting to be explored to accurately use exosomes as cancer biomarkers [108].

In summary, the research avenues for the near future in the field of exosomes in cancer liquid biopsy are multiple, wide, and very exciting.

Funding: This research was funded by SPANISH MINISTRY OF ECONOMY AND INNOVATION AND FONDO DE INVESTIGACIÓN SANITARIA-FONDO EUROPEO DE DESARROLLO REGIONAL (PI19/00098).

Conflicts of Interest: The authors declare no conflict of interest.

References

1. Chronopoulos, A.; Kalluri, R. Emerging role of bacterial extracellular vesicles in cancer. *Oncogene* **2020**, *39*, 6951–6960. [CrossRef]
2. Mathivanan, S.; Ji, H.; Simpson, R.J. Exosomes: Extracellular organelles important in intercellular communication. *J. Proteom.* **2010**, *73*, 1907–1920. [CrossRef] [PubMed]
3. Kowal, J.; Tkach, M.; Théry, C. Biogenesis and secretion of exosomes. *Curr. Opin. Cell Biol.* **2014**, *29*, 116–125. [CrossRef] [PubMed]
4. Simons, M.; Raposo, G. Exosomes—Vesicular carriers for intercellular communication. *Curr. Opin. Cell Biol.* **2009**, *21*, 575–581. [CrossRef] [PubMed]
5. Runz, S.; Keller, S.; Rupp, C.; Stoeck, A.; Issa, Y.; Koensgen, D.; Mustea, A.; Sehouli, J.; Kristiansen, G.; Altevogt, P. Malignant ascites-derived exosomes of ovarian carcinoma patients contain CD24 and EpCAM. *Gynecol. Oncol.* **2007**, *107*, 563–571. [CrossRef]
6. Keerthikumar, S.; Chisanga, D.; Ariyaratne, D.; Al Saffar, H.; Anand, S.; Zhao, K.; Samuel, M.; Pathan, M.; Jois, M.; Chilamkurti, N.; et al. ExoCarta: A Web-Based Compendium of Exosomal Cargo. *J. Mol. Biol.* **2016**, *428*, 688–692. [CrossRef]
7. Yokoi, A.; Villar-Prados, A.; Oliphint, P.A.; Zhang, J.; Song, X.; DeHoff, P.; Morey, R.; Liu, J.; Roszik, J.; Clise-Dwyer, K.; et al. Mechanisms of nuclear content loading to exosomes. *Sci. Adv.* **2019**, *5*, 1–17. [CrossRef]
8. Picca, A.; Guerra, F.; Calvani, R.; Coelho-Junior, H.J.; Bossola, M.; Landi, F.; Bernabei, R.; Bucci, C.; Marzetti, E. Generation and Release of Mitochondrial-Derived Vesicles in Health, Aging and Disease. *J. Clin. Med.* **2020**, *9*, 1440. [CrossRef]
9. Lázaro-Ibáñez, E.; Lässer, C.; Shelke, G.V.; Crescitelli, R.; Jang, S.C.; Cvjetkovic, A.; García-Rodríguez, A.; Lötvall, J. DNA analysis of low- and high-density fractions defines heterogeneous subpopulations of small extracellular vesicles based on their DNA cargo and topology. *J. Extracell. Vesicles* **2019**, *8*. [CrossRef]
10. Kalluri, R. The biology and function of exosomes in cancer. *J. Clin. Invest.* **2016**, *126*, 1208–1215. [CrossRef]
11. Fitts, C.A.; Ji, N.; Li, Y.; Tan, C. Exploiting Exosomes in Cancer Liquid Biopsies and Drug Delivery. *Adv. Healthc. Mater.* **2019**, *8*, 1–8. [CrossRef]
12. Brock, G.; Castellanos-Rizaldos, E.; Hu, L.; Coticchia, C.; Skog, J. Liquid biopsy for cancer screening, patient stratification and monitoring. *Transl. Cancer Res.* **2015**, *4*, 280–290. [CrossRef]
13. Salehi, M.; Sharifi, M. Exosomal miRNAs as novel cancer biomarkers: Challenges and opportunities. *J. Cell. Physiol.* **2018**, *233*, 6370–6380. [CrossRef]
14. Thind, A.; Wilson, C. Exosomal miRNAs as cancer biomarkers and therapeutic targets. *J. Extracell. Vesicles* **2016**, *5*, 1–11. [CrossRef]
15. Balaj, L.; Lessard, R.; Dai, L.; Cho, Y.-J.; Pomeroy, S.L.; Breakefield, X.O.; Skog, J. Tumour microvesicles contain retrotransposon elements and amplified oncogene sequences. *Nat. Commun.* **2011**, *2*, 180. [CrossRef] [PubMed]
16. Thakur, B.K.; Zhang, H.; Becker, A.; Matei, I.; Huang, Y.; Costa-Silva, B.; Zheng, Y.; Hoshino, A.; Brazier, H.; Xiang, J.; et al. Double-stranded DNA in exosomes: A novel biomarker in cancer detection. *Cell Res.* **2014**, *24*, 766–769. [CrossRef]
17. Maguire, G. Exosomes: Smart nanospheres for drug delivery naturally produced by stem cells. In *Fabrication and Self Assembly of Nanobiomaterials*; Grumezescu, A., Ed.; Elsevier: Amsterdam, The Netherlands, 2016; pp. 179–209.
18. Mathivanan, S.; Lim, J.W.E.; Tauro, B.J.; Ji, H.; Moritz, R.L.; Simpson, R.J. Proteomics analysis of A33 immunoaffinity-purified exosomes released from the human colon tumor cell line LIM1215 reveals a tissue-specific protein signature. *Mol. Cell. Proteomics* **2010**, *9*, 197–208. [CrossRef]
19. Heitzer, E.; Haque, I.S.; Roberts, C.E.S.; Speicher, M.R. Current and future perspectives of liquid biopsies in genomics-driven oncology. *Nat. Rev. Genet.* **2019**, *20*, 71–88. [CrossRef] [PubMed]
20. Ciravolo, V.; Huber, V.; Ghedini, G.C.; Venturelli, E.; Bianchi, F.; Campiglio, M.; Morelli, D.; Villa, A.; Mina, P.D.; Menard, S.; et al. Potential role of HER2-overexpressing exosomes in countering trastuzumab-based therapy. *J. Cell. Physiol.* **2012**, *227*, 658–667. [CrossRef] [PubMed]
21. Adamczyk, K.A.; Klein-Scory, S.; Tehrani, M.M.; Warnken, U.; Schmiegel, W.; Schnölzer, M.; Schwarte-Waldhoff, I. Characterization of soluble and exosomal forms of the EGFR released from pancreatic cancer cells. *Life Sci.* **2011**, *89*, 304–312. [CrossRef]
22. Wen, S.W.; Lima, L.G.; Lobb, R.J.; Norris, E.L.; Hastie, M.L.; Krumeich, S.; Möller, A. Breast Cancer-Derived Exosomes Reflect the Cell-of-Origin Phenotype. *Proteomics* **2019**, *19*, e1800180. [CrossRef]
23. Hoshino, A.; Costa-Silva, B.; Shen, T.-L.; Rodrigues, G.; Hashimoto, A.; Tesic Mark, M.; Molina, H.; Kohsaka, S.; Di Giannatale, A.; Ceder, S.; et al. Tumour exosome integrins determine organotropic metastasis. *Nature* **2015**, *527*, 329–335. [CrossRef] [PubMed]
24. Rodrigues, G.; Hoshino, A.; Kenific, C.M.; Matei, I.R.; Steiner, L.; Freitas, D.; Kim, H.S.; Oxley, P.R.; Scandariato, I.; Casanova-Salas, I.; et al. Tumour exosomal CEMIP protein promotes cancer cell colonization in brain metastasis. *Nat. Cell Biol.* **2019**, *21*, 1403–1412. [CrossRef] [PubMed]
25. Valencia, K.; Luis-Ravelo, D.; Bovy, N.; Antón, I.; Martínez-Canarias, S.; Zandueta, C.; Ormazábal, C.; Struman, I.; Tabruyn, S.; Rebmann, V.; et al. MiRNA cargo within exosome-like vesicle transfer influences metastatic bone colonization. *Mol. Oncol.* **2014**, *8*, 689–703. [CrossRef]
26. García-Silva, S.; Benito-Martín, A.; Sánchez-Redondo, S.; Hernández-Barranco, A.; Ximénez-Embún, P.; Nogués, L.; Mazariegos, M.S.; Brinkmann, K.; López, A.A.; Meyer, L.; et al. Use of extracellular vesicles from lymphatic drainage as surrogate markers of melanoma progression and BRAFV600E mutation. *J. Exp. Med.* **2019**, *216*, 1061–1070. [CrossRef] [PubMed]
27. Szczepanski, M.J.; Szajnik, M.; Welsh, A.; Whiteside, T.L.; Boyiadzis, M. Blast-derived microvesicles in sera from patients with acute myeloid leukemia suppress natural killer cell function via membrane-associated transforming growth factor-β1. *Haematologica* **2011**, *96*, 1302–1309. [CrossRef]

28. Bebelman, M.P.; Smit, M.J.; Pegtel, D.M.; Baglio, S.R. Biogenesis and function of extracellular vesicles in cancer. *Pharmacol. Ther.* **2018**, *188*, 1–11. [CrossRef]
29. Chiu, Y.-J.; Cai, W.; Shih, Y.-R.V.; Lian, I.; Lo, Y.-H. A Single-Cell Assay for Time Lapse Studies of Exosome Secretion and Cell Behaviors. *Small* **2016**, *12*, 3658–3666. [CrossRef]
30. Melo, S.A.; Luecke, L.B.; Kahlert, C.; Fernandez, A.F.; Gammon, S.T.; Kaye, J.; LeBleu, V.S.; Mittendorf, E.A.; Weitz, J.; Rahbari, N.; et al. Glypican-1 identifies cancer exosomes and detects early pancreatic cancer. *Nature* **2015**, *523*, 177–182. [CrossRef]
31. Reclusa, P.; Verstraelen, P.; Taverna, S.; Gunasekaran, M.; Pucci, M.; Pintelon, I.; Claes, N.; de Miguel-Pérez, D.; Alessandro, R.; Bals, S.; et al. Improving extracellular vesicles visualization: From static to motion. *Sci. Rep.* **2020**, *10*, 1–9. [CrossRef]
32. You, S.; Barkalifa, R.; Chaney, E.J.; Tu, H.; Park, J.; Sorrells, J.E.; Sun, Y.; Liu, Y.Z.; Yang, L.; Chen, D.Z.; et al. Label-free visualization and characterization of extracellular vesicles in breast cancer. *Proc. Natl. Acad. Sci. USA* **2019**, *116*, 24012–24018. [CrossRef] [PubMed]
33. Mathieu, M.; Martin-Jaular, L.; Lavieu, G.; Théry, C. Specificities of secretion and uptake of exosomes and other extracellular vesicles for cell-to-cell communication. *Nat. Cell Biol.* **2019**, *21*, 9–17. [CrossRef]
34. Hanahan, D.; Weinberg, R.A. Hallmarks of cancer: The next generation. *Cell* **2011**, *144*, 646–674. [CrossRef]
35. Ribatti, D.; Tamma, R.; Annese, T. Epithelial-Mesenchymal Transition in Cancer: A Historical Overview. *Transl. Oncol.* **2020**, *13*, 100773. [CrossRef] [PubMed]
36. Blackwell, R.; Foreman, K.; Gupta, G. The Role of Cancer-Derived Exosomes in Tumorigenicity & Epithelial-to-Mesenchymal Transition. *Cancers* **2017**, *9*, 105. [CrossRef]
37. Xia, Y.; Wei, K.; Hu, L.-Q.; Zhou, C.-R.; Lu, Z.-B.; Zhan, G.-S.; Pan, X.-L.; Pan, C.-F.; Wang, J.; Wen, W.; et al. Exosome-mediated transfer of miR-1260b promotes cell invasion through Wnt/β-catenin signaling pathway in lung adenocarcinoma. *J. Cell. Physiol.* **2020**, *235*, 6843–6853. [CrossRef] [PubMed]
38. Lu, C.; Shan, Z.; Hong, J.; Yang, L. MicroRNA-92a promotes epithelial-mesenchymal transition through activation of PTEN/PI3K/AKT signaling pathway in non-small cell lung cancer metastasis. *Int. J. Oncol.* **2017**, *51*, 235–244. [CrossRef] [PubMed]
39. Hu, Y.; Yang, C.; Yang, S.; Cheng, F.; Rao, J.; Wang, X. miR-665 promotes hepatocellular carcinoma cell migration, invasion, and proliferation by decreasing Hippo signaling through targeting PTPRB. *Cell Death Dis.* **2018**, *9*, 954. [CrossRef] [PubMed]
40. He, M.; Qin, H.; Poon, T.C.W.; Sze, S.-C.; Ding, X.; Co, N.N.; Ngai, S.-M.; Chan, T.-F.; Wong, N. Hepatocellular carcinoma-derived exosomes promote motility of immortalized hepatocyte through transfer of oncogenic proteins and RNAs. *Carcinogenesis* **2015**, *36*, 1008–1018. [CrossRef] [PubMed]
41. Yu, F.; Chen, B.; Dong, P.; Zheng, J. HOTAIR Epigenetically Modulates PTEN Expression via MicroRNA-29b: A Novel Mechanism in Regulation of Liver Fibrosis. *Mol. Ther.* **2017**, *25*, 205–217. [CrossRef]
42. Nabet, B.Y.; Qiu, Y.; Shabason, J.E.; Wu, T.J.; Yoon, T.; Kim, B.C.; Benci, J.L.; DeMichele, A.M.; Tchou, J.; Marcotrigiano, J.; et al. Exosome RNA Unshielding Couples Stromal Activation to Pattern Recognition Receptor Signaling in Cancer. *Cell* **2017**, *170*, 352–366.e13. [CrossRef]
43. Gutiérrez-Vázquez, C.; Villarroya-Beltri, C.; Mittelbrunn, M.; Sánchez-Madrid, F. Transfer of extracellular vesicles during immune cell-cell interactions. *Immunol. Rev.* **2013**, *251*, 125–142. [CrossRef]
44. Wieckowski, E.U.; Visus, C.; Szajnik, M.; Szczepanski, M.J.; Storkus, W.J.; Whiteside, T.L. Tumor-derived microvesicles promote regulatory T cell expansion and induce apoptosis in tumor-reactive activated CD8+ T lymphocytes. *J. Immunol.* **2009**, *183*, 3720–3730. [CrossRef]
45. Whiteside, T.L. Immune modulation of T-cell and NK (natural killer) cell activities by TEXs (tumour-derived exosomes). *Biochem. Soc. Trans.* **2013**, *41*, 245–251. [CrossRef]
46. Clayton, A.; Mitchell, J.P.; Court, J.; Mason, M.D.; Tabi, Z. Human tumor-derived exosomes selectively impair lymphocyte responses to interleukin-2. *Cancer Res.* **2007**, *67*, 7458–7466. [CrossRef] [PubMed]
47. Muller-Haegele, S.; Muller, L.; Whiteside, T.L. Immunoregulatory activity of adenosine and its role in human cancer progression. *Expert Rev. Clin. Immunol.* **2014**, *10*, 897–914. [CrossRef]
48. Schuler, P.J.; Saze, Z.; Hong, C.-S.; Muller, L.; Gillespie, D.G.; Cheng, D.; Harasymczuk, M.; Mandapathil, M.; Lang, S.; Jackson, E.K.; et al. Human CD4+ CD39+ regulatory T cells produce adenosine upon co-expression of surface CD73 or contact with CD73+ exosomes or CD73+ cells. *Clin. Exp. Immunol.* **2014**, *177*, 531–543. [CrossRef] [PubMed]
49. Kim, J.W.; Wieckowski, E.; Taylor, D.D.; Reichert, T.E.; Watkins, S.; Whiteside, T.L. Fas ligand-positive membranous vesicles isolated from sera of patients with oral cancer induce apoptosis of activated T lymphocytes. *Clin. Cancer Res.* **2005**, *11*, 1010–1020. [PubMed]
50. Czystowska, M.; Szczepanski, M.J.; Szajnik, M.; Quadrini, K.; Brandwein, H.; Hadden, J.W.; Whiteside, T.L. Mechanisms of T-cell protection from death by IRX-2: A new immunotherapeutic. *Cancer Immunol. Immunother.* **2011**, *60*, 495–506. [CrossRef]
51. Hoffmann, T.K.; Dworacki, G.; Tsukihiro, T.; Meidenbauer, N.; Gooding, W.; Johnson, J.T.; Whiteside, T.L. Spontaneous apoptosis of circulating T lymphocytes in patients with head and neck cancer and its clinical importance. *Clin. Cancer Res.* **2002**, *8*, 2553–2562.
52. Poggio, M.; Hu, T.; Pai, C.-C.; Chu, B.; Belair, C.D.; Chang, A.; Montabana, E.; Lang, U.E.; Fu, Q.; Fong, L.; et al. Suppression of Exosomal PD-L1 Induces Systemic Anti-tumor Immunity and Memory. *Cell* **2019**, *177*, 414–427.e13. [CrossRef]

53. Mrizak, D.; Martin, N.; Barjon, C.; Jimenez-Pailhes, A.-S.; Mustapha, R.; Niki, T.; Guigay, J.; Pancré, V.; de Launoit, Y.; Busson, P.; et al. Effect of nasopharyngeal carcinoma-derived exosomes on human regulatory T cells. *J. Natl. Cancer Inst.* **2015**, *107*, 363. [CrossRef]
54. Clayton, A.; Mitchell, J.P.; Court, J.; Linnane, S.; Mason, M.D.; Tabi, Z. Human tumor-derived exosomes down-modulate NKG2D expression. *J. Immunol.* **2008**, *180*, 7249–7258. [CrossRef]
55. Valenti, R.; Huber, V.; Iero, M.; Filipazzi, P.; Parmiani, G.; Rivoltini, L. Tumor-released microvesicles as vehicles of immunosuppression. *Cancer Res.* **2007**, *67*, 2912–2915. [CrossRef]
56. Torralba, D.; Baixauli, F.; Villarroya-Beltri, C.; Fernández-Delgado, I.; Latorre-Pellicer, A.; Acín-Pérez, R.; Martín-Cófreces, N.B.; Jaso-Tamame, Á.L.; Iborra, S.; Jorge, I.; et al. Priming of dendritic cells by DNA-containing extracellular vesicles from activated T cells through antigen-driven contacts. *Nat. Commun.* **2018**, *9*, 2658. [CrossRef] [PubMed]
57. Gabrilovich, D.I.; Nagaraj, S. Myeloid-derived suppressor cells as regulators of the immune system. *Nat. Rev. Immunol.* **2009**, *9*, 162–174. [CrossRef] [PubMed]
58. Altevogt, P.; Bretz, N.P.; Ridinger, J.; Utikal, J.; Umansky, V. Novel insights into exosome-induced, tumor-associated inflammation and immunomodulation. *Semin. Cancer Biol.* **2014**, *28*, 51–57. [CrossRef] [PubMed]
59. Xiang, X.; Poliakov, A.; Liu, C.; Liu, Y.; Deng, Z.B.; Wang, J.; Cheng, Z.; Shah, S.V.; Wang, G.J.; Zhang, L.; et al. Induction of myeloid-derived suppressor cells by tumor exosomes. *Int. J. Cancer* **2009**, *124*, 2621–2633. [CrossRef]
60. Filipazzi, P.; Bürdek, M.; Villa, A.; Rivoltini, L.; Huber, V. Recent advances on the role of tumor exosomes in immunosuppression and disease progression. *Semin. Cancer Biol.* **2012**, *22*, 342–349. [CrossRef]
61. Chennakrishnaiah, S.; Meehan, B.; D'Asti, E.; Montermini, L.; Lee, T.-H.; Karatzas, N.; Buchanan, M.; Tawil, N.; Choi, D.; Divangahi, M.; et al. Leukocytes as a reservoir of circulating oncogenic DNA and regulatory targets of tumor-derived extracellular vesicles. *J. Thromb. Haemost.* **2018**, *16*, 1800–1813. [CrossRef] [PubMed]
62. Ying, X.; Wu, Q.; Wu, X.; Zhu, Q.; Wang, X.; Jiang, L.; Chen, X.; Wang, X. Epithelial ovarian cancer-secreted exosomal miR-222-3p induces polarization of tumor-associated macrophages. *Oncotarget* **2016**, *7*, 43076–43087. [CrossRef] [PubMed]
63. Gabrusiewicz, K.; Li, X.; Wei, J.; Hashimoto, Y.; Marisetty, A.L.; Ott, M.; Wang, F.; Hawke, D.; Yu, J.; Healy, L.M.; et al. Glioblastoma stem cell-derived exosomes induce M2 macrophages and PD-L1 expression on human monocytes. *Oncoimmunology* **2018**, *7*, e1412909. [CrossRef] [PubMed]
64. Capello, M.; Vykoukal, J.V.; Katayama, H.; Bantis, L.E.; Wang, H.; Kundnani, D.L.; Aguilar-Bonavides, C.; Aguilar, M.; Tripathi, S.C.; Dhillon, D.S.; et al. Exosomes harbor B cell targets in pancreatic adenocarcinoma and exert decoy function against complement-mediated cytotoxicity. *Nat. Commun.* **2019**, *10*. [CrossRef]
65. Zhou, W.; Fong, M.Y.; Min, Y.; Somlo, G.; Liu, L.; Palomares, M.R.; Yu, Y.; Chow, A.; O'Connor, S.T.F.; Chin, A.R.; et al. Cancer-secreted miR-105 destroys vascular endothelial barriers to promote metastasis. *Cancer Cell* **2014**, *25*, 501–515. [CrossRef] [PubMed]
66. Yokoi, A.; Yoshioka, Y.; Yamamoto, Y.; Ishikawa, M.; Ikeda, S.-I.; Kato, T.; Kiyono, T.; Takeshita, F.; Kajiyama, H.; Kikkawa, F.; et al. Malignant extracellular vesicles carrying MMP1 mRNA facilitate peritoneal dissemination in ovarian cancer. *Nat. Commun.* **2017**, *8*, 14470. [CrossRef] [PubMed]
67. Zhao, H.; Yang, L.; Baddour, J.; Achreja, A.; Bernard, V.; Moss, T.; Marini, J.C.; Tudawe, T.; Seviour, E.G.; San Lucas, F.A.; et al. Tumor microenvironment derived exosomes pleiotropically modulate cancer cell metabolism. *Elife* **2016**, *5*, e10250. [CrossRef] [PubMed]
68. Luga, V.; Zhang, L.; Viloria-Petit, A.M.; Ogunjimi, A.A.; Inanlou, M.R.; Chiu, E.; Buchanan, M.; Hosein, A.N.; Basik, M.; Wrana, J.L. Exosomes mediate stromal mobilization of autocrine Wnt-PCP signaling in breast cancer cell migration. *Cell* **2012**, *151*, 1542–1556. [CrossRef]
69. Peinado, H.; Alečković, M.; Lavotshkin, S.; Matei, I.; Costa-Silva, B.; Moreno-Bueno, G.; Hergueta-Redondo, M.; Williams, C.; García-Santos, G.; Ghajar, C.M.; et al. Melanoma exosomes educate bone marrow progenitor cells toward a pro-metastatic phenotype through MET. *Nat. Med.* **2012**, *18*, 883–891. [CrossRef]
70. Becker, A.; Thakur, B.K.; Weiss, J.M.; Kim, H.S.; Peinado, H.; Lyden, D. Extracellular Vesicles in Cancer: Cell-to-Cell Mediators of Metastasis. *Cancer Cell* **2016**, *30*, 836–848. [CrossRef]
71. Costa-Silva, B.; Aiello, N.M.; Ocean, A.J.; Singh, S.; Zhang, H.; Thakur, B.K.; Becker, A.; Hoshino, A.; Mark, M.T.; Molina, H.; et al. Pancreatic cancer exosomes initiate pre-metastatic niche formation in the liver. *Nat. Cell Biol.* **2015**, *17*, 816–826. [CrossRef]
72. Plebanek, M.P.; Angeloni, N.L.; Vinokour, E.; Li, J.; Henkin, A.; Martinez-Marin, D.; Filleur, S.; Bhowmick, R.; Henkin, J.; Miller, S.D.; et al. Pre-metastatic cancer exosomes induce immune surveillance by patrolling monocytes at the metastatic niche. *Nat. Commun.* **2017**, *8*, 1319. [CrossRef]
73. Webber, J.; Steadman, R.; Mason, M.D.; Tabi, Z.; Clayton, A. Cancer exosomes trigger fibroblast to myofibroblast differentiation. *Cancer Res.* **2010**, *70*, 9621–9630. [CrossRef]
74. Sansone, P.; Savini, C.; Kurelac, I.; Chang, Q.; Amato, L.B.; Strillacci, A.; Stepanova, A.; Iommarini, L.; Mastroleo, C.; Daly, L.; et al. Packaging and transfer of mitochondrial DNA via exosomes regulate escape from dormancy in hormonal therapy-resistant breast cancer. *Proc. Natl. Acad. Sci. USA* **2017**, *114*, E9066–E9075. [CrossRef]
75. Amintas, S.; Vendrely, V.; Dupin, C.; Buscail, L.; Laurent, C.; Bournet, B.; Merlio, J.-P.; Bedel, A.; Moreau-Gaudry, F.; Boutin, J.; et al. Next-Generation Cancer Biomarkers: Extracellular Vesicle DNA as a Circulating Surrogate of Tumor DNA. *Front. Cell Dev. Biol.* **2021**, *8*, 1–9. [CrossRef]

76. Zhang, P.; Zhou, X.; He, M.; Shang, Y.; Tetlow, A.L.; Godwin, A.K.; Zeng, Y. Ultrasensitive detection of circulating exosomes with a 3D-nanopatterned microfluidic chip. *Nat. Biomed. Eng.* **2019**, *3*, 438–451. [CrossRef] [PubMed]
77. Iliescu, F.; Vrtačnik, D.; Neuzil, P.; Iliescu, C. Microfluidic Technology for Clinical Applications of Exosomes. *Micromachines* **2019**, *10*, 392. [CrossRef]
78. Zhang, P.; Zhou, X.; Zeng, Y. Multiplexed immunophenotyping of circulating exosomes on nano-engineered ExoProfile chip towards early diagnosis of cancer. *Chem. Sci.* **2019**, *10*, 5495–5504. [CrossRef]
79. Zhu, Q.; Heon, M.; Zhao, Z.; He, M. Microfluidic engineering of exosomes: Editing cellular messages for precision therapeutics. *Lab Chip* **2018**, *18*, 1690–1703. [CrossRef] [PubMed]
80. Tayebi, M.; Zhou, Y.; Tripathi, P.; Chandramohanadas, R.; Ai, Y. Exosome Purification and Analysis Using a Facile Microfluidic Hydrodynamic Trapping Device. *Anal. Chem.* **2020**, *92*, 10733–10742. [CrossRef] [PubMed]
81. Shen, M.; Di, K.; He, H.; Xia, Y.; Xie, H.; Huang, R.; Liu, C.; Yang, M.; Zheng, S.; He, N.; et al. Progress in exosome associated tumor markers and their detection methods. *Mol. Biomed.* **2020**, *1*, 1–25. [CrossRef]
82. Shin, H.; Oh, S.; Hong, S.; Kang, M.; Kang, D.; Ji, Y.G.; Choi, B.H.; Kang, K.W.; Jeong, H.; Park, Y.; et al. Early-Stage Lung Cancer Diagnosis by Deep Learning-Based Spectroscopic Analysis of Circulating Exosomes. *ACS Nano* **2020**, *14*, 5435–5444. [CrossRef]
83. Théry, C.; Witwer, K.W.; Aikawa, E.; Alcaraz, M.J.; Anderson, J.D.; Andriantsitohaina, R.; Antoniou, A.; Arab, T.; Archer, F.; Atkin-Smith, G.K.; et al. Minimal information for studies of extracellular vesicles 2018 (MISEV2018): A position statement of the International Society for Extracellular Vesicles and update of the MISEV2014 guidelines. *J. Extracell. Vesicles* **2018**, *7*. [CrossRef] [PubMed]
84. Li, M.; Zou, X.; Xia, T.; Wang, T.; Liu, P.; Zhou, X.; Wang, S.; Zhu, W. A five-miRNA panel in plasma was identified for breast cancer diagnosis. *Cancer Med.* **2019**, *8*, 7006–7017. [CrossRef] [PubMed]
85. Cui, Y.; Xu, H.-F.; Liu, M.-Y.; Xu, Y.-J.; He, J.-C.; Zhou, Y.; Cang, S.-D. Mechanism of exosomal microRNA-224 in development of hepatocellular carcinoma and its diagnostic and prognostic value. *World J. Gastroenterol.* **2019**, *25*, 1890–1898. [CrossRef]
86. Sun, S.; Chen, H.; Xu, C.; Zhang, Y.; Zhang, Q.; Chen, L.; Ding, Q.; Deng, Z. Exosomal miR-106b serves as a novel marker for lung cancer and promotes cancer metastasis via targeting PTEN. *Life Sci.* **2020**, *244*, 117297. [CrossRef] [PubMed]
87. Wang, X.; Jiang, X.; Li, J.; Wang, J.; Binang, H.; Shi, S.; Duan, W.; Zhao, Y.; Zhang, Y. Serum exosomal miR-1269a serves as a diagnostic marker and plays an oncogenic role in non-small cell lung cancer. *Thorac. Cancer* **2020**, *11*, 3436–3447. [CrossRef]
88. Su, Y.Y.; Sun, L.; Guo, Z.R.; Li, J.C.; Bai, T.T.; Cai, X.X.; Li, W.H.; Zhu, Y.F. Upregulated expression of serum exosomal miR-375 and miR-1307 enhance the diagnostic power of CA125 for ovarian cancer. *J. Ovarian Res.* **2019**, *12*, 6. [CrossRef]
89. Zhang, S.; Du, L.; Wang, L.; Jiang, X.; Zhan, Y.; Li, J.; Yan, K.; Duan, W.; Zhao, Y.; Wang, L.; et al. Evaluation of serum exosomal LncRNA-based biomarker panel for diagnosis and recurrence prediction of bladder cancer. *J. Cell. Mol. Med.* **2019**, *23*, 1396–1405. [CrossRef]
90. Zhang, R.; Xia, Y.; Wang, Z.; Zheng, J.; Chen, Y.; Li, X.; Wang, Y.; Ming, H. Serum long non coding RNA MALAT-1 protected by exosomes is up-regulated and promotes cell proliferation and migration in non-small cell lung cancer. *Biochem. Biophys. Res. Commun.* **2017**, *490*, 406–414. [CrossRef]
91. Hao, Y.-X.; Li, Y.-M.; Ye, M.; Guo, Y.-Y.; Li, Q.-W.; Peng, X.-M.; Wang, Q.; Zhang, S.-F.; Zhao, H.-X.; Zhang, H.; et al. KRAS and BRAF mutations in serum exosomes from patients with colorectal cancer in a Chinese population. *Oncol. Lett.* **2017**, *13*, 3608–3616. [CrossRef]
92. García-Romero, N.; Carrión-Navarro, J.; Esteban-Rubio, S.; Lázaro-Ibáñez, E.; Peris-Celda, M.; Alonso, M.M.; Guzmán-De-Villoria, J.; Fernández-Carballal, C.; de Mendivil, A.O.; García-Duque, S.; et al. DNA sequences within glioma-derived extracellular vesicles can cross the intact blood-brain barrier and be detected in peripheral blood of patients. *Oncotarget* **2017**, *8*, 1416–1428. [CrossRef]
93. Castellanos-Rizaldos, E.; Grimm, D.G.; Tadigotla, V.; Hurley, J.; Healy, J.; Neal, P.L.; Sher, M.; Venkatesan, R.; Karlovich, C.; Raponi, M.; et al. Exosome-Based Detection of EGFR T790M in Plasma from Non-Small Cell Lung Cancer Patients. *Clin. Cancer Res.* **2018**, *24*, 2944–2950. [CrossRef] [PubMed]
94. Krug, A.K.; Enderle, D.; Karlovich, C.; Priewasser, T.; Bentink, S.; Spiel, A.; Brinkmann, K.; Emenegger, J.; Grimm, D.G.; Castellanos-Rizaldos, E.; et al. Improved EGFR mutation detection using combined exosomal RNA and circulating tumor DNA in NSCLC patient plasma. *Ann. Oncol. Off. J. Eur. Soc. Med. Oncol.* **2018**, *29*, 700–706. [CrossRef] [PubMed]
95. Park, J.; Lee, C.; Eom, J.S.; Kim, M.-H.; Cho, Y.-K. Detection of EGFR Mutations Using Bronchial Washing-Derived Extracellular Vesicles in Patients with Non-Small-Cell Lung Carcinoma. *Cancers* **2020**, *12*, 2822. [CrossRef]
96. Hur, J.Y.; Kim, H.J.; Lee, J.S.; Choi, C.-M.; Lee, J.C.; Jung, M.K.; Pack, C.G.; Lee, K.Y. Extracellular vesicle-derived DNA for performing EGFR genotyping of NSCLC patients. *Mol. Cancer* **2018**, *17*, 15. [CrossRef]
97. Zocco, D.; Bernardi, S.; Novelli, M.; Astrua, C.; Fava, P.; Zarovni, N.; Carpi, F.M.; Bianciardi, L.; Malavenda, O.; Quaglino, P.; et al. Isolation of extracellular vesicles improves the detection of mutant DNA from plasma of metastatic melanoma patients. *Sci. Rep.* **2020**, *10*, 15745. [CrossRef] [PubMed]
98. Allenson, K.; Castillo, J.; San Lucas, F.A.; Scelo, G.; Kim, D.U.; Bernard, V.; Davis, G.; Kumar, T.; Katz, M.; Overman, M.J.; et al. High prevalence of mutant KRAS in circulating exosome-derived DNA from early-stage pancreatic cancer patients. *Ann. Oncol. Off. J. Eur. Soc. Med. Oncol.* **2017**, *28*, 741–747. [CrossRef]

99. Bernard, V.; Kim, D.U.; San Lucas, F.A.; Castillo, J.; Allenson, K.; Mulu, F.C.; Stephens, B.M.; Huang, J.; Semaan, A.; Guerrero, P.A.; et al. Circulating Nucleic Acids Are Associated with Outcomes of Patients with Pancreatic Cancer. *Gastroenterology* **2019**, *156*, 108–118.e4. [CrossRef]
100. Foroni, C.; Zarovni, N.; Bianciardi, L.; Bernardi, S.; Triggiani, L.; Zocco, D.; Venturella, M.; Chiesi, A.; Valcamonico, F.; Berruti, A. When Less Is More: Specific Capture and Analysis of Tumor Exosomes in Plasma Increases the Sensitivity of Liquid Biopsy for Comprehensive Detection of Multiple Androgen Receptor Phenotypes in Advanced Prostate Cancer Patients. *Biomedicines* **2020**, *8*, 131. [CrossRef]
101. Vagner, T.; Spinelli, C.; Minciacchi, V.R.; Balaj, L.; Zandian, M.; Conley, A.; Zijlstra, A.; Freeman, M.R.; Demichelis, F.; De, S.; et al. Large extracellular vesicles carry most of the tumour DNA circulating in prostate cancer patient plasma. *J. Extracell. Vesicles* **2018**, *7*, 1505403. [CrossRef] [PubMed]
102. Cordonnier, M.; Nardin, C.; Chanteloup, G.; Derangere, V.; Algros, M.; Arnould, L.; Garrido, C.; Aubin, F.; Gobbo, J. Tracking the evolution of circulating exosomal-PD-L1 to monitor melanoma patients. *J. Extracell. Vesicles* **2020**, *9*, 1710899. [CrossRef]
103. Global Liquid Biopsy Market Outlook to 2020. Available online: https://www.prnewswire.com/news-releases/global-liquid-biopsy-market-outlook-to-2020-300220260.html (accessed on 27 April 2021).
104. FDA Approves Blood Tests That Can Help Guide Cancer Treatment. Available online: https://www.cancer.gov/news-events/cancer-currents-blog/2020/fda-guardant-360-foundation-one-cancer-liquid-biopsy (accessed on 27 April 2021).
105. Cancer "Liquid Biopsy" Blood Test Gets Expanded FDA Approval. Available online: https://www.cancer.gov/news-events/cancer-currents-blog/2020/fda-foundation-one-cancer-liquid-biopsy-expanded-approval (accessed on 27 April 2021).
106. Teijeira, Á.; Garasa, S.; Gato, M.; Alfaro, C.; Migueliz, I.; Cirella, A.; de Andrea, C.; Ochoa, M.C.; Otano, I.; Etxeberria, I.; et al. CXCR1 and CXCR2 Chemokine Receptor Agonists Produced by Tumors Induce Neutrophil Extracellular Traps that Interfere with Immune Cytotoxicity. *Immunity* **2020**, *52*, 856–871.e8. [CrossRef] [PubMed]
107. Fernando, M.R.; Jiang, C.; Krzyzanowski, G.D.; Ryan, W.L. New evidence that a large proportion of human blood plasma cell-free DNA is localized in exosomes. *PLoS ONE* **2017**, *12*, e0183915. [CrossRef] [PubMed]
108. Dlamini, Z.; Francies, F.Z.; Hull, R.; Marima, R. Artificial intelligence (AI) and big data in cancer and precision oncology. *Comput. Struct. Biotechnol. J.* **2020**, *18*, 2300–2311. [CrossRef] [PubMed]

Review

Serum Exosomes and Their miRNA Load—A Potential Biomarker of Lung Cancer

Mateusz Smolarz and Piotr Widlak *

Maria Skłodowska-Curie National Research Institute of Oncology, Gliwice Branch, 44-101 Gliwice, Poland; mateusz.smolarz@io.gliwice.pl
* Correspondence: piotr.widlak@io.gliwice.pl; Tel.: +48-32-2789672

Simple Summary: Exosomes are an emerging source of cancer biomarkers. Molecular components of serum-derived exosomes have been addressed in several reports in the context of biomarkers for early detection of lung cancer. However, despite the promising results of pilot studies, the clinical applicability of such biomarkers has not been validated yet. In this review, the diagnostic potential of miRNA content of serum-derived exosomes is presented. Moreover, potential target genes and signaling pathways affected by miRNA present in lung cancer signatures are discussed.

Abstract: Early detection of lung cancer in screening programs is a rational way to reduce mortality associated with this malignancy. Low-dose computed tomography, a diagnostic tool used in lung cancer screening, generates a relatively large number of false-positive results, and its complementation with molecular biomarkers would greatly improve the effectiveness of such programs. Several biomarkers of lung cancer based on different components of blood, including miRNA signatures, were proposed. However, only a few of them have been positively validated in the context of early cancer detection yet, which imposes a constant need for new biomarker candidates. An emerging source of cancer biomarkers are exosomes and other types of extracellular vesicles circulating in body fluids. Hence, different molecular components of serum/plasma-derived exosomes were tested and showed different levels in lung cancer patients and healthy individuals. Several studies focused on the miRNA component of these vesicles. Proposed signatures of exosome miRNA had promising diagnostic value, though none of them have yet been clinically validated. These signatures involved a few dozen miRNA species overall, including a few species that recurred in different signatures. It is worth noting that all these miRNA species have cancer-related functions and have been associated with lung cancer progression. Moreover, a few of them, including known oncomirs miR-17, miR-19, miR-21, and miR-221, appeared in multiple miRNA signatures of lung cancer based on both the whole serum/plasma and serum/plasma-derived exosomes.

Keywords: biomarkers; exosome; extracellular vesicles; lung cancer; miRNA; plasma; serum

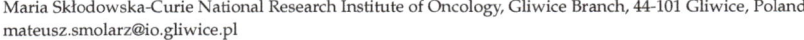

1. Introduction

Lung cancer is among the major cancer-related public health problem responsible for about a quarter of cancer-related deaths worldwide. Overall, the lung cancer five-year survival rate (below 20%) is much lower than other leading cancer sites, such as colorectal (about 65%), breast (about 90%), and prostate (about 95%). Though the risk and incidence of lung cancer are slightly higher among men, this malignancy is becoming the major cause of cancer-related death also in women. The majority of lung cancer cases are diagnosed at advanced stages and have unfavorable prognoses (the average five-year survival of about 10–15%). However, in the case of the disease detected at the early stages, the prognosis is much better (the average five-year survival varies between 65 and 85%). Thus, in addition to primary prevention (i.e., tobacco smoking control), screening for early detection was proposed as a promising strategy to reduce lung cancer mortality [1,2]. Several screening tools have been investigated during the past decades, but only one, the low-dose computed tomography (LD-CT), has found an application in clinical practice. Originally, the results of the National Lung Screening Trial (NLST) showed that compared to chest X-ray examination, the LD-CT screening was associated with over 20% reduction of lung cancer-specific mortality in a high-risk group of subjects defined by their smoking status and age [3]. The potential of LD-CT screening programs to reduce lung cancer mortality was further confirmed by other studies [2], including the Dutch–Belgian NELSON trial [4] and the Danish Lung Cancer Screening Trial (DLCST) [5]. It is estimated that the use of LD-CT allows for earlier detection of lung cancers in about 12,000 people a year, which is about 8% of deaths annually due to this disease. It is worth noting, however, that LD-CT allows detecting abnormalities in 20–40% of people undergoing this examination, but as much as 95% of results could be false-positive [6]. Hence, due to the low specificity of LD-CT (positive predictive value of only 3.8% in the NLST), the vast majority of patients with screen-detected chest abnormalities are subjected to further expensive and potentially harmful diagnostic procedures, such as transthoracic or bronchoscopic biopsy or surgery. It is estimated that about 75% of patients unnecessarily underwent diagnostic workup, including 25% subjected to invasive procedures [7]. Hence, there is an urgent need for clinical and molecular tests supporting CT-based screening for the detection of lung cancer to reduce "over-diagnosis" and decrease the costs. Such test(s) could either pre-select individuals for LD-CT examination or discriminate between benign and malignant chest abnormalities detected by LD-CT [8,9].

Potential biomarkers for early lung cancer can be found in various biological fluids; however, blood is the richest and most readily available source [10,11]. Candidates for such biomarkers include serum proteins, free nucleic acids, and metabolites [11,12]. Several works reported serum/plasma proteins, which levels are associated with the risk of lung cancer [13]. Another candidate for the biomarker of lung cancer is circulating free DNA (cfDNA) [14] and circulating tumor cells (CTC) [15]. More recently, serum metabolites and lipids have emerged as another class of potential biomarkers in lung cancer [16,17]. Several other review papers could be suggested that cover this well-researched field [11–13,18–21]. However, though numerous biomarker candidates have been proposed only a few of them have been positively validated in the proper clinical settings. The main reason was the lack of sensitivity and analytical reproducibility, which in turn led to the elimination of potential candidates from further stages of biomarker testing [9,12]. Moreover, none of the tested biomarkers increased the actual number of detected early lung cancer cases yet [18,20,22]. Currently, only two molecular tests are used in clinical practice to help in the diagnosis of indeterminate pulmonary nodules detected by CT. One of them is the autoantigen-based EarlyCDT-Lung test, which enables the classification of indeterminate nodules with a positive predictive value (PPV) >70% [23]. Another test is the XL2 test, which combines the clinical probability of cancer score with the level of two plasma proteins: LG3BP and C163A [24]. Hence, the identification of the reliable molecular biomarker that could be used for the early detection of lung cancer remains a timely and vital issue.

The purpose of this literature review is to summarize current data on the emerging biomarker of early lung cancer-circulating serum exosomes and their microRNA cargo.

2. Micro RNA Signatures of Lung Cancer

In the search for a lung cancer biomarker, there were numerous studies focused on microRNAs (miRNAs). It is a class of small endogenous non-coding RNAs of 18–24 nucleotides responsible for the regulation of target genes. More than 2500 mature miRNAs have been described in humans yet [25–27]. miRNA is transcribed in the cell nucleus with the participation of RNA polymerase II resulting in pri-miRNA, which is processed by the Drosh/DGCR8 enzyme complex to precursor miRNA (pre-miRNA). The resulting pre-miRNA is transported from the nucleus to the cytoplasm involving Exportin-5, where it is processed by Dicer nuclease to form miRNA duplexes or mature miRNA. Usually, a less-thermostable 5′-terminus strand is packed to the protein complex (RISC), whose main component is a protein from the Argonaut family (AGO), while the second strand is degraded. The RISC complex then recognizes the target mRNA and binds at the 3'UTR position: mRNA degradation occurs in the case of perfect miRNA/mRNA matching, while translation repression in the case of incomplete alignment. Thus, by silencing target mRNAs, miRNAs affect many critical cellular processes such as cell proliferation, apoptosis, differentiation, and metabolism [27,28].

The composition of miRNA component of tissues (so-called miR-ome) could be affected by different pathological conditions; hence, the diagnostic and prognostic values of miRNA signatures have been addressed in many studies [29–34]. miRNA is resistant to RNase digestion, boiling, extended storage, extreme pH, and multiple freezing and thawing cycles [35]. Moreover, miRNA is considered to be more stable than other classes of RNA in blood and other biofluids. However, it should be noted that during the analysis of free circulating miRNA in human blood, miRNA molecules released by cancer cells and other classes of "normal" cells (platelets, red blood cells, and endothelial cells) are co-purified and co-analyzed [36]. Nevertheless, miRNA circulating in the blood and present in the isolated serum (i.e., the liquid fraction of blood remaining after removal of the clot followed coagulation) or plasma (i.e., the liquid fraction of blood remaining after removal of cell components without coagulation), is an emerging source of disease biomarkers including lung cancer.

Several studies addressed circulating miRNA as potential molecular signatures to be used for the diagnosis of lung cancer. Numerous papers have been published since 2011 that described signatures of serum/plasma miRNA, which enabled the differentiation between lung cancer patients and healthy individuals. Some of these reports described single miRNA, yet most of them proposed multi-component panels up to 24 plasma miRNAs [37] or 34 serum miRNAs [38]. Examples of such studies are listed in Table 1. Proposed lung cancer signatures involved about 100 miRNA species overall, which (according to our literature review) included 39 miRNA species that recurred in more than one signature. However, only four miRNA species were included in more than five signatures, namely, miR-21 (11 signatures), miR-148b (8 signatures), miR-126, and miR-486–5p (seven signatures). Hence, the overlap among different signatures was relatively low, which putatively reflected different clinical characteristics of lung cancer patients and their ethnic/genetic backgrounds as well as different analytical approaches used in different studies. Nevertheless, we analyzed a subset of 39 miRNA species that appeared in multiple lung cancer signatures in the search for their target genes and associated biological functions; the bioinformatics tool miRSystem (version 20160513) was used [39]. Among the biological processes associated with this subset of miRNAs and statistically overrepresented, several pathways were involved in cancer development, including the MAPK signaling, FGFR signaling, transport of glucose, apoptosis, and antigen processing/presentation. This subset included several known "oncomirs", exemplified by miR-21, which will be discussed in detail below. Furthermore, among the genes hypothetically targeted by the highest number of miRs from this subset were a few genes with putative cancer-related functions, exemplified by *IFI30*, *PLA2G10*, *FGF6*, *ZBTB16*, and *CORO1A*. *IFI30* encodes a lysosomal thiol reductase involved in the processing of MHC class II-restricted antigen, which was reported in the development of melanoma [40]. *PLA2G10* encodes a phospholipase A2 family member involved in the production of inflammatory lipid mediators (e.g., prostaglandins), which was reported in the progression of breast cancer [41]. *FGF6* encodes a fibroblast growth factor (FGF) family member involved in tumor growth [42]. *ZBTB16* encodes a Krueppel C2H2 zinc finger family member involved in the regulation of cell cycle, apoptosis, and the AKT/Foxo3a pathway [43]. *CORO1A* encodes a WD-repeat protein family member involved in the cell cycle progression, apoptosis, and signal transduction [44]. Hence, cancer-related functions of miRNA species present in the proposed lung cancer signatures provide additional validation of their putative diagnostic importance.

In conclusion, circulating miRNA appears a forward-looking diagnostic tool in the detection of lung cancer. Proposed signatures revealed promising sensitivity and specificity, which usually reached 80–90%. Still, their actual diagnostic reproducibility requires further validation and clinical testing [25,35,45–47]. Further, none of the proposed miRNA signatures have yet been conclusively validated in the prospective clinical studies. Nevertheless, three registered clinical trials are currently ongoing that include validation of the serum/plasma miRNA signatures of early lung cancer. The BIOMILD study (NCT02247453) sponsored by the Fondazione IRCCS Istituto Nazionale dei Tumori (Milano) is aimed at the validation of the Plasma miR Signature Classifier [37]. The COSMOS study (NCT01248806) sponsored by the European Institute of Oncology involves validation of the miR-Test [48] in the context of lung cancer screening. Moreover, a smaller study sponsored by Hummingbird Diagnostics (NCT03452514) is aimed at the validation of the commercial HMBDx microRNA Test in a group of participants of the LD-CT lung cancer screening. However, all these clinical trials are still running, and no conclusions are available yet (the planned completion date of these studies is 2021).

Table 1. Examples of serum/plasma miRNAs as biomarkers of lung cancer.

Biofluid	miRNA Signature	Size of Groups	Diagnostic Value	Reference
Plasma	miR-21, miR-126, miR-210, miR-486	Control: 29 Cases: 29 (Stage I–IV)	AUC = 0.86 SEN = 75% SPE = 85%	[49]
Plasma	miR-21, miR-335	Control: 38 Cases: 36 (Stage I)	AUC = 0.86 SEN = 72% SPE = 81%	[50]
Plasma	miR-21, miR-486	Control: 46 Cases: 54 (Stage I–III)	AUC = 0.90 SEN = 87% SPE = 87%	[51]
Plasma	miR-21, miR-145, miR-155	Control: 92 Cases: 96 (Stage I–IV)	AUC = 0.85 SEN = 69% SPE = 78%	[52]
Plasma	miR-101, miR-106a, miR-126, miR-133a, miR-140-3p, miR-140-5p, miR-142-3p, miR-145, miR-148a, miR-15b, miR-16, miR-17, miR-197, miR-19b, miR-21, miR-221, miR-28-3p, miR-30b, miR-30c, miR-320, miR-451, miR-486-5p, miR-660, and miR-92a (Plasma miR Signature Classifier; MSC)	Control: 870 Cases: 69 (Stage I–III)	SEN = 87% SPE = 81%	[37]
Plasma	miR-182, miR-183, miR-210, miR-126	Control: 40 Cases: 112 (Stage I–III)	AUC = 0.97 SEN = 81% SPE = 100%	[53]
Plasma	miR-145, miR-20a, miR-21, miR-223	Control: 83 Cases: 129 (Stage I–II)	AUC = 0.90 SEN = 82% SPE = 90%	[54]
Plasma	miR-19b, miR-21, miR-221, miR-409, miR-425, miR-584	Control: 124 Cases: 141 (Stage I–IV)	AUC = 0.84 SEN = 73% SPE = 80%	[55]
Serum	miR-92, miR-484, miR-486, miR-328, miR-191, miR-376a, miR-342, miR-331, miR-30c, miR-28, miR-98, miR-17, miR-26b, miR-374, miR-30b, miR-26a, miR-142, miR-103, miR-126, let-7a, let-7d, let-7b, miR-32, miR-133b, miR-566, miR-432, miR-223, miR-29a, miR-148a, miR-142, miR-22, miR-148b, miR-140, miR-139	Control: 69 Cases: 95 (Stage I–IV)	AUC = 0.89 SEN = 71% SPE = 90%	[38]
Serum	miR-15b, miR-27b	Control: 95 Cases: 85 (Stage I–IV)	AUC = 0.98 SEN = 100% SPE = 84%	[56]

Table 1. Cont.

Biofluid	miRNA Signature	Size of Groups	Diagnostic Value	Reference
Serum	miR-92a-3p, miR-30b-5p, miR-191-5p, miR-484, miR-328-3p, miR-30c-5p, miR-374a-5p, let-7d-5p, miR-331-3p, miR-29a-3p, miR-148a-3p, miR-223-3p, miR-140-5p (miR-Test)	Control: 984 Cases: 48 (Stage I–III)	AUC = 0.85 SEN = 72% SPE = 77%	[48]
Serum	miR-193b, miR-301, miR-141, miR-200b	Control: 45 Cases: 154 (Stage I–III)	AUC = 0.99 SEN = 97% SPE = 96%	[57]
Serum	miR-483, miR-193a, miR-25, miR-214, miR-7	Control: 63 Cases: 63 (Stage I–IV)	AUC = 0.82 SEN = 89% SPE = 68%	[58]
Serum	miR-152, miR-148a, miR-148b, miR-21	Control: 70 Cases: 70 (Stage I–IV)	AUC = 0.97 SEN = 96% SPE = 91%	[59]
Serum	miR-15b, miR-16, miR-20a	Control: 58 Cases: 94 (Stage I–III)	AUC = 0.93 SEN = 86% SPE = 91%	[60]
Serum	miR-429, miR-205, miR-200b, miR-203, miR-12, miR-34b	Control: 74 Cases: 138 (Stage I–IV)	AUC = 0.89 SEN = 88% SPE = 71%	[61]
Serum	miR-141, miR-193b, miR200b, miR-301	Control: 185 Cases: 213 (Stage I–IV)	AUC = 0.92 SEN = 91% SPE = 78%	[62]
Serum	miR-1268b, miR-6075	Control: 2178 Cases: 1566 (Stage I–IV)	AUC = 0.99 SEN = 99% SPE = 99%	[63]

AUC—Area Under the Receiver Operating Characteristic (ROC) Curve; SEN—Sensitivity; SPE—Specificity.

3. Exosomes, an Emerging Type of Liquid Biopsy

Exosomes are membrane-enclosed nanovesicles (30–150 nm) of endosomal origin. Exosomes arise as a result of the concavity of the plasma membrane inward, resulting in the formation of an early endosome. The early endosome matures into the late endosome, which then transforms into a multivesicular body (MVB) that could attach to the plasma membrane from inside and release exosomes into the extracellular space [64,65] (Figure 1). Exosomes can be detected in various biological fluids such as urine, cerebrospinal fluid, saliva, blood, and its derivatives (serum and plasma). Exosomes are secreted by all types of cells, either non-tumorigenic and cancerous. These vesicles are enclosed by a double film of symmetrically distributed lipids containing several tetraspanins and other membrane proteins involved in the formation of MVB (CD9, CD63, CD81, TSG101, and Alix). However, the full set of proteins present in the exosome cargo (involving thousands of different cellular proteins) is variable and reflects the current phenotype of the parent cell. Except for proteins and lipids, exosomes also contain different classes of nucleic acids (single-stranded RNA, long non-coding RNA, and microRNA) and metabolites, whose composition is also regulated by the state of the cell [64,66,67].

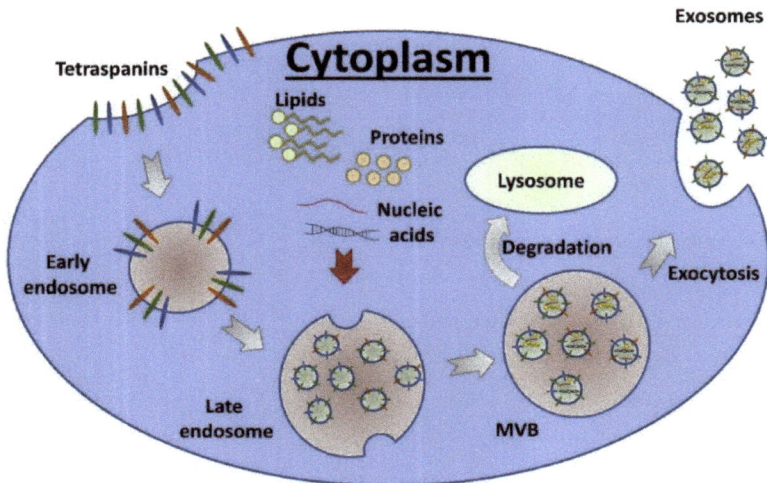

Figure 1. Biogenesis of exosomes.

In general, exosomes are involved in many aspects of cell-to-cell communication working in both paracrine and endocrine modes. In the case of exosomes from "normal" (non-tumorogenic) cells, their role in immunity, coagulation, angiogenesis, spermatogenesis, and various physiological processes in the central nervous system has been confirmed. In the case of tumor-derived exosomes (TEX), several lines of evidence indicate their association with immunomodulation, pre-metastatic niche formation, tumor growth, resistance to the treatment, and drug removal from cells [68]. TEX are signal mediators and promote disease development by participating in processes such as angiogenesis, metastasis, and many others [66,68–70]. TEX are released into the bloodstream so they can reach distant organs and modify the phenotype of many different cell types. This ability of TEX depends on their bioactive cargo, which differs from the content of exosomes released by "normal" cells and corresponds to the malignant phenotype of cancer cells [71]. Several review papers focused on the functional role of TEX have already been published, including a few recent ones [68,70,72,73].

Exosomes released by lung cancer cells were reported to be involved in tumor promotion, immunomodulation, and remodeling of the tumor microenvironment, also in the context of metastatic niche [66,69]. TEX secreted by lung cancer cells contain several proteins involved in tumor development, including CD91, Galectin-9, LRG1, EGFR, and Wnt5b [53,70,73–76]. Several studies also addressed the functional importance of non-coding RNA present in TEX released by lung cancer cells. For example, miR-103a present in TEX directly affected the polarization of macrophages by reducing PTEN protein expression, which in turn led to the accumulation of tumor-promoting factors such as IL10, CCL2, and VEGF-A [70,77]. Moreover, miR-21 present in TEX promoted tumor growth by increasing the permeability of blood vessels and the accumulation of hypoxia-induced factor-1α (HIF-1α) under both normoxic and hypoxic conditions [78]. Other miRNAs present in TEX secreted by lung cancer cells (e.g., miR-9, miR-126, miR-122, and miR-210) could also participate in the process of angiogenesis of neoplastic blood vessels [73,74,79–82]. Long non-coding RNAs (lncRNAs) are another group of nucleic acids present in TEX secreted from lung cancer cells. It has been reported that several such lncRNAs (MALAT1, AK126698, SCAL1, and HOTAIR) are associated with the anti-apoptotic activity, resistance to cisplatin, protection of cells against oxidative stress, and increased migration proliferation and invasiveness [74,79,83,84].

The molecular composition of TEX reflects that of parental cancer cells. Therefore, TEX present in blood and other biofluids are an emerging type of liquid biopsy, considered a gold mine of potential cancer markers [26,72,85–87]. It should be emphasized, however, that exosomes represent only a subset of the heterogeneous group of extracellular vesicles (EV) that also include microvesicles (also known as ectosomes; 250–1000 nm) and apoptotic bodies (>1000 nm) formed by outward budding ("blebbing") of the plasma membrane. The term "exosomes" should be reserved for vesicles of endosomal origin that form via MVB. However, due to the limitations of current methods used for the isolation of EV the adequate discrimination between various EV subsets is not feasible. Therefore, to avoid possible misconceptions, a simplified nomenclature has been recently proposed that distinguishes small EV (i.e., <200 nm) and medium/large EV (>200 nm). A class of small EV (sEV) consists mostly of exosomes, yet other types of EV, e.g., small microvesicles, could also copurify with this fraction [88]; in this review, the terms "exosome" and "sEV" are used interchangeably for simplicity. Moreover, sEV present in blood and other biofluids represent a complex mixture of vesicles released by different types of cells. It is estimated that TEX represent about 20–60% of sEV present in the plasma of cancer patients while the remaining exosomes and other sEV present in this specimen are released by "normal" non-cancerous types of cells (e.g., platelets, immune cells, and endothelial cells) [89]. However, due to current limitations of methods allowing purification of specific TEX from body fluids [90], the mixture of different sEV that could be isolated from serum or plasma remains a feasible material in the search of cancer markers. Nevertheless, even such heterogeneous material is a promising source of biomarkers for the detection of lung cancer, which is discussed below.

4. Serum Exosomes as Potential Lung Cancer Biomarkers

Exosomes are secreted by various cells. However, the concentration of exosomes is much higher in the blood of cancer patients, including lung cancer, compared to healthy individuals. Recent reports indicate that the concentration of vesicles in the blood of cancer patients may reach 10^9 vesicles/mL of blood [71]. The above observations have been confirmed in many types of cancers, including prostate cancer, ovarian cancer, breast cancer, pancreatic ductal adenocarcinoma, hepatocellular carcinoma, and breast cancer [91–95]. Increased levels of vesicles in the blood of cancer patients correlate with a worse prognosis. The molecular cargo of exosomes is the primary source of cancer biomarkers. However, apart from a different molecular cargo, TEX may have a different morphology than exosomes secreted by "normal" cells. Exosomes isolated from the serum of patients diagnosed with pancreatic cancer had a significantly smaller size compared to exosomes isolated from healthy people [91]. Similar observations were made with the use of atomic force microscopy in the case of exosomes present in patients with oral cancer [96]. Hence, the number, composition, and morphology of exosomes can be an important diagnostic cancer biomarker, though no specific data regarding lung cancer patients is available yet.

Different molecular components of exosomes existing in body fluids (serum, plasma, and saliva) of patients with lung cancer have been tested in the search for a biomarker of this malignancy [85,97–102]. Identified biomarker candidates include different classes of molecules-nucleic acids, proteins, and metabolites. Results of these studies (except for exosome miRNA discussed in the subsequent paragraph) are listed in Table 2. A few signatures of lung cancer have been proposed based on proteins present in serum/plasma-derived exosomes [86,103–107]. Moreover, several studies have proposed long non-coding RNAs and circular RNAs present in serum-derived exosomes as lung cancer biomarkers [84,108–111]. Furthermore, different levels of several phospholipids (phosphatidylcholines and sphingomyelins), triglycerides, and cholesterol esters present in the exosome membrane have been observed in plasma-derived exosomes in lung cancer patients and healthy controls [112]. Different diagnostic performance of proposed signatures was reported (Area Under the ROC Curve, AUC, was in the range 0.70 to 0.90), yet the observed difference could be attributed to differences in the statistical methodology. Nevertheless, though some of these biomarker candidates are promising, their actual diagnostic performance has not yet been validated in the proper clinical settings.

Table 2. Potential exosome biomarkers of lung cancer.

Biofluid/EV Isolation	Size of Groups	Proposed Biomarker	Analytic. Method	Diagnostic Value	Reference
Serum/UC TEM, NTA, WB	Control: 46 Cases: 125 (Stage I–IV)	AHSG, ECM1 proteins	MS	AUC = 0.80 SEN = 54% SPE = 89%	[104]
Serum/IMA	Control: 10 Cases: 26 (Stage III–IV)	CD91	MS	AUC = 0.72 SEN = 60% SPE = 89%	[105]
Plasma/UC TEM, NTA, WB	Control: 15 Cases: 13 (Stage I–II)	SRGN, TPM3, THBS1, HUWE1 proteins	MS	AUC = 0.90 SEN = 81% SPE = 82%	[106]
Serum/UC TEM, NTA, WB	Control: 90 Cases: 183 (Stage I–IV)	LPS-binding protein (LBP)	ELISA	AUC = 0.71 SEN = 65% SPE = 76%	[107]
Plasma/EV array	Control: 150 Cases: 431 (Stage I–IV)	CD151, Tspan8, NYESO1, HER2, CD171, EGFRvIII SFTPD, Flotilin1, CD142, Mucin16	EV array	AUC = 0.74 SEN = 71% SPE = 69%	[103]
Serum/PRE TEA, NTA	Control: 150 Cases: 150 (Stage I–IV)	lncRNA (TBILA)	qPCR	AUC = 0.78 SEN = 65% SPE = 81%	[108]
Serum/PRE TEA, NTA	Control: 150 Cases: 150 (Stage I–IV)	lncRNA (AGAP2-AS1)	qPCR	AUC = 0.73 SEN = 67% SPE = 73%	[108]
Serum/PRE TEM, NTA, WB	Control: 64 Cases: 72 (Stage I–IV)	lncRNA (DLX6-AS1)	qPCR	AUC = 0.81 SEN = 78% SPE = 86%	[109]
Serum/PRE TEM, NTA, WB	Control: 30 Cases: 77 (Stage I–IV)	lncRNA (MALAT-1)	qPCR	AUC = 0.70 SEN = 60% SPE = 81%	[85]
Serum/PRE TEM, NTA, WB	Control: 40 Cases: 64 (Stage I–IV)	lncRNA (GAS5)	qPCR	AUC = 0.86 SEN = 86% SPE = 70%	[110]
Serum/PRE WB	Control: 30 Cases: 120 (Stage I–IV)	circular RNA (circRNA-002178)	qPCR	AUC = 0.99 SEN = 99% SPE = 100%	[111]
Plasma/UC	Control: 39 Cases: 44 (Stage I–II)	PC(32:0), PC(34:2), PC(36:1)/(36:2)/(36:3), PC(38:3)/(38:5)/(38:6), LPC(12:0), LPC(16:0), SM(34:1), SM(42:2), TG(52:5), TG(54:6), CE(20:4)	MS	AUC = 0.85 SEN = 77% SPE = 72%	[112]

sEV's isolation and characterization methods: UC—Ultracentrifugation; PRE—Precipitation; IMA—Immunoaffinity; TEM—Transmission Electron Microscopy; NTA—Nanoparticle Tracking Analysis; WB—Western Blot; MS—mass spectrometry; qPCR—quantitative real-time PCR; AUC—Area Under the ROC Curve; SEN—Sensitivity; SPE—Specificity.

5. Exosome miRNA as a Biomarker of Lung Cancer

The miRNA content of serum/plasma-derived exosomes is another promising source of lung cancer biomarkers addressed in several papers. Two analytical methods of miRNA detection dominate in these studies—quantitative PCR and next-generation sequencing.

However, many different approaches were applied to isolate and characterize sEV from serum or plasma; hence, different classes of vesicles could be studied in different reports. The representative papers are summarized in Table 3. Some of these studies tested the diagnostic performance of miRNA signatures, which resulted in AUC values that ranged between 0.71 and 0.98. However, none of these signatures have yet been validated in an independent study. Furthermore, none of them have been studied in the context of lung cancer screening. Analyzed groups had different sizes and represented different clinical characteristics and ethnic/genetic backgrounds. Therefore, different miRNA signatures of serum/plasma exosomes proposed to discriminate lung cancer patients from healthy controls should be compared with caution.

Table 3. Potential sEV miRNA biomarkers of lung cancer.

Biofluid/EV Isolation	miRNA Signature	Size of Groups	Diagnostic Value	Reference
Plasma/PRE	miR-378a, miR-379, miR-139-5p, miR-200b-5p	Control: 25 Cases: 80 (Stage I)	AUC = 0.91 SEN = 98% SPE = 72%	[113]
Plasma/PRE WB, TEM	miR-30b, miR-30c, miR-103, miR-122, miR-195, miR-203, miR-221, miR-222	Control: 6 Cases: 12 (Stage -)	-	[114]
Plasma/PRE	miR-19-3p, miR-21-5p, miR-221-3p	Control: 14 Cases: 18 (Stage I–IV)	-	[55]
Plasma/PRE WB, NTA, TEM	miR-23b-3p, miR-10b-5p, miR-21-5p	Control: 10 Cases: 10 (Stage I–IV)	AUC = 0.91 SEN = 82% SPE = 85%	[115]
Plasma/PRE WB, NTA, TEM	miR-451a, miR-194-5p, miR-486-5p	Control: 149 Cases: 434 (Stage I-IV)	AUC = 0.97 SEN = 95% SPE = 71%	[36]
Plasma/PRE WB, NTA, TEM	miR-185-5p, miR-32-5p, miR-140-3p, let-7f-5p	Control: 20 Cases:79 (Stage I–III)	AUC = 0.91 SEN = 59% SPE = 100%	[116]
Plasma/SEC + IMA	miR-17-3p, miR-21, miR-106a, miR-146, miR-155, miR-191, miR-192, miR-203, miR-205, miR-210, miR-212, miR-214	Control: 8 Cases: 28 (Stage I–IV)	-	[117]
Plasma/IMA	let-7f, miR-20b, miR-30e-3p, miR-223, miR-301	Control: 48 Cases:78 (Stage I–IV)	-	[118]
Plasma/UC + IMA WB, NTA	let-7b-5p, let-7e-5p, miR-24-5p, miR-21-5p	Control: 13 Cases: 47 (Stage I)	AUC = 0.90 SEN = 80% SPE = 92%	[119]
Plasma/UC TEM	miR-21, miR-4257	Control: 30 Cases: 195 (Stage I–III)	-	[120]
Plasma/SEC	miR-411-5p	Control: 7 Cases: 19 (Stage -)	-	[121]
Serum/PRE	miR-451a, miR-486-5p, miR-363-3p, miR-660-5p, miR-15b-5p, miR-25-3p, miR-16-2-3p	Control: 10 Cases: 20 (Stage I–IV)	AUC = 0.98 SEN = 100% SPE = 90%	[122]
Serum/PRE WB, NTA, TEM	miR-17-5p	Control: 137 Cases: 172 (Stage I–III)	AUC = 0.74 SEN = 67% SPE = 77%	[123]
Serum/PRE WB, NTA, TEM	miR-146a-5p, miR-486-5p	Control: 80 Cases: 48 (Stage I–II)	AUC = 0.90 SEN = 83% SPE = 90%	[124]

Table 3. *Cont.*

Biofluid/EV Isolation	miRNA Signature	Size of Groups	Diagnostic Value	Reference
Serum/PRE	miR-216b	Control: 60 Cases: 105 (Stage I–IV)	AUC = 0.84 SEN = 87% SPE = 75%	[125]
Serum/PRE WB, TEM	miR-106b	Control: 72 Cases: 72 (Stage I–IV)	-	[126]
Serum/PRE	106a-5p, miR-20a-5p, miR-93-5p	Control: 36 Cases: 34 (Stage I–III)	AUC = 0.83	[127]
Serum/PRE WB, NTA, TEM	miR-210-5p, miR-1269a, miR-205-5p, miR-9-3p	Control: 150 Cases: 148 (Stage I–III)	AUC = 0.74 SEN = 81% SPE = 61%	[128]
Serum/PRE WB, NTA, TEM	miR-1290	Control: 40 Cases: 70 (Stage I–IV)	AUC = 0.94 SEN = 80% SPE = 97%	[129]
Serum/PRE	miR-378	Control: 60 Cases: 103 (Stage I–IV)	AUC = 0.84 SEN = 78% SPE = 82%	[130]
Serum/PRE WB, TEM	miR-7977, miR-98-3p	Control: 65 Cases: 65 (Stage I–IV)	AUC = 0.82 SEN = 81% SPE = 75%	[131]
Serum/UC WB, NTA, TEM	miR-126	Control: 31 Cases: 45 (Stage I–III)	AUC = 0.84 SEN = 90% SPE = 86%	[132]
Serum/UC WB, NTA, TEM	miR-21-5p, miR-126-3p, miR-140-5p	Control: 16 Cases: 23 (Stage I–IV)	-	[133]
Serum/UC WB, NTA, TEM	miR-620	Control: 231 Cases: 235 (Stage I–IV)	AUC = 0.71 SEN = 63% SPE = 68%	[134]
Serum/UC WB, NTA, TEM	miR-5684, miR-125b-5p	Control: 312 Cases: 330 (Stage I–IV)	AUC = 0.74 SEN = 81% SPE = 61%	[135]
Serum/UC WB, NTA, TEM	miR-20b-5p, miR-3187-5p	Control: 30 Cases: 380 (Stage 0–I)	AUC = 0.84	[136]

sEV's isolation and characterization methods: UC—Ultracentrifugation; PRE—Precipitation; IMA—Immunoaffinity; SEC—Size Exclusion Chromatography; TEM—Transmission Electron Microscopy; NTA—Nanoparticle Tracking Analysis; WB—Western Blot; AUC—Area Under the ROC Curve; SEN—Sensitivity; SPE—Specificity.

According to current literature research, proposed lung cancer exosome signatures involved above 60 miRNA species overall, and 14 miRNA species appeared in more than one signature. This included miR-21 (seven signatures), miR-221 (three signatures), and miR-486-5p (three signatures). Figure 2 illustrates miRNA species present in lung cancer signatures, detected in either whole serum/plasma or serum/plasma-derived exosomes, which were included in more than one signature. There were nine miRNA species, namely, miR-17, miR-19, miR-21, miR-221, miR-451, miR-486-5p, miR-126, miR-140, and miR-210, which appeared in both whole serum/plasma and exosome-based signatures. Functions associated with this interesting subset of miRNAs are discussed below.

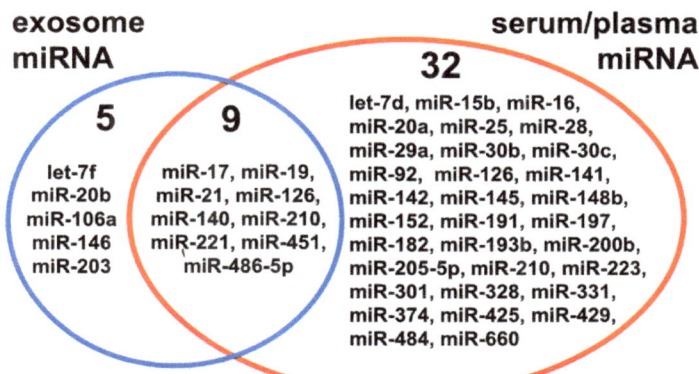

Figure 2. MicroRNA species present in lung cancer signatures. Showed are components present in at least 2 signatures identified in either whole serum/plasma or serum/plasma-derived exosomes (small extracellular vesicles).

Shared components of the whole serum/plasma-based and exosome-based lung cancer signatures contain several oncomirs, i.e., miRNAs with known cancer-related functions. These include miR-17 and miR-19 belonging to the miR-17-92 cluster, which is regulated by MYC. The miR-17-92 cluster is a unique oncomir due to the polycistronic miRNA transcript, which allows obtaining six individual miRNAs involved in many cancer-associated processes: miR-17, miR-18a, miR-19a, miR-20a, miR-19b-1, and miR-92a-1 [137]. A high level of miR-17 and miR-19 induces cell proliferation, while the deletion is lethal (it causes lung and lymphoid cell developmental defects) [138]. miR-17 suppresses the expression of the E2F1 transcription factor, shifting the cellular balance in favor of increased proliferation. In lung cancer, overexpression of miR-17 and miR-19 affects the expression of *HIF1A*, *PTEN*, *BCL2L11*, *CDKNA*, and *TSP1*, enhancing tumor growth by increasing the permeability of blood vessels, inducing hypoxia, increasing proliferation, inhibiting apoptosis, and stimulating tumor cell migration [139,140]. miR-21 is another oncomir frequently overexpressed in cancer cells, one of the first miRNAs identified in mammals. Among the targets of miR-21 are tumor suppressor genes such as *PTEN*, *RHOB*, and *TP63*. Further, miR-21 blocks AKT and MAPK signaling pathways via inhibition of several phosphatases. As a result of miR-21 overexpression, the action of tumor suppressors is blocked, causing the development of many cancers such as lung, ovarian, breast, brain, and many others [141]. In lung cancer, overexpression of miR-21 is associated with increased cell proliferation, angiogenesis, cell invasion, and metastasis, as well as chemo- and radioresistance [142]. The inhibition of miR-21 resulted in the induction of apoptosis (due to inhibiting the PI3K/Akt/NF-KB signaling pathway and increased caspase activity) as well as impeded the migration and invasiveness of NSCLC cells [143]. miR-21 is involved in modulating the tumor microenvironment by targeting *PTEN* in the stromal compartment, which is mediated by miR-21-containing TEX [144]. Another oncogenic miRNA found in TEX is miR-221. miR-221 inhibits p27 tumor suppressor, which causes the transition from G1 to S phase and acceleration of cell division [145]. Among miR-221 targets is also *CD117*, a known proto-oncogene that regulates cell survival, migration, and differentiation. Overexpression of miR-221 induces proliferation and migration of tumor cells as well as tumor angiogenesis via the Wnt/β-catenin signaling pathway and has been shown to promote the chemoresistance of lung cancer cells by activating the PTEN/Akt pathway [146]. miR-210 is also an important factor in the development of lung cancer, whose level increases in NSCLS tissues and is associated with a worse prognosis [147]. The action of miR-210 involves the regulation of HIF-1, ATG7, LC3, and Beclin-1 [148].

Other miRNA observed in multiple lung cancer signatures are putative tumor suppressors. In lung cancer, a decreased level of miR-451 correlates with poor prognosis [149]. Functionally, decreased expression of miR-451 increases drug resistance and accelerates the epithelial-mesenchymal transition due to *MYC* overexpression, which is a miR-451 target [150]. Moreover, miR-451 targets several genes involved in the inflammation and stress response pathways that modulate the tumor microenvironment, including *PSMB8*, *NOS2*, and *CARF* [151]. Another component of exosome lung cancer signature is miR-126, which level was reduced in cancer patients. Overexpression of miR-126 inhibits cancer cell proliferation, colony formation, migration, invasion, induces cell cycle arrest, and apoptosis via targeting *ITGA6* gene [152]. Another characteristic component of serum-derived exosomes is miR-140 involved in carcinogenesis and tumor progression, which level is significantly lowered in tumors. Overexpression of miR-140 is associated with inhibition of proliferation, migration, and invasion of NSCLC cells via targeting of *ATP8A1* and *IGF1R* genes [153,154]. Another miRNA shared by serum and exosome-based signatures is miR-486-5p. This miRNA, one of the most abundant miRNAs in the peripheral blood, plays an important role in the development of many cancers. Overexpression of miR-486-5p increases cell proliferation by regulating the PTEN/PI3K/AKT pathway [155]. On the other hand, however, decreased levels of miRNA-486-5p in NSCLC tissues correlated with increased drug resistance and a worse prognosis [156]. Moreover, overexpression of miR-486-5p inhibits the development of lung cancer due to the suppression of *GAB2* [157]. Further, decreased level of miRNA-486-5p correlates with KIAA1199 protein overexpression, which in turn results in increased cancer proliferation and poor prognosis [158].

Interesting cancer-related features could be attributed also to five miRNA species detected only in exosome-based signatures of lung cancer, namely, let-7f, miR-146, miR-203, miR-106a, and miR-20b. Let-7f belongs to the let-7 (lethal-7) family, which consists of 12 members that regulate cell cycle and cell proliferation by affecting RAS, cyclin A2, CDC34, Aurora A and B kinases, E2F5, CDK8, and HMGA2 [159]. Decreased expression of let-7 is observed in different tumor tissues [160]. Increased expression of let-7f is associated with inhibition of proliferation, migration, and invasion of neoplastic cells, including lung cancer cells, while its decreased expression was observed in metastatic cells [161]. miR-146 is involved in the regulation of inflammation [162]. The overexpression of miR-146 is associated with increased survival and migration of NSCLC cells via suppressing *TRAF6* [163]. Further, increased expression of miR-146 in lung cancer cells lowers the level of claudin-12, which in turn leads to activation of the Wnt/β-catenin and PI3K/AKT/MAPK signaling pathways resulting in the increased viability and migration, as well as resistance to cisplatin and inhibition of apoptosis [164]. Another oncogenic miRNA observed in exosomes of lung cancer patients is miR-106, which increased expression correlates lymph node metastases, drug resistance, and poor prognosis [165]. Increased level of miR-106 decreased expression of *BTG3*, which in turn promotes proliferation and inhibits apoptosis [166]. The expression of miR-20b is also significantly higher in lung cancer cells. miR-20b contributes to the development of NSCLC by inhibiting APC via the canonical Wnt signaling pathway [167]. Moreover, similar to miR-106, miR-20b directly targets *BTG3* [168]. The last miRNA detected in multiple lung cancer signatures is miR-203, which is a putative tumor suppressor. High expression of miR-203 inhibits the proliferation and invasiveness of lung cancer cells through negative regulation of survivin [169]. Moreover, increased expression of miR-203 inhibits *RGS17* oncogene, which results in reduced cell proliferation through the cAMP-PKA-CREB pathway [170]. Furthermore, miR-203 acts as a suppressor of the SRC/Ras/ERK pathway by inhibiting the expression of *SRC* oncogene, resulting in the suppression of proliferation and migration of lung cancer cells [171].

Furthermore, to search systemically for genes regulated by 14 miRNA species that recurred in sEV-based signatures of lung cancer (Figure 2), the miRTarBase database of experimentally validated interactions between miRNA and genes [172] was analyzed. This returned the set of about 600 genes, which functions were analyzed using the FunRich functional enrichment analysis tool [173]. The set comprised of 390 genes associated with lung cancer, including several ones responsible for clinical features of this cancer (e.g., *KRAS, EGFR, CASP8, PIK3CA, ERBB2, FASLG, RB1, MYD88*, and *TP53*). Among molecular functions and biological processes associated with this set of genes, several terms potentially involved in cancer development and progression were significantly over-represented, which is summarized in Table 4. Moreover, over-represented biological pathways associated with the most numerous subsets of genes were outweighed by signaling pathways associated with inflammation, immune response, cell growth, cell-to-cell communication, and cancer.

Table 4. Functions associated with genes regulated by exosome miRNAs common in lung cancer signatures.

Molecular Function	No. of Genes	Fold Enrichment	FDR
Transcription factor activity	72	2.64	<0.00001
Receptor activity	32	2.73	0.00003
Protein serine/threonine kinase activity	28	2.87	0.00005
Transmembrane receptor protein tyrosine kinase activity	11	6.06	0.00008
Receptor signaling complex scaffold activity	28	2.68	0.00010
Receptor binding	16	3.83	0.00016
Protein-tyrosine kinase activity	8	6.50	0.00077
Transcription regulator activity	47	1.74	0.00451
GTPase activity	18	2.50	0.00868
Kinase regulator activity	5	6.71	0.01639
Biological Process	**No. of Genes**	**Fold Enrichment**	**FDR**
Signal transduction	240	1.88	<0.00001
Cell communication	223	1.85	<0.00001
Regulation of nucleotide and nucleic acid metabolism	144	1.57	<0.00001
Apoptosis	26	3.64	<0.00001
Regulation of cell growth	5	7.35	0.01662

No. of genes—number of genes connected to specific term among 600 genes in the whole set; FDR—corrected p-value of the hypergeometric test for the significance of over-representation.

6. Conclusions

MicroRNA component of serum/plasma is an attractive source of cancer biomarkers, and several miRNA signatures of lung cancer have been proposed. Though none of them is applied in clinical practice yet, a few are currently tested in prospective clinical trials aimed at validation of their applicability in the early detection of lung cancer and/or diagnosis of the indeterminate pulmonary nodules. Among other potential biomarkers of early lung cancer are exosomes (or rather small extracellular vesicle, sEV) circulating in the blood. Several molecular components of sEV, including proteins, lipids, and noncoding RNAs, have been reported to have different levels in vesicles isolated from lung cancer patients and healthy individuals. The largest number of published reports that address this issue focus on the miRNA component of vesicles. Proposed signatures of exosome miRNA have promising diagnostic value (AUC in the 0.75–0.95 range), yet none of them has been validated in the context of the early detection of lung cancer. These signatures involve a few dozen miRNA species overall, including 14 miRNA (so far) that recurred in different signatures. It is worth noting that all these miRNA species have cancer-related functions and have been associated with lung cancer progression, which further confirms their diagnostic importance. Importantly, a few miRNA species, including known oncomirs miR-17, miR-19, and miR-21, appear in multiple miRNA signatures of lung cancer that are based on both the whole serum/plasma and serum/plasma-derived exosomes. However, one should note, that due to barely standardized methods of sEV isolation, the analysis of exosome miRNA content represents a diagnostic challenge. Therefore, the direct comparison of a diagnostic value of miRNA signature based on the serum/plasma-derived sEV and the whole specimen is desired, which is not available yet.

Author Contributions: Writing—original draft preparation, M.S.; writing—review and editing, P.W. All authors have read and agreed to the published version of the manuscript.

Funding: This work was supported by the National Science Centre, Poland, Grant 2017/27/B/NZ7/01833 (to P.W.).

Institutional Review Board Statement: Not applicable.

Informed Consent Statement: Not applicable.

Data Availability Statement: Not applicable.

Conflicts of Interest: The authors declare no conflict of interest.

References

1. Blandin Knight, S.; Crosbie, P.A.; Balata, H.; Chudziak, J.; Hussell, T.; Dive, C. Progress and prospects of early detection in lung cancer. *Open Biol.* **2017**, *7*, 170070. [CrossRef]
2. Kauczor, H.U.; Baird, A.M.; Blum, T.G.; Bonomo, L.; Bostantzoglou, C.; Burghuber, O.; Čepická, B.; Comanescu, A.; Couraud, S.; Devaraj, A.; et al. ESR/ERS statement paper on lung cancer screening. *Eur. Radiol.* **2020**, *30*, 3277–3294. [CrossRef] [PubMed]
3. Aberle, D.R.; Adams, A.M.; Berg, C.D.; Black, W.C.; Clapp, J.D.; Fagerstrom, R.M.; Gareen, I.F.; Gatsonis, C.; Marcus, P.M.; Sicks, J.D. Reduced lung-cancer mortality with low-dose computed tomographic screening. *N. Engl. J. Med.* **2011**, *365*, 395–409. [CrossRef] [PubMed]
4. de Koning, H.J.; van der Aalst, C.M.; de Jong, P.A.; Scholten, E.T.; Nackaerts, K.; Heuvelmans, M.A.; Lammers, J.J.; Weenink, C.; Yousaf-Khan, U.; Horeweg, N.; et al. Reduced Lung-Cancer Mortality with Volume CT Screening in a Randomized Trial. *N. Engl. J. Med.* **2020**, *382*, 503–513. [CrossRef] [PubMed]
5. Wille, M.M.; Dirksen, A.; Ashraf, H.; Saghir, Z.; Bach, K.S.; Brodersen, J.; Clementsen, P.F.; Hansen, H.; Larsen, K.R.; Mortensen, J.; et al. Results of the Randomized Danish Lung Cancer Screening Trial with Focus on High-Risk Profiling. *Am. J. Respir. Crit. Care Med.* **2016**, *193*, 542–551. [CrossRef]
6. Priola, A.M.; Priola, S.M.; Giaj-Levra, M.; Basso, E.; Veltri, A.; Fava, C.; Cardinale, L. Clinical implications and added costs of incidental findings in an early detection study of lung cancer by using low-dose spiral computed tomography. *Clin. Lung Cancer* **2013**, *14*, 139–148. [CrossRef] [PubMed]
7. Rzyman, W.; Jelitto-Gorska, M.; Dziedzic, R.; Biadacz, I.; Ksiazek, J.; Chwirot, P.; Marjanski, T. Diagnostic work-up and surgery in participants of the Gdansk lung cancer screening programme, the incidence of surgery for non-malignant conditions. *Interact. Cardiovasc. Thorac. Surg.* **2013**, *17*, 969–973. [CrossRef]
8. Hasan, N.; Kumar, R.; Kavuru, M.S. Lung cancer screening beyond low-dose computed tomography: The role of novel biomarkers. *Lung* **2014**, *192*, 639–648. [CrossRef]
9. Atwater, T.; Massion, P.P. Biomarkers of risk to develop lung cancer in the new screening era. *Ann. Transl. Med.* **2016**, *4*, 158. [CrossRef]
10. Chu, G.C.W.; Lazare, K.; Sullivan, F. Serum and blood based biomarkers for lung cancer screening, a systematic review. *BMC Cancer* **2018**, *18*, 181. [CrossRef]
11. Ostrin, E.J.; Sidransky, D.; Spira, A.; Hanash, S.M. Biomarkers for Lung Cancer Screening and Detection. *Cancer Epidemiol. Biomark. Prev.* **2020**, *29*, 2411–2415. [CrossRef] [PubMed]
12. Hassanein, M.; Callison, J.C.; Callaway-Lane, C.; Aldrich, M.C.; Grogan, E.L.; Massion, P.P. The state of molecular biomarkers for the early detection of lung cancer. *Cancer Prev. Res.* **2012**, *5*, 992–1006. [CrossRef]
13. Zamay, T.N.; Zamay, G.S.; Kolovskaya, O.S.; Zukov, R.A.; Petrova, M.M.; Gargaun, A.; Berezovski, M.V.; Kichkailo, A.S. Current and Prospective Protein Biomarkers of Lung Cancer. *Cancers* **2017**, *9*, 155. [CrossRef]
14. Leighl, N.B.; Page, R.D.; Raymond, V.M.; Daniel, D.B.; Divers, S.G.; Reckamp, K.L.; Villalona-Calero, M.A.; Dix, D.; Odegaard, J.I.; Lanman, R.B.; et al. Clinical Utility of Comprehensive Cell-free DNA Analysis to Identify Genomic Biomarkers in Patients with Newly Diagnosed Metastatic Non-small Cell Lung Cancer. *Clin. Cancer Res.* **2019**, *25*, 4691–4700. [CrossRef]
15. Maly, V.; Maly, O.; Kolostova, K.; Bobek, V. Circulating Tumor Cells in Diagnosis and Treatment of Lung Cancer. *In Vivo* **2019**, *33*, 1027–1037. [CrossRef] [PubMed]
16. Lee, K.B.; Ang, L.; Yau, W.P.; Seow, W.J. Association between Metabolites and the Risk of Lung Cancer, A Systematic Literature Review and Meta-Analysis of Observational Studies. *Metabolites* **2020**, *10*, 362. [CrossRef] [PubMed]
17. Ros-Mazurczyk, M.; Jelonek, K.; Marczyk, M.; Binczyk, F.; Pietrowska, M.; Polanska, J.; Dziadziuszko, R.; Jassem, J.; Rzyman, W.; Widlak, P. Serum lipid profile discriminates patients with early lung cancer from healthy controls. *Lung Cancer* **2017**, *112*, 69–74. [CrossRef] [PubMed]
18. Seijo, L.M.; Peled, N.; Ajona, D.; Boeri, M.; Field, J.K.; Sozzi, G.; Pio, R.; Zulueta, J.J.; Spira, A.; Massion, P.P.; et al. Biomarkers in Lung Cancer Screening: Achievements, Promises, and Challenges. *J. Thorac. Oncol.* **2019**, *14*, 343–357. [CrossRef] [PubMed]
19. Duffy, M.J.; O'Byrne, K. Tissue and Blood Biomarkers in Lung Cancer, A Review. *Adv. Clin. Chem.* **2018**, *86*, 1–21. [CrossRef]
20. Vargas, A.J.; Harris, C.C. Biomarker development in the precision medicine era, lung cancer as a case study. *Nat. Rev. Cancer* **2016**, *16*, 525–537. [CrossRef]

21. Dama, E.; Melocchi, V.; Colangelo, T.; Cuttano, R.; Bianchi, F. Deciphering the Molecular Profile of Lung Cancer: New Strategies for the Early Detection and Prognostic Stratification. *J. Clin. Med.* **2019**, *8*, 108. [CrossRef] [PubMed]
22. Mazzone, P.J.; Sears, C.R.; Arenberg, D.A.; Gaga, M.; Gould, M.K.; Massion, P.P.; Nair, V.S.; Powell, C.A.; Silvestri, G.A.; Vachani, A.; et al. ATS Assembly on Thoracic Oncology. Evaluating Molecular Biomarkers for the Early Detection of Lung Cancer, When Is a Biomarker Ready for Clinical Use? An Official American Thoracic Society Policy Statement. *Am. J. Respir. Crit. Care Med.* **2017**, *196*, e15–e29. [CrossRef] [PubMed]
23. Massion, P.P.; Healey, G.F.; Peek, L.J.; Fredericks, L.; Sewell, H.F.; Murray, A.; Robertson, J.F. Autoantibody Signature Enhances the Positive Predictive Power of Computed Tomography and Nodule-Based Risk Models for Detection of Lung Cancer. *J. Thorac. Oncol.* **2017**, *12*, 578–584. [CrossRef]
24. Silvestri, G.A.; Tanner, N.T.; Kearney, P.; Vachani, A.; Massion, P.P.; Porter, A.; Springmeyer, S.C.; Fang, K.C.; Midthun, D.; Mazzone, P.J.; et al. Assessment of Plasma Proteomics Biomarker's Ability to Distinguish Benign From Malignant Lung Nodules, Results of the PANOPTIC (Pulmonary Nodule Plasma Proteomic Classifier) Trial. *Chest* **2018**, *154*, 491–500. [CrossRef] [PubMed]
25. Han, Y.; Li, H. miRNAs as biomarkers and for the early detection of non-small cell lung cancer (NSCLC). *J. Thorac. Dis.* **2018**, *10*, 3119–3131. [CrossRef]
26. Rijavec, E.; Coco, S.; Genova, C.; Rossi, G.; Longo, L.; Grossi, F. Liquid Biopsy in Non-Small Cell Lung Cancer: Highlights and Challenges. *Cancers* **2019**, *12*, 17. [CrossRef] [PubMed]
27. Iqbal, M.A.; Arora, S.; Prakasam, G.; Calin, G.A.; Syed, M.A. MicroRNA in lung cancer: Role, mechanisms, pathways and therapeutic relevance. *Mol. Aspects Med.* **2019**, *70*, 3–20. [CrossRef] [PubMed]
28. Vishnoi, A.; Rani, S. MiRNA Biogenesis and Regulation of Diseases: An Overview. *Methods Mol. Biol.* **2017**, *1509*, 1–10. [CrossRef]
29. Liu, G.; Li, B. Role of miRNA in transformation from normal tissue to colorectal adenoma and cancer. *J. Cancer Res. Ther.* **2019**, *15*, 278–285. [CrossRef]
30. Abreu, F.B.; Liu, X.; Tsongalis, G.J. miRNA analysis in pancreatic cancer: The Dartmouth experience. *Clin. Chem. Lab. Med.* **2017**, *55*, 755–762. [CrossRef]
31. Ghafouri-Fard, S.; Shoorei, H.; Taheri, M. miRNA profile in ovarian cancer. *Exp. Mol. Pathol.* **2020**, *113*, 104381. [CrossRef] [PubMed]
32. Balacescu, O.; Dumitrescu, R.G.; Marian, C. MicroRNAs Role in Prostate Cancer. *Methods Mol. Biol.* **2018**, *1856*, 103–117. [CrossRef]
33. Wang, N.; Guo, W.; Song, X.; Liu, L.; Niu, L.; Song, X.; Xie, L. Tumor-associated exosomal miRNA biomarkers to differentiate metastatic vs. nonmetastatic non-small cell lung cancer. *Clin. Chem. Lab. Med.* **2020**, *58*, 1535–1545. [CrossRef] [PubMed]
34. Montani, F.; Bianchi, F. Circulating Cancer Biomarkers: The Macro-revolution of the Micro-RNA. *EBioMedicine* **2016**, *5*, 4–6. [CrossRef] [PubMed]
35. He, W.J.; Li, W.H.; Jiang, B.; Wang, Y.F.; Xia, Y.X.; Wang, L. MicroRNAs level as an initial screening method for early-stage lung cancer, a bivariate diagnostic random-effects meta-analysis. *Int. J. Clin. Exp. Med.* **2015**, *8*, 12317–12326.
36. Yao, B.; Qu, S.; Hu, R.; Gao, W.; Jin, S.; Liu, M.; Zhao, Q. A panel of miRNAs derived from plasma extracellular vesicles as novel diagnostic biomarkers of lung adenocarcinoma. *FEBS Open Bio.* **2019**, *9*, 2149–2158. [CrossRef]
37. Sozzi, G.; Boeri, M.; Rossi, M.; Verri, C.; Suatoni, P.; Bravi, F.; Roz, L.; Conte, D.; Grassi, M.; Sverzellati, N.; et al. Clinical utility of a plasma-based miRNA signature classifier within computed tomography lung cancer screening, a correlative MILD trial study. *J. Clin. Oncol.* **2014**, *32*, 768–773. [CrossRef] [PubMed]
38. Bianchi, F.; Nicassio, F.; Marzi, M.; Belloni, E.; Dall'olio, V.; Bernard, L.; Pelosi, G.; Maisonneuve, P.; Veronesi, G.; Di Fiore, P.P. A serum circulating miRNA diagnostic test to identify asymptomatic high-risk individuals with early stage lung cancer. *EMBO Mol. Med.* **2011**, *3*, 495–503. [CrossRef]
39. Lu, T.P.; Lee, C.Y.; Tsai, M.H.; Chiu, Y.C.; Hsiao, C.K.; Lai, L.C.; Chuang, E.Y. miRSystem, an integrated system for characterizing enriched functions and pathways of microRNA targets. *PLoS ONE* **2012**, *7*, e42390. [CrossRef] [PubMed]
40. Nguyen, J.; Bernert, R.; In, K.; Kang, P.; Sebastiao, N.; Hu, C.; Hastings, K.T. Gamma-interferon-inducible lysosomal thiol reductase is upregulated in human melanoma. *Melanoma Res.* **2016**, *26*, 125–137. [CrossRef]
41. Pucer, A.; Brglez, V.; Payré, C.; Pungerčar, J.; Lambeau, G.; Petan, T. Group X secreted phospholipase A(2) induces lipid droplet formation and prolongs breast cancer cell survival. *Mol. Cancer* **2013**, *12*, 111. [CrossRef]
42. Coulier, F.; Batoz, M.; Marics, I.; de Lapeyrière, O.; Birnbaum, D. Putative structure of the FGF6 gene product and role of the signal peptide. *Oncogene* **1991**, *6*, 1437–1444.
43. Li, J.X.; Zhang, Z.F.; Wang, X.B.; Yang, E.Q.; Dong, L.; Meng, J. PLZF regulates apoptosis of leukemia cells by regulating AKT/Foxo3a pathway. *Eur. Rev. Med. Pharmacol. Sci.* **2019**, *23*, 6411–6418. [CrossRef]
44. Kim, G.Y.; Kim, H.; Lim, H.J.; Park, H.Y. Coronin 1A depletion protects endothelial cells from TNFα-induced apoptosis by modulating p38β expression and activation. *Cell. Signal.* **2015**, *27*, 1688–1693. [CrossRef]
45. Wang, H.; Wu, S.; Zhao, L.; Zhao, J.; Liu, J.; Wang, Z. Clinical use of microRNAs as potential non-invasive biomarkers for detecting non-small cell lung cancer, a meta-analysis. *Respirology* **2015**, *20*, 56–65. [CrossRef] [PubMed]
46. Shen, Y.; Wang, T.; Yang, T.; Hu, Q.; Wan, C.; Chen, L.; Wen, F. Diagnostic value of circulating microRNAs for lung cancer, a meta-analysis. *Genet. Test. Mol. Biomark.* **2013**, *17*, 359–366. [CrossRef] [PubMed]
47. Jiang, M.; Li, X.; Quan, X.; Li, X.; Zhou, B. Clinically Correlated MicroRNAs in the Diagnosis of Non-Small Cell Lung Cancer, A Systematic Review and Meta-Analysis. *Biomed. Res. Int.* **2018**, *2018*, 5930951. [CrossRef] [PubMed]

48. Montani, F.; Marzi, M.J.; Dezi, F.; Dama, E.; Carletti, R.M.; Bonizzi, G.; Bertolotti, R.; Bellomi, M.; Rampinelli, C.; Maisonneuve, P.; et al. miR-Test, a blood test for lung cancer early detection. *J. Natl. Cancer Inst.* **2015**, *107*, djv063. [CrossRef]
49. Shen, J.; Liu, Z.; Todd, N.W.; Zhang, H.; Liao, J.; Yu, L.; Guarnera, M.A.; Li, R.; Cai, L.; Zhan, M.; et al. Diagnosis of lung cancer in individuals with solitary pulmonary nodules by plasma microRNA biomarkers. *BMC Cancer* **2011**, *11*, 374. [CrossRef]
50. Ma, J.; Li, N.; Guarnera, M.; Jiang, F. Quantification of Plasma miRNAs by Digital PCR for Cancer Diagnosis. *Biomark. Insights* **2013**, *8*, 127–136. [CrossRef] [PubMed]
51. Mozzoni, P.; Banda, I.; Goldoni, M.; Corradi, M.; Tiseo, M.; Acampa, O.; Balestra, V.; Ampollini, L.; Casalini, A.; Carbognani, P.; et al. Plasma and EBC microRNAs as early biomarkers of non-small-cell lung cancer. *Biomarkers* **2013**, *18*, 679–686. [CrossRef]
52. Tang, D.; Shen, Y.; Wang, M.; Yang, R.; Wang, Z.; Sui, A.; Jiao, W.; Wang, Y. Identification of plasma microRNAs as novel noninvasive biomarkers for early detection of lung cancer. *Eur. J. Cancer Prev.* **2013**, *22*, 540–548. [CrossRef]
53. Wang, Y.; Yi, J.; Chen, X.; Zhang, Y.; Xu, M.; Yang, Z. The regulation of cancer cell migration by lung cancer cell-derived exosomes through TGF-β and IL-10. *Oncol. Lett.* **2016**, *11*, 1527–1530. [CrossRef]
54. Zhang, H.; Mao, F.; Shen, T.; Luo, Q.; Ding, Z.; Qian, L.; Huang, J. Plasma miR-145.; miR-20a.; miR-21 and miR-223 as novel biomarkers for screening early-stage non-small cell lung cancer. *Oncol. Lett.* **2017**, *13*, 669–676. [CrossRef] [PubMed]
55. Zhou, X.; Wen, W.; Shan, X.; Zhu, W.; Xu, J.; Guo, R.; Cheng, W.; Wang, F.; Qi, L.W.; Chen, Y.; et al. A six-microRNA panel in plasma was identified as a potential biomarker for lung adenocarcinoma diagnosis. *Oncotarget* **2017**, *8*, 6513–6525. [CrossRef]
56. Hennessey, P.T.; Sanford, T.; Choudhary, A.; Mydlarz, W.W.; Brown, D.; Adai, A.T.; Ochs, M.F.; Ahrendt, S.A.; Mambo, E.; Califano, J.A. Serum microRNA biomarkers for detection of non-small cell lung cancer. *PLoS ONE* **2012**, *7*, e32307. [CrossRef]
57. Nadal, E.; Truini, A.; Nakata, A.; Lin, J.; Reddy, R.M.; Chang, A.C.; Ramnath, N.; Gotoh, N.; Beer, D.G.; Chen, G. A Novel Serum 4-microRNA Signature for Lung Cancer Detection. *Sci. Rep.* **2015**, *5*, 12464. [CrossRef]
58. Wang, C.; Ding, M.; Xia, M.; Chen, S.; Van Le, A.; Soto-Gil, R.; Shen, Y.; Wang, N.; Wang, J.; Gu, W.; et al. A Five-miRNA Panel Identified From a Multicentric Case-control Study Serves as a Novel Diagnostic Tool for Ethnically Diverse Non-small-cell Lung Cancer Patients. *EBioMedicine* **2015**, *2*, 1377–1385. [CrossRef]
59. Yang, J.S.; Li, B.J.; Lu, H.W.; Chen, Y.; Lu, C.; Zhu, R.X.; Liu, S.H.; Yi, Q.T.; Li, J.; Song, C.H. Serum miR-152.; miR-148a.; miR-148b.; and miR-21 as novel biomarkers in non-small cell lung cancer screening. *Tumour. Biol.* **2015**, *36*, 3035–3042. [CrossRef]
60. Fan, L.; Qi, H.; Teng, J.; Su, B.; Chen, H.; Wang, C.; Xia, Q. Identification of serum miRNAs by nano-quantum dots microarray as diagnostic biomarkers for early detection of non-small cell lung cancer. *Tumour. Biol.* **2016**, *37*, 7777–7784. [CrossRef] [PubMed]
61. Halvorsen, A.R.; Bjaanæs, M.; LeBlanc, M.; Holm, A.M.; Bolstad, N.; Rubio, L.; Peñalver, J.C.; Cervera, J.; Mojarrieta, J.C.; López-Guerrero, J.A.; et al. A unique set of 6 circulating microRNAs for early detection of non-small cell lung cancer. *Oncotarget* **2016**, *7*, 37250–37259. [CrossRef] [PubMed]
62. Yang, X.; Su, W.; Chen, X.; Geng, Q.; Zhai, J.; Shan, H.; Guo, C.; Wang, Z.; Fu, H.; Jiang, H.; et al. Validation of a serum 4-microRNA signature for the detection of lung cancer. *Transl. Lung Cancer Res.* **2019**, *8*, 636–648. [CrossRef]
63. Asakura, K.; Kadota, T.; Matsuzaki, J.; Yoshida, Y.; Yamamoto, Y.; Nakagawa, K.; Takizawa, S.; Aoki, Y.; Nakamura, E.; Miura, J.; et al. A miRNA-based diagnostic model predicts resectable lung cancer in humans with high accuracy. *Commun. Biol.* **2020**, *3*, 134. [CrossRef] [PubMed]
64. Pegtel, D.M.; Gould, S.J. Exosomes. *Annu. Rev. Biochem.* **2019**, *88*, 487–514. [CrossRef]
65. Kalluri, R.; LeBleu, V.S. The biology, function, and biomedical applications of exosomes. *Science* **2020**, *367*, eaau6977. [CrossRef]
66. Kalluri, R. The biology and function of exosomes in cancer. *J. Clin. Investig.* **2016**, *126*, 1208–1215. [CrossRef]
67. Zhang, L.; Yu, D. Exosomes in cancer development, metastasis, and immunity. *Biochim. Biophys. Acta Rev. Cancer* **2019**, *1871*, 455–468. [CrossRef]
68. Whiteside, T.L. The effect of tumor-derived exosomes on immune regulation and cancer immunotherapy. *Future Oncol.* **2017**, *13*, 2583–2592. [CrossRef]
69. Kim, D.H.; Kim, H.; Choi, Y.J.; Kim, S.Y.; Lee, J.E.; Sung, K.J.; Sung, Y.H.; Pack, C.G.; Jung, M.K.; Han, B.; et al. Exosomal PD-L1 promotes tumor growth through immune escape in non-small cell lung cancer. *Exp. Mol. Med.* **2019**, *51*, 1–13. [CrossRef] [PubMed]
70. Alipoor, S.D.; Mortaz, E.; Varahram, M.; Movassaghi, M.; Kraneveld, A.D.; Garssen, J.; Adcock, I.M. The Potential Biomarkers and Immunological Effects of Tumor-Derived Exosomes in Lung Cancer. *Front. Immunol.* **2018**, *9*, 819. [CrossRef]
71. Meng, X.; Jinchang, P.; Shifang, S.; Zhaohui, G. Circulating exosomes and their cargos in blood as novel biomarkers for cancer. *Transl. Cancer Res.* **2018**, *7*, 1. [CrossRef]
72. Whiteside, T.L. The potential of tumor-derived exosomes for noninvasive cancer monitoring, an update. *Expert. Rev. Mol. Diagn.* **2018**, *18*, 1029–1040. [CrossRef] [PubMed]
73. Zheng, H.; Zhan, Y.; Liu, S.; Lu, J.; Luo, J.; Feng, J.; Fan, S. The roles of tumor-derived exosomes in non-small cell lung cancer and their clinical implications. *J. Exp. Clin. Cancer Res.* **2018**, *37*, 226. [CrossRef]
74. Liu, S.; Zhan, Y.; Luo, J.; Feng, J.; Lu, J.; Zheng, H.; Wen, Q.; Fan, S. Roles of exosomes in the carcinogenesis and clinical therapy of non-small cell lung cancer. *Biomed. Pharmacother.* **2019**, *111*, 338–346. [CrossRef] [PubMed]
75. Li, Z.; Zeng, C.; Nong, Q.; Long, F.; Liu, J.; Mu, Z.; Chen, B.; Wu, D.; Wu, H. Exosomal Leucine-Rich-Alpha2-Glycoprotein 1 Derived from Non-Small-Cell Lung Cancer Cells Promotes Angiogenesis via TGF-β Signal Pathway. *Mol. Ther. Oncolytics* **2019**, *14*, 313–322. [CrossRef]

76. Harada, T.; Yamamoto, H.; Kishida, S.; Kishida, M.; Awada, C.; Takao, T.; Kikuchi, A. Wnt5b-associated exosomes promote cancer cell migration and proliferation. *Cancer Sci.* **2017**, *108*, 42–52. [CrossRef] [PubMed]
77. Baig, M.S.; Roy, A.; Rajpoot, S.; Liu, D.; Savai, R.; Banerjee, S.; Kawada, M.; Faisal, S.M.; Saluja, R.; Saqib, U.; et al. Tumor-derived exosomes in the regulation of macrophage polarization. *Inflamm. Res.* **2020**, *69*, 435–451. [CrossRef]
78. Dong, C.; Liu, X.; Wang, H.; Li, J.; Dai, L.; Li, J.; Xu, Z. Hypoxic non-small-cell lung cancer cell-derived exosomal miR-21 promotes resistance of normoxic cell to cisplatin. *Oncol. Targets Ther.* **2019**, *12*, 1947–1956. [CrossRef] [PubMed]
79. Chen, R.; Xu, X.; Qian, Z.; Zhang, C.; Niu, Y.; Wang, Z.; Sun, J.; Zhang, X.; Yu, Y. The biological functions and clinical applications of exosomes in lung cancer. *Cell. Mol. Life Sci.* **2019**, *76*, 4613–4633. [CrossRef]
80. Li, Y.; Yin, Z.; Fan, J.; Zhang, S.; Yang, W. The roles of exosomal miRNAs and lncRNAs in lung diseases. *Signal Transduct. Target. Ther.* **2019**, *4*, 47. [CrossRef]
81. Fan, J.; Xu, G.; Chang, Z.; Zhu, L.; Yao, J. miR-210 transferred by lung cancer cell-derived exosomes may act as proangiogenic factor in cancer-associated fibroblasts by modulating JAK2/STAT3 pathway. *Clin. Sci.* **2020**, *134*, 807–825. [CrossRef]
82. Yuwen, D.L.; Sheng, B.B.; Liu, J.; Wenyu, W.; Shu, Y.Q. MiR-146a-5p level in serum exosomes predicts therapeutic effect of cisplatin in non-small cell lung cancer. *Eur. Rev. Med. Pharmacol. Sci.* **2017**, *21*, 2650–2658.
83. Sun, T.; Kalionis, B.; Lv, G.; Xia, S.; Gao, W. Role of Exosomal Noncoding RNAs in Lung Carcinogenesis. *Biomed. Res. Int.* **2015**, *2015*, 125807. [CrossRef] [PubMed]
84. Zhang, R.; Xia, Y.; Wang, Z.; Zheng, J.; Chen, Y.; Li, X.; Wang, Y.; Ming, H. Serum long non coding RNA MALAT-1 protected by exosomes is up-regulated and promotes cell proliferation and migration in non-small cell lung cancer. *Biochem. Biophys. Res. Commun.* **2017**, *490*, 406–414. [CrossRef] [PubMed]
85. Reclusa, P.; Taverna, S.; Pucci, M.; Durendez, E.; Calabuig, S.; Manca, P.; Serrano, M.J.; Sober, L.; Pauwels, P.; Russo, A.; et al. Exosomes as diagnostic and predictive biomarkers in lung cancer. *J. Thorac. Dis.* **2017**, *9*, S1373–S1382. [CrossRef]
86. Cui, S.; Cheng, Z.; Qin, W.; Jiang, L. Exosomes as a liquid biopsy for lung cancer. *Lung Cancer* **2018**, *116*, 46–54. [CrossRef] [PubMed]
87. Zheng, H.; Wu, X.; Yin, J.; Wang, S.; Li, Z.; You, C. Clinical applications of liquid biopsies for early lung cancer detection. *Am. J. Cancer Res.* **2019**, *9*, 2567–2579.
88. Théry, C.; Witwer, K.W.; Aikawa, E.; Alcaraz, M.J.; Anderson, J.D.; Andriantsitohaina, R.; Antoniou, A.; Arab, T.; Archer, F.; Atkin-Smith, G.K.; et al. Minimal information for studies of extracellular vesicles 2018 (MISEV2018), a position statement of the International Society for Extracellular Vesicles and update of the MISEV2014 guidelines. *J. Extracell. Vesicles* **2018**, *7*, 1535750. [CrossRef]
89. Theodoraki, M.N.; Hoffmann, T.K.; Whiteside, T.L. Separation of plasma-derived exosomes into CD3(+) and CD3(-) fractions allows for association of immune cell and tumour cell markers with disease activity in HNSCC patients. *Clin. Exp. Immunol.* **2018**, *192*, 271–283. [CrossRef]
90. Taylor, D.D.; Shah, S. Methods of isolating extracellular vesicles impact down-stream analyses of their cargoes. *Methods* **2015**, *87*, 3–10. [CrossRef]
91. Melo, S.A.; Luecke, L.B.; Kahlert, C.; Fernandez, A.F.; Gammon, S.T.; Kaye, J.; LeBleu, V.S.; Mittendorf, E.A.; Weitz, J.; Rahbari, N.; et al. Glypican-1 identifies cancer exosomes and detects early pancreatic cancer. *Nature* **2015**, *523*, 177–182. [CrossRef]
92. Turay, D.; Khan, S.; Diaz Osterman, C.J.; Curtis, M.P.; Khaira, B.; Neidigh, J.W.; Mirshahidi, S.; Casiano, C.A.; Wall, N.R. Proteomic Profiling of Serum-Derived Exosomes from Ethnically Diverse Prostate Cancer Patients. *Cancer Investig.* **2016**, *34*, 1–11. [CrossRef] [PubMed]
93. Arbelaiz, A.; Azkargorta, M.; Krawczyk, M.; Santos-Laso, A.; Lapitz, A.; Perugorria, M.J.; Erice, O.; Gonzalez, E.; Jimenez-Agüero, R.; Lacasta, A.; et al. Serum extracellular vesicles contain protein biomarkers for primary sclerosing cholangitis and cholangiocarcinoma. *Hepatology* **2017**, *66*, 1125–1143. [CrossRef]
94. Fang, S.; Tian, H.; Li, X.; Jin, D.; Li, X.; Kong, J.; Yang, C.; Yang, X.; Lu, Y.; Luo, Y.; et al. Clinical application of a microfluidic chip for immunocapture and quantification of circulating exosomes to assist breast cancer diagnosis and molecular classification. *PLoS ONE* **2017**, *12*, e0175050. [CrossRef]
95. Meng, X.; Müller, V.; Milde-Langosch, K.; Trillsch, F.; Pantel, K.; Schwarzenbach, H. Diagnostic and prognostic relevance of circulating exosomal miR-373.; miR-200a.; miR-200b and miR-200c in patients with epithelial ovarian cancer. *Oncotarget* **2016**, *7*, 16923–16935. [CrossRef]
96. Zlotogorski-Hurvitz, A.; Dayan, D.; Chaushu, G.; Salo, T.; Vered, M. Morphological and molecular features of oral fluid-derived exosomes, oral cancer patients versus healthy individuals. *J. Cancer Res. Clin. Oncol.* **2016**, *142*, 101–110. [CrossRef]
97. Santarpia, M.; Liguori, A.; D'Aveni, A.; Karachaliou, N.; Gonzalez-Cao, M.; Daffinà, M.G.; Lazzari, C.; Altavilla, G.; Rosell, R. Liquid biopsy for lung cancer early detection. *J. Thorac. Dis.* **2018**, *10*, S882–S897. [CrossRef]
98. Taverna, S.; Giallombardo, M.; Gil-Bazo, I.; Carreca, A.P.; Castiglia, M.; Chacártegui, J.; Araujo, A.; Alessandro, R.; Pauwels, P.; Peeters, M.; et al. Exosomes isolation and characterization in serum is feasible in non-small cell lung cancer patients, critical analysis of evidence and potential role in clinical practice. *Oncotarget* **2016**, *7*, 28748–28760. [CrossRef]
99. Sun, Y.; Liu, S.; Qiao, Z.; Shang, Z.; Xia, Z.; Niu, X.; Qian, L.; Zhang, Y.; Fan, L.; Cao, C.X.; et al. Systematic comparison of exosomal proteomes from human saliva and serum for the detection of lung cancer. *Anal. Chim. Acta* **2017**, *982*, 84–95. [CrossRef]
100. Fortunato, O.; Gasparini, P.; Boeri, M.; Sozzi, G. Exo-miRNAs as a New Tool for Liquid Biopsy in Lung Cancer. *Cancers* **2019**, *11*, 888. [CrossRef]

101. Srivastava, A.; Amreddy, N.; Razaq, M.; Towner, R.; Zhao, Y.D.; Ahmed, R.A.; Munshi, A.; Ramesh, R. Exosomes as Theranostics for Lung Cancer. *Adv. Cancer Res.* **2018**, *139*, 1–33. [CrossRef] [PubMed]
102. Cao, B.; Wang, P.; Gu, L.; Liu, J. Use of four genes in exosomes as biomarkers for the identification of lung adenocarcinoma and lung squamous cell carcinoma. *Oncol. Lett.* **2021**, *21*, 249. [CrossRef]
103. Sandfeld-Paulsen, B.; Jakobsen, K.R.; Bæk, R.; Folkersen, B.H.; Rasmussen, T.R.; Meldgaard, P.; Varming, K.; Jørgensen, M.M.; Sorensen, B.S. Exosomal Proteins as Diagnostic Biomarkers in Lung Cancer. *J. Thorac. Oncol.* **2016**, *11*, 1701–1710. [CrossRef] [PubMed]
104. Niu, L.; Song, X.; Wang, N.; Xue, L.; Song, X.; Xie, L. Tumor-derived exosomal proteins as diagnostic biomarkers in non-small cell lung cancer. *Cancer Sci.* **2019**, *110*, 433–442. [CrossRef]
105. Ueda, K.; Ishikawa, N.; Tatsuguchi, A.; Saichi, N.; Fujii, R.; Nakagawa, H. Antibody-coupled monolithic silica microtips for highthroughput molecular profiling of circulating exosomes. *Sci. Rep.* **2014**, *4*, 6232. [CrossRef] [PubMed]
106. Vykoukal, J.; Sun, N.; Aguilar-Bonavides, C.; Katayama, H.; Tanaka, I.; Fahrmann, J.F.; Capello, M.; Fujimoto, J.; Aguilar, M.; Wistuba, I.I.; et al. Plasma-derived extracellular vesicle proteins as a source of biomarkers for lung adenocarcinoma. *Oncotarget* **2017**, *8*, 95466–95480. [CrossRef] [PubMed]
107. Wang, N.; Song, X.; Liu, L.; Niu, L.; Wang, X.; Song, X.; Xie, L. Circulating exosomes contain protein biomarkers of metastatic non-small-cell lung cancer. *Cancer Sci.* **2018**, *109*, 1701–1709. [CrossRef] [PubMed]
108. Tao, Y.; Tang, Y.; Yang, Z.; Wu, F.; Wang, L.; Yang, L.; Lei, L.; Jing, Y.; Jiang, X.; Jin, H.; et al. Exploration of Serum Exosomal LncRNA TBILA and AGAP2-AS1 as Promising Biomarkers for Diagnosis of Non-Small Cell Lung Cancer. *Int. J. Biol. Sci.* **2020**, *16*, 471–482. [CrossRef]
109. Zhang, X.; Guo, H.; Bao, Y.; Yu, H.; Xie, D.; Wang, X. Exosomal long non-coding RNA DLX6-AS1 as a potential diagnostic biomarker for non-small cell lung cancer. *Oncol. Lett.* **2019**, *18*, 5197–5204. [CrossRef]
110. Li, C.; Lv, Y.; Shao, C.; Chen, C.; Zhang, T.; Wei, Y.; Fan, H.; Lv, T.; Liu, H.; Song, Y. Tumor-derived exosomal lncRNA GAS5 as a biomarker for early-stage non-small-cell lung cancer diagnosis. *J. Cell. Physiol.* **2019**, *234*, 20721–20727. [CrossRef] [PubMed]
111. Wang, J.; Zhao, X.; Wang, Y.; Ren, F.; Sun, D.; Yan, Y.; Kong, X.; Bu, J.; Liu, M.; Xu, S. circRNA-002178 act as a ceRNA to promote PDL1/PD1 expression in lung adenocarcinoma. *Cell Death Dis.* **2020**, *11*, 32. [CrossRef]
112. Fan, T.W.M.; Zhang, X.; Wang, C.; Yang, Y.; Kang, W.Y.; Arnold, S.; Higashi, R.M.; Liu, J.; Lane, A.N. Exosomal lipids for classifying early and late stage non-small cell lung cancer. *Anal. Chim. Acta* **2018**, *1037*, 256–264. [CrossRef]
113. Cazzoli, R.; Buttitta, F.; Di Nicola, M.; Malatesta, S.; Marchetti, A.; Rom, W.N.; Pass, H.I. microRNAs derived from circulating exosomes as noninvasive biomarkers for screening and diagnosing lung cancer. *J. Thorac. Oncol.* **2013**, *8*, 1156–1162. [CrossRef]
114. Giallombardo, M.; Chacártegui, B.J.; Castiglia, M.; Van Der Steen, N.; Mertens, I.; Pauwels, P.; Peeters, M.; Rolfo, C. Exosomal miRNA Analysis in Non-small Cell Lung Cancer (NSCLC) Patients' Plasma Through qPCR, A Feasible Liquid Biopsy Tool. *J. Vis. Exp.* **2016**, *111*, 53900. [CrossRef] [PubMed]
115. Liu, Q.; Yu, Z.; Yuan, S.; Xie, W.; Li, C.; Hu, Z.; Xiang, Y.; Wu, N.; Wu, L.; Bai, L.; et al. Circulating exosomal microRNAs as prognostic biomarkers for non-small-cell lung cancer. *Oncotarget* **2017**, *8*, 13048–13058. [CrossRef]
116. Zhang, J.T.; Qin, H.; Man Cheung, F.K.; Su, J.; Zhang, D.D.; Liu, S.Y.; Li, X.F.; Qin, J.; Lin, J.T.; Jiang, B.Y.; et al. Plasma extracellular vesicle microRNAs for pulmonary ground-glass nodules. *J. Extracell. Vesicles* **2019**, *8*, 1663666. [CrossRef] [PubMed]
117. Rabinowits, G.; Gerçel-Taylor, C.; Day, J.M.; Taylor, D.D.; Kloecker, G.H. Exosomal microRNA, a diagnostic marker for lung cancer. *Clin. Lung Cancer* **2009**, *10*, 42–46. [CrossRef]
118. Silva, J.; García, V.; Zaballos, Á.; Provencio, M.; Lombardía, L.; Almonacid, L.; García, J.M.; Domínguez, G.; Peña, C.; Diaz, R.; et al. Vesicle-related microRNAs in plasma of nonsmall cell lung cancer patients and correlation with survival. *Eur. Respir. J.* **2011**, *37*, 617–623. [CrossRef]
119. Jin, X.; Chen, Y.; Chen, H.; Fei, S.; Chen, D.; Cai, X.; Liu, L.; Lin, B.; Su, H.; Zhao, L.; et al. Evaluation of Tumor-Derived Exosomal miRNA as Potential Diagnostic Biomarkers for Early-Stage Non-Small Cell Lung Cancer Using Next-Generation Sequencing. *Clin. Cancer Res.* **2017**, *23*, 5311–5319. [CrossRef] [PubMed]
120. Dejima, H.; Iinuma, H.; Kanaoka, R.; Matsutani, N.; Kawamura, M. Exosomal microRNA in plasma as a non-invasive biomarker for the recurrence of non-small cell lung cancer. *Oncol. Lett.* **2017**, *13*, 1256–1263. [CrossRef] [PubMed]
121. Nigita, G.; Distefano, R.; Veneziano, D.; Romano, G.; Rahman, M.; Wang, K.; Pass, H.; Croce, C.M.; Acunzo, M.; Nana-Sinkam, P. Tissue and exosomal miRNA editing in Non-Small Cell Lung Cancer. *Sci. Rep.* **2018**, *8*, 10222. [CrossRef]
122. Poroyko, V.; Mirzapoiazova, T.; Nam, A.; Mambetsariev, I.; Mambetsariev, B.; Wu, X.; Husain, A.; Vokes, E.E.; Wheeler, D.L.; Salgia, R. Exosomal miRNAs species in the blood of small cell and non-small cell lung cancer patients. *Oncotarget* **2018**, *9*, 19793–19806. [CrossRef]
123. Zhang, Y.; Yin, Y.; Li, S. Detection of circulating exosomal miR-17-5p serves as a novel non-invasive diagnostic marker for non-small cell lung cancer patients. *Pathol. Res. Pract.* **2019**, *215*, 152466. [CrossRef]
124. Wu, Q.; Yu, L.; Lin, X.; Zheng, Q.; Zhang, S.; Chen, D.; Pan, X.; Huang, Y. Combination of Serum miRNAs with Serum Exosomal miRNAs in Early Diagnosis for Non-Small-Cell Lung Cancer. *Cancer Manag. Res.* **2020**, *12*, 485–495. [CrossRef]
125. Liu, W.; Liu, J.; Zhang, Q.; Wei, L. Downregulation of serum exosomal miR-216b predicts unfavorable prognosis in patients with non-small cell lung cancer. *Cancer Biomark.* **2020**, *27*, 113–120. [CrossRef]
126. Sun, S.; Chen, H.; Xu, C.; Zhang, Y.; Zhang, Q.; Chen, L.; Ding, Q.; Deng, Z. Exosomal miR-106b serves as a novel marker for lung cancer and promotes cancer metastasis via targeting PTEN. *Life Sci.* **2020**, *244*, 117297. [CrossRef] [PubMed]

127. Zhang, L.; Shan, X.; Wang, J.; Zhu, J.; Huang, Z.; Zhang, H.; Zhou, X.; Cheng, W.; Shu, Y.; Zhu, W.; et al. A three-microRNA signature for lung squamous cell carcinoma diagnosis in Chinese male patients. *Oncotarget* **2017**, *8*, 86897–86907. [CrossRef] [PubMed]
128. Wang, X.; Jiang, X.; Li, J.; Wang, J.; Binang, H.; Shi, S.; Duan, W.; Zhao, Y.; Zhang, Y. Serum exosomal miR-1269a serves as a diagnostic marker and plays an oncogenic role in non-small cell lung cancer. *Thorac. Cancer* **2020**, *11*, 3436–3447. [CrossRef] [PubMed]
129. Wu, Y.; Wei, J.; Zhang, W.; Xie, M.; Wang, X.; Xu, J. Serum Exosomal miR-1290 is a Potential Biomarker for Lung Adenocarcinoma. *Oncol. Targets Ther.* **2020**, *13*, 7809–7818. [CrossRef] [PubMed]
130. Zhang, Y.; Xu, H. Serum exosomal miR-378 upregulation is associated with poor prognosis in non-small-cell lung cancer patients. *J. Clin. Lab. Anal.* **2020**, *34*, e23345. [CrossRef]
131. Chen, L.; Cao, P.; Huang, C.; Wu, Q.; Chen, S.; Chen, F. Serum exosomal miR-7977 as a novel biomarker for lung adenocarcinoma. *J. Cell. Biochem.* **2020**, *121*, 3382–3391. [CrossRef]
132. Grimolizzi, F.; Monaco, F.; Leoni, F.; Bracci, M.; Staffolani, S.; Bersaglieri, C.; Gaetani, S.; Valentino, M.; Amati, M.; Rubini, C.; et al. Exosomal miR-126 as a circulating biomarker in non-small-cell lung cancer regulating cancer progression. *Sci. Rep.* **2017**, *7*, 15277. [CrossRef]
133. Feng, M.; Zhao, J.; Wang, L.; Liu, J. Upregulated Expression of Serum Exosomal microRNAs as Diagnostic Biomarkers of Lung Adenocarcinoma. *Ann. Clin. Lab. Sci.* **2018**, *48*, 712–718. [PubMed]
134. Tang, Y.; Zhang, Z.; Song, X.; Yu, M.; Niu, L.; Zhao, Y.; Wang, L.; Song, X.; Xie, L. Tumor-Derived Exosomal miR-620 as a Diagnostic Biomarker in Non-Small-Cell Lung Cancer. *J. Oncol.* **2020**, *2020*, 6691211. [CrossRef]
135. Zhang, Z.; Tang, Y.; Song, X.; Xie, L.; Zhao, S.; Song, X. Tumor-Derived Exosomal miRNAs as Diagnostic Biomarkers in Non-Small Cell Lung Cancer. *Front. Oncol.* **2020**, *10*, 560025. [CrossRef] [PubMed]
136. Zhang, Z.J.; Song, X.G.; Xie, L.; Wang, K.Y.; Tang, Y.Y.; Yu, M.; Feng, X.D.; Song, X.R. Circulating serum exosomal miR-20b-5p and miR-3187-5p as efficient diagnostic biomarkers for early-stage non-small cell lung cancer. *Exp. Biol. Med.* **2020**, *245*, 1428–1436. [CrossRef] [PubMed]
137. Olive, V.; Li, Q.; He, L. mir-17-92, a polycistronic oncomir with pleiotropic functions. *Immunol. Rev.* **2013**, *253*, 158–166. [CrossRef] [PubMed]
138. Ventura, A.; Young, A.G.; Winslow, M.M.; Lintault, L.; Meissner, A.; Erkeland, S.J.; Newman, J.; Bronson, R.T.; Crowley, D.; Stone, J.R.; et al. Targeted deletion reveals essential and overlapping functions of the miR-17 through 92 family of miRNA clusters. *Cell* **2008**, *132*, 875–886. [CrossRef]
139. Mogilyansky, E.; Rigoutsos, I. The miR-17/92 cluster, a comprehensive update on its genomics; genetics; functions and increasingly important and numerous roles in health and disease. *Cell Death Differ.* **2013**, *20*, 1603–1614. [CrossRef]
140. Zhang, X.; Li, Y.; Qi, P.; Ma, Z. Biology of MiR-17-92 Cluster and Its Progress in Lung Cancer. *Int. J. Med. Sci.* **2018**, *15*, 1443–1448. [CrossRef]
141. Feng, Y.H.; Tsao, C.J. Emerging role of microRNA-21 in cancer. *Biomed. Rep.* **2016**, *5*, 395–402. [CrossRef]
142. Bica-Pop, C.; Cojocneanu-Petric, R.; Magdo, L.; Raduly, L.; Gulei, D.; Berindan-Neagoe, I. Overview upon miR-21 in lung cancer, focus on NSCLC. *Cell. Mol. Life Sci.* **2018**, *75*, 3539–3551. [CrossRef]
143. Zhou, B.; Wang, D.; Sun, G.; Mei, F.; Cui, Y.; Xu, H. Effect of miR-21 on Apoptosis in Lung Cancer Cell Through Inhibiting the PI3K/Akt/NF-κB Signaling Pathway in Vitro and in Vivo. *Cell. Physiol. Biochem.* **2018**, *46*, 999–1008. [CrossRef]
144. Marin, I.; Ofek, E.; Bar, J.; Prisant, N.; Perelman, M.; Avivi, C.; Lavy-Shahaf, G.; Onn, A.; Katz, R.; Barshack, I. MiR-21; EGFR and PTEN in non-small cell lung cancer, an in situ hybridisation and immunohistochemistry study. *J. Clin. Pathol.* **2020**, *73*, 636–641. [CrossRef] [PubMed]
145. Yin, G.; Zhang, B.; Li, J. miR-221-3p promotes the cell growth of non-small cell lung cancer by targeting p27. *Mol. Med. Rep.* **2019**, *20*, 604–612. [CrossRef] [PubMed]
146. Zheng, J.; Dong, L.; Hu, X.; Xiao, Y.; Wu, Q.; Wang, Y.; Zhu, Y. MicroRNA-221-3p promotes proliferation and invasion in non-small cell lung cancer via targeting Axin2 to regulate Wnt/β-catenin signaling pathway. *Res. Sq.* **2020**, *1*, 1–20. [CrossRef]
147. He, R.Q.; Cen, W.L.; Cen, J.M.; Cen, W.N.; Li, J.Y.; Li, M.W.; Gan, T.Q.; Hu, X.H.; Chen, G. Clinical Significance of miR-210 and its Prospective Signaling Pathways in Non-Small Cell Lung Cancer: Evidence from Gene Expression Omnibus and the Cancer Genome Atlas Data Mining with 2763 Samples and Validation via Real-Time Quantitative PCR. *Cell. Physiol. Biochem.* **2018**, *46*, 925–952. [CrossRef]
148. Ju, S.; Liang, Z.; Li, C.; Ding, C.; Xu, C.; Song, X.; Zhao, J. The effect and mechanism of miR-210 in down-regulating the autophagy of lung cancer cells. *Pathol. Res. Pract.* **2019**, *215*, 453–458. [CrossRef] [PubMed]
149. Goto, A.; Tanaka, M.; Yoshida, M.; Umakoshi, M.; Nanjo, H.; Shiraishi, K.; Saito, M.; Kohno, T.; Kuriyama, S.; Konno, H.; et al. The low expression of miR-451 predicts a worse prognosis in non-small cell lung cancer cases. *PLoS ONE* **2017**, *12*, e0181270. [CrossRef]
150. Chen, D.; Huang, J.; Zhang, K.; Pan, B.; Chen, J.; De, W.; Wang, R.; Chen, L. MicroRNA-451 induces epithelial-mesenchymal transition in docetaxel-resistant lung adenocarcinoma cells by targeting proto-oncogene c-Myc. *Eur. J. Cancer* **2014**, *50*, 3050–3067. [CrossRef] [PubMed]
151. Li, L.; Gao, R.; Yu, Y.; Kaul, Z.; Wang, J.; Kalra, R.S.; Zhang, Z.; Kaul, S.C.; Wadhwa, R. Tumor suppressor activity of miR-451, Identification of CARF as a new target. *Sci. Rep.* **2018**, *8*, 375. [CrossRef]

152. Li, M.; Wang, Q.; Zhang, X.; Yan, N.; Li, X. Exosomal miR-126 blocks the development of non-small cell lung cancer through the inhibition of ITGA6. *Cancer Cell Int.* **2020**, *20*, 574. [CrossRef]
153. Dong, W.; Yao, C.; Teng, X.; Chai, J.; Yang, X.; Li, B. MiR-140-3p suppressed cell growth and invasion by downregulating the expression of ATP8A1 in non-small cell lung cancer. *Tumour. Biol.* **2016**, *37*, 2973–2985. [CrossRef] [PubMed]
154. Yuan, Y.; Shen, Y.; Xue, L.; Fan, H. miR-140 suppresses tumor growth and metastasis of non-small cell lung cancer by targeting insulin-like growth factor 1 receptor. *PLoS ONE* **2013**, *8*, e73604. [CrossRef]
155. Gao, Z.J.; Yuan, W.D.; Yuan, J.Q.; Yuan, K.; Wang, Y. miR-486-5p functions as an oncogene by targeting PTEN in non-small cell lung cancer. *Pathol. Res. Pract.* **2018**, *214*, 700–705. [CrossRef] [PubMed]
156. Jin, X.; Pang, W.; Zhang, Q.; Huang, H. MicroRNA-486-5p improves nonsmall-cell lung cancer chemotherapy sensitivity and inhibits epithelial-mesenchymal transition by targeting twinfilin actin binding protein 1. *J. Int. Med. Res.* **2019**, *47*, 3745–3756. [CrossRef]
157. Yu, S.; Geng, S.; Hu, Y. miR-486-5p inhibits cell proliferation and invasion through repressing GAB2 in non-small cell lung cancer. *Oncol. Lett.* **2018**, *16*, 3525–3530. [CrossRef]
158. Wang, A.; Zhu, J.; Li, J.; Du, W.; Zhang, Y.; Cai, T.; Liu, T.; Fu, Y.; Zeng, Y.; Liu, Z.; et al. Downregulation of KIAA1199 by miR-486-5p suppresses tumorigenesis in lung cancer. *Cancer Med.* **2020**, *9*, 5570–5586. [CrossRef]
159. Su, J.L.; Chen, P.S.; Johansson, G.; Kuo, M.L. Function and regulation of let-7 family microRNAs. *Microrna* **2012**, *1*, 34–39. [CrossRef] [PubMed]
160. Gilles, M.E.; Slack, F.J. Let-7 microRNA as a potential therapeutic target with implications for immunotherapy. *Expert. Opin. Ther. Targets* **2018**, *22*, 929–939. [CrossRef]
161. Chiu, S.C.; Chung, H.Y.; Cho, D.Y.; Chan, T.M.; Liu, M.C.; Huang, H.M.; Li, T.Y.; Lin, J.Y.; Chou, P.C.; Fu, R.H.; et al. Therapeutic potential of microRNA let-7, tumor suppression or impeding normal stemness. *Cell Transplant.* **2014**, *23*, 459–469. [CrossRef]
162. Sonkoly, E.; Ståhle, M.; Pivarcsi, A. MicroRNAs and immunity, novel players in the regulation of normal immune function and inflammation. *Semin. Cancer Biol.* **2008**, *18*, 131–140. [CrossRef]
163. Liu, X.; Liu, B.; Li, R.; Wang, F.; Wang, N.; Zhang, M.; Bai, Y.; Wu, J.; Liu, L.; Han, D.; et al. miR-146a-5p Plays an Oncogenic Role in NSCLC via Suppression of TRAF6. *Front. Cell Dev. Biol.* **2020**, *8*, 847. [CrossRef]
164. Sun, X.; Cui, S.; Fu, X.; Liu, C.; Wang, Z.; Liu, Y. MicroRNA-146-5p promotes proliferation.; migration and invasion in lung cancer cells by targeting claudin-12. *Cancer Biomark.* **2019**, *25*, 89–99. [CrossRef]
165. Li, Y.; Tian, J.; Guo, Z.J.; Zhang, Z.B.; Xiao, C.Y.; Wang, X.C. Expression of microRNAs-106b in nonsmall cell lung cancer. *J. Cancer Res. Ther.* **2018**, *14*, S295–S298. [CrossRef]
166. Wei, K.; Pan, C.; Yao, G.; Liu, B.; Ma, T.; Xia, Y.; Jiang, W.; Chen, L.; Chen, Y. MiR-106b-5p Promotes Proliferation and Inhibits Apoptosis by Regulating BTG3 in Non-Small Cell Lung Cancer. *Cell. Physiol. Biochem.* **2017**, *44*, 1545–1558. [CrossRef]
167. Ren, T.; Fan, X.X.; Wang, M.F.; Duan, F.G.; Wei, C.L.; Li, R.Z.; Jiang, Z.B.; Wang, Y.W.; Yao, X.J.; Chen, M.W.; et al. miR-20b promotes growth of non-small cell lung cancer through a positive feedback loop of the Wnt/β-catenin signaling pathway. *Int. J. Oncol.* **2020**, *56*, 470–479. [CrossRef]
168. Peng, L.; Li, S.; Li, Y.; Wan, M.; Fang, X.; Zhao, Y.; Zuo, W.; Long, D.; Xuan, Y. Regulation of BTG3 by microRNA-20b-5p in non-small cell lung cancer. *Oncol. Lett.* **2019**, *18*, 137–144. [CrossRef]
169. Jin, J.; Deng, J.; Wang, F.; Xia, X.; Qiu, T.; Lu, W.; Li, X.; Zhang, H.; Gu, X.; Liu, Y.; et al. The expression and function of microRNA-203 in lung cancer. *Tumour. Biol.* **2013**, *34*, 349–357. [CrossRef]
170. Chi, Y.; Jin, Q.; Liu, X.; Xu, L.; He, X.; Shen, Y.; Zhou, Q.; Zhang, J.; Jin, M. miR-203 inhibits cell proliferation.; invasion.; and migration of non-small-cell lung cancer by downregulating RGS17. *Cancer Sci.* **2017**, *108*, 2366–2372. [CrossRef]
171. Wang, N.; Liang, H.; Zhou, Y.; Wang, C.; Zhang, S.; Pan, Y.; Wang, Y.; Yan, X.; Zhang, J.; Zhang, C.Y.; et al. miR-203 suppresses the proliferation and migration and promotes the apoptosis of lung cancer cells by targeting SRC. *PLoS ONE* **2014**, *9*, e105570. [CrossRef] [PubMed]
172. Huang, H.Y.; Lin, Y.C.; Li, J.; Huang, K.Y.; Shrestha, S.; Hong, H.C.; Tang, Y.; Chen, Y.G.; Jin, C.N.; Yu, Y.; et al. miRTarBase 2020: Updates to the experimentally validated microRNA-target interaction database. *Nucleic Acids Res.* **2020**, *48*, D148–D154. [CrossRef]
173. Pathan, M.; Keerthikumar, S.; Ang, C.S.; Gangoda, L.; Quek, C.Y.; Williamson, N.A.; Mouradov, D.; Sieber, O.M.; Simpson, R.J.; Salim, A.; et al. An open access standalone functional enrichment and interaction network analysis tool. *Proteomics* **2015**, *15*, 2597–2601. [CrossRef]

Review

Exosomes: Their Role in Pathogenesis, Diagnosis and Treatment of Diseases

Houssam Aheget [1,2,*], Loubna Mazini [3], Francisco Martin [1], Boutaïna Belqat [2], Juan Antonio Marchal [4,5,6,7] and Karim Benabdellah [1,*]

1. GENYO Centre for Genomics and Oncological Research, Genomic Medicine Department, Pfizer/University of Granada/Andalusian Regional Government, Health Sciences Technology Park, Av. de la Illustration 114, 18016 Granada, Spain; francisco.martin@genyo.es
2. Department of Biology, Faculty of Sciences, University Abdelmalek Essaâdi, Tétouan 93000, Morocco; bbelqat@uae.ac.ma
3. Center of Biological and Medical Sciences (CIAM), Mohammed VI Polytechnic University, Ben-Guerir 43152, Morocco; loubna.MAZINI@um6p.ma
4. Biomedical Research Institute (ibs. GRANADA), 18012 Granada, Spain; jmarchal@ugr.es
5. Biopathology and Regenerative Medicine Institute (IBIMER), Centre for Biomedical Research (CIBM), University of Granada, 18016 Granada, Spain
6. Department of Human Anatomy and Embryology, Faculty of Medicine, University of Granada, 18016 Granada, Spain
7. Excellence Research Unit Modeling Nature (MNat), University of Granada, 18016 Granada, Spain
* Correspondence: houssam.aheget@genyo.es (H.A.); karim.benabdel@genyo.es (K.B.)

Simple Summary: The aim of this review is to provide an overview of the current scientific evidence concerning the role played by exosomes in the pathogenesis, diagnosis and treatment of diseases. The potential use of exosomes as delivery vectors for small-molecule therapeutic agents will be discussed. In addition, a special emphasis will be placed on the involvement of exosomes in oncological diseases, as well as to their potential therapeutic application as liquid biopsy tools mainly in cancer diagnosis. A better understanding of exosome biology could improve the results of clinical interventions using exosomes as therapeutic agents.

Abstract: Exosomes are lipid bilayer particles released from cells into their surrounding environment. These vesicles are mediators of near and long-distance intercellular communication and affect various aspects of cell biology. In addition to their biological function, they play an increasingly important role both in diagnosis and as therapeutic agents. In this paper, we review recent literature related to the molecular composition of exosomes, paying special attention to their role in pathogenesis, along with their application as biomarkers and as therapeutic tools. In this context, we analyze the potential use of exosomes in biomedicine, as well as the limitations that preclude their wider application.

Keywords: exosomes; molecular composition; cancer pathogenesis; diagnostics; therapeutics

1. Introduction

Membrane-bound and heterogeneous extracellular vesicles (EVs) were initially considered anecdotal examples of cell debris or apoptotic bodies released by the majority of cells [1]. EVs are now regarded as key diagnostic tools [2–4] and therapeutic agents [5]. EVs facilitate communication processes between near and distant cells. In addition, these vesicles can be grouped into two major categories: (a) microvesicles (MVs; 100–1000 nm), considered to be functional liposomes composed of molecules such as nucleic acids, proteins and functional lipids surrounded by a lipid bilayer and (b) exosomes (EXOs; 30–150 nm) (Figure 1) [6], which differ from MVs in their size, protein composition, buoyant density, release mechanism and potential physiological role [7–10]. In this review, we will focus mainly on exosomes, with particular emphasis on their composition. We will discuss their potential role in signaling under both physiological and different pathological conditions. Special attention will be paid to the therapeutic role of exosomes as delivery vectors, as well as their potential use as biomarkers and in clinical interventions.

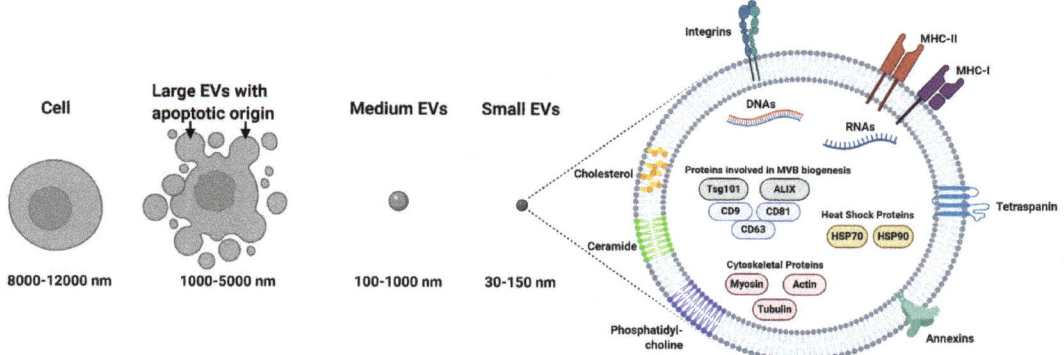

Figure 1. Sizes of most common cell particles: small extracellular vesicles (EVs) are 30 nm to 150 nm in size, and medium-sized EVs are in the 100 nm to 1000 nm range, while large EVs of apoptotic origin are typically 50 nm to 5000 nm in diameter [11]. The release of small EVs or exosomes differs from migracytosis, which involves the translocation of cytoplasmic material to migrasomes, followed by their release when the retraction fibers break [12]. In addition, the uptake of small EVs may have an effect on recipient cells different from that of multivesicular body (MVB)-like EVs, whose release may lead to a relatively delayed effect on the microenvironment [13]. Molecular composition of exosomes: exosomes are surrounded by a phospholipid bilayer and contain numerous molecules, including proteins, lipids, DNA and several types of RNA. (MHC: major histocompatibility complex).

2. Pathological Functions of Exosomes

Exosomes are known to transfer bioactive cargo between donor and recipient cells, ensuring pleiotropic functions in intercellular communication. They are also considered to be an important factor in tumor pathogenesis and immunosuppression [14]. They generate an intricate network of interactions that inhibit the immune system by delivering similar contents of tumor cells to immune cells and also impair natural killer cell activation and induce effector T cell apoptosis [15]. These vesicles have been reported to use autocrine and paracrine signaling pathways to regulate cell characteristics, to modulate their microenvironment and to boost their effects [16]. In addition, exosomes can act as external stimuli and modify the biological phenotype of recipient cells by changing their RNA expression and activating their receptors. Interestingly, cancer cells exchange exosomes with stromal cells in order to create a protumor microenvironment and to increase tumor invasion and proliferation [17].

On the other hand, these vesicles facilitate the interneuronal transmission of pathogenic proteins that are responsible for several neurodegenerative diseases, such as Parkinson's disease (PD) and Alzheimer's disease (AD) [18]. The exosomal transfer of p-tau and $A\beta1\text{-}42$ between cells and body fluids is potentially involved in the slow progress of AD. Moreover, early detection of these neurodegenerative proteins could lead to successful treatments and longer survival [19]. Thus, the key protein involved in PD pathology α-synuclein is secreted via a calcium-dependent mechanism and transported by exosomes, leading to cell death in recipient cells [20]. Exosomes have also been reported to release cellular prion protein (PrPc) and prion protein scrapie (PrPsc) to the extracellular environment, thereby contributing to the pathological spread of infectious prions [21].

2.1. Tumor Pathogenesis

Tumor cells influence both their surrounding microenvironment and distant organs where they can promote angiogenesis, proliferation and cancer metastasis. Exosomes, which are considerably involved in cancer growth and metastatic spread, are considered the main cause of the paracrine effect on recipient cells (Figure 2). The regulation of oncogene expression and abnormal transformations might also result from different initiation factor effects. Eukaryotic translation initiation factor 3 (eIF3) bridges the 43S preinitiation complex and eIF4F-bound mRNA to control protein synthesis, and their aberrant expressed subunits are associated with different cancers [22]. The transforming growth factor beta (TGF-β) signaling pathway, another cancer initiation and progression factor, acts through its central mediator SMAD4 by disrupting DNA damage responses and repair mechanisms, thus enhancing their genomic instability [23]. This signaling is also targeted by the migration inhibitory factor (MIF) to induce the fibronectin production necessary for the remodeling of the premetastatic niche [24]. Additionally, TGF-β is reported to increase fibroblast growth factor-2 (FGF-2) production and mesenchymal stem cell (MSC) differentiation into myofibroblasts to trigger cancer proliferation and invasiveness [25,26]. In the tumor environment, the production of hypoxia inducible factor-1 (HIF-1) plays a crucial role in cancer initiation and progression. Consequently, hypoxia induces HIF-1 stabilization, and its nuclear translocation fosters the expression of genes such as vascular endothelial growth factor (VEGF), hepatocyte growth factor (HGF) and the Met protooncogene [27]. The oncogenes Kristen rat sarcoma 2 viral oncogene homolog (KRAS), epidermal growth factor (EGF) and SRC are transferred by exosomes to recipient tumor cells to promote tumor invasion [28]. To ensure the tumor evasion of immune surveillance, exosomes also release programmed death-ligand 1 (PD-L1) [29].

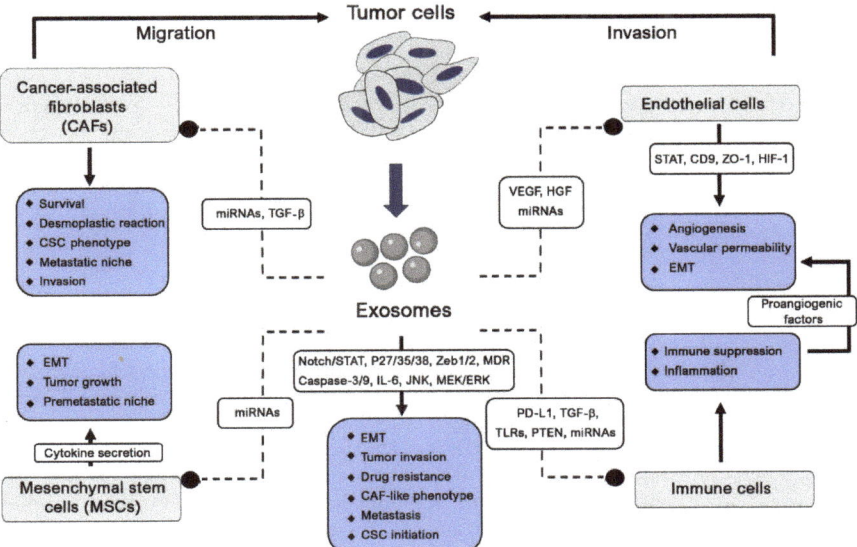

Figure 2. Roles of tumor-derived exosomes in cancer pathogenesis. Cancer stem cells: CSC. Epithelial-to-mesenchymal transition: EMT. microRNAs: miRNAs. Transforming growth factor beta: TGF-β. Signal transducer and activator of transcription: Stat. zinc finger E-box-binding homeobox: Zeb. Multidrug resistance: MDR. Interleukin 6: IL-6. Jun N-terminal kinases: JNK. Mitogen-activated protein kinase kinase: MEK. Extracellular signal-regulated kinases: ERK. Vascular endothelial growth factor: VEGF. Hepatocyte growth factor: HGF. Programmed death-ligand 1: PD-L1. Toll-like receptors: TLRs. Phosphatase and tensin homolog: PTEN. Cluster of differentiation 9: CD9. Zonula occludens-1: ZO-1. Hypoxia inducible factor-1: HIF-1).

Heat shock proteins (HSPs), which are associated with stress conditions, are key regulators of exosome formation and release [30,31] and are involved in antitumor activity in a murine model in a major histocompatibility complex (MHC)-independent manner [32]. Furthermore, the P53 protein is mutated or lost in the majority of cancer types and also modulates many surveillance pathways [33,34]. This protein modulates the transcription of different genes, including TSAP6 and CHMP4C, thus promoting exosome production [35]. These transcriptional signals are involved in cell communication and immune activation [36]. Another tumor suppresser, the phosphatase and tensin homolog (PTEN) protein secreted in exosomes, presents phosphatase activity in target cells, resulting in the activation of the apoptosis cascade and suppression of cell proliferation through interactions with Notch signaling [37,38].

Exosomes containing noncoding RNAs (long noncoding and microRNAs (lnc- and miRNAs)) are associated with many cellular mechanisms [39,40]. MiRNAs were first identified in human serum and later in biological fluids such as saliva, urine and breast milk, thus confirming their role in cell-to-cell communication [7,41]. By modulating mRNA translation in target cells, exosome-associated miRNAs can improve and suppress cellular unbalance, development and tumorigenesis [42].

MiRNA secreted by nontumor cells can affect various cancer-associated mechanisms. Tumors not only contain cancerous cells but also vascular, immune and cancer-associated fibroblastic cells, as well as an extracellular matrix (ECM), all nontumor cells involved in cellular communication and signaling and sustain neighboring tumor cell growth and metastasis [43]. These normal cells secrete tumor-suppressive miRNAs in their EVs to competitively overcome the anarchic growth of their neighbors, a system failure that might initiate cancer [44]. This was observed in prostate cancer, where miRNA-143 acts as a death signal, inducing growth inhibition through a cell-competitive process [45]. Table 1 summarizes the commonly reported miRNAs and lnc-miRNAs found in cancer pathogenesis. Cancer is a multifactorial process in which different miRNAs are secreted by different cells belonging to the tumor microenvironment, resulting in intercellular communication and a single pathway, causing initiation and progression of the disease, angiogenesis, metastasis and drug resistance. In contrast, a single miRNA could be a key modulator of different signaling systems in multiple intercellular networks in recipient cells, thereby modifying their destination and signaling pathway, thus promoting tumorigenesis. Lnc-RNAs, which are highly expressed in exosomes, play a crucial role in the microenvironment by transferring cell signaling and by modulating gene expression [46,47].

2.1.1. Cancer Initiation

Cancer is a genetic and irreversible change due to the activation of specific oncogenes, inactivation of tumor-suppressive genes or other genes involved in genome stability. The evolution of these cancer cells is the result of dual interactions between cancer cells and their surrounding microenvironments. Inflammation is considered the driving initiator of tumor development. Exosome integrins are reported to upregulate S100 proinflammatory molecules, probably by activating and phosphorylating SRCs [48]. Additionally, tumor cells induce the secretion of inflammatory factors, including VEGF, tumor necrosis factor-α (TNF-α), TGF-β and interleukins, to stimulate myeloid cells and immune cells to migrate, thus amplifying inflammatory factor secretion [49]. The immune response is prevented later after the programmed death receptor (PD-1) expressed on activated T and B cells and macrophages binds to its ligand, PD-L1, inducing T-cell apoptosis and inhibiting T-cell activation and proliferation [50,51]. Tumor-associated macrophages (TAMs), T-regulators and myeloid-derived suppressor cells are also recruited to the tumor to inhibit the immune response [52,53]. This immune suppression phase is followed by the improvement of angiogenesis and vascular permeability. In this case, MSCs and endothelial cell interactions mediated by Akt phosphorylation lead to the formation of a vascular microenvironment [54]. By expressing E-cadherin and carbonic anhydrase-9 (CA-9) on their surfaces, exosomes also promote angiogenesis [55,56]. Additionally, integrins (ITGs) present on their surfaces determine organotropism, and their different expressions are organ-specific [48]. These ITGs colocalize specifically with ECM components (laminin or fibronectin) whose composition is modulated by fibroblasts and endothelial cells, suggesting that exosomes drive the colonization of tumor cells by remodeling the stromal microenvironment of the target organ. Mesenchymal stem cells (MSCs) are part of the tumor microenvironment [57], where they are educated and transformed through the release of exosomes into tumor-supportive myofibroblasts, leading to cancer initiation [58].

Likewise, cancer cell-derived exosomes from multiple myeloma (MM) cells are reported to transfer miRNAs to MSCs to initiate cancer, which, in turn, activates cytokine secretion, tumor growth and migration [59]. This mutual intercellular communication is of primordial importance in initiating tumorigenesis in different organs. Tumor cells can also inhibit or decrease antitumorigenic miRNA activity, leading to cancer initiation [60]. The release of miRNA-202-3p by exosomes into the microenvironment negatively regulates its antitumorigenic target [61] (Table 1). From an alternative perspective, cancer-associated fibroblasts (CAFs), which are mostly present in the cancer microenvironment, could induce tumor development and progression. These cells, which secrete miRNAs such as miRNA-21, miRNA-378e and miRNA-143, significantly increase the stemness of breast cancer cells and their epithelial–mesenchymal phenotype [62]. In addition, infiltrating monocytes play an important role in tumor cell progression, as they are driven to differentiate into M2 tumor-associated macrophages (TAMs) by the derived exosomes miRNA-103a and miRNA-203, leading to the secretion of fibroblastic and proangiogenic growth factors [63,64].

On the other hand, exosome lncRNA-p21 is reported to suppress prostate cancer initiation and the expression of genes transcriptionally regulated by P53 [65]. P53 expression is also stimulated by lnc-RNA-MEG3 to inhibit cell proliferation in lung cancer [66]. Lnc-RNA-GAS5 represses antiapoptotic genes when binding to the DNA-binding domain of the glucocorticoid receptor to prevent prostate cancer initiation [67].

Other lnc-RNAs are reported to favor tumor progression by regulating or silencing different miRNAs involved in cancer initiation repression. LncRNA-HOTAIR, which is associated with poor prognosis in urothelial bladder cancer, sponges miRNA-205, thus facilitating tumor initiation and progression [68]. Similarly, lncRNA-MALAT1 is reported to modulate EMT and to promote cervical cancer cell growth and invasion [69,70]. LncRNA-MONC and MIR100HG are both expressed in acute megakaryoblastic leukemia blasts and act as oncogenes associated with tumor development [71]. LncRNA-RoR is a stress-responsive lncRNA in hepatocellular cancer, preventing the activation of cellular stress and miRNA-145 sponging, which can also promote the expression of hypoxia-inducible genes associated with cell growth, apoptosis, angiogenesis, differentiation and survival [72]. Another lnc-RNA, lncRNA DANCR, has been reported to sponge miRNA-33a-5p and to increase EMT, cell proliferation and migration in gliomas [73].

2.1.2. Tumor Angiogenesis

The formation of tumor-associated vessels might be mediated by the sprouting of tumors surrounding pre-existing vessels or the newly recruited endothelial progenitor cells from bone marrow [60]. Exosome-derived miRNA-21 is reported to increase vascular endothelial growth factor (VEGF) levels (Table 1), the key player in angiogenesis, which facilitates endothelial cell migration and new blood vessel formation [74,75]. STAT proteins are also targeted by miRNA-9, whereby tumor neovascularization is strongly activated [76–78]. Another miRNA, miRNA-135b, transferred to endothelial cells by multiple myeloma cell-derived exosomes, inhibits hypoxia-inducible factor 1 (HIF-1) and promotes angiogenesis [79].

Angiogenesis is an important mediator of tumor progression through the induction of protumoral TAMs when monocytes incorporate miRNA-203-derived exosomes secreted by colorectal cancer cells [63] and miRNA-103a-derived exosomes from lung cancer [64]. This mechanism underlies the spread of cancer through the polarization of tumor-suppressive and proangiogenic macrophages. Exosomes also mediate the endothelial cell phenotype in CD90+ liver cancer cells through lnc-RNA H19 and promote angiogenesis and cell-to-cell adhesion [80].

2.1.3. Tumor Metastasis

Since 1989, when Steven Paget introduced the concept of "seed and soil" in relation to tumor progression and metastasis, a great body of literature has been developed, with a better understanding of the mechanisms underlying tumor growth and metastasis [81]. The spread of tumor cells was proposed as the result of the interaction and cooperation between cancer cells (seed) and the host organ (soil) [82]. The metastatic process was later identified as including several stages, such as intravasation, extravasation, tumor latency and the development of micrometastasis and macrometastasis. However, the preferential target organs (soil) may be prepared for metastatic deposits through the development of premetastatic niches that facilitate tumor cell homing, colonization and growth. The primary tumor (seed) plays a key role in the development of premetastatic niches by producing soluble factors, inducing bone marrow-derived hematopoietic cell migration to the premetastatic niche. The primary tumor also secretes exosomes, thus modulating the tumor microenvironment in the premetastatic niches. EMT and mesenchymal-to-epithelial transition (MET) enable migratory phenotypes and seed behaviors. EMT enables tumor cells to enter the circulation and seeding at distant sites [83], where MET is responsible for colonization and metastasis [84]. Moreover, premetastatic niche formation is associated with the composition of molecular and cellular components undergoing four stages to support tumor growth and metastasis. In the primary phase, the primary tumor, which is affected by the uncontrolled proliferation and secretion of exosomes or other tumor-derived secreted factors (TDSFs), becomes hypoxic and inflammatory. Bone marrow-derived (BMD) immune/suppressive cells are prepared and mobilized to form an immature premetastatic niche at a distant organ or at another site of the same organ [85]. In the second licensing phase, BMDCs are continuously recruited in the secondary site in response to exosomes and TDSFs, and their interactions with the distant microenvironment lead to their maturation and preparation for tumor cell colonization. Apart from these BMDCs, bone marrow mesenchymal stem cells (BM-MSCs), which have been identified in different studies, are recruited by the evolving tumor microenvironment as a major source of cancer-associated fibroblasts (CAFs) that boost tumor cell survival [86–88]. The activation of integrins, chemokines and the ECM plays a key role in this organotropism by enabling seeding and colonization in the secondary licensed site [48]. ECM remodeling, as well as the presence of interleukin-1 β (IL-1β) and myeloid-derived suppresser cells, result in the EMT profile of tumor cells [89,90]. The mature and fertile premetastatic niche is colonized by the tumor cells that can undergo latency if the niche microenvironment is not yet suitable during the initiation phase. In the case of a well-prepared niche, seeding and colonization with tumor cells lead to the formation of micrometastases. In the final progression phase, premetastatic niche hosting and support of more migratory tumor cells induce growth, expansion and progression to form macrometastases.

From another perspective, cancer stem cells (CSCs), also known as cancer-initiating cells, have the ability to self-renew and to regenerate the different cell subpopulations constituting the tumor [91], with evidence showing that few tumor cells can form a tumor and accomplish metastasis [91,92]. CSCs from metastatic breast cancer show significantly higher tumorigenic and metastatic capacities than low metastatic cells [93]. Althogh autophagy, whose contribution to tumor progression and metastasis remains controversial is considered to be another seed-type factor, some evidence has demonstrated its involvement in tumor invasion, colonization [94,95], in EMT [96] and CSC viability [95,97,98]. Tumor cells can also disseminate and metastasize in distant sites; however, a lag between both these processes can occur, with tumor cells entering a dormant state for long periods before giving rise to metastasis months or several years after the primary tumor treatment [99]. When these residual tumor cells, whose reactivation appears to be regulated by microenvironmental factors in certain organs, enter a dormant state, they become immune to therapy.

According to Paget, soil factors may first be represented by the primary tumor microenvironment and some molecules providing primary seed-to-soil signaling to enhance the invasive properties of tumor cells [100,101]. In different cancers, TAMs have been shown to induce tumor cell invasiveness through exosome-derived oncogenic miRNA-223, CCL18 and CCL19 [102,103]. MSCs promote cell motility through CCL5 signaling and endothelial cells by modulating oxygenation and tumor perfusion [104]. Besides promoting tumor growth and angiogenesis, CAFs also secrete SDF-1 to induce tumor cell motility and invasion [105]. Additionally, secondary soil, which plays a critical role in influencing cancer metastasis, is composed of many factors and cell types in the metastatic environment (distant organ microenvironment). In each cancer type, microenvironment-derived factors promote specific signaling, leading to tumor migration, cell adhesion, growth and metastasis by enabling tumor cells to enter the niche.

Invasive features are commonly associated with morphological changes in EMT migration, cytoskeleton organization, motility, the basal membrane and extracellular matrix (ECM) remodeling. Exosomes have emerged as potential regulators of the EMT promotion of tumor invasion and spread. Given that EMT is reversible, mesenchymal-to-epithelial transition (MET) might enable cancer cells to adopt an epithelial profile and capacity and, thus, transmigrate to distant sites, promoting metastasis [106]. The miRNA-200 family (miRNA-200a, -200b, -200c, -429 and -141) has the ability to regulate this epithelial cancer cell phenotype by inhibiting the expression of Zeb1 and Zeb2 gene repressors [107,108]. Being the principal component of the tumor microenvironment, fibroblasts play a crucial role in tumor progression. Their reprogramming into cancer-associated fibroblasts (CAFs) occurs after miRNA-105 and miRNA-155 induction in breast cancer and pancreatic cancer, respectively [109,110].

In addition, exosomes carrying different miRNAs have been shown to display migratory and metastatic behaviors leading to distant tumors [111]. By disrupting the vascular endothelial barrier, miRNA-939 and miRNA-105 are reported to increase its permeability through the VE-cadherin gene and by targeting the tight junction protein ZO-1, respectively [112,113]. In exosomes derived from breast cancer, miRNA-10b, with its higher enrichment levels, also promotes cell invasion [114]. The blood–brain barrier (BBB) is another aspect of tumor cell invasion, in which the modulation of permeability is the key feature of brain metastasis. BBB degradation is caused by miRNA-181c, which downregulates PDPK1 gene expression [115].

Glucose uptake suppression by nontumor cells has also been reported to increase nutrient availability in the premetastatic niche via high-secretion miRNA-122, as observed in breast cancer patients with metastatic progression [116,117].

2.1.4. Tumor Immunity

Exosomes have been reported to regulate adaptive immunity in different organs through the cytokines and miRNAs they secrete [118]. Their involvement in tumor immunity can range from the regulation of tumor antigens to tumor immunity polarization [119,120]. However, the most commonly reported involvement of exosomes in immune responses relates to antitumor supportive activity and to their role in preventing immune surveillance. Tumor exosomes inhibit bone marrow dendritic cell (DC) differentiation via the modulation of interleukin-6 (IL-6) expression [121]. The regulatory factor-X-associated protein (RFXAP), a key transcription factor for the MHC-II gene, is downregulated by pancreatic cancer-secreted exosomes containing miRNA-212-3p, leading to the inhibition of MHC class II expression and CD4+ T-cell inactivation [122]. On the other hand, T-cell apoptosis can be induced via the Fas ligand [123], while cytotoxic natural killer (NK) cell activity can be inhibited via the downregulation of NK group 2 member D by tumor exosomes [124]. Regulatory T cells are induced by exosome-derived transforming growth factor β-1 (TGF-β) or miRNA-214 in order to downregulate the phosphatase and tensin homolog (PTEN) and to increase IL-10 secretion, leading to tumor growth [125,126].

On the other hand, tumor cells can evade immunosuppression responses by upregulating the surface expression of PD-L1 and by inactivating T cells. After binding to its receptor PD-1, the Sh2p-driven dephosphorylation of the T cell receptor and its coreceptor CD28 occurs, resulting in the suppression of the antigen-driven activation of T cells [127]. The level of PD-L1 in blood cancer patients is related to their pathoclinical features. Poggio et al. have also demonstrated the differential expression of exosomal PD-L1 in prostate cancer and melanoma cell lines [128].

Cancer cells release exosomes expressing PD-L1, which binds PD-1 through its extracellular domain on CD8 T cells in a concentration-dependent manner [53,128–130]. This PD-L1 secretion can be significantly amplified in tumor cells and in exosomes in response to interferon gamma (IFN-γ) [128,131]. Exosomal PD-L1 levels, which correlate with tumor size, have been reported to be significantly higher in the plasma of melanoma patients as compared to healthy donors. Breast and lung cancer cells also exhibit immunosuppressive exosomal PD-L1. Physical interactions were identified with exosomal PD-L1 and activated PD-1+ CD8 T cells, leading to the inhibition of their proliferation by reducing the expression of Ki-67 and Granzyme B, cytokine production and cytotoxicity through the inhibition of IFN-γ, IL-2 and TNF-α [129]. Using a preclinical model of prostate cancer, the TRAMP-C2 model, the cluster regulatory interspaced short palindromic repeats (CRISPR)/Cas9-mediated deletion of *Rab27a* and *PD1l*, thus inducing exosomal PD-L1 loss, has proven that exosomal PD-L1 is involved in in vivo tumor growth, even at distant sites [128]. Additionally, in the absence of exosomes or PD-L1, the CD8 T-cell fraction increases in lymph nodes relative to wild-type animals and decreases the exhaustion marker Tim 3 characterizing cell subpopulations and increases the Granzyme B marker. Thus, exposure to exosomal PD-L1-deficient tumor cells or the use of anti-PD-L1 antibodies, considered to be new antitumor therapeutic targets, suppresses tumor growth. Moreover, antibodies against PD-L1 and PD-1 have been demonstrated to be efficient in treating many cancer types.

Known to express different toll-like receptors (TLRs), DCs and MSCs are expected to interact with miRNAs to modulate immunity under normal and tumor conditions. Tumor exosomes release miRNA-21 and miRNA-29a, which are considered TLR family ligands in immune cells and act as key regulators of immune responses associated with prometastatic microenvironments [132]. Pancreatic cancer-derived exosomes regulate TLR4 secretion and the production of cytokines such as TNF-α and IL-12 in DCs through miRNA-203 transfer [133]. DC-derived exosomes are reported to activate T and B cells, thus facilitating the presentation of tumor antigens released by cancer cell-derived exosomes [134]. Additionally, this activation of T and B cells might be amplified by mast cells when DC differentiation is induced [135].

2.1.5. Cancer Drug Resistance

Tumor cells often display resistance, hampering tumor treatments aimed at decrease inter- and intracellular drug concentrations. This resistance can be the result of different mechanisms due to genetic or phenotypic changes termed intrinsic resistance or to extrinsic resistance involving the effect of the tumor microenvironment (TME) [136]. In the TME, endothelial cells, fibroblasts and immune cells interact to support tumor growth and progression, where homotypic or heterotypic exosome transfers are regarded as potent effectors [136–138].

Tumor cells presenting cancer predisposition display multidrug resistance (MDR), which is related to the increase in the expression of drug transporters from the adenosine triphosphate (ATP)-binding cassette transporter (ABC) family [139]. These transporters are present in more than 50% of cancer-presenting MDR phenotypes or can be induced by chemotherapy [140] and encoded by multidrug resistance protein 1 gene (*MDR1* or *ABCB1*) for the p-glycoprotein or the *ABCG2* gene for the breast cancer resistant protein

(BCRP) [141]. Additionally, these transporters are able to transfer drug resistance through exosomes to sensitive cells [142–144]. On the other hand, by reversing their orientation in the exosome membrane, the transporters can drive drugs from donor cells into exosomes for sequestration [143–145]. Acidification of the tumor microenvironment appears to promote drug sequestration by increasing the expression of H+-ATPases [146]. Exosomes can also act as sponges by presenting on their surface bait targets for drug molecules such as CD20 to trap the anti-CD20 rituximab [147].

Exosomes are also reported to mediate irradiation resistance by interacting with the cell cycle and DNA repair. Stroma-derived exosomes are reported to induce tumor cell dormancy through their recruitment in the G0 phase and a CSC phenotype, thus increasing chemoresistance [148]. When exosomes were derived from MSCs, a CSC phenotype was improved in tumor cells [149,150]. Exosomes can also mediate antiapoptosis in donor cells by decreasing the intracellular levels of proapoptotic proteins by releasing caspase-3 and -9 [151,152]. Besides decreasing these proapoptotic proteins, exosomes prevent apoptosis in recipient cells by stimulating antiapoptotic pathways mediated by IL-6, CD41, p38 and p53 and JNK, Raf/MEK/ERK and Akt [152–154]. IL-6, activin A and granulocyte-colony stimulating factor (G-CSF) have been shown to induce a CSC phenotype in lung carcinoma cells by stimulating their de-differentiation [155].

Inducing DNA damage repair is triggered by exosomes to induce tumor cell survival after exposure to genotoxic irradiation. Furthermore, irradiation increases tumor cell exosome release [156]. In breast cancer exosomes, the phosphorylation of ataxia telangiectasia mutated (ATM) kinase, Histone H2AX and checkpoint kinase 1 (ChK1) increases in recipient cells, leading to DNA damage repair responses [157]. DNA double-stranded break repair, induced by tumor cell exosomes to increase irradiation therapy, can occur in response to irradiation [156–158]. Exosomes derived from irradiated tumor cells can adopt a migratory profile to escape from the irradiated site, leading to an increase in irradiation resistance [159].

Cancer-associated fibroblasts (CAFs), which are largely regarded as the principal component of tumors and supportive cells, provide a nursing niche and actively regulate the survival and proliferation of cancer cells [137,138]. CAFs affect cross-interactions between the stroma and tumor to activate tumor-supportive mechanisms [160,161]. One of these mechanisms is related to the decrease in drug penetrance in the tumor microenvironment due to a desmoplastic reaction [162]. After exposure to chemotherapy, CAFs contribute to therapy resistance through the significant increase in exosome release. In response to gemcitabine exposure, these exosomes increase the chemoresistance-inducing factor SNAIL in recipient epithelial cells, leading to proliferation and resistance of pancreatic ductal adenocarcinoma [163]. In breast cancer, fibroblast-derived exosomes induce a CSC phenotype through Notch3/STAT1 [164], where, in lung cancer, these fibroblasts create a nursing microenvironment around aldehyde dehydrogenase 1-positive CSCs to resist chemotherapy [165].

Therapy resistance mediated by the CSC phenotype is closely related to EMT. Exosomes are actually regarded as the main inducers of EMT [166,167], and cross-interactions between EMT, CSCs, resistance and exosomes appear to be responsible for increasing CSC markers in breast cancer biopsies after chemotherapy [168]. Moreover, this EMT confers cell plasticity on CSCs and CAFs. However, CAFs and CAF-like phenotypes may release cancer-supportive signals after exposure to different chemotherapies, as well as to a single ablative dose of radiotherapy [138,161,169].

Increasing evidence demonstrates that miRNA-derived exosomes are involved in drug resistance in different cancers. Breast cancer exosome-derived miRNA-221/222 has been reported to increase tamoxifen resistance by reducing the target gene expression of P27 and Era [170]. Transferred by monocytes, miRNA-155 has been reported to target telomerase activity and telomere length through TERF1 in neuroblastoma cells, leading to enhanced chemotherapy resistance. The authors cited above also report that miRNA-21 is involved in ovarian cancer chemoresistance, which suppresses cell apoptosis by binding to its target, APAF1 [171]. In addition, multidrug resistance protein 1 (MRP-1) is reported to be overexpressed in the promyelocytic leukemia HL60 cell line [172]. Nevertheless, cancer cells might target other adaptation mechanisms to escape chemotherapy; for example, in breast cancer, exosome-derived miRNA-9-5p, miRNA-195-5p and miRNA-203a-3p trigger the expression of stemness-associated genes, including Notch1, SRY-box transcription factor 9 (SOX9), SOX2, NANOG and octamer-binding transcription factor 4 (OCT4), leading to a cancer stem-like cell phenotype [173].

In pancreatic cancer, overexpression of reactive oxygen species (ROS)-detoxifying genes superoxide dismutase 2 (SOD2) and catalase (CAT) and downregulation of gemcitabine-metabolizing enzyme deoxycytidine kinase (DCK) confers cellular chemoresistance through exosome-derived miRNA-155 [174]. Another nc-miRNA associated with cellular stress, lncRNA-RoR, has been reported to act as a mediator of cell-to-cell communication in hepatocellular cancer, which elevates miRNA TGF levels in recipient cells, resulting in chemoresistance [47].

Table 1. Roles and mechanisms of microRNAs (miRNAs) and long noncoding (lnc)-miRNAs reported in cancer pathogenesis.

Exosome Components	Cancer Type	Cell Function	Induced Mechanism	Reference
miRNA-202-3p	Chronic lymphoblastic leukemia (CLL)	Inhibits cancer initiation	Discarded by tumor cells in extracellular vesicles (EVs)	[61]
miRNA-19b miRNA-20a	Acute myeloid leukemia (AML)	Multidrug resistance	Transfer of multidrug resistance protein-1 (MRP-1)	[172]
miRNA-126	Chronic myelogenous leukemia (CML)	Leukemic stem cell quiescence and leukemia growth	Not defined	[109]
LncRNA-MONC miRNA 100HG	Acute megacaryobastic leukemia	Tumor growth	Oncogenes	[71]
miRNA-103a	Lung cancer	Cancer progression and angiogenesis	Decreased phosphatase and tensin homolog (PTEN) and M2 polarization of protumoral macrophages	[64]
miRNA-21	Lung cancer	Modulates immunity, promotes angiogenesis	Increase in ligands of long terminal repeats (LTRs) in immune cells, vascular endothelial growth factor (VEGF) levels	[28,55] [75,132]
miRNA-21	Ovarian cancer	Suppresses apoptosis (drug resistance)	Binding to apoptotic protease activating factor 1 (APAF1)	[175]
miRNA-21	Glioblastoma	Priming tumor microenvironment	Microglial cell reprograming	[176]
miRNA-21	Esophageal squamous cell carcinoma	Cancer cell migration and invasion	Activator of cancer-associated fibroblasts (CAFs), cancer cell migration	[177]
miRNA-21	Breast cancer	Tumor progression	Cancer cell stemness and epithelial-mesenchymal transition (EMT), induction of proinflammatory and pro-tumorigenic monocyte profile	[62,178]
miRNA-9	Breast cancer	Promotes angiogenesis tumor metastasis	Janus kinase-signal transducer and activator of transcription (JAK-STAT) activation Induction of CAFs	[78,179]
miRNA-9-5p miRNA-195-5p miRNA-203a-3p	Breast cancer	Stimulate cancer stem-like line phenotype	Transcription factor one cut homeobox 2 (ONECUT2)	[173]
miRNA-939 miRNA-105	Breast cancer	Destruction of endothelial barrier	Downregulation of vascular endothelial (VE)-cadherin, tight junction protein Zonula occludens-1 (ZO-1)	[113,133]
miRNA-105	Breast cancer	Tumor growth	CAF mediation of metabolic reprograming	[180]
miRNA-10b	Breast cancer	Cell invasion	Suppression target genes homeobox D10 (HOXD10) and Kruppel-like factor 4 (KLF4)	[114]

Table 1. Cont.

Exosome Components	Cancer Type	Cell Function	Induced Mechanism	Reference
miRNA-200 miRNA-122	Breast cancer	Promote metastasis	Mesenchymal-to-epithelial transition (MET) regulation process, glucose metabolism reprogramming	[108,116]
miRNA-181c	Breast cancer and metastatic brain cancer	BBB destruction Brain metastasis	Downregulation of gene phosphoinositide dependent protein kinase 1 (PDPK1)	[115]
miRNA-221/222	Breast cancer	Drug resistance	Reduction in expression of target genes P27 and ERa	[170]
miRNA-222/223	Breast cancer	Breast cancer cell dormancy in bone marrow and drug resistance	Not defined	[148]
miRNA-143	Breast cancer	Promotion of cancer cell stemness and EMT phenotype	Not defined	[62]
miRNA-143	Prostate cancer	Inhibition of cell growth	Induce death signaling between normal and cancer cells	[45]
miRNA-203 miRNA-212-3p	Pancreatic cancer	Immune dysfunction, immune tolerance	Toll-like receptor 4 (TLR4) regulation, downregulation of regulatory factor X-associated protein (RFXAP) expression	[122,133]
miRNA-155	Pancreatic cancer	Chemoresistance Tumor invasion and progression	Promotion of reactive oxygen species (ROS) detoxification Reprograming of normal fibroblasts into CAFs via tumor protein P53 inducible nuclear protein 1 (TP53INP1)	[110,174]
miRNA-21/155	Neuroblastoma	Resistance to chemotherapy	Crosstalk with miRNA-21 Activation of toll-like receptor 8/nuclear factor Kappa B (TLR8/NFKB) and telomeric repeat binding factor 1 (TERF1) axis	[171]
LncRNA DANCR	Glioma	Tumor progression and malignancy	Sponging miRNA-33a-5p	[73]
miRNA-146a	Multiple myeloma	Tumor cell growth	Increased cytokine and chemokine secretion	[59]
miRNA-24-3p	Nansopharyngeal carcinoma (NPC)	Tumor growth	Target fibroblast growth factor 11 (FGF11) to suppress T cells	[181]
Let-7 family	Gastric cancer	Suppression of cancer initiation	Not defined	[182]
miRNA-15b-3a	Gastric cancer	Tumor progression	Restraining dynein light chain Tctex-type 1 (DYNLT1)/caspase-3/Caspase-9 signaling pathway	[183]
miRNA-203	Colorectal cancer	Metastasis	Differentiation of monocytes to M2 tumor-associated macrophages	[63]
miRNA-210	Hepatocellular carcinoma	Angiogenesis	Inhibition of Mothers against decapentaplegic homolog 4 (SMAD4) and Signal transducer and activator of transcription 6 (STAT6) secretion by endothelial cells	[109]

Table 1. *Cont.*

Exosome Components	Cancer Type	Cell Function	Induced Mechanism	Reference
miRNA-103	Hepatocellular carcinoma	Increase in vascular permeability	Inhibition of VE-cadherin, P120-catenin and zonula occludens 1 expression	[184]
LncRNA-RoR	Hepatocellular cancer	Tumor growth	Sponge miRNA-145 and promote hypoxia-inducible factor (HIF)	[72]
LncRNA-HOTAIR	Urotheral bladder cancer	Tumor initiation and progression	Sponge miRNA-205	[68]
LncRNA-MALAT1	Cervical cancer	Tumor invasion	Modulation of epithelial-to-mesenchymal transition (EMT)	[69,70]

2.2. Neurodegenerative Disease

In the central nervous system (CNS), close interactions between neurons, microglia, astrocytes and oligodendrocytes facilitate nerve homeostasis, cellular communication and signal transduction by secreting exosomes, which, however, also leads to the transfer of abnormal mediators [185]. These exosomes, which are released into the extracellular microenvironment, have, in recent years, led to increased interest in the pathophysiology of neurodegenerative diseases associated with aging and increasing life expectancy. Alzheimer's disease (AD), frontotemporal dementia, Parkinson's disease (PD), Huntington's disease (HD), multiple sclerosis (MS) and amyotrophic lateral sclerosis (ALS) have been the subject of intense study focused on different aspects of these diseases, including their physiology, etiopathology, diagnosis and biomarkers, as well as emerging treatments [16,186]. These pathologies are characterized by protein aggregates and the formation of inclusion bodies in specific sites in the brain due to neuronal cell death. The impairment of the quality control mechanisms of these proteins resulting from age-related external stress induces the transmission of these aggregates to other aggregate-free cells in the brain [186]. Recently, exosomes have been identified as potential new biomarkers of great interest in synaptic transmission and nerve regeneration. Additionally, some evidence shows that they are involved in pathogenesis and could play a role in the advanced treatment of neurodegenerative diseases. These exosomes, which act as key mediators in intercellular communication, have recently been observed to be involved in age-related neurodegenerative diseases, leading to cognitive impairment due to their ability to transmigrate the blood–brain barrier (BBB) and to transfer pathological protein aggregates such as amyloidβ (Aβ), tau and α-synuclein proteins to distant brain cells [187]. Cancer cell-derived exosomes can reach the CNS by destroying the BBB and transferring to neural cells. miRNA-181-c has been shown to activate actin mislocalization, enabling exosomes to be transferred to the periphery of the CNS [115]. There is also evidence that exosomes have the ability to cross the BBB in the opposite direction. Hematopoietic cells are reported to transfer their exosomes to Purkinje cells in the brain, leading to a modification in gene expression via the inflammatory pathway [188]. Moreover, exosomes are involved in nerve injuries associated with infectious agents. Prion proteins might be taken up in the infected cells and then delivered to target cells through membrane fusion after secretion in the extracellular fluid [21], suggesting that they play a role in spreading the infectious disease in the brain.

AD is the first common neurogenerative disease in which affected neurons probably secrete tau protein in the exosomes released, thus contributing to the spread and progression of tauopathy due to tau protein hyperphosphorylation [189]. Wang et al. have demonstrated that neuron depolarization leads to the release of exosome-derived tau, whose trans-synaptic transmission is confirmed by its trans-neuronal and microglial transfer [190,191]. Exosomes effectively spread within interconnected neurons and transfer Aβ and tau proteins through an endosomal pathway and axonal transport [192]. The exosomal hyperphosphorylated tau (p-tau) protein and the extracellular senile plaque containing the Aβ peptide result in neuron degeneration and the secretion of proinflammatory cytokines by microglia and astrocytes, thus altering the BBB and causing AD [193]. Rajendran et al. reported that exosome-derived p-tau protein concentrations increase significantly in the mild/severe stages (Braak stages 3–6) of AD, as compared to patients in the early stages (Braak stages 0–2), suggesting that exosomes play a crucial role in the abnormal processing of tau in the cerebrospinal fluid (CSF) in early onset AD [194,195]. On the other hand, Aβ is transported by exosomes to be degraded by lysosomes in normal settings, and the disruption of this clearance could lead to their accumulation in exosomes and AD spread [196]. Similarly, this lysosomal dysfunction has been observed in relation to exosomal α-synuclein release and transmission [197]. Disruption of the secretory pathway of neurons is another pathological mechanism leading to AD, in which the neuroprotective signal peptide sequence targeted by cystatin C is downregulated in exosomes [198]. The soluble amyloid protein precursor (APP) is thus decreased and associated with the involvement of Aβ aggregates [199]. Exosomes from activated astrocytes have also been observed in the pathogenesis of AD by targeting the inflammatory and proapoptotic pathways [200,201]. Astrocytic damage is related to Aβ senile plaques through the activation of prostate apoptosis response 4 (PAR-4) [202,203], while the exosome secretion of PAR-4/ceramide results in neuroprotective astrocyte apoptosis and AD induction [204].

The neurons are likely to modulate myelin biogenesis by regulating the secretion of oligodendroglia-derived exosomes, whereby myelin sheaths are slowed down during CNS development [205]. These exosomes contain myelin proteins and RNAs involved in promoting myelination [206,207]. Their impact is not restricted to a positive effect on myelination through an increase in neuron resistance to stress and their enhanced growth but might also be involved in repairing damaged myelin sheaths [101].

In an immunological setting, exosomes from astrocytes, microglia, platelets, leukocytes and endothelial cells have been demonstrated to secrete metalloproteinase (MMP)-14 and caspase 1 following stimulation by proinflammatory cytokines in MS. These enzymes facilitate lymphocyte and myeloid cell transmigration to CNS by inducing the disintegration of the BBB [208,209]. In addition, endothelial-derived exosomes transfer the ICAM-1 receptor for integrin Mac-1 to monocytes, thus increasing their transmigration through the barrier [210]. Furthermore, activated T lymphocytes are involved in this immunological cascade by releasing exosomes containing larger amounts of chemokine CCL5, which facilitates their adhesion to brain microvessel endothelium cells [211]. This suggests that exosome generation by the neural and immune cell network is of great importance in MS pathogenesis.

Exosome cargo is also transferred outside the CNS. In MS, serum-derived exosomes have been found to contain three myelin proteins: the myelin basic protein, the proteolipid protein and the myelin oligodendrocyte glycoprotein (MOG). Some evidence indicates that MOG content is strongly associated with MS, which modulates anti-myelin immune reactions in both relapsing-remitting MS (RRMS) and secondary progressive MS (SPMS) patients [212]. Significant sphingomyelinase enzymatic activity has recently been found in MS patient-derived exosomes, resulting in decreasing levels of different sphingomyelins in their CSF, which is associated with axonal damage and neuronal dysfunction [213].

Dopaminergic neuron degeneration in substantia nigra, the formation of intracytoplasmic Lewy bodies in other surviving neurons and the abnormal accumulation of α-synuclein are related to the occurrence of PD [214,215]. In addition, α-synucleins control synaptic transmission and vesicle release [216], where Lewy bodies indicate pathological α-synuclein aggregation in neurons and glial cells [217], which propagate according to a prion-like pattern [218]. Some evidence indicates that exosomes are involved in PD by transporting α-synucleins to lysosomes for degradation, which might then be accumulated and released into the intercellular space, resulting in cytotoxicity [219,220]. The coaggregation of α-synuclein with Aβ and the protein tau has also been reported, thus accelerating the neuropathology and cognitive decline [221,222].

Although protein aggregation is a major cause of neurodegenerative disease, exosome-derived miRNAs play a key role in controlling protein levels by regulating their mRNAs [223,224]. Differential miRNA expression is closely associated with AD, PD, ALS, MS and HD [225–231]. In MS, different miRNAs have been identified in serum-derived exosomes, whose signatures appear to be indicative of disease subtypes. MiRNA-15b-5p, miRNA-451a, miRNA-30b-5p and miRNA-342-3p have been identified in RRMS patients, while miRNA-127-3p, miRNA-370-3p, miRNA-409-3p and miRNA-432-5p have been found in SPMS patients [232]. Given the T-cell-mediated autoimmune nature of MS, various studies have reported the involvement of miRNAs in CNS immunomodulation. Exosomal miRNA let-7i was found to increase in MS patients and to suppress T-reg cell induction by targeting insulin-like growth factor 1 receptor (IGF1R) and TGF-β receptor 1 (TGF-β R1), leading to autoimmune modulation [233]. Exosomal miRNA let-7 can also activate TLR7 in neuronal cells and trigger inflammation, causing neuronal death [234,235]. On the other hand, Winkler et al. have suggested that neurons activate TLR7 proteins present in endosomes and the uptake of exosomes containing miRNA let-7, thus inducing cell degeneration [236]. In the CNS, TLRs are widely expressed in different cell types, whose crosstalk with miRNAs is associated with immune damage, causing inflammation and neurodegenerative diseases. Additionally, the pathogenesis of MS is related to an increase in miRNA-326 secretion from T-cell-derived exosomes in RRMSs, thus targeting TH17 differentiation and maturation [237].

In AD, Aβ and the hyperphosphorylated tau protein are individually regulated by the APP gene. Increased APP activity results in higher Aβ levels, which negatively impacts synaptic function and neuron degeneration [238]. Various studies have reported that miRNA-16; miRNA-101; miRNA-193b; miRNA-200b and the miRNA-20a family (miRNA-20a, -106b and -17-5p) downregulate APP expression [239–241]. On the other hand, the post-transcriptional protein tau is targeted by miRNA-34a by combining with the 3′-untranslated region (UTR) of microtubule-associated protein tau (MAPT), which inhibits its endogenous expression and leads to AD neuron degeneration [242,243].

The α-synuclein protein characterizing PD pathogenesis has been found to be overexpressed, with a recent study reporting that the α-synuclein gene (SNCA) combines its 3′-UTR mRNA with miRNA-7, resulting in the inhibition of transcription and protein expression. In PD, given the decrease in miRNA-7 expression, α-synuclein was found to be toxic to dopamine neurons [244,245]. In addition, the blood plasma of patients is enriched in miRNA-4639-5p as a result of the post-transcriptional downregulation of the DJ-1 gene, given that the decrease in DJ-1 protein levels causes severe oxidative stress and neuron death [230].

3. Exosome Composition

Exosomes contain numerous molecules, including proteins, lipids, metabolites, mRNA and microRNA [246], as well as genomic and mitochondrial DNA [247,248]. Other forms of RNA, including transfer, ribosomal, small nucleolar and long noncoding RNA (lncRNA), have also been identified [249] (Figure 1). These can be transferred from host to recipient cells in order to regulate cellular functions [250–252]. In addition, the ExoCarta, EVpedia and Vesiclepedia exosome databases provide detailed information regarding the molecular content of exosomes [253]. The composition of exosomes is a tightly regulated process that is influenced by environmental factors such as cell activation and stress conditions [254]. Exosomes secreted by the same cells are expected to have a similar protein, lipid and nucleic acid composition. However, the molecular composition of exosomes has recently been shown to be non-cell type-dependent and differs even when the exosomes originate from the same parental cells [255–257]. On the other hand, some cargos are common to exosomes of different origins [258]. Novel methods and technologies, including high-resolution flow cytometry [259], laser tweezer Raman spectroscopy (LTRS) [257], ultracentrifugation [260] and immunocapturing [261], have recently been developed in order to differentiate features of exosomes such as exosomal heterogeneity [262].

3.1. Nucleic Acids

Exosomes contain nucleic acids, including messenger RNA (mRNA), microRNA (miRNA) and other noncoding RNAs, which can be transferred between cells and possibly regulate gene expression in recipient cells [263]. Exosomes released from cancer patients have been found to contain fragments of single-stranded DNA and double-stranded genomic DNA, including all chromosomes [264,265]. These vesicles also excrete harmful DNA from cells in order to maintain cellular homeostasis [266]. Exosomal RNA content is a subset of cellular RNA and, in some cases, may differ significantly from that of its parent cell. However, other RNAs are ubiquitous among all types of exosomes regardless of their cell of origin due to their specific targeting in multivesicular bodies (MVBs) during biogenesis [267], indicating that specific RNAs are actively sorted into exosomes. In addition, miRNA packaging in EVs is different from that of the parent cell and is particularly influenced by external stimuli. As exosomal miRNAs play a prominent role in disease progression, induce angiogenesis and facilitate metastasis in cancers [112,268], they can be used as potential noninvasive biomarkers of disease states [269,270].

Koppers-Lalic and colleagues have suggested that post-transcriptional modifications, notably 3′-end adenylation and uridylation, have opposite effects that may contribute, at least in part, to directing ncRNA sorting towards EVs, given the overrepresentation of 3′-end-adenylated miRNAs and 3′-end-uridylated miRNAs in human B cells and their secreted exosomes, respectively [271]. Dicer and Ago2, key components of miRNA processing, have been found to be functionally present in exosomes [272]. A tetranucleotide sequence is also present in miRNAs that controls their localization in exosomes. In fact, the protein heterogeneous nuclear ribonucleoprotein A2B1 (hnRNPA2B1) specifically binds exosomal miRNAs through the recognition of this sequence and controls their loading into exosomes [273]. Similarly, the synaptotagmin-binding cytoplasmic RNA-interacting protein (SYNCRIP) can control miRNA sorting in exosomes. This protein binds directly to miRNAs enriched in exosomes that share a similar sequence or hEXO motif. This motif, whose introduction into a poorly exported miRNA improves its exosomal loading, can regulate miRNA localization [274].

Exosomes produced by endothelial cells promote angiogenesis in vivo in a small RNA-dependent manner. Exosomes produced by human breast cancer cell lines MDA-MB-231 and MDA-MB-436 contain various classes of RNA, such as small nucleolar RNAs (snoRNAs), ribosomal RNAs (rRNAs), transfer RNAs (tRNAs), microRNAs (miRNAs) and yRNAs, with the major class of RNA being fragmented rRNAs, particularly rRNA subunits 28S and 18S [275]. On the other hand, tRNAs are the most common RNA species found in exosomes derived from human adipose- and bone marrow-derived mesenchymal stem cells (MSCs). More than 50% of total small RNAs are tRNAs in adipose-derived exosomes (ASC), while tRNAs account for 23–25% of the total small RNA content in bone marrow (BMSC) exosomes [276]. Similarly, exosomes isolated from urine contain high concentrations of rRNAs (40–60%) and tRNAs (20–50%), followed by mRNAs (5–15%) and miRNAs (5–10%), while serum-derived exosomes are enriched with miRNAs (30–75%), mRNAs (10–20%) and tRNAs (20–30%) [277]. As tRNAs can bind to argonaut proteins and recognize mRNA targets similar to miRNAs, tRNAs may play a major role in RNA silencing [278]. Furthermore, vault RNAs (vRNAs) have been reported to play an important role by mediating the drug-resistant phenotype of malignant cells, suggesting that vRNAs may be involved in the sequestration of chemotherapeutic compounds. On this basis, mitoxantrone has the ability to bind to vRNAs, which potentially sequesters the drug and prevents it from reaching the target site [279].

3.2. Proteins

Exosome protein contents have been well-identified using a wide variety of proteomic techniques. High-throughput proteomic analyses have revealed the presence of proteins involved in cell structure, motility and adhesion, such as actins, myosin, radixin, tubulins, integrins, and cell surface receptors, including epidermal growth factor receptors (EGFRs), platelet-derived growth factor receptor beta (PDGFRB) proteins and plasminogen activator urokinase receptors (PLAURs), as well as signaling proteins, transcription factors and metabolic enzymes [280,281]. In addition, ExoCarta has indicated the presence of over 4000 proteins in exosomes. Exosomal protein composition can vary between different cell types and under different culture conditions. Ingenuity pathway analysis (IPA) has identified the presence of 157 proteins in placenta mesenchymal stem cell (PlaMsc)-derived exosomes exposed to 1% O_2. On the other hand, 34 and 37 individual proteins were found to be present in PlaMSC-3%O_2 and PlaMSC-8%O_2 exosomes, respectively. More proteins associated with vascular endothelial growth factor (VEGF), actin cytoskeleton, growth hormone and clathrin-mediated endocytosis signaling in exosomes have been reported to be isolated from pMSC exposed to 1% O_2 as compared to 3% or 8% O_2, possibly leading to an increase in the exosome uptake of target cells [282]. As characterized by matrix-assisted laser desorption ionization time-of-flight (MALDI-TOF) analysis, MHC-I, together with heat shock proteins HSC70 and HSP90, annexins, PV-1 and developmental endothelial locus-1 (DEL-1), were found to be present in exosomes derived from human mesothelioma cells [283].

Certain molecular markers commonly found in exosomes are essential for the overall biological and pharmacological effects of exosomes. Heat shock proteins HSP70 and HSP90 are molecular chaperones, and tumor susceptibility gene 101 (TSG101) is involved in multivesicular body (MVB) biogenesis. Moreover, tetraspanin and integrin proteins such as CD63, CD9, CD81 and CD82 are pivotal for cell targeting and adhesion, while Rab GTPases, annexins and flotillin are involved in membrane transport and fusion [284]. Different α and β chains of integrins ($\alpha 4\beta 1$, $\alpha M\beta 2$, $\alpha L\beta 2$ and $\beta 2$); A33 antigen and P-selectin; ICAM1/CD54 and cell-surface peptidases CD26 and CD13 are also present in exosomes [285]. Interestingly, given their competition with membrane MHC-II for T-cell receptor binding on $CD4^+$ T cells, soluble MHC-II (sMHC-II) proteins play a prominent role in immune response suppression and the maintenance of tolerance [286].

As the protein composition of exosomes is not identical to that of the parent cell, there are two major protein sorting pathways: the dependent and independent endosomal sorting complexes for transport (ESCRT). ESCRT are composed of four multimeric complexes, ESCRT-0 to ESCRT-III. Baietti and colleagues showed that cytoplasmic adaptor syntenin interacts directly with ALIX, which, in turn, binds to ESCRT-III and is involved in the sorting of syndecan membrane proteins in exosomes [287]. On the other hand, other studies have indicated that proteins can also be packaged into MVBs without the involvement of ESCRT or ubiquitin. Intraluminal vesicle (ILV) formation and melanosomal protein (Pmel17) sorting continue following the disruption of the Hrs/ESCRT function, suggesting that Pmel may be sorted into ILVs independently of Hrs/ESCRT machinery [288]. In addition, the features of protein Sna3p enable its selective inclusion in invaginating vesicles independently of ubiquitin [289]. Intriguingly, Lin et al. found that many ribosomal proteins are secreted by exosomes that are derived from embryonic fibroblasts in sirtuin 6 knockout mice, indicating that SIRT6 affects the sorting of many proteins to exosomes [290].

Le Pecq and colleagues showed that dendritic cell-derived exosomes (dexosomes) induce strong antitumor activity by displaying antigens to $CD8^+$ and $CD4^+$ T cells. In addition, this form of immunotherapy is well-tolerated in patients with advanced non-small cell lung cancer (NSCLC), thus rendering dexosomes a viable acellular alternative to dendritic cells (DC) for use in cancer vaccinations in preclinical and clinical studies [291,292]. Some highly potent proteins in MSC-derived exosomes have the potential to improve cardiac function after myocardial infarction (MI), including growth factors such as fibroblast growth factor 1 (FGF1) and neuregulin-1 (NRG1), involved in cardiac development and regeneration in an MI rat model [293]. In addition, cardiac-specific human fibroblast growth factor 1 (FGF-1) is also associated with enhanced postischemic hemodynamic recovery and the attenuation of reperfusion-induced myocardial cell necrosis during ischemia reperfusion (IR) [294]. Macrophage colony-stimulating factor (M-CSF) increases vascular endothelial growth factor (VEGF) production from cardiomyocytes, protects cardiomyocytes and myotubes from cell death and enhances cardiac function after ischemic injury [295]. Hill et al. demonstrated that glial growth factor 2 (GGF2) improves cardiac function in rats with MI-induced systolic dysfunction [296]. Similarly, chronic leukemia inhibitory factor (LIF) treatment has a positive effect on systolic heart function, suggesting that LIF may have a therapeutic role in preventing or repairing myocardium injury [297].

3.3. Lipids

The effects of exosomes are not only mediated by their nucleic acid and protein content, but exosomal lipids, in particular, can also modulate their bioactivity and vesicle stability. Understanding the biological and pharmacological effects of exosomal lipids can improve our knowledge of exosome biogenesis and will help to develop efficient exosome-based therapeutics [262].

Exosomes are a heterogeneous population of extracellular vesicles (EVs) with different surface-expressed molecular patterns, thus providing an additional tool for their identification. The lipid composition of exosomes, which accounts for their unique rigidity, differs from that of the parent cell's plasma membrane, partly because exosomes also contain lipids from the Golgi apparatus. These vesicles are also rich in cholesterol, ceramide and other sphingolipids, as well as phosphoglycerides with long saturated fatty acyl chains [298]. In this regard, B-cell-derived exosomes are rich in ceramides [299], whose role in the budding of exosome vesicles into MVBs has also been reported [298]. On the other hand, exosomes secreted from oligodendrocytes are highly rich in phosphatidylcholine (40%), phosphatidylserine (25%) and phosphatidylethanolamine (20%) but contain only 2.2% cholesterol [300].

Exosomes from mast and dendritic cells have increased levels of phosphatidylethanolamines, which have a higher rate of flipping between the two leaflets of the exosome bilayer than in cellular membranes [301]. Interestingly, exosomes are able to deliver prostaglandins to the target cells and carry prostaglandins bound to the exosomal membrane with potentially enhanced biological activity rather than the soluble form of prostaglandins [302]. Recent studies have shown that exosomes may affect the lipid composition of recipient cells, specifically cholesterol and sphingomyelin, thus modulating cell homeostasis [303]. Beloribi-Djefaflia and colleagues suggested that exosomal lipids contribute to tumor progression and drug resistance in Mia-PaCa-2 cells [304]. Finally, ceramide-enriched exosomes have been shown to induce astrocyte apoptosis, potentially contributing to the progression of Alzheimer's disease [204].

4. Applications of Exosomes in Biomedicine
4.1. Exosomes as Biomarkers

Exosomes are now regarded as new players in regenerative medicine due to their therapeutic capacity and their potential as noninvasive biomarkers for early diagnosis; the evaluation of treatment efficacy and monitoring of the progression of cancer, neurodegenerative, metabolic and infectious diseases [5,305]. They offer a simple method for the molecular analysis of biofluids that reduces invasive surgery and promotes more precise medical interventions. Several clinical trials have been launched for both early screening and accurate diagnosis to reduce mortality rates and to increase recovery rates (Table 2). The molecular content of exosomes reflects the origin and pathophysiological conditions of releasing cells, suggesting that the analysis of exosomal markers is a highly specific and sensitive method that could potentially replace invasive biopsies. In addition, their small volume, specific biological information, strong permeability through body tissue barriers, abundance and long half-lives in all biological fluids make these biomarkers highly attractive targets for clinical diagnostic applications. In addition to nucleic acids, exosomal proteins have been found to be potential biomarkers for a variety of pathologies, including cancer, as well as a number of noncancer diseases in different organs, such the central nervous system [195,197], the kidneys [306,307], liver [308] and lungs [309].

4.1.1. Exosomes for Cancer Diagnosis

Several types of cancer have long been known to shed exosomes into the blood. Fortunately, recent technological advances have enabled the capture and analysis of these cancer-derived exosomes to be improved upon, making them valuable diagnostic tools. RNAs, including mRNAs, lncRNAs, circular RNAs (circRNAs) and miRNAs, DNA, proteins and lipids, have been extensively used as cancer biomarkers (Figure 3).

DNA. Exosomes produced by several cancer types have been reported to contain DNA. These vesicles carry either long double-stranded DNA fragments [310] or single-stranded DNA [264]. Some studies have revealed the presence of double-stranded DNA in exosomes secreted by human carcinoma and murine melanoma, suggesting its potential use in the early clinical detection of cancer [248]. Similarly, Kahlert and coworkers detected the predominance of double-stranded DNA in pancreatic cancer-derived exosomes, as well as similar genomic mutations among exosomes and parental cancer cells [265]. On the other hand, Balaj et al. identified single-stranded DNA in medulloblastoma-derived exosomes, thus illustrating its promising potential use in cancer diagnosis and therapy [264].

Table 2. Representative table showing a selection of concluded and ongoing clinical trials (Clinical trial.gov; November 2020) utilizing exosomes as biomarkers mainly for cancer diagnosis and, to a lesser extent, for the early detection of other diseases. The clinical work involving the diagnostic application of exosomes has not yet been published or made available for peer review.

CT Identification	Aim of Study	Source of Exosomes	Associated Markers	Promoted by
NCT04182893	Identification of benign and malignant pulmonary nodules	Blood and alveolar lavage fluid	Exosomal RNA	Shanghai Chest Hospital, Shanghai, China
NCT04499794	Study of exosome EML4-ALK fusion in NSCLC clinical diagnosis	Plasma	EML4-ALK fusion	Cancer Hospital Chinese Academy of Medical Sciences, Beijing, China
NCT03032913	Onco-exosome quantification in diagnosis of pancreatic cancer	Blood	Onco-exosomes	CHU de Bordeaux, Bordeaux, France
NCT04529915	Early diagnosis of lung cancer using blood plasma-derived exosomes	Blood	Exosomal proteins	Korea University Guro Hospital, Seoul, Republic of Korea
NCT04394572	Identification of new diagnostic protein markers for colorectal cancer	Blood	Exosomal proteins	CHU Reims, Reims, France
NCT04155359	Diagnosis of bladder cancer in hematuria patients	Urine	sncRNAs	Integrated Medical Professionals, Farmingdale, New York, United States
NCT03974204	Analysis of exosomes in cerebrospinal fluid for breast cancer patients	Cerebrospinal fluid and blood	Exosomal proteins	Centre Hospitalier Régional Universitaire de Lille, Lille, Hauts-de-France, France
NCT03830619	Exosomal long noncoding RNAs as potential biomarkers for lung cancer diagnosis	Plasma	Exosomal lncRNAs	Union Hospital, Tongji Medical College, Huazhong University of Science and Technology, Wuhan, Hubei, China
NCT03562715	Role of exosomal miRNAs 136, 494 and 495 in pre-eclampsia diagnosis	Blood	miRNAs 136, 494 and 495	Cairo University, Cairo, Egypt
NCT03415984	Estimation of age-related macular degeneration (ARMD) prevalence in Parkinson's patients	Not defined	Pro-inflammatory components	Fondation Ophtalmologique Adolphe de Rothschild, Paris, France
NCT04523389	Analysis of extracellular vesicle contents as biomarkers in colorectal cancer patients	Blood	miRNAs	CHU Dijon Bourgogne, Dijon, France
NCT03110877	Evaluation of circulating exosomal RNA as biomarker for lung metastases of primary high-grade osteosarcoma	Blood	Exosomal RNA	Ruijin Hospital, Shanghai, China
NCT04258735	Genomic analysis of metastatic breast cancer patients	Blood	ctDNA and RNA	Sungkyunkwan University School of Medicine, Seoul, Republic of Korea

Table 2. Cont.

CT Identification	Aim of Study	Source of Exosomes	Associated Markers	Promoted by
NCT04053855	Evaluation of urinary exosomes presence from clear cell renal cell carcinoma	Urine	CD9,CD63,CD81,CA9 and VGEFR2	CHU Saint-Etienne, Saint-Étienne, France
NCT04315753	assessment of exosomes' role in improving lung cancer management and early detection	Blood	Exosomal antigens	Istituto Clinico Humanitas, Rozzano, Milano, Italy
NCT04459182	evaluation of miRNA in exosomes in obese and OSA patients with endothelial dysfunction	Not defined	miRNA	University Hospital, Angers, France
NCT04556916	Early detection of prostate cancer	Blood	Exosomes	University Hospital, Montpellier, France
NCT02464930	Evaluation of microRNA expression in blood and cytology for detecting Barrett's esophagus and associated neoplasia	Bile and serum	miRNAs 192-5p, 215-5p and 194-5p	Kansas City Veterans Affairs Medical Center, Kansas City, Missouri, United States
NCT03800121	Study of exosomes in monitoring patients with sarcoma	Blood	Exosomal RNA and proteins	Centre Georges François Leclerc, Dijon, France
NCT04154332	Defining the functional role of exosomes in the development of preeclampsia leading to cardiovascular remodeling	Urine and Blood	Exosome abnormalities	University of Alabama, Birmingham, Alabama, United States
NCT03102268	Characterization of cholangiocarcinoma-derived exosomal ncRNAs	Plasma	Non-coding RNAs	Second Affiliated Hospital of Nanjing Medical University, Nanjing, Jiangsu, China
NCT03581435	Study of circulating exosome proteomics in gallbladder carcinoma patients	Blood	Exosomal proteins	Xinhua Hospital, Shanghai, China
NCT03738319	Analysis of non-coding RNAs in epithelia ovarian cancer	Blood	miRNA and lncRNA	Peking Union Medical College Hospital, Beijing, China
NCT03911999	Investigating relationship between urinary exosomes and aggressiveness of prostate cancer	Urine	Exosomal miRNA	Prince of Wales Hospital, Hong Kong, Hong Kong
NCT04120272	Search for biomarkers for early detection and prevention of delirium	Urine and blood	miRNA	College of Nursing, Yonsei University, Seoul, Republic of Korea

Table 2. Cont.

CT Identification	Aim of Study	Source of Exosomes	Associated Markers	Promoted by
NCT04053855	Evaluation of urinary exosomes presence from clear cell renal cell carcinoma	Urine	CD9,CD63,CD81,CA9 and VGEFR2	CHU Saint-Etienne, Saint-Étienne, France
NCT03419000	Evaluation of microRNAs as biomarkers of respiratory dysfunction in refractory epilepsy	Blood	miRNAs	Hospices Civils de Lyon, Bron, France
NCT04534647	Assessment of correlation between serological and urinary biomarkers and systemic lupus erythematosus	Blood and urine	Exosomes	Liga Panamericana de Asociaciones de Reumatologia (PANLAR), Rosario, Argentina
NCT02147418	Exosome testing as screening modality for human papillomavirus-positive oropharyngeal squamous cell carcinoma	Saliva	Exosomal proteins	New Mexico Cancer Care Alliance, Albuquerque, New Mexico, United States

Messenger RNAs (mRNAs). Increased levels of epidermal growth factor receptor variant type III (EGFRvIII) mRNA have been detected in the serum exosomes of glioblastoma patients, suggesting its use as a new glioblastoma diagnosis method instead of surgery [270]. Exosome Diagnostics, Inc. (Waltham, MA, USA) have developed methods for detecting one or more biomarkers in urine microvesicles in order to aid the diagnosis, monitoring and treatment of diseases such as cancer, especially prostate gland-related pathologies. Biomarkers, which are mRNAs of one or more isoforms of a large group of genes, facilitate the detection of prostate cancer by determining the fusion between SLC45A3 and BRAF genes in urinary microvesicles [311]. Recently, Dong and coworkers found that exosomal serum membrane type 1-matrix metalloproteinase (MT1-MMP) mRNA increases significantly in gastric cancer (GC) patients, which correlates with the tumor, lymph node and metastasis (TNM) stage and lymphatic metastasis. These findings indicate that exosomal MT1-MMP mRNA can be utilized as a biomarker for GC diagnosis and early treatment [312]. Similarly, exosomal heterogeneous nuclear ribonucleoprotein H1 (hnRNPH1) mRNA levels, which are remarkably higher in hepatocellular carcinoma (HCC) patients than in other groups, are associated with the Child-Pugh and TNM stage classification, portal vein tumor emboli and lymph node metastasis. This confirms that exosomal serum hnRNPH1 mRNA could be an effective marker of HCC [313]. Esophageal cancer-related gene-4 (Ecrg4) has been shown to be a tumor suppressor in several studies. Mao and colleagues have reported that serum exosomes contain higher levels of ECRG4 mRNA in healthy individuals than in their cancer counterparts, thus showing that exosomal ECRG4 mRNA can be used for cancer detection [314].

MicroRNAs (miRNAs) are small noncoding, double-stranded RNA molecules that degrade complementary mRNA sequences in target cells in order to inhibit protein translation. These molecules are reported to be abnormally expressed in several types of cancer, suggesting their role in the pathogenesis of human cancer [315]. Eight miRNAs, previously shown to be diagnostic markers of ovarian cancer, have been reported to be present at similar levels in biopsy specimens of ovarian cancer and circulating exosomes isolated from the same ovarian cancer patients [316]. With respect to lung tumors, Rabinowits and coworkers found similar miRNA patterns in plasma exosomes and tumor biopsies from lung adenocarcinoma patients. However, miRNA levels in lung cancer patients and control subjects differed significantly, indicating that circulating exosomal miRNA could be useful for lung adenocarcinoma screening tests [269]. Hepatocellular carcinoma (HCC) is a primary liver malignancy and a leading cause of cancer-related mortality worldwide. Exosomal miRNA-210 secreted by hepatocellular carcinoma cells is reported to promote angiogenesis by targeting SMAD4 and STAT6 in endothelial cells. Therefore, exosomal miRNA-210 could be used as a therapeutic target in anti-HCC therapy [109]. In this regard, circulating miRNAs in serum exosomes have potential as novel biomarkers for predicting hepatocellular carcinoma recurrence following liver transplantation [317]. In addition, Takeshita and colleagues reported that the sensitivity and specificity of serum miRNA-1246 in an esophageal squamous cell cancer (ESCC) diagnosis are 71.3% and 73.9%, respectively. Serum miRNA-1246, which closely correlates with the tumor, lymph node and metastasis (TNM) stage, has been shown to be a strong independent risk indicator of poor survival rates. Intriguingly, miRNA-1246 levels were found to be elevated in serum exosomes from ESCC patients but not in ESCC tissue samples, suggesting that exosomal serum miRNA-1246 could be a valuable diagnostic and prognostic biomarker of ESCC [318]. Circulating exosomal miRNA-17-5p and miRNA-92a-3p were found to be upregulated in colorectal cancer (CRC) patients. Their expression levels correlated closely with metastasis and chemotherapy resistance [319]. Moreover, exosomal miRNA-320d has been identified as a promising blood-based biomarker for distinguishing metastatic from nonmetastatic diseases in the serum of CRC patients. Therefore, these noninvasive biomarkers may have great potential to predict the clinical behavior of CRC and to monitor tumor metastasis [320,321]. Mitchell et al. reported that circulating miRNA-141 levels are strong diagnostic markers of prostate cancer [322]. Furthermore, exosomal serum miRNA-141 and miRNA-375 have been found to correlate with tumor progression in prostate cancer [323]. The enrichment of the let-7 miRNA family in exosomes from AZ-P7a cells may reflect their oncogenic characteristics, including tumorigenesis and metastasis, suggesting that AZ-P7a cells release let-7 miRNAs via exosomes into the extracellular environment to maintain their oncogenesis [182].

Long noncoding RNAs (lncRNAs). Exosomes also contain lncRNAs, now characterized as potential diagnostic and prognostic biomarkers for a wide range of pathologies. These functional RNAs, which are longer than 200 nucleotides, do not code for proteins but, rather, bind to a variety of nucleic acids and proteins as a means to regulate gene expression at the transcriptional and/or translational level. Colon cancer-associated transcript 2 (CCAT2), a novel lncRNA transcript encompassing the rs6983267 SNP, is significantly upregulated in CRC tissues as compared to adjacent noncancerous tissues. The higher expression levels of CCAT2 are associated with a greater depth of local invasion, positive lymph node metastasis and advanced TNM stage [324]. Moreover, exosomal lncRNA and miRNA-217 are differentially expressed in the serum of colorectal carcinoma patients and correlate with tumor classifications (T3/T4), advanced clinical stages (III/IV) and lymph node or distant metastasis [325]. LncRNA 91H is known to play a prominent role in tumor development by enhancing tumor cell migration and invasion through the modification of heterogeneous nuclear ribonucleoprotein K (HNRNPK) protein expression. In addition, CRC patients with high lncRNA 91H expression demonstrate a higher risk of tumor recurrence and metastasis [326]. Interestingly, exosomes from healthy donors carry a significant amount of HOTTIP (HOXA distal transcript antisense RNA) transcripts in comparison to CRC patients, with a significant statistical correlation between low exosomal HOTTIP levels and poor overall survival rates. Therefore, lncRNA *HOTTIP* could be a viable biomarker for CRC patients to predict the postsurgical survival time [327]. Exosomal serum lncRNA HOTAIR (Hox transcript antisense intergenic RNA) and miRNA-21 expression levels were higher in patients with lymph node metastasis than those without. In addition, exosomal HOTAIR and miRNA-21 achieved a sensitivity and specificity of 94.2% and 73.5%, respectively, in differentiating the malignant from benign laryngeal disease, suggesting that the combined evaluation of their serum expression levels may be a valuable biomarker of laryngeal squamous cell carcinoma [328].

Proteins. Exosomal protein signatures have also been used as potent alternative diagnostic markers of cancer. The epidermal growth factor receptor (EGFR) localized to exosome membranes has been found to be a possible marker for lung cancer diagnosis [329]. In this regard, Jakobsen and coworkers reported that the EGFR is highly expressed on the exosomal surface by analyzing the extracellular vesicles secreted by lung cancer cells [330], indicating that the EGFR is a promising biomarker for diagnosing non-small cell lung cancer (NSCLC). The epidermal growth factor receptor variant type III (EGFRvIII) transcript was detected in serum exosomes from 25 spongioblastoma patients but was not found in serum exosomes from 30 normal control individuals. Therefore, exosomal EGFRvIII may provide diagnostic information for glioblastoma patients [270]. Similarly, Graner et al. reported that brain tumor exosomes can escape from the blood–brain barrier, with potential systemic and distal signaling and immune consequences, and that serum exosomes from brain tumor patients possess EGFR, EGFRvIII and TGF-beta [331]. A microfluidic chip was used to analyze exosomal protein types in the blood circulation of spongioblastoma patients. In this regard, Shao and colleagues found that circulating exosomes contain EGFR-VII, EGFR, PDPN and IDH1, which can be used to analyze primary tumor mutations and to indicate drug efficacy [332]. Urinary exosomal proteins have also been investigated as potential biomarkers for prostate and bladder cancers. Nilsson et al. showed that urinary exosomes in prostate cancer patients express prostate-specific antigen (PSA), prostate cancer gene-3 (PCA-3), transmembrane serine protease 2-erythroblast transformation-specific (ETS) transcription factor family member-related gene fusion (TMPRSS2-ERG) and other prostate cancer-related markers, indicating the potential for the diagnosis and monitoring of cancer patients [333]. In this respect, Chen and colleagues found that 24 urinary exosomal proteins presented at significantly different levels in hernia (control) and bladder cancer patients. In particular, they revealed the strong association of TACSTD2 with bladder cancer and the potential of human urinary exosomes in noninvasive cancer diagnosis [334]. CD24, found in the MVB cytoplasm, is released into the extracellular environment via exosomes and is associated with the poor prognosis of ovarian carcinomas [335]. Logozzi and colleagues found that plasma CD63+ exosome levels are significantly higher in melanoma patients as compared to healthy control individuals [336]. This team recently showed that plasmatic exosomes from prostate cancer patients overexpress carbonic anhydrase IX (CA IX), as well as CA IX-related activity. In addition, CA IX expression correlated with intraluminal acidity in the plasmatic exosomes of these cancer patients [337]. The acidic microenvironment was reported to induce an upregulation of both the expression and activity of CA IX in cancer-derived exosomes, along with an increase in their production levels [338]. Finally, leucine-rich alpha-2-glycoprotein 1 (LRG1) expression levels were found to be higher in the urinary exosomes and lung tissue of NSCLC patients as compared to healthy individuals, indicating that LRG1 may be a candidate biomarker for noninvasive NSCLC diagnosis [309].

Lipids. Exosome lipidomics show great potential for the identification of suitable markers for cancer diagnosis. Recently, using an untargeted high-resolution mass spectrometry approach, our research group identified similarities between structural lipids differentially expressed in cancer stem cell (CSC)-derived exosomes and those derived from patients with malignant melanoma (MM) [339]. Our results showed significant metabolomic differences between exosomes derived from MM CSCs and those from differentiated tumor cells and, also, between serum-derived exosomes from patients with MM (MMPs) and those from healthy controls (HCs). We detected metabolites from different lipid classes, such as glycerophosphoglycerols, glycerophosphoserines, triacylglycerols and glycerophosphocholines. Interestingly, we found that PC 16:0/0:0 glycerophosphocholine expression was lower in both CSCs and MMPs in comparison with differentiated tumor cells and HCs, respectively, while lysophospholipid sphingosine 1-phosphate (S1P) levels were found to be lower in serum-derived exosomes from MMP patients than from HCs. These results indicate the importance of structural lipids detected in exosomes as biomarkers in the early detection of cancer and their potential in the determination of aggressiveness and therapeutic monitoring [339].

4.1.2. Use of Exosomes for Molecular Diagnostics of Neurodegenerative Diseases

Recent evidence indicates the potential involvement of exosomes in the nervous system and highlights their role in transcription regulation, neurogenesis and plasticity [340]. Several central nervous system (CNS) cell types, such as neurons and glial cells, are known to communicate intercellularly by releasing EVs. However, these vesicles could also play a role in the development of neurodegenerative diseases. Parkinson's disease (PD) is a progressive neurodegenerative disorder that mostly affects the motor system. Proteomic profiling was used to differentially identify proteins expressed in serum exosomes from PD patients and healthy individuals [341]. In addition, Fraser and colleagues identified leucine-rich repeat kinase 2 (LRRK2) as a biomarker in urinary exosomes from PD patients that predicts the risk of the development of this disease among LRRK2 mutation carriers [342]. The aggregation of α-synuclein may play an important role in PD pathology. Exosomes have been shown to be able to transfer the α-synuclein protein to neighboring normal cells, thus possibly exacerbating PD pathogenesis [197].

Alzheimer's disease (AD), another neurodegenerative disorder, is now regarded as the most common casue of dementia. The early detection of exosome-associated tau, which is present in human cerebrospinal fluid (CSF) samples and is phosphorylated at Thr-181 (AT270), would be helpful for AD diagnosis [194]. In this regard, the T-tau, P-tau and neurofilament light (NFL) biomarkers could be used to differentiate effectively between AD patients and healthy subjects [343]. Exosomal lipids could also be used as promising biomarkers for AD diagnosis. In this respect, 10 lipids from plasma were able to predict phenoconversion to AD within a two-to-three-year timeframe with over 90% accuracy [344].

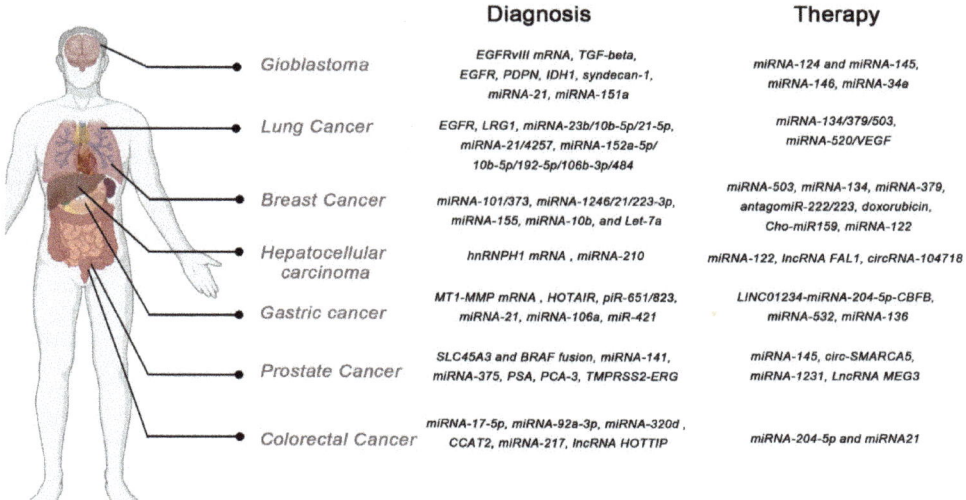

Figure 3. Exosome-associated molecules used for diagnosis and therapy. For instance, epidermal growth factor receptor variant type III (EGFRvIII) is associated with the classical glioblastoma (GBM) subtype [345]. MicroRNA (miRNA)-124 has been reported to enhance the chemosensitivity of GBM cells to temozolomide and to decrease GBM cell migration [346]. In addition, the delivery of miRNA-34a results in the inhibition of GBM cell proliferation, invasion, migration and tumorgenesis both in vivo and in vitro [347]. Lung cancer was also detected using exosomal biomarkers. In this context, Liu et al. found that miRNA-23b/10b-5p/21-5p were good candidates for its diagnosis [348], while Dejima and coworkers considered miRNA-21/4257/451a reliable biomarkers [349,350]. Other miRNAs such as homo sapiens (hsa)-miRNA-320d/320c/320b were suggested as potential biomarkers [351]. On the other hand, exosome miRNA-101/373 serum levels were found to be linked to aggressive breast carcinomas [352]. Other authors recommend miRNA-1246/21/223-3p as potential indicators of breast cancer [353,354]. Therapeutic quantities of doxorubicin (Dox) and cholesterol-modified miRNA 159 (Cho-miRNA-159) were delivered to triple-negative breast cancer (TNBC) cells and exhibited improved anticancer effects [355]. In addition, miRNA-204-5p and miRNA-21 efficiently inhibited cancer cell proliferation and increased chemosensitivity by specifically suppressing their target genes in human colorectal cancer cells [356,357]. Adipose-derived stromal cells (ASCs) were shown to be able to promote prostate cancer cell apoptosis via exosomal miRNA-145 through the caspase-3/7 pathway [358,359].

4.2. Use of Exosomes as Therapeutic Agents

In many studies, exosomes have been used as delivery vectors for small-molecule therapeutic agents, as they are capable of traveling from one cell to another and of conveying their cargo in a biologically active form, thus acting as attractive gene and drug delivery vehicles [360]. Cancer cells internalize a significantly larger percentage of exosomes as compared to normal cells. HEK293 and MSC exosomes were therefore effectively used as delivery vectors to transport PLK-1 small interfering RNA (siRNA) to bladder cancer cells in vitro, resulting in the selective gene silencing of PLK1 [361]. In addition, the internalization of exosomes in tumor cells is ten times greater than that of liposomes of comparable size due to their lipid composition and surface proteins, indicating the superior specificity of exosomes for cancer targeting [362]. Furthermore, exosomes offer several advantages over standard delivery vehicles. For example, exosomes are able to cross biological barriers, such as the blood–brain barrier (BBB), have poor immunogenicity and can be cell-specific [363]. Therefore, exosomes could be next-generation nontoxic delivery tools that combine nanoparticle sizes with high capacity levels, making them powerful vectors for the treatment of a variety of pathologies [364].

Doxorubicin-loaded exosomes are transported to tumor tissues and reduce tumor growth in mice without any adverse effects observed from this equipotent free drug [365]. Tian and coworkers used mouse immature dendritic cells (imDCs) for exosome production due to their lack of immunostimulatory markers. Purified imDC-derived exosomes were gently mixed with doxorubicin (DOX) in an electroporation buffer and then examined under a transmission electron microscope to verify the recovery of their plasma membrane. After loading the therapeutic cargo, these vesicles successfully delivered DOX to the targeted cell nucleus, leading to the inhibition of tumor growth without overt toxicity [366]. In another study, exosomes derived from a brain endothelial cell line, bEND.3, were loaded with DOX and used to deliver the anticancer drug across the blood–brain barrier (BBB) for the treatment of brain cancer in a zebrafish model [367]. The membrane vesicles mediated the autonomous intercellular migration of anticancer agents through multiple cancer cell layers and enabled hydrophobic and hydrophilic compounds to significantly penetrate both spheroids and in vivo tumors, thereby enhancing their therapeutic efficacy [368]. Interestingly, chemotherapeutic agents epirubicin and paclitaxel increased miR-503 levels in exosomes released from human umbilical vein endothelial cells (HUVECs) as compared to control conditions and were demonstrated to induce antitumor responses during breast cancer chemotherapy [369].

Exosomes also have the potential to deliver oligonucleotides, such as mRNA, miRNA and various noncoding RNAs, as well as mitochondrial and genomic DNA, to other cells, thus offering considerable advantages as ideal delivery systems for gene therapy [370]. As with the incorporation of genetic material into living cells, Alvarez-Erviti and colleagues used electroporation to deliver short interfering siRNA analogs to the brain in mice via exosomes [363]. In addition, Wahlgren and coworkers used plasma exosomes as gene delivery vectors to transport exogenous siRNA to human blood cells. The vesicles successfully delivered the administered siRNA to monocytes and lymphocytes, leading to robust gene silencing of mitogen-activated protein kinase 1, thus suggesting exosomes as a new generation of drug carriers that enable the development of safe and effective gene therapies [371]. Similarly, Kamerkar et al. demonstrated a technique for the direct and specific targeting of oncogenic KRAS in tumors using electroporated MSC-derived exosomes with siRNA. This treatment suppressed cancer in multiple mouse models of pancreatic cancer and significantly increased overall survival rates [372]. The same method was used to load exosomes with miRNA to the epidermal growth factor receptor (EGFR) expressed in breast cancer cells, indicating that exosomes can be used therapeutically to target EGFR-expressing cancerous tissues with nucleic acid drugs [373]. Finally, endothelial cells treated with chemotherapeutic agents are reported to release more exosomes that contain miRNA-503. Given that miRNA-503 is downregulated in exosomes released from endothelial cells cultured under tumoral conditions, the introduction of miRNA-503 into breast cancer cells altered their proliferative and metastatic capacities by inhibiting both CCND2 and CCND3 [369].

Lee and colleagues demonstrated that exosomes derived from MSCs deliver specific miRNA mimics (miRNA-124 and miRNA-145) and decrease glioma cell migration and the stem cell properties of cancer cells, providing an efficient route of therapeutic miRNA delivery in vivo [374]. In addition, the intratumoral injection of exosomes derived from miRNA-146-expressing MSCs results in a considerable reduction in glioma xenograft development in a rat brain tumor model and decreases cell growth and invasion, suggesting that the export of specific therapeutic miRNA into MSC exosomes represents an effective treatment strategy for malignant glioma [375]. O'Brien and coworkers engineered EVs loaded with miRNA-134, which is substantially downregulated in breast cancer tissue as compared to healthy tissue. It has been demonstrated that miRNA-134-enriched EVs reduce STAT5B and Hsp90 levels in target breast cancer cells, as well as cellular migration and invasion, and enhance the sensitivity of these cancer cells to anti-Hsp90 drugs [376]. Similarly, MSC-derived exosomes encapsulated with miRNA-379 were administered in breast cancer therapy in vivo. The results of this study show that miRNA-379-enriched EVs are potent tumor suppressors with an exciting potential as an innovative therapy for metastatic breast cancer [377]. Bovy et al. identified miRNA-503, whose expression levels are downregulated in exosomes released from endothelial cells cultured under tumoral conditions. Endothelial cells are able to transfer miRNA-503 via exosomes to breast cancer cells, thus impairing their growth and altering their proliferative capacity [369]. Breast cancer cells prime MSCs to secrete exosomes containing distinct miRNA contents, which promotes quiescence in a subset of cancer cells and confers drug resistance. According to this study, a novel therapeutic approach to target dormant breast cancer cells based on the systemic administration of MSCs loaded with antagomiRNA-222/223 resulted in the chemosensitization of cancer cells and increased survival rates [148].

Shtam et al. introduced two different anti-RAD51 and -RAD52 siRNAs into Henrietta Lacks (HeLa) cell-derived exosomes. These exosomes effectively delivered siRNA into the recipient cancer cells and caused strong RAD51 knockdown, providing additional evidence of the ability to use human exosomes as vectors in cancer therapy [378]. In a similar study, Shimbo and coworkers found that the transfer of miRNA-143 by means of MSC-derived exosomes decreases in the in vitro migration of osteosarcoma cells [379]. In addition, miRNA-122-transfected adipose tissue-derived MSCs (AMSCs) can effectively generate miRNA-122-encapsulated exosomes, which can mediate miRNA-122 communication between AMSCs and hepatocellular carcinoma (HCC) cells, thereby elevating tumor cell sensitivity to chemotherapeutic agents through the alteration of miRNA-122 target gene expression in HCC cells [380]. Usman and colleagues have described a strategy for generating large-scale amounts of exosomes for the delivery of RNA drugs, including antisense oligonucleotides (ASOs). They chose human red blood cells (RBCs), which are devoid of DNA, for EV production. RBC EVs were demonstrated to deliver therapeutic ASOs in order to effectively antagonize oncogenic micro-RNAs (oncomiRNAs) and to suppress tumorigenesis [381]. Exosomes could potentially deliver therapeutic proteins to recipient cells, with a recent study reporting the feasibility of using exosomes as biocompatible vectors that could improve the targeting and delivery of therapeutic proteins to specific cells in diseased tissues [382]. In addition, Haney et al. used a new method to treat Parkinson's disease (PD). In fact, catalase-loaded exosomes produce a potent neuroprotective effect on both in vitro and in mouse brains following intranasal administration. This result demonstrates the capacity of exosomes to load fully functional proteins and to treat specific disorders [383]. Several approaches have envisaged the utilization of specific conserved domains in order to enhance the loading of proteins. For instance, Sterzenbach and colleagues exploited late-domain (L-Domain) proteins and ESCRT machinery pathways to load Cre recombinase into exosomes. This protein was successfully delivered to neurons through a nasal route, a well-characterized noninvasive method to deliver exogenous proteins to the brain via exosomes [384]. Human ubiquitin was also used as a sorting sequence to deliver diverse proteins into exosomes such as EGFP and nHER2. Interestingly, C-terminal–ubiquitin fusion may act as an efficient signal sequence of antigenic proteins into exosomes, which could support the use of exosomes as vaccines [385].

5. Conclusions

A considerable number of physiological and pathological processes are undoubtedly governed or, at least, modulated by the intervention of exosomes. This places exosomes in a privileged position and optimizes their use as a potential tool in clinical applications for both diagnosis and therapy. Despite groundbreaking improvements, a number of limitations and challenges remain with regards to transforming exosome applications into clinical therapies. Further exploration of the molecular composition and function of exosomes, along with an appropriate cell source for exosome production according to the intended therapeutic use, will undoubtedly enhance the final outcome of any clinical applications using these membrane vesicles. Taking into account the low biofluid volumes available for diagnosis application, standard and highly effective exosome isolation, purification, characterization and manipulation methods need to be developed to make these vesicles a clinical reality. Furthermore, the loading of exosomes without altering their functional efficacy and the natural characteristics of the donor cell are crucial for oncological research and their development. Finally, with research in exosome biology in its infancy, further studies to evaluate the possible impacts of exosomes in major preclinical models are required to assess the safety/toxicology issues and to ensure their safe and effective use in therapeutic settings.

Author Contributions: H.A. and L.M.: manuscript writing, H.A. and K.B.: figure artwork, F.M. and B.B.: manuscript review and J.A.M. and K.B.: manuscript writing and final approval of the manuscript. All authors have read and agreed to the published version of the manuscript.

Funding: This study was funded by the Spanish ISCIII Health Research Fund and the European Regional Development Fund (FEDER) through research grants PI12/01097, PI15/02015, PI18/00337 (F.M.), PIE16-00045 (J.A.M.), DTS19/00145 (J.A.M.) and PI18/00330 (K.B.). The study was also supported by the Ministry of Science, Innovation and Universities (MICIU, RTI2018-101309-B-C22, FEDER funds), by the Consejería de Economía, Conocimiento, Empresas y Universidad de la Junta de Andalucía (SOMM17/6109/UGR, FEDER Funds), the Chair of "Doctors Galera-Requena in cancer stem cell research" (CMC-CTS963) and the Junta de Andalucia Health and Families Department (CARTPI-0001-201). The CECEyU and CSyF of the Junta de Andalucía FEDER/European Cohesion Fund (FSE) provided the following research grants: 2016000073391-TRA, 2016000073332-TRA, PI-57069 and PAIDI-Bio326 (F.M.) and PI-0014-2016 (K.B.). K.B. was also on a Nicolas Monardes Regional Ministry of Health contract (0006/2018). H.A. held a Research Excellence PhD Fellowship (2UAE2020) from the National Center for Scientific and Technical Research (CNRST). L.M. was supported by the Mohammed VI Polytechnic University of Morocco.

Institutional Review Board Statement: Not applicable as this is a review article of the availableliterature and not a research study involving human participants.

Informed Consent Statement: Not applicable as this is a review article of the availablelitera-ture and not a research study involving human participants.

Data Availability Statement: The data presented in this study are available on request from the corresponding authors.

Acknowledgments: Figures were created using Biorender.com. We also wish to thank Michael O'Shea for proofreading the review and María Tristán-Manzano for her help in generating the figures.

Conflicts of Interest: The authors declare no conflict of interest. The funders had no role in the design of the study; in the collection, analyses or interpretation of data; in the writing of the manuscript or in the decision to publish the results.

References

1. Ratajczak, M.Z.; Ratajczak, J. Extracellular microvesicles/exosomes: Discovery, disbelief, acceptance, and the future? *Leukemia* **2020**, *34*, 3126–3136. [CrossRef] [PubMed]
2. Ferreira, B.; Caetano, J.; Barahona, F.; Lopes, R.; Carneiro, E.; Costa-Silva, B.; Joao, C. Liquid biopsies for multiple myeloma in a time of precision medicine. *J. Mol. Med.* **2020**, *98*, 513–525. [CrossRef] [PubMed]
3. Heitzer, E.; Haque, I.S.; Roberts, C.E.S.; Speicher, M.R. Current and future perspectives of liquid biopsies in genomics-driven oncology. *Nat. Rev. Genet.* **2019**, *20*, 71–88. [CrossRef] [PubMed]
4. Wong, S.Q.; Dawson, S.J. Combining liquid biopsies and PET-CT for early cancer detection. *Nat. Med.* **2020**, *26*, 1010–1011. [CrossRef] [PubMed]
5. Aheget, H.; Tristán-Manzano, M.; Mazini, L.; Cortijo-Gutierrez, M.; Galindo-Moreno, P.; Herrera, C.; Martin, F.; Marchal, J.A.; Benabdellah, K. Exosome: A new player in translational nanomedicine. *J. Clin. Med.* **2020**, *9*, 2380. [CrossRef]
6. Kalluri, R. The biology and function of exosomes in cancer. *J. Clin. Investig.* **2016**, *126*, 1208–1215. [CrossRef]
7. Tkach, M.; Thery, C. Communication by Extracellular Vesicles: Where We Are and Where We Need to Go. *Cell* **2016**, *164*, 1226–1232. [CrossRef]
8. Raposo, G.; Stoorvogel, W. Extracellular vesicles: Exosomes, microvesicles, and friends. *J. Cell Biol.* **2013**, *200*, 373–383. [CrossRef]
9. Santucci, L.; Bruschi, M.; Del Zotto, G.; Antonini, F.; Ghiggeri, G.M.; Panfoli, I.; Candiano, G. Biological surface properties in extracellular vesicles and their effect on cargo proteins. *Sci. Rep.* **2019**, *9*, 13048. [CrossRef]

10. Yanez-Mo, M.; Siljander, P.R.; Andreu, Z.; Zavec, A.B.; Borras, F.E.; Buzas, E.I.; Buzas, K.; Casal, E.; Cappello, F.; Carvalho, J.; et al. Biological properties of extracellular vesicles and their physiological functions. *J. Extracell. Vesicles* **2015**, *4*, 27066. [CrossRef]
11. Théry, C.; Witwer, K.W.; Aikawa, E.; Alcaraz, M.J.; Anderson, J.D.; Andriantsitohaina, R.; Antoniou, A.; Arab, T.; Archer, F.; Atkin-Smith, G.K. Minimal information for studies of extracellular vesicles 2018 (MISEV2018): A position statement of the International Society for Extracellular Vesicles and update of the MISEV2014 guidelines. *J. Extracell. Vesicles* **2018**, *7*, 1535750. [CrossRef] [PubMed]
12. Ma, L.; Li, Y.; Peng, J.; Wu, D.; Zhao, X.; Cui, Y.; Chen, L.; Yan, X.; Du, Y.; Yu, L. Discovery of the migrasome, an organelle mediating release of cytoplasmic contents during cell migration. *Cell Res.* **2015**, *25*, 24–38. [CrossRef] [PubMed]
13. Valcz, G.; Buzás, E.I.; Kittel, Á.; Krenács, T.; Visnovitz, T.; Spisák, S.; Török, G.; Homolya, L.; Zsigrai, S.; Kiszler, G. En bloc release of MVB-like small extracellular vesicle clusters by colorectal carcinoma cells. *J. Extracell. Vesicles* **2019**, *8*, 1596668. [CrossRef] [PubMed]
14. Malm, T.; Loppi, S.; Kanninen, K.M. Exosomes in Alzheimer's disease. *Neurochem. Int.* **2016**, *97*, 193–199. [CrossRef]
15. Li, K.; Chen, Y.; Li, A.; Tan, C.; Liu, X. Exosomes play roles in sequential processes of tumor metastasis. *Int. J. Cancer* **2019**, *144*, 1486–1495. [CrossRef]
16. Zhang, L.; Yu, D. Exosomes in cancer development, metastasis, and immunity. *Biochim. Biophys. Acta Rev. Cancer* **2019**, *1871*, 455–468. [CrossRef]
17. Nabet, B.Y.; Qiu, Y.; Shabason, J.E.; Wu, T.J.; Yoon, T.; Kim, B.C.; Benci, J.L.; DeMichele, A.M.; Tchou, J.; Marcotrigiano, J. Exosome RNA unshielding couples stromal activation to pattern recognition receptor signaling in cancer. *Cell* **2017**, *170*, 352–366.e313. [CrossRef]
18. Hill, A.F. Extracellular vesicles and neurodegenerative diseases. *J. Neurosci.* **2019**, *39*, 9269–9273. [CrossRef]
19. Jiang, L.; Dong, H.; Cao, H.; Ji, X.; Luan, S.; Liu, J. Exosomes in pathogenesis, diagnosis, and treatment of Alzheimer's disease. *Medical Sci. Monit.* **2019**, *25*, 3329. [CrossRef]
20. Emmanouilidou, E.; Melachroinou, K.; Roumeliotis, T.; Garbis, S.D.; Ntzouni, M.; Margaritis, L.H.; Stefanis, L.; Vekrellis, K. Cell-produced α-synuclein is secreted in a calcium-dependent manner by exosomes and impacts neuronal survival. *J. Neurosci.* **2010**, *30*, 6838–6851. [CrossRef]
21. Fevrier, B.; Vilette, D.; Archer, F.; Loew, D.; Faigle, W.; Vidal, M.; Laude, H.; Raposo, G. Cells release prions in association with exosomes. *Proc. Natl. Acad. Sci. USA* **2004**, *101*, 9683–9688. [CrossRef] [PubMed]
22. Yin, Y.; Long, J.; Sun, Y.; Li, H.; Jiang, E.; Zeng, C.; Zhu, W. The function and clinical significance of eIF3 in cancer. *Gene* **2018**, *673*, 130–133. [CrossRef] [PubMed]
23. Zhao, M.; Mishra, L.; Deng, C.X. The role of TGF-β/SMAD4 signaling in cancer. *Int. J. Biol. Sci.* **2018**, *14*, 111–123. [CrossRef] [PubMed]
24. Peinado, H.; Aleckovic, M.; Lavotshkin, S.; Matei, I.; Costa-Silva, B.; Moreno-Bueno, G.; Hergueta-Redondo, M.; Williams, C.; Garcia-Santos, G.; Ghajar, C.; et al. Melanoma exosomes educate bone marrow progenitor cells toward a pro-metastatic phenotype through MET. *Nat. Med.* **2012**, *18*, 883–891. [CrossRef]
25. Webber, J.; Steadman, R.; Mason, M.D.; Tabi, Z.; Clayton, A. Cancer exosomes trigger fibroblast to myofibroblast differentiation. *Cancer Res.* **2010**, *70*, 9621–9630. [CrossRef]
26. Chowdhury, R.; Webber, J.P.; Gurney, M.; Mason, M.D.; Tabi, Z.; Clayton, A. Cancer exosomes trigger mesenchymal stem cell differentiation into pro-angiogenic and pro-invasive myofibroblasts. *Oncotarget* **2015**, *6*, 715–731. [CrossRef]
27. Pezzuto, A.; Carico, E. Role of HIF-1 in Cancer Progression: Novel Insights. A Review. *Curr. Mol. Med.* **2018**, *18*, 343–351. [CrossRef]
28. Demory Beckler, M.; Higginbotham, J.N.; Franklin, J.L.; Ham, A.J.; Halvey, P.J.; Imasuen, I.E.; Whitwell, C.; Li, M.; Liebler, D.C.; Coffey, R.J. Proteomic analysis of exosomes from mutant KRAS colon cancer cells identifies intercellular transfer of mutant KRAS. *Mol. Cell. Proteom.* **2013**, *12*, 343–355. [CrossRef]
29. Yang, Y.; Li, C.-W.; Chan, L.-C.; Wei, Y.; Hsu, J.-M.; Xia, W.; Cha, J.-H.; Hou, J.; Hsu, J.L.; Sun, L. Exosomal PD-L1 harbors active defense function to suppress T cell killing of breast cancer cells and promote tumor growth. *Cell Res.* **2018**, *28*, 862–864. [CrossRef]
30. Lancaster, G.I.; Febbraio, M.A. Exosome-Dependent trafficking of HSP70 A novel secretory pathway for cellular stress proteins. *J. Biol. Chem.* **2005**, *280*, 23349–23355. [CrossRef]
31. Clayton, A.; Turkes, A.; Navabi, H.; Mason, M.D.; Tabi, Z. Induction of heat shock proteins in B-cell exosomes. *J. Cell Sci.* **2005**, *118*, 3631–3638. [CrossRef] [PubMed]
32. Cho, J.A.; Lee, Y.S.; Kim, S.H.; Ko, J.K.; Kim, C.W. MHC independent anti-tumor immune responses induced by Hsp70-enriched exosomes generate tumor regression in murine models. *Cancer Lett.* **2009**, *275*, 256–265. [CrossRef] [PubMed]
33. Mukhopadhyay, U.K.; Mak, A.S. p53: Is the guardian of the genome also a suppressor of cell invasion? *Cell Cycle* **2009**, *8*, 2481. [CrossRef] [PubMed]
34. Muller, P.A.; Vousden, K.H. p53 mutations in cancer. *Nat. Cell Biol.* **2013**, *15*, 2–8. [CrossRef]
35. Yu, X.; Riley, T.; Levine, A.J. The regulation of the endosomal compartment by p53 the tumor suppressor gene. *FEBS J.* **2009**, *276*, 2201–2212. [CrossRef]
36. Lespagnol, A.; Duflaut, D.; Beekman, C.; Blanc, L.; Fiucci, G.; Marine, J.C.; Vidal, M.; Amson, R.; Telerman, A. Exosome secretion, including the DNA damage-induced p53-dependent secretory pathway, is severely compromised in TSAP6/Steap3-null mice. *Cell Death Differ.* **2008**, *15*, 1723–1733. [CrossRef]

37. Putz, U.; Howitt, J.; Doan, A.; Goh, C.P.; Low, L.H.; Silke, J.; Tan, S.S. The tumor suppressor PTEN is exported in exosomes and has phosphatase activity in recipient cells. *Sci. Signal.* **2012**, *5*, ra70. [CrossRef]
38. Ristorcelli, E.; Beraud, E.; Mathieu, S.; Lombardo, D.; Verine, A. Essential role of Notch signaling in apoptosis of human pancreatic tumoral cells mediated by exosomal nanoparticles. *Int. J. Cancer* **2009**, *125*, 1016–1026. [CrossRef]
39. Hinger, S.A.; Cha, D.J.; Franklin, J.L.; Higginbotham, J.N.; Dou, Y.; Ping, J.; Shu, L.; Prasad, N.; Levy, S.; Zhang, B.; et al. Diverse Long RNAs Are Differentially Sorted into Extracellular Vesicles Secreted by Colorectal Cancer Cells. *Cell Rep.* **2018**, *25*, 715–725e714. [CrossRef]
40. Sork, H.; Corso, G.; Krjutskov, K.; Johansson, H.J.; Nordin, J.Z.; Wiklander, O.P.B.; Lee, Y.X.F.; Westholm, J.O.; Lehtio, J.; Wood, M.J.A.; et al. Heterogeneity and interplay of the extracellular vesicle small RNA transcriptome and proteome. *Sci. Rep.* **2018**, *8*, 10813. [CrossRef]
41. Montecalvo, A.; Larregina, A.T.; Shufesky, W.J.; Stolz, D.B.; Sullivan, M.L.; Karlsson, J.M.; Baty, C.J.; Gibson, G.A.; Erdos, G.; Wang, Z.; et al. Mechanism of transfer of functional microRNAs between mouse dendritic cells via exosomes. *Blood* **2011**, *119*, 756–766. [CrossRef] [PubMed]
42. Bronisz, A.; Godlewski, J.; Chiocca, E.A. Extracellular Vesicles and MicroRNAs: Their Role in Tumorigenicity and Therapy for Brain Tumors. *Cell Mol. Neurobiol.* **2016**, *36*, 361–376. [CrossRef] [PubMed]
43. Hanahan, D.; Coussens, L.M. Accessories to the crime: Functions of cells recruited to the tumor microenvironment. *Cancer Cell* **2012**, *21*, 309–322. [CrossRef] [PubMed]
44. Kosaka, M.; Kang, M.R.; Yang, G.; Li, L.C. Targeted p21WAF1/CIP1 activation by RNAa inhibits hepatocellular carcinoma cells. *Nucleic Acid Ther.* **2012**, *22*, 335–343. [CrossRef]
45. Che, Y.; Shi, X.; Shi, Y.; Jiang, X.; Ai, Q.; Shi, Y.; Gong, F.; Jiang, W. Exosomes derived from miR-143-overexpressing MSCs inhibit cell migration and invasion in human prostate cancer by downregulating TFF3. *Mol. Ther. Nucleic Acids* **2019**, *18*, 232–244. [CrossRef]
46. Ahadi, A.; Brennan, S.; Kennedy, P.J.; Hutvagner, G.; Tran, N. Long non-coding RNAs harboring miRNA seed regions are enriched in prostate cancer exosomes. *Sci. Rep.* **2016**, *6*, 24922. [CrossRef]
47. Takahashi, K.; Yan, I.K.; Haga, H.; Patel, T. Modulation of hypoxia-signaling pathways by extracellular linc-RoR. *J. Cell Sci.* **2014**, *127*, 1585–1594. [CrossRef]
48. Hoshino, A.; Costa-Silva, B.; Shen, T.L.; Rodrigues, G.; Hashimoto, A.; Tesic Mark, M.; Molina, H.; Kohsaka, S.; Di Giannatale, A.; Ceder, S.; et al. Tumour exosome integrins determine organotropic metastasis. *Nature* **2015**, *527*, 329–335. [CrossRef]
49. Deng, J.; Galipeau, J. Reprogramming of B cells into regulatory cells with engineered fusokines. *Infect. Disord. Drug Targets* **2012**, *12*, 248–254. [CrossRef]
50. Dong, H.; Strome, S.E.; Salomao, D.R.; Tamura, H.; Hirano, F.; Flies, D.B.; Roche, P.C.; Lu, J.; Zhu, G.; Tamada, K.; et al. Tumor-associated B7-H1 promotes T-cell apoptosis: A potential mechanism of immune evasion. *Nat. Med.* **2002**, *8*, 793–800. [CrossRef]
51. Gordon, S.R.; Maute, R.L.; Dulken, B.W.; Hutter, G.; George, B.M.; McCracken, M.N.; Gupta, R.; Tsai, J.M.; Sinha, R.; Corey, D.; et al. PD-1 expression by tumour-associated macrophages inhibits phagocytosis and tumour immunity. *Nature* **2017**, *545*, 495–499. [CrossRef] [PubMed]
52. Liu, Y.; Huang, L.; Guan, X.; Li, H.; Zhang, Q.Q.; Han, C.; Wang, Y.J.; Wang, C.; Zhang, Y.; Qu, C.; et al. ER-α36, a novel variant of ERα, is involved in the regulation of Tamoxifen-sensitivity of glioblastoma cells. *Steroids* **2016**, *111*, 127–133. [CrossRef] [PubMed]
53. Ludwig, S.; Floros, T.; Theodoraki, M.N.; Hong, C.S.; Jackson, E.K.; Lang, S.; Whiteside, T.L. Suppression of Lymphocyte Functions by Plasma Exosomes Correlates with Disease Activity in Patients with Head and Neck Cancer. *Clin. Cancer Res.* **2017**, *23*, 4843–4854. [CrossRef] [PubMed]
54. Pasquier, J.; Thawadi, H.A.; Ghiabi, P.; Abu-Kaoud, N.; Maleki, M.; Guerrouahen, B.S.; Vidal, F.; Courderc, B.; Ferron, G.; Martinez, A.; et al. Microparticles mediated cross-talk between tumoral and endothelial cells promote the constitution of a pro-metastatic vascular niche through Arf6 up regulation. *Cancer Microenviron.* **2014**, *7*, 41–59. [CrossRef] [PubMed]
55. Horie, K.; Kawakami, K.; Fujita, Y.; Sugaya, M.; Kameyama, K.; Mizutani, K.; Deguchi, T.; Ito, M. Exosomes expressing carbonic anhydrase 9 promote angiogenesis. *Biochem. Biophys. Res. Commun.* **2017**, *492*, 356–361. [CrossRef] [PubMed]
56. Tang, M.K.S.; Yue, P.Y.K.; Ip, P.P.; Huang, R.L.; Lai, H.C.; Cheung, A.N.Y.; Tse, K.Y.; Ngan, H.Y.S.; Wong, A.S.T. Soluble E-cadherin promotes tumor angiogenesis and localizes to exosome surface. *Nat. Commun.* **2018**, *9*, 2270. [CrossRef]
57. Pittenger, M.F.; Mackay, A.M.; Beck, S.C.; Jaiswal, R.K.; Douglas, R.; Mosca, J.D.; Moorman, M.A.; Simonetti, D.W.; Craig, S.; Marshak, D.R. Multilineage potential of adult human mesenchymal stem cells. *Science* **1999**, *284*, 143–147. [CrossRef]
58. Li, X.; Wang, S.; Zhu, R.; Li, H.; Han, Q.; Zhao, R.C. Lung tumor exosomes induce a pro-inflammatory phenotype in mesenchymal stem cells via NFκB-TLR signaling pathway. *J. Hematol. Oncol.* **2016**, *9*, 42. [CrossRef]
59. De Veirman, K.; Wang, J.; Xu, S.; Leleu, X.; Himpe, E.; Maes, K.; De Bruyne, E.; Van Valckenborgh, E.; Vanderkerken, K.; Menu, E.; et al. Induction of miR-146a by multiple myeloma cells in mesenchymal stromal cells stimulates their pro-tumoral activity. *Cancer Lett.* **2016**, *377*, 17–24. [CrossRef]
60. Dhondt, B.; Rousseau, Q.; De Wever, O.; Hendrix, A. Function of extracellular vesicle-associated miRNAs in metastasis. *Cell Tissue Res.* **2016**, *365*, 621–641. [CrossRef]
61. Farahani, M.; Rubbi, C.; Liu, L.; Slupsky, J.R.; Kalakonda, N. CLL Exosomes Modulate the Transcriptome and Behaviour of Recipient Stromal Cells and Are Selectively Enriched in miR-202-3p. *PLoS ONE* **2015**, *10*, e0141429. [CrossRef] [PubMed]

62. Donnarumma, E.; Fiore, D.; Nappa, M.; Roscigno, G.; Adamo, A.; Iaboni, M.; Russo, V.; Affinito, A.; Puoti, I.; Quintavalle, C.; et al. Cancer-Associated fibroblasts release exosomal microRNAs that dictate an aggressive phenotype in breast cancer. *Oncotarget* **2017**, *8*, 19592–19608. [CrossRef] [PubMed]
63. Takano, Y.; Masuda, T.; Iinuma, H.; Yamaguchi, R.; Sato, K.; Tobo, T.; Hirata, H.; Kuroda, Y.; Nambara, S.; Hayashi, N.; et al. Circulating exosomal microRNA-203 is associated with metastasis possibly via inducing tumor-associated macrophages in colorectal cancer. *Oncotarget* **2017**, *8*, 78598–78613. [CrossRef] [PubMed]
64. Hsu, Y.L.; Hung, J.Y.; Chang, W.A.; Jian, S.F.; Lin, Y.S.; Pan, Y.C.; Wu, C.Y.; Kuo, P.L. Hypoxic Lung-Cancer-Derived Extracellular Vesicle MicroRNA-103a Increases the Oncogenic Effects of Macrophages by Targeting PTEN. *Mol. Ther.* **2018**, *26*, 568–581. [CrossRef]
65. Huarte, M.; Guttman, M.; Feldser, D.; Garber, M.; Koziol, M.J.; Kenzelmann-Broz, D.; Khalil, A.M.; Zuk, O.; Amit, I.; Rabani, M.; et al. A large intergenic noncoding RNA induced by p53 mediates global gene repression in the p53 response. *Cell* **2010**, *142*, 409–419. [CrossRef]
66. Hewson, C.; Morris, K.V. Form and Function of Exosome-Associated Long Non-coding RNAs in Cancer. *Curr. Top. Microbiol. Immunol.* **2016**, *394*, 41–56.
67. Kino, T.; Hurt, D.E.; Ichijo, T.; Nader, N.; Chrousos, G.P. Noncoding RNA gas5 is a growth arrest- and starvation-associated repressor of the glucocorticoid receptor. *Sci. Signal.* **2010**, *3*, ra8. [CrossRef]
68. Sun, X.; Du, P.; Yuan, W.; Du, Z.; Yu, M.; Yu, X.; Hu, T. Long non-coding RNA HOTAIR regulates cyclin J via inhibition of microRNA-205 expression in bladder cancer. *Cell Death Dis.* **2015**, *6*, e1907. [CrossRef]
69. Sun, R.; Qin, C.; Jiang, B.; Fang, S.; Pan, X.; Peng, L.; Liu, Z.; Li, W.; Li, Y.; Li, G. Down-Regulation of MALAT1 inhibits cervical cancer cell invasion and metastasis by inhibition of epithelial-mesenchymal transition. *Mol. Biosyst.* **2016**, *12*, 952–962. [CrossRef]
70. Yang, L.; Bai, H.S.; Deng, Y.; Fan, L. High MALAT1 expression predicts a poor prognosis of cervical cancer and promotes cancer cell growth and invasion. *Eur. Rev. Med. Pharmacol. Sci.* **2015**, *19*, 3187–3193.
71. Emmrich, S.; Streltsov, A.; Schmidt, F.; Thangapandi, V.R.; Reinhardt, D.; Klusmann, J.H. LincRNAs MONC and MIR100HG act as oncogenes in acute megakaryoblastic leukemia. *Mol. Cancer* **2014**, *13*, 171. [CrossRef]
72. Nath, B.; Szabo, G. Hypoxia and hypoxia inducible factors: Diverse roles in liver diseases. *Hepatology* **2011**, *55*, 622–633. [CrossRef]
73. Yang, J.X.; Sun, Y.; Gao, L.; Meng, Q.; Yang, B.Y. Long non-coding RNA DANCR facilitates glioma malignancy by sponging miR-33a-5p. *Neoplasma* **2018**, *65*, 790–798. [CrossRef] [PubMed]
74. Bolat, F.; Kayaselcuk, F.; Nursal, T.Z.; Yagmurdur, M.C.; Bal, N.; Demirhan, B. Microvessel density, VEGF expression, and tumor-associated macrophages in breast tumors: Correlations with prognostic parameters. *J. Exp. Clin. Cancer Res.* **2006**, *25*, 365–372.
75. Liu, Y.; Luo, F.; Wang, B.; Li, H.; Xu, Y.; Liu, X.; Shi, L.; Lu, X.; Xu, W.; Lu, L.; et al. STAT3-Regulated exosomal miR-21 promotes angiogenesis and is involved in neoplastic processes of transformed human bronchial epithelial cells. *Cancer Lett.* **2015**, *370*, 125–135. [CrossRef] [PubMed]
76. Leong, H.; Mathur, P.S.; Greene, G.L. Green tea catechins inhibit angiogenesis through suppression of STAT3 activation. *Breast Cancer Res. Treat.* **2009**, *117*, 505–515. [CrossRef]
77. Dong, Y.; Lu, B.; Zhang, X.; Zhang, J.; Lai, L.; Li, D.; Wu, Y.; Song, Y.; Luo, J.; Pang, X.; et al. Cucurbitacin E, a tetracyclic triterpenes compound from Chinese medicine, inhibits tumor angiogenesis through VEGFR2-mediated Jak2-STAT3 signaling pathway. *Carcinogenesis* **2010**, *31*, 2097–2104. [CrossRef] [PubMed]
78. Zhuang, G.; Wu, X.; Jiang, Z.; Kasman, I.; Yao, J.; Guan, Y.; Oeh, J.; Modrusan, Z.; Bais, C.; Sampath, D.; et al. Tumour-Secreted miR-9 promotes endothelial cell migration and angiogenesis by activating the JAK-STAT pathway. *EMBO J.* **2012**, *31*, 3513–3523. [CrossRef] [PubMed]
79. Umezu, T.; Tadokoro, H.; Azuma, K.; Yoshizawa, S.; Ohyashiki, K.; Ohyashiki, J.H. Exosomal miR-135b shed from hypoxic multiple myeloma cells enhances angiogenesis by targeting factor-inhibiting HIF-1. *Blood* **2014**, *124*, 3748–3757. [CrossRef] [PubMed]
80. Conigliaro, A.; Costa, V.; Lo Dico, A.; Saieva, L.; Buccheri, S.; Dieli, F.; Manno, M.; Raccosta, S.; Mancone, C.; Tripodi, M.; et al. CD90+ liver cancer cells modulate endothelial cell phenotype through the release of exosomes containing H19 lncRNA. *Mol. Cancer* **2015**, *14*, 155. [CrossRef] [PubMed]
81. Paget, S. The distribution of secondary growths in cancer of the breast. *Cancer Metastasis Rev.* **1989**, *8*, 98–101. [CrossRef]
82. Akhtar, M.; Haider, A.; Rashid, S.; Al-Nabet, A. Paget's "Seed and Soil" Theory of Cancer Metastasis: An Idea Whose Time has Come. *Adv. Anatom. Pathol.* **2019**, *26*, 69–74. [CrossRef] [PubMed]
83. Chaffer, C.L.; Morris, M.J. The feeding response to melanin-concentrating hormone is attenuated by antagonism of the NPY Y(1)-receptor in the rat. *Endocrinology* **2002**, *143*, 191–197. [CrossRef] [PubMed]
84. Ocana, O.H.; Corcoles, R.; Fabra, A.; Moreno-Bueno, G.; Acloque, H.; Vega, S.; Barrallo-Gimeno, A.; Cano, A.; Nieto, M.A. Metastatic colonization requires the repression of the epithelial-mesenchymal transition inducer Prrx1. *Cancer Cell* **2012**, *22*, 709–724. [CrossRef] [PubMed]
85. Liu, Y.; Cao, X. Immunosuppressive cells in tumor immune escape and metastasis. *J. Mol. Med.* **2016**, *94*, 509–522. [CrossRef]
86. Bergfeld, S.A.; Blavier, L.; DeClerck, Y.A. Bone marrow-derived mesenchymal stromal cells promote survival and drug resistance in tumor cells. *Mol. Cancer Ther.* **2014**, *13*, 962–975. [CrossRef]

87. Camorani, S.; Hill, B.S.; Fontanella, R.; Greco, A.; Gramanzini, M.; Auletta, L.; Gargiulo, S.; Albanese, S.; Lucarelli, E.; Cerchia, L.; et al. Inhibition of Bone Marrow-Derived Mesenchymal Stem Cells Homing Towards Triple-Negative Breast Cancer Microenvironment Using an Anti-PDGFRβ Aptamer. *Theranostics* **2017**, *7*, 3595–3607. [CrossRef]
88. Luo, J.; Ok Lee, S.; Liang, L.; Huang, C.K.; Li, L.; Wen, S.; Chang, C. Infiltrating bone marrow mesenchymal stem cells increase prostate cancer stem cell population and metastatic ability via secreting cytokines to suppress androgen receptor signaling. *Oncogene* **2014**, *33*, 2768–2778. [CrossRef]
89. Cui, T.X.; Kryczek, I.; Zhao, L.; Zhao, E.; Kuick, R.; Roh, M.H.; Vatan, L.; Szeliga, W.; Mao, Y.; Thomas, D.G.; et al. Myeloid-derived suppressor cells enhance stemness of cancer cells by inducing microRNA101 and suppressing the corepressor CtBP2. *Immunity* **2013**, *39*, 611–621. [CrossRef]
90. Weyemi, U.; Redon, C.E.; Sethi, T.K.; Burrell, A.S.; Jailwala, P.; Kasoji, M.; Abrams, N.; Merchant, A.; Bonner, W.M. Twist1 and Slug mediate H2AX-regulated epithelial-mesenchymal transition in breast cells. *Cell Cycle* **2016**, *15*, 2398–2404. [CrossRef]
91. Brabletz, T.; Hlubek, F.; Spaderna, S.; Schmalhofer, O.; Hiendlmeyer, E.; Jung, A.; Kirchner, T. Invasion and metastasis in colorectal cancer: Epithelial-mesenchymal transition, mesenchymal-epithelial transition, stem cells and β-catenin. *Cells Tissues Organs* **2005**, *179*, 56–65. [CrossRef]
92. Al-Hajj, M.; Wicha, M.S.; Benito-Hernandez, A.; Morrison, S.J.; Clarke, M.F. Prospective identification of tumorigenic breast cancer cells. *Proc. Natl. Acad. Sci. USA* **2003**, *100*, 3983–3988. [CrossRef]
93. Okuda, H.; Kobayashi, A.; Xia, B.; Watabe, M.; Pai, S.K.; Hirota, S.; Xing, F.; Liu, W.; Pandey, P.R.; Fukuda, K.; et al. Hyaluronan synthase HAS2 promotes tumor progression in bone by stimulating the interaction of breast cancer stem-like cells with macrophages and stromal cells. *Cancer Res.* **2012**, *72*, 537–547. [CrossRef]
94. Lock, R.; Kenific, C.M.; Leidal, A.M.; Salas, E.; Debnath, J. Autophagy-Dependent production of secreted factors facilitates oncogenic RAS-driven invasion. *Cancer Discov.* **2014**, *4*, 466–479. [CrossRef] [PubMed]
95. Peng, Y.F.; Shi, Y.H.; Shen, Y.H.; Ding, Z.B.; Ke, A.W.; Zhou, J.; Qiu, S.J.; Fan, J. Promoting colonization in metastatic HCC cells by modulation of autophagy. *PLoS ONE* **2013**, *8*, e74407. [CrossRef] [PubMed]
96. Zhu, H.; Wang, D.; Zhang, L.; Xie, X.; Wu, Y.; Liu, Y.; Shao, G.; Su, Z. Upregulation of autophagy by hypoxia-inducible factor-1α promotes EMT and metastatic ability of CD133+ pancreatic cancer stem-like cells during intermittent hypoxia. *Oncol. Rep.* **2014**, *32*, 935–942. [CrossRef] [PubMed]
97. Bellodi, C.; Lidonnici, M.R.; Hamilton, A.; Helgason, G.V.; Soliera, A.R.; Ronchetti, M.; Galavotti, S.; Young, K.W.; Selmi, T.; Yacobi, R.; et al. Targeting autophagy potentiates tyrosine kinase inhibitor-induced cell death in Philadelphia chromosome-positive cells, including primary CML stem cells. *J. Clin. Investig.* **2009**, *119*, 1109–1123. [CrossRef]
98. Zhu, H.; Wang, D.; Liu, Y.; Su, Z.; Zhang, L.; Chen, F.; Zhou, Y.; Wu, Y.; Yu, M.; Zhang, Z.; et al. Role of the Hypoxia-inducible factor-1 α induced autophagy in the conversion of non-stem pancreatic cancer cells into CD133+ pancreatic cancer stem-like cells. *Cancer Cell Int.* **2013**, *13*, 119. [CrossRef]
99. Husemann, Y.; Klein, C.A. The analysis of metastasis in transgenic mouse models. *Transgenic Res.* **2009**, *18*, 1–5. [CrossRef]
100. Joyce, J.A.; Pollard, J.W. Microenvironmental regulation of metastasis. *Nat. Rev. Cancer* **2009**, *9*, 239–252. [CrossRef]
101. Pollard, J.W. Trophic macrophages in development and disease. *Nat. Rev. Immunol.* **2009**, *9*, 259–270. [CrossRef] [PubMed]
102. Chen, J.; Yao, Y.; Gong, C.; Yu, F.; Su, S.; Liu, B.; Deng, H.; Wang, F.; Lin, L.; Yao, H.; et al. CCL18 from tumor-associated macrophages promotes breast cancer metastasis via PITPNM3. *Cancer Cell* **2011**, *19*, 541–555. [CrossRef] [PubMed]
103. Yang, M.; Chen, J.; Su, F.; Yu, B.; Lin, L.; Liu, Y.; Huang, J.D.; Song, E. Microvesicles secreted by macrophages shuttle invasion-potentiating microRNAs into breast cancer cells. *Mol. Cancer* **2011**, *10*, 117. [CrossRef] [PubMed]
104. Mazzone, M.; Dettori, D.; de Oliveira, R.L.; Loges, S.; Schmidt, T.; Jonckx, B.; Tian, Y.M.; Lanahan, A.A.; Pollard, P.; de Almodovar, C.R.; et al. Heterozygous deficiency of PHD2 restores tumor oxygenation and inhibits metastasis via endothelial normalization. *Cell* **2009**, *136*, 839–851. [CrossRef] [PubMed]
105. Orimo, A.; Gupta, P.B.; Sgroi, D.C.; Arenzana-Seisdedos, F.; Delaunay, T.; Naeem, R.; Carey, V.J.; Richardson, A.L.; Weinberg, R.A. Stromal fibroblasts present in invasive human breast carcinomas promote tumor growth and angiogenesis through elevated SDF-1/CXCL12 secretion. *Cell* **2005**, *121*, 335–348. [CrossRef] [PubMed]
106. Tsai, J.H.; Yang, J. Epithelial-Mesenchymal plasticity in carcinoma metastasis. *Genes Dev.* **2013**, *27*, 2192–2206. [CrossRef]
107. Park, S.M.; Gaur, A.B.; Lengyel, E.; Peter, M.E. The miR-200 family determines the epithelial phenotype of cancer cells by targeting the E-cadherin repressors ZEB1 and ZEB2. *Genes Dev.* **2008**, *22*, 894–907. [CrossRef]
108. Le, M.T.; Hamar, P.; Guo, C.; Basar, E.; Perdigao-Henriques, R.; Balaj, L.; Lieberman, J. miR-200-containing extracellular vesicles promote breast cancer cell metastasis. *J. Clin. Investig.* **2014**, *124*, 5109–5128. [CrossRef]
109. Lin, X.J.; Fang, J.H.; Yang, X.J.; Zhang, C.; Yuan, Y.; Zheng, L.; Zhuang, S.M. Hepatocellular Carcinoma Cell-Secreted Exosomal MicroRNA-210 Promotes Angiogenesis In Vitro and In Vivo. *Mol. Ther. Nucleic Acids* **2018**, *11*, 243–252. [CrossRef]
110. Pang, W.; Su, J.; Wang, Y.; Feng, H.; Dai, X.; Yuan, Y.; Chen, X.; Yao, W. Pancreatic cancer-secreted miR-155 implicates in the conversion from normal fibroblasts to cancer-associated fibroblasts. *Cancer Sci.* **2015**, *106*, 1362–1369. [CrossRef]
111. Zomer, A.; Maynard, C.; Verweij, F.J.; Kamermans, A.; Schafer, R.; Beerling, E.; Schiffelers, R.M.; de Wit, E.; Berenguer, J.; Ellenbroek, S.I.J.; et al. In Vivo imaging reveals extracellular vesicle-mediated phenocopying of metastatic behavior. *Cell* **2015**, *161*, 1046–1057. [CrossRef] [PubMed]
112. Zhou, W.; Fong, M.Y.; Min, Y.; Somlo, G.; Liu, L.; Palomares, M.R.; Yu, Y.; Chow, A.; O'Connor, S.T.F.; Chin, A.R. Cancer-secreted miR-105 destroys vascular endothelial barriers to promote metastasis. *Cancer Cell* **2014**, *25*, 501–515. [CrossRef] [PubMed]

113. Di Modica, M.; Regondi, V.; Sandri, M.; Iorio, M.V.; Zanetti, A.; Tagliabue, E.; Casalini, P.; Triulzi, T. Breast cancer-secreted miR-939 downregulates VE-cadherin and destroys the barrier function of endothelial monolayers. *Cancer Lett.* **2016**, *384*, 94–100. [CrossRef] [PubMed]
114. Singh, R.; Pochampally, R.; Watabe, K.; Lu, Z.; Mo, Y.Y. Exosome-Mediated transfer of miR-10b promotes cell invasion in breast cancer. *Mol. Cancer* **2014**, *13*, 256. [CrossRef]
115. Tominaga, N.; Kosaka, N.; Ono, M.; Katsuda, T.; Yoshioka, Y.; Tamura, K.; Lotvall, J.; Nakagama, H.; Ochiya, T. Brain metastatic cancer cells release microRNA-181c-containing extracellular vesicles capable of destructing blood-brain barrier. *Nat. Commun.* **2015**, *6*, 6716. [CrossRef]
116. Fong, M.Y.; Zhou, W.; Liu, L.; Alontaga, A.Y.; Chandra, M.; Ashby, J.; Chow, A.; O'Connor, S.T.; Li, S.; Chin, A.R.; et al. Breast-Cancer-Secreted miR-122 reprograms glucose metabolism in premetastatic niche to promote metastasis. *Nat. Cell Biol.* **2015**, *17*, 183–194. [CrossRef]
117. Wu, X.; Somlo, G.; Yu, Y.; Palomares, M.R.; Li, A.X.; Zhou, W.; Chow, A.; Yen, Y.; Rossi, J.J.; Gao, H.; et al. De novo sequencing of circulating miRNAs identifies novel markers predicting clinical outcome of locally advanced breast cancer. *J. Transl. Med.* **2012**, *10*, 42. [CrossRef]
118. Thery, C.; Duban, L.; Segura, E.; Veron, P.; Lantz, O.; Amigorena, S. Indirect activation of naive CD4+ T cells by dendritic cell-derived exosomes. *Nat. Immunol.* **2002**, *3*, 1156–1162. [CrossRef]
119. Clayton, A.; Mason, M.D. Exosomes in tumour immunity. *Curr. Oncol.* **2009**, *16*, 46–49. [CrossRef]
120. Greening, D.W.; Gopal, S.K.; Xu, R.; Simpson, R.J.; Chen, W. Exosomes and their roles in immune regulation and cancer. *Semin. Cell Dev. Biol.* **2015**, *40*, 72–81. [CrossRef]
121. Yu, S.; Liu, C.; Su, K.; Wang, J.; Liu, Y.; Zhang, L.; Li, C.; Cong, Y.; Kimberly, R.; Grizzle, W.E.; et al. Tumor exosomes inhibit differentiation of bone marrow dendritic cells. *J. Immunol.* **2007**, *178*, 6867–6875. [CrossRef] [PubMed]
122. Ding, G.; Zhou, L.; Qian, Y.; Fu, M.; Chen, J.; Xiang, J.; Wu, Z.; Jiang, G.; Cao, L. Pancreatic cancer-derived exosomes transfer miRNAs to dendritic cells and inhibit RFXAP expression via miR-212-3p. *Oncotarget* **2015**, *6*, 29877–29888. [CrossRef] [PubMed]
123. Andreola, G.; Rivoltini, L.; Castelli, C.; Huber, V.; Perego, P.; Deho, P.; Squarcina, P.; Accornero, P.; Lozupone, F.; Lugini, L.; et al. Induction of lymphocyte apoptosis by tumor cell secretion of FasL-bearing microvesicles. *J. Exp. Med.* **2002**, *195*, 1303–1316. [CrossRef]
124. Clayton, A.; Mitchell, J.P.; Court, J.; Linnane, S.; Mason, M.D.; Tabi, Z. Human tumor-derived exosomes down-modulate NKG2D expression. *J. Immunol.* **2008**, *180*, 7249–7258. [CrossRef]
125. Clayton, A.; Mitchell, J.P.; Court, J.; Mason, M.D.; Tabi, Z. Human tumor-derived exosomes selectively impair lymphocyte responses to interleukin-2. *Cancer Res.* **2007**, *67*, 7458–7466. [CrossRef]
126. Yin, Y.; Cai, X.; Chen, X.; Liang, H.; Zhang, Y.; Li, J.; Wang, Z.; Zhang, W.; Yokoyama, S.; Wang, C.; et al. Tumor-Secreted miR-214 induces regulatory T cells: A major link between immune evasion and tumor growth. *Cell Res.* **2014**, *24*, 1164–1180. [CrossRef] [PubMed]
127. Hui, E.; Cheung, J.; Zhu, J.; Su, X.; Taylor, M.J.; Wallweber, H.A.; Sasmal, D.K.; Huang, J.; Kim, J.M.; Mellman, I.; et al. T cell costimulatory receptor CD28 is a primary target for PD-1-mediated inhibition. *Science* **2017**, *355*, 1428–1433. [CrossRef]
128. Poggio, M.; Hu, T.; Pai, C.C.; Chu, B.; Belair, C.D.; Chang, A.; Montabana, E.; Lang, U.E.; Fu, Q.; Fong, L.; et al. Suppression of Exosomal PD-L1 Induces Systemic Anti-Tumor Immunity and Memory. *Cell* **2019**, *177*, 414–427.e413. [CrossRef]
129. Chen, G.; Huang, A.C.; Zhang, W.; Zhang, G.; Wu, M.; Xu, W.; Yu, Z.; Yang, J.; Wang, B.; Sun, H.; et al. Exosomal PD-L1 contributes to immunosuppression and is associated with anti-PD-1 response. *Nature* **2018**, *560*, 382–386. [CrossRef]
130. Theodoraki, M.N.; Yerneni, S.S.; Hoffmann, T.K.; Gooding, W.E.; Whiteside, T.L. Clinical Significance of PD-L1(+) Exosomes in Plasma of Head and Neck Cancer Patients. *Clin. Cancer Res.* **2018**, *24*, 896–905. [CrossRef]
131. Chen, L.; Han, X. Anti-PD-1/PD-L1 therapy of human cancer: Past, present, and future. *J. Clin. Investig.* **2015**, *125*, 3384–3391. [CrossRef] [PubMed]
132. Fabbri, M.; Paone, A.; Calore, F.; Galli, R.; Gaudio, E.; Santhanam, R.; Lovat, F.; Fadda, P.; Mao, C.; Nuovo, G.J.; et al. MicroRNAs bind to toll-like receptors to induce prometastatic inflammatory response. *Proc. Natl. Acad. Sci. USA* **2012**, *109*, E2110–E2116. [CrossRef] [PubMed]
133. Zhou, M.; Chen, J.; Zhou, L.; Chen, W.; Ding, G.; Cao, L. Pancreatic cancer derived exosomes regulate the expression of TLR4 in dendritic cells via miR-203. *Cell Immunol.* **2014**, *292*, 65–69. [CrossRef]
134. Andre, F.; Schartz, N.E.; Chaput, N.; Flament, C.; Raposo, G.; Amigorena, S.; Angevin, E.; Zitvogel, L. Tumor-Derived exosomes: A new source of tumor rejection antigens. *Vaccine* **2002**, *20*, A28–A31. [CrossRef]
135. Skokos, D.; Botros, H.G.; Demeure, C.; Morin, J.; Peronet, R.; Birkenmeier, G.; Boudaly, S.; Mecheri, S. Mast cell-derived exosomes induce phenotypic and functional maturation of dendritic cells and elicit specific immune responses in vivo. *J. Immunol.* **2003**, *170*, 3037–3045. [CrossRef] [PubMed]
136. Li, I.; Nabet, B.Y. Exosomes in the tumor microenvironment as mediators of cancer therapy resistance. *Mol. Cancer* **2019**, *18*, 32. [CrossRef] [PubMed]
137. Azmi, A.S.; Bao, B.; Sarkar, F.H. Exosomes in cancer development, metastasis, and drug resistance: A comprehensive review. *Cancer Metastasis Rev.* **2013**, *32*, 623–642. [CrossRef]

138. Valcz, G.; Buzas, E.I.; Sebestyen, A.; Krenacs, T.; Szallasi, Z.; Igaz, P.; Molnar, B. Extracellular Vesicle-Based Communication May Contribute to the Co-Evolution of Cancer Stem Cells and Cancer-Associated Fibroblasts in Anti-Cancer Therapy. *Cancers* **2020**, *12*, 2324. [CrossRef]
139. Steinbichler, T.B.; Dudas, J.; Skvortsov, S.; Ganswindt, U.; Riechelmann, H.; Skvortsova, I.I. Therapy resistance mediated by exosomes. *Mol. Cancer* **2019**, *18*, 58. [CrossRef]
140. Moitra, K.; Lou, H.; Dean, M. Multidrug efflux pumps and cancer stem cells: Insights into multidrug resistance and therapeutic development. *Clin. Pharmacol. Ther.* **2011**, *89*, 491–502. [CrossRef]
141. Januchowski, R.; Sterzynska, K.; Zaorska, K.; Sosinska, P.; Klejewski, A.; Brazert, M.; Nowicki, M.; Zabel, M. Analysis of MDR genes expression and cross-resistance in eight drug resistant ovarian cancer cell lines. *J. Ovarian Res.* **2016**, *9*, 65. [CrossRef] [PubMed]
142. Bebawy, M.; Combes, V.; Lee, E.; Jaiswal, R.; Gong, J.; Bonhoure, A.; Grau, G.E. Membrane microparticles mediate transfer of P-glycoprotein to drug sensitive cancer cells. *Leukemia* **2009**, *23*, 1643–1649. [CrossRef] [PubMed]
143. Corcoran, C.; Rani, S.; O'Brien, K.; O'Neill, A.; Prencipe, M.; Sheikh, R.; Webb, G.; McDermott, R.; Watson, W.; Crown, J.; et al. Docetaxel-Resistance in prostate cancer: Evaluating associated phenotypic changes and potential for resistance transfer via exosomes. *PLoS ONE* **2012**, *7*, e50999. [CrossRef] [PubMed]
144. Sousa, D.; Lima, R.T.; Vasconcelos, M.H. Intercellular Transfer of Cancer Drug Resistance Traits by Extracellular Vesicles. *Trends Mol. Med.* **2015**, *21*, 595–608. [CrossRef]
145. Gong, J.; Luk, F.; Jaiswal, R.; George, A.M.; Grau, G.E.; Bebawy, M. Microparticle drug sequestration provides a parallel pathway in the acquisition of cancer drug resistance. *Eur. J. Pharmacol.* **2013**, *721*, 116–125. [CrossRef]
146. Federici, C.; Petrucci, F.; Caimi, S.; Cesolini, A.; Logozzi, M.; Borghi, M.; D'Ilio, S.; Lugini, L.; Violante, N.; Azzarito, T.; et al. Exosome release and low pH belong to a framework of resistance of human melanoma cells to cisplatin. *PLoS ONE* **2014**, *9*, e88193. [CrossRef]
147. Aung, T.; Chapuy, B.; Vogel, D.; Wenzel, D.; Oppermann, M.; Lahmann, M.; Weinhage, T.; Menck, K.; Hupfeld, T.; Koch, R.; et al. Exosomal evasion of humoral immunotherapy in aggressive B-cell lymphoma modulated by ATP-binding cassette transporter A3. *Proc. Natl. Acad. Sci. USA* **2011**, *108*, 15336–15341. [CrossRef]
148. Bliss, S.A.; Sinha, G.; Sandiford, O.A.; Williams, L.M.; Engelberth, D.J.; Guiro, K.; Isenalumhe, L.L.; Greco, S.J.; Ayer, S.; Bryan, M. Mesenchymal stem cell–derived exosomes stimulate cycling quiescence and early breast cancer dormancy in bone marrow. *Cancer Res.* **2016**, *76*, 5832–5844. [CrossRef]
149. Akhter, M.Z.; Sharawat, S.K.; Kumar, V.; Kochat, V.; Equbal, Z.; Ramakrishnan, M.; Kumar, U.; Mathur, S.; Kumar, L.; Mukhopadhyay, A. Aggressive serous epithelial ovarian cancer is potentially propagated by EpCAM(+)CD45(+) phenotype. *Oncogene* **2018**, *37*, 2089–2103. [CrossRef]
150. Ji, R.; Zhang, B.; Zhang, X.; Xue, J.; Yuan, X.; Yan, Y.; Wang, M.; Zhu, W.; Qian, H.; Xu, W. Exosomes derived from human mesenchymal stem cells confer drug resistance in gastric cancer. *Cell Cycle* **2015**, *14*, 2473–2483. [CrossRef]
151. Boing, A.N.; Stap, J.; Hau, C.M.; Afink, G.B.; Ris-Stalpers, C.; Reits, E.A.; Sturk, A.; van Noorden, C.J.; Nieuwland, R. Active caspase-3 is removed from cells by release of caspase-3-enriched vesicles. *Biochim. Biophys. Acta* **2013**, *1833*, 1844–1852. [CrossRef] [PubMed]
152. Wang, J.; Zhang, X.; Wei, P.; Zhang, J.; Niu, Y.; Kang, N.; Zhang, Y.; Zhang, W.; Xing, N. Livin, Survivin and Caspase 3 as early recurrence markers in non-muscle-invasive bladder cancer. *World J. Urol.* **2014**, *32*, 1477–1484. [CrossRef] [PubMed]
153. Manier, S.; Sacco, A.; Leleu, X.; Ghobrial, I.M.; Roccaro, A.M. Bone marrow microenvironment in multiple myeloma progression. *J. Biomed. Biotechnol.* **2012**, *2012*, 157496. [CrossRef]
154. Xu, F.H.; Sharma, S.; Gardner, A.; Tu, Y.; Raitano, A.; Sawyers, C.; Lichtenstein, A. Interleukin-6-induced inhibition of multiple myeloma cell apoptosis: Support for the hypothesis that protection is mediated via inhibition of the JNK/SAPK pathway. *Blood* **1998**, *92*, 241–251. [CrossRef] [PubMed]
155. Rodrigues, C.F.D.; Serrano, E.; Patricio, M.I.; Val, M.M.; Albuquerque, P.; Fonseca, J.; Gomes, C.M.F.; Abrunhosa, A.J.; Paiva, A.; Carvalho, L.; et al. Stroma-Derived IL-6, G-CSF and Activin-A mediated dedifferentiation of lung carcinoma cells into cancer stem cells. *Sci. Rep.* **2018**, *8*, 11573. [CrossRef] [PubMed]
156. Hazawa, M.; Tomiyama, K.; Saotome-Nakamura, A.; Obara, C.; Yasuda, T.; Gotoh, T.; Tanaka, I.; Yakumaru, H.; Ishihara, H.; Tajima, K. Radiation increases the cellular uptake of exosomes through CD29/CD81 complex formation. *Biochem. Biophys. Res. Commun.* **2014**, *446*, 1165–1171. [CrossRef] [PubMed]
157. Dutta, S.; Warshall, C.; Bandyopadhyay, C.; Dutta, D.; Chandran, B. Interactions between exosomes from breast cancer cells and primary mammary epithelial cells leads to generation of reactive oxygen species which induce DNA damage response, stabilization of p53 and autophagy in epithelial cells. *PLoS ONE* **2014**, *9*, e97580. [CrossRef]
158. Mutschelknaus, L.; Peters, C.; Winkler, K.; Yentrapalli, R.; Heider, T.; Atkinson, M.J.; Moertl, S. Exosomes Derived from Squamous Head and Neck Cancer Promote Cell Survival after Ionizing Radiation. *PLoS ONE* **2016**, *11*, e0152213. [CrossRef]
159. Arscott, W.T.; Tandle, A.T.; Zhao, S.; Shabason, J.E.; Gordon, I.K.; Schlaff, C.D.; Zhang, G.; Tofilon, P.J.; Camphausen, K.A. Ionizing radiation and glioblastoma exosomes: Implications in tumor biology and cell migration. *Transl. Oncol.* **2013**, *6*, 638–648. [CrossRef]
160. Peiris-Pages, M.; Sotgia, F.; Lisanti, M.P. Chemotherapy induces the cancer-associated fibroblast phenotype, activating paracrine Hedgehog-GLI signalling in breast cancer cells. *Oncotarget* **2015**, *6*, 10728–10745. [CrossRef]

161. Wang, Z.; Tang, Y.; Tan, Y.; Wei, Q.; Yu, W. Cancer-Associated fibroblasts in radiotherapy: Challenges and new opportunities. *Cell Commun. Signal.* **2019**, *17*, 47. [CrossRef] [PubMed]
162. McMillin, D.W.; Negri, J.M.; Mitsiades, C.S. The role of tumour-stromal interactions in modifying drug response: Challenges and opportunities. *Nat. Rev. Drug Discov.* **2013**, *12*, 217–228. [CrossRef] [PubMed]
163. Richards, K.E.; Zeleniak, A.E.; Fishel, M.L.; Wu, J.; Littlepage, L.E.; Hill, R. Cancer-Associated fibroblast exosomes regulate survival and proliferation of pancreatic cancer cells. *Oncogene* **2017**, *36*, 1770–1778. [CrossRef] [PubMed]
164. Boelens, M.C.; Wu, T.J.; Nabet, B.Y.; Xu, B.; Qiu, Y.; Yoon, T.; Azzam, D.J.; Twyman-Saint Victor, C.; Wiemann, B.Z.; Ishwaran, H.; et al. Exosome transfer from stromal to breast cancer cells regulates therapy resistance pathways. *Cell* **2014**, *159*, 499–513. [CrossRef]
165. Su, S.; Chen, J.; Yao, H.; Liu, J.; Yu, S.; Lao, L.; Wang, M.; Luo, M.; Xing, Y.; Chen, F.; et al. CD10(+)GPR77(+) Cancer-Associated Fibroblasts Promote Cancer Formation and Chemoresistance by Sustaining Cancer Stemness. *Cell* **2018**, *172*, 841–856.e816. [CrossRef]
166. Mitra, A.; Mishra, L.; Li, S. EMT, CTCs and CSCs in tumor relapse and drug-resistance. *Oncotarget* **2015**, *6*, 10697–10711. [CrossRef]
167. Steinbichler, T.B.; Dudas, J.; Skvortsov, S.; Ganswindt, U.; Riechelmann, H.; Skvortsova, I.I. Therapy resistance mediated by cancer stem cells. *Semin. Cancer Biol.* **2018**, *53*, 156–167. [CrossRef]
168. Johansson, J.; Landgren, M.; Fernell, E.; Lewander, T.; Venizelos, N. Decreased binding capacity (Bmax) of muscarinic acetylcholine receptors in fibroblasts from boys with attention-deficit/hyperactivity disorder (ADHD). *Atten. Defic. Hyperact. Disord.* **2013**, *5*, 267–271. [CrossRef]
169. Grinde, M.T.; Vik, J.; Camilio, K.A.; Martinez-Zubiaurre, I.; Hellevik, T. Ionizing radiation abrogates the pro-tumorigenic capacity of cancer-associated fibroblasts co-implanted in xenografts. *Sci. Rep.* **2017**, *7*, 46714. [CrossRef]
170. Wei, Y.; Lai, X.; Yu, S.; Chen, S.; Ma, Y.; Zhang, Y.; Li, H.; Zhu, X.; Yao, L.; Zhang, J. Exosomal miR-221/222 enhances tamoxifen resistance in recipient ER-positive breast cancer cells. *Breast Cancer Res. Treat.* **2014**, *147*, 423–431. [CrossRef]
171. Challagundla, K.B.; Wise, P.M.; Neviani, P.; Chava, H.; Murtadha, M.; Xu, T.; Kennedy, R.; Ivan, C.; Zhang, X.; Vannini, I.; et al. Exosome-Mediated transfer of microRNAs within the tumor microenvironment and neuroblastoma resistance to chemotherapy. *J. Natl. Cancer Inst.* **2015**, *107*. [CrossRef] [PubMed]
172. Bouvy, C.; Wannez, A.; Laloy, J.; Chatelain, C.; Dogne, J.M. Transfer of multidrug resistance among acute myeloid leukemia cells via extracellular vesicles and their microRNA cargo. *Leukemia Res.* **2017**, *62*, 70–76. [CrossRef] [PubMed]
173. Shen, M.; Dong, C.; Ruan, X.; Yan, W.; Cao, M.; Pizzo, D.; Wu, X.; Yang, L.; Liu, L.; Ren, X.; et al. Chemotherapy-Induced Extracellular Vesicle miRNAs Promote Breast Cancer Stemness by Targeting ONECUT2. *Cancer Res.* **2019**, *79*, 3608–3621. [CrossRef] [PubMed]
174. Patel, G.K.; Khan, M.A.; Bhardwaj, A.; Srivastava, S.K.; Zubair, H.; Patton, M.C.; Singh, S.; Khushman, M.; Singh, A.P. Exosomes confer chemoresistance to pancreatic cancer cells by promoting ROS detoxification and miR-155-mediated suppression of key gemcitabine-metabolising enzyme, DCK. *Br. J. Cancer* **2017**, *116*, 609–619. [CrossRef]
175. Au Yeung, C.L.; Co, N.N.; Tsuruga, T.; Yeung, T.L.; Kwan, S.Y.; Leung, C.S.; Li, Y.; Lu, E.S.; Kwan, K.; Wong, K.K.; et al. Exosomal transfer of stroma-derived miR21 confers paclitaxel resistance in ovarian cancer cells through targeting APAF1. *Nat. Commun.* **2016**, *7*, 11150. [CrossRef]
176. Abels, E.R.; Maas, S.L.N.; Nieland, L.; Wei, Z.; Cheah, P.S.; Tai, E.; Kolsteeg, C.J.; Dusoswa, S.A.; Ting, D.T.; Hickman, S.; et al. Glioblastoma-Associated Microglia Reprogramming Is Mediated by Functional Transfer of Extracellular miR-21. *Cell Rep.* **2019**, *28*, 3105–3119.e3107. [CrossRef]
177. Nouraee, N.; Van Roosbroeck, K.; Vasei, M.; Semnani, S.; Samaei, N.M.; Naghshvar, F.; Omidi, A.A.; Calin, G.A.; Mowla, S.J. Expression, tissue distribution and function of miR-21 in esophageal squamous cell carcinoma. *PLoS ONE* **2013**, *8*, e73009. [CrossRef]
178. Momen-Heravi, F.; Bala, S. Extracellular vesicles in oral squamous carcinoma carry oncogenic miRNA profile and reprogram monocytes via NF-κB pathway. *Oncotarget* **2018**, *9*, 34838–34854. [CrossRef]
179. Baroni, S.; Romero-Cordoba, S.; Plantamura, I.; Dugo, M.; D'Ippolito, E.; Cataldo, A.; Cosentino, G.; Angeloni, V.; Rossini, A.; Daidone, M.G.; et al. Exosome-Mediated delivery of miR-9 induces cancer-associated fibroblast-like properties in human breast fibroblasts. *Cell Death Dis.* **2016**, *7*, e2312. [CrossRef]
180. Yan, W.; Wu, X.; Zhou, W.; Fong, M.Y.; Cao, M.; Liu, J.; Liu, X.; Chen, C.H.; Fadare, O.; Pizzo, D.P.; et al. Cancer-Cell-Secreted exosomal miR-105 promotes tumour growth through the MYC-dependent metabolic reprogramming of stromal cells. *Nat. Cell Biol.* **2018**, *20*, 597–609. [CrossRef]
181. Ye, S.B.; Zhang, H.; Cai, T.T.; Liu, Y.N.; Ni, J.J.; He, J.; Peng, J.Y.; Chen, Q.Y.; Mo, H.Y.; Jun, C.; et al. Exosomal miR-24-3p impedes T-cell function by targeting FGF11 and serves as a potential prognostic biomarker for nasopharyngeal carcinoma. *J. Pathol.* **2016**, *240*, 329–340. [CrossRef] [PubMed]
182. Ohshima, K.; Inoue, K.; Fujiwara, A.; Hatakeyama, K.; Kanto, K.; Watanabe, Y.; Muramatsu, K.; Fukuda, Y.; Ogura, S.-I.; Yamaguchi, K. Let-7 microRNA family is selectively secreted into the extracellular environment via exosomes in a metastatic gastric cancer cell line. *PLoS ONE* **2010**, *5*, e13247. [CrossRef] [PubMed]
183. Wei, S.; Peng, L.; Yang, J.; Sang, H.; Jin, D.; Li, X.; Chen, M.; Zhang, W.; Dang, Y.; Zhang, G. Exosomal transfer of miR-15b-3p enhances tumorigenesis and malignant transformation through the DYNLT1/Caspase-3/Caspase-9 signaling pathway in gastric cancer. *J. Exp. Clin. Cancer Res.* **2020**, *39*, 32. [CrossRef]

184. Fang, J.H.; Zhang, Z.J.; Shang, L.R.; Luo, Y.W.; Lin, Y.F.; Yuan, Y.; Zhuang, S.M. Hepatoma cell-secreted exosomal microRNA-103 increases vascular permeability and promotes metastasis by targeting junction proteins. *Hepatology* **2018**, *68*, 1459–1475. [CrossRef]
185. Fruhbeis, C.; Frohlich, D.; Kramer-Albers, E.M. Emerging roles of exosomes in neuron-glia communication. *Front. Physiol.* **2012**, *3*, 119. [CrossRef]
186. Howitt, J.; Hill, A.F. Exosomes in the Pathology of Neurodegenerative Diseases. *J. Biol. Chem.* **2016**, *291*, 26589–26597. [CrossRef] [PubMed]
187. D'Anca, M.; Fenoglio, C.; Serpente, M.; Arosio, B.; Cesari, M.; Scarpini, E.A.; Galimberti, D. Exosome Determinants of Physiological Aging and Age-Related Neurodegenerative Diseases. *Front. Aging Neurosci.* **2019**, *11*, 232. [CrossRef]
188. Ridder, K.; Keller, S.; Dams, M.; Rupp, A.K.; Schlaudraff, J.; Del Turco, D.; Starmann, J.; Macas, J.; Karpova, D.; Devraj, K.; et al. Extracellular vesicle-mediated transfer of genetic information between the hematopoietic system and the brain in response to inflammation. *PLoS Biol.* **2014**, *12*, e1001874. [CrossRef]
189. Perez, M.; Avila, J.; Hernandez, F. Propagation of Tau via Extracellular Vesicles. *Front. Neurosci.* **2019**, *13*, 698. [CrossRef] [PubMed]
190. Guix, F.X.; Corbett, G.T.; Cha, D.J.; Mustapic, M.; Liu, W.; Mengel, D.; Chen, Z.; Aikawa, E.; Young-Pearse, T.; Kapogiannis, D.; et al. Detection of Aggregation-Competent Tau in Neuron-Derived Extracellular Vesicles. *Int. J. Mol. Sci.* **2018**, *19*, 663. [CrossRef]
191. Wang, Y.; Balaji, V.; Kaniyappan, S.; Kruger, L.; Irsen, S.; Tepper, K.; Chandupatla, R.; Maetzler, W.; Schneider, A.; Mandelkow, E.; et al. The release and trans-synaptic transmission of Tau via exosomes. *Mol. Neurodegener.* **2017**, *12*, 5. [CrossRef] [PubMed]
192. Polanco, J.C.; Li, C.; Durisic, N.; Sullivan, R.; Gotz, J. Exosomes taken up by neurons hijack the endosomal pathway to spread to interconnected neurons. *Acta Neuropathologica Commun.* **2018**, *6*, 10. [CrossRef] [PubMed]
193. Querfurth, H.W.; LaFerla, F.M. Alzheimer's disease. *N. Engl. J. Med.* **2010**, *362*, 329–344. [CrossRef] [PubMed]
194. Saman, S.; Kim, W.; Raya, M.; Visnick, Y.; Miro, S.; Saman, S.; Jackson, B.; McKee, A.C.; Alvarez, V.E.; Lee, N.C. Exosome-associated tau is secreted in tauopathy models and is selectively phosphorylated in cerebrospinal fluid in early Alzheimer disease. *J. Biol. Chem.* **2012**, *287*, 3842–3849. [CrossRef]
195. Rajendran, L.; Honsho, M.; Zahn, T.R.; Keller, P.; Geiger, K.D.; Verkade, P.; Simons, K. Alzheimer's disease β-amyloid peptides are released in association with exosomes. *Proc. Natl. Acad. Sci. USA* **2006**, *103*, 11172–11177. [CrossRef]
196. Yuyama, K.; Sun, H.; Mitsutake, S.; Igarashi, Y. Sphingolipid-Modulated exosome secretion promotes clearance of amyloid-beta by microglia. *J. Biol. Chem.* **2012**, *287*, 10977–10989. [CrossRef]
197. Alvarez-Erviti, L.; Seow, Y.; Schapira, A.H.; Gardiner, C.; Sargent, I.L.; Wood, M.J.; Cooper, J.M. Lysosomal dysfunction increases exosome-mediated α-synuclein release and transmission. *Neurobiol. Dis.* **2011**, *42*, 360–367. [CrossRef]
198. Sundelof, J.; Giedraitis, V.; Irizarry, M.C.; Sundstrom, J.; Ingelsson, E.; Ronnemaa, E.; Arnlov, J.; Gunnarsson, M.D.; Hyman, B.T.; Basun, H.; et al. Plasma β amyloid and the risk of Alzheimer disease and dementia in elderly men: A prospective, population-based cohort study. *Arch. Neurol.* **2008**, *65*, 256–263. [CrossRef]
199. Ghidoni, R.; Paterlini, A.; Albertini, V.; Stoppani, E.; Binetti, G.; Fuxe, K.; Benussi, L.; Agnati, L.F. A window into the heterogeneity of human cerebrospinal fluid Aβ peptides. *J. Biomed. Biotechnol.* **2011**, *2011*, 697036. [CrossRef]
200. Fuller, S.; Steele, M.; Munch, G. Activated astroglia during chronic inflammation in Alzheimer's disease—Do they neglect their neurosupportive roles? *Mutat. Res.* **2009**, *690*, 40–49. [CrossRef]
201. Kobayashi, K.; Hernandez, L.D.; Galan, J.E.; Janeway, C.A., Jr.; Medzhitov, R.; Flavell, R.A. IRAK-M is a negative regulator of Toll-like receptor signaling. *Cell* **2002**, *110*, 191–202. [CrossRef]
202. Bieberich, E.; MacKinnon, S.; Silva, J.; Noggle, S.; Condie, B.G. Regulation of cell death in mitotic neural progenitor cells by asymmetric distribution of prostate apoptosis response 4 (PAR-4) and simultaneous elevation of endogenous ceramide. *J. Cell Biol.* **2003**, *162*, 469–479. [CrossRef] [PubMed]
203. Bieberich, E. Integration of glycosphingolipid metabolism and cell-fate decisions in cancer and stem cells: Review and hypothesis. *Glycoconj. J.* **2004**, *21*, 315–327. [CrossRef] [PubMed]
204. Wang, G.; Dinkins, M.; He, Q.; Zhu, G.; Poirier, C.; Campbell, A.; Mayer-Proschel, M.; Bieberich, E. Astrocytes secrete exosomes enriched with proapoptotic ceramide and prostate apoptosis response 4 (PAR-4) potential mechanism of apoptosis induction in Alzheimer disease (AD). *J. Biol. Chem.* **2012**, *287*, 21384–21395. [CrossRef] [PubMed]
205. Bakhti, M.; Winter, C.; Simons, M. Inhibition of myelin membrane sheath formation by oligodendrocyte-derived exosome-like vesicles. *J. Biol. Chem.* **2010**, *286*, 787–796. [CrossRef]
206. Frohlich, D.; Kuo, W.P.; Fruhbeis, C.; Sun, J.J.; Zehendner, C.M.; Luhmann, H.J.; Pinto, S.; Toedling, J.; Trotter, J.; Kramer-Albers, E.M. Multifaceted effects of oligodendroglial exosomes on neurons: Impact on neuronal firing rate, signal transduction and gene regulation. *Philos. Trans. R. Soc. Lond. B Biol. Sci.* **2014**, *369*. [CrossRef]
207. Fruhbeis, C.; Frohlich, D.; Kuo, W.P.; Amphornrat, J.; Thilemann, S.; Saab, A.S.; Kirchhoff, F.; Mobius, W.; Goebbels, S.; Nave, K.A.; et al. Neurotransmitter-Triggered transfer of exosomes mediates oligodendrocyte-neuron communication. *PLoS Biol.* **2013**, *11*, e1001604. [CrossRef]
208. Hakulinen, J.; Sankkila, L.; Sugiyama, N.; Lehti, K.; Keski-Oja, J. Secretion of active membrane type 1 matrix metalloproteinase (MMP-14) into extracellular space in microvesicular exosomes. *J. Cell Biochem.* **2008**, *105*, 1211–1218. [CrossRef]
209. Sarkar, S.; Yong, V.W. Inflammatory cytokine modulation of matrix metalloproteinase expression and invasiveness of glioma cells in a 3-dimensional collagen matrix. *J. Neurooncol.* **2009**, *91*, 157–164. [CrossRef]

210. Jy, W.; Minagar, A.; Jimenez, J.J.; Sheremata, W.A.; Mauro, L.M.; Horstman, L.L.; Bidot, C.; Ahn, Y.S. Endothelial microparticles (EMP) bind and activate monocytes: Elevated EMP-monocyte conjugates in multiple sclerosis. *Front. Biosci.* **2004**, *9*, 3137–3144. [CrossRef]
211. Quandt, J.; Dorovini-Zis, K. The β chemokines CCL4 and CCL5 enhance adhesion of specific CD4+ T cell subsets to human brain endothelial cells. *J. Neuropathol. Exp. Neurol.* **2004**, *63*, 350–362. [CrossRef] [PubMed]
212. Galazka, G.; Mycko, M.P.; Selmaj, I.; Raine, C.S.; Selmaj, K.W. Multiple sclerosis: Serum-Derived exosomes express myelin proteins. *Mult. Scler.* **2017**, *24*, 449–458. [CrossRef] [PubMed]
213. Pieragostino, D.; Cicalini, I.; Lanuti, P.; Ercolino, E.; di Ioia, M.; Zucchelli, M.; Zappacosta, R.; Miscia, S.; Marchisio, M.; Sacchetta, P.; et al. Enhanced release of acid sphingomyelinase-enriched exosomes generates a lipidomics signature in CSF of Multiple Sclerosis patients. *Sci. Rep.* **2018**, *8*, 3071. [CrossRef] [PubMed]
214. Hirsch, E.; Ruberg, M.; Portier, M.M.; Dardenne, M.; Agid, Y. Characterization of two antigens in parkinsonian Lewy bodies. *Brain Res.* **1988**, *441*, 139–144. [CrossRef]
215. Harraz, M.M.; Eacker, S.M.; Wang, X.; Dawson, T.M.; Dawson, V.L. MicroRNA-223 is neuroprotective by targeting glutamate receptors. *Proc. Natl. Acad. Sci. USA* **2012**, *109*, 18962–18967. [CrossRef]
216. Fortin, D.L.; Nemani, V.M.; Voglmaier, S.M.; Anthony, M.D.; Ryan, T.A.; Edwards, R.H. Neural activity controls the synaptic accumulation of α-synuclein. *J. Neurosci.* **2005**, *25*, 10913–10921. [CrossRef]
217. Goedert, M. The significance of tau and α-synuclein inclusions in neurodegenerative diseases. *Curr. Opin. Genet. Dev.* **2001**, *11*, 343–351. [CrossRef]
218. Prusiner, S.B.; Woerman, A.L.; Mordes, D.A.; Watts, J.C.; Rampersaud, R.; Berry, D.B.; Patel, S.; Oehler, A.; Lowe, J.K.; Kravitz, S.N.; et al. Evidence for α-synuclein prions causing multiple system atrophy in humans with parkinsonism. *Proc. Natl. Acad. Sci. USA* **2015**, *112*, E5308–E5317. [CrossRef]
219. Danzer, K.M.; Kranich, L.R.; Ruf, W.P.; Cagsal-Getkin, O.; Winslow, A.R.; Zhu, L.; Vanderburg, C.R.; McLean, P.J. Exosomal cell-to-cell transmission of alpha synuclein oligomers. *Mol. Neurodegener.* **2012**, *7*, 42. [CrossRef]
220. Grozdanov, V.; Danzer, K.M. Release and uptake of pathologic α-synuclein. *Cell Tissue Res.* **2018**, *373*, 175–182. [CrossRef] [PubMed]
221. Ishizawa, T.; Mattila, P.; Davies, P.; Wang, D.; Dickson, D.W. Colocalization of tau and α-synuclein epitopes in Lewy bodies. *J. Neuropathol. Exp. Neurol.* **2003**, *62*, 389–397. [CrossRef] [PubMed]
222. Clinton, L.K.; Blurton-Jones, M.; Myczek, K.; Trojanowski, J.Q.; LaFerla, F.M. Synergistic Interactions between Aβ, τ, and α-synuclein: Acceleration of neuropathology and cognitive decline. *J. Neurosci.* **2010**, *30*, 7281–7289. [CrossRef] [PubMed]
223. Li, D.; Li, Y.P.; Li, Y.X.; Zhu, X.H.; Du, X.G.; Zhou, M.; Li, W.B.; Deng, H.Y. Effect of Regulatory Network of Exosomes and microRNAs on Neurodegenerative Diseases. *Chin. Med. J.* **2018**, *131*, 2216–2225. [CrossRef] [PubMed]
224. Kalani, A.; Tyagi, A.; Tyagi, N. Exosomes: Mediators of neurodegeneration, neuroprotection and therapeutics. *Mol. Neurobiol.* **2013**, *49*, 590–600. [CrossRef] [PubMed]
225. Leidinger, P.; Backes, C.; Deutscher, S.; Schmitt, K.; Mueller, S.C.; Frese, K.; Haas, J.; Ruprecht, K.; Paul, F.; Stahler, C.; et al. A blood based 12-miRNA signature of Alzheimer disease patients. *Genome Biol.* **2013**, *14*, R78. [CrossRef]
226. Zuccato, C.; Ciammola, A.; Rigamonti, D.; Leavitt, B.R.; Goffredo, D.; Conti, L.; MacDonald, M.E.; Friedlander, R.M.; Silani, V.; Hayden, M.R.; et al. Loss of huntingtin-mediated BDNF gene transcription in Huntington's disease. *Science* **2001**, *293*, 493–498. [CrossRef]
227. Hoss, A.G.; Labadorf, A.; Beach, T.G.; Latourelle, J.C.; Myers, R.H. microRNA Profiles in Parkinson's Disease Prefrontal Cortex. *Front. Aging Neurosci.* **2016**, *8*, 36. [CrossRef]
228. Reed, E.R.; Latourelle, J.C.; Bockholt, J.H.; Bregu, J.; Smock, J.; Paulsen, J.S.; Myers, R.H. MicroRNAs in CSF as prodromal biomarkers for Huntington disease in the PREDICT-HD study. *Neurology* **2017**, *90*, e264–e272. [CrossRef]
229. Si, Y.; Cui, X.; Crossman, D.K.; Hao, J.; Kazamel, M.; Kwon, Y.; King, P.H. Muscle microRNA signatures as biomarkers of disease progression in amyotrophic lateral sclerosis. *Neurobiol. Dis.* **2018**, *114*, 85–94. [CrossRef]
230. Chen, J.J.; Zhao, B.; Zhao, J.; Li, S. Potential Roles of Exosomal MicroRNAs as Diagnostic Biomarkers and Therapeutic Application in Alzheimer's Disease. *Neural Plast.* **2017**, *2017*, 7027380. [CrossRef]
231. Cardo, L.F.; Coto, E.; de Mena, L.; Ribacoba, R.; Moris, G.; Menendez, M.; Alvarez, V. Profile of microRNAs in the plasma of Parkinson's disease patients and healthy controls. *J. Neurol.* **2013**, *260*, 1420–1422. [CrossRef] [PubMed]
232. Ebrahimkhani, S.; Vafaee, F.; Young, P.E.; Hur, S.S.J.; Hawke, S.; Devenney, E.; Beadnall, H.; Barnett, M.H.; Suter, C.M.; Buckland, M.E. Exosomal microRNA signatures in multiple sclerosis reflect disease status. *Sci. Rep.* **2017**, *7*, 14293. [CrossRef] [PubMed]
233. Kimura, K.; Hohjoh, H.; Fukuoka, M.; Sato, W.; Oki, S.; Tomi, C.; Yamaguchi, H.; Kondo, T.; Takahashi, R.; Yamamura, T. Circulating exosomes suppress the induction of regulatory T cells via let-7i in multiple sclerosis. *Nat. Commun.* **2018**, *9*, 17. [CrossRef] [PubMed]
234. Properzi, F.; Ferroni, E.; Poleggi, A.; Vinci, R. The regulation of exosome function in the CNS: Implications for neurodegeneration. *Swiss Med. Wkly.* **2015**, *145*, w14204. [CrossRef]
235. Lehmann, S.M.; Kruger, C.; Park, B.; Derkow, K.; Rosenberger, K.; Baumgart, J.; Trimbuch, T.; Eom, G.; Hinz, M.; Kaul, D.; et al. An unconventional role for miRNA: Let-7 activates Toll-like receptor 7 and causes neurodegeneration. *Nat. Neurosci.* **2012**, *15*, 827–835. [CrossRef]

236. Winkler, C.W.; Taylor, K.G.; Peterson, K.E. Location is everything: Let-7b microRNA and TLR7 signaling results in a painful TRP. *Sci. Signal.* **2014**, *7*, pe14. [CrossRef]
237. Azimi, M.; Ghabaee, M.; Naser Moghadasi, A.; Izad, M. Altered Expression of miR-326 in T Cell-derived Exosomes of Patients with Relapsing-Remitting Multiple Sclerosis. *Iran J. Allergy Asthma Immunol.* **2019**, *18*, 108–113. [CrossRef]
238. Nicolas, M.; Hassan, B.A. Amyloid precursor protein and neural development. *Development* **2014**, *141*, 2543–2548. [CrossRef]
239. Vilardo, E.; Barbato, C.; Ciotti, M.; Cogoni, C.; Ruberti, F. MicroRNA-101 regulates amyloid precursor protein expression in hippocampal neurons. *J. Biol. Chem.* **2010**, *285*, 18344–18351. [CrossRef]
240. Liu, C.G.; Wang, J.L.; Li, L.; Wang, P.C. MicroRNA-384 regulates both amyloid precursor protein and β-secretase expression and is a potential biomarker for Alzheimer's disease. *Int. J. Mol. Med.* **2014**, *34*, 160–166. [CrossRef]
241. Zhang, B.; Chen, C.F.; Wang, A.H.; Lin, Q.F. MiR-16 regulates cell death in Alzheimer's disease by targeting amyloid precursor protein. *Eur. Rev. Med. Pharmacol. Sci.* **2015**, *19*, 4020–4027. [PubMed]
242. Dickson, J.R.; Kruse, C.; Montagna, D.R.; Finsen, B.; Wolfe, M.S. Alternative polyadenylation and miR-34 family members regulate tau expression. *J. Neurochem.* **2013**, *127*, 739–749. [CrossRef] [PubMed]
243. Santa-Maria, I.; Alaniz, M.E.; Renwick, N.; Cela, C.; Fulga, T.A.; Van Vactor, D.; Tuschl, T.; Clark, L.N.; Shelanski, M.L.; McCabe, B.D.; et al. Dysregulation of microRNA-219 promotes neurodegeneration through post-transcriptional regulation of tau. *J. Clin. Investig.* **2015**, *125*, 681–686. [CrossRef] [PubMed]
244. Du, J.J.; Chen, S.D. Current Nondopaminergic Therapeutic Options for Motor Symptoms of Parkinson's Disease. *Chin. Med. J.* **2017**, *130*, 1856–1866. [CrossRef] [PubMed]
245. McMillan, K.J.; Murray, T.K.; Bengoa-Vergniory, N.; Cordero-Llana, O.; Cooper, J.; Buckley, A.; Wade-Martins, R.; Uney, J.B.; O'Neill, M.J.; Wong, L.F.; et al. Loss of MicroRNA-7 Regulation Leads to α-Synuclein Accumulation and Dopaminergic Neuronal Loss In Vivo. *Mol. Ther.* **2017**, *25*, 2404–2414. [CrossRef] [PubMed]
246. Valadi, H.; Ekström, K.; Bossios, A.; Sjöstrand, M.; Lee, J.J.; Lötvall, J.O. Exosome-Mediated transfer of mRNAs and microRNAs is a novel mechanism of genetic exchange between cells. *Nat. Cell Biol.* **2007**, *9*, 654–659. [CrossRef]
247. Guescini, M.; Genedani, S.; Stocchi, V.; Agnati, L.F. Astrocytes and Glioblastoma cells release exosomes carrying mtDNA. *J. Neural Transm.* **2010**, *117*, 1. [CrossRef]
248. Thakur, B.K.; Zhang, H.; Becker, A.; Matei, I.; Huang, Y.; Costa-Silva, B.; Zheng, Y.; Hoshino, A.; Brazier, H.; Xiang, J. Double-stranded DNA in exosomes: A novel biomarker in cancer detection. *Cell Res.* **2014**, *24*, 766–769. [CrossRef]
249. Huang, X.; Yuan, T.; Tschannen, M.; Sun, Z.; Jacob, H.; Du, M.; Liang, M.; Dittmar, R.L.; Liu, Y.; Liang, M. Characterization of human plasma-derived exosomal RNAs by deep sequencing. *BMC Genom.* **2013**, *14*, 319. [CrossRef]
250. Aliotta, J.M.; Pereira, M.; Johnson, K.W.; de Paz, N.; Dooner, M.S.; Puente, N.; Ayala, C.; Brilliant, K.; Berz, D.; Lee, D. Microvesicle entry into marrow cells mediates tissue-specific changes in mRNA by direct delivery of mRNA and induction of transcription. *Exp. Hematol.* **2010**, *38*, 233–245. [CrossRef]
251. Camussi, G.; Deregibus, M.C.; Bruno, S.; Cantaluppi, V.; Biancone, L. Exosomes/Microvesicles as a mechanism of cell-to-cell communication. *Kidney Int.* **2010**, *78*, 838–848. [CrossRef] [PubMed]
252. Luo, S.-S.; Ishibashi, O.; Ishikawa, G.; Ishikawa, T.; Katayama, A.; Mishima, T.; Takizawa, T.; Shigihara, T.; Goto, T.; Izumi, A. Human villous trophoblasts express and secrete placenta-specific microRNAs into maternal circulation via exosomes. *Biol. Reprod.* **2009**, *81*, 717–729. [CrossRef] [PubMed]
253. He, C.; Zheng, S.; Luo, Y.; Wang, B. Exosome theranostics: Biology and translational medicine. *Theranostics* **2018**, *8*, 237. [CrossRef] [PubMed]
254. De Jong, O.G.; Verhaar, M.C.; Chen, Y.; Vader, P.; Gremmels, H.; Posthuma, G.; Schiffelers, R.M.; Gucek, M.; van Balkom, B.W. Cellular stress conditions are reflected in the protein and RNA content of endothelial cell-derived exosomes. *J. Extracell. Vesicles* **2012**, *1*, 18396. [CrossRef] [PubMed]
255. Laulagnier, K.; Vincent-Schneider, H.; Hamdi, S.; Subra, C.; Lankar, D.; Record, M. Characterization of exosome subpopulations from RBL-2H3 cells using fluorescent lipids. *Blood Cells Mol. Dis.* **2005**, *35*, 116–121. [CrossRef]
256. Oksvold, M.P.; Kullmann, A.; Forfang, L.; Kierulf, B.; Li, M.; Brech, A.; Vlassov, A.V.; Smeland, E.B.; Neurauter, A.; Pedersen, K.W. Expression of B-cell surface antigens in subpopulations of exosomes released from B-cell lymphoma cells. *Clin. Ther.* **2014**, *36*, 847–862.e841. [CrossRef]
257. Smith, Z.J.; Lee, C.; Rojalin, T.; Carney, R.P.; Hazari, S.; Knudson, A.; Lam, K.; Saari, H.; Ibañez, E.L.; Viitala, T. Single exosome study reveals subpopulations distributed among cell lines with variability related to membrane content. *J. Extracell. Vesicles* **2015**, *4*, 28533. [CrossRef]
258. Mathivanan, S.; Ji, H.; Simpson, R.J. Exosomes: Extracellular organelles important in intercellular communication. *J. Proteom.* **2010**, *73*, 1907–1920. [CrossRef]
259. Van Der Vlist, E.J.; Nolte, E.N.; Stoorvogel, W.; Arkesteijn, G.J.; Wauben, M.H. Fluorescent labeling of nano-sized vesicles released by cells and subsequent quantitative and qualitative analysis by high-resolution flow cytometry. *Nat. Protoc.* **2012**, *7*, 1311–1326. [CrossRef]
260. Pospichalova, V.; Svoboda, J.; Dave, Z.; Kotrbova, A.; Kaiser, K.; Klemova, D.; Ilkovics, L.; Hampl, A.; Crha, I.; Jandakova, E. Simplified protocol for flow cytometry analysis of fluorescently labeled exosomes and microvesicles using dedicated flow cytometer. *J. Extracell. Vesicles* **2015**, *4*, 25530. [CrossRef]

261. Tauro, B.J.; Greening, D.W.; Mathias, R.A.; Mathivanan, S.; Ji, H.; Simpson, R.J. Two distinct populations of exosomes are released from LIM1863 colon carcinoma cell-derived organoids. *Mol. Cell. Proteom.* **2013**, *12*, 587–598. [CrossRef] [PubMed]
262. Ferguson, S.W.; Nguyen, J. Exosomes as therapeutics: The implications of molecular composition and exosomal heterogeneity. *J. Control. Release* **2016**, *228*, 179–190. [CrossRef] [PubMed]
263. Pegtel, D.M.; Cosmopoulos, K.; Thorley-Lawson, D.A.; van Eijndhoven, M.A.; Hopmans, E.S.; Lindenberg, J.L.; de Gruijl, T.D.; Würdinger, T.; Middeldorp, J.M. Functional delivery of viral miRNAs via exosomes. *Proc. Natl. Acad. Sci. USA* **2010**, *107*, 6328–6333. [CrossRef] [PubMed]
264. Balaj, L.; Lessard, R.; Dai, L.; Cho, Y.-J.; Pomeroy, S.L.; Breakefield, X.O.; Skog, J. Tumour microvesicles contain retrotransposon elements and amplified oncogene sequences. *Nat. Commun.* **2011**, *2*, 1–9. [CrossRef]
265. Kahlert, C.; Melo, S.A.; Protopopov, A.; Tang, J.; Seth, S.; Koch, M.; Zhang, J.; Weitz, J.; Chin, L.; Futreal, A. Identification of double-stranded genomic DNA spanning all chromosomes with mutated KRAS and p53 DNA in the serum exosomes of patients with pancreatic cancer. *J. Biol. Chem.* **2014**, *289*, 3869–3875. [CrossRef] [PubMed]
266. Takahashi, A.; Okada, R.; Nagao, K.; Kawamata, Y.; Hanyu, H.; Yoshimoto, S.; Takasugi, M.; Watanabe, S.; Kanemaki, M.T.; Obuse, C. Exosomes maintain cellular homeostasis by excreting harmful DNA from cells. *Nat. Commun.* **2017**, *8*, 15287. [CrossRef]
267. Gibbings, D.J.; Ciaudo, C.; Erhardt, M.; Voinnet, O. Multivesicular bodies associate with components of miRNA effector complexes and modulate miRNA activity. *Nat. Cell Biol.* **2009**, *11*, 1143–1149. [CrossRef]
268. Png, K.J.; Halberg, N.; Yoshida, M.; Tavazoie, S.F. A microRNA regulon that mediates endothelial recruitment and metastasis by cancer cells. *Nature* **2012**, *481*, 190–194. [CrossRef]
269. Rabinowits, G.; Gerçel-Taylor, C.; Day, J.M.; Taylor, D.D.; Kloecker, G.H. Exosomal microRNA: A diagnostic marker for lung cancer. *Clin. Lung Cancer* **2009**, *10*, 42–46. [CrossRef]
270. Skog, J.; Würdinger, T.; Van Rijn, S.; Meijer, D.H.; Gainche, L.; Curry, W.T.; Carter, B.S.; Krichevsky, A.M.; Breakefield, X.O. Glioblastoma microvesicles transport RNA and proteins that promote tumour growth and provide diagnostic biomarkers. *Nat. Cell Biol.* **2008**, *10*, 1470–1476. [CrossRef]
271. Koppers-Lalic, D.; Hackenberg, M.; Bijnsdorp, I.V.; van Eijndhoven, M.A.; Sadek, P.; Sie, D.; Zini, N.; Middeldorp, J.M.; Ylstra, B.; de Menezes, R.X. Nontemplated nucleotide additions distinguish the small RNA composition in cells from exosomes. *Cell Rep.* **2014**, *8*, 1649–1658. [CrossRef] [PubMed]
272. Tran, N. Cancer exosomes as miRNA factories. *Trends Cancer* **2016**, *2*, 329–331. [CrossRef] [PubMed]
273. Villarroya-Beltri, C.; Gutiérrez-Vázquez, C.; Sánchez-Cabo, F.; Pérez-Hernández, D.; Vázquez, J.; Martin-Cofreces, N.; Martinez-Herrera, D.J.; Pascual-Montano, A.; Mittelbrunn, M.; Sánchez-Madrid, F. Sumoylated hnRNPA2B1 controls the sorting of miRNAs into exosomes through binding to specific motifs. *Nat. Commun.* **2013**, *4*, 2980. [CrossRef] [PubMed]
274. Santangelo, L.; Giurato, G.; Cicchini, C.; Montaldo, C.; Mancone, C.; Tarallo, R.; Battistelli, C.; Alonzi, T.; Weisz, A.; Tripodi, M. The RNA-binding protein SYNCRIP is a component of the hepatocyte exosomal machinery controlling microRNA sorting. *Cell Rep.* **2016**, *17*, 799–808. [CrossRef]
275. Jenjaroenpun, P.; Kremenska, Y.; Nair, V.M.; Kremenskoy, M.; Joseph, B.; Kurochkin, I.V. Characterization of RNA in exosomes secreted by human breast cancer cell lines using next-generation sequencing. *PeerJ* **2013**, *1*, e201. [CrossRef]
276. Baglio, S.R.; Rooijers, K.; Koppers-Lalic, D.; Verweij, F.J.; Lanzón, M.P.; Zini, N.; Naaijkens, B.; Perut, F.; Niessen, H.W.; Baldini, N. Human bone marrow-and adipose-mesenchymal stem cells secrete exosomes enriched in distinctive miRNA and tRNA species. *Stem Cell Res. Ther.* **2015**, *6*, 127. [CrossRef]
277. Li, M.; Zeringer, E.; Barta, T.; Schageman, J.; Cheng, A.; Vlassov, A.V. Analysis of the RNA content of the exosomes derived from blood serum and urine and its potential as biomarkers. *Philos. Trans. R. Soc. B Biol. Sci.* **2014**, *369*, 20130502. [CrossRef]
278. Kumar, P.; Anaya, J.; Mudunuri, S.B.; Dutta, A. Meta-Analysis of tRNA derived RNA fragments reveals that they are evolutionarily conserved and associate with AGO proteins to recognize specific RNA targets. *BMC Biol.* **2014**, *12*, 78. [CrossRef]
279. Gopinath, S.C.; Wadhwa, R.; Kumar, P.K. Expression of noncoding vault RNA in human malignant cells and its importance in mitoxantrone resistance. *Mol. Cancer Res.* **2010**, *8*, 1536–1546. [CrossRef]
280. Kim, H.-S.; Choi, D.-Y.; Yun, S.J.; Choi, S.-M.; Kang, J.W.; Jung, J.W.; Hwang, D.; Kim, K.P.; Kim, D.-W. Proteomic analysis of microvesicles derived from human mesenchymal stem cells. *J. Proteome Res.* **2012**, *11*, 839–849. [CrossRef]
281. Simpson, R.J.; Jensen, S.S.; Lim, J.W. Proteomic profiling of exosomes: Current perspectives. *Proteomics* **2008**, *8*, 4083–4099. [CrossRef] [PubMed]
282. Salomon, C.; Ryan, J.; Sobrevia, L.; Kobayashi, M.; Ashman, K.; Mitchell, M.; Rice, G.E. Exosomal signaling during hypoxia mediates microvascular endothelial cell migration and vasculogenesis. *PLoS ONE* **2013**, *8*, e68451. [CrossRef] [PubMed]
283. Hegmans, J.P.; Bard, M.P.; Hemmes, A.; Luider, T.M.; Kleijmeer, M.J.; Prins, J.-B.; Zitvogel, L.; Burgers, S.A.; Hoogsteden, H.C.; Lambrecht, B.N. Proteomic analysis of exosomes secreted by human mesothelioma cells. *Am. J. Pathol.* **2004**, *164*, 1807–1815. [CrossRef]
284. Simons, M.; Raposo, G. Exosomes–Vesicular carriers for intercellular communication. *Curr. Opin. Cell Biol.* **2009**, *21*, 575–581. [CrossRef] [PubMed]
285. Théry, C.; Zitvogel, L.; Amigorena, S. Exosomes: Composition, biogenesis and function. *Nat. Rev. Immunol.* **2002**, *2*, 569–579. [CrossRef]
286. Bakela, K.; Kountourakis, N.; Aivaliotis, M.; Athanassakis, I. Soluble MHC-II proteins promote suppressive activity in CD 4+ T cells. *Immunology* **2015**, *144*, 158–169. [CrossRef]

287. Baietti, M.F.; Zhang, Z.; Mortier, E.; Melchior, A.; Degeest, G.; Geeraerts, A.; Ivarsson, Y.; Depoortere, F.; Coomans, C.; Vermeiren, E. Syndecan–Syntenin–ALIX regulates the biogenesis of exosomes. *Nat. Cell Biol.* **2012**, *14*, 677–685. [CrossRef]
288. Theos, A.C.; Truschel, S.T.; Tenza, D.; Hurbain, I.; Harper, D.C.; Berson, J.F.; Thomas, P.C.; Raposo, G.; Marks, M.S. A lumenal domain-dependent pathway for sorting to intralumenal vesicles of multivesicular endosomes involved in organelle morphogenesis. *Dev. Cell* **2006**, *10*, 343–354. [CrossRef]
289. Reggiori, F.; Pelham, H.R. Sorting of proteins into multivesicular bodies: Ubiquitin-dependent and-independent targeting. *EMBO J.* **2001**, *20*, 5176–5186. [CrossRef]
290. Zhang, X.; Khan, S.; Jiang, H.; Antonyak, M.A.; Chen, X.; Spiegelman, N.A.; Shrimp, J.H.; Cerione, R.A.; Lin, H. Identifying the functional contribution of the defatty-acylase activity of SIRT6. *Nat. Chem. Biol.* **2016**, *12*, 614–620. [CrossRef]
291. Hsu, D.-H.; Paz, P.; Villaflor, G.; Rivas, A.; Mehta-Damani, A.; Angevin, E.; Zitvogel, L.; Le Pecq, J.-B. Exosomes as a tumor vaccine: Enhancing potency through direct loading of antigenic peptides. *J. Immunother.* **2003**, *26*, 440–450. [CrossRef] [PubMed]
292. Morse, M.A.; Garst, J.; Osada, T.; Khan, S.; Hobeika, A.; Clay, T.M.; Valente, N.; Shreeniwas, R.; Sutton, M.A.; Delcayre, A. A phase I study of dexosome immunotherapy in patients with advanced non-small cell lung cancer. *J. Transl. Med.* **2005**, *3*, 9. [CrossRef] [PubMed]
293. Formiga, F.R.; Pelacho, B.; Garbayo, E.; Imbuluzqueta, I.; Díaz-Herráez, P.; Abizanda, G.; Gavira, J.J.; Simón-Yarza, T.; Albiasu, E.; Tamayo, E. Controlled delivery of fibroblast growth factor-1 and neuregulin-1 from biodegradable microparticles promotes cardiac repair in a rat myocardial infarction model through activation of endogenous regeneration. *J. Control. Release* **2014**, *173*, 132–139. [CrossRef] [PubMed]
294. Palmen, M.; Daemen, M.J.; De Windt, L.J.; Willems, J.; Dassen, W.R.; Heeneman, S.; Zimmermann, R.; Van Bilsen, M.; Doevendans, P.A. Fibroblast growth factor-1 improves cardiac functional recovery and enhances cell survival after ischemia and reperfusion: A fibroblast growth factor receptor, protein kinase C, and tyrosine kinase-dependent mechanism. *J. Am. Coll. Cardiol.* **2004**, *44*, 1113–1123. [CrossRef]
295. Okazaki, T.; Ebihara, S.; Asada, M.; Yamanda, S.; Saijo, Y.; Shiraishi, Y.; Ebihara, T.; Niu, K.; Mei, H.; Arai, H. Macrophage colony-stimulating factor improves cardiac function after ischemic injury by inducing vascular endothelial growth factor production and survival of cardiomyocytes. *Am. J. Pathol.* **2007**, *171*, 1093–1103. [CrossRef]
296. Hill, M.F.; Patel, A.V.; Murphy, A.; Smith, H.M.; Galindo, C.L.; Pentassuglia, L.; Peng, X.; Lenneman, C.G.; Odiete, O.; Friedman, D.B. Intravenous glial growth factor 2 (GGF2) isoform of neuregulin-1β improves left ventricular function, gene and protein expression in rats after myocardial infarction. *PLoS ONE* **2013**, *8*, e55741. [CrossRef]
297. Zgheib, C.; Zouein, F.A.; Kurdi, M.; Booz, G.W. Chronic treatment of mice with leukemia inhibitory factor does not cause adverse cardiac remodeling but improves heart function. *Eur. Cytokine Netw.* **2012**, *23*, 191. [CrossRef]
298. Trajkovic, K.; Hsu, C.; Chiantia, S.; Rajendran, L.; Wenzel, D.; Wieland, F.; Schwille, P.; Brügger, B.; Simons, M. Ceramide triggers budding of exosome vesicles into multivesicular endosomes. *Science* **2008**, *319*, 1244–1247. [CrossRef]
299. Wubbolts, R.; Leckie, R.S.; Veenhuizen, P.T.; Schwarzmann, G.; Möbius, W.; Hoernschemeyer, J.; Slot, J.-W.; Geuze, H.J.; Stoorvogel, W. Proteomic and biochemical analyses of human B cell-derived exosomes Potential implications for their function and multivesicular body formation. *J. Biol. Chem.* **2003**, *278*, 10963–10972. [CrossRef]
300. Krämer-Albers, E.M.; Bretz, N.; Tenzer, S.; Winterstein, C.; Möbius, W.; Berger, H.; Nave, K.A.; Schild, H.; Trotter, J. Oligodendrocytes secrete exosomes containing major myelin and stress-protective proteins: Trophic support for axons? *Proteom. Clin. Appl.* **2007**, *1*, 1446–1461. [CrossRef]
301. Laulagnier, K.; Motta, C.; Hamdi, S.; Roy, S.; Fauvelle, F.; Pageaux, J.-F.; Kobayashi, T.; Salles, J.-P.; Perret, B.; Bonnerot, C. Mast cell-and dendritic cell-derived exosomes display a specific lipid composition and an unusual membrane organization. *Biochem. J.* **2004**, *380*, 161–171. [CrossRef] [PubMed]
302. Subra, C.; Grand, D.; Laulagnier, K.; Stella, A.; Lambeau, G.; Paillasse, M.; De Medina, P.; Monsarrat, B.; Perret, B.; Silvente-Poirot, S. Exosomes account for vesicle-mediated transcellular transport of activatable phospholipases and prostaglandins. *J. Lipid Res.* **2010**, *51*, 2105–2120. [CrossRef] [PubMed]
303. Mashouri, L.; Yousefi, H.; Aref, A.R.; Mohammad Ahadi, A.; Molaei, F.; Alahari, S.K. Exosomes: Composition, biogenesis, and mechanisms in cancer metastasis and drug resistance. *Mol. Cancer* **2019**, *18*, 75. [CrossRef] [PubMed]
304. Beloribi-Djefaflia, S.; Siret, C.; Lombardo, D. Exosomal lipids induce human pancreatic tumoral MiaPaCa-2 cells resistance through the CXCR4-SDF-1α signaling axis. *Oncoscience* **2015**, *2*, 15. [CrossRef] [PubMed]
305. Urbanelli, L.; Magini, A.; Buratta, S.; Brozzi, A.; Sagini, K.; Polchi, A.; Tancini, B.; Emiliani, C. Signaling pathways in exosomes biogenesis, secretion and fate. *Genes* **2013**, *4*, 152–170. [CrossRef]
306. Zhou, H.; Cheruvanky, A.; Hu, X.; Matsumoto, T.; Hiramatsu, N.; Cho, M.E.; Berger, A.; Leelahavanichkul, A.; Doi, K.; Chawla, L.S. Urinary exosomal transcription factors, a new class of biomarkers for renal disease. *Kidney Int.* **2008**, *74*, 613–621. [CrossRef]
307. Zhou, H.; Pisitkun, T.; Aponte, A.; Yuen, P.S.; Hoffert, J.D.; Yasuda, H.; Hu, X.; Chawla, L.; Shen, R.-F.; Knepper, M.A. Exosomal Fetuin-A identified by proteomics: A novel urinary biomarker for detecting acute kidney injury. *Kidney Int.* **2006**, *70*, 1847–1857. [CrossRef]
308. Welker, M.-W.; Reichert, D.; Susser, S.; Sarrazin, C.; Martinez, Y.; Herrmann, E.; Zeuzem, S.; Piiper, A.; Kronenberger, B. Soluble serum CD81 is elevated in patients with chronic hepatitis C and correlates with alanine aminotransferase serum activity. *PLoS ONE* **2012**, *7*, e30796. [CrossRef]

309. Li, Y.; Zhang, Y.; Qiu, F.; Qiu, Z. Proteomic identification of exosomal LRG1: A potential urinary biomarker for detecting NSCLC. *Electrophoresis* **2011**, *32*, 1976–1983. [CrossRef]
310. Lee, T.H.; Chennakrishnaiah, S.; Audemard, E.; Montermini, L.; Meehan, B.; Rak, J. Oncogenic ras-driven cancer cell vesiculation leads to emission of double-stranded DNA capable of interacting with target cells. *Biochem. Biophys. Res. Commun.* **2014**, *451*, 295–301. [CrossRef]
311. Russo, L. Urine Biomarkers. WO2013028788A1, 28 February 2013.
312. Dong, Z.; Sun, X.; Xu, J.; Han, X.; Xing, Z.; Wang, D.; Ge, J.; Meng, L.; Xu, X. Serum Membrane Type 1-Matrix Metalloproteinase (MT1-MMP) mRNA Protected by Exosomes as a Potential Biomarker for Gastric Cancer. *Medical Sci. Monit.* **2019**, *25*, 7770. [CrossRef] [PubMed]
313. Xu, H.; Dong, X.; Chen, Y.; Wang, X. Serum exosomal hnRNPH1 mRNA as a novel marker for hepatocellular carcinoma. *Clin. Chem. Lab. Med.* **2018**, *56*, 479–484. [CrossRef] [PubMed]
314. Mao, L.; Li, X.; Gong, S.; Yuan, H.; Jiang, Y.; Huang, W.; Sun, X.; Dang, X. Serum exosomes contain ECRG 4 mRNA that suppresses tumor growth via inhibition of genes involved in inflammation, cell proliferation, and angiogenesis. *Cancer Gene Ther.* **2018**, *25*, 248–259. [CrossRef] [PubMed]
315. Bracci, L.; Lozupone, F.; Parolini, I. The role of exosomes in colorectal cancer disease progression and response to therapy. *Cytokine Growth Factor Rev.* **2020**, *51*, 84–91. [CrossRef] [PubMed]
316. Taylor, D.D.; Gercel-Taylor, C. MicroRNA signatures of tumor-derived exosomes as diagnostic biomarkers of ovarian cancer. *Gynecol. Oncol.* **2008**, *110*, 13–21. [CrossRef] [PubMed]
317. Sugimachi, K.; Matsumura, T.; Hirata, H.; Uchi, R.; Ueda, M.; Ueo, H.; Shinden, Y.; Iguchi, T.; Eguchi, H.; Shirabe, K. Identification of a bona fide microRNA biomarker in serum exosomes that predicts hepatocellular carcinoma recurrence after liver transplantation. *Br. J. Cancer* **2015**, *112*, 532–538. [CrossRef]
318. Takeshita, N.; Hoshino, I.; Mori, M.; Akutsu, Y.; Hanari, N.; Yoneyama, Y.; Ikeda, N.; Isozaki, Y.; Maruyama, T.; Akanuma, N. Serum microRNA expression profile: miR-1246 as a novel diagnostic and prognostic biomarker for oesophageal squamous cell carcinoma. *Br. J. Cancer* **2013**, *108*, 644–652. [CrossRef]
319. Fu, F.; Jiang, W.; Zhou, L.; Chen, Z. Circulating exosomal miR-17-5p and miR-92a-3p predict pathologic stage and grade of colorectal cancer. *Trans. Oncol.* **2018**, *11*, 221–232. [CrossRef]
320. De Miguel Pérez, D.; Martínez, A.R.; Palomo, A.O.; Ureña, M.D.; Puche, J.L.G.; Remacho, A.R.; Hernandez, J.E.; Acosta, J.A.L.; Sánchez, F.G.O.; Serrano, M.J. Extracellular vesicle-miRNAs as liquid biopsy biomarkers for disease identification and prognosis in metastatic colorectal cancer patients. *Sci. Rep.* **2020**, *10*, 1–13. [CrossRef]
321. Tang, Y.; Zhao, Y.; Song, X.; Song, X.; Niu, L.; Xie, L. Tumor-Derived exosomal miRNA-320d as a biomarker for metastatic colorectal cancer. *J. Clin. Lab. Anal.* **2019**, *33*, e23004. [CrossRef]
322. Mitchell, P.S.; Parkin, R.K.; Kroh, E.M.; Fritz, B.R.; Wyman, S.K.; Pogosova-Agadjanyan, E.L.; Peterson, A.; Noteboom, J.; O'Briant, K.C.; Allen, A. Circulating microRNAs as stable blood-based markers for cancer detection. *Proc. Natl. Acad. Sci. USA* **2008**, *105*, 10513–10518. [CrossRef] [PubMed]
323. Brase, J.C.; Johannes, M.; Schlomm, T.; Fälth, M.; Haese, A.; Steuber, T.; Beissbarth, T.; Kuner, R.; Sültmann, H. Circulating miRNAs are correlated with tumor progression in prostate cancer. *Int. J. Cancer* **2011**, *128*, 608–616. [CrossRef] [PubMed]
324. Wang, L.; Duan, W.; Yan, S.; Xie, Y.; Wang, C. Circulating long non-coding RNA colon cancer-associated transcript 2 protected by exosome as a potential biomarker for colorectal cancer. *Biomed. Pharmacother.* **2019**, *113*, 108758. [CrossRef] [PubMed]
325. Yu, B.; Du, Q.; Li, H.; Liu, H.-Y.; Ye, X.; Zhu, B.; Zhai, Q.; Li, X.-X. Diagnostic potential of serum exosomal colorectal neoplasia differentially expressed long non-coding RNA (CRNDE-p) and microRNA-217 expression in colorectal carcinoma. *Oncotarget* **2017**, *8*, 83745. [CrossRef] [PubMed]
326. Gao, T.; Liu, X.; He, B.; Nie, Z.; Zhu, C.; Zhang, P.; Wang, S. Exosomal lncRNA 91H is associated with poor development in colorectal cancer by modifying HNRNPK expression. *Cancer Cell Int.* **2018**, *18*, 11. [CrossRef] [PubMed]
327. Oehme, F.; Krahl, S.; Gyorffy, B.; Muessle, B.; Rao, V.; Greif, H.; Ziegler, N.; Lin, K.; Thepkaysone, M.-L.; Polster, H. Low level of exosomal long non-coding RNA HOTTIP is a prognostic biomarker in colorectal cancer. *RNA Biol.* **2019**, *16*, 1339–1345. [CrossRef] [PubMed]
328. Wang, J.; Zhou, Y.; Lu, J.; Sun, Y.; Xiao, H.; Liu, M.; Tian, L. Combined detection of serum exosomal miR-21 and HOTAIR as diagnostic and prognostic biomarkers for laryngeal squamous cell carcinoma. *Med. Oncol.* **2014**, *31*, 148. [CrossRef]
329. Yamashita, T.; Kamada, H.; Kanasaki, S.; Maeda, Y.; Nagano, K.; Abe, Y.; Inoue, M.; Yoshioka, Y.; Tsutsumi, Y.; Katayama, S. Epidermal growth factor receptor localized to exosome membranes as a possible biomarker for lung cancer diagnosis. *Die Pharm. Int. J. Pharm. Sci.* **2013**, *68*, 969–973.
330. Jakobsen, K.R.; Paulsen, B.S.; Bæk, R.; Varming, K.; Sorensen, B.S.; Jørgensen, M.M. Exosomal proteins as potential diagnostic markers in advanced non-small cell lung carcinoma. *J. Extracell. Vesicles* **2015**, *4*, 26659. [CrossRef]
331. Graner, M.W.; Alzate, O.; Dechkovskaia, A.M.; Keene, J.D.; Sampson, J.H.; Mitchell, D.A.; Bigner, D.D. Proteomic and immunologic analyses of brain tumor exosomes. *FASEB J.* **2009**, *23*, 1541–1557. [CrossRef]
332. Shao, H.; Chung, J.; Balaj, L.; Charest, A.; Bigner, D.D.; Carter, B.S.; Hochberg, F.H.; Breakefield, X.O.; Weissleder, R.; Lee, H. Protein typing of circulating microvesicles allows real-time monitoring of glioblastoma therapy. *Nat. Med.* **2012**, *18*, 1835. [CrossRef] [PubMed]

333. Nilsson, J.; Skog, J.; Nordstrand, A.; Baranov, V.; Mincheva-Nilsson, L.; Breakefield, X.; Widmark, A. Prostate cancer-derived urine exosomes: A novel approach to biomarkers for prostate cancer. *Br. J. Cancer* **2009**, *100*, 1603–1607. [CrossRef] [PubMed]
334. Chen, C.-L.; Lai, Y.-F.; Tang, P.; Chien, K.-Y.; Yu, J.-S.; Tsai, C.-H.; Chen, H.-W.; Wu, C.-C.; Chung, T.; Hsu, C.-W. Comparative and targeted proteomic analyses of urinary microparticles from bladder cancer and hernia patients. *J. Proteome Res.* **2012**, *11*, 5611–5629. [CrossRef] [PubMed]
335. Runz, S.; Keller, S.; Rupp, C.; Stoeck, A.; Issa, Y.; Koensgen, D.; Mustea, A.; Sehouli, J.; Kristiansen, G.; Altevogt, P. Malignant ascites-derived exosomes from ovarian carcinoma patients contain CD24 and EpCAM. *Gynecol. Oncol.* **2007**, *107*, 563–571. [CrossRef] [PubMed]
336. Logozzi, M.; De Milito, A.; Lugini, L.; Borghi, M.; Calabro, L.; Spada, M.; Perdicchio, M.; Marino, M.L.; Federici, C.; Iessi, E. High levels of exosomes expressing CD63 and caveolin-1 in plasma of melanoma patients. *PLoS ONE* **2009**, *4*, e5219. [CrossRef] [PubMed]
337. Logozzi, M.; Mizzoni, D.; Capasso, C.; Del Prete, S.; Di Raimo, R.; Falchi, M.; Angelini, D.F.; Sciarra, A.; Maggi, M.; Supuran, C.T. Plasmatic exosomes from prostate cancer patients show increased carbonic anhydrase IX expression and activity and low pH. *J. Enzym. Inhib. Med. Chem.* **2020**, *35*, 280–288. [CrossRef]
338. Logozzi, M.; Capasso, C.; Di Raimo, R.; Del Prete, S.; Mizzoni, D.; Falchi, M.; Supuran, C.T.; Fais, S. Prostate cancer cells and exosomes in acidic condition show increased carbonic anhydrase IX expression and activity. *J. Enzym. Inhib. Med. Chem.* **2019**, *34*, 272–278. [CrossRef]
339. Palacios-Ferrer, J.L.; García-Ortega, M.B.; Gallardo-Gómez, M.; García, M.Á.; Díaz, C.; Boulaiz, H.; Valdivia, J.; Jurado, J.M.; Almazan-Fernandez, F.M.; Arias-Santiago, S. Metabolomic profile of cancer stem cell-derived exosomes from patients with malignant melanoma. *Mol. Oncol.* **2020**. [CrossRef]
340. Delpech, J.C.; Herron, S.; Botros, M.B.; Ikezu, T. Neuroimmune Crosstalk through Extracellular Vesicles in Health and Disease. *Trends Neurosci.* **2019**, *42*, 361–372. [CrossRef]
341. Tomlinson, P.R.; Zheng, Y.; Fischer, R.; Heidasch, R.; Gardiner, C.; Evetts, S.; Hu, M.; Wade-Martins, R.; Turner, M.R.; Morris, J. Identification of distinct circulating exosomes in Parkinson's disease. *Ann. Clin. Transl. Neurol.* **2015**, *2*, 353–361. [CrossRef]
342. Fraser, K.B.; Moehle, M.S.; Alcalay, R.N.; West, A.B. Urinary LRRK2 phosphorylation predicts parkinsonian phenotypes in G2019S LRRK2 carriers. *Neurology* **2016**, *86*, 994–999. [CrossRef] [PubMed]
343. Olsson, B.; Lautner, R.; Andreasson, U.; Öhrfelt, A.; Portelius, E.; Bjerke, M.; Hölttä, M.; Rosén, C.; Olsson, C.; Strobel, G. CSF and blood biomarkers for the diagnosis of Alzheimer's disease: A systematic review and meta-analysis. *Lancet Neurol.* **2016**, *15*, 673–684. [CrossRef]
344. Mapstone, M.; Cheema, A.K.; Fiandaca, M.S.; Zhong, X.; Mhyre, T.R.; MacArthur, L.H.; Hall, W.J.; Fisher, S.G.; Peterson, D.R.; Haley, J.M. Plasma phospholipids identify antecedent memory impairment in older adults. *Nat. Med.* **2014**, *20*, 415–418. [CrossRef] [PubMed]
345. Roth, P.; Weller, M. Challenges to targeting epidermal growth factor receptor in glioblastoma: Escape mechanisms and combinatorial treatment strategies. *Neuro Oncol.* **2014**, *16*, 14–19. [CrossRef] [PubMed]
346. Sharif, S.; Ghahremani, M.H.; Soleimani, M. Delivery of Exogenous miR-124 to Glioblastoma Multiform Cells by Wharton's Jelly Mesenchymal Stem Cells Decreases Cell Proliferation and Migration, and Confers Chemosensitivity. *Stem Cell Rev. Rep.* **2018**, *14*, 236–246. [CrossRef] [PubMed]
347. Wang, B.; Wu, Z.H.; Lou, P.Y.; Chai, C.; Han, S.Y.; Ning, J.F.; Li, M. Human bone marrow-derived mesenchymal stem cell-secreted exosomes overexpressing microRNA-34a ameliorate glioblastoma development via down-regulating MYCN. *Cell. Oncol.* **2019**, *42*, 783–799. [CrossRef]
348. Liu, Q.; Yu, Z.; Yuan, S.; Xie, W.; Li, C.; Hu, Z.; Xiang, Y.; Wu, N.; Wu, L.; Bai, L.; et al. Circulating exosomal microRNAs as prognostic biomarkers for non-small-cell lung cancer. *Oncotarget* **2017**, *8*, 13048–13058. [CrossRef]
349. Kanaoka, R.; Iinuma, H.; Dejima, H.; Sakai, T.; Uehara, H.; Matsutani, N.; Kawamura, M. Usefulness of Plasma Exosomal MicroRNA-451a as a Noninvasive Biomarker for Early Prediction of Recurrence and Prognosis of Non-Small Cell Lung Cancer. *Oncology* **2018**, *94*, 311–323. [CrossRef] [PubMed]
350. Dejima, H.; Iinuma, H.; Kanaoka, R.; Matsutani, N.; Kawamura, M. Exosomal microRNA in plasma as a non-invasive biomarker for the recurrence of non-small cell lung cancer. *Oncol. Lett.* **2017**, *13*, 1256–1263. [CrossRef] [PubMed]
351. Peng, X.X.; Yu, R.; Wu, X.; Wu, S.Y.; Pi, C.; Chen, Z.H.; Zhang, X.C.; Gao, C.Y.; Shao, Y.W.; Liu, L.; et al. Correlation of plasma exosomal microRNAs with the efficacy of immunotherapy in EGFR/ALK wild-type advanced non-small cell lung cancer. *J. Immunother. Cancer* **2020**, *8*, e000376. [CrossRef] [PubMed]
352. Eichelser, C.; Stuckrath, I.; Muller, V.; Milde-Langosch, K.; Wikman, H.; Pantel, K.; Schwarzenbach, H. Increased serum levels of circulating exosomal microRNA-373 in receptor-negative breast cancer patients. *Oncotarget* **2014**, *5*, 9650–9663. [CrossRef] [PubMed]
353. Hannafon, B.N.; Trigoso, Y.D.; Calloway, C.L.; Zhao, Y.D.; Lum, D.H.; Welm, A.L.; Zhao, Z.J.; Blick, K.E.; Dooley, W.C.; Ding, W.Q. Plasma exosome microRNAs are indicative of breast cancer. *Breast Cancer Res.* **2016**, *18*, 90. [CrossRef] [PubMed]
354. Yoshikawa, M.; Iinuma, H.; Umemoto, Y.; Yanagisawa, T.; Matsumoto, A.; Jinno, H. Exosome-Encapsulated microRNA-223-3p as a minimally invasive biomarker for the early detection of invasive breast cancer. *Oncol. Lett.* **2018**, *15*, 9584–9592. [CrossRef]

355. Gong, C.; Tian, J.; Wang, Z.; Gao, Y.; Wu, X.; Ding, X.; Qiang, L.; Li, G.; Han, Z.; Yuan, Y.; et al. Functional exosome-mediated co-delivery of doxorubicin and hydrophobically modified microRNA 159 for triple-negative breast cancer therapy. *J. Nanobiotechnol.* **2019**, *17*, 93. [CrossRef] [PubMed]
356. Yao, S.; Yin, Y.; Jin, G.; Li, D.; Li, M.; Hu, Y.; Feng, Y.; Liu, Y.; Bian, Z.; Wang, X. Exosome-mediated delivery of miR-204-5p inhibits tumor growth and chemoresistance. *Cancer Med.* **2020**, *9*, 5989–5998. [CrossRef] [PubMed]
357. Liang, G.; Zhu, Y.; Ali, D.J.; Tian, T.; Xu, H.; Si, K.; Sun, B.; Chen, B.; Xiao, Z. Engineered exosomes for targeted co-delivery of miR-21 inhibitor and chemotherapeutics to reverse drug resistance in colon cancer. *J. Nanobiotechnol.* **2020**, *18*, 10. [CrossRef] [PubMed]
358. Dai, J.; Escara-Wilke, J.; Keller, J.M.; Jung, Y.; Taichman, R.S.; Pienta, K.J.; Keller, E.T. Primary prostate cancer educates bone stroma through exosomal pyruvate kinase M2 to promote bone metastasis. *J. Exp. Med.* **2019**, *216*, 2883–2899. [CrossRef]
359. Takahara, K.; Ii, M.; Inamoto, T.; Nakagawa, T.; Ibuki, N.; Yoshikawa, Y.; Tsujino, T.; Uchimoto, T.; Saito, K.; Takai, T. microRNA-145 mediates the inhibitory effect of adipose tissue-derived stromal cells on prostate cancer. *Stem. Cells Dev.* **2016**, *25*, 1290–1298. [CrossRef]
360. Lässer, C. Exosomes in diagnostic and therapeutic applications: Biomarker, vaccine and RNA interference delivery vehicle. *Exp. Opin. Biol. Ther.* **2015**, *15*, 103–117. [CrossRef]
361. Greco, K.A.; Franzen, C.A.; Foreman, K.E.; Flanigan, R.C.; Kuo, P.C.; Gupta, G.N. PLK-1 silencing in bladder cancer by siRNA delivered with exosomes. *Urology* **2016**, *91*, 241.e1–241.e7. [CrossRef]
362. Smyth, T.J.; Redzic, J.S.; Graner, M.W.; Anchordoquy, T.J. Examination of the specificity of tumor cell derived exosomes with tumor cells in vitro. *Biochimica et Biophysica Acta Biomembr.* **2014**, *1838*, 2954–2965. [CrossRef] [PubMed]
363. Alvarez-Erviti, L.; Seow, Y.; Yin, H.; Betts, C.; Lakhal, S.; Wood, M.J. Delivery of siRNA to the mouse brain by systemic injection of targeted exosomes. *Nat. Biotechnol.* **2011**, *29*, 341–345. [CrossRef] [PubMed]
364. Sun, D.; Zhuang, X.; Zhang, S.; Deng, Z.-B.; Grizzle, W.; Miller, D.; Zhang, H.-G. Exosomes are endogenous nanoparticles that can deliver biological information between cells. *Adv. Drug Deliv. Rev.* **2013**, *65*, 342–347. [CrossRef] [PubMed]
365. Jang, S.C.; Kim, O.Y.; Yoon, C.M.; Choi, D.-S.; Roh, T.-Y.; Park, J.; Nilsson, J.; Lotvall, J.; Kim, Y.-K.; Gho, Y.S. Bioinspired exosome-mimetic nanovesicles for targeted delivery of chemotherapeutics to malignant tumors. *ACS Nano* **2013**, *7*, 7698–7710. [CrossRef]
366. Tian, Y.; Li, S.; Song, J.; Ji, T.; Zhu, M.; Anderson, G.J.; Wei, J.; Nie, G. A doxorubicin delivery platform using engineered natural membrane vesicle exosomes for targeted tumor therapy. *Biomaterials* **2014**, *35*, 2383–2390. [CrossRef]
367. Yang, T.; Martin, P.; Fogarty, B.; Brown, A.; Schurman, K.; Phipps, R.; Yin, V.P.; Lockman, P.; Bai, S. Exosome delivered anticancer drugs across the blood-brain barrier for brain cancer therapy in Danio rerio. *Pharm. Res.* **2015**, *32*, 2003–2014. [CrossRef]
368. Lee, J.; Kim, J.; Jeong, M.; Lee, H.; Goh, U.; Kim, H.; Kim, B.; Park, J.-H. Liposome-Based engineering of cells to package hydrophobic compounds in membrane vesicles for tumor penetration. *Nano Lett.* **2015**, *15*, 2938–2944. [CrossRef]
369. Bovy, N.; Blomme, B.; Frères, P.; Dederen, S.; Nivelles, O.; Lion, M.; Carnet, O.; Martial, J.A.; Noël, A.; Thiry, M. Endothelial exosomes contribute to the antitumor response during breast cancer neoadjuvant chemotherapy via microRNA transfer. *Oncotarget* **2015**, *6*, 10253. [CrossRef]
370. Didiot, M.-C.; Hall, L.M.; Coles, A.H.; Haraszti, R.A.; Godinho, B.M.; Chase, K.; Sapp, E.; Ly, S.; Alterman, J.F.; Hassler, M.R. Exosome-Mediated delivery of hydrophobically modified siRNA for huntingtin mRNA silencing. *Mol. Ther.* **2016**, *24*, 1836–1847. [CrossRef]
371. Wahlgren, J.; Karlson, T.D.L.; Brisslert, M.; Vaziri, S.F.; Telemo, E.; Sunnerhagen, P.; Valadi, H. Plasma exosomes can deliver exogenous short interfering RNA to monocytes and lymphocytes. *Nucleic Acids Res.* **2012**, *40*, e130. [CrossRef]
372. Kamerkar, S.; LeBleu, V.S.; Sugimoto, H.; Yang, S.; Ruivo, C.F.; Melo, S.A.; Lee, J.J.; Kalluri, R. Exosomes facilitate therapeutic targeting of oncogenic KRAS in pancreatic cancer. *Nature* **2017**, *546*, 498–503. [CrossRef] [PubMed]
373. Ohno, S.-I.; Takanashi, M.; Sudo, K.; Ueda, S.; Ishikawa, A.; Matsuyama, N.; Fujita, K.; Mizutani, T.; Ohgi, T.; Ochiya, T. Systemically injected exosomes targeted to EGFR deliver antitumor microRNA to breast cancer cells. *Mol. Ther.* **2013**, *21*, 185–191. [CrossRef] [PubMed]
374. Lee, H.K.; Finniss, S.; Cazacu, S.; Bucris, E.; Ziv-Av, A.; Xiang, C.; Bobbitt, K.; Rempel, S.A.; Hasselbach, L.; Mikkelsen, T. Mesenchymal stem cells deliver synthetic microRNA mimics to glioma cells and glioma stem cells and inhibit their cell migration and self-renewal. *Oncotarget* **2013**, *4*, 346. [CrossRef] [PubMed]
375. Katakowski, M.; Buller, B.; Zheng, X.; Lu, Y.; Rogers, T.; Osobamiro, O.; Shu, W.; Jiang, F.; Chopp, M. Exosomes from marrow stromal cells expressing miR-146b inhibit glioma growth. *Cancer Lett.* **2013**, *335*, 201–204. [CrossRef] [PubMed]
376. O'Brien, K.; Lowry, M.C.; Corcoran, C.; Martinez, V.G.; Daly, M.; Rani, S.; Gallagher, W.M.; Radomski, M.W.; MacLeod, R.A.; O'Driscoll, L. miR-134 in extracellular vesicles reduces triple-negative breast cancer aggression and increases drug sensitivity. *Oncotarget* **2015**, *6*, 32774. [CrossRef] [PubMed]
377. O'brien, K.; Khan, S.; Gilligan, K.; Zafar, H.; Lalor, P.; Glynn, C.; O'Flatharta, C.; Ingoldsby, H.; Dockery, P.; De Bhulbh, A. Employing mesenchymal stem cells to support tumor-targeted delivery of extracellular vesicle (EV)-encapsulated microRNA-379. *Oncogene* **2018**, *37*, 2137–2149. [CrossRef]
378. Shtam, T.A.; Kovalev, R.A.; Varfolomeeva, E.Y.; Makarov, E.M.; Kil, Y.V.; Filatov, M.V. Exosomes are natural carriers of exogenous siRNA to human cells in vitro. *Cell Commun. Signal.* **2013**, *11*, 1–10. [CrossRef]

379. Shimbo, K.; Miyaki, S.; Ishitobi, H.; Kato, Y.; Kubo, T.; Shimose, S.; Ochi, M. Exosome-Formed synthetic microRNA-143 is transferred to osteosarcoma cells and inhibits their migration. *Biochem. Biophys. Res. Commun.* **2014**, *445*, 381–387. [CrossRef]
380. Lou, G.; Song, X.; Yang, F.; Wu, S.; Wang, J.; Chen, Z.; Liu, Y. Exosomes derived from miR-122-modified adipose tissue-derived MSCs increase chemosensitivity of hepatocellular carcinoma. *J. Hematol. Oncol.* **2015**, *8*, 1–11. [CrossRef]
381. Usman, W.M.; Pham, T.C.; Kwok, Y.Y.; Vu, L.T.; Ma, V.; Peng, B.; San Chan, Y.; Wei, L.; Chin, S.M.; Azad, A. Efficient RNA drug delivery using red blood cell extracellular vesicles. *Nat. Commun.* **2018**, *9*, 1–15. [CrossRef]
382. Meyer, C.; Losacco, J.; Stickney, Z.; Li, L.; Marriott, G.; Lu, B. Pseudotyping exosomes for enhanced protein delivery in mammalian cells. *Int. J. Nanomed.* **2017**, *12*, 3153. [CrossRef] [PubMed]
383. Haney, M.J.; Klyachko, N.L.; Zhao, Y.; Gupta, R.; Plotnikova, E.G.; He, Z.; Patel, T.; Piroyan, A.; Sokolsky, M.; Kabanov, A.V. Exosomes as drug delivery vehicles for Parkinson's disease therapy. *J. Control. Release* **2015**, *207*, 18–30. [CrossRef] [PubMed]
384. Sterzenbach, U.; Putz, U.; Low, L.-H.; Silke, J.; Tan, S.-S.; Howitt, J. Engineered exosomes as vehicles for biologically active proteins. *Mol. Ther.* **2017**, *25*, 1269–1278. [CrossRef] [PubMed]
385. Cheng, Y.; Schorey, J.S. Targeting soluble proteins to exosomes using a ubiquitin tag. *Biotechnol. Bioeng.* **2016**, *113*, 1315–1324. [CrossRef]

Review

Urine as a Source of Liquid Biopsy for Cancer

Masanori Oshi [1,2,†], Vijayashree Murthy [1,†], Hideo Takahashi [3], Michelle Huyser [1], Maiko Okano [4], Yoshihisa Tokumaru [1,5], Omar M. Rashid [6,7], Ryusei Matsuyama [2], Itaru Endo [2] and Kazuaki Takabe [1,2,3,8,9,10,*]

1. Breast Surgery, Department of Surgical Oncology, Roswell Park Comprehensive Cancer Center, Buffalo, NY 14263, USA; Masanori.Oshi@RoswellPark.org (M.O.); vijayashree.murthy@roswellpark.org (V.M.); Michelle.Huyser@RoswellPark.org (M.H.); Yoshihisa.Tokumaru@RoswellPark.org (Y.T.)
2. Department of Gastroenterological Surgery, Yokohama City University School of Medicine, Yokohama 236-0004, Japan; ryusei@yokohama-cu.ac.jp (R.M.); endoit@yokohama-cu.ac.jp (I.E.)
3. Hepatopancreatobiliary (HPB), Surgical Oncology, Department of Surgery, Mount Sinai South Nassau, One South Central Avenue, Valley Stream, NY 11580, USA; hideo.takahashi@snch.org
4. Department of Breast Surgery, Fukushima Medical University, Fukushima 960-1295, Japan; tentekomaikocco@hotmail.com
5. Department of Surgical Oncology, Graduate School of Medicine, Gifu University, 1-1 Yanagido, Gifu 501-1194, Japan
6. Michael and Dianne Bienes Comprehensive Cancer Center, Department of Surgery, Holy Cross Hospital, Fort Lauderdale, FL 33308, USA; omarmrashidmdjd@gmail.com
7. Department of Surgery, Massachusetts General Hospital, Boston, MA 02114, USA
8. Department of Surgery, University at Buffalo Jacobs School of Medicine and Biomedical Sciences, The State University of New York, Buffalo, NY 14263, USA
9. Department of Breast Surgery and Oncology, Tokyo Medical University, Tokyo 160-8402, Japan
10. Department of Surgery, Graduate School of Medical and Dental Sciences, Niigata University, Niigata 951-8510, Japan
* Correspondence: kazuaki.takabe@roswellpark.org
† These authors equally contributed to this work.

Simple Summary: Tissue biopsy is essential for diagnosis and characterization of a tumor. Recently circulating tumor cells and other tumor-derived nucleic acid can be detected from blood, which is called liquid biopsy. Now this concept has been expanded to many other body fluids including urine. Urine is the least invasive method to obtain a liquid biopsy and can be done anywhere, which allows longitudinal repeated sampling. Here, we review the latest update on urine liquid biopsy in urological and non-urological cancers.

Abstract: Tissue biopsy is the gold standard for diagnosis and morphological and immunohistochemical analyses to characterize cancer. However, tissue biopsy usually requires an invasive procedure, and it can be challenging depending on the condition of the patient and the location of the tumor. Even liquid biopsy analysis of body fluids such as blood, saliva, gastric juice, sweat, tears and cerebrospinal fluid may require invasive procedures to obtain samples. Liquid biopsy can be applied to circulating tumor cells (CTCs) or nucleic acids (NAs) in blood. Recently, urine has gained popularity due to its less invasive sampling, ability to easily repeat samples, and ability to follow tumor evolution in real-time, making it a powerful tool for diagnosis and treatment monitoring in cancer patients. With the development and advancements in extraction methods of urinary substances, urinary NAs have been found to be closely related to carcinogenesis, metastasis, and therapeutic response, not only in urological cancers but also in non-urological cancers. This review mainly highlights the components of urine liquid biopsy and their utility and limitations in oncology, especially in non-urological cancers.

Keywords: liquid biopsy; urine; urine liquid biopsy; DNA; mRNA; microRNA; sncRNA

1. Introduction

Radiological evaluation followed by biopsy for assessment of tumor tissue and pathological confirmation are the main investigatory methods for cancer diagnosis and treatment planning. Depending upon the location of the tumor, invasive biopsies can be painful with risk of complication and associated costs. This is particularly the case when lesions are in vital organs or close to major vessels, thus making biopsy very challenging to access. Treatment decisions are often made based on the pathological profile of the primary tumors, which may or may not be the same genomic clone of the metastatic tumor. It is well known that treatment effect is different between primary and metastatic tumors [1]. Cancer cells proliferate continuously, through clonal evolution, to adapt to new environments and exhibit clonal selection by selective pressure from a tumor microenvironment (TME) or treatment [2]. It is now known that a bulk tumor may consist of multiple clones of the same cancer cells (with different molecular and phenotypical profiles) that have a different cancer biology and clonal evolution in response to treatment, also known as intratumor heterogeneity. Intratumor heterogeneity is a key challenge in cancer treatment, requiring real-time assessment of tumor genomic information for precision medicine. Tissue biopsy often takes samples from only a small part of a bulk tumor and thus may not capture the entire spatial diversity of tumor heterogeneity [3]. Although multi-region sequential biopsy can be performed in order to address intratumor heterogeneity [4,5], it may be impractical in clinical practice and limited to the number of samples that can be tolerated by the patient. At the present time, cancer surveillance and assessment of treatment effect is dependent on imaging studies. However, they can only capture morphology of the tumor as a snapshot at a specific time and location, which does not correspond to the whole characteristics or function of the tumor. Multiple follow-up visits with imaging studies and possible biopsies significantly reduce patient compliance and quality of life, and it may be cost-prohibitive. In order to avoid the shortcomings of current imaging and tissue biopsy modalities but capture tumor heterogeneity, a non-invasive method to monitor tumor-wide genomic information during tumor progression or treatment responses is needed.

Liquid biopsy can be an answer to these challenges. Body fluids contain large amounts of substances secreted from cells after they are utilized in intercellular communication or released upon cell death. They include metabolites, proteins, and nucleic acids which may reflect the changes or abnormalities of cells in the body. Liquid biopsy refers to a process of obtaining tumor-derived materials from body fluids. It is a non-invasive investigative modality suited for repetitive assessment of tumor-related substances for assessing changes in gene expression patterns and to study the genomic profile of the tumor. Liquid biopsy (regardless of blood or urine) measures cells or nucleic acids, either secreted out of the tumor or brushed out of the tumor after being destroyed. Thus, liquid biopsy involves sampling from the entire tumor and not a specific area of a bulk tumor. First, in regard to tumor heterogeneity, blood or urine are expected to contain materials secreted or released from all cells and its quantity is expected to be reflective of the amount in the bulk of a tumor. Second, since blood or urine capture the secretome from cells, it is expected to capture the function of the cells. Changes in circulating materials reflect overall changes in the TME, such as stromal interactions between the cancer and immune cells [6]. Theoretically, liquid biopsy has the possibility to capture everything from the cells in the TME and is not spatially or longitudinally limited. Finally, based upon the homeostasis of the body, anything produced and secreted by a neoplasm should be an excess to the body and should be excreted via the urine; thus, urine is theoretically an ideal medium to detect a neoplasm-derived material.

Because of these advantages, liquid biopsy is expected to become a powerful tool in oncology not only for diagnosis and prognosis, but also for surveillance and assessing therapeutic effects. Liquid biopsy initially referred only to circulating tumor cells (CTCs) (although with a short half-life) but now extends to other components released by tumors like cell-free circulating nucleic acids (NAs) such as DNA, messenger RNA (mRNA), microRNA (miRNA), non-coding RNA, extracellular vesicles (EVs or exosomes) and tumor educated platelets (TEPs). Liquid biopsy corresponds to tumor burden and measurement of ctDNA appears to be even more beneficial in the metastatic setting (with levels < 5 CTCs per 7.5 cc correlating to better progression free survival and overall survival) [7]. For the same reason, high false negative rates (FNR) are seen when lower levels of tumor-derived products are seen in body fluids. For example, cell-free DNA (cfDNA) can be poor in quality secondary to inflammation or infections that result in high false positive rates. Droplet digital polymerase chain reaction (ddPCR) has become one of the most sensitive methods for detection of somatic mutations by improving cfDNA extraction methods, thereby optimizing the yield of cfDNA [8–11]. Although liquid biopsy can be performed with various body fluids, such as blood [12], CSF [13,14], pleural fluid [15,16], gastric juice [17,18], or saliva [19,20], we will focus on urine, as it can be easily and non-invasively collected without use of any special techniques or instruments and in copious amounts. Table 1 illustrates the specifics of standard tissue biopsy and its comparison with blood and urine as liquid biopsy. While several reviews have been published on urine as a source of liquid biopsy for cancer, most of them have mainly focused on genitourinary cancers. Chen et al. focused on urine liquid biopsy technologies and its use in cancer, glomerular disease, and tuberculosis [21], while Yu et al. focused on prostate and bladder cancer [22], and Hentschel et al. on bladder, prostate, and cervix cancer [23]. This review seeks to highlight the components of urine liquid biopsy and its utility and limitations in oncology, mainly in non-urological cancers.

Table 1. Advantages and disadvantages of standard tissue biopsy versus blood liquid biopsy versus urine liquid biopsy.

	Standard Tissue Biopsy	Blood Liquid Biopsy	Urine Liquid Biopsy
Components	Cell structure, grade, stromal and immune cells, Lymphovascular invasion, DNA seq, RNA seq, gene signatures	CTCs, cell free nucleic acids, exosomes, tumor educated platelets	Cell free DNA, urinary mRNA, miRNA, lnc RNA, other snc RNA, exosomes
Advantages	• Standard of care • Standard technique, low FNR • Histological information and immunohistochemical profiling excellent	• Minimally invasive procedure • Early detection and molecular profile assessment • Intratumor heterogeneity • Real time monitoring of cancer evolution • Corelates with tumor burden • Identifies genetic markers of treatment and treatment resistance • DNA fresh and not modified by storage technique • Quick turnaround testing time for ctDNA • ctDNA more beneficial in metastatic setting	• Noninvasive procedure • Early detection and molecular profile assessment • Intratumor heterogeneity • Large quantities available and centrifuged for concentrates • High DNA yield • Identifies genetic markers of treatment and treatment resistance • Good for longitudinal follow up • ucfDNA can potentially help in localizing "cancer of unknown primary"
Disadvantages	• Invasive procedure, involves patient risk • Lacks assessment of intratumor heterogeneity • Time period of analysis fixed • Repetitive invasive biopsies cumbersome • Early detection of cancer not possible • DNA quality highly variable in FFPE • Variable quantity of DNA based on sampling methods, high risk of DNA degradation	• Investigational setting • High FNR • Lack of standardized technique for cfDNA and cellular genomic DNA • ctDNA quality and extraction methods. • Short half life of CTCs (1–2.4 h) in peripheral blood	• Investigational setting • No histological assessment • Effect of hydration status and medications • ucfDNA integrity sensitivity and specificity issues • Artifacts from microchip analysis • Variations in assay protocols/sample handling • Measurement of urinary RNAs challenging • Lack of large multicenter studies

CTC: Circulating tumor cells; FNR: False negative rate; lnc RNA: Long non-coding RNA, sncRNA: small non-coding RNA; ucf DNA: urine cell-free DNA; RT-PCR: Reverse transcription polymerase chain reaction; FFPE: Formalin fixed paraffin embedded; ctDNA: circulating tumor DNA.

2. Urine Liquid Biopsy Components

Urine is a biological fluid consisting of organic and inorganic compounds; salts; cells like leucocytes, renal cells, urothelial cells, prostate cells, and exfoliated tumor cells; and tumor cell-free nucleic acids. Tumor-derived DNA, mRNA, and miRNA can be obtained via whole urine sample, centrifugation to obtain urine sediment, or filtration to obtain urinary supernatant and cells [24]. With recent technological advances, it has become possible to extract and analyze minute amounts of NAs from body fluids. It is easy to collect large amounts of urine for larger samples of urinary NAs. Urinary NAs are expected to provide very useful clinical information that may reflect tumor heterogeneity.

2.1. Urinary DNAs

While the exact mechanism of origin of circulating tumor DNA (ctDNA) and its filtration by the kidneys remains unclear, some hypotheses of origin include: (i) from dying cells, exfoliated either in urine (bladder and prostatic cells) or from circulation (ii) from CTCs and (iii) via active release [25]. Urine contains DNA as a result of renal clearance of blood. Only molecular substances smaller than 6.4 nm in diameter and molecular weight not greater than 70 kDa can pass through the lumen of a nephron [26]. This corresponds to about 100 base pair (bp) DNA in size [27]. Since the size of a mononucleosome exceeds the size of the nephron barrier pores, they cannot pass through the nephron. Only protein and NAs can pass through and are excreted in the urine. Many studies of urine liquid biopsy have reported the correlation between urinary DNAs and urological malignancies, such as cancers of the bladder [28,29], prostate [30], and kidney [31], as a result of directly shedding breakdown products in the urine. Isolating DNA fragments in urine is technically easier than blood since urine contains less protein [32]. On the other hand, NA-hydrolyzing enzymes that breakdown DNAs are easily activated in urine. DNA hydrolase deoxyribonuclease I and II (DNase I and II) are present both in urine and blood and are more active in urine. DNase I is a major DNA hydrolase released from the pancreas. The amount of DNase II is less than DNase I in urine although it is more potent.

DNA methylation changes, which are considered one of the primary events in carcinogenesis, can be identified by DNA-sodium bisulfite in the urine. This method selectively deaminates unmethylated cytosines to uracil but methylated forms of cytosines escape the bisulfite reaction, allowing them to be analyzed by polymerase chain reaction (PCR)-based technology to target specific functional locations like CpG islands where methylation genes are expressed. However, there is great variability in its sensitivity and specificity. GSTP1 methylation is a biomarker for prostate cancer [33] and ONECUT2 (One Cut Homeobox 2) is for upper ureteral carcinoma [34]. However, due to its low sensitivity, DNA methylation is recommended only in combination with other biomarkers.

Urinary cell-free DNA (ucfDNA) originates directly from dying cells exfoliated in urine and gives important information regarding DNA derived from cancer cells and is considered to be more representative than the tissue biopsy of a tumor [35]. There are no standard protocols for isolation or detection of ucfDNA to date, but it can be detected by conventional PCR-based assays or by using commercially available kits [36]. Recently, next generation sequencing (NGS) has been used for better sensitivity [37]. ucfDNA has mainly seen utility in urological cancers and was first described by Sidransky et al. in 1991 with the presence of p53 mutations in the urine sediment of patients with muscle invasive bladder cancer [38]. ucfDNA has also been investigated for EGFR mutation in non-small cell lung cancer [39], with elevated levels seen in Stage III and IV. Elevated levels of ucfDNA p53 mutations have been demonstrated by Lin et al. in hepatocellular carcinoma and could be potentially explored for screening [40]. Su et al. reported that KRAS mutation was detected in higher incidence in urine compared to serum (35%) or plasma (40%) among patients with colorectal cancer or colonic polyps [41]. KRAS gene G12/13 mutation has also been found in ucfDNA by the NGS approach of patients with colorectal cancer [42].

2.2. RNA-Based Biomarkers

Several types of RNA are present and measurable in the supernatant of the urine. RNA molecules are biochemically unstable and sensitive to heavy metal ions, alkaline pH, and RNA-hydrolyzing enzymes. There are abundant RNA hydrolases in urine, such as RNA-hydrolyzing enzyme (RNase II) and Ribonuclease I, which hydrolyze both RNAs and DNAs. Despite this mechanism, mRNAs are still detectable in the urine because they are somewhat protected from degradation by extracellular vesicles, ribonucleoproteins, and lipoproteins [43]. Through alternate splicing of mRNA, many genes generate different isoforms of protein products in cancer. Thus, mRNA, being a protein coding transcript, represents a good biomarker for establishing the correlation between information in DNA and proteins. Several methods to isolate urinary mRNA have been described, including the QIAamp Circulating Nucleic acid kit (Qiagen) [44], RNeasy kit (Qiagen [45]), and Quick-RNA MicroPrep Kit (Zymo Research) [46], and miRNeasy kit (Qiagen) [47]. After isolation of mRNA, molecular biology methods such as quantitative-PCR, droplet digital PCR, or Next Generation Sequencing are required to search or determine NAs. Currently, the Xpert BC Monitor test [48] and 2-Gene mRNA Urine test [49] are used for bladder cancer and prostate cancer, respectively. Since these kits have been used predominantly for urological cancer, further studies are needed to expand their application for non-urological cancers.

2.2.1. Urinary microRNAs

Compared with mRNA that can be easily degraded by RNA-hydrolyzing enzymes, microRNA is more resistant to nucleases and remains relatively stable in urine [50]. MicroRNA are a class of short single strand RNAs (22–24 nucleotides in length) and are involved in cell proliferation, differentiation, stress response, inflammation, and cell death [51–58]. They epigenetically inhibit the translation of target mRNA into proteins [59]. They are known to play roles in different mechanisms of cancer progression, including carcinogenesis, angiogenesis, and metastasis [51]. MicroRNA is encapsulated and bound to RNA-binding protein, which stabilize it to the point that it withstands several cycles of freeze and thaw and remains stable at room temperature for long periods of time. They can be evaluated in different fractions such as non-centrifuged urine, urine sediment, supernatant, and as part of exosomes [60]. Since some microRNAs released from cancer cells are also highly expressed in activated T-cells, some suggest that monitoring circulating microRNA released from the host immune cells can be used as a biomarker in predicting cancer progression [61]. Furthermore, given the possible association between circulating microRNA and cancer immunity, studies on circulating microRNA are expected to lead to the future development of new therapeutic agents through immunomodulation. MicroRNAs represent a new source of reliable biomarkers that can be diagnostic, prognostic, and predictive during therapy of cancer patients and has been widely studied in prostate [62], renal [63], and urothelial carcinoma [64]. MicroRNA can be quantified by reverse transcription-PCR (RT-PCR), Northern blotting, in situ hybridization, gene expression microarray, or NGS technology but also with commercially available isolation kits including the miRNeasy Mini kit (Qiagen) [65], ZR urine RNA isolation kit (Zymo Research) [66] for bladder cancer; Acid phenol–chloroform plus Silica columns (BioSilica Ltd.) [67], Urine Exfoliated Cell and Bacteria RNA Purification Kit (Norgen) [68] for prostate cancer; TRIZOL reagent (Invitrogen) [69] and miRNeasy Serum/Plasma kit (Qiagen) [70] for gastric cancer.

2.2.2. Long Non-Coding RNAs (lncRNAs)

Long non-coding RNAs (lncRNAs) are transcripts with length greater than 200 nucleotides encoding no protein and are gene regulators involved in many biological functions and dysregulated in various cancers [71]. Expression of lncRNAs is associated with a broad range of cellular processes, such as cell growth, survival, migration, invasion, and differentiation [72]. More recently, studies have investigated their possible role as biomarkers in cancer by highlighting the role of lncRNAs in carcinogenesis through impairment of cell cycle arrest and apoptosis [73]. Many lncRNAs are exosome-derived in urine and have been found to be more protected by RNAse activity. The gold standard method for lncRNA detection is quantitative RT-PCR [74]. Prostate cancer antigen 3 (PCA3) was the first lncRNA identified in 1999 mapped on chromosome 9q21–22 and found to be overexpressed in greater than 95% of prostate cancers [75]. The human urothelial carcinoma–associated 1 (UCA1), a 2314-bp lncRNA located on human chromosome 19, has been found to be upregulated in many cancers, such as hepatocellular cancer [76], colorectal cancer [77], gastric cancer [78], esophageal squamous cell carcinoma [79], and epithelial ovarian cancer [80].

2.2.3. Other Urinary Small Non-Coding RNAs (sncRNAs)

Small non-coding RNAs are usually shorter in length by about 18–200 nucleotides. While mRNAs are highly susceptible to nucleases, sncRNAs, which are smaller in size, form stable complexes in urine, making them more resistant to nuclease [81]. They include small nuclear RNA (snRNA), small nucleolar RNA (snoRNA), Piwi-interacting RNA (piRNA) and tRNA-derived small RNA (tsRNA). They have diverse roles, which in conjunction with other molecules involve gene regulation through RNA interference or RNA modification. SncRNAs circulate as part of nucleoprotein complexes or membrane-coated microparticles such as exosomes [82]. Their role as a biomarker of cancers remains unclear [83].

3. Utility of Urine for Liquid Biopsy

While urine is a relatively cell-free biofluid, it contains large numbers of complex substances, including protein, circulating NAs (DNAs and RNAs), and extracellular vesicles (EVs). Since the yield and sensitivity of urine cfDNAs are comparable to blood cfDNAs, attention has been directed to urine sampling as an alternative body fluid source in lieu of blood to monitor clinical course and follow-up therapeutic effects [84]. Genomic abnormalities detected from urine NAs are shown to be useful in both urologic and non-urologic cancers. It has been shown that the sensitivity of cfDNA/ctDNA in urine is comparable to blood among patients with multiple cancers, such as urothelial carcinoma [85], breast cancer [84], colon cancer [41], and lung cancer [37]. One of the major advantages of using urine is its non-invasive nature of collection compared to tissue or blood, especially in patients requiring repeated sampling to monitor cancer progression and/or therapeutic effects [86]. Urine can be collected in large quantities, which solves one of the major problems with tissue biopsy or other liquid biopsy materials that often suffer from a limited number of samples. It is more patient-friendly since the collection of urine can be done anywhere as opposed to access to other body fluid or tissue which needs to be done in clinics or hospitals. Even in clinic settings, obtaining sufficient blood draws can be a challenge in some populations including geriatric patients, intravenous drug abusers, or anyone with thin veins [87]. Sampling cerebrospinal fluid (CSF) or gastric juice is even more invasive, and sophisticated techniques are required for their collection. Therefore, liquid biopsy using urine is expected to significantly reduce labor and cost as well as patients' pain. Due to these advantages, urine liquid biopsy has been investigated for cancer screening, monitoring of cancer progression or recurrence, and the efficacy of chemo and radiation therapy.

3.1. Urinary Liquid Biopsy for Urological Cancers

Most of the studies regarding urine liquid biopsy have been performed on urological cancers, since many of the substances secreted from urological cancer are likely to drain directly into the urinary tract [27,88]. First morning urine contains the highest number of cells and cellular debris from the urological tract exfoliated in urine at night [89]. Both low-molecular-weight DNA (<100 bp) and high-molecular-weight DNA ($\geq$1 kbp) can be detected in urine [84]. Urinary protein biomarkers for early detection of prostate cancer and bladder cancer have already been established and approved by the FDA, such as Nuclear Matrix Protein 22 (NMP22), Urovysion Fluorescence In Situ Hybridization (FISH), and Prostate Cancer gene 3 (PCA3) [90]. As a matter of fact, several tests based on urine liquid biopsy have been already included in the National Comprehensive Cancer

Network (NCCN) Guidelines for Prostate Cancer Early Detection since 2020. These tests are Mi-Prostate scores that include measurements of PCA3 and TMPRSS2:ERG fusion gene expression in the urine, IntelliScore and SelectMDx, which may reduce the number of unnecessary biopsies [91]. In addition to ctDNA/cfDNA, the other NAs, such as mRNA [44,92], lncRNA [93], microRNA, piRNA [94], and circRNA [95], have been reported to be useful as biomarkers in urological cancers. The first commercial exosome-based prostate Intelliscore test for prostate cancer became available in 2016 [96]. Several urinary lncRNAs, such as FR0348383, MALAT1, and DD3 (PCA3), have been reported as better biomarkers in prostate cancer compared to serum prostate-specific antigens (PSA) [97,98]. Given its quality and accuracy, detection of urinary PCA3 has been approved by the FDA as a diagnostic tool for prostate cancer [98]. PCA3 levels have also been associated with tumor volume burden and extracapsular extension and provide prognostic information before a radical prostatectomy [99]. Urothelial carcinoma associated 1 (UCA1) is one of the most well studied genes in bladder cancer, and urinary lncRNA of UCA1 was often detected in patients with bladder cancer [100,101]. Currently, there are two clinical trials evaluating urine as a source for liquid biopsy. NCT04432909 is a prospective multi-center, single-blinded study to evaluate the utility of UroCAD for urothelial carcinoma diagnosis and follow-up in 500 participants (https://clinicaltrials.gov/ct2/show/NCT04432909, accessed on 19 May 2021). Patients with urothelial carcinoma prior to resection are compared with the patients being treated for other diseases but without any tumor to determine the sensitivity and specificity of UroCAD analysis, which will be compared with cytology and FISH. Another trial was reported at the American Society of Clinical Oncology Genitourinary (ASCO-GU) 2021 meeting by Zhang et al. from Shanghai, China, which is a prospective clinical trial that compares blood and urine liquid biopsy using PredicineCARE NGS 152 gene assay with the gold standard of tissue biopsy in 59 treatment-naïve bladder cancer patients. The mutation profiles of urine samples (sensitivity of 86.7%) were found to be very similar with tissue biopsy compared to blood liquid biopsy samples (sensitivity of 10.3%). At this point, we were unable to identify any current ongoing clinical trials in non-urological cancers. With increasing evidence, it is conceivable that detection of not only DNA and miRNA but also oncogenic lncRNAs in urine might enable early cancer diagnosis and can be promising therapeutic targets for patients with genitourinary cancer.

3.2. Urinary Liquid Biopsy for Non-Urological Cancers

There are numerous studies identifying common mutations in each type of cancer, such as EGFR mutation in lung cancer, that guide us in assigning a cell of origin to a biomarker like cfDNA. We have summarized these molecules detected in urine and the cancer type in Table 2. In addition, given the strength of liquid biopsy in longitudinal follow-up, we may discover a unique/novel biomarker for a particular patient of a particular cancer type that may become a strategy in the future. It is speculated that urinary RNAs may be also associated with clinical outcomes in patients with various types of cancers [102,103].

Table 2. Application of urine liquid biopsy in non-urological cancers.

Study, Reference Number	Cancer Type	Early Stage, Advanced or Metastatic	No of Patients	Molecules Assessed	Methodology/ Quantitative Analysis	Clinical Application of Urine Biopsy	Sensitivity in Urine
Reckamp [37]	NSCLC	Advanced Stage	63	ctDNA for EGFR T790M mutation	ddPCR, NGS	Predictive response to Rociletinib (EGFR TKI)	75%
Husain [104]	NSCLC	Advanced Stage	8	ctDNA for EGFR T790M mutation	ddPCR, NGS	Predictive response to Osimertinib (III generation EGFR TKI)	86%

Table 2. Cont.

Study, Reference Number	Cancer Type	Early Stage, Advanced or Metastatic	No of Patients	Molecules Assessed	Methodology/ Quantitative Analysis	Clinical Application of Urine Biopsy	Sensitivity in Urine
Wu [105]	NSCLC	Advanced Stage & Metastatic	50	TP53 and EGFR mutation	PCR, NGS	Detection of driver gene alterations	60%
Liu [106]	NSCLC	Early stage	74	DNA methylation	Methylation specific PCR	Early detection after incidental finding of nodule on CT chest	73%
Zhang [107]	Breast	Early stage	200	ctDNA for PIK3CA	ddPCR	Prognostic and predictive	77%
Ritter [108]	Endometrial & Ovarian	Early stage	10	MiR-10b-5p	RT-qPCR, Human miRNA V21.0 microarray	Early detection	50%
Kao [69]	Gastric	Early stage	50	MiR-21-5p	Quantitative stem loop RT-PCR	Predictive	NA
Iwasaki [70]	Gastric	Early stage	197	MiR-6807-5p MiR-6856-5p	miRNeasy kit (Qiagen), miRNA microarray	Early detection and Prognostic	63.4%
Su [41]	Colorectal	Advanced stage	20	cfDNA KRAS mutation	RT-PCR	Early detection	95%

NSCLC: Non-small cell lung cancer; ddPCR: Droplet digital polymerase chain reaction; NGS: Next generation sequencing; RT-PCR: Reverse transcription polymerase chain reaction; ctDNA: circulating tumor DNA.

3.2.1. Urine Liquid Biopsy in Lung Cancers

Conventional tissue biopsies are particularly cumbersome and carry potential risk for significant morbidity to lung cancer patients since they can cause pneumothorax and significant bleeding within the airway. Various urine liquid biopsy components have been investigated for patients with non-small cell lung cancer (NSCLC) and have been reported to reduce costs by improving detection of EGFR T790M mutations and reducing the complications associated with tissue biopsy [109]. Reckamp et al. studied 63 patients with advanced EGFR-mutant NSCLC and found that the sensitivities of tissue, plasma, and urine were 73%, 82%, and 75%, respectively, for T790M detection in these complementary specimens [36]. They also found a significant decrease in T790M MAF in urine in patients treated with Rociletinib (an EGFR tyrosine kinase inhibitor (TKI)), highlighting a potential of using urine for follow-up. These findings were confirmed in another study by Husain et al. who found that early kinetics of ctDNA in the urine of eight patients treated with Osimertinib, a third generation anti-EGFR TKI, correlated with tumor response [104]. These studies demonstrate that urine testing successfully identifies EGFR mutations in patients with advanced stage/metastatic NSCLC and has high concordance with tumor tissue and plasma and can be used as a viable approach for assessing EGFR mutation status. In advanced NSCLC, Wu et al. demonstrated a good correlation and complementarity between genomic profiles of cfDNA extracted from plasma, sputum, and urine compared to tissue [105]. In early-stage NSCLC, the analysis of DNA methylation at cancer-specific loci in urine were shown to help characterize nodules after screening via computed tomography (CT) [106]. Thus, various studies have demonstrated the utility of urine liquid biopsy not only as a diagnostic but also a prognostic and predictive marker in NSCLC.

3.2.2. Urine Liquid Biopsy in Breast, Gynecological, and Gastrointestinal Cancers

Some have investigated the role of urine liquid biopsy in early breast cancer. In a prospective study, Zhang et al. compared serum and urine ctDNA levels using a droplet digital PCR (ddPCR) technique of 200 breast cancer patients and healthy volunteers [107]. The authors found 3.5-fold higher levels of ctDNA as well as wild-type PIK3CA genotype in early breast cancer patients compared to healthy volunteers. These results demonstrate that urinary ctDNA is capable of discerning between healthy populations while providing early disease detection, especially in high-risk individuals. Zhang et al. also evaluated a decline of urinary ctDNA following initial treatment and found a 6.8-fold decrease [107].

Among the tested 10 microRNAs, miR-10b-5p was identified as a candidate biomarker for endometrial and ovarian cancer [108]. It was found to be elevated in patients with endometrial cancer compared to healthy women; however, its relevance in ovarian cancer remains unelucidated. MiR-200c-3p was found to be enriched in the urine of endometrial cancer patients, paving the way for the development of a non-invasive biomarker for early detection [110]. Abnormal lncRNA UCA1 expression has been linked to adverse clinicopathological characteristics including lymph node metastasis, chemoresistance, and poor overall survival in both cancers [80,111].

Identification of biomarkers for gastric cancer still remains a challenge. The most frequently used tumor markers include CEA, CA19-9, CA72-4, CA50, pepsinogen, and alfa fetoprotein, however their sensitivity and specificity are poor and hence not specific to a diagnosis of gastric cancer. Hung et al. reported that miR-376c promotes gastric cancer cell proliferation and migration, and it was increased in urine and plasma of gastric cancer patients [112]. Kao et al. detected miR-21-5p levels in the urine of gastric cancer patients pre- and post-op at one and three months and found that its levels consistently decreased following gastric surgery [69]. MiR-6807-5p and miR-6856-5p were also found to be significantly increased in the urine of gastric cancer patients but fell to almost non-detectable levels following gastric resection [70]. These results appear promising for both early detection and prognosis of patients with gastric cancer.

KRAS mutations were detected from the cfDNAs in the urine of advanced colorectal cancer patients. This was the first reported urinary cfDNA as a biomarker in a non-urological cancer, proving that the kidney barrier in humans is permeable to DNA molecules large enough to be analyzed by standard genetic technologies [113]. Su YH et al. compared the concentration of DNA in different body fluids and found that it was similar in urine compared to serum, but it was significantly lower in plasma than in either urine or serum ($p < 0.05$). They also reported that when DNA was derived from 10 µL of body fluid in each mutation assay, the mutated KRAS DNA detection was comparable among serum, plasma, and urine. However, in patients with colorectal cancer, when a larger amount of body fluid (200 µL) was used the detection rate of the KRAS gene in urine was significantly higher (95%) than in serum (35%) or plasma (40%). These findings suggest that inhibitory factors (such as DNase) in serum and plasma might be less abundant in urine, and that urine does not usually contain large molecules, such as protein, that can interfere with PCR amplification compared to blood or serum, as they are filtered by the kidneys [41].

4. Limitations of Urinary Liquid Biopsy

By definition, urine is generated by the kidney and there are many components that may not get filtered in the urine compared to blood. Therefore, urine liquid biopsy has been more intensively studied in genitourinary cancers and is one of the major limitations in non-urological cancers. Since urine is a dynamic body fluid, concentrations change with hydration status, renal pathology, urine volume, and effect of medications. Hence, concentrations will most likely not be reliable with a high degree of variability within the urine composition and will require an absolute amount or centrifugation. Therefore, measuring 24 h urine volumes would be the gold standard to assess hydration status. Measuring creatinine ratios or specific gravity remain as other possibilities and potentially more feasible alternatives. Despite being a useful tool for diagnosis, prognosis, and a predictive marker for treatment response, a major limitation of urine cfDNA-based tests includes lack of specificity. Increased levels of cfDNA are seen in non-malignant conditions such as trauma, inflammation, pregnancy, autoimmune conditions like lupus, and infections such as tuberculosis. Due to these very reasons, cfDNA-based tests lack application in the clinical setting [36]. Mutation rates in individuals may be influenced by environmental and physiological factors [114], and spontaneous mutations known to typically contribute to cancer development can occur with increasing age but may not directly cause cancer. There is also less abundance of mRNA in urine, a lack of stable targeted molecules, along with possible contamination of cellular RNA during sample preparation [115]. Thus, utilization of urine liquid biopsy in pre-symptomatic stages may yield false positive results and overdiagnosis of cancer. With regards to methodology, microchip analysis is an efficient method to screen for urine biomarkers; however, challenges when applying this method include repetitive sequences in the discovery phase miRNAs when designing probes or primers (due to short length of nucleotides), which may result in artifacts [108] masking the results of microchip analysis. Varied analytical methods include NGS, RT-PCR, and microarray, which can lead to aberrational findings [116]. Variations exist in assay protocols and sample handling despite the same analytical method performed. More importantly, why certain specific RNA extraction kits are used for detection of biomarkers in different studies depending on the cancer site, remains elusive. Lack of large multicenter studies remain the major reason for precluding its adoption in clinical practice.

While there are several limitations to urine as liquid biopsy, it can also be used to our advantage. The biggest advantage of urine is that an unlimited amount can be collected. Instead of an absolute value, urine samples can be collected as a set quantity per day and quantified as a fraction of the total quantity (especially in patients suffering from excessive diuresis). Ideally, it would be beneficial to confirm the presence of sufficient amounts and quality of ctDNA to identify the most appropriate ctDNA quantification methods to maintain uniformity and improve the sensitivity of ctDNA detection to anticipate drug resistances by urine biopsy. In addition to looking into ctDNA, we can look into smaller nucleic acids such as messenger RNA, micro-RNA, circular RNA, transfer RNA, or even RNA in exosomes. Analyzing exosomes in the future can become an important strategy as cells communicate through exosomes. More recently, with newer tools like SiRe NGS panel testing [12] or the TargetPlex FFPE Direct DNA library preparation kit [117] being applied to patient blood and tissue samples with advanced-stage NSCLC, we cannot help but speculate that these more cost-effective methods may gain more widespread application in urine liquid biopsies in the future. Considering the positive effects on biomarker studies and beyond, we hope that funding bodies will take steps to complement the current emphasis on these novel studies and support programs for reproduction studies of existing findings to validate their clinical utility.

5. Conclusions and Future Perspectives

Changes in genomic and genetic material in the urine potentially precede changes in imaging and can detect minimal tumor burden of urological and non-urological cancers. There still remains a need for standardized methods and normalization procedures. Despite the non-invasive nature of sample collection and its potential benefits, this newer urine-based approach still requires large-scale research for validation by large cohorts prospectively. Although a promising innovation, an important question that remains to be answered is whether urine biomarkers offer better profiling for disease recurrence and whether urine biomarker elevation-driven interventions translate into better outcomes.

Author Contributions: Conceptualization, K.T. and M.O. (Masanori Oshi); methodology, K.T., M.O. (Masanori Oshi), V.M., H.T., M.O. (Maiko Okano), Y.T., O.M.R., R.M., I.E. and K.T.; formal analysis, V.M. and M.O. (Masanori Oshi), writing—original draft preparation, V.M., M.O. (Masanori Oshi) and M.H., writing—review and editing, K.T. and M.O. (Masanori Oshi), supervision, K.T., project administration, K.T. All authors have read and agreed to the published version of the manuscript.

Funding: This work was supported by US National Institutes of Health/National Cancer Institute grant 5T32CA108456 to V.M., R01CA160688, R01CA250412, R37CA248018, US Department of Defense BCRP grant W81XWH-19-1-0674, as well as the Edward K. Duch Foundation and Paul & Helen Ellis Charitable Trust to K.T., and US National Cancer Institute cancer center support grant P30-CA016056 to Roswell Park Comprehensive Cancer Center.

Conflicts of Interest: The authors declare no conflict of interest.

References

1. Johnson, B.E.; Mazor, T.; Hong, C.; Barnes, M.; Aihara, K.; McLean, C.Y.; Fouse, S.D.; Yamamoto, S.; Ueda, H.; Tatsuno, K.; et al. Mutational analysis reveals the origin and therapy-driven evolution of recurrent glioma. *Science* **2014**, *343*, 189–193. [CrossRef]
2. McGranahan, N.; Swanton, C. Clonal Heterogeneity and Tumor Evolution: Past, Present, and the Future. *Cell* **2017**, *168*, 613–628. [CrossRef]
3. Sun, R.; Hu, Z.; Sottoriva, A.; Graham, T.A.; Harpak, A.; Ma, Z.; Fischer, J.M.; Shibata, D.; Curtis, C. Between-region genetic divergence reflects the mode and tempo of tumor evolution. *Nat. Genet.* **2017**, *49*, 1015–1024. [CrossRef]
4. Yan, T.; Cui, H.; Zhou, Y.; Yang, B.; Kong, P.; Zhang, Y.; Liu, Y.; Wang, B.; Cheng, Y.; Li, J.; et al. Multi-region sequencing unveils novel actionable targets and spatial heterogeneity in esophageal squamous cell carcinoma. *Nat. Commun.* **2019**, *10*, 1670. [CrossRef]
5. Hu, X.; Fujimoto, J.; Ying, L.; Fukuoka, J.; Ashizawa, K.; Sun, W.; Reuben, A.; Chow, C.W.; McGranahan, N.; Chen, R.; et al. Multi-region exome sequencing reveals genomic evolution from preneoplasia to lung adenocarcinoma. *Nat. Commun.* **2019**, *10*, 2978. [CrossRef]
6. Rapisuwon, S.; Vietsch, E.E.; Wellstein, A. Circulating biomarkers to monitor cancer progression and treatment. *Comput. Struct. Biotechnol. J.* **2016**, *14*, 211–222. [CrossRef]
7. Cristofanilli, M.; Budd, G.T.; Ellis, M.J.; Stopeck, A.; Matera, J.; Miller, M.C.; Reuben, J.M.; Doyle, G.V.; Allard, W.J.; Terstappen, L.W.; et al. Circulating tumor cells, disease progression, and survival in metastatic breast cancer. *N. Engl. J. Med.* **2004**, *351*, 781–791. [CrossRef]
8. Takeshita, T.; Yamamoto, Y.; Yamamoto-Ibusuki, M.; Inao, T.; Sueta, A.; Fujiwara, S.; Omoto, Y.; Iwase, H. Prognostic role of PIK3CA mutations of cell-free DNA in early-stage triple negative breast cancer. *Cancer Sci.* **2015**, *106*, 1582–1589. [CrossRef]
9. Takeshita, T.; Yamamoto, Y.; Yamamoto-Ibusuki, M.; Inao, T.; Sueta, A.; Fujiwara, S.; Omoto, Y.; Iwase, H. Clinical significance of monitoring ESR1 mutations in circulating cell-free DNA in estrogen receptor positive breast cancer patients. *Oncotarget* **2016**, *7*, 32504–32518. [CrossRef]
10. Takeshita, T.; Yamamoto, Y.; Yamamoto-Ibusuki, M.; Tomiguchi, M.; Sueta, A.; Murakami, K.; Iwase, H. Clinical significance of plasma cell-free DNA mutations in PIK3CA, AKT1, and ESR1 gene according to treatment lines in ER-positive breast cancer. *Mol. Cancer* **2018**, *17*, 67. [CrossRef]
11. Zmrzljak, U.P.; Košir, R.; Krivokapić, Z.; Radojković, D.; Nikolić, A. Detection of Somatic Mutations with ddPCR from Liquid Biopsy of Colorectal Cancer Patients. *Genes* **2021**, *12*, 289. [CrossRef] [PubMed]
12. Nacchio, M.; Sgariglia, R.; Gristina, V.; Pisapia, P.; Pepe, F.; De Luca, C.; Migliatico, I.; Clery, E.; Greco, L.; Vigliar, E.; et al. KRAS mutations testing in non-small cell lung cancer: The role of Liquid biopsy in the basal setting. *J. Thorac. Dis.* **2020**, *12*, 3836–3843. [CrossRef]
13. Pentsova, E.I.; Shah, R.H.; Tang, J.; Boire, A.; You, D.; Briggs, S.; Omuro, A.; Lin, X.; Fleisher, M.; Grommes, C.; et al. Evaluating Cancer of the Central Nervous System Through Next-Generation Sequencing of Cerebrospinal Fluid. *J. Clin. Oncol.* **2016**, *34*, 2404–2415. [CrossRef] [PubMed]
14. Miller, A.M.; Shah, R.H.; Pentsova, E.I.; Pourmaleki, M.; Briggs, S.; Distefano, N.; Zheng, Y.; Skakodub, A.; Mehta, S.A.; Campos, C.; et al. Tracking tumour evolution in glioma through liquid biopsies of cerebrospinal fluid. *Nature* **2019**, *565*, 654–658. [CrossRef]
15. Song, Z.; Wang, W.; Li, M.; Liu, J.; Zhang, Y. Cytological-negative pleural effusion can be an alternative liquid biopsy media for detection of EGFR mutation in NSCLC patients. *Lung Cancer* **2019**, *136*, 23–29. [CrossRef]
16. Villatoro, S.; Mayo-de-Las-Casas, C.; Jordana-Ariza, N.; Viteri-Ramírez, S.; Garzón-Ibañez, M.; Moya-Horno, I.; García-Peláez, B.; González-Cao, M.; Malapelle, U.; Balada-Bel, A.; et al. Prospective detection of mutations in cerebrospinal fluid, pleural effusion, and ascites of advanced cancer patients to guide treatment decisions. *Mol. Oncol.* **2019**, *13*, 2633–2645. [CrossRef]
17. Deng, K.; Lin, S.; Zhou, L.; Geng, Q.; Li, Y.; Xu, M.; Na, R. Three aromatic amino acids in gastric juice as potential biomarkers for gastric malignancies. *Anal. Chim. Acta* **2011**, *694*, 100–107. [CrossRef] [PubMed]
18. Shao, Y.; Ye, M.; Jiang, X.; Sun, W.; Ding, X.; Liu, Z.; Ye, G.; Zhang, X.; Xiao, B.; Guo, J. Gastric juice long noncoding RNA used as a tumor marker for screening gastric cancer. *Cancer* **2014**, *120*, 3320–3328. [CrossRef]
19. Cheng, J.; Nonaka, T.; Wong, D.T.W. Salivary Exosomes as Nanocarriers for Cancer Biomarker Delivery. *Materials* **2019**, *12*, 654. [CrossRef]
20. Tang, H.; Wu, Z.; Zhang, J.; Su, B. Salivary lncRNA as a potential marker for oral squamous cell carcinoma diagnosis. *Mol. Med. Rep.* **2013**, *7*, 761–766. [CrossRef]
21. Chen, C.K.; Liao, J.; Li, M.S.; Khoo, B.L. Urine biopsy technologies: Cancer and beyond. *Theranostics* **2020**, *10*, 7872–7888. [CrossRef]

22. Yu, W.; Hurley, J.; Roberts, D.; Chakrabortty, S.K.; Enderle, D.; Noerholm, M.; Breakefield, X.O.; Skog, J.K. Exosome-based liquid biopsies in cancer: Opportunities and challenges. *Ann. Oncol.* **2021**, *32*, 466–477. [CrossRef]
23. Hentschel, A.E.; van den Helder, R.; van Trommel, N.E.; van Splunter, A.P.; van Boerdonk, R.A.A.; van Gent, M.; Nieuwenhuijzen, J.A.; Steenbergen, R.D.M. The Origin of Tumor DNA in Urine of Urogenital Cancer Patients: Local Shedding and Transrenal Excretion. *Cancers* **2021**, *13*, 535. [CrossRef]
24. Bosschieter, J.; Bach, S.; Bijnsdorp, I.V.; Segerink, L.I.; Rurup, W.F.; van Splunter, A.P.; Bahce, I.; Novianti, P.W.; Kazemier, G.; van Moorselaar, R.J.A.; et al. A protocol for urine collection and storage prior to DNA methylation analysis. *PLoS ONE* **2018**, *13*, e0200906. [CrossRef]
25. Stroun, M.; Lyautey, J.; Lederrey, C.; Mulcahy, H.E.; Anker, P. Alu repeat sequences are present in increased proportions compared to a unique gene in plasma/serum DNA: Evidence for a preferential release from viable cells? *Ann. N. Y. Acad. Sci.* **2001**, *945*, 258–264. [CrossRef]
26. Simkin, M.; Abdalla, M.; El-Mogy, M.; Haj-Ahmad, Y. Differences in the quantity of DNA found in the urine and saliva of smokers versus nonsmokers: Implications for the timing of epigenetic events. *Epigenomics* **2012**, *4*, 343–352. [CrossRef] [PubMed]
27. Bryzgunova, O.E.; Laktionov, P.P. Extracellular Nucleic Acids in Urine: Sources, Structure, Diagnostic Potential. *Acta Naturae* **2015**, *7*, 48–54. [CrossRef] [PubMed]
28. Utting, M.; Werner, W.; Dahse, R.; Schubert, J.; Junker, K. Microsatellite analysis of free tumor DNA in urine, serum, and plasma of patients: A minimally invasive method for the detection of bladder cancer. *Clin. Cancer Res.* **2002**, *8*, 35–40. [PubMed]
29. Patel, K.M.; van der Vos, K.E.; Smith, C.G.; Mouliere, F.; Tsui, D.; Morris, J.; Chandrananda, D.; Marass, F.; van den Broek, D.; Neal, D.E.; et al. Association of Plasma and Urinary Mutant DNA With Clinical Outcomes In Muscle Invasive Bladder Cancer. *Sci. Rep.* **2017**, *7*, 5554. [CrossRef] [PubMed]
30. Casadio, V.; Calistri, D.; Salvi, S.; Gunelli, R.; Carretta, E.; Amadori, D.; Silvestrini, R.; Zoli, W. Urine cell-free DNA integrity as a marker for early prostate cancer diagnosis: A pilot study. *BioMed. Res. Int.* **2013**, *2013*, 270457. [CrossRef]
31. Cairns, P. Detection of promoter hypermethylation of tumor suppressor genes in urine from kidney cancer patients. *Ann. N. Y. Acad. Sci.* **2004**, *1022*, 40–43. [CrossRef]
32. Su, Y.H.; Wang, M.; Brenner, D.E.; Ng, A.; Melkonyan, H.; Umansky, S.; Syngal, S.; Block, T.M. Human urine contains small, 150 to 250 nucleotide-sized, soluble DNA derived from the circulation and may be useful in the detection of colorectal cancer. *J. Mol. Diagn.* **2004**, *6*, 101–107. [CrossRef]
33. Bryzgunova, O.E.; Morozkin, E.S.; Yarmoschuk, S.V.; Vlassov, V.V.; Laktionov, P.P. Methylation-specific sequencing of GSTP1 gene promoter in circulating/extracellular DNA from blood and urine of healthy donors and prostate cancer patients. *Ann. N. Y. Acad. Sci.* **2008**, *1137*, 222–225. [CrossRef]
34. Xu, Y.; Ma, X.; Ai, X.; Gao, J.; Liang, Y.; Zhang, Q.; Ma, T.; Mao, K.; Zheng, Q.; Wang, S.; et al. A Urine-Based Liquid Biopsy Method for Detection of Upper Tract Urinary Carcinoma. *Front. Oncol.* **2020**, *10*, 597486. [CrossRef]
35. Pasic, M.D.; Samaan, S.; Yousef, G.M. Genomic medicine: New frontiers and new challenges. *Clin. Chem* **2013**, *59*, 158–167. [CrossRef]
36. Salvi, S.; Martignano, F.; Molinari, C.; Gurioli, G.; Calistri, D.; De Giorgi, U.; Conteduca, V.; Casadio, V. The potential use of urine cell free DNA as a marker for cancer. *Expert Rev. Mol. Diagn.* **2016**, *16*, 1283–1290. [CrossRef] [PubMed]
37. Reckamp, K.L.; Melnikova, V.O.; Karlovich, C.; Sequist, L.V.; Camidge, D.R.; Wakelee, H.; Perol, M.; Oxnard, G.R.; Kosco, K.; Croucher, P.; et al. A Highly Sensitive and Quantitative Test Platform for Detection of NSCLC EGFR Mutations in Urine and Plasma. *J. Thorac. Oncol.* **2016**, *11*, 1690–1700. [CrossRef] [PubMed]
38. Sidransky, D.; Von Eschenbach, A.; Tsai, Y.C.; Jones, P.; Summerhayes, I.; Marshall, F.; Paul, M.; Green, P.; Hamilton, S.R.; Frost, P.; et al. Identification of p53 gene mutations in bladder cancers and urine samples. *Science* **1991**, *252*, 706–709. [CrossRef] [PubMed]
39. Chen, S.; Zhao, J.; Cui, L.; Liu, Y. Urinary circulating DNA detection for dynamic tracking of EGFR mutations for NSCLC patients treated with EGFR-TKIs. *Clin. Transl. Oncol.* **2017**, *19*, 332–340. [CrossRef]
40. Lin, S.Y.; Dhillon, V.; Jain, S.; Chang, T.T.; Hu, C.T.; Lin, Y.J.; Chen, S.H.; Chang, K.C.; Song, W.; Yu, L.; et al. A locked nucleic acid clamp-mediated PCR assay for detection of a p53 codon 249 hotspot mutation in urine. *J. Mol. Diagn.* **2011**, *13*, 474–484. [CrossRef] [PubMed]
41. Su, Y.H.; Wang, M.; Brenner, D.E.; Norton, P.A.; Block, T.M. Detection of mutated K-ras DNA in urine, plasma, and serum of patients with colorectal carcinoma or adenomatous polyps. *Ann. N. Y. Acad. Sci.* **2008**, *1137*, 197–206. [CrossRef]
42. Fujii, T.; Barzi, A.; Sartore-Bianchi, A.; Cassingena, A.; Siravegna, G.; Karp, D.D.; Piha-Paul, S.A.; Subbiah, V.; Tsimberidou, A.M.; Huang, H.J.; et al. Mutation-Enrichment Next-Generation Sequencing for Quantitative Detection of KRAS Mutations in Urine Cell-Free DNA from Patients with Advanced Cancers. *Clin. Cancer Res.* **2017**, *23*, 3657–3666. [CrossRef]
43. Gunasekaran, P.M.; Luther, J.M.; Byrd, J.B. For what factors should we normalize urinary extracellular mRNA biomarkers? *Biomol. Detect. Quantif.* **2019**, *17*, 100090. [CrossRef] [PubMed]
44. Kim, W.T.; Kim, Y.H.; Jeong, P.; Seo, S.P.; Kang, H.W.; Kim, Y.J.; Yun, S.J.; Lee, S.C.; Moon, S.K.; Choi, Y.H.; et al. Urinary cell-free nucleic acid IQGAP3: A new non-invasive diagnostic marker for bladder cancer. *Oncotarget* **2018**, *9*, 14354–14365. [CrossRef]
45. Urquidi, V.; Netherton, M.; Gomes-Giacoia, E.; Serie, D.; Eckel-Passow, J.; Rosser, C.J.; Goodison, S. Urinary mRNA biomarker panel for the detection of urothelial carcinoma. *Oncotarget* **2016**, *7*, 38733–38740. [CrossRef] [PubMed]
46. Guo, J.; Yang, J.; Zhang, X.; Feng, X.; Zhang, H.; Chen, L.; Johnson, H.; Persson, J.L.; Xiao, K. A Panel of Biomarkers for Diagnosis of Prostate Cancer Using Urine Samples. *Anticancer Res.* **2018**, *38*, 1471–1477. [CrossRef]

47. Woo, H.K.; Park, J.; Ku, J.Y.; Lee, C.H.; Sunkara, V.; Ha, H.K.; Cho, Y.K. Urine-based liquid biopsy: Non-invasive and sensitive AR-V7 detection in urinary EVs from patients with prostate cancer. *Lab Chip* **2018**, *19*, 87–97. [CrossRef]
48. Pichler, R.; Fritz, J.; Tulchiner, G.; Klinglmair, G.; Soleiman, A.; Horninger, W.; Klocker, H.; Heidegger, I. Increased accuracy of a novel mRNA-based urine test for bladder cancer surveillance. *BJU Int.* **2018**, *121*, 29–37. [CrossRef]
49. Haese, A.; Trooskens, G.; Steyaert, S.; Hessels, D.; Brawer, M.; Vlaeminck-Guillem, V.; Ruffion, A.; Tilki, D.; Schalken, J.; Groskopf, J.; et al. Multicenter Optimization and Validation of a 2-Gene mRNA Urine Test for Detection of Clinically Significant Prostate Cancer before Initial Prostate Biopsy. *J. Urol.* **2019**, *202*, 256–263. [CrossRef]
50. Mall, C.; Rocke, D.M.; Durbin-Johnson, B.; Weiss, R.H. Stability of miRNA in human urine supports its biomarker potential. *Biomark. Med.* **2013**, *7*, 623–631. [CrossRef] [PubMed]
51. Tokumaru, Y.; Takabe, K.; Yoshida, K.; Akao, Y. Effects of MIR143 on rat sarcoma signaling networks in solid tumors: A brief overview. *Cancer Sci.* **2020**, *111*, 1076–1083. [CrossRef] [PubMed]
52. Kim, S.Y.; Kawaguchi, T.; Yan, L.; Young, J.; Qi, Q.; Takabe, K. Clinical Relevance of microRNA Expressions in Breast Cancer Validated Using the Cancer Genome Atlas (TCGA). *Ann. Surg. Oncol.* **2017**, *24*, 2943–2949. [CrossRef]
53. Kawaguchi, T.; Yan, L.; Qi, Q.; Peng, X.; Gabriel, E.M.; Young, J.; Liu, S.; Takabe, K. Overexpression of suppressive microRNAs, miR-30a and miR-200c are associated with improved survival of breast cancer patients. *Sci. Rep.* **2017**, *7*, 15945. [CrossRef]
54. Young, J.; Kawaguchi, T.; Yan, L.; Qi, Q.; Liu, S.; Takabe, K. Tamoxifen sensitivity-related microRNA-342 is a useful biomarker for breast cancer survival. *Oncotarget* **2017**, *8*, 99978–99989. [CrossRef]
55. Kawaguchi, T.; Yan, L.; Qi, Q.; Peng, X.; Edge, S.B.; Young, J.; Yao, S.; Liu, S.; Otsuji, E.; Takabe, K. Novel MicroRNA-Based Risk Score Identified by Integrated Analyses to Predict Metastasis and Poor Prognosis in Breast Cancer. *Ann. Surg Oncol.* **2018**, *25*, 4037–4046. [CrossRef]
56. Sporn, J.C.; Katsuta, E.; Yan, L.; Takabe, K. Expression of MicroRNA-9 is Associated with Overall Survival in Breast Cancer Patients. *J. Surg. Res.* **2019**, *233*, 426–435. [CrossRef]
57. Tokumaru, Y.; Katsuta, E.; Oshi, M.; Sporn, J.C.; Yan, L.; Le, L.; Matsuhashi, N.; Futamura, M.; Akao, Y.; Yoshida, K.; et al. High Expression of miR-34a Associated with Less Aggressive Cancer Biology but Not with Survival in Breast Cancer. *Int. J. Mol. Sci.* **2020**, *21*, 3045. [CrossRef] [PubMed]
58. Tokumaru, Y.; Asaoka, M.; Oshi, M.; Katsuta, E.; Yan, L.; Narayanan, S.; Sugito, N.; Matsuhashi, N.; Futamura, M.; Akao, Y.; et al. High Expression of microRNA-143 is Associated with Favorable Tumor Immune Microenvironment and Better Survival in Estrogen Receptor Positive Breast Cancer. *Int. J. Mol. Sci.* **2020**, *21*, 3213. [CrossRef]
59. Bandini, E.; Fanini, F. MicroRNAs and Androgen Receptor: Emerging Players in Breast Cancer. *Front. Genet.* **2019**, *10*, 203. [CrossRef] [PubMed]
60. Bandini, E. Urinary microRNA and mRNA in Tumors. *Methods Mol. Biol* **2021**, *2292*, 57–72. [CrossRef] [PubMed]
61. Bayraktar, R.; Bertilaccio, M.T.S.; Calin, G.A. The Interaction Between Two Worlds: MicroRNAs and Toll-Like Receptors. *Front. Immunol.* **2019**, *10*, 1053. [CrossRef]
62. Lewis, H.; Lance, R.; Troyer, D.; Beydoun, H.; Hadley, M.; Orians, J.; Benzine, T.; Madric, K.; Semmes, O.J.; Drake, R.; et al. miR-888 is an expressed prostatic secretions-derived microRNA that promotes prostate cell growth and migration. *Cell Cycle* **2014**, *13*, 227–239. [CrossRef]
63. Di Meo, A.; Batruch, I.; Brown, M.D.; Yang, C.; Finelli, A.; Jewett, M.A.S.; Diamandis, E.P.; Yousef, G.M. Identification of Prognostic Biomarkers in the Urinary Peptidome of the Small Renal Mass. *Am. J. Pathol.* **2019**, *189*, 2366–2376. [CrossRef] [PubMed]
64. Yamada, Y.; Enokida, H.; Kojima, S.; Kawakami, K.; Chiyomaru, T.; Tatarano, S.; Yoshino, H.; Kawahara, K.; Nishiyama, K.; Seki, N.; et al. MiR-96 and miR-183 detection in urine serve as potential tumor markers of urothelial carcinoma: Correlation with stage and grade, and comparison with urinary cytology. *Cancer Sci.* **2011**, *102*, 522–529. [CrossRef] [PubMed]
65. Andreu, Z.; Otta Oshiro, R.; Redruello, A.; López-Martín, S.; Gutiérrez-Vázquez, C.; Morato, E.; Marina, A.I.; Olivier Gómez, C.; Yáñez-Mó, M. Extracellular vesicles as a source for non-invasive biomarkers in bladder cancer progression. *Eur. J. Pharm. Sci.* **2017**, *98*, 70–79. [CrossRef] [PubMed]
66. Hofbauer, S.L.; de Martino, M.; Lucca, I.; Haitel, A.; Susani, M.; Shariat, S.F.; Klatte, T. A urinary microRNA (miR) signature for diagnosis of bladder cancer. *Urol. Oncol.* **2018**, *36*, 531.e531–531.e538. [CrossRef]
67. Lekchnov, E.A.; Amelina, E.V.; Bryzgunova, O.E.; Zaporozhchenko, I.A.; Konoshenko, M.Y.; Yarmoschuk, S.V.; Murashov, I.S.; Pashkovskaya, O.A.; Gorizkii, A.M.; Zheravin, A.A.; et al. Searching for the Novel Specific Predictors of Prostate Cancer in Urine: The Analysis of 84 miRNA Expression. *Int. J. Mol. Sci.* **2018**, *19*, 4088. [CrossRef]
68. Guelfi, G.; Cochetti, G.; Stefanetti, V.; Zampini, D.; Diverio, S.; Boni, A.; Mearini, E. Next Generation Sequencing of urine exfoliated cells: An approach of prostate cancer microRNAs research. *Sci. Rep.* **2018**, *8*, 7111. [CrossRef] [PubMed]
69. Kao, H.W.; Pan, C.Y.; Lai, C.H.; Wu, C.W.; Fang, W.L.; Huang, K.H.; Lin, W.C. Urine miR-21-5p as a potential non-invasive biomarker for gastric cancer. *Oncotarget* **2017**, *8*, 56389–56397. [CrossRef]
70. Iwasaki, H.; Shimura, T.; Yamada, T.; Okuda, Y.; Natsume, M.; Kitagawa, M.; Horike, S.I.; Kataoka, H. A novel urinary microRNA biomarker panel for detecting gastric cancer. *J. Gastroenterol.* **2019**, *54*, 1061–1069. [CrossRef]
71. Kapranov, P.; Cheng, J.; Dike, S.; Nix, D.A.; Duttagupta, R.; Willingham, A.T.; Stadler, P.F.; Hertel, J.; Hackermüller, J.; Hofacker, I.L.; et al. RNA maps reveal new RNA classes and a possible function for pervasive transcription. *Science* **2007**, *316*, 1484–1488. [CrossRef]

72. Wilusz, J.E.; Sunwoo, H.; Spector, D.L. Long noncoding RNAs: Functional surprises from the RNA world. *Genes Dev.* **2009**, *23*, 1494–1504. [CrossRef]
73. Wang, F.; Li, X.; Xie, X.; Zhao, L.; Chen, W. UCA1, a non-protein-coding RNA up-regulated in bladder carcinoma and embryo, influencing cell growth and promoting invasion. *FEBS Lett.* **2008**, *582*, 1919–1927. [CrossRef]
74. Shi, T.; Gao, G.; Cao, Y. Long Noncoding RNAs as Novel Biomarkers Have a Promising Future in Cancer Diagnostics. *Dis. Markers 2016*, *2016*, 9085195. [CrossRef] [PubMed]
75. Bussemakers, M.J.; van Bokhoven, A.; Verhaegh, G.W.; Smit, F.P.; Karthaus, H.F.; Schalken, J.A.; Debruyne, F.M.; Ru, N.; Isaacs, W.B. DD3: A new prostate-specific gene, highly overexpressed in prostate cancer. *Cancer Res.* **1999**, *59*, 5975–5979. [PubMed]
76. Wang, F.; Ying, H.Q.; He, B.S.; Pan, Y.Q.; Deng, Q.W.; Sun, H.L.; Chen, J.; Liu, X.; Wang, S.K. Upregulated lncRNA-UCA1 contributes to progression of hepatocellular carcinoma through inhibition of miR-216b and activation of FGFR1/ERK signaling pathway. *Oncotarget* **2015**, *6*, 7899–7917. [CrossRef] [PubMed]
77. Bian, Z.; Jin, L.; Zhang, J.; Yin, Y.; Quan, C.; Hu, Y.; Feng, Y.; Liu, H.; Fei, B.; Mao, Y.; et al. LncRNA-UCA1 enhances cell proliferation and 5-fluorouracil resistance in colorectal cancer by inhibiting miR-204-5p. *Sci Rep.* **2016**, *6*, 23892. [CrossRef]
78. Zheng, Q.; Wu, F.; Dai, W.Y.; Zheng, D.C.; Zheng, C.; Ye, H.; Zhou, B.; Chen, J.J.; Chen, P. Aberrant expression of UCA1 in gastric cancer and its clinical significance. *Clin. Transl. Oncol.* **2015**, *17*, 640–646. [CrossRef]
79. Li, J.Y.; Ma, X.; Zhang, C.B. Overexpression of long non-coding RNA UCA1 predicts a poor prognosis in patients with esophageal squamous cell carcinoma. *Int. J. Clin. Exp. Pathol.* **2014**, *7*, 7938–7944.
80. Yang, Y.; Jiang, Y.; Wan, Y.; Zhang, L.; Qiu, J.; Zhou, S.; Cheng, W. UCA1 functions as a competing endogenous RNA to suppress epithelial ovarian cancer metastasis. *Tumour Biol.* **2016**, *37*, 10633–10641. [CrossRef]
81. Li, M.; Zeringer, E.; Barta, T.; Schageman, J.; Cheng, A.; Vlassov, A.V. Analysis of the RNA content of the exosomes derived from blood serum and urine and its potential as biomarkers. *Philos. Trans. R. Soc. Lond. B Biol. Sci.* **2014**, *369*. [CrossRef]
82. Santosh, B.; Varshney, A.; Yadava, P.K. Non-coding RNAs: Biological functions and applications. *Cell Biochem. Funct.* **2015**, *33*, 14–22. [CrossRef]
83. Zeuschner, P.; Linxweiler, J.; Junker, K. Non-coding RNAs as biomarkers in liquid biopsies with a special emphasis on extracellular vesicles in urological malignancies. *Expert Rev. Mol. Diagn.* **2020**, *20*, 151–167. [CrossRef]
84. Bryzgunova, O.E.; Skvortsova, T.E.; Kolesnikova, E.V.; Starikov, A.V.; Rykova, E.Y.; Vlassov, V.V.; Laktionov, P.P. Isolation and comparative study of cell-free nucleic acids from human urine. *Ann. N. Y. Acad. Sci.* **2006**, *1075*, 334–340. [CrossRef] [PubMed]
85. de Almeida, E.F.; Abdalla, T.E.; Arrym, T.P.; de Oliveira Delgado, P.; Wroclawski, M.L.; da Costa Aguiar Alves, B.; de, S.G.F.; Azzalis, L.A.; Alves, S.; Tobias-Machado, M.; et al. Plasma and urine DNA levels are related to microscopic hematuria in patients with bladder urothelial carcinoma. *Clin. Biochem.* **2016**, *49*, 1274–1277. [CrossRef] [PubMed]
86. Lu, T.; Li, J. Clinical applications of urinary cell-free DNA in cancer: Current insights and promising future. *Am. J. Cancer Res.* **2017**, *7*, 2318–2332.
87. Woelfel, I.A.; Takabe, K. Successful intravenous catheterization by medical students. *J. Surg. Res.* **2016**, *204*, 351–360. [CrossRef]
88. Casadio, V.; Calistri, D.; Tebaldi, M.; Bravaccini, S.; Gunelli, R.; Martorana, G.; Bertaccini, A.; Serra, L.; Scarpi, E.; Amadori, D.; et al. Urine cell-free DNA integrity as a marker for early bladder cancer diagnosis: Preliminary data. *Urol. Oncol.* **2013**, *31*, 1744–1750. [CrossRef]
89. Casadio, V.; Salvi, S. Urinary Cell-Free DNA Integrity Analysis. *Methods Mol. Biol.* **2021**, *2292*, 17–22. [CrossRef] [PubMed]
90. Calistri, D.; Casadio, V.; Bravaccini, S.; Zoli, W.; Amadori, D. Urinary biomarkers of non-muscle-invasive bladder cancer: Current status and future potential. *Expert Rev. Anticancer Ther.* **2012**, *12*, 743–752. [CrossRef]
91. NCCN Guidelines for Prostate Cancer Early Detection. 2021. Available online: https://www.nccn.org/professionals/physician_gls/pdf/prostate_detection.pdf (accessed on 19 May 2021).
92. Kim, W.T.; Jeong, P.; Yan, C.; Kim, Y.H.; Lee, I.S.; Kang, H.W.; Kim, Y.J.; Lee, S.C.; Kim, S.J.; Kim, Y.T.; et al. UBE2C cell-free RNA in urine can discriminate between bladder cancer and hematuria. *Oncotarget* **2016**, *7*, 58193–58202. [CrossRef] [PubMed]
93. Qu, L.; Ding, J.; Chen, C.; Wu, Z.J.; Liu, B.; Gao, Y.; Chen, W.; Liu, F.; Sun, W.; Li, X.F.; et al. Exosome-Transmitted lncARSR Promotes Sunitinib Resistance in Renal Cancer by Acting as a Competing Endogenous RNA. *Cancer Cell* **2016**, *29*, 653–668. [CrossRef]
94. Iliev, R.; Fedorko, M.; Machackova, T.; Mlcochova, H.; Svoboda, M.; Pacik, D.; Dolezel, J.; Stanik, M.; Slaby, O. Expression Levels of PIWI-interacting RNA, piR-823, Are Deregulated in Tumor Tissue, Blood Serum and Urine of Patients with Renal Cell Carcinoma. *Anticancer Res.* **2016**, *36*, 6419–6423. [CrossRef] [PubMed]
95. Chen, X.; Chen, R.X.; Wei, W.S.; Li, Y.H.; Feng, Z.H.; Tan, L.; Chen, J.W.; Yuan, G.J.; Chen, S.L.; Guo, S.J.; et al. PRMT5 Circular RNA Promotes Metastasis of Urothelial Carcinoma of the Bladder through Sponging miR-30c to Induce Epithelial-Mesenchymal Transition. *Clin. Cancer Res.* **2018**, *24*, 6319–6330. [CrossRef] [PubMed]
96. McKiernan, J.; Donovan, M.J.; O'Neill, V.; Bentink, S.; Noerholm, M.; Belzer, S.; Skog, J.; Kattan, M.W.; Partin, A.; Andriole, G.; et al. A Novel Urine Exosome Gene Expression Assay to Predict High-grade Prostate Cancer at Initial Biopsy. *JAMA Oncol.* **2016**, *2*, 882–889. [CrossRef] [PubMed]
97. Sanguedolce, F.; Cormio, A.; Brunelli, M.; D'Amuri, A.; Carrieri, G.; Bufo, P.; Cormio, L. Urine TMPRSS2: ERG Fusion Transcript as a Biomarker for Prostate Cancer: Literature Review. *Clin. Genitourin. Cancer* **2016**, *14*, 117–121. [CrossRef]

98. Groskopf, J.; Aubin, S.M.; Deras, I.L.; Blase, A.; Bodrug, S.; Clark, C.; Brentano, S.; Mathis, J.; Pham, J.; Meyer, T.; et al. APTIMA PCA3 molecular urine test: Development of a method to aid in the diagnosis of prostate cancer. *Clin. Chem.* **2006**, *52*, 1089–1095. [CrossRef]
99. Whitman, E.J.; Groskopf, J.; Ali, A.; Chen, Y.; Blase, A.; Furusato, B.; Petrovics, G.; Ibrahim, M.; Elsamanoudi, S.; Cullen, J.; et al. PCA3 score before radical prostatectomy predicts extracapsular extension and tumor volume. *J. Urol.* **2008**, *180*, 1975–1978; discussion 1978–1979. [CrossRef]
100. Srivastava, A.K.; Singh, P.K.; Rath, S.K.; Dalela, D.; Goel, M.M.; Bhatt, M.L. Appraisal of diagnostic ability of UCA1 as a biomarker of carcinoma of the urinary bladder. *Tumour. Biol.* **2014**, *35*, 11435–11442. [CrossRef] [PubMed]
101. Wang, Z.; Wang, X.; Zhang, D.; Yu, Y.; Cai, L.; Zhang, C. Long non-coding RNA urothelial carcinoma-associated 1 as a tumor biomarker for the diagnosis of urinary bladder cancer. *Tumour Biol.* **2017**, *39*, 1010428317709990. [CrossRef]
102. Jin, P.C.; Gou, B.; Qian, W. Urinary markers in treatment monitoring of lung cancer patients with bone metastasis. *Int. J. Biol. Markers* **2019**, *34*, 243–250. [CrossRef] [PubMed]
103. Wang, X.; Meng, Q.; Wang, C.; Li, F.; Zhu, Z.; Liu, S.; Shi, Y.; Huang, J.; Chen, S.; Li, C. Investigation of transrenal KRAS mutation in late stage NSCLC patients correlates with disease progression. *Biomarkers* **2017**, *22*, 654–660. [CrossRef] [PubMed]
104. Husain, H.; Melnikova, V.O.; Kosco, K.; Woodward, B.; More, S.; Pingle, S.C.; Weihe, E.; Park, B.H.; Tewari, M.; Erlander, M.G.; et al. Monitoring Daily Dynamics of Early Tumor Response to Targeted Therapy by Detecting Circulating Tumor DNA in Urine. *Clin. Cancer Res.* **2017**, *23*, 4716–4723. [CrossRef] [PubMed]
105. Wu, Z.; Yang, Z.; Li, C.S.; Zhao, W.; Liang, Z.X.; Dai, Y.; Zhu, Q.; Miao, K.L.; Cui, D.H.; Chen, L.A. Differences in the genomic profiles of cell-free DNA between plasma, sputum, urine, and tumor tissue in advanced NSCLC. *Cancer Med.* **2019**, *8*, 910–919. [CrossRef] [PubMed]
106. Liu, B.; Ricarte Filho, J.; Mallisetty, A.; Villani, C.; Kottorou, A.; Rodgers, K.; Chen, C.; Ito, T.; Holmes, K.; Gastala, N.; et al. Detection of Promoter DNA Methylation in Urine and Plasma Aids the Detection of Non-Small Cell Lung Cancer. *Clin. Cancer Res.* **2020**, *26*, 4339–4348. [CrossRef] [PubMed]
107. Zhang, J.; Zhang, X.; Shen, S. Treatment and relapse in breast cancer show significant correlations to noninvasive testing using urinary and plasma DNA. *Future Oncol.* **2020**, *16*, 849–858. [CrossRef]
108. Ritter, A.; Hirschfeld, M.; Berner, K.; Jaeger, M.; Grundner-Culemann, F.; Schlosser, P.; Asberger, J.; Weiss, D.; Noethling, C.; Mayer, S.; et al. Discovery of potential serum and urine-based microRNA as minimally-invasive biomarkers for breast and gynecological cancer. *Cancer Biomark.* **2020**, *27*, 225–242. [CrossRef] [PubMed]
109. Sands, J.; Li, Q.; Hornberger, J. Urine circulating-tumor DNA (ctDNA) detection of acquired EGFR T790M mutation in non-small-cell lung cancer: An outcomes and total cost-of-care analysis. *Lung Cancer* **2017**, *110*, 19–25. [CrossRef] [PubMed]
110. Srivastava, A.; Moxley, K.; Ruskin, R.; Dhanasekaran, D.N.; Zhao, Y.D.; Ramesh, R. A Non-invasive Liquid Biopsy Screening of Urine-Derived Exosomes for miRNAs as Biomarkers in Endometrial Cancer Patients. *AAPS J.* **2018**, *20*, 82. [CrossRef]
111. Zhang, L.; Cao, X.; Zhang, L.; Zhang, X.; Sheng, H.; Tao, K. UCA1 overexpression predicts clinical outcome of patients with ovarian cancer receiving adjuvant chemotherapy. *Cancer Chemother. Pharmacol.* **2016**, *77*, 629–634. [CrossRef]
112. Hung, P.S.; Chen, C.Y.; Chen, W.T.; Kuo, C.Y.; Fang, W.L.; Huang, K.H.; Chiu, P.C.; Lo, S.S. miR-376c promotes carcinogenesis and serves as a plasma marker for gastric carcinoma. *PLoS ONE* **2017**, *12*, e0177346. [CrossRef] [PubMed]
113. Botezatu, I.; Serdyuk, O.; Potapova, G.; Shelepov, V.; Alechina, R.; Molyaka, Y.; Ananév, V.; Bazin, I.; Garin, A.; Narimanov, M.; et al. Genetic analysis of DNA excreted in urine: A new approach for detecting specific genomic DNA sequences from cells dying in an organism. *Clin. Chem* **2000**, *46*, 1078–1084. [CrossRef] [PubMed]
114. Maltoni, R.; Casadio, V.; Ravaioli, S.; Foca, F.; Tumedei, M.M.; Salvi, S.; Martignano, F.; Calistri, D.; Rocca, A.; Schirone, A.; et al. Cell-free DNA detected by "liquid biopsy" as a potential prognostic biomarker in early breast cancer. *Oncotarget* **2017**, *8*, 16642–16649. [CrossRef]
115. Yuan, T.; Huang, X.; Woodcock, M.; Du, M.; Dittmar, R.; Wang, Y.; Tsai, S.; Kohli, M.; Boardman, L.; Patel, T.; et al. Plasma extracellular RNA profiles in healthy and cancer patients. *Sci. Rep.* **2016**, *6*, 19413. [CrossRef]
116. Zendjabil, M.; Favard, S.; Tse, C.; Abbou, O.; Hainque, B. The microRNAs as biomarkers: What prospects? *C. R. Biol.* **2017**, *340*, 114–131. [CrossRef]
117. Malapelle, U.; Pepe, F.; Pisapia, P.; Sgariglia, R.; Nacchio, M.; Barberis, M.; Bilh, M.; Bubendorf, L.; Büttner, R.; Cabibi, D.; et al. TargetPlex FFPE-Direct DNA Library Preparation Kit for SiRe NGS panel: An international performance evaluation study. *J. Clin. Pathol.* **2021**. [CrossRef]

Review

Cancer Epigenetic Biomarkers in Liquid Biopsy for High Incidence Malignancies

Cora Palanca-Ballester [1,†], Aitor Rodriguez-Casanova [2,3,†], Susana Torres [4,5,6,†], Silvia Calabuig-Fariñas [4,5,6,7], Francisco Exposito [4,8,9], Diego Serrano [8,9], Esther Redin [4,8,9], Karmele Valencia [4,8,10], Eloisa Jantus-Lewintre [4,5,6,11], Angel Diaz-Lagares [2,4], Luis Montuenga [4,8,9], Juan Sandoval [1,*,‡] and Alfonso Calvo [4,8,9,*,‡]

1. Biomarkers and Precision Medicine (UBMP) and Epigenomics Unit, IIS, La Fe, 46026 Valencia, Spain; cora_palanca@iislafe.es
2. Cancer Epigenomics, Translational Medical Oncology (Oncomet), Health Research Institute of Santiago (IDIS), University Clinical Hospital of Santiago (CHUS/SERGAS), 15706 Santiago de Compostela, Spain; aitorrodriguez@me.com (A.R.-C.); angel.diaz.lagares@sergas.es (A.D.-L.)
3. Roche-CHUS Joint Unit, Translational Medical Oncology Group (Oncomet), Health Research Institute of Santiago (IDIS), 15706 Santiago de Compostela, Spain
4. CIBERONC, ISCIII, 28029 Madrid, Spain; susana.torres@alu.umh.es (S.T.); calabuix_sil@gva.es (S.C.-F.); fexposito@unav.es (F.E.); eredin@alumni.unav.es (E.R.); kvalencia@external.unav.es (K.V.); jantus_elo@gva.es (E.J.-L.); lmontuenga@unav.es (L.M.)
5. Molecular Oncology Laboratory, Fundación Hospital General Universitario de Valencia, 46014 Valencia, Spain
6. TRIAL Mixed Unit, Centro de Investigación Príncipe Felipe-Fundación para la Investigación del Hospital General Universitario de Valencia, 46014 Valencia, Spain
7. Department of Pathology, Universitat de València, 46010 Valencia, Spain
8. DISNA and Program in Solid Tumors, Center for Applied Medical Research (CIMA), 31008 Pamplona, Spain; dserrano@unav.es
9. Department of Pathology, Anatomy and Physiology, School of Medicine, University of Navarra, 31008 Pamplona, Spain
10. Department of Biochemistry and Genetics, School of Sciences, University of Navarra, 31008 Pamplona, Spain
11. Department of Biotechnology, Universitat Politècnica de València, 46022 Valencia, Spain
* Correspondence: epigenomica@iislafe.es (J.S.); acalvo@unav.es (A.C.)
† These authors contributed equally and should be considered as first authors.
‡ Co-senior authors.

Simple Summary: Apart from genetic changes, cancer is characterized by epigenetic alterations, which indicate modifications in the DNA (such as DNA methylation) and histones (such as methylation and acetylation), as well as gene expression regulation by non-coding (nc)RNAs. These changes can be used in biological fluids (liquid biopsies) for diagnosis, prognosis and prediction of cancer drug response. Although these alterations are not widely used as biomarkers in the clinical practice yet, increasing number of commercial kits and clinical trials are expected to prove that epigenetic changes are able to offer valuable information for cancer patients.

Abstract: Early alterations in cancer include the deregulation of epigenetic events such as changes in DNA methylation and abnormal levels of non-coding (nc)RNAs. Although these changes can be identified in tumors, alternative sources of samples may offer advantages over tissue biopsies. Because tumors shed DNA, RNA, and proteins, biological fluids containing these molecules can accurately reflect alterations found in cancer cells, not only coming from the primary tumor, but also from metastasis and from the tumor microenvironment (TME). Depending on the type of cancer, biological fluids encompass blood, urine, cerebrospinal fluid, and saliva, among others. Such samples are named with the general term "liquid biopsy" (LB). With the advent of ultrasensitive technologies during the last decade, the identification of actionable genetic alterations (i.e., mutations) in LB is a common practice to decide whether or not targeted therapy should be applied. Likewise, the analysis of global or specific epigenetic alterations may also be important as biomarkers for diagnosis, prognosis, and even for cancer drug response. Several commercial kits that assess the DNA promoter methylation of single genes or gene sets are available, with some of them being tested as biomarkers for diagnosis in clinical trials. From the tumors with highest incidence, we can stress the relevance of DNA methylation changes in the following genes found in LB: *SHOX2* (for lung cancer); *RASSF1A*, *RARB2*, and *GSTP1* (for lung, breast, genitourinary and colon cancers); and *SEPT9* (for colon cancer). Moreover, multi-cancer high-throughput methylation-based tests are now commercially available. Increased levels of the microRNA miR21 and several miRNA- and long ncRNA-signatures can also be indicative biomarkers in LB. Therefore, epigenetic biomarkers are attractive and may have a clinical value in cancer. Nonetheless, validation, standardization, and demonstration of an added value over the common clinical practice are issues needed to be addressed in the transfer of this knowledge from "bench to bedside".

Keywords: epigenetic biomarkers; cancer; DNA methylation; micro-RNAs

1. Introduction

Aberrant epigenetic changes are recognized as one of the key events leading to carcinogenesis [1]. Cancer cells harbor global epigenetic abnormalities in addition to genetic alterations. The use of "omic" techniques in recent years has allowed us to get a comprehensive view of the extensive reprograming that occurs in the epigenetic machinery of cancer cells. These epigenetic changes include DNA methylation, histone modifications, nucleosome positioning, and de-regulation of non-coding RNAs, mainly micro-RNAs (miRNAs) [2].

The most widely studied epigenetic modification and the one closer to be transferred to the clinic as a cancer biomarker is DNA methylation. This modification is the result of the addition of a methyl group at the 5'-carbon of the pyrimidine ring of a cytosine followed by a guanine (CpG), which impedes gene transcription. Cancer is characterized by global DNA hypomethylation and focal hypermethylation of certain genes such as tumor suppressor genes [3] or miRNAs, whose silencing promotes tumor growth [4]. Hypomethylation takes place mainly in repetitive regions of the genome and has been shown to facilitate genomic instability and DNA damage [5].

Both genetic and epigenetic alterations identified in cancer can be used as biomarkers for diagnosis, prognosis, and prediction of drug response. Although assessment of biomarkers in tumor specimens may offer direct information about genetic and epigenetic alterations, the amount of tissue obtained from advanced tumors is often insufficient and may not reflect the whole tumor heterogeneity. To overcome these inconveniences, an alternative option to tissue samples has emerged in the last years, known as liquid biopsy (LB). LB is an non-invasive method that allows for the analysis of different biomarkers in fluids such as blood, saliva, bronchoalveolar lavage (BAL), cerebrospinal fluid (CSF), urine, or other body fluids [6]. These samples are easily obtained and may pick up DNA/RNA/proteins coming from both the primary tumor and the different metastatic sites, representing tumor heterogeneity and clonal evolution. In LB, we can find circulating tumor DNA (ctDNA), circulating tumor RNA (ctRNA), circulating tumor cells (CTCs), and extracellular vesicles (EVs) that may contain RNA, proteins, and DNA. Figure 1 graphically depicts the possible contribution of epigenetic biomarkers (free or vesicle-enclosed methylated DNA and ncRNAs) isolated in LB, in conjunction with clinical data, for patient's stratification, prognosis, and prediction of response to therapy [6].

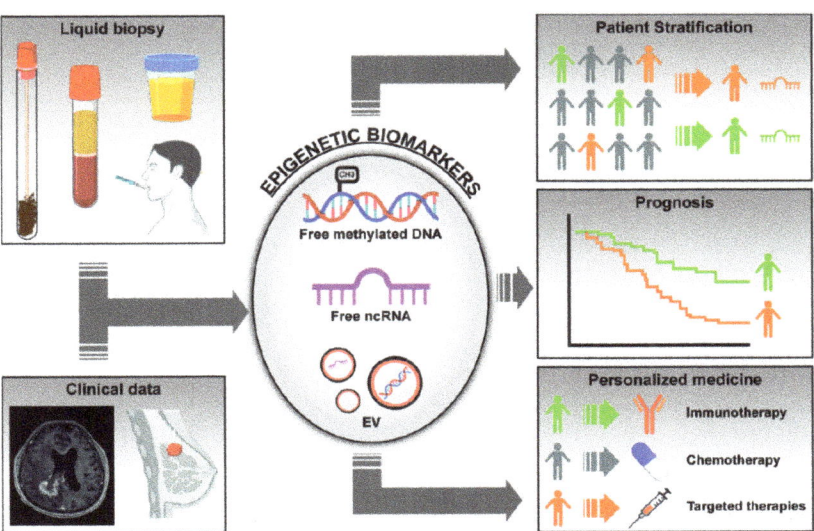

Figure 1. Scheme representing the utility of epigenetic changes found in liquid biopsies as biomarkers for cancer diagnosis, patient's stratification, prognosis, and response to treatments. Changes in DNA methylation of gene promoters and abnormal levels of non-coding RNAs (ncRNAs) can be found in fluids as free molecules or inside extracellular vesicles (EVs). Integration of these biological markers with clinical and radiological data may help in the management of cancer patients, in particular in the field of screening and diagnosis.

The clinical value of identifying actionable genetic mutations in LB (mainly blood) to treat patients with targeted therapy has been widely proven. However, regarding epigenetic changes, translation of these potential biomarkers into the clinic still lags far behind the genetic biomarkers. Although with some exceptions, rather than prediction of response to drugs, epigenetic biomarkers could be particularly useful as diagnostic and prognostic indicators, with numerous commercially available tests already developed to detect changes in DNA methylation levels [7]. The performance of diagnostic test is commonly evaluated in terms of sensitivity, specificity, and the area under the ROC curve (AUC). Sensitivity is defined as the percentage of positive cases that is correctly identified and specificity as the percentage of negative cases that is correctly identified. The AUC, which takes into account both sensitivity and specificity, defines diagnostic accuracy and is optimal when values are closer to 1. The fact that epigenetic changes are found early in carcinogenesis and that DNA methylation is stable in ctDNA, makes this epigenetic modification an excellent potential cancer diagnostic biomarker in LB.

The term "epigenetic", considered as any change in gene expression that does not permanently affect the DNA, may also include gene expression regulation by non-coding (nc)RNAs and histone modifications [8]. The aberrantly expressed ncRNAs may be promising therapeutic targets as well as cancer biomarkers. ncRNAs are the principal regulators of key molecular and cellular processes such as RNA splicing, gene regulation, proliferation, and apoptosis. ncRNAs can be classified into two groups based on their length and their roles: housekeeping ncRNAs and regulatory ncRNAs, which in turn include small ncRNAs and long ncRNAs (Figure 2). Circulating RNA species can be found free in fluids or inside EVs, where they are protected from degradation. EVs can be classified into three main types according to their size and biogenesis: exosomes, microvesicles, and apoptotic bodies [9]. Approximately 70% of studies have assessed exosomes as the source of choice for ncRNA when evaluating biomarkers [9]. Yuan et al. analyzed the RNA content in exosomes and estimated that mature miRNAs spanned 40.4%, piwi-interacting RNAs 40%, pseudo-genes 3.7%, lncRNAs 2.4%, tRNAs 2.1%, and mRNAs 2.1% of the total RNA [10]. From the different species of ncRNAs, microRNAs stand out as potential epigenetic markers in fluids, although implementation in the clinic encounters several difficulties such as RNA instability and variability of the methodologies used [11]. In general, RNA is less stable than DNA and proteins, and in particular, some regulatory lncRNAs show short half-lives [12]. Besides, there are many factors that may influence RNA stability in body fluids such as hemolysis in plasma/serum samples, which are a major cause of variation in miRNA levels [13]. Another possible pitfall when analyzing tumor miRNAs in liquid biopsy is their unknown cellular origin and the masking effect from ncRNAs released by non-tumor cells. Several studies postulate that the material contained in exosomes derived from the tumor microenvironment (TME) can also contribute to the characterization of the tumor, and as a consequence, TME-derived exosomes could be a good source for biomarkers [14].

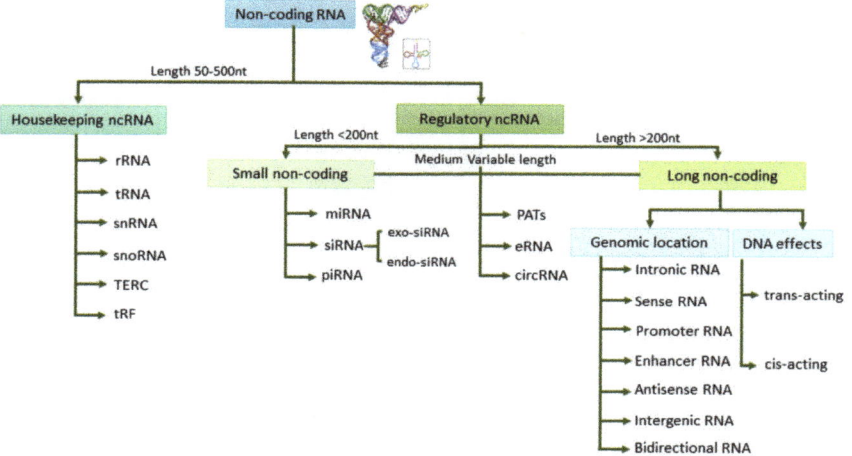

Figure 2. Non-coding (nc)RNAs classification into different groups based on their length and their regulatory roles. Small non coding RNA (sncRNA), long non coding RNA (lncRNA), ribosomal RNA (rRNA), transfer RNA (tRNA), small nuclear (snRNA), small nucleolar (snoRNA), telomerase RNA component (TERC), tRNA-Derived Fragments (tRF) and tRNA halves (tiRNA), microRNA (miRNA), small interfering RNA (siRNA), piwi-interacting RNA (piRNA), promoter-associated transcripts (PATs), enhancer RNA (eRNA), circular RNA (circRNA), and long non-coding RNA (lncRNA).

Changes in histone modification have also been identified in circulation in cancer patients and are another source of epigenetic biomarkers. Nonetheless, due to the complexity of modifications, we will not cover it in our review. Information on this issue has been comprehensively reviewed in a recent study [15].

In this review, we address the most relevant evidence (according to authors' criteria) on epigenetic biomarkers in LB, with special emphasis on tumors with high incidence. We summarize the data about biomarkers currently registered on the market as well as novel emerging candidates.

2. Types of Biological Fluids for Epigenetic Analysis

In cancer patients, ctDNA can harbor the same mutational and epigenetic traits as the corresponding tumor [16]. A common and convenient source of LB is blood, but certain tumors are characterized by shedding low amounts of DNA into the blood (i.e., brain, kidney, bladder, prostate, thyroid, or head and neck cancers). In these cases, since their tumor location allows a direct communication with other body fluids, it could be more informative to use alternative samples for the analysis of biomarkers [17].

In head and neck cancer, saliva is an attractive non-invasive sample for screening, diagnosis, and monitoring due to its simple collection and low cost. Salivary nucleic acids have been used, for example, for the identification and validation of DNA methylation and miRNAs, demonstrating their utility in several clinical contexts (recently reviewed by [18]). Airway-derived fluids such as bronchial aspirates/lavages and sputum samples, have proven to be accurate tools for the early detection of tumors arising in the respiratory system [7]. Pleural effusion is also a very informative biological sample for biomarker assessment in lung cancer (LuCa) patients. It is well established that the *EGFR* mutational status can be reliably determined in ctDNA from pleural effusions to predict response to EGFR-tyrosine kinase inhibitors (TKIs) [19]. In contrast, the information on epigenetic biomarkers in pleural fluid is more limited and the clinical value of such biomarkers has to be clearly determined, but recent reports are finding possible association with prognosis [20].

Urine is a bona-fide source of epigenetic biomarkers in the case of genitourinary cancers. DNA hypermethylation has been described as one of the earliest and most frequent aberrations in prostate cancer (PrCa), and the detection of methylation patterns in urinary ctDNA has shown to be clinically meaningful [21]. In bladder cancer (BdCa), promising results from urine-based tests that measure DNA methylation patterns have been described [22]. The potential of miRNAs as biomarkers (individually or in combination) has also been demonstrated for BdCa, showing high sensitivity and specificity [23].

In the case of central nervous system (CNS) tumors and due to the existence of the blood–brain barrier, ctDNA in CSF seems to be a better source than blood [24]. A high concordance between methylation patterns in CSF and matched tumor samples has been reported, indicating the potential use of this biofluid for epigenetic biomarker analysis [22,25].

3. Technologies for Epigenetic Assays in Liquid Biopsy

Several techniques have been described for the analysis of epigenetic alterations in CTCs, free circulating nucleic acids and exosomes [7,26,27]. According to the number of targets analyzed, these technologies can be divided into (a) single-locus or multiplexed assays; and (b) genome-wide approaches, which are mainly based on microarrays and next-generation sequencing (NGS).

3.1. DNA Methylation

The analyses of DNA methylation in ctDNA by single-locus or multiplexed assays are mainly amplification-based methods such as methylation specific PCR (MSP) or digital droplet PCR (ddPCR). MSP is a classical method that encompasses Methylight and MethylQuant assays. MSP detects a small amount of ctDNA among a considerable number of circulating free DNA (cfDNA) [23]. In Methylight assays, the DNA methylation level is analyzed by comparing the fluorescence of specific probes between methylated and unmethylated molecules [28,29]. ddPCR is an ultrasensitive and quantitative method that is useful for the discovery of clinical biomarkers in samples with a low amount of cfDNA. This method is based on a PCR that is conducted in water-oil emulsion droplets where a single DNA molecule can be amplified inside each droplet, thus avoiding the mask effect of the non-target DNA [26,28]. Moreover, there are other approaches that combine techniques used for ctDNA studies such as BEAming technology, epityper epigenetic analysis, and methylation sensitive high-resolution technology (MS-HRT). BEAming technology is a method that combines ddPCR and flow cytometry for the analysis of ctDNA [30,31]. Epityper epigenetic analysis combines specific enzymatic cleavage with mass spectrometry (MALDI-TOF-MS) [32]. MS-HRT compares the melting profiles of sequences that present differences in their base compositions [7,26].

Regarding genome-wide assays, the methylation analysis of ctDNA can be performed by different techniques such as Infinium DNA methylation EPIC array (EPIC). EPIC is considered the gold standard method for DNA methylation assays due to its cost-effectiveness and its ability to examine more than 850,000 CpG sites [33].

3.2. Non-Coding RNAs

The expression of ncRNAs can be evaluated in LB at targeted-specific level using amplification-based methods such as reverse transcription quantitative PCR (RT-qPCR) and ddPCR. Of note, the use of ddPCR represents a highly sensitive method to quantify the expression of specific transcripts in LB [34–36]. In addition, there are other targeted methods such as peptide nucleic acids (PNAs)-based fluorogenic biosensors [37] and the NanoString nCounter platform that allow for the detection of ncRNA expression levels without the need of previous amplification [38]. In particular, the NanoString nCounter platform is able to analyze a large panel of miRNAs in several types of biological fluids including plasma and urine [39]. Other targeted approaches have also been developed for the detection of miRNAs in CTCs such as in situ hybridization (ISH) with locked-nucleic-acid (LNA) probes [40] and methods based on signal amplification in microfluidic droplets for single-cell analysis of multiple miRNAs [41].

The expression of ncRNAs can also be detected in LB at transcriptomic level with the use of NGS (RNA-Seq) or microarrays. Although both technologies allow for the analysis of ncRNA transcripts in LB [42,43], unlike microarrays, RNA-Seq does not require prior knowledge of the target transcripts and shows higher sensitivity for the detection of ncRNAs [42,43]. A summary of the types of technologies, applications, and advantages/disadvantages can be found in Table S1.

4. Epigenetic Biomarkers in Lung Cancer

Lung cancer (LuCa) is currently the second most commonly diagnosed cancer and the leading cause of cancer-related deaths worldwide among women and men. Globally, there were an estimated 2.2 million lung cancer cases and 1.8 million deaths in 2018, accounting for approximately a third of all cancer cases and deaths [44]. LuCa is one of the most aggressive tumor types, with a 5-year survival rate that varies globally but remains consistently low, not exceeding 19% [45]. There are two main types of LuCa: non-small cell lung cancer (NSCLC, ~85% cases) and small cell lung cancer (SCLC, ~15% cases). NSCLC is subdivided in three main histological subtypes: adenocarcinoma (LUAD) (~40% of NSCLC cases), squamous cell carcinoma (LUSC) (~30% of NSCLC cases), and large-cell carcinoma (~10–15% of NSCLC cases).

Despite breakthroughs in LuCa treatments in the last few decades, which have gradually improved patient's outcome, the mortality rate is still considerably behind that observed for other prevalent types such as breast or colon cancer. A major factor is the late diagnosis and, consequently, its late-stage presentation. In recent years, increased interest has been directed toward the use of imaging techniques and biomarkers for screening and early detection. In randomized trials, the use of low-dose computed tomography (LDCT) in populations at risk has shown a significant reduction in lung cancer mortality [46]. However, there are still several questions regarding LDCT, which require further research. Examples are clarification of cost-effectiveness in different populations; characterization of detected nodules with indeterminate risk level; the small but significant percentage of false-positive cases and the potential harms associated with unnecessary invasive interventions (biopsies or even surgeries) of these cases; and the potential tools to optimize risk assessment, to recommend for screening only those individuals with higher risk, not only based on age and smoking exposure [47]. It is then possible that the use of LB-based molecular biomarkers in screening programs might help LDCTs in identifying NSCLC. Biological fluids that can be used as a source for biomarkers in LuCa include blood, bronchial aspirates (BAS), bronchial lavages (BAL), sputum or pleural effusions, for analysis of ctDNA, exosomes, and CTCs.

4.1. DNA Methylation

Among DNA methylation biomarkers in LuCa diagnosis, *SHOX2* hypermethylation is clearly the best studied epigenetic alteration. *SHOX2* hypermethylation was first described by Schmidt et al. using bronchial fluid aspirates during bronchoscopy, showing 68%

sensitivity and 95% specificity [48]. Other studies [49–51] have later validated the diagnostic potential of *SHOX2* methylation status in plasma and pleural effusions. The EpiProLung® assay is the only commercial test specifically designed for LuCa diagnosis. This test is based on a PCR assay that analyzes methylation of *SHOX2* and *PTGER4* in blood, with a sensitivity of 78% and specificity of 96%, and an area under the ROC curve (AUC) of 0.73. This gene combination has also been tested in lavage fluid samples.

Other genes have been found to be differentially methylated in plasma samples when comparing LuCa patients and healthy controls including *DCLK1* (49% sensitivity and 91% specificity) [52], *SEPT9* (44% sensitivity and 92% specificity) [53], *RASSF1A* and *RARB2* (87% sensitivity and 75% specificity) [54]. Hulbert et al. demonstrated that analyzing the DNA methylation status of different genes such as *TAC1* (86% sensitivity and 75% specificity), *HOXA7* (63% sensitivity and 92% specificity) and *SOX17* (84% sensitivity and 88% specificity) allowed for the detection of LuCa in sputum samples with a global sensitivity of 98% and specificity of 71% [55,56]. Interestingly this group has also published that methylation analysis of *CDO1*, *TAC1*, *HOXA9*, and *SOX17* in urine (as well as in plasma) can be useful as an adjunct to LDCT screening [57]. Recently, our group has developed an epigenetic model identified through epigenomic analysis by which the DNA methylation status of four genes (*BCAT1*, *CDO1*, *TRIM58*, and *ZNF177*) in BAS/BAL/sputum samples was able to discriminate between NSCLC patients (even at early stages) and controls (82% sensitivity and 76% specificity, AUC, ~0.9) [55,58]. We have also described that *TMPRSS4* hypomethylation can be used as a diagnostic tool in early stages, with an AUC of 0.72 (52% sensitivity and 91% specificity) for BAL and 0.73 (90% sensitivity and 65% specificity) for plasma [59].

Through genome-wide DNA methylation assays, Hsu et al. detected a multiple epigenetic panel in tumor samples by studying the methylation status of genes *CDH13*, *BLU*, *FHIT*, *RASSF1A*, and *RARB*, whose diagnostic potential was also validated in plasma samples with a sensitivity of 73% and a specificity of 82% [60]. Similarly, Ostrow et al. validated in plasma a group of four genes (*DCC*, *Kif1a*, *NISCH*, *RARB*) that was previously found in tumors, which discriminated between LuCa patients and tumor-free individuals, with a sensitivity of 73% and specificity of 71% [61]. In addition, Ooki et al. described a serum-based gene signature, previously identified in tumors from TCGA (*MARCH11*, *HOXA9*, *CDO1*, *UNCX*, *PTGDR*, and *AJAP1*) that was able to differentiate stage I NSCLC patients from the controls, with 72.1% sensitivity and 71.4% specificity [62].

Unlike for diagnosis, only a few studies have described the association between DNA methylation status and outcome or response to drugs using ctDNA [63,64]. Hypermethylation of *SHP1P2* in plasma was associated with reduced progression-free survival (PFS) in advanced NSCLC [65]. DNA methylation of the gene panel *SOX17*, *BMRS1*, and *DCLK1* in plasma had a negative impact on survival [52,66,67], whereas *SFN* methylation in serum samples was associated with improved survival [68]. Salazar et al. described that patients with unmethylated *CHFR* had an improved survival when treated with second-line EGFR TKIs [69]. Additionally, increased plasma ctDNA methylation levels of *RASSF1A* and *APC* within 24 h after chemotherapy administration was found to be associated with good response to cisplatin [70]. In addition, using plasma samples, prolonged survival has been observed in patients with low *SHOX2* promoter methylation after chemotherapy/radiotherapy [48].

4.2. ncRNAs

ncRNAs are also becoming a valuable tool for the early detection of LuCa. miRNAs, the most widely studied type of ncRNA, provide promising biomarkers for the diagnosis and prognosis of LuCa [71]. For instance, miR-1285 was significantly decreased while miR-324-3p was significantly increased in plasma of stage I LUSC patients in contrast to healthy donors (AUC 0.85 and 0.79, respectively) [72]. Chen et al. described 10 miRNAs (miR-20a, miR-222, miR-221, miR-320, miR-152, miR-145, miR-223, miR-199a-5p, miR-24, miR-25) able to discriminate NSCLC patients from healthy controls, with high sensitivity (92.5%)

and specificity (90%) rates (AUC 0.97) [73]. miR-21, the most commonly studied miRNA in LuCa, has been found consistently upregulated in both serum and plasma samples and may serve as a diagnostic biomarker of early-stage NSCLC. Yu et al. reported that miR-21 was suitable for diagnosis, with 69% sensitivity and 71.9% specificity [74]. A multicenter study was performed with a total of 3102 participants to investigate the potential use of circulating miRNAs as diagnostic biomarkers in LuCa. Results reported that a 14-miRNA signature might be useful to discriminate patients with early-stage lung cancer (stage I or II) from healthy individuals. Specifically, miR-374b-5p differentiated patients with early-stage LuCa from those without cancer, with an AUC of 0.83 [75]. Some groups have studied miRNA precursors as diagnostic biomarkers in LuCa. Powrózek et al. reported that miRNA-944 precursors distinguished SCC from ADC with 78.6% sensitivity and 91.7% specificity (AUC = 0.77), and pri-miRNA-3662 discriminated SCC from ADC with 57.1% sensitivity and 90% specificity (AUC = 0.845). Both markers allowed to distinguish stage I-IIIA NSCLC from healthy individuals with 75.7% sensitivity and 82.3% specificity (AUC = 0.898) [76].

Some ncRNAs have also been proposed as prognostic biomarkers for LuCa. A recent study using a cohort of 182 patients with resected early-stage NSCLC reported that, among 84 circulating microRNAs, only miR-126-3p had an independent prognostic value in SCC patients [77]. Moreover, Yanaihara et al. showed that high expression of precursor has-mir-155 could be an independent poor prognosis biomarker in ADC patients [78]. Increasing evidence shows that lncRNA can also act as biomarkers for prognosis. Xie et al. reported in a cohort of 460 patients that low serum levels of SOX2OT and ANRIL were associated with higher overall survival (OS) rate. Multivariate analysis revealed that SOX2OT could be an independent prognostic factor for NSCLC [79]. Yung-Hung Luo et al. studied the correlation between clinicopathological characteristics and circRNAs using plasma from a cohort of 231 LuCa patients (65 had stage I–II and 166 stage III–IV) and 41 healthy donors. They reported that higher levels of circ_0000190 were correlated with larger primary tumor size, advanced stage, extrathoracic metastasis, and poor survival [80].

miR-21 has been identified as a key miRNA in the regulation of acquired resistance to EGFR-TKIs in NSCLC, and high serum levels of this miRNA have been found to be significantly increased at the time of EGFR-TKI progression when compared to those observed before treatment [81]. Wang et al. also demonstrated that patients who were resistant to EGFR-TKIs had higher levels of circulating miR-21, miR-27a, and miR-218 than patients who were sensitive [82]. Jinshuo Fan et al. found that NSCLC patients who were responsive to ICIs (immune checkpoint inhibitors) had increased levels of a signature composed of miR-27a, miR-28, miR-34a, miR-93, miR-106b, miR-138-5p, miR-181a, miR-193a-3p, miR-200, and miR-424 compared to non-responders. Moreover, patients with high levels of this signature showed improved PFS and OS than those where levels were low [83]. Recently, expression of circ_0000190 has been found to be correlated with PD-L1 expression and response to immunotherapy in NSCLC [80].

5. Epigenetic Biomarkers in Genitourinary Cancers

The most prevalent tumors of the genitourinary (GU) tract are prostate cancer (PrCa), bladder cancer (BdCa), and renal cell carcinoma (RCC) [84]. PrCa is the second most commonly diagnosed cancer and the sixth leading cause of cancer death among men worldwide [85]. PrCa diagnosis has not evolved significantly since the 1980s, when blood levels of prostate-specific antigen (PSA) were first introduced as a follow-up marker for recurrent tumors, and, subsequently, for early detection in combination with digital rectal examination (DRE) [77,86]. PSA for PrCa screening has low positive predictive value (~30%), potentially driving to over-diagnosis and over-treatment. This highlights the need of more accurate biomarkers that are alternative or complementary to PSA for screening and diagnosis [87]. In the case of BdCa, there was an estimated number of 550,000 cases and 200,000 deaths (2.1% of all cancer deaths) in 2018 [88]. The 5-year survival rates (~77%) have remained mostly unchanged since the 90s [89]. Although less frequent, RCC accounts for ~2% of all diagnosed cancers, but its incidence has more than doubled over the past fifty years, being the tenth most common neoplasm in men [84,90]. RCC diagnosis is mostly an accidental finding and represents 1.8% of all cancer deaths worldwide [90]. The therapeutic options for RCC have increased tremendously in recent years, but biomarkers of response to these drugs are still lacking.

5.1. DNA Methylation

A common trait of GU tumors is the possibility of using urine as a biological fluid for the analysis of CTCs and ctDNA [7]. Commercial epigenetic-based kits for the detection of PrCa and Bdca in both urine and blood samples are currently available [7]. Unfortunately, epigenetic markers in liquid biopsies from RCC patients are underdeveloped, as this is one of the tumor types with less ctDNA shedding into biological fluids [91–93]. Among DNA methylation biomarkers in GU cancers, *GSTP1* hypermethylation is by far the most frequently described epigenetic alteration, especially in PrCa patients [94]. Many authors have described its utility for PrCa detection [95,96] showing much higher specificity (~90%) than PSA (~30%), although sensitivity was similar for both PSA levels and *GSTP1* methylation [87,95]. Matched assessment of ctDNA *GSTP1* in urine and plasma samples revealed that urinary analysis outperforms plasma for diagnostic purposes [96,97]. It is worth mentioning that the DNA methylation analysis of multi-gene panels in serum including *GSTP1*, *RASSF1A*, and *RARB* have increased the diagnostic coverage of *GSTP1* alone [98] for PrCa. Similar strategies have also been used for BdCa detection (with 100% sensitivity) in a multi-gene panel that assessed *CDKN2A, ARF, MGMT,* and *GSTP1* [86]. In the case of RCC, methylation analysis of serum cfDNA using *GSTP1* alone or in combination with either *APC, p14ARF, p16, RARB, RASSF1, TIMP3,* or *PTGS2* has been shown to provide a high accuracy of detection (AUC ranging from 0.73 to 0.75; 95% IC 0.50–0.84) [99]. Independently, Hoque et al. measured *GSTP1* together with *CDH1, APC, MGMT, RASSF1A, p16, RARB2,* and *ARF* methylation for RCC detection using urine and plasma samples, showing that at least one gene was hypermethylated in 88% and 67% of the patient's urine sediments and plasma, respectively. [100].

Other gene panels that do not include *GSTP1* are also under study for PrCa and BdCa detection. Analysis of *MCAM, ERalpha,* and *ERbeta* showed 75% sensitivity and 70% specificity for early PrCa detection [101]. Similarly, *ST6GALNAC3, ZNF660, CCDC181,* and *HAPLN3* detected PrCa patients with up to 100% specificity and 67% sensitivity [102]. With respect to BdCa, methylation status of several genes are reliable alone or in combination using ctDNA in serum: *CDH13* [103], *PCDH10* [104], and *PCDH17* [105]. Additionally, dual combinations such as *PCDH17* and *POU4F2* (93.96% sensitivity, 90% specificity) [106] or *NID2* and *TWIST1* (90% sensitivity and 93% specificity) [107] have been proven to be accurate for the detection of BdCa patients using urine samples [108]. Several commercial tests that include epigenetic and non-epigenetic biomarkers are now available for the diagnosis of BdCa using urine or blood (AssureMDx®, Bladder CARE®, Bladder EPICHECK®). In RCC, a panel of genes that act as Wnt antagonists can serve as biomarkers for diagnosis, staging, and prognosis using serum ctDNA [109]. Notably, Vitale Nuzzo P et al. used cell-free methylated DNA immunoprecipitation and high-throughput sequencing (cfMeDIP–seq) as a highly sensitive assay capable of detecting and discriminating early-stage RCC from other tumor types and healthy controls in plasma (AUC 0.9) and urine (AUC 0.86) [110]. Taking into account the different studies related to diagnosis in PrCa, BdCa, and RCC, hypermethylation of *RASSF1A, APC, RARB2,* and *ARF* [111,112] seem to be the most consolidated biomarkers to use in plasma, serum, and/or urine.

The potential of DNA methylation analysis in LB related to progression and therapy response is an area of intense study. Sunami et al. reported that methylation of *GSTP1, RASSF1A,* and *RARB2* associated with PrCa's Gleason score and serum PSA; in addition, *GSTP1* and *RARB2* were associated with the disease's advanced stage [98]. Likewise, it has recently been shown that hypermethylation of *APC, GSTP1,* and *RARB2* in urine sediments correlated with shorter RFS and higher PrCa grade [113]. Additionally, in the case of urine, dual assessment of *GSTP1* and *APC* discriminates between low-risk and aggressive PrCa [114]. Interestingly, Zhao et al. showed that monitoring *GSTP1, APC, CRIP3,* and *HOXD8* methylation was useful for noninvasive prediction of PrCa aggressive disease in patients on active surveillance [115]. Indeed, the same group of authors developed a PrCa urinary epigenetic assay (ProCUrE®) with diagnostic and prognostic purposes based on the optimized measurement of *GSTP1* and *HOXD3* gene methylation [116]. In the case of BdCa, *PCDH10* and *PCDH17* hypermethylation were independent predictors of cancer survival and correlated with higher stage and grade [104,105], as described for *NID2* and *TWIST1*, which were able to discriminate between different patient's BdCa grades [108]. Finally, a multigene panel useful for BdCa recurrence surveillance has been developed, which included *EOMES, HOXA9, POU4F2, TWIST1, VIM,* and *ZNF154* [117]. A number of registered clinical trials (in some cases using commercial tests) for screening or recurrence purposes have been initiated in the case of PrCa and BdCa (https://www.clinicaltrials.gov/ (accessed on 30 December 2020)) (Table S2).

5.2. ncRNAs

In terms of ncRNA, several studies have shown their possible role as biomarkers in LB. Yu et al. recently designed a 4-lncRNA panel of urinary biomarkers (UCA1-201, HOTAIR, HYMA1, and MALAT1) for the diagnosis of non-muscle invasive bladder cancer (NMIBC) [118]. This signature confirmed the presence of tumor in a validation cohort of 140 NMIBC patients. A different study identified a 7-miRNA panel providing high diagnostic accuracy in BdCa using urine samples (miR-7-5p, miR-22-3p, miR-29a-3p, miR-126-5p, miR-200a-3p, miR-375, and miR-423-5p) [119]. Urquidi et al. described a sensitivity of 87% and a specificity of 100% using a different 25-miRNA urine signature [120]. An interesting approach integrated the expression of the mRNA HYAL1 together with two miRNAs (miR-96 and miR-210) and one lncRNA (UCA1) in an urine diagnostic panel that achieved a sensitivity of 100% and a specificity of 89.5% [121]. Interestingly, the lncRNA UCA1 increases cisplatin resistance in BdCa [122]. In the case of PrCa, ncRNA profiling could be a powerful tool to complement PSA screening. A recent study found a robust diagnostic model in serum using two different miRNAs (miR-17-3p and miR-1185-2-3p) with an associated 90% sensitivity and 90% specificity [123]. Serum detection of PSA in combination with miR-103a-3p and let-7a-5p detected PrCa cases better than PSA alone [124]. Serum miR-106b, miR-141-3p, miR-21, and miR-375 have also been combined in a panel with AUC of 0.86 [125].

6. Epigenetic Biomarkers in Breast Cancer

Breast cancer (BrCa) is the most common neoplasm diagnosed in females worldwide, with an incidence of 11.7% of all women cancer cases [44]. Screening based on imaging is key for the early detection and better prognosis of this disease, with mammograms being the most frequent technique recommended. However, the breast cancer nodules do not always exhibit pathognomonic characteristics, which can prevent the radiologist from performing a biopsy, or in other cases, generate false positives. Other limitations of this technique include the possible cumulative radiation exposure and over-diagnosis [126–128]. For these reasons among others, the search for potential non-invasive biomarkers in BrCa is needed.

6.1. DNA Methylation

Multiple studies have explored epigenetic alterations in ctDNA from BrCa patients that could serve for diagnosis, prognosis, classification of BrCa subtypes, and prediction of response to therapies. One of the most frequently studied markers has been the hypermethylation of *RASSF1A* [129–133], which discriminates between healthy individuals and BrCa patients and acts as a poor prognosis indicator [134]. Moreover, this aberration predicts the response to tamoxifen or neoadjuvant chemotherapy [135]. Other methylated targets found in plasma from BrCa patients with a diagnostic value encompass

SOX17, CST6, APC, DAK-K, MASPIN, HIC-1, HIN-1, RARB, RARbeta2, GSTP1, BRCA1, and *KIF1A* [32,129–132,136–139]. Among these targets, the hypermethylation of *RASSF1A, BRCA1, RARB,* and *RARB2,* in estrogen receptor+ (ER+) and progesterone receptor (PR+) breast tumors and plasmas, were validated as indicators of poor prognosis in at least two independent studies [140]. Furthermore, Fujita et al. showed that the simultaneous detection of *RASSF1A, RARB,* and *GSTP1* methylation (93% specificity) was strongly correlated with poor outcome [141]. *SOX17* methylation is another independent prognostic factor (HR: 4.737; 95% CI: 2.088–10.747) and its methylation status in ctDNA from plasma samples was found to correlate (70.9% concordance) with that observed in CTCs from matched BrCa patients [142,143]. *PITX2* hypermethylation in plasma has also been reported as another indicator of poor OS (HR: 3.4; 95% CI: 1.2–9.8) [144]. Interestingly, *PITX2* hypermethylation also predicted the response to anthracycline-based therapy [145].

In relation to ER status, Martinez-Galan et al. demonstrated methylation of *ER* and *ESR1* promoters in plasma from ER- patients [146]. In contrast, *PTPRO* methylation was found as a prognostic factor (HR: 3.66; 95% CI: 1.371–9.784) in ER+ positive BrCa patients but not in ER− [141]. Some gene panels have been designed to simultaneously analyze several gene methylation patterns in BrCa serum/plasma. For instance, Visvanathan et al. developed a panel of 10 genes including *RASSF1A*, whose methylation index predicted worse PFS (HR: 1.79; CI 95%: 1.23–2.60, and OS (HR: 1.75; 95% CI: 1.21–2.54) in metastatic BrCa patients [147]. Although evaluation of the methylation status in these genes is promising, there is currently only one commercial kit available specifically for BrCa, which tests for *PITX2* methylation in paraffin samples (Therascreen®, Qiagen, Frankfurt, Germany) [134].

6.2. ncRNAs

The potential value of ncRNAs as BrCa biomarkers in serum or plasma (either in exosomes or as cfRNA) has also been reported. Exosomal miR-21 has been widely shown to be a diagnostic biomarker, with sensitivity and specificity in pooled studies of ~75% and ~85%, respectively, and an AUC of 0.93 [148]. There are hundreds of studies proposing miRNAs as diagnostic and prognostic biomarkers in BrCa, but they need validation. We summarize here some recent relevant publications. In BrCa plasma samples, combination of four miRNAs (miR-1246, miR-206, miR-24, miR-373) distinguished BrCa from healthy individuals with 98% sensitivity, 96% specificity, and accuracy of 97% [149]. A panel composed of exosomal miR-142-5p, miR-320a, and miR-4433b-5p isolated from a BrCa patient's serum differentiated patients from their control counterparts with 93.33% sensitivity, 68.75% specificity, and AUC of 0.83 [135]. Furthermore, the combination of miR-142-5p and miR-320a discriminated luminal A subtype from healthy donors with 100% sensitivity, 93.80% specificity, and AUC of 0.94. Interestingly, decreased expression of miR-142-5p and miR-150-5p were significantly associated with more advanced tumor grades (grade III), while the decreased expression of miR-142-5p and miR-320a was associated with a larger tumor size [135]. Additionally, circulating miR-30b-5b has been recently reported to act as a BrCa prognostic factor [150].

Serum miRNA profiles may be useful for the diagnosis of axillary lymph node metastasis before surgery in a less-invasive manner than sentinel lymph node biopsy. A model that includes a combination of two miRNAs (miR-629-3p and miR-4710) and three clinicopathological factors (T stage, lymphovascular invasion, and ultrasound findings) showed an optimal diagnostic potential, with 88% sensitivity, 69% specificity, and accuracy of 0.86 [151].

There are also data that correlate ncRNA levels in serum to treatment response. For example, a set of exosomal miRNAs (miR-185, miR-4283, miR-5008 and miR-3613, miR-1302, miR-4715, and miR-3144) that target pathways of immune maturation predicted poor neoadjuvant chemotherapy response prior to surgery [152]. Similarly, lncH19 levels in the plasma of BrCa patients have also been reported to predict response to neoadjuvant chemotherapy [153].

7. Epigenetic Biomarkers in Colorectal Cancer

Colorectal cancer (CRC) is the third most common cancer worldwide. This tumor represents approximately 10% of all diagnosed cancer cases, with approximately 1.8 million new cases estimated in 2018. It is important to note that CRC is responsible for approximately 9% of all cancer deaths, being the second leading cause of cancer mortality [154]. In CRC, screening strategies have been shown to be effective to detect early CRC and precancerous lesions, and to reduce its morbidity and mortality. Among the detection strategies, the fecal immunochemical test (FIT) represents a non-invasive and cost-effective assay for detecting the presence of fecal hemoglobin. This is currently the most commonly used method for CRC screening, with an overall sensitivity and specificity for detection of 79% and 94%, respectively. However, the ability of this assay to detect advanced precancerous lesions is limited, showing 24% sensitivity and 95% specificity [155]. After a positive result for FIT, colonoscopy is the gold standard diagnostic technique for CRC detection. However, it is an invasive method that needs bowel preparation and sedation, presenting certain risk of complications for the patients [156]. In this context, the use of epigenetic biomarkers such as DNA methylation in stool samples might provide a non-invasive and the most cost-effective approach in population-based screening for both CRC and precancerous lesions [157]. Thus, for example, the simultaneous methylation analysis of *SEPT9* and *SDC2* (ColoDefense® test) in stool samples was able to obtain a sensitivity of 66.7% for advanced adenoma (AA) and 92.3% for CRC, with a specificity of 93.2% [158].

7.1. DNA Methylation

Among the most frequently studied epigenetic biomarkers in ctDNA for CRC, the methylation of *SEPTIN9* (*SEPT9*) stands out for screening and early detection [144,159,160]. The EpiproColon® test was the first commercially available FDA-approved test for the detection of *SEPT9* methylation in plasma by real-time PCR [161,162]. In addition to blood samples, methylation of this gene has also been analyzed in stool, showing a 35.9% improvement in detecting pre-tumoral stages (AA) and 7.9% in identifying early CRC tumors, in comparison with the plasma test [163]. The use of ColoDefense® in blood enabled the detection of AA and CRC, with an overall sensitivity of 88.9% and a specificity of 92.8% [164]. Similarly, other studies have proposed the analysis of the methylation of several genes in plasma as circulating epigenetic biomarkers able to discriminate between healthy controls and patients with AA or CRC [165,166]. In addition, approaches based on methylation microarrays [33] and NGS [167] have been used to identify epigenetic biomarkers in ctDNA for cancer detection.

Regarding prognosis, hypermethylation of the *P16* promoter in ctDNA has been associated with poor OS [168]. Additionally, hypermethylation of *HPP1* and *HLTF* indicates a poor prognosis and high mortality [169], and hypermethylation of *RARB* and *RASSF1A* was associated with the aggressiveness of the disease [170] in patients with CRC. Methylation of ctDNA can also be used to monitor tumor burden and evaluate the therapeutic response of patients [171,172], correlating better than classical biomarkers such as carcinoembryonic antigen (CEA) and carbohydrate antigen (CA)-19-9. For example, the analysis of the methylation status of the 2-gene panel *BCAT1/IKZF1* in plasma showed higher sensitivity for detecting CRC recurrence than CEA, with an odds ratio of 14.4 (95% CI: 5–39) and 6.9 (95% CI: 2–22), respectively [173]. Similarly, the plasma methylation of *SEPT9*, *DCC*, *BOLL*, and *SRFP2* showed stronger correlation with tumor burden than CEA and CA-19-9 [172].

7.2. ncRNAs

Circulating levels of ncRNAs have also shown utility as biomarkers in the management of CRC. Circulating miR-21 levels in blood and saliva allow for the detection of CRC [174,175]. In addition, miRNA signatures evaluated in fluids can be useful for discriminating between healthy controls, patients with adenomas, and patients with CRC with high sensitivity and specificity. In particular, the plasma levels of miR-601 and miR-760 showed an AUC of 0.68 with 72.1% sensitivity and 62.1% specificity, which can discriminate between AA and healthy donors. In addition, this panel of miRNAs was able to differentiate CRC from the control samples with an AUC of 0.79, a sensitivity of 83.3%, and a specificity of 69.1% [176]. Another study has recently identified a signature of six miRNAs (miRNA19a, miRNA19b, miRNA15b, miRNA29a, miRNA335, and miRNA18a) with an AUC of 0.92, a sensitivity of 85%, and a specificity of 90% that is able to detect CRC and AA in comparison to healthy individuals [157]. Regarding prognosis, high levels of circulating miR-210 and miR-141 are associated with shorter survival [165,177], while high levels of miR-23b are associated with longer survival [178]. Besides, levels of different miRNAs in blood may be useful for the early detection of recurrence [179] and evaluation of therapy response in CRC patients [180]. High plasma levels of the lncRNA HOTAIR have shown utility for the detection of CRC and association with a worse prognosis and higher mortality [181]. Of note, other studies have analyzed different combinations of circulating lncRNAs as diagnostic biomarkers, which were useful for the detection of adenomas and CRC [166,182,183].

8. Epigenetic Biomarkers in Other Tumor Types and Multi-Cancer Tests

In addition to common tumor types, epigenetic alterations may also be detected in LB from other less frequent malignancies, showing clinical utility as tumor biomarkers. In cutaneous melanoma, where the use of circulating epigenetic biomarkers has been proposed as a non-invasive tool for tumor detection, promoter hypermethylation of *RASSF1A* has been described in plasma samples as a diagnostic indicator, with the ability of discriminating between melanoma patients and healthy individuals, showing a good diagnostic accuracy with an AUC of 0.90 [184]. Besides, the detection of hypermethylated *RASSF1A* in serum before treatment was able to predict the prognosis and clinical response to drugs in advanced melanoma patients [185]. In a recent pilot study using NGS and machine learning, Bustos et al. were able to identify a circulating miRNA signature (miR-4649-3p, miR-615-3p, and miR-1234-3p) associated with the response to ICIs in advanced melanoma patients, suggesting that circulating miRNAs could enable real-time monitoring of patients receiving this type of treatment [167]. The plasmatic levels of other ncRNAs such as lncRNAs (IGF2AS, anti-Peg11, MEG3, Zeb2NAT) were also found to be associated with prognosis and therapy response in *BRAF*-mutant advanced melanoma patients treated with the BRAF inhibitor vemurafenib [186]. Similar to melanoma, the blood-based analysis of DNA methylation and ncRNAs has shown utility as circulating epigenetic biomarkers for other tumors including pancreatic cancer [187], ovarian and endometrial carcinomas [188,189], and brain tumors [190], among others.

In brain tumors such as glioblastoma, promoter hypermethylation of several genes (*MGMT*, *p16INK4a*, *TIMP-3*, and *THBS1*) has been detected at high frequencies in serum and CSF. In glioblastoma, hypermethylation of *MGMT* is associated with response to temozolamide [191]. The methylation status of *MGMT* and *THBS1* in CSF was also able to independently predict PFS of glioblastoma patients [22]. Similar to methylation, the circulating microRNA profiling of CSF has also been proposed as a good approach for the non-invasive detection (miR-30e, miR-140, let-7b, mR-10a, and miR-21-3p) and prognosis (miR-10b and miR-196b) of glioblastoma patients [192]. In other tumor types such as oral cancer, the analysis of epigenetic biomarkers in saliva has been explored. In this sense, the promoter hypermethylation of different types of genes (e.g., *RASSF1A*, *p16 INK4a*, *TIMP3*, and *PCQAP/MED15*) and the expression levels of miRNAs in saliva have been detected in association with oral tumors [18,193]. Thus, the study of epigenetic biomarkers in saliva has been proposed as an easily accessible LB sample for oral cancer detection.

The recent application of NGS has allowed for the development of sensitive epigenetic assays for the detection of both common and less-frequent tumors. Thus, using NGS and machine learning, Liu et al., in a very large clinical trial including individuals with (n = 2482) and without cancer (n = 4207), recently developed a classifier based on the methylation of cfDNA, assessing >100,000 methylation sites in plasma for the sensitive detection of more than 50 tumor types [169]. This multi-cancer approach was useful across all stages of the disease, and also for the identification of the tissue of origin with high accuracy, which could be relevant for the treatment and follow-up of the patients. The assay is going to be commercialized by the Biotech Company GRAIL. PanSeer® is an NGS-based assay that is able to detect cancer in asymptomatic individuals, years before standard diagnosis [194].

Table 1 shows a list of the top methylated genes and ncRNAs identified in LB from cancer patients, with an emerging role as biomarkers. This list has been established based on the number of studies and robustness of the genes/ncRNAs published and/or inclusion in commercial tests.

Table 1. Top methylated genes/signatures and ncRNAs identified in liquid biopsies from cancer patients, with an emerging role as biomarkers. BAL: bronchoalveolar lavage; BAS: bronchoalveolar aspirate; CSF: cerebrospinal fluid.

Epigenetic Alteration	Gene Name(s)/Epigenetic Kit	Type of Liquid Biopsy	Intended Use	Reference
	LUNG CANCER			
DNA methylation	-SHOX2/PTGER4 (EpiProLung)®	Blood	Diagnosis	[66]
	-Gene sets including RASFF1A and other genes	BAL/sputum	Diagnosis	[54–57,62]
	-BCAT1/CDO1/TRIM58/ZNF177	BAS/BAL/sputum	Diagnosis	[58]
ncRNAs	-miR21	Blood	Diagnosis	[74]
	-Several miRNA signatures	Blood	Diagnosis	[72,73]
	GENITOURINARY CANCERS			
DNA methylation	-Gene sets including GSTP1, RASFF1A, APC, ARF and RARB2	Urine, blood	Diagnosis	[94,96]
	-Several gene sets (AssureMDx®, Bladder CARE®, Bladder EPICHECK®)	Urine, blood	Diagnosis	[103,108]
	-GSTP1 and HOXD3 (ProCUrE, Prostate cancer)	Urine	Diagnosis	[116]
ncRNAs	-Several miRNA and lncRNA signatures	Blood	Diagnosis	[118–120]
	BREAST CANCER			
DNA methylation	-Gene sets including GSTP1, RASFF1A, BRCA1 and RARB2	Blood	Diagnosis	[129–133,136–139]
	-PITX2	Blood	Prognosis/response	[144,145]
ncRNAs	-miR21	Blood	Diagnosis	[148]
	-Several miRNA signatures	Blood	Diagnosis	[149,151]
	COLORECTAL CANCER			
DNA methylation	-SEPT9 (EpiProColon®)	Stool, blood,	Diagnosis	[161,162]
	-SEPT9 and SDC2 (ColoDefense®)	Stool, blood	Diagnosis	[158]
	-p16, RASFF1A, RARB2	Blood	Diagnosis, prognosis	[168–171]
	-BCAT1 and IKZF1	Blood	Diagnosis	[172]
ncRNAs	-miR21	Blood, saliva	Diagnosis	[174,175]
	-Several miRNA signatures	Blood	Diagnosis	[176,177]
	OTHER CANCER TYPES AND MULTI-CANCER BIOMARKERS			
DNA methylation	-RASFF1A (melanoma)	Blood	Diagnosis	[184]
	-MGMT (glioblastoma)	CSF	Response to therapy	[191]
	-RASFF1A, p16, TIMP3 (oral cancer)	Blood, saliva	Diagnosis, prognosis	[193]
	-DNA methylation signature PanSeer®	Blood	Diagnosis	[194]
	-DNA methylation signature GRAIL®	Blood	Diagnosis	[169]
ncRNAs	-Several miRNA signatures	Blood	Diagnosis	[186–189]

9. Epigenetic Biomarkers in Cancer: Translation to the Clinic

The use of non-invasive epigenetic biomarkers is considered as a promising option in oncology. However, these biomarkers (with few exceptions) have not successfully reached clinical practice yet. Progress in the path to translation will be made provided clinical value is added to the current management of patients. These are some of the difficulties to take into consideration for clinical translation [195]:

(1) Clinical value and confirmatory results: confirmed clinical evidence in prospective trials is critical for medical professionals and regulatory agencies.
(2) Performance and affordability: it is essential to develop a commercial product with demonstrated good performance, affordable price, and is easy to use.
(3) Pre-analytical issues: preservation of the sample, storage time, and temperature, etc., have to be extensively studied.
(4) Technical barriers: when using some of the epigenetic techniques, there may be a technical barrier, particularly for advanced procedures such as mass spectrometry or next generation sequencing (NGS).
(5) Training: formation on new epigenetic platforms and interpretation of the results is needed, especially for the "omic" epigenetic technologies.
(6) Global regulation: establishing a global harmonization of regulation would facilitate translating an epigenetic assay into the clinic.

Overall, the continuous technological development and commercialization activity of epigenetic kits would lead to an innovative and competitive environment that will result in significant benefits for the clinical practice in the near future.

10. Conclusions and Future Perspectives

Evaluation of epigenetic biomarkers in LB is an emerging field in oncology that may help in cancer screening, diagnosis, identification of tumor subtypes as well as in the prediction of response to therapy and outcome. LB offers the opportunity of evaluating tumor markers using non-invasive methods and may represent better tumor heterogeneity and evolution. While the evaluation of actionable mutations in LB has a demonstrated clinical value, the use of epigenetic alterations (with few exceptions) has not reached clinical practice yet. Among the different fields where epigenetic changes may play a role as biomarkers, we envision that screening and diagnosis are the areas closer to the clinic. Current screening tests such as mammography, analysis of occult blood in feces and colonoscopy are routinely performed to detect BrCa and CRC, respectively. Nonetheless, over-diagnosis and false positives are of concern. In the case of PrCa, blood levels of PSA lack diagnostic accuracy and for LuCa, LDCT is not a common practice yet. Therefore, epigenetic biomarkers in LB could be of great value in screening and diagnosis for these cancer types. Moreover, the development of platforms that analyze thousands of methylation alterations in blood has been shown to be highly valuable in screening for multiple cancers.

Some epigenetic commercial tests have been developed and are currently being evaluated in clinical trials. These tests are designed for individual cancer types or as multi-cancer diagnostic tools; some others include both DNA methylation and mutational assays in the same kit. With constant information being provided by "omic" techniques for both DNA methylation and ncRNAs, new potential sources of epigenetic markers will be introduced and tested. However, the path to clinical translation is long and costly and thus the identified epigenetic biomarkers need to offer an added value over the established clinical practice and to attract investment for their development.

The discovery of new gene/signature candidates can also face several issues. For example, in the case of blood, studies show the need to use large amounts of plasma or serum to evaluate DNA methylation (1–4 mL) in comparison with protein-based techniques that can use much lower amounts (10–100 µL). This limitation could be solved with the introduction of new ultrasensitive techniques. The discovery of novel aberrantly methylated genes using "omic" platforms may also need specialized technicians and bioinformaticians to analyze the data correctly. In addition, these technologies are expensive and could be outsourced at reference hospitals.

Supplementary Materials: The following are available online at https://www.mdpi.com/article/10.3390/cancers13123016/s1, Table S1: Technologies for epigenetic assays in liquid biopsy, Table S2: Clinical trials using epigenetic biomarkers in PrCa, BdCa and RCC.

Author Contributions: Conceptualization: A.C., J.S., E.J.-L., A.D.-L., L.M.; Writing: original draft preparation, C.P.-B., A.R.-C., S.T., S.C.-F., F.E., D.S., E.R., K.V., E.J.-L., A.D.-L., J.S., A.C., review and editing, A.C., J.S., E.J.-L., A.D.-L., L.M. All authors have read and agreed to the published version of the manuscript.

Funding: A.C. and L.M. were funded by FIMA (Foundation for Applied Medical Research), ISCIII-Fondo de Investigación Sanitaria-Fondo Europeo de Desarrollo Regional "Una manera de hacer Europa" (PI19/00230, to AC; PI19/00098 to LMM), CIBERONC CB16/12/00443, AECC, and Ramón Areces Foundations (all to L.M.). J.S. was funded by FIS grant (PI19/00572) from the FEDER, FSE, Carlos III Health Institute (ISCIII) and INNVA1/2020/71 de la Línea 1 de Valorización y Transferencia a las empresas, de la Agencia Valenciana de la Innovación. E.J-L. was supported by the Spanish Health Institute Carlos III (ISCII, Fondo de Investigación Sanitaria: CB16/12/00350 and PI18/00266). S.T. was supported by the Generalitat Valenciana and Fondo Social Europeo, fellowship ACIF/2018/275. A.D-L. was funded by a contract Juan Rodés from Spanish Health Institute Carlos III (ISCIII) (JR17/00016) and by a research grant (PI18/00307) co-funded by ISCIII and the European Regional Development Fund (FEDER). A.R.-C. was supported by the Roche-CHUS Joint Unit (IN853B2018/03) funded by GAIN, Consellería de Economía, Emprego e Industria". Fellowship support: E.R. "FPU, Spanish Ministry of Education"; D.S., "Juan de la Cierva-Incorporacion, Spanish Ministry of Science and Innovation".

Institutional Review Board Statement: Non-applicable.

Informed Consent Statement: Non-applicable.

Data Availability Statement: Non-applicable.

Conflicts of Interest: Authors declare no conflict of interests.

References

1. Alvarez, H.; Opalinska, J.; Zhou, L.; Sohal, D.; Fazzari, M.J.; Yu, Y.; Montagna, C.; Montgomery, E.A.; Canto, M.; Dunbar, K.B.; et al. Widespread hypomethylation occurs early and synergizes with gene amplification during esophageal carcinogenesis. *PLoS Genet.* **2011**, *7*, e1001356. [CrossRef]
2. Sharma, G.; Dua, P.; Agarwal, S.M. A Comprehensive review of dysregulated MiRNAs involved in cervical cancer. *Curr. Genom.* **2014**, *15*, 310–323. [CrossRef] [PubMed]
3. Ross, J.P.; Rand, K.N.; Molloy, P.L. Hypomethylation of repeated DNA sequences in cancer. *Epigenomics* **2010**, *2*, 245–269. [CrossRef]
4. Karimzadeh, M.R.; Pourdavoud, P.; Ehtesham, N.; Qadbeigi, M.; Asl, M.M.; Alani, B.; Mosallaei, M.; Pakzad, B. Regulation of DNA methylation machinery by Epi-MiRNAs in human cancer: Emerging new targets in cancer therapy. *Cancer Gene Ther.* **2020**. [CrossRef] [PubMed]
5. Mutirangura, A. Is global hypomethylation a nidus for molecular pathogenesis of age-related noncommunicable diseases? *Epigenomics* **2019**, *11*, 577–579. [CrossRef] [PubMed]
6. Amelio, I.; Bertolo, R.; Bove, P.; Buonomo, O.C.; Candi, E.; Chiocchi, M.; Cipriani, C.; Di Daniele, N.; Ganini, C.; Juhl, H.; et al. Liquid biopsies and cancer omics. *Cell Death Discov.* **2020**, *6*, 131. [CrossRef] [PubMed]
7. Locke, W.J.; Guanzon, D.; Ma, C.; Liew, Y.J.; Duesing, K.R.; Fung, K.Y.C.; Ross, J.P. DNA methylation cancer biomarkers: Translation to the clinic. *Front. Genet.* **2019**. [CrossRef]
8. Baylin, S.B.; Jones, P.A. Epigenetic determinants of cancer. *Cold Spring Harb. Perspect. Biol.* **2016**, *8*. [CrossRef]
9. Doyle, L.M.; Wang, M.Z. Overview of extracellular vesicles, their origin, composition, purpose, and methods for exosome isolation and analysis. *Cells* **2019**, *8*, 727. [CrossRef]
10. Yuan, T.; Huang, X.; Woodcock, M.; Du, M.; Dittmar, R.; Wang, Y.; Tsai, S.; Kohli, M.; Boardman, L.; Patel, T.; et al. Plasma extracellular RNA profiles in healthy and cancer patients. *Sci. Rep.* **2016**. [CrossRef]
11. Ayupe, A.C.; Reis, E.M. Evaluating the stability of MRNAs and noncoding RNAs. In *Enhancer RNAs*; Methods in Molecular Biology Book Series; Humana Press: Clifton, NJ, USA, 2017; Volume 1468, pp. 139–153. [CrossRef]

12. Tani, H.; Mizutani, R.; Salam, K.A.; Tano, K.; Ijiri, K.; Wakamatsu, A.; Isogai, T.; Suzuki, Y.; Akimitsu, N. Genome-Wide Determination of RNA Stability reveals hundreds of short-lived noncoding transcripts in mammals. *Genome Res.* **2012**, *22*, 947–956. [CrossRef]
13. McDonald, J.S.; Milosevic, D.; Reddi, H.V.; Grebe, S.K.; Algeciras-Schimnich, A. Analysis of circulating MicroRNA: Preanalytical and analytical challenges. *Clin. Chem.* **2011**, *57*, 833–840. [CrossRef]
14. Li, I.; Nabet, B.Y. Exosomes in the tumor microenvironment as mediators of cancer therapy resistance. *Mol. Cancer* **2019**, *18*, 32. [CrossRef]
15. McAnena, P.; Brown, J.A.L.; Kerin, M.J. Circulating nucleosomes and nucleosome modifications as biomarkers in cancer. *Cancers* **2017**, *9*, 5. [CrossRef]
16. Bettegowda, C.; Sausen, M.; Leary, R.J.; Kinde, I.; Wang, Y.; Agrawal, N.; Bartlett, B.R.; Wang, H.; Luber, B.; Alani, R.M.; et al. Detection of circulating tumor DNA in early- and late-stage human malignancies. *Sci. Transl. Med.* **2014**, *6*, 224ra24. [CrossRef]
17. Ponti, G.; Manfredini, M.; Tomasi, A. Non-blood sources of cell-free DNA for cancer molecular profiling in clinical pathology and oncology. *Crit. Rev. Oncol. Hematol.* **2019**, *141*, 36–42. [CrossRef]
18. Rapado-González, Ó.; López-López, R.; López-Cedrún, J.L.; Triana-Martínez, G.; Muinelo-Romay, L.; Suárez-Cunqueiro, M.M. Cell-free MicroRNAs as potential oral cancer biomarkers: From diagnosis to therapy. *Cells* **2019**, *8*, 1653. [CrossRef]
19. Kimura, H.; Fujiwara, Y.; Sone, T.; Kunitoh, H.; Tamura, T.; Kasahara, K.; Nishio, K. High sensitivity detection of epidermal growth factor receptor mutations in the pleural effusion of non-small cell lung cancer patients. *Cancer Sci.* **2006**, *97*, 642–648. [CrossRef]
20. O'Reilly, E.; Tuzova, A.V.; Walsh, A.L.; Russell, N.M.; O'Brien, O.; Kelly, S.; Dhomhnallain, O.N.; DeBarra, L.; Dale, C.M.; Brugman, R.; et al. EpiCaPture: A Urine DNA methylation test for early detection of aggressive prostate cancer. *JCO Precis. Oncol.* **2019**, *3*. [CrossRef]
21. Van Kessel, K.E.M.; Beukers, W.; Lurkin, I.; Ziel-van der Made, A.; van der Keur, K.A.; Boormans, J.L.; Dyrskjøt, L.; Márquez, M.; Ørntoft, T.F.; Real, F.X.; et al. Validation of a DNA methylation-mutation urine assay to select patients with hematuria for cystoscopy. *J. Urol.* **2017**, *197 Pt 1*, 590–595. [CrossRef]
22. Liu, B.-L.; Cheng, J.-X.; Zhang, W.; Zhang, X.; Wang, R.; Lin, H.; Huo, J.-L.; Cheng, H. Quantitative detection of multiple gene promoter hypermethylation in tumor tissue, serum, and cerebrospinal fluid predicts prognosis of malignant gliomas. *Neuro Oncol.* **2010**, *12*, 540–548. [CrossRef] [PubMed]
23. Shivapurkar, N.; Gazdar, A.F. DNA methylation based biomarkers in non-invasive cancer screening. *Curr. Mol. Med.* **2010**, *10*, 123–132. [CrossRef] [PubMed]
24. Alix-Panabières, C.; Pantel, K. Clinical applications of circulating tumor cells and circulating tumor DNA as liquid biopsy. *Cancer Discov.* **2016**, *6*, 479–491. [CrossRef] [PubMed]
25. Wang, Z.; Jiang, W.; Wang, Y.; Guo, Y.; Cong, Z.; DU, F.; Song, B. mgmt promoter methylation in serum and cerebrospinal fluid as a tumor-specific biomarker of glioma. *Biomed. Rep.* **2015**, *3*, 543–548. [CrossRef] [PubMed]
26. Pajares, M.J.; Palanca-Ballester, C.; Urtasun, R.; Alemany-Cosme, E.; Lahoz, A.; Sandoval, J. Methods for analysis of specific DNA methylation status. *Methods* **2021**, *187*, 3–12. [CrossRef] [PubMed]
27. Bao-Caamano, A.; Rodriguez-Casanova, A.; Diaz-Lagares, A. Epigenetics of circulating tumor cells in breast cancer. *Adv. Exp. Med. Biol.* **2020**, *1220*, 117–134. [CrossRef]
28. Han, X.; Wang, J.; Sun, Y. Circulating tumor DNA as biomarkers for cancer detection. *Genom. Proteom. Bioinform.* **2017**, *15*, 59–72. [CrossRef]
29. Eads, C.A.; Danenberg, K.D.; Kawakami, K.; Saltz, L.B.; Blake, C.; Shibata, D.; Danenberg, P.V.; Laird, P.W. MethyLight: A high-throughput assay to measure DNA methylation. *Nucleic Acids Res.* **2000**, *28*, E32. [CrossRef]
30. Wan, J.C.M.; Massie, C.; Garcia-Corbacho, J.; Mouliere, F.; Brenton, J.D.; Caldas, C.; Pacey, S.; Baird, R.; Rosenfeld, N. Liquid biopsies come of age: Towards implementation of circulating tumour DNA. *Nat. Rev. Cancer* **2017**, *17*, 223–238. [CrossRef]
31. Li, M.; Chen, W.-D.; Papadopoulos, N.; Goodman, S.N.; Bjerregaard, N.C.; Laurberg, S.; Levin, B.; Juhl, H.; Arber, N.; Moinova, H.; et al. Sensitive digital quantification of DNA methylation in clinical samples. *Nat. Biotechnol.* **2009**, *27*, 858–863. [CrossRef] [PubMed]
32. Radpour, R.; Barekati, Z.; Kohler, C.; Lv, Q.; Bürki, N.; Diesch, C.; Bitzer, J.; Zheng, H.; Schmid, S.; Zhong, X.Y. Hypermethylation of tumor suppressor genes involved in critical regulatory pathways for developing a blood-based test in breast cancer. *PLoS ONE* **2011**, *6*, e16080. [CrossRef]
33. Gallardo-Gómez, M.; Moran, S.; Páez de la Cadena, M.; Martínez-Zorzano, V.S.; Rodríguez-Berrocal, F.J.; Rodríguez-Girondo, M.; Esteller, M.; Cubiella, J.; Bujanda, L.; Castells, A.; et al. A new approach to epigenome-wide discovery of non-invasive methylation biomarkers for colorectal cancer screening in circulating cell-free DNA using pooled samples. *Clin. Epigenet.* **2018**, *10*, 53. [CrossRef]
34. Solé, C.; Tramonti, D.; Schramm, M.; Goicoechea, I.; Armesto, M.; Hernandez, L.I.; Manterola, L.; Fernandez-Mercado, M.; Mujika, K.; Tuneu, A.; et al. The circulating transcriptome as a source of biomarkers for melanoma. *Cancers* **2019**, *11*, 70. [CrossRef]
35. Cojocneanu, R.; Braicu, C.; Raduly, L.; Jurj, A.; Zanoaga, O.; Magdo, L.; Irimie, A.; Muresan, M.-S.; Ionescu, C.; Grigorescu, M.; et al. Plasma and tissue specific MiRNA expression pattern and functional analysis associated to colorectal cancer patients. *Cancers* **2020**, *12*, 843. [CrossRef]

36. Gasparello, J.; Papi, C.; Allegretti, M.; Giordani, E.; Carboni, F.; Zazza, S.; Pescarmona, E.; Romania, P.; Giacomini, P.; Scapoli, C.; et al. A Distinctive MicroRNA (MiRNA) signature in the blood of colorectal cancer (CRC) patients at surgery. *Cancers* **2020**, *12*, 2410. [CrossRef]
37. Metcalf, G.A.D.; Shibakawa, A.; Patel, H.; Sita-Lumsden, A.; Zivi, A.; Rama, N.; Bevan, C.L.; Ladame, S. Amplification-free detection of circulating microrna biomarkers from body fluids based on fluorogenic oligonucleotide-templated reaction between engineered peptide-nucleic acid probes: Application to prostate cancer diagnosis. *Anal. Chem.* **2016**, *88*, 8091–8098. [CrossRef]
38. Shukla, N.; Yan, I.K.; Patel, T. Multiplexed Detection and Quantitation of Extracellular Vesicle RNA Expression Using NanoString. *Methods Mol. Biol.* **2018**, *1740*, 177–185. [CrossRef]
39. Armstrong, D.A.; Green, B.B.; Seigne, J.D.; Schned, A.R.; Marsit, C.J. MicroRNA molecular profiling from matched tumor and bio-fluids in bladder cancer. *Mol. Cancer* **2015**, *14*, 194. [CrossRef]
40. Ortega, F.G.; Lorente, J.A.; Garcia Puche, J.L.; Ruiz, M.P.; Sanchez-Martin, R.M.; de Miguel-Pérez, D.; Diaz-Mochon, J.J.; Serrano, M.J. MiRNA in situ hybridization in circulating tumor cells—MishCTC. *Sci. Rep.* **2015**, *5*, 9207. [CrossRef]
41. Li, L.; Lu, M.; Fan, Y.; Shui, L.; Xie, S.; Sheng, R.; Si, H.; Li, Q.; Wang, Y.; Tang, B. High-throughput and ultra-sensitive single-cell profiling of multiple MicroRNAs and identification of human cancer. *Chem. Commun.* **2019**, *55*, 10404–10407. [CrossRef] [PubMed]
42. Hurd, P.J.; Nelson, C.J. Advantages of next-generation sequencing versus the microarray in epigenetic research. *Brief. Funct. Genom. Proteom.* **2009**, *8*, 174–183. [CrossRef] [PubMed]
43. Wang, Y.-M.; Trinh, M.P.; Zheng, Y.; Guo, K.; Jimenez, L.A.; Zhong, W. Analysis of circulating non-coding RNAs in a non-invasive and cost-effective manner. *Trends Anal. Chem. TRAC* **2019**, *117*, 242–262. [CrossRef] [PubMed]
44. Sung, H.; Ferlay, J.; Siegel, R.L.; Laversanne, M.; Soerjomataram, I.; Jemal, A.; Bray, F. Global Cancer Statistics 2020: GLOBOCAN estimates of incidence and mortality worldwide for 36 cancers in 185 countries. *CA Cancer J. Clin.* **2021**. [CrossRef]
45. Siegel, R.L.; Miller, K.D.; Jemal, A. Cancer statistics, 2019. *CA Cancer J. Clin.* **2019**, *69*, 7–34. [CrossRef]
46. Krist, A.H.; Davidson, K.W.; Mangione, C.M.; Barry, M.J.; Cabana, M.; Caughey, A.B.; Davis, E.M.; Donahue, K.E.; Doubeni, C.A.; Kubik, M.; et al. Screening for lung cancer: US preventive services task force recommendation statement. *JAMA* **2021**, *325*, 962–970. [CrossRef]
47. Seijo, L.M.; Peled, N.; Ajona, D.; Boeri, M.; Field, J.K.; Sozzi, G.; Pio, R.; Zulueta, J.J.; Spira, A.; Massion, P.P.; et al. Biomarkers in lung cancer screening: Achievements, promises, and challenges. *J. Thorac. Oncol.* **2019**, *14*, 343–357. [CrossRef]
48. Schmidt, B.; Beyer, J.; Dietrich, D.; Bork, I.; Liebenberg, V.; Fleischhacker, M. Quantification of cell-free MSHOX2 plasma DNA for therapy monitoring in advanced stage non-small cell (NSCLC) and small-cell lung cancer (SCLC) patients. *PLoS ONE* **2015**, *10*, e0118195. [CrossRef]
49. Kneip, C.; Schmidt, B.; Seegebarth, A.; Weickmann, S.; Fleischhacker, M.; Liebenberg, V.; Field, J.K.; Dietrich, D. SHOX2 DNA methylation is a biomarker for the diagnosis of lung cancer in plasma. *J. Thorac. Oncol.* **2011**, *6*, 1632–1638. [CrossRef]
50. Konecny, M.; Markus, J.; Waczulikova, I.; Dolesova, L.; Kozlova, R.; Repiska, V.; Novosadova, H.; Majer, I. The Value of SHOX2 methylation test in peripheral blood samples used for the differential diagnosis of lung cancer and other lung disorders. *Neoplasma* **2016**, *63*, 246–253. [CrossRef]
51. Ilse, P.; Biesterfeld, S.; Pomjanski, N.; Fink, C.; Schramm, M. SHOX2 DNA methylation is a tumour marker in pleural effusions. *Cancer Genom. Proteom.* **2013**, *10*, 217–223.
52. Powrózek, T.; Krawczyk, P.; Nicoś, M.; Kuźnar-Kamińska, B.; Batura-Gabryel, H.; Milanowski, J. Methylation of the DCLK1 promoter region in circulating free DNA and its prognostic value in lung cancer patients. *Clin. Transl. Oncol.* **2016**, *18*, 398–404. [CrossRef]
53. Powrózek, T.; Krawczyk, P.; Kucharczyk, T.; Milanowski, J. Septin 9 promoter region methylation in free circulating DNA-potential role in noninvasive diagnosis of lung cancer: Preliminary report. *Med. Oncol.* **2014**, *31*, 917. [CrossRef]
54. Ponomaryova, A.A.; Rykova, E.Y.; Cherdyntseva, N.V.; Skvortsova, T.E.; Dobrodeev, A.Y.; Zav'yalov, A.A.; Bryzgalov, L.O.; Tuzikov, S.A.; Vlassov, V.V.; Laktionov, P.P. Potentialities of aberrantly methylated circulating DNA for diagnostics and post-treatment follow-up of lung cancer patients. *Lung Cancer* **2013**, *81*, 397–403. [CrossRef]
55. Hulbert, A.; Jusue-Torres, I.; Stark, A.; Chen, C.; Rodgers, K.; Lee, B.; Griffin, C.; Yang, A.; Huang, P.; Wrangle, J.; et al. Early detection of lung cancer using DNA promoter hypermethylation in plasma and sputum. *Clin. Cancer Res.* **2017**, *23*, 1998–2005. [CrossRef]
56. Liu, D.; Peng, H.; Sun, Q.; Zhao, Z.; Yu, X.; Ge, S.; Wang, H.; Fang, H.; Gao, Q.; Liu, J.; et al. The indirect efficacy comparison of DNA methylation in sputum for early screening and auxiliary detection of lung cancer: A meta-analysis. *Int. J. Environ. Res. Public Health* **2017**, *14*, 679. [CrossRef]
57. Liu, B.; Ricarte Filho, J.; Mallisetty, A.; Villani, C.; Kottorou, A.; Rodgers, K.; Chen, C.; Ito, T.; Holmes, K.; Gastala, N.; et al. Detection of promoter DNA methylation in urine and plasma aids the detection of non-small cell lung cancer. *Clin. Cancer Res.* **2020**, *26*, 4339–4348. [CrossRef]
58. Diaz-Lagares, A.; Mendez-Gonzalez, J.; Hervas, D.; Saigi, M.; Pajares, M.J.; Garcia, D.; Crujerias, A.B.; Pio, R.; Montuenga, L.M.; Zulueta, J.; et al. A novel epigenetic signature for early diagnosis in lung cancer. *Clin. Cancer Res.* **2016**, *22*, 3361–3371. [CrossRef]
59. Exposito, F.; Villalba, M.; Redrado, M.; de Aberasturi, A.L.; Cirauqui, C.; Redin, E.; Guruceaga, E.; de Andrea, C.; Vicent, S.; Ajona, D.; et al. Targeting of TMPRSS4 sensitizes lung cancer cells to chemotherapy by impairing the proliferation machinery. *Cancer Lett.* **2019**, *453*, 21–33. [CrossRef]

60. Hsu, H.-S.; Chen, T.-P.; Hung, C.-H.; Wen, C.-K.; Lin, R.-K.; Lee, H.-C.; Wang, Y.-C. Characterization of a multiple epigenetic marker panel for lung cancer detection and risk assessment in plasma. *Cancer* **2007**, *110*, 2019–2026. [CrossRef]
61. Ostrow, K.L.; Hoque, M.O.; Loyo, M.; Brait, M.; Greenberg, A.; Siegfried, J.M.; Grandis, J.R.; Gaither Davis, A.; Bigbee, W.L.; Rom, W.; et al. Molecular analysis of plasma DNA for the early detection of lung cancer by quantitative methylation-specific PCR. *Clin. Cancer Res.* **2010**, *16*, 3463–3472. [CrossRef]
62. Ooki, A.; Maleki, Z.; Tsay, J.-C.J.; Goparaju, C.; Brait, M.; Turaga, N.; Nam, H.-S.; Rom, W.N.; Pass, H.I.; Sidransky, D.; et al. A Panel of novel detection and prognostic methylated dna markers in primary non-small cell lung cancer and serum DNA. *Clin. Cancer Res.* **2017**, *23*, 7141–7152. [CrossRef]
63. Constâncio, V.; Nunes, S.P.; Henrique, R.; Jerónimo, C. DNA methylation-based testing in liquid biopsies as detection and prognostic biomarkers for the four major cancer types. *Cells* **2020**, *9*, 624. [CrossRef]
64. Lissa, D.; Robles, A.I. Methylation analyses in liquid biopsy. *Transl. Lung Cancer Res.* **2016**, *5*, 492–504. [CrossRef]
65. Vinayanuwattikun, C.; Sriuranpong, V.; Tanasanvimon, S.; Chantranuwat, P.; Mutirangura, A. Epithelial-Specific Methylation marker: A potential plasma biomarker in advanced non-small cell lung cancer. *J. Thorac. Oncol.* **2011**, *6*, 1818–1825. [CrossRef]
66. Balgkouranidou, I.; Chimonidou, M.; Milaki, G.; Tsaroupa, E.G.; Kakolyris, S.; Welch, D.R.; Georgoulias, V.; Lianidou, E.S. Breast cancer metastasis suppressor-1 promoter methylation in cell-free DNA provides prognostic information in non-small cell lung cancer. *Br. J. Cancer* **2014**, *110*, 2054–2062. [CrossRef]
67. Balgkouranidou, I.; Chimonidou, M.; Milaki, G.; Tsaroucha, E.; Kakolyris, S.; Georgoulias, V.; Lianidou, E. SOX17 promoter methylation in plasma circulating tumor DNA of patients with non-small cell lung cancer. *Clin. Chem. Lab. Med.* **2016**, *54*, 1385–1393. [CrossRef]
68. Ramirez, J.L.; Rosell, R.; Taron, M.; Sanchez-Ronco, M.; Alberola, V.; de Las Peñas, R.; Sanchez, J.M.; Moran, T.; Camps, C.; Massuti, B.; et al. 14-3-3sigma methylation in pretreatment serum circulating DNA of cisplatin-plus-gemcitabine-treated advanced non-small-cell lung cancer patients predicts survival: The Spanish lung cancer group. *J. Clin. Oncol.* **2005**, *23*, 9105–9112. [CrossRef] [PubMed]
69. Salazar, F.; Molina, M.A.; Sanchez-Ronco, M.; Moran, T.; Ramirez, J.L.; Sanchez, J.M.; Stahel, R.; Garrido, P.; Cobo, M.; Isla, D.; et al. First-line therapy and methylation status of CHFR in serum influence outcome to chemotherapy versus EGFR tyrosine kinase inhibitors as second-line therapy in stage IV non-small-cell lung cancer patients. *Lung Cancer* **2011**, *72*, 84–91. [CrossRef] [PubMed]
70. Wang, H.; Zhang, B.; Chen, D.; Xia, W.; Zhang, J.; Wang, F.; Xu, J.; Zhang, Y.; Zhang, M.; Zhang, L.; et al. Real-time monitoring efficiency and toxicity of chemotherapy in patients with advanced lung cancer. *Clin. Epigenet.* **2015**, *7*, 119. [CrossRef] [PubMed]
71. Shen, J.; Wang, S.; Zhang, Y.-J.; Kappil, M.A.; Chen Wu, H.; Kibriya, M.G.; Wang, Q.; Jasmine, F.; Ahsan, H.; Lee, P.-H.; et al. Genome-wide aberrant DNA methylation of MicroRNA host genes in hepatocellular carcinoma. *Epigenetics* **2012**, *7*, 1230–1237. [CrossRef]
72. Gao, X.; Wang, Y.; Zhao, H.; Wei, F.; Zhang, X.; Su, Y.; Wang, C.; Li, H.; Ren, X. Plasma MiR-324-3p and MiR-1285 as Diagnostic and prognostic biomarkers for early stage lung squamous cell carcinoma. *Oncotarget* **2016**, *7*, 59664–59675. [CrossRef]
73. Chen, X.; Hu, Z.; Wang, W.; Ba, Y.; Ma, L.; Zhang, C.; Wang, C.; Ren, Z.; Zhao, Y.; Wu, S.; et al. Identification of ten serum MicroRNAs from a genome-wide serum MicroRNA expression profile as novel noninvasive biomarkers for nonsmall cell lung cancer diagnosis. *Int. J. Cancer* **2012**, *130*, 1620–1628. [CrossRef]
74. Yu, H.; Guan, Z.; Cuk, K.; Zhang, Y.; Brenner, H. Circulating MicroRNA biomarkers for lung cancer detection in east Asian populations. *Cancers* **2019**, *11*, 415. [CrossRef]
75. Fehlmann, T.; Kahraman, M.; Ludwig, N.; Backes, C.; Galata, V.; Keller, V.; Geffers, L.; Mercaldo, N.; Hornung, D.; Weis, T.; et al. Evaluating the use of circulating MicroRNA profiles for lung cancer detection in symptomatic patients. *JAMA Oncol.* **2020**, *6*, 714–723. [CrossRef]
76. Powrózek, T.; Kuźnar-Kamińska, B.; Dziedzic, M.; Mlak, R.; Batura-Gabryel, H.; Sagan, D.; Krawczyk, P.; Milanowski, J.; Małecka-Massalska, T. The diagnostic role of plasma circulating precursors of MiRNA-944 and MiRNA-3662 for non-small cell lung cancer detection. *Pathol. Res. Pract.* **2017**, *213*, 1384–1387. [CrossRef]
77. Ulivi, P.; Petracci, E.; Marisi, G.; Baglivo, S.; Chiari, R.; Billi, M.; Canale, M.; Pasini, L.; Racanicchi, S.; Vagheggini, A.; et al. Prognostic role of circulating MiRNAs in early-stage non-small cell lung cancer. *J. Clin. Med.* **2019**, *8*, 131. [CrossRef]
78. Yanaihara, N.; Caplen, N.; Bowman, E.; Seike, M.; Kumamoto, K.; Yi, M.; Stephens, R.M.; Okamoto, A.; Yokota, J.; Tanaka, T.; et al. Unique MicroRNA molecular profiles in lung cancer diagnosis and prognosis. *Cancer Cell* **2006**, *9*, 189–198. [CrossRef]
79. Xie, Y.; Zhang, Y.; Du, L.; Jiang, X.; Yan, S.; Duan, W.; Li, J.; Zhan, Y.; Wang, L.; Zhang, S.; et al. Circulating long noncoding RNA act as potential novel biomarkers for diagnosis and prognosis of non-small cell lung cancer. *Mol. Oncol.* **2018**, *12*, 648–658. [CrossRef]
80. Luo, Y.-H.; Yang, Y.-P.; Chien, C.-S.; Yarmishyn, A.A.; Ishola, A.A.; Chien, Y.; Chen, Y.-M.; Huang, T.-W.; Lee, K.-Y.; Huang, W.-C.; et al. Plasma level of circular RNA Hsa_circ_0000190 correlates with tumor progression and poor treatment response in advanced lung cancers. *Cancers* **2020**, *12*, 1740. [CrossRef]
81. Li, B.; Ren, S.; Li, X.; Wang, Y.; Garfield, D.; Zhou, S.; Chen, X.; Su, C.; Chen, M.; Kuang, P.; et al. MiR-21 Overexpression is associated with acquired resistance of EGFR-TKI in non-small cell lung cancer. *Lung Cancer* **2014**, *83*, 146–153. [CrossRef]

82. Wang, S.; Su, X.; Bai, H.; Zhao, J.; Duan, J.; An, T.; Zhuo, M.; Wang, Z.; Wu, M.; Li, Z.; et al. Identification of plasma MicroRNA profiles for primary resistance to EGFR-TKIs in advanced non-small cell lung cancer (NSCLC) patients with EGFR activating mutation. *J. Hematol. Oncol.* **2015**, *8*, 127. [CrossRef]
83. Fan, J.; Yin, Z.; Xu, J.; Wu, F.; Huang, Q.; Yang, L.; Jin, Y.; Yang, G. Circulating MicroRNAs predict the response to Anti-PD-1 therapy in non-small cell lung cancer. *Genomics* **2020**, *112*, 2063–2071. [CrossRef]
84. International Agency for Research on Cancer (IARC). W.H.O. GLOBOCAN. 2020. Available online: https://gco.iarc.fr/today/home (accessed on 15 June 2021).
85. Culp, M.B.B.; Soerjomataram, I.; Efstathiou, J.A.; Bray, F.; Jemal, A. Recent global patterns in prostate cancer incidence and mortality rates. *Eur. Urol.* **2020**, 38–52. [CrossRef]
86. Hoque, M.O.; Begum, S.; Topaloglu, O.; Chatterjee, A.; Rosenbaum, E.; Van Criekinge, W.; Westra, W.H.; Schoenberg, M.; Zahurak, M.; Goodman, S.N.; et al. Quantitation of promoter methylation of multiple genes in urine DNA and bladder cancer detection. *J. Natl. Cancer Inst.* **2006**, *98*, 996–1004. [CrossRef]
87. Porzycki, P.; Ciszkowicz, E. Modern biomarkers in prostate cancer diagnosis. *Cent. Eur. J. Urol.* **2020**, 300–306. [CrossRef]
88. Saginala, K.; Barsouk, A.; Aluru, J.S.; Rawla, P.; Padala, S.A.; Barsouk, A. Epidemiology of bladder cancer. *Med. Sci.* **2020**, *8*, 15. [CrossRef]
89. Berdik, C.; Ashour, M. Unlocking Bladder Cancer. *Nature* **2017**, S34–S35. [CrossRef] [PubMed]
90. Padala, S.A.; Barsouk, A.; Thandra, K.C.; Saginala, K.; Mohammed, A.; Vakiti, A.; Rawla, P.; Barsouk, A. Epidemiology of renal cell carcinoma. *World J. Oncol.* **2020**, *11*, 79–87. [CrossRef] [PubMed]
91. Wu, P.; Cao, Z.; Wu, S. New progress of epigenetic biomarkers in urological cancer. *Dis. Markers* **2016**. [CrossRef] [PubMed]
92. Lasseigne, B.N.; Brooks, J.D. The Role of DNA Methylation in renal cell carcinoma. *Mol. Diagn. Ther.* **2018**, 431–442. [CrossRef]
93. Zill, O.A.; Banks, K.C.; Fairclough, S.R.; Mortimer, S.A.; Vowles, J.V.; Mokhtari, R.; Gandara, D.R.; Mack, P.C.; Odegaard, J.I.; Nagy, R.J.; et al. The landscape of actionable genomic alterations in cell-free circulating tumor DNA from 21,807 advanced cancer patients. *Clin. Cancer Res.* **2018**, *24*, 3528–3538. [CrossRef]
94. Gurioli, G.; Martignano, F.; Salvi, S.; Costantini, M.; Gunelli, R.; Casadio, V. GSTP1 methylation in cancer: A liquid biopsy biomarker? *Clin. Chem. Lab. Med.* **2018**, 705–717. [CrossRef]
95. Wu, T.; Giovannucci, E.; Welge, J.; Mallick, P.; Tang, W.Y.; Ho, S.M. Measurement of GSTP1 promoter methylation in body fluids may complement PSA screening: A meta-analysis. *Br. J. Cancer* **2011**, *105*, 65–73. [CrossRef]
96. Payne, S.R.; Serth, J.; Schostak, M.; Kamradt, J.; Strauss, A.; Thelen, P.; Model, F.; Day, J.K.; Liebenberg, V.; Morotti, A.; et al. DNA methylation biomarkers of prostate cancer: Confirmation of candidates and evidence urine is the most sensitive body fluid for non-invasive detection. *Prostate* **2009**, *69*, 1257–1269. [CrossRef]
97. Goessl, C.; Müller, M.; Heicappell, R.; Krause, H.; Miller, K. DNA-Based detection of prostate cancer in blood, urine, and ejaculates. *Ann. N. Y. Acad. Sci.* **2001**, *945*, 51–58. [CrossRef]
98. Sunami, E.; Shinozaki, M.; Higano, C.S.; Wollman, R.; Dorff, T.B.; Tucker, S.J.; Martinez, S.R.; Singer, F.R.; Hoon, D.S.B. Multimarker circulating DNA assay for assessing blood of prostate cancer patients. *Clin. Chem.* **2009**, *55*, 559–567. [CrossRef]
99. Hauser, S.; Tobias, Z.; Fechner, G.; Lummen, G.; Muller, S.C.; Ellinger, J. Serum DNA hypermethylation in patients with kidney cancer: Results of a prospective study. *Anticancer Res.* **2013**, *33*, 4651–4656.
100. Hoque, M.O.; Begum, S.; Topaloglu, O.; Jeronimo, C.; Mambo, E.; Westra, W.H.; Califano, J.A.; Sidransky, D. Quantitative detection of promoter hypermethylation of multiple genes in the tumor, urine, and serum DNA of patients with renal cancer. *Cancer Res.* **2004**, *64*, 5511–5517. [CrossRef]
101. Brait, M.; Banerjee, M.; Maldonado, L.; Ooki, A.; Loyo, M.; Guida, E.; Izumchenko, E.; Mangold, L.; Humphreys, E.; Rosenbaum, E.; et al. Promoter methylation of MCAM, ERa and ERβ in serum of early stage prostate cancer patients. *Oncotarget* **2017**, *8*, 15431–15440. [CrossRef]
102. Haldrup, C.; Pedersen, A.L.; Øgaard, N.; Strand, S.H.; Høyer, S.; Borre, M.; Ørntoft, T.F.; Sørensen, K.D. Biomarker potential of ST6GALNAC3 and ZNF660 promoter hypermethylation in prostate cancer tissue and liquid biopsies. *Mol. Oncol.* **2018**, *12*, 545–560. [CrossRef]
103. Lin, Y.L.; Sun, G.; Liu, X.Q.; Li, W.P.; Ma, J.G. Clinical significance of CDH13 promoter methylation in serum samples from patients with bladder transitional cell carcinoma. *J. Int. Med. Res.* **2011**, *39*, 179–186. [CrossRef] [PubMed]
104. Lin, Y.L.; Li, Z.G.; He, Z.K.; Guan, T.Y.; Ma, J.Q. Clinical and prognostic significance of protocadherin-10 (PCDH10) promoter methylation in bladder cancer. *J. Int. Med. Res.* **2012**, *40*, 2117–2123. [CrossRef] [PubMed]
105. Luo, Z.G.; Li, Z.G.; Gui, S.L.; Chi, B.J.; Ma, J.G. Protocadherin-17 promoter methylation in serum-derived dna is associated with poor prognosis of bladder cancer. *J. Int. Med. Res.* **2014**, *42*, 35–41. [CrossRef] [PubMed]
106. Wang, Y.; Yu, Y.; Ye, R.; Zhang, D.; Li, Q.; An, D.; Fang, L.; Lin, Y.; Hou, Y.; Xu, A.; et al. An epigenetic biomarker combination of PCDH17 and POU4F2 detects bladder cancer accurately by methylation analyses of urine sediment DNA in Han Chinese. *Oncotarget* **2016**, *7*, 2754–2764. [CrossRef]
107. Fantony, J.J.; Longo, T.A.; Gopalakrishna, A.; Owusu, R.; Lance, R.S.; Foo, W.C.; Inman, B.A.; Abern, M.R. Urinary NID2 and TWIST1 methylation to augment conventional urine cytology for the detection of bladder cancer. *Cancer Biomark.* **2017**, *18*, 381–387. [CrossRef]

108. Hermanns, T.; Savio, A.J.; Olkhov-Mitsel, E.; Mari, A.; Wettstein, M.S.; Saba, K.; Bhindi, B.; Kuk, C.; Poyet, C.; Wild, P.J.; et al. A Noninvasive urine-based methylation biomarker panel to detect bladder cancer and discriminate cancer grade. *Urol. Oncol. Semin. Orig. Investig.* **2020**, *38*, 603.e1–603.e7. [CrossRef]
109. Urakami, S.; Shiina, H.; Enokida, H.; Hirata, H.; Kawamoto, K.; Kawakami, T.; Kikuno, N.; Tanaka, Y.; Majid, S.; Nakagawa, M.; et al. Wnt antagonist family genes as biomarkers for diagnosis, staging, and prognosis of renal cell carcinoma using tumor and serum DNA. *Clin. Cancer Res.* **2006**, *12*, 6989–6997. [CrossRef]
110. Nuzzo, P.V.; Berchuck, J.E.; Korthauer, K.; Spisak, S.; Nassar, A.H.; Abou Alaiwi, S.; Chakravarthy, A.; Shen, S.Y.; Bakouny, Z.; Boccardo, F.; et al. Detection of renal cell carcinoma using plasma and urine cell-free DNA methylomes. *Nat. Med.* **2020**, *26*, 1041–1043. [CrossRef]
111. Vener, T.; Derecho, C.; Baden, J.; Wang, H.; Rajpurohit, Y.; Skelton, J.; Mehrotra, J.; Varde, S.; Chowdary, D.; Stallings, W.; et al. Development of a multiplexed urine assay for prostate cancer diagnosis. *Clin. Chem.* **2008**, *54*, 874–882. [CrossRef]
112. Kawamoto, K.; Enokida, H.; Gotanda, T.; Kubo, H.; Nishiyama, K.; Kawahara, M.; Nakagawa, M. P16INK4a and P14ARF methylation as a potential biomarker for human bladder cancer. *Biochem. Biophys. Res. Commun.* **2006**, *339*, 790–796. [CrossRef]
113. Moreira-Barbosa, C.; Barros-Silva, D.; Costa-Pinheiro, P.; Torres-Ferreira, J.; Constâncio, V.; Freitas, R.; Oliveira, J.; Antunes, L.; Henrique, R.; Jerónimo, C. Comparing diagnostic and prognostic performance of two-gene promoter methylation panels in tissue biopsies and urines of prostate cancer patients. *Clin. Epigenet.* **2018**, *10*. [CrossRef]
114. Jatkoe, T.A.; Karnes, R.J.; Freedland, S.J.; Wang, Y.; Le, A.; Baden, J. A urine-based methylation signature for risk stratification within low-risk prostate cancer. *Br. J. Cancer* **2015**, *112*, 802–808. [CrossRef]
115. Zhao, F.; Olkhov-Mitsel, E.; van der Kwast, T.; Sykes, J.; Zdravic, D.; Venkateswaran, V.; Zlotta, A.R.; Loblaw, A.; Fleshner, N.E.; Klotz, L.; et al. Urinary DNA methylation biomarkers for noninvasive prediction of aggressive disease in patients with prostate cancer on active surveillance. *J. Urol.* **2017**, *197*, 335–341. [CrossRef]
116. Zhao, F.; Olkhov-Mitsel, E.; Kamdar, S.; Jeyapala, R.; Garcia, J.; Hurst, R.; Hanna, M.Y.; Mills, R.; Tuzova, A.V.; O'Reilly, E.; et al. A urine-based DNA methylation assay, ProCUrE, to identify clinically significant prostate cancer 11 medical and health sciences 1112 oncology and carcinogenesis. *Clin. Epigenet.* **2018**, *10*. [CrossRef]
117. Reinert, T.; Borre, M.; Christiansen, A.; Hermann, G.G.; Ørntoft, T.F.; Dyrskjøt, L. Diagnosis of bladder cancer recurrence based on urinary levels of EOMES, HOXA9, POU4F2, TWIST1, VIM, and ZNF154 hypermethylation. *PLoS ONE* **2012**, *7*, e46297. [CrossRef]
118. Yu, X.; Wang, R.; Han, C.; Wang, Z.; Jin, X. A panel of urinary long non-coding RNAs differentiate bladder cancer from urocystitis. *J. Cancer* **2020**, *11*, 781–787. [CrossRef]
119. Du, L.; Jiang, X.; Duan, W.; Wang, R.; Wang, L.; Zheng, G.; Yan, K.; Wang, L.; Li, J.; Zhang, X.; et al. Cell-Free MicroRNA expression signatures in urine serve as novel noninvasive biomarkers for diagnosis and recurrence prediction of bladder cancer. *Oncotarget* **2017**, *8*, 40832–40842. [CrossRef]
120. Urquidi, V.; Netherton, M.; Gomes-Giacoia, E.; Serie, D.J.; Eckel-Passow, J.; Rosser, C.J.; Goodison, S. A MicroRNA biomarker panel for the non-invasive detection of bladder cancer. *Oncotarget* **2016**, *7*, 86290–86299. [CrossRef]
121. Eissa, S.; Matboli, M.; Essawy, N.O.E.; Kotb, Y.M. Integrative functional genetic-epigenetic approach for selecting genes as urine biomarkers for bladder cancer diagnosis. *Tumour Biol.* **2015**, *36*, 9545–9552. [CrossRef]
122. Fan, Y.; Shen, B.; Tan, M.; Mu, X.; Qin, Y.; Zhang, F.; Liu, Y. Long Non-Coding RNA UCA1 increases chemoresistance of bladder cancer cells by regulating wnt signaling. *FEBS J.* **2014**, *281*, 1750–1758. [CrossRef]
123. Urabe, F.; Matsuzaki, J.; Yamamoto, Y.; Kimura, T.; Hara, T.; Ichikawa, M.; Takizawa, S.; Aoki, Y.; Niida, S.; Sakamoto, H.; et al. Large-scale circulating MicroRNA profiling for the liquid biopsy of prostate cancer. *Clin. Cancer Res.* **2019**, *25*, 3016–3025. [CrossRef] [PubMed]
124. Mello-Grand, M.; Gregnanin, I.; Sacchetto, L.; Ostano, P.; Zitella, A.; Bottoni, G.; Oderda, M.; Marra, G.; Munegato, S.; Pardini, B.; et al. Circulating MicroRNAs combined with PSA for accurate and non-invasive prostate cancer detection. *Carcinogenesis* **2019**, *40*, 246–253. [CrossRef] [PubMed]
125. Porzycki, P.; Ciszkowicz, E.; Semik, M.; Tyrka, M. Combination of three MiRNA (MiR-141, MiR-21, and MiR-375) as potential diagnostic tool for prostate cancer recognition. *Int. Urol. Nephrol.* **2018**, *50*, 1619–1626. [CrossRef] [PubMed]
126. Bayo, J.; Castaño, M.A.; Rivera, F.; Navarro, F. Analysis of blood markers for early breast cancer diagnosis. *Clin. Transl. Oncol.* **2018**, *20*, 467–475. [CrossRef]
127. Melnikow, J.; Fenton, J.J.; Whitlock, E.P.; Miglioretti, D.L.; Weyrich, M.S.; Thompson, J.H.; Shah, K. Supplemental screening for breast cancer in women with dense breasts: A systematic review for the U.S. preventive services task force. *Ann. Intern. Med.* **2016**, *164*, 268–278. [CrossRef]
128. Miller, A.B.; Wall, C.; Baines, C.J.; Sun, P.; To, T.; Narod, S.A. Twenty five year follow-up for breast cancer incidence and mortality of the canadian national breast screening study: Randomised screening trial. *BMJ* **2014**, *348*, g366. [CrossRef]
129. Dulaimi, E.; Hillinck, J.; Ibanez de Caceres, I.; Al-Saleem, T.; Cairns, P. Tumor suppressor gene promoter hypermethylation in serum of breast cancer patients. *Clin. Cancer Res.* **2004**, *10 18 Pt 1*, 6189–6193. [CrossRef]
130. Hoque, M.O.; Feng, Q.; Toure, P.; Dem, A.; Critchlow, C.W.; Hawes, S.E.; Wood, T.; Jeronimo, C.; Rosenbaum, E.; Stern, J.; et al. Detection of aberrant methylation of four genes in plasma DNA for the detection of breast cancer. *J. Clin. Oncol.* **2006**, *24*, 4262–4269. [CrossRef]

131. Skvortsova, T.E.; Rykova, E.Y.; Tamkovich, S.N.; Bryzgunova, O.E.; Starikov, A.V.; Kuznetsova, N.P.; Vlassov, V.V.; Laktionov, P.P. Cell-Free and cell-bound circulating DNA in breast tumours: DNA quantification and analysis of tumour-related gene methylation. *Br. J. Cancer* **2006**, *94*, 1492–1495. [CrossRef]
132. Yamamoto, N.; Nakayama, T.; Kajita, M.; Miyake, T.; Iwamoto, T.; Kim, S.J.; Sakai, A.; Ishihara, H.; Tamaki, Y.; Noguchi, S. Detection of aberrant promoter methylation of GSTP1, RASSF1A, and RARβ2 in serum DNA of patients with breast cancer by a newly established one-step methylation-specific PCR assay. *Breast Cancer Res. Treat.* **2012**, *132*, 165–173. [CrossRef]
133. Han, Z.-H.; Xu, C.-S.; Han, H.; Wang, C.; Lin, S.-G. Value of the level of methylation of RASSF1A and WIF-1 in tissue and serum in neoadjuvant chemotherapeutic assessment for advanced breast cancer. *Oncol. Lett.* **2017**, *14*, 4499–4504. [CrossRef]
134. Taryma-Leśniak, O.; Sokolowska, K.E.; Wojdacz, T.K. Current status of development of methylation biomarkers for in vitro diagnostic IVD applications. *Clin. Epigenet.* **2020**, *12*, 100. [CrossRef]
135. Ozawa, P.M.M.; Vieira, E.; Lemos, D.S.; Souza, I.L.M.; Zanata, S.M.; Pankievicz, V.C.; Tuleski, T.R.; Souza, E.M.; Wowk, P.F.; de Andrade Urban, C.; et al. Identification of MiRNAs enriched in extracellular vesicles derived from serum samples of breast cancer patients. *Biomolecules* **2020**, *10*, 150. [CrossRef]
136. Schwarzenbach, H.; Pantel, K. Circulating DNA as biomarker in breast cancer. *Breast Cancer Res.* **2015**, *17*, 136. [CrossRef]
137. Chimonidou, M.; Kallergi, G.; Georgoulias, V.; Welch, D.R.; Lianidou, E.S. Breast cancer metastasis suppressor-1 promoter methylation in primary breast tumors and corresponding circulating tumor cells. *Mol. Cancer Res.* **2013**, *11*, 1248–1257. [CrossRef]
138. Chimonidou, M.; Tzitzira, A.; Strati, A.; Sotiropoulou, G.; Sfikas, C.; Malamos, N.; Georgoulias, V.; Lianidou, E. CST6 promoter methylation in circulating cell-free DNA of breast cancer patients. *Clin. Biochem.* **2013**, *46*, 235–240. [CrossRef]
139. Sharma, G.; Mirza, S.; Parshad, R.; Srivastava, A.; Gupta, S.D.; Pandya, P.; Ralhan, R. Clinical significance of maspin promoter methylation and loss of its protein expression in invasive ductal breast carcinoma: Correlation with VEGF-A and MTA1 expression. *Tumour Biol.* **2011**, *32*, 23–32. [CrossRef]
140. De Ruijter, T.C.; van der Heide, F.; Smits, K.M.; Aarts, M.J.; van Engeland, M.; Heijnen, V.C.G. Prognostic DNA methylation markers for hormone receptor breast cancer: A systematic review. *Breast Cancer Res. BCR* **2020**, *22*, 13. [CrossRef]
141. Huang, Y.-T.; Li, F.-F.; Ke, C.; Li, Z.; Li, Z.-T.; Zou, X.-F.; Zheng, X.-X.; Chen, Y.-P.; Zhang, H. PTPRO promoter methylation is predictive of poorer outcome for HER2-positive breast cancer: Indication for personalized therapy. *J. Transl. Med.* **2013**, *11*, 245. [CrossRef]
142. Chimonidou, M.; Strati, A.; Malamos, N.; Georgoulias, V.; Lianidou, E.S. sox17 promoter methylation in circulating tumor cells and matched cell-free DNA isolated from plasma of patients with breast cancer. *Clin. Chem.* **2013**, *59*, 270–279. [CrossRef]
143. Fu, D.; Ren, C.; Tan, H.; Wei, J.; Zhu, Y.; He, C.; Shao, W.; Zhang, J. Sox17 promoter methylation in plasma DNA is associated with poor survival and can be used as a prognostic factor in breast cancer. *Medicine* **2015**, *94*, e637. [CrossRef] [PubMed]
144. Church, T.R.; Wandell, M.; Lofton-Day, C.; Mongin, S.J.; Burger, M.; Payne, S.R.; Castaños-Vélez, E.; Blumenstein, B.A.; Rösch, T.; Osborn, N.; et al. Prospective evaluation of methylated SEPT9 in plasma for detection of asymptomatic colorectal cancer. *Gut* **2014**, *63*, 317–325. [CrossRef] [PubMed]
145. Absmaier, M.; Napieralski, R.; Schuster, T.; Aubele, M.; Walch, A.; Magdolen, V.; Dorn, J.; Gross, E.; Harbeck, N.; Noske, A.; et al. PITX2 DNA-methylation predicts response to anthracycline-based adjuvant chemotherapy in triple-negative breast cancer patients. *Int. J. Oncol.* **2018**, *52*, 755–767. [CrossRef] [PubMed]
146. Martínez-Galán, J.; Torres-Torres, B.; Núñez, M.I.; López-Peñalver, J.; Del Moral, R.; Ruiz De Almodóvar, J.M.; Menjón, S.; Concha, A.; Chamorro, C.; Ríos, S.; et al. ESR1 gene promoter region methylation in free circulating DNA and its correlation with estrogen receptor protein expression in tumor tissue in breast cancer patients. *BMC Cancer* **2014**, *14*, 59. [CrossRef]
147. Visvanathan, K.; Fackler, M.S.; Zhang, Z.; Lopez-Bujanda, Z.A.; Jeter, S.C.; Sokoll, L.J.; Garrett-Mayer, E.; Cope, L.M.; Umbricht, C.B.; Euhus, D.M.; et al. Monitoring of serum dna methylation as an early independent marker of response and survival in metastatic breast cancer: TBCRC 005 prospective biomarker study. *J. Clin. Oncol.* **2017**, *35*, 751–758. [CrossRef]
148. Shi, J. Considering exosomal MiR-21 as a biomarker for cancer. *J. Clin. Med.* **2016**, *5*, 42. [CrossRef]
149. Jang, J.Y.; Kim, Y.S.; Kang, K.N.; Kim, K.H.; Park, Y.J.; Kim, C.W. Multiple MicroRNAs as biomarkers for early breast cancer diagnosis. *Mol. Clin. Oncol.* **2021**, *14*, 31. [CrossRef]
150. Estevão-Pereira, H.; Lobo, J.; Salta, S.; Amorim, M.; Lopes, P.; Cantante, M.; Reis, B.; Antunes, L.; Castro, F.; Palma de Sousa, S.; et al. Overexpression of circulating MiR-30b-5p identifies advanced breast cancer. *J. Transl. Med.* **2019**, *17*, 435. [CrossRef]
151. Shiino, S.; Matsuzaki, J.; Shimomura, A.; Kawauchi, J.; Takizawa, S.; Sakamoto, H.; Aoki, Y.; Yoshida, M.; Tamura, K.; Kato, K.; et al. Serum MiRNA-based prediction of axillary lymph node metastasis in breast cancer. *Clin. Cancer Res.* **2019**, *25*, 1817–1827. [CrossRef]
152. Salvador-Coloma, C.; Santaballa, A.; Sanmartín, E.; Calvo, D.; García, A.; Hervás, D.; Cordón, L.; Quintas, G.; Ripoll, F.; Panadero, J.; et al. Immunosuppressive profiles in liquid biopsy at diagnosis predict response to neoadjuvant chemotherapy in triple-negative breast cancer. *Eur. J. Cancer* **2020**, *139*, 119–134. [CrossRef]
153. Özgür, E.; Ferhatoğlu, F.; Şen, F.; Saip, P.; Gezer, U. Circulating LncRNA H19 may be a useful marker of response to neoadjuvant chemotherapy in breast cancer. *Cancer Biomark. Sect. Dis. Markers* **2020**, *27*, 11–17. [CrossRef]
154. Bray, F.; Ferlay, J.; Soerjomataram, I.; Siegel, R.L.; Torre, L.A.; Jemal, A. Global cancer statistics 2018: GLOBOCAN estimates of incidence and mortality worldwide for 36 cancers in 185 countries. *CA Cancer J. Clin.* **2018**, *68*, 394–424. [CrossRef]
155. Song, L.-L.; Li, Y.-M. Current noninvasive tests for colorectal cancer screening: An overview of colorectal cancer screening tests. *World J. Gastrointest. Oncol.* **2016**, *8*, 793–800. [CrossRef]

156. Nee, J.; Chippendale, R.Z.; Feuerstein, J.D. Screening for colon cancer in older adults: Risks, benefits, and when to stop. *Mayo Clin. Proc.* **2020**, *95*, 184–196. [CrossRef]
157. Herreros-Villanueva, M.; Duran-Sanchon, S.; Martín, A.C.; Pérez-Palacios, R.; Vila-Navarro, E.; Marcuello, M.; Diaz-Centeno, M.; Cubiella, J.; Diez, M.S.; Bujanda, L.; et al. Plasma MicroRNA signature validation for early detection of colorectal cancer. *Clin. Transl. Gastroenterol.* **2019**, *10*, e00003. [CrossRef]
158. Zhao, G.; Liu, X.; Liu, Y.; Li, H.; Ma, Y.; Li, S.; Zhu, Y.; Miao, J.; Xiong, S.; Fei, S.; et al. Aberrant DNA methylation of SEPT9 and SDC2 in stool specimens as an integrated biomarker for colorectal cancer early detection. *Front. Genet.* **2020**, *11*, 643. [CrossRef]
159. Lofton-Day, C.; Model, F.; Devos, T.; Tetzner, R.; Distler, J.; Schuster, M.; Song, X.; Lesche, R.; Liebenberg, V.; Ebert, M.; et al. DNA methylation biomarkers for blood-based colorectal cancer screening. *Clin. Chem.* **2008**, *54*, 414–423. [CrossRef]
160. Wang, Y.; Chen, P.-M.; Liu, R.-B. Advance in plasma SEPT9 gene methylation assay for colorectal cancer early detection. *World J. Gastrointest. Oncol.* **2018**, *10*, 15–22. [CrossRef]
161. Pickhardt, P.J. Emerging stool-based and blood-based non-invasive DNA tests for colorectal cancer screening: The importance of cancer prevention in addition to cancer detection. *Abdom. Radiol.* **2016**, *41*, 1441–1444. [CrossRef]
162. Issa, I.A.; Noureddine, M. Colorectal cancer screening: An updated review of the available options. *World J. Gastroenterol.* **2017**, *23*, 5086–5096. [CrossRef]
163. Liu, Y.; Zhao, G.; Miao, J.; Li, H.; Ma, Y.; Liu, X.; Li, S.; Zhu, Y.; Xiong, S.; Zheng, M.; et al. Performance comparison between plasma and stool methylated SEPT9 tests for detecting colorectal cancer. *Front. Genet.* **2020**, *11*, 324. [CrossRef] [PubMed]
164. Zhao, G.; Li, H.; Yang, Z.; Wang, Z.; Xu, M.; Xiong, S.; Li, S.; Wu, X.; Liu, X.; Wang, Z.; et al. Multiplex methylated DNA testing in plasma with high sensitivity and specificity for colorectal cancer screening. *Cancer Med.* **2019**, *8*, 5619–5628. [CrossRef] [PubMed]
165. Wang, W.; Qu, A.; Liu, W.; Liu, Y.; Zheng, G.; Du, L.; Zhang, X.; Yang, Y.; Wang, C.; Chen, X. Circulating MiR-210 as a diagnostic and prognostic biomarker for colorectal cancer. *Eur. J. Cancer Care* **2017**, *26*. [CrossRef] [PubMed]
166. Shi, J.; Li, X.; Zhang, F.; Zhang, C.; Guan, Q.; Cao, X.; Zhu, W.; Zhang, X.; Cheng, Y.; Ou, K.; et al. Circulating LncRNAs associated with occurrence of colorectal cancer progression. *Am. J. Cancer Res.* **2015**, *5*, 2258–2265.
167. Bustos, M.A.; Gross, R.; Rahimzadeh, N.; Cole, H.; Tran, L.T.; Tran, K.D.; Takeshima, L.; Stern, S.L.; O'Day, S.; Hoon, D.S.B. A pilot study comparing the efficacy of lactate dehydrogenase levels versus circulating cell-free MicroRNAs in monitoring responses to checkpoint inhibitor immunotherapy in metastatic melanoma patients. *Cancers* **2020**, *12*, 3361. [CrossRef]
168. Xing, X.-B.; Cai, W.-B.; Luo, L.; Liu, L.-S.; Shi, H.-J.; Chen, M.-H. The prognostic value of P16 hypermethylation in cancer: A meta-analysis. *PLoS ONE* **2013**, *8*, e66587. [CrossRef]
169. Liu, M.C.; Oxnard, G.R.; Klein, E.A.; Swanton, C.; Seiden, M.V. Sensitive and specific multi-cancer detection and localization using methylation signatures in cell-free DNA. *Ann. Oncol.* **2020**, *31*, 745–759. [CrossRef]
170. Rasmussen, S.L.; Krarup, H.B.; Sunesen, K.G.; Johansen, M.B.; Stender, M.T.; Pedersen, I.S.; Madsen, P.H.; Thorlacius-Ussing, O. The prognostic efficacy of Cell-Free DNA hypermethylation in colorectal cancer. *Oncotarget* **2018**, *9*, 7010–7022. [CrossRef]
171. Barault, L.; Amatu, A.; Siravegna, G.; Ponzetti, A.; Moran, S.; Cassingena, A.; Mussolin, B.; Falcomatà, C.; Binder, A.M.; Cristiano, C.; et al. Discovery of methylated circulating DNA biomarkers for comprehensive non-invasive monitoring of treatment response in metastatic colorectal cancer. *Gut* **2018**, *67*, 1995–2005. [CrossRef]
172. Bhangu, J.S.; Beer, A.; Mittlböck, M.; Tamandl, D.; Pulverer, W.; Schönthaler, S.; Taghizadeh, H.; Stremitzer, S.; Kaczirek, K.; Gruenberger, T.; et al. Circulating free methylated tumor DNA markers for sensitive assessment of tumor burden and early response monitoring in patients receiving systemic chemotherapy for colorectal cancer liver metastasis. *Ann. Surg.* **2018**, *268*, 894–902. [CrossRef]
173. Young, G.P.; Pedersen, S.K.; Mansfield, S.; Murray, D.H.; Baker, R.T.; Rabbitt, P.; Byrne, S.; Bambacas, L.; Hollington, P.; Symonds, E.L. A cross-sectional study comparing a blood test for methylated BCAT1 and IKZF1 tumor-derived DNA with CEA for detection of recurrent colorectal cancer. *Cancer Med.* **2016**, *5*, 2763–2772. [CrossRef]
174. Peng, Q.; Zhang, X.; Min, M.; Zou, L.; Shen, P.; Zhu, Y. The clinical role of MicroRNA-21 as a promising biomarker in the diagnosis and prognosis of colorectal cancer: A systematic review and meta-analysis. *Oncotarget* **2017**, *8*, 44893–44909. [CrossRef]
175. Sazanov, A.A.; Kiselyova, E.V.; Zakharenko, A.A.; Romanov, M.N.; Zaraysky, M.I. Plasma and saliva MiR-21 expression in colorectal cancer patients. *J. Appl. Genet.* **2017**, *58*, 231–237. [CrossRef]
176. Wang, Q.; Huang, Z.; Ni, S.; Xiao, X.; Xu, Q.; Wang, L.; Huang, D.; Tan, C.; Sheng, W.; Du, X. Plasma MiR-601 and MiR-760 are novel biomarkers for the early detection of colorectal cancer. *PLoS ONE* **2012**, *7*, e44398. [CrossRef]
177. Cheng, H.; Zhang, L.; Cogdell, D.E.; Zheng, H.; Schetter, A.J.; Nykter, M.; Harris, C.C.; Chen, K.; Hamilton, S.R.; Zhang, W. Circulating plasma MiR-141 is a novel biomarker for metastatic colon cancer and predicts poor prognosis. *PLoS ONE* **2011**, *6*, e17745. [CrossRef]
178. Kou, C.-H.; Zhou, T.; Han, X.-L.; Zhuang, H.-J.; Qian, H.-X. Downregulation of Mir-23b in plasma is associated with poor prognosis in patients with colorectal cancer. *Oncol. Lett.* **2016**, *12*, 4838–4844. [CrossRef]
179. Yuan, Z.; Baker, K.; Redman, M.W.; Wang, L.; Adams, S.V.; Yu, M.; Dickinson, B.; Makar, K.; Ulrich, N.; Böhm, J.; et al. Dynamic plasma MicroRNAs Are biomarkers for prognosis and early detection of recurrence in colorectal cancer. *Br. J. Cancer* **2017**, *117*, 1202–1210. [CrossRef]
180. Hansen, T.F.; Carlsen, A.L.; Heegaard, N.H.H.; Sørensen, F.B.; Jakobsen, A. Changes in circulating MicroRNA-126 during treatment with chemotherapy and bevacizumab predicts treatment response in patients with metastatic colorectal cancer. *Br. J. Cancer* **2015**, *112*, 624–629. [CrossRef]

181. Svoboda, M.; Slyskova, J.; Schneiderova, M.; Makovicky, P.; Bielik, L.; Levy, M.; Lipska, L.; Hemmelova, B.; Kala, Z.; Protivankova, M.; et al. HOTAIR long non-coding RNA is a negative prognostic factor not only in primary tumors, but also in the blood of colorectal cancer patients. *Carcinogenesis* **2014**, *35*, 1510–1515. [CrossRef]
182. Liu, H.; Ye, D.; Chen, A.; Tan, D.; Zhang, W.; Jiang, W.; Wang, M.; Zhang, X. A pilot study of new promising non-coding RNA diagnostic biomarkers for early-stage colorectal cancers. *Clin. Chem. Lab. Med.* **2019**, *57*, 1073–1083. [CrossRef]
183. Xu, W.; Zhou, G.; Wang, H.; Liu, Y.; Chen, B.; Chen, W.; Lin, C.; Wu, S.; Gong, A.; Xu, M. Circulating LncRNA SNHG11 as a novel biomarker for early diagnosis and prognosis of colorectal cancer. *Int. J. Cancer* **2020**, *146*, 2901–2912. [CrossRef]
184. Salvianti, F.; Orlando, C.; Massi, D.; De Giorgi, V.; Grazzini, M.; Pazzagli, M.; Pinzani, P. Tumor-related methylated cell-free DNA and circulating tumor cells in melanoma. *Front. Mol. Biosci.* **2015**, *2*, 76. [CrossRef]
185. Mori, T.; O'Day, S.J.; Umetani, N.; Martinez, S.R.; Kitago, M.; Koyanagi, K.; Kuo, C.; Takeshima, T.-L.; Milford, R.; Wang, H.-J.; et al. Predictive utility of circulating methylated DNA in serum of melanoma patients receiving biochemotherapy. *J. Clin. Oncol.* **2005**, *23*, 9351–9358. [CrossRef]
186. Kolenda, T.; Rutkowski, P.; Michalak, M.; Kozak, K.; Guglas, K.; Ryś, M.; Galus, Ł.; Woźniak, S.; Ługowska, I.; Gos, A.; et al. Plasma LncRNA expression profile as a prognostic tool in BRAF-mutant metastatic melanoma patients treated with BRAF inhibitor. *Oncotarget* **2019**, *10*, 3879–3893. [CrossRef]
187. Manoochehri, M.; Wu, Y.; Giese, N.A.; Strobel, O.; Kutschmann, S.; Haller, F.; Hoheisel, J.D.; Moskalev, E.A.; Hackert, T.; Bauer, A.S. SST gene hypermethylation acts as a pan-cancer marker for pancreatic ductal adenocarcinoma and multiple other tumors: Toward its use for blood-based diagnosis. *Mol. Oncol.* **2020**, *14*, 1252–1267. [CrossRef]
188. Giannopoulou, L.; Kasimir-Bauer, S.; Lianidou, E.S. Liquid biopsy in ovarian cancer: Recent advances on circulating tumor cells and circulating tumor DNA. *Clin. Chem. Lab. Med.* **2018**, *56*, 186–197. [CrossRef]
189. Wang, L.; Chen, Y.-J.; Xu, K.; Xu, H.; Shen, X.-Z.; Tu, R.-Q. Circulating MicroRNAs as a fingerprint for endometrial endometrioid adenocarcinoma. *PLoS ONE* **2014**, *9*, e110767. [CrossRef]
190. Nassiri, F.; Chakravarthy, A.; Feng, S.; Shen, S.Y.; Nejad, R.; Zuccato, J.A.; Voisin, M.R.; Patil, V.; Horbinski, C.; Aldape, K.; et al. Detection and discrimination of intracranial tumors using plasma cell-free DNA methylomes. *Nat. Med.* **2020**, *26*, 1044–1047. [CrossRef]
191. Hegi, M.E.; Diserens, A.-C.; Gorlia, T.; Hamou, M.-F.; de Tribolet, N.; Weller, M.; Kros, J.M.; Hainfellner, J.A.; Mason, W.; Mariani, L.; et al. MGMT gene silencing and benefit from temozolomide in glioblastoma. *N. Engl. J. Med.* **2005**, *352*, 997–1003. [CrossRef]
192. Kopkova, A.; Sana, J.; Machackova, T.; Vecera, M.; Radova, L.; Trachtova, K.; Vybihal, V.; Smrcka, M.; Kazda, T.; Slaby, O.; et al. Cerebrospinal fluid MicroRNA signatures as diagnostic biomarkers in brain tumors. *Cancers* **2019**, *11*, 1546. [CrossRef]
193. Lim, Y.; Wan, Y.; Vagenas, D.; Ovchinnikov, D.A.; Perry, C.F.L.; Davis, M.J.; Punyadeera, C. Salivary DNA Methylation panel to diagnose HPV-positive and HPV-negative head and neck cancers. *BMC Cancer* **2016**, *16*, 749. [CrossRef] [PubMed]
194. Chen, X.; Gole, J.; Gore, A.; He, Q.; Lu, M.; Min, J.; Yuan, Z.; Yang, X.; Jiang, Y.; Zhang, T.; et al. Non-invasive early detection of cancer four years before conventional diagnosis using a blood test. *Nat. Commun.* **2020**, *11*, 3475. [CrossRef] [PubMed]
195. García-Giménez, J.L.; Seco-Cervera, M.; Tollefsbol, T.O.; Romá-Mateo, C.; Peiró-Chova, L.; Lapunzina, P.; Pallardó, F.V. Epigenetic biomarkers: Current strategies and future challenges for their use in the clinical laboratory. *Crit. Rev. Clin. Lab. Sci.* **2017**, *54*, 529–550. [CrossRef] [PubMed]

Commentary

Rheumatoid Factor: A Novel Determiner in Cancer History

Alessio Ugolini [1,2] and Marianna Nuti [1,*]

1 Department of Experimental Medicine, "Sapienza" University of Rome, Viale Regina Elena 324, 00161 Rome, Italy; alessio.ugolini@moffitt.org
2 Department of Immunology, H. Lee Moffitt Cancer Center, Tampa, FL 33612, USA
* Correspondence: marianna.nuti@uniroma1.it

Simple Summary: Rheumatoid factors are autoantibodies that characterize different autoimmune diseases, in particular rheumatoid arthritis, but that can also be found in the sera of the general healthy population. They have been mainly studied in the context of autoimmune diseases, but some evidence have suggested an association between their presence and the predisposition to develop cancer as well as a facilitation of cancer growth and progression in oncologic patients. In this review, for the first time we thus analyze and discuss the possible roles that these autoantibodies can assume in tumor history, from determiners of a heightened susceptibility of developing cancer to drivers of a reduced response to immunotherapies.

Abstract: The possible interplay between autoimmunity and cancer is a topic that still needs to be deeply explored. Rheumatoid factors are autoantibodies that are able to bind the constant regions (Fc) of immunoglobulins class G (IgGs). In physiological conditions, their production is a transient event aimed at contributing to the elimination of pathogens as well as limiting a redundant immune response by facilitating the clearance of antibodies and immune complexes. Their production can become persistent in case of different chronic infections or diseases, being for instance a fundamental marker for the diagnosis and prognosis of rheumatoid arthritis. Their presence is also associated with aging. Some studies highlighted how elevated levels of rheumatoid factors (RFs) in the blood of patients are correlated with an increased cancer risk, tumor recurrence, and load and with a reduced response to anti-tumor immunotherapies. In line with their physiological roles, RFs showed in different works the ability to impair in vitro anti-cancer immune responses and effector functions, suggesting their potential immunosuppressive activity in the context of tumor immunity. Thus, the aim of this review is to investigate the emerging role of RFs as determiners of cancer faith.

Keywords: rheumatoid factor; autoimmunity; autoantibodies; cancer; biomarker; predictive biomarker; prognostic biomarker; cancer progression; cancer development; immunotherapy; cancer susceptibility; tumor recurrence; tumor load

1. Introduction

The link between autoimmunity and cancer is considered a hot topic since the relationship existing between these two conditions is still to be clarified. Rheumatoid factors (RFs) are autoantibodies with different isotypes and affinities, which bind the constant regions (Fc) of immunoglobulins class G (IgG). RFs were initially discovered in sera of patients with rheumatoid arthritis (RA), and they are still considered a fundamental marker for the diagnosis and prediction of the prognosis of these patients [1,2]. Later, it was highlighted that RFs are not crucial for the development of the arthritis and, above all, that they are not specific only for RA [3].

In fact, high levels of RFs can be found in the sera of patients with other diseases (both autoimmune and non-autoimmune) in the same way as in healthy subjects [4].

2. Rheumatoid Factors Isotypes and Affinities

RFs mainly belong to the immunoglobulins of class M (IgM), class G (IgG), and class A (IgA) isotypes; rarely, also class E (IgE) and class D (IgD) RFs can be detected. While physiological RFs are mainly of the IgM isotype, are polyreactive, have low affinity, and show a reduced usage of the V gene (encoding for the variable region of the antibody), pathological RFs can belong to IgM, IgA, IgG, IgD, and IgE classes and show a high affinity and a wide usage of the V gene, thus indicating that pathological RFs are the result of an immune response against a specific antigen [5–9].

RFs can bind to the Fc of all four subclasses of IgG. In different studies mapping the RFs' binding sites, it resulted that the affinity for the IgG1 subclass was the highest, whereas the affinity for the IgG3 subclass was the most variable among the different sera samples that were tested [10,11].

3. Physiological Rheumatoid Factor Production and Its Presence in Different Clinical Conditions

In peripheral blood of healthy subjects, researchers found a B-cells repertoire that was able to secrete RFs (RF^+ B-cells) [12–14]. This population seems to be anergic in subjects that are RF-seronegative, while it requires a specific activation pattern to start synthetizing RFs [15].

Whereas in pathological conditions, such as RA, the chronic presence of RFs in patient sera is due to the production carried out by terminally differentiated plasma cells in the absence of a specific stimulus [16,17], in physiological conditions, the RFs production is a transient event that results from an initiating stimulus capable of activating the B-cells repertoire [15]. This initiating stimulus can be represented by an infection (bacterial, viral, or parasitic) or by an active immunization [18–23]. The activation of this repertoire of RF^+ B-cells is due to their interaction with T-helper cells, which react against a foreign antigen during a secondary immune response [24,25]. In fact, it was proven that activated T-cells are strong inducers of RF^+ B-cells and, therefore, of physiological RFs production [12–15]. This is coherent with the RFs physiological role: on one side, they seem essential in fighting pathogens by contributing to the formation and clearance of immune complexes (thanks to IgM and IgG RFs capability at forming bigger immune complexes that can both bind the complement and be phagocytosed) and because RF^+ B-cells can act as antigen-presenting cells (APCs); on the other side, RFs are also important in limiting a redundant immune response against pathogens by destroying the antibodies produced in excess [26–29].

3.1. RFs in Patients with Non-Autoimmune Conditions

As outlined above, RFs production is essential in protecting the host against infections in an inflammatory milieu. This is why high levels of RFs can be detected in patients with different types of infections and chronic diseases. Conversely to the RFs found in RA, those detected during infections are not damaging and are usually transient [4]. They are also, as physiological RFs, polyreactive and low-affinity IgMs that show a reduced usage of V gene [6,7,9]. If the infection evolves in a chronic disease, also the RFs circulating levels can become persistent. RFs can indeed be found in the sera of 40–50% of patients with HCV infection, reaching even 76% in some studies [30]. This is probably due to a continuous stimulation and activation of an immune response triggered by the presence of the virus in HCV patients. Since HCV infection nowadays has reached a high prevalence in a large number of countries, it has become the first cause of high RFs levels in sera [30,31].

3.2. RFs in the General Healthy Population

High levels of RFs can be detected in the general healthy population, with a worldwide variability in prevalence: for example, RFs positivity in sera show the highest prevalence in North American Indian tribes (up to 30%), while in young Caucasians it is up to 4% [32–37]. As physiological RFs, those detected in the healthy population are not damaging, are usually transient, and are polyreactive and low-affinity IgM, showing a reduced usage of the V gene [4,6,7,9]. The transient production of these physiological RFs can be the result of any kind of infection [18–23,38]. Polyreactive IgM RFs can be persistently found in 18% of presumably healthy aging subjects, suggesting that their chronic production could be an age-related immune deregulation phenomenon [39–41], whereas in other individuals, the reason of their presence can still not be identified [4].

3.3. RFs in Rheumatoid Arthritis and Other Autoimmune Diseases

In the sera of RA patients, IgM is the most frequent RFs isotype detected, which is followed by IgG and IgA and, very rarely, also IgE and IgD. Most (70–90%) RA patients are RF-positive; three isotypes of RFs, IgM, IgG, and IgA, are detected in up to 52% of patients with RA [4,5,8,42–45]. RFs can also be found in the sera of patients with other autoimmune systemic syndromes, such as Sjogren's syndrome (SS), mixed cryoglobulinemia, systemic lupus erythematosus (SLE), mixed connective tissue disease, polymyositis, and dermatomyositis. They can be detected also in 10% of patients with Waldenstrom's macroglobulinemia (a rare plasma cell cancer). Patients with SS (up to 60%) and type II and type III of mixed cryoglobulinemia (often HCV-related) show the highest RFs titers [26,43,46–49].

4. Rheumatoid Factor and Cancer History

During the years, the presence of circulating RFs was almost exclusively correlated with the diagnosis and prognosis of RA and other autoimmune diseases. The role of RFs in cancer was poorly investigated.

The presence of RFs can be detected in the blood of 10–20% of cancer patients [50], reaching 26% in non-small lung cancer (NSCLC) patients [51]. The higher prevalence of RF positivity in cancer patients compared to the general healthy population can be surely explained by the older age of subjects affected by cancer, since RFs production is associated with aging; however, it could also suggest a possible association between RF positivity and cancer. In this scenario, its production could be the result of a regulatory-skewing B-cells activation.

This association was investigated for the first time in the Reykjavik area (Iceland) where, starting from 1967, a general health survey was conducted [52,53]. Then, the women that were tested positive for RF were divided in groups based on RF titers and followed up until 1974: of the four women who died during this observation period, three belonged to the group with the highest RF titers; all of these three women had been diagnosed with cancer (two mammalian cancer and one lung cancer). Thus, it was suggested that high blood concentrations of RFs in healthy subjects might be associated with an increased risk of developing cancer [52,53].

After this study, other publications showed how the presence of high RF titers in patient sera were associated with an increased cancer risk, tumor recurrence, and tumor load if compared with patients that were RF-negative [54–60].

In particular, in a longitudinal study conducted in 2016 including 2331 patients with early RA, Ajeganova et al. [54] studied the presence of RFs, anticitrullinated protein antibodies (ACPA), and anticarbamylated protein (anti-CarP) antibodies in relation to all causes of mortality. Interestingly, they found that the presence of RF, differently from the other autoantibodies, was associated with an increased number of neoplasm-related deaths [54].

In a cohort study made in 2017 and involving 295,837 RA-free participants, Ahn et al. clearly showed how cancer mortality risk was significantly greater in healthy adults that were positive for RF when compared with those that were RF-negative; moreover, they also demonstrated that cancer mortality risk was even higher in subjects with RF titers greater than 100 IU/mL than in those with RF-negativity, suggesting a dose-dependent effect [55].

Finally, a retrospective study conducted in our laboratory brought clear evidence of how the IgM–RF positivity is a strong predictive factor for the development of NSCLC patients' early progression in response to the treatment with anti-PD-1 immune checkpoint inhibitors (ICIs) [51,61]. IgM-RF also correlates with a negative prognosis in terms of both overall survival (OS) and progression-free survival (PFS) in metastatic NSCLC patients in treatment with an anti-PD-1 ICI, with the worst outcome shown by patients with titers greater than 50 IU/mL [51,61].

Taken together, all these studies strongly suggest a facilitation of cancer growth and progression in patients that are positive for RFs. The mechanism lying behind this phenomenon still remains unknown, but some in vitro experiments highlighted an association between the presence of RFs and an altered anti-tumor immunity.

Indeed, it was pointed out that RF preparations are able to impair the tumor-specific in vitro cytotoxicity of cancer patients' lymphoid cells [62,63]; IgM preparations lacking of RF anti-IgG activity, used as control, did not block the cytotoxicity, indicating that the impairing effect was the result of the specific RFs activity. The pre-incubation of lymphoid effector cells with the RF preparations inhibited their cytotoxic action, whereas pre-incubation of the tumor target cells with RF preparations before cytotoxic lymphocytes were added had no effect, thus indicating that the observed phenomenon was mediated by a direct effect on lymphoid effector cells [62].

Another important evidence supporting the suppressive activity of RFs on lymphoid effector cells directed against tumor cells came out when it was shown that RF can have a blocking effect in antibody-dependent cell-mediated cytotoxicity (ADCC) [64,65], which is an important mechanism underlying the killing of tumor cells exerted by lymphoid cells.

In addition, Giuliano et al. [66] demonstrated that RFs are able to affect the melanoma patients' humoral immune response directed against membrane antigens of melanoma cells in vitro. They indeed showed that the presence of RF in Indirect Membrane Immunofluorescence (IMI) assays increases the IgM reactivity detection, while in the Immune Adherence (IA) assays, its presence reduces the detection of anti-membrane antibodies, thus suggesting that the presence of RF prevents the binding of anti-tumor antibodies to their target antigens on cancer cells [66].

Interestingly, in 2013, Jones at al. [67] found that RF can inhibit Rituximab effector function. Rituximab is an IgG1 monoclonal antibody directed against the receptor CD20, expressed by B-cells, which uses the complement-dependent cytotoxicity (CDC) and other mechanisms to eliminate pathogenic B-cells [68]. It is used in the treatment of some B-cell neoplasms, RA, and other autoimmune diseases [69,70]. In this study, Jones et al. demonstrated that RF inhibits Rituximab-mediated CDC. Since RF does not block the interaction between Rituximab and B-cells, it seems plausible that RF impairs this effector function through the recognition and the binding of Rituximab Fc, which mediates the CDC. Supporting this, they demonstrated that RF can also inhibit the trogocytosis, which is an FcγR-dependent effect [67].

In accordance with these observations, in a recent work, we showed that IgM-RF is not able to prevent the engagement between the drug Nivolumab (an IgG4 monoclonal antibody) and its target receptor PD-1 on T-cells [51]. Instead, IgM-RF is able to bind preferentially naïve and central memory CD4$^+$ and CD8$^+$ T-cells, leading to an impaired in vitro migration of these T-cell subsets in response to the CCL19 cytokine [51,61]. Moreover, RF-positive NSCLC patients showed a significant reduction of the CD137$^+$ T-cells, which identify the tumor-specific effector T-cell population [71]. This suggests that the dysfunctional recirculation of naïve and central memory T-cells due to the presence of IgM-RF can lead to an impaired expansion of the tumor-directed effector T-cell population, consequently resulting in the failure of the anti-PD-1 treatments that relies on tumor-specific effector T-cells in order to be effective [51] (Figure 1).

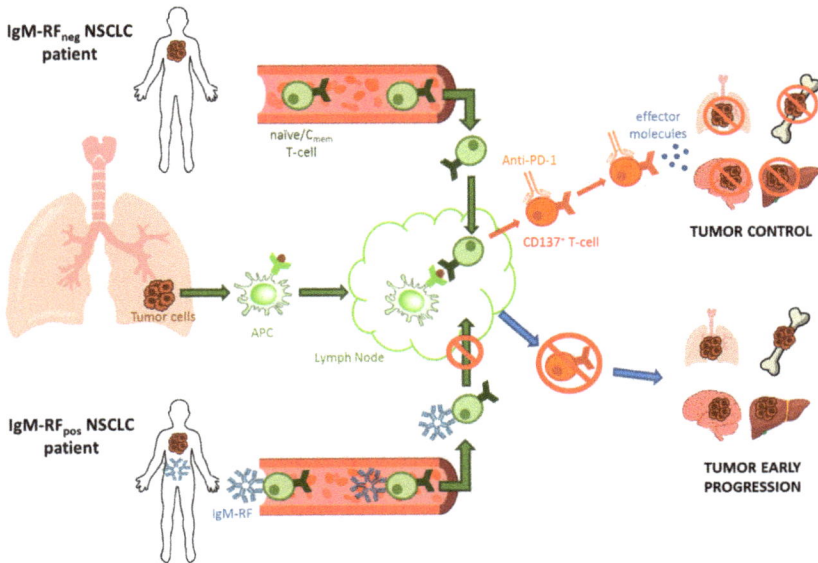

Figure 1. Schematic representation of the suggested effect of immunoglobulin class M (IgM)-rheumatoid factor (RF) in limiting naïve and central memory T-cells recirculation in non-small lung cancer (NSCLC) patients, leading to an impaired CD137+ tumor-directed T-cells expansion and a consequent failure of immune-based immunotherapies.

5. Conclusions

Altogether, these findings suggest that high titers of RF in patients' sera could inhibit the anti-tumor immunity in different ways (Figure 2).

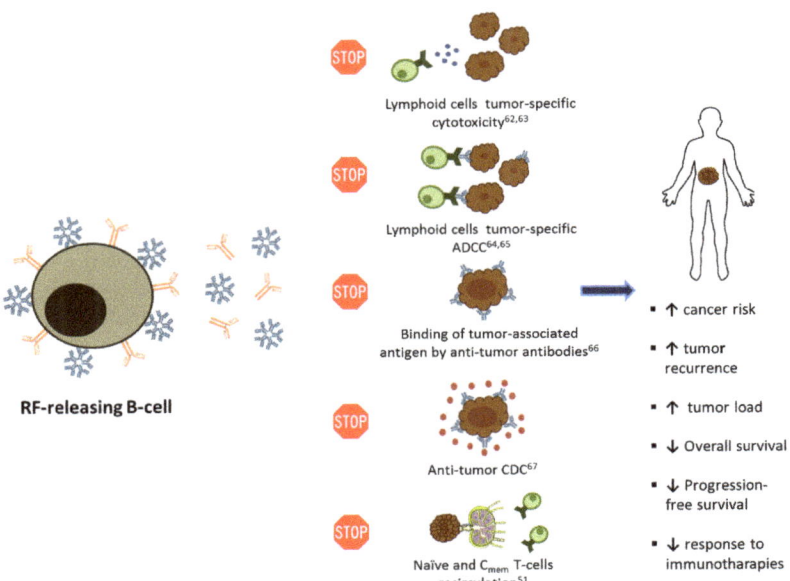

Figure 2. Schematic representation of the different in vitro immune-suppressive activities of RFs in the context of anti-tumor responses with their effects on cancer patients.

In fact, RFs physiological role of limiting a redundant immune response against pathogens seems coherent with their potential immunosuppressive activity within anti-tumor immune responses. On one side, in healthy conditions, it was pointed out how RFs are able to facilitate the clearance of immune complexes and antibodies. Similarly, in cancer, their ability to bind antibodies results in interfering with the anti-tumor effect of both endogenous and therapeutic antibodies. Indeed, it has been demonstrated that RFs can hamper the interaction between tumor-directed antibodies and antigens on cancer cells' surface as well as impair antibodies-mediated anti-tumor effector functions, such as CDC and ADCC. On the other side, the physiological effect exerted by RFs on effector T-cells has not been clarified yet. However, from different works presented in this review, it emerged that in the context of anti-cancer immune responses, the presence of RFs is able to impair lymphocytes-mediated anti-tumor cytotoxicity and the recirculation of naïve and central memory T-cells, leading to a reduced expansion and effectiveness of tumor-specific effector T-cells. Therefore, in the light of these results, it is reliable to assume that RFs' presence in the blood of cancer patients could facilitate cancer growth and progression.

However, although some evidence of the pro-tumor role of RFs was provided in vitro, in vivo data are still lacking.

Thus, the altered anti-tumor immunity due to the presence of RFs may lead both to an increased risk of developing cancer in RF-positive subjects and to a failure of immune-based anti-tumor therapies, in terms of a higher amount of early progressions and a reduced overall survival and progression-free survival rate following immunotherapies. Different studies have indeed demonstrated the association between RF positivity and the predisposition to develop cancer, the increased tumor recurrence and tumor load, the worse prognosis, and eventually the failure of immune checkpoint inhibitors therapies in cancer patients.

These findings opened the possibility of using RFs as both prognostic and predictive biomarkers in cancer patients, which looks promising from two different perspectives. First, being now clear that RFs can be used as an indicator of a heightened susceptibility of developing cancer as well as of an increased tumor recurrence, subjects that are positive for RFs could be monitored more strictly so as to increase early cancer diagnosis. Second, since the presence of RFs in cancer patients correlates with a reduced OS and PFS and a lack of response to anti-cancer immunotherapies, the RFs' dosage could be introduced in oncologic patients together with other parameters in order to improve their risk stratification and help adjust the therapeutic plan based on the single patient's characteristics. Since RF dosage is a routinely performed and already standardized test, this seems to be a easily practicable and promising opportunity to aid clinicians that are struggling to predict patients faith and, above all, to increase the number of responder patients to anti-cancer therapies.

Finally, the higher prevalence of RFs in cancer patients compared to healthy adults could certainly be referred to the older age of subjects affected by cancer, but it could also be more evidence of the association between RF positivity and cancer risk. Indeed, even if direct proof is lacking, we would like to take into consideration the hypothesis that RF secretion could be the result of a regulatory-skewing B-cells activation due to the chronic inflammation carried out by the presence of the tumor itself and, in this scenario, it could serve as a further mechanism exerted by the tumor in its inflammatory milieu in order to escape from the immune surveillance.

Other studies will be required to further clarify the role of RFs in cancer. Nevertheless, the aim of this review is to open a new chapter in the study of cancer history, a chapter in which RF can be considered as a novel determiner of cancer susceptibility and a novel predictive and prognostic biomarker of a negative outcome in cancer patients that should now be taken into account when stratifying oncologic patients.

Author Contributions: Conceptualization, visualization, review, and editing: A.U. and M.N.; original draft preparation: A.U. All authors have read and agreed to the published version of the manuscript.

Funding: This study was supported by MIUR-"Sapienza" (M.N., C26H15Y42B).

Institutional Review Board Statement: Not applicable.

Informed Consent Statement: Not applicable.

Conflicts of Interest: The authors declare that the research was conducted in the absence of any commercial or financial relationships that could be construed as potential conflicts of interest.

References

1. Van Zeben, D.; Hazes, J.M.W.; Zwinderman, A.H.; Cats, A.; Van Der Voort, E.A.M.; Breedveld, F.C.; Van Zeben, D.; Hazes, J.M.W.; Cats, A.; Van Der Voort, E.A.M.; et al. Clinical significance of rheumatoid factors in early rheumatoid arthritis: Results of a follow up study. *Ann. Rheum. Dis.* **1992**, *51*, 1029–1035. [CrossRef] [PubMed]
2. Steiner, G.; Smolen, J.S. Autoantibodies in rheumatoid arthritis: Prevalence and clinical significance. In *Rheumatoid Arthritis*; Oxford University Press: Oxford, UK, 2006; pp. 193–198. ISBN 9780198566304.
3. Dörner, T.; Egerer, K.; Feist, E.; Burmester, G.R. Rheumatoid factor revisited. *Curr. Opin. Rheumatol.* **2004**, *16*, 246–253. [CrossRef] [PubMed]
4. Ingegnoli, F.; Castelli, R.; Gualtierotti, R. Rheumatoid factors: Clinical applications. *Dis. Markers* **2013**, *35*, 727–734. [CrossRef] [PubMed]
5. Vaughan, J.H. Pathogenetic concepts and origins of rheumatoid factor in rheumatoid arthritis. *Arthritis Rheum.* **1993**, *36*, 1–6. [CrossRef] [PubMed]
6. Carson, D.A.; Chen, P.P.; Fox, R.I.; Kipps, T.J.; Jirik, F.; Goldfien, R.D.; Silverman, G.; Radoux, V.; Fong, S. Rheumatoid factor and immune networks. *Annu. Rev. Immunol.* **1987**, *5*, 109–126. [CrossRef]
7. Carson, D.A.; Chen, P.P.; Kipps, T.J. New roles for rheumatoid factor. *J. Clin. Investig.* **1991**, *87*, 379–383. [CrossRef]
8. Bos, W.H.; van de Stadt, L.A.; Sohrabian, A.; Rönnelid, J.; Van Schaardenburg, D. Development of anti-citrullinated protein antibody and rheumatoid factor isotypes prior to the onset of rheumatoid arthritis. *Arthritis Res. Ther.* **2014**, *16*, 1–2. [CrossRef]
9. Dresser, D.W.; Popham, A.M. Induction of an IgM anti-(bovine)-IgG response in mice by bacterial lipopolysaccharide. *Nature* **1976**, *264*, 552–554. [CrossRef]
10. Falkenburg, W.J.J.; Van Schaardenburg, D.; Ooijevaar-De Heer, P.; Wolbink, G.; Rispens, T. IgG subclass specificity discriminates restricted IgM rheumatoid factor responses from more mature anti-citrullinated protein antibody-associated or isotype-switched IgA responses. *Arthritis Rheumatol.* **2015**, *67*, 3124–3134. [CrossRef]
11. Bonagura, V.R.; Artandi, S.E.; Agostino, N.; Tao, M.H.; Morrison, S.L. Mapping rheumatoid factor binding sites using genetically engineered, chimeric IgG antibodies. *DNA Cell Biol.* **1992**, *11*, 245–252. [CrossRef]
12. Hirohata, S.; Inoue, T.; Ito, K. Frequency analysis of human peripheral blood B cells producing autoantibodies: Differential activation requirements of precursors for B cells producing IgM-RF and anti-DNA antibody. *Cell. Immunol.* **1991**, *138*, 445–455. [CrossRef]
13. He, X.; Goronzy, J.J.; Weyand, C.M. The repertoire of rheumatoid factor-producing B cells in normal subjects and patients with rheumatoid arthritis. *Arthritis Rheum.* **1993**, *36*, 1061–1069. [CrossRef] [PubMed]
14. Carson, D.A.; Chen, P.P.; Kipps, T.J.; Roudier, J.; Silverman, G.J.; Tighe, H. Regulation of rheumatoid factor synthesis. *Clin. Exp. Rheumatol.* **1989**, *7*, S69–S73. [PubMed]
15. Van Esch, W.J.; Reparon-Schuijt, C.C.; Levarht, E.W.; Van Kooten, C.; Breedveld, F.C.; Verweij, C.L. Differential requirements for induction of total immunoglobulin and physiological rheumatoid factor production by human peripheral blood B cells. *Clin. Exp. Immunol.* **2001**, *123*, 496–504. [CrossRef]
16. Reparon-Schuijt, C.C.; Van Esch, W.J.E.; Van Kooten, C.; Levarht, E.W.N.; Breedveld, F.C.; Verweij, C.L. Functional analysis of rheumatoid factor-producing B cells from the synovial fluid of rheumatoid arthritis patients. *Arthritis Rheum.* **1998**, *41*, 2211–2220. [CrossRef]
17. Reparon-Schuijt, C.C.; Van Esch, W.J.E.; Van Kooten, C.; Rozier, B.C.D.; Levarht, E.W.N.; Breedveld, F.C.; Verweij, C.L. Regulation of synovial B cell survival nrheumatoid arthritis by vascular cell adhesion molecule 1 (CD106) expressed on fibroblast-like synoviocytes. *Arthritis Rheum.* **2000**, *43*, 1115–1121. [CrossRef]
18. Meyer, M.P.; Malan, A.F. Rheumatoid factor in congenital syphilis. *Genitourin. Med.* **1989**, *65*, 304–307. [CrossRef]
19. Svec, K.H.; Dingle, J.H. The Occurence of Rheumatoid Factor in Association with Antibody Response to Influenza A2(Asian) Virus. *Arthritis Rheum.* **1965**, *8*, 524–529. [CrossRef]
20. Carson, D.A.; Bayer, A.S.; Eisenberg, R.A.; Lawrance, S.; Theofilopoulos, A. IgG rheumatoid factor in subacute bacterial endocarditis: Relationship to IgM rheumatoid factor and circulating immune complexes. *Clin. Exp. Immunol.* **1978**, *31*, 100–103.
21. Harboe, M. Rheumatoid factors in leprosy and parasitic diseases. *Scand. J. Rheumatol.* **1988**, *17*, 309–313. [CrossRef]
22. Slaughter, L.; Carson, D.A.; Jensen, F.C.; Holbrook, T.L.; Vaughan, J.H. In vitro effects of Epstein-Barr virus on peripheral blood mononuclear cells from patients with rheumatoid arthritis and normal subjects. *J. Exp. Med.* **1978**, *148*, 1429–1434. [CrossRef] [PubMed]
23. Williams, R.C. Rheumatoid factors in subacute bacterial endocarditis and other infectious diseases. *Scand. J. Rheumatol. Suppl.* **1988**, *75*, 300–308. [CrossRef] [PubMed]
24. Roosnek, E.; Lanzavecchia, A. Efficient and selective presentation of antigen-antibody complexes by rheumatoid factor B cells. *J. Exp. Med.* **1991**, *173*, 487–489. [CrossRef] [PubMed]

25. Tighe, H.; Chen, P.P.; Tucker, R.; Kipps, T.J.; Roudier, J.; Jirik, F.R.; Carson, D.A. Function of B cells expressing a human immunoglobulin M rheumatoid factor autoantibody in transgenic mice. *J. Exp. Med.* **1993**, *177*, 109–118. [CrossRef]
26. Newkirk, M.M. Rheumatoid factors: Host resistance or autoimmunity? *Clin. Immunol.* **2002**, *104*, 1–13. [CrossRef]
27. Westwood, O.M.R.; Nelson, P.N.; Hay, F.C. Rheumatoid factors: What's new? *Rheumatology* **2006**, *45*, 379–385. [CrossRef]
28. Sinclair, N.R.S.; Panoskaltsis, A. Immunoregulation by Fc signals. A mechanism for self-nonself discrimination. *Immunol. Today* **1987**, *8*, 76–79. [CrossRef]
29. Van Snick, J.L.; Van Roost, E.; Markowetz, B.; Cambiaso, C.L.; Masson, P.L. Enhancement by IgM rheumatoid factor of in vitro ingestion by macrophages and in vivo clearance of aggregated IgG or antigen-antibody complexes. *Eur. J. Immunol.* **1978**, *8*, 279–285. [CrossRef]
30. Palazzi, C.; Buskila, D.; D'Angelo, S.; D'Amico, E.; Olivieri, I. Autoantibodies in patients with chronic hepatitis C virus infection: Pitfalls for the diagnosis of rheumatic diseases. *Autoimmun. Rev.* **2012**, *11*, 659–663. [CrossRef]
31. Charles, E.D.; Orloff, M.I.M.; Nishiuchi, E.; Marukian, S.; Rice, C.M.; Dustin, L.B. Somatic hypermutations confer rheumatoid factor activity in hepatitis C virus-associated mixed cryoglobulinemia. *Arthritis Rheum.* **2013**, *65*, 2430–2440. [CrossRef]
32. Børretzen, M.; Chapman, C.; Natvig, J.B.; Thompson, K.M. Differences in mutational patterns between rheumatoid factors in health and disease are related to variable heavy chain family and germ-line gene usage. *Eur. J. Immunol.* **1997**, *27*, 735–741. [CrossRef] [PubMed]
33. Simard, J.F.; Holmqvist, M. Rheumatoid factor positivity in the general population. *BMJ* **2012**, *345*, e5841. [CrossRef] [PubMed]
34. Tasliyurt, T.; Kisacik, B.; Kaya, S.U.; Yildirim, F.; Pehlivan, Y.; Kutluturk, F.; Ozyurt, H.; Sahin, S.; Onat, A.M. The frequency of antibodies against cyclic citrullinated peptides and rheumatoid factor in healthy population: A field study of rheumatoid arthritis from northern turkey. *Rheumatol. Int.* **2013**, *33*, 939–942. [CrossRef] [PubMed]
35. Newkirk, M.M. Rheumatoid factors: What do they tell us? *J. Rheumatol.* **2002**, *29*, 2034–2040.
36. Jacobsson, L.T.H.; Knowler, W.C.; Pillemer, S.; Hanson, R.L.; Pettitt, D.J.; Nelson, R.G.; Puente, A.; Del Mccance, D.R.; Charles, M.A.; Bennett, P.H. Rheumatoid arthritis and mortality. A longitudinal study in pima indians. *Arthritis Rheum.* **1993**, *36*, 1045–1053. [CrossRef] [PubMed]
37. Korpilähde, T.; Heliövaara, M.; Kaipiainen-Seppänen, O.; Knekt, P.; Aho, K. Regional differences in Finland in the prevalence of rheumatoid factor in the presence and absence of arthritis. *Ann. Rheum. Dis.* **2003**, *62*, 353–355. [CrossRef]
38. Dresser, D.W. Most IgM-producing cells in the mouse secrete auto-antibodies (rheumatoid factor). *Nature* **1978**, *274*, 480–483. [CrossRef]
39. Goodwin, J.S.; Searles, R.P.; Tung, K.S.K. Immunological responses of a healthy elderly population. *Clin. Exp. Immunol.* **1982**, *48*, 403–410.
40. Van Schaardenburg, D.; Lagaay, A.M.; Breedveld, F.C.; Hijmans, W.; Vandebroucke, J.P. Rheumatoid arthritis in a population of persons aged 85 years and over. *Rheumatology* **1993**, *32*, 104–108. [CrossRef]
41. Ursum, J.; Bos, W.H.; van de Stadt, R.J.; Dijkmans, B.A.C.; van Schaardenburg, D. Different properties of ACPA and IgM-RF derived from a large dataset: Further evidence of two distinct autoantibody systems. *Arthritis Res. Ther.* **2009**, *11*. [CrossRef]
42. Schroeder, H.W.; Cavacini, L.; Schroeder, H.W., Jr.; Cavacini, L.; Schroeder, H.W.; Cavacini, L. Structure and function of immunoglobulins. *J. Allergy Clin. Immunol.* **2010**, *125*, S41–S52. [CrossRef] [PubMed]
43. Meek, B.; Kelder, J.C.; Claessen, A.M.E.; van Houte, A.J.; ter Borg, E.J. Rheumatoid factor isotype and Ro epitope distribution in primary Sjögren syndrome and rheumatoid arthritis with keratoconjunctivitis sicca. *Rheumatol. Int.* **2018**, *38*, 1487–1493. [CrossRef] [PubMed]
44. Sieghart, D.; Platzer, A.; Studenic, P.; Alasti, F.; Grundhuber, M.; Swiniarski, S.; Horn, T.; Haslacher, H.; Blüml, S.; Smolen, J.; et al. Determination of autoantibody isotypes increases the sensitivity of serodiagnostics in rheumatoid arthritis. *Front. Immunol.* **2018**, *9*. [CrossRef] [PubMed]
45. Carson, D.A.; Pasquali, J.L.; Tsoukas, C.D.; Fong, S.; Slovin, S.F.; Lawrance, S.K.; Slaughter, L.; Vaughan, J.H. Physiology and pathology of rheumatoid factors. *Springer Semin. Immunopathol.* **1981**, *4*, 161–179. [CrossRef]
46. Seligmann, M.; Brouet, J.C. Antibody activity of human myeloma globulins. *Semin. Hematol.* **1973**, *10*, 163–177.
47. Shmerling, R.H.; Delbanco, T.L. The rheumatoid factor: An analysis of clinical utility. *Am. J. Med.* **1991**, *91*, 528–534. [CrossRef]
48. Diaz-Lopez, C.; Geli, C.; Corominas, H.; Malat, N.; Diaz-Torner, C.; Llobet, J.M.; De La Serna, A.R.; Laiz, A.; Moreno, M.; Vazquez, G. Are there clinical or serological differences between male and female patients with primary Sjogren's syndrome? *J. Rheumatol.* **2004**, *31*, 1352–1355.
49. Sansonno, D.; Lauletta, G.; Nisi, L.; Gatti, P.; Pesola, F.; Pansini, N.; Dammacco, F. Non-enveloped HCV core protein as constitutive antigen of cold-precipitable immune complexes in type II mixed cryoglobulinaemia. *Clin. Exp. Immunol.* **2003**, *133*, 275–282. [CrossRef]
50. Hurri, L.; Perttala, Y. Observations on non-specific Waaler-Rose and latex reactions in cancer patients. *Ann. Med. Intern. Fenn.* **1965**, *54*, 181–183.
51. Ugolini, A.; Zizzari, I.G.; Ceccarelli, F.; Botticelli, A.; Colasanti, T.; Strigari, L.; Rughetti, A.; Rahimi, H.; Conti, F.; Valesini, G.; et al. IgM-Rheumatoid factor confers primary resistance to anti-PD-1 immunotherapies in NSCLC patients by reducing CD137+T-cells. *EBioMedicine* **2020**, *62*, 103098. [CrossRef]
52. Allander, E.; Björnsson, O.J.; Kolbeinsson, A.; Olafsson, O.; Sigfússon, N.; Thorsteinsson, J. Rheumatoid factor in Iceland: A population study. *Int. J. Epidemiol.* **1972**, *1*, 211–223. [CrossRef] [PubMed]

53. Thorsteinsson, J.; Björnsson, O.J.; Kolbeinsson, A.; Allander, E.; Sigfússon, N.; Olafsson, O. A population study of rheumatoid factor in Iceland. A 5-year follow-up of 50 women with rheumatoid factor (RF). *Ann. Clin. Res.* **1975**, *7*, 183–194. [PubMed]
54. Ajeganova, S.; Humphreys, J.H.; Verheul, M.K.; Van Steenbergen, H.W.; Van Nies, J.A.B.; Svensson, B.; Hafström, I.J.; Huizinga, T.W.; Trouw, L.A.; Verstappen, S.M.M.; et al. Factor are associated with increased mortality but with different causes of death in patients with rheumatoid arthritis: A longitudinal study in three European cohorts. *Ann. Rheum. Dis.* **2016**, *75*. [CrossRef] [PubMed]
55. Ahn, J.K.; Hwang, J.; Chang, Y.; Ryu, S. Rheumatoid factor positivity increases all-cause and cancer mortality: A cohort study. *Rheumatol. Int.* **2017**. [CrossRef]
56. Gupta, N.P.; Malaviya, A.N.; Singh, S.M. Rheumatoid factor: Correlation with recurrence in transitional cell carcinoma of the bladder. *J. Urol.* **1979**, *121*, 417–418. [CrossRef]
57. Schattner, A.; Shani, A.; Talpaz, M.; Bentwich, Z. Rheumatoid factors in the sera of patients with gastrointestinal carcinoma. *Cancer* **1983**, *52*. [CrossRef]
58. Lochman, I.; Tvrdik, J.; Lochmanova, A.; Machálek, J.; Vrtná, L.; Kyselá, T.; Beska, F.; Repisták, J.; Konrád, B.; Cesaný, P. Antiimmunoglobulins of rheumatoid factor (RF) type in prediction of melanoma patients. *Neoplasma* **1986**, *33*, 737–741.
59. Turnbull, A.R.; Turner, D.T.; Fraser, J.D.; Lloyd, R.S.; Lang, C.J.; Wright, R. Autoantibodies in early breast cancer: A stage-related phenomenon? *Br. J. Cancer* **1978**, *38*, 461–463. [CrossRef]
60. Lewis, M.G.; Hartman, D.; Jerry, L.M. Antibodies and anti-antibodies in human malignancy: An expression of deranged immune regulation. *Ann. N. Y. Acad. Sci.* **1976**, *276*, 316–327. [CrossRef]
61. Ugolini, A.; Zizzari, I.G.; Ceccarelli, F.; Botticelli, A.; Colasanti, T.; Strigari, L.; Rughetti, A.; Rahimi, H.; Conti, F.; Valesini, G.; et al. 4P IgM-rheumatoid factor as a novel biomarker for a reduced survival in anti-PD-1 treated NSCLC patients through the decrease of CD137+ T-cells. *Ann. Oncol.* **2020**, *31*, S1418, [CrossRef]
62. Saksela, E.; Pyrhøonen, S.; Timonen, T.; Teppo, A.-M.; Wager, O.; Penttinen, K. Blocking effect of rheumatoid factor on the in vitro cytotoxicity of lymphoid cells from carcinoma patients. *Scand. J. Immunol.* **1976**, *5*, 1075–1080. [CrossRef]
63. Pyrhönen, S.; Timonen, T.; Heikkinen, A.; Penttinen, K.; Alftan, O.; Saksela, E.; Wager, O. Rheumatoid factor as an indicator of serum blocking activity and tumour recurrences in bladder tumours. *Eur. J. Cancer* **1976**, *12*, 87–94. [CrossRef]
64. Isturiz, M.A.; Maria, M.M.; Pizzi, A.M.; Manni, J.A. Antibody-dependent cell-mediated cytotoxicity in rheumatoid arthritis. Effect of rheumatoid serum fractions on normal lymphocytes. *Arthritis Rheum.* **1976**, *19*, 725–730. [CrossRef]
65. Hallberg, T. In vitro cytotoxicity of human lymphocytes for sensitized chicken erythrocytes is inhibited by sera from rheumatoid arthritis patients. *Scand. J. Immunol.* **1972**, *1*, 329–338. [CrossRef] [PubMed]
66. Giuliano, A.E.; Irie, R.; Morton, D.L. Rheumatoid factor in melanoma patients: Alterations of humoral tumor immunity in vitro. *Cancer* **1979**, *43*, 1624–1629. [CrossRef]
67. Jones, J.D.; Shyu, I.; Newkirk, M.M.; Rigby, W.F.C. A rheumatoid factor paradox: Inhibition of rituximab effector function. *Arthritis Res. Ther.* **2013**, *15*. [CrossRef]
68. Glennie, M.J.; French, R.R.; Cragg, M.S.; Taylor, R.P. Mechanisms of killing by anti-CD20 monoclonal antibodies. *Mol. Immunol.* **2007**, *44*, 3823–3837. [CrossRef]
69. Johnson, P.W.M.; Glennie, M.J. Rituximab: Mechanisms and applications. *Br. J. Cancer* **2001**, *85*, 1619–1623. [CrossRef]
70. King, K.M.; Younes, A. Rituximab: Review and clinical applications focusing on non-Hodgkin's lymphoma. *Expert Rev. Ant Cancer Ther.* **2001**, *1*, 177–186. [CrossRef]
71. Ugolini, A.; Nuti, M. CD137+ T-Cells: Protagonists of the Immunotherapy Revolution. *Cancers* **2021**, *13*, 456. [CrossRef]

Review

Prostate Cancer Liquid Biopsy Biomarkers' Clinical Utility in Diagnosis and Prognosis

Milena Matuszczak [1], Jack A. Schalken [2] and Maciej Salagierski [1,*]

[1] Department of Urology, Collegium Medicum, University of Zielona Góra, 65-046 Zielona Góra, Poland; matuszczakmilena@gmail.com
[2] Department of Urology, Radboud University Medical Centre, 6525 GA Nijmegen, The Netherlands; J.Schalken@uro.umcn.nl
* Correspondence: m.salagierski@cm.uz.zgora.pl

Simple Summary: In prostate cancer, overdiagnosis and overtreatment is a common problem for clinicians. Accurate diagnosis and prognosis are essential to avoid unnecessary biopsy and to increases the effectiveness of treatment. A new, easy-to-use and non-invasive test based on liquid biopsy biomarkers such as Progensa PCA3, MyProstateScore, ExoDx, SelectMDx, PHI, 4K, Stockholm3 and ConfirmMDx have been developed to improve diagnosis, prognosis and to help guide the decision-making process. This article provides an overview of the above-mentioned commercial tests. The performance and financial aspects of the tests have been compared using available studies. Then the application of biomarker tests as an adjunct to multiparametric MRI in the diagnosis, prognosis and monitoring of prostate cancer has been discussed.

Abstract: Prostate cancer (PCa) is the most common cancer in men worldwide. The current gold standard for diagnosing PCa relies on a transrectal ultrasound-guided systematic core needle biopsy indicated after detection changes in a digital rectal examination (DRE) and elevated prostate-specific antigen (PSA) level in the blood serum. PSA is a marker produced by prostate cells, not just cancer cells. Therefore, an elevated PSA level may be associated with other symptoms such as benign prostatic hyperplasia or inflammation of the prostate gland. Due to this marker's low specificity, a common problem is overdiagnosis, which leads to unnecessary biopsies and overtreatment. This is associated with various treatment complications (such as bleeding or infection) and generates unnecessary costs. Therefore, there is no doubt that the improvement of the current procedure by applying effective, sensitive and specific markers is an urgent need. Several non-invasive, cost-effective, high-accuracy liquid biopsy diagnostic biomarkers such as Progensa PCA3, MyProstateScore ExoDx, SelectMDx, PHI, 4K, Stockholm3 and ConfirmMDx have been developed in recent years. This article compares current knowledge about them and their potential application in clinical practice.

Keywords: cancer biomarkers; prostate cancer; liquid biopsy; prognosis; diagnosis; early detection

1. Introduction: Prostate Cancer Diagnosis

Prostate cancer (PCa) is the most common cancer in men and the second most common cause of mortality in this population in the United States, with 191,930 new cases and 33,330 deaths in 2020 [1]. Globally, there are approximately 1,276,106 new cases and 358,989 deaths each year [2]. The lifetime risk of being diagnosed with prostate cancer is estimated to be 1 in 9 men, while the risk of death is, fortunately, not as high at around 2% [1].

There is an emerging role for liquid biopsy in PCa, which has excellent potential in preoperative medicine. It is a minimally invasive procedure, analysing even small numbers of targets, which allows its usefulness in screening, diagnosis, prognosis, follow-up and therapeutic management [3]. This review compares the diagnostic and prognostic utility of prostate cancer tests. Good clinical outcomes can be achieved by accurate diagnosis

followed by acute treatment or active surveillance in patients with disease located within the gland. There is an unmet clinical need for non-invasive, easily performed diagnostic tests to assess whether a prostate biopsy is indicated. The EAU 2020 guidelines [4] recommend mpMRI before the first biopsy in men with a clinical suspicion of prostate cancer (PCa). Indeed, when mpMRI shows lesions suspicious for PCa (i.e., PI-RADS $\geq$ 3), targeted biopsy (TBx) and systemic biopsy (SBx) are recommended in patients who have not had a previous biopsy. It therefore represents an important diagnostic tool, and its combination with biomarkers further improves the accuracy of the initial diagnosis of PCa.

The traditional diagnosis of PCa (Figure 1) is based on the assessment of serum prostate-specific antigen (PSA) levels, digital rectal examination (DRE), followed by biopsy under the guidance of transrectal ultrasonography (TRUS). In screening programmes, high PSA levels, despite a normal DRE, lead to the diagnosis of PCa in more than 60% of asymptomatic patients. Serum PSA levels are commonly used for detection, risk stratification and monitoring of PCa [5]; unfortunately, it results in a high number of unnecessary biopsies and detection of asymptomatic cancers with low clinical risk [6]. The reason may be that PSA has a low positive predictive value (~30%) and poor specificity, being organ rather than cancer-specific. This highlights the need to develop more precise methods to identify clinically relevant PCa, such as liquid biopsy-derived biomarkers.

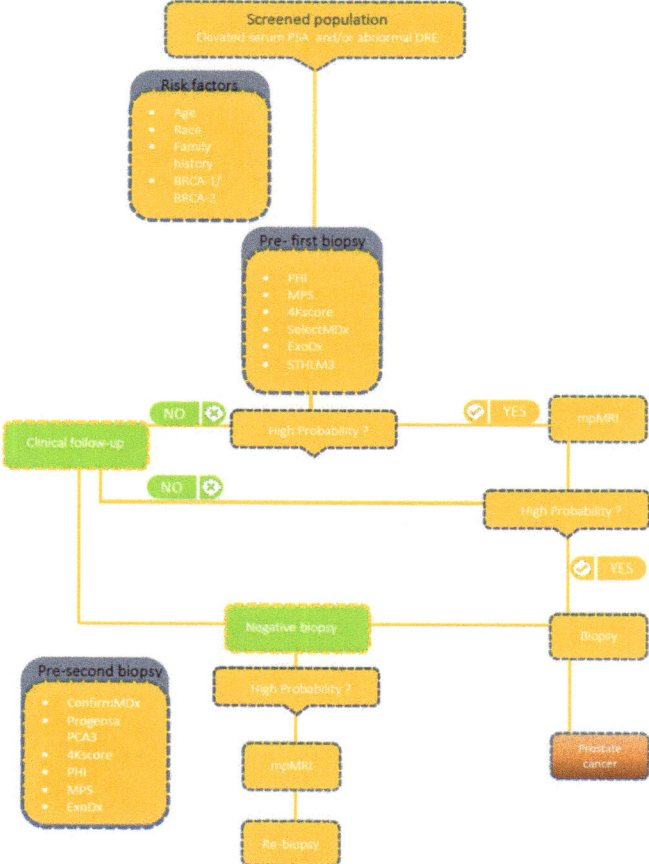

Figure 1. Suggested workflow for utilisation of prostate cancer biomarkers.

As prostate cancer is a heterogeneous disease, urologists, after identifying the presence of disease during the baseline assessment, focus primarily on assessing the risk group. Risk groups have been classified since 2014 using a classification system with five distinct Grade Groups based on modified Gleason score groups. Group 1 = Gleason score ≤ 6, Group 2 = Gleason score 3 + 4 = 7, Group 3 = Gleason score 4 + 3 = 7, Group 4 = Gleason score 4 + 4 = 8, Group 5 = Gleason score 9 and 10.

Currently, the gold-standard test to confirm all of the above clinical situations is the histopathological result of a prostate biopsy.

Unfortunately, this invasive procedure is painful, expensive and may pose a risk of complications (e.g., infection or sepsis). Furthermore, the procedure is prone to significant sampling error. It is therefore important to avoid unnecessary biopsies [7,8].

Liquid biopsy biomarkers are proving to be a promising new diagnostic and prognostic approach to help optimise the pre-biopsy decision and stratify whether the patient requires treatment or can be monitored under active surveillance.

2. Material and Methods

A literature review was performed by searching MEDLINE/PubMed, Google Scholar and and CrossRef electronic databases to identify articles published from January 2000 to October 2020 whose methods included commercially available prognostic and diagnostic prostate cancer liquid biopsy biomarkers or contain information about the characteristics of a relevant biomarker. The search terms included ConfirmMDx, ExoDx, MiPS, PCA3, PHI, SelectMDx, Stockholm3, 4Kscore and and prostate cancer liquid biopsy using search terms database = specific—medical subject headings terms in various combinations appropriate to the research objective. Articles on biomarkers not available in clinical practice or studies based on less than 40 patients were excluded.

3. Urine Biomarkers

Urine is obtained non-invasively and contains fluid excreted from the prostate gland, which may contain products from prostate cancer cells [9]. For many urinary biomarkers, performing a DRE is crucial as it increases the excretion of fluid from the prostate. To date, four tests are available with proven clinical utility.

3.1. Prostate Cancer Antigen 3 (PCA3)

The Progensa PCA3 test (Hologic Inc., Marlborough, MA, USA) is a test that measures PSA messenger RNA (mRNA) and PCA3 mRNA detectable in the first catch urine sample after DRE.

Prostate cancer antigen 3 (PCA3, previously called "DD3") is a long, non-coding RNA (lncRNA) that is overexpressed in 95% of prostate cancers [9]. The test is based on the fact that 60–100 times more PCA3 gene mRNA is detected in prostate cancer cells compared to non-cancerous prostate tissue.

The PCA3 score is calculated using the Progensa PCA3 method. The test result represents the PCA3/mRNA PSA $\times 1000$ ratio [10].

It is the first urine biomarker test to be approved in 2006 by the European Union, Canada [11] and in 2012 by the FDA. The FDA recommends its use in men ≥ 50 years old to support repeat biopsy decision-making in whom one or more previous prostate biopsies have been negative and for whom repeat biopsy is recommended based on current standards of care [12]. However, some clinical studies [13,14] report the benefits of using the test as early as the first biopsy.

Although the FDA recommends a cut-off PCA3 score = 25, many studies [12,13] suggest a cut-off score of 35 as a more optimal cut-off point. Establishing a cut-off point appears to be of vital importance.

A study [12] evaluated different PCA3 score cut-off points: 10 and 35. For these values, the sensitivity was 87% and 58%, respectively, and the specificity was 28% and 72%, respectively. The results showed that a PCA3 score cut-off of 35 could provide an optimal balance between sensitivity (58%) and specificity (72%) for the diagnosis of PCa and was superior to PSA (Table 1).

Although the study [12] demonstrated high sensitivity and specificity, the ability to improve prostate cancer detection was not shown. For this reason, Wei et al. conducted a prospective validation trial on 859 men [14] to assess whether the PCA3 score could improve the PPV for initial biopsy and NPV for repeat biopsy. The results were PPV = 80% for detecting any PCa at initial biopsy and NPV = 88% at repeat biopsy. This showed that at initial biopsy, a PCA3 score > 60 increases the likelihood of detecting PCa, and at repeat biopsy, a PCA3 score < 20 indicates a low risk of detecting PCa at biopsy [14].

A systematic review and meta-analysis of studies (with a threshold of 35) [13] yielded the following overall values: AUC = 0.734, sensitivity 69% and specificity 65%. These results support the greater clinical utility of cut-off point = 35 than 25 (FDA approved).

Determining the best cut-off value is controversial, especially for primary biopsy—the available studies are very heterogeneous. Several have highlighted that PCA3 does not perform well at a single threshold, showing a high NPV below the low cut-off and a high PPV above the high cut-off, with a grey zone in between—reflecting prostate cancer specificity [14].

Roobol et al., in a publication [15], highlight men with a PCA3 score $\geq$ 100 and no PCa in a biopsy. This study combines data from the initial and re-biopsies that provided a PPV of 52.2% in men with PCA3 $\geq$ 100, resulting in almost 50% unexplained high results. To date, there is no explanation why PCA3 scores can be excessively high despite the absence of biopsy-detectable PCa.

Publications [16–18] do not show a relationship between PCA3 value and prostate cancer aggressiveness (Gleason score). A high PCA3 level, due to its low specificity, does not help assess prognostic parameters and is therefore of low utility in clinical practice, as it does not provide an answer to how to proceed with the patient. For this reason, to detect patients who require rapid and radical treatment, it is reasonable to use newer, more sensitive and specific diagnostic tools, e.g., SelectMDx (MDxHealth, Inc., Irvine, CA, USA).

Numerous studies [14,19] have shown that the diagnostic value of the test increases when adding other predictors (i.e., age, PSA value, DRE result or prostate volume). Therefore, the producer recommends its use in combination with standard diagnostic parameters [20].

To determine the clinical utility of the PCA3 test in African Americans, Feibus et al. conducted a study [21] (Table 2) on a racially diverse group of men, where 60% of the participants were African American. They demonstrated that the PCA3 test in African Americans also improves the ability to predict the presence of any prostate cancer and high malignancy.

Ochiai et al. [22] (Table 2) examined the diagnostic utility of PCA 3 in Japanese men undergoing prostate biopsy. They achieved a similar diagnostic value to that obtained in men in Europe and the USA. The PCA3 score for men with prostate cancer was significantly higher than for men with negative biopsy results. Furthermore, they showed that also in Japanese men, PCA3 was significantly better than PSA in predicting PCa.

The reported clinical utility of the study mentioned above on the Japanese population and the desire to verify the promising reports of Shen et al. [23] (Table 2) on a small group of Chinese men (prostate cancer patient group ($n = 35$), BPH patient group ($n = 64$)), inspired other researchers to study the Chinese population. Wang et al. conducted a study on a cohort of 500 Chinese men [24] (Table 2). This study showed a moderate improvement in diagnostic accuracy using PCA3 during the initial prostate biopsy. In patients qualified for initial biopsy (PSA $\geq$ 4 and/or suspicious DRE), the Progensa test was not used, but the RC-PCR-based PCA3 test was used. The values obtained were sufficient to distinguish positive from negative prostate biopsy results but were not correlated with PCa aggressiveness.

In a study [25] (Table 2) involving Latino Americans, results were comparable to those obtained for other populations, indicating its potential use in Latino Americans with persistently elevated PSA and previous negative biopsies.

Table 1. Predictive capacity of prostate cancer after negative biopsy biomarkers.

Commercial Product	Biomarkers	Purpose	Indication	Cohort	Avoid Biopsies	Specimen	FDA Approved	Method	Predictive Capacity	Ref.
PCA3	lncRNA PCA3, PSA mRNA	Predicts the presence of malignancy. Supports initial biopsy decisions by enhancing diagnostic value. Determines whether repeat biopsy is needed after an initially negative biopsy.	Diagnosis: repeat biopsy Prognosis:*	n = 233 [12] n = 859 [14] n = 1072 [19] n = 351 [26] n = 3073 [27]	For PCA3 score < 20 and PSA < 4 ng/mL 8% of men would have avoided biopsies, while 9% of cancer (non-HG) have been underdiagnosed. For only PCA3 score < 20 46% biopsies would have been avoided, while missing 12% of cancers (3% HG [14])	First catch (20–30 mL) post-DRE urine	Yes	Urine specimens were held at 2 °C to 8 °C and processed within 4 h by mixing with an equal volume of detergent-based stabilisation buffer (Gen-Probe®Hologic, San Diego, CA, USA) to lyse the prostate cells and stabilise the RNA. Samples were stored at −70 °C until testing and batch shipped on ice packs if needed. PCA3 and PSA mRNA were isolated from processed urine samples by capturing magnetic microparticles and amplified by transcription-assisted amplification. Products were detected with chemiluminescent DNA probes using a hybridisation protection assay [12]. Statistical analyses were performed by the data coordinating centre using SAS version 9.2 (SAS Institute, Cary, NC, USA) [14]. NCSS 2004 (NCSS Inc., Kaysville, UT, USA) was used for the analysis [19]. Data analysis was performed using Statistical Package for Social Sciences version 12.0.1 (SPSS, Chicago, IL, USA) [26].	AUC = 0.68 Se. = 58% Sp. = 72% [12] For detection of any cancer, PPV = 80% for initial prostate biopsy, and for repeat prostate biopsy NPV = 88% Se. = 76% Sp. = 52% [14] AUC = 0.693 Se. = 48.4% Sp. = 78.6% [19] AUC = 0.72 Sp. = 61% Sp. = 74% [26] AUC = 0.697 for prediction PCa AUC = 0.682 for HG PCa NPV = 0.67 PPV = 0.62 Se. = 53% Sp. = 75% [27]	[12,14,19,26,27]
ConfirmMDx	Hypermethylation of GSTP1, APC and RASSF1 genes, PSA	Screening patients at risk of HG PCa after an initial negative biopsy. It is clinically validated for detection of PCa in tissue from PCa-negative biopsies. Helps to distinguish true negative biopsies from those with possible undetected cancer, and decide when to re-biopsy.	Diagnosis: repeat biopsy	n = 498 [28] n = 350 [29] n = 803 [30] n = 211 [31] **	30% of repeat biopsies can be safely avoided [30]	Tissue from prostate biopsy	No	All men underwent two consecutive biopsies: a negative index biopsy and then negative or positive rebiopsy. DNA was extracted and processed from fixed, paraffin-embedded blocks of prostate biopsy core tissue. In histologically negative prostate biopsy core tissues, epigenetic analyses were performed in a randomised, blinded fashion profiled for GSTP1, APC RASSF1 against the reference ACTB gene using methylation-specific PCR (MSP). In DOCUMENT (The Detection Of Cancer Using Methylated Events in Negative Tissue) [29] and studies [30,31] for direct comparison with the MATLOC (Methylation Analysis To Locate Occult Cancer) [28] cohort, the previously determined analytical gene cutoff values for determining methylation status were identical. All statistical analyses, including logistic regression and cross-validation, were performed in R software (R Foundation for Statistical Computing, Vienna, Austria).	NPV = 90% Se. = 68% Sp. = 64% for any PCa [28] AUC = 0.628 NPV = 88% Se. = 62% Sp. = 64% [29] AUC = 0.742 NPV = 96% for HG NPV = 89.2% for all cancers PPV = 28.2% for any cancer [30] NPV = 78.8% PPV = 53.6% for detection of all PCa Se. = 74.1% Sp. = 60.0% [31] NPV = 94.2% PPV = 19.4% for detection of GS ≥ 7 PCa Se. = 77.8% Sp. = 52.7% [31]	[28–31]

*—Prognostic value of PCA3 is controversial. **—group of African Americans. Abbreviations: APC- adenomatous polyposis coli, ACTB-beta-actin (reference gene) AUC—area under the curve, GS—Gleason score, GSTP1- glutathione s-transferase pi 1, HG- high grade, lncRNA—long non-coding RNA, mRNA—messenger RNA, NPV—negative predictive value, PCa—prostate cancer, PCA3—PCa antigen 3, PPV—positive predictive value, PSA—prostate-specific antigen, RASSF1—ras association domain family member 1, Ref—references, Se.—sensitivity, Sp—specificity.

Table 2. Evaluation of biomarkers in multiethnic populations.

Commercial Product	African-Americans	Japanese Men	Chinese Men	Latino American
PCA3	In a study [21] on a racially diverse group of men, 60% of the participants were African-American. It demonstrated that the PCA3 test also in African-Americans improves the ability to predict the presence of any prostate cancer and high malignancy.	Study [22] examined the diagnostic utility of PCA 3 in Japanese men undergoing prostate biopsy. They achieved a similar diagnostic value to that obtained in men in Europe and the USA.	Studies [23,24] showed the utility of PCA3 in Chinese men.	In a study [25] involving Latino Americans, results were comparable to those obtained in other publications for other populations indicating its potential use in Latino Americans with persistently elevated PSA and previous negative biopsies.
ConfirmMDx	Study [31] showed no significant differences in sensitivity and specificity between this test and previously described validation studies involving predominantly Caucasian populations and indicates usefulness for African Americans in risk stratification after an initially negative biopsy.	No data about these ethnic groups were found.		
PHI	To assess the ability of PHI to detect Gleason grade 2-5 (GGG) PCa in African Americans, 158 patients with elevated PSA levels and 135 controls were recruited [32]. Results indicate that PHI ≥ 28.0 can be safely used to avoid unnecessary biopsies in African Americans.	In a study [33] involving a European (n = 503) and Asian (n = 1652) population, more biopsies were avoided in the Asian group (56% vs. 40%). This study also identified the need to establish differential cut-off points for diverse ethnic groups. The authors of the publication recommended cut-off points for csPCa: PHI > 40 for European men and PHI > 30 for men of Asian origin.		No data about these ethnic groups were found.
4Kscore	The study [34] included 366 men, 205 of whom were African American. The results of the study showed an AUC = 0.81 in predicting aggressive prostate cancer in this population, therefore the 4Kscore can be used to guide biopsy decisions also in this ethnic group.	A multiethnic group study [35] (African Americans, Japanese, Latinos, Native Hawaiians, and Whites) confirmed the 4Kscore's accuracy to discriminate benign from malignant cases and indolent from aggressive tumors.		
Mi-Prostate Score	It is unknown what the cut-off values should be and what the diagnostic and prognostic accuracy. There is a lack of studies on African-American, Asian or Latino American populations.			
ExoDx Prostate IntelliScore	It is unknown what the cut-off values should be and what the diagnostic and prognostic accuracy. There is a lack of studies on African-American, Asian or Latino American populations.			
SelectMDx	It is unknown what the cut-off values should be and what the diagnostic and prognostic accuracy. There is a lack of studies on African-American, Asian or Latino American populations.			
Stockholm3 Model	This test was evaluated only on men from an ethnically homogeneous population (Stockholm County, Sweden).			

PCA3 shows more significant diagnostic and prognostic potential when combined with other biomarkers, such as TMPRSS2 fusion: ERG [36] hK2, PSA [37] and PSAD [38]. Currently, some researchers are making efforts to develop more precise detection methods for PCA3 [39,40].

3.2. Mi-Prostate Score (MiPS)

MyProstateScore (MPS, LynxDx, Ann Arbor, MI, USA (previously known as MiPS—Michigan Prostate Score)) is an algorithm that measures mRNA, PCA3 and TMPRSS-ERG (abnormal fusion of TMPRSS2 and ERG (T2-ERG)) expression in urine from the first collection after DRE and serum PSA.

More than 50% of patients with PCa have an ERG gene fusion with TMPRSS2 [41]. The presence of this translocation has been shown to be associated with poor patient prognosis—an increased risk of recurrence and mortality from PCa.

The test is indicated for men with suspicious PSA levels who are being considered for initial or repeat biopsy. The test result validates the need for biopsy and predicts the risk of high-grade prostate cancer (GS > 7) in a diagnostic needle biopsy [42–44]. Values range from 0 to 100, and the higher the score, the greater the risk of aggressive cancer.

In 2013, Salami et al. [44] showed that the MPS test was significantly more accurate than any single variable (TMPRSS2-ERG AUC = 0.77 compared with 0.65 for PCA3 and 0.72 for serum PSA alone), AUC was 0.88, with specificity and sensitivity of 90% and 80%, respectively (Table 3).

A pivotal study published by Tomlins et al. [43] in 2016 indicated the high diagnostic value of MPS, AUC = 0.751 for detecting PCa on biopsy and AUC = 0.772 for detecting clinically significant PCa (defined as Gleason ≥ 7), which was significantly better than for PSA alone (AUC = 0.651) (Table 3).

In a prospective study [45] involving 1077 men, MPS was shown to increase the detection of aggressive prostate cancers compared with PSA alone. When the cut-off point was set at 95% sensitivity, the specificity of detecting HG PCa increased from 18% (PSA alone) to 39%. The authors further demonstrated that if biopsies were performed in patients with positive urine PCA3 (score > 20) or T2-ERG (score > 8) or with serum PSA > 10 ng/mL, 42% of unnecessary biopsies could be avoided.

In a study including a validation cohort of 1525 men, the MPS test was confirmed to improve the detection of csPCa. The authors also intended to set a threshold to exclude GG cancer ≥ 2. An MPS threshold of ≤ 10 was recommended. At this value, sensitivity (96%) and NPV (97%) were obtained, avoiding 32% of unnecessary biopsies while missing 3.7% of GG cancer cases ≥ 2 (Table 3) [46].

3.3. ExoDx Prostate ® (IntelliScore) (EPI)

The ExoDx Prostate (IntelliScore) (EPI, Exosome Diagnostics, Waltham, MA, USA) assesses the exosomal RNA expression of three genes (ERG, PCA3 and SPDEF) involved in the initiation and progression of PCa. ExoDx prostate is a test performed from a urine sample that does not require prior DRE testing. Exosomal RNA is derived from exosomes, which are small membrane vesicles secreted by several cell types, including immune and cancer cells [47]. The high potential of exosomes as biomarkers is due to their structure—a lipid bilayer protects the contents from degradation by proteases.

The test scores range from 1 to 100 and a cut-off point of 15.6 indicates men at increased risk of HG PCa ($\geq$GG2) at subsequent biopsy, making the test helpful in validating the need for biopsy in men at risk. The test is recommended for men aged $\geq$50 years who are in the PSA "grey zone" (2–10 ng/mL) to distinguish a benign (HG1; when the test value < 15.6) from high-grade PCa (HG2 $\geq$ 15.6) [48].

Table 3. Predictive capacity of prostate cancer prebiopsy biomarkers.

Commercial Product	Biomarkers	Purpose	Indication	Cohort	Avoid Biopsies	Specimen	FDA Approved	Method	Predictive Capacity	Ref.
PHI	tPSA, fPSA, p2PSA	Estimates the probability of a diagnosis of all grades PCa and csPCa (GS ≥ 7). Indicates the need for a biopsy, reduces the number of unnecessary ones and continues to follow up. Reduces overdiagnosis and overtreatment.	Diagnosis: initial biopsy, repeat biopsy. Prognosis.	n = 893 [49]. Two independent cohorts n = 561 and n = 395 [50] n = 769 [51] n = 350 [52] n = 658 [53] n = 1652 Asian men and n = 503 European men [33] n = 531 [54] n = 16,762 [55].	A total of 26% of unnecessary biopsies [49]. In the primary cohort, avoided 41% of unnecessary biopsies. In the validation cohort, avoided 36% of unnecessary biopsies while missing only 2.5% of high-grade PCa [50] Among Asian men at 90% sensitivity for HG PCa and cut-off > 30, 56% of biopsies and 33% of GS 6 diagnoses could have been avoided [33]. Among European men at 90% sensitivity for HG PCa and cut-off > 40, 40% of biopsies and 31% of GS 6 diagnoses could have been avoided [33].	blood serum	Yes	Specimens were analysed at the EDRN Biomarker Reference Laboratory at Johns Hopkins University. Serum was stored at −80 °C before testing. Prebiopsy measurements of total PSA, fPSA and p2PSA were performed using an Access 2 automated immunoassay analyser (Beckman Coulter Inc., CA, US). Technologists performing the assays were blinded to prostate biopsy results. PHI was calculated using the equation (p2PSA/fPSA) × √(PSA) [50] Statistical analysis was conducted by using SAS, version 9.3 and R, version 3.1.0 (R Foundation for Statistical Computing, Vienna, Austria) [50].	AUC = 0.703 Se. = 95% Sp. = 16% [49] AUC = 0.815 for detecting aggressive PCa Se. = 95% Sp. = 36% [50] AUC = 0.73 [51] For prediction of GS ≥ 7 Se. = 90.8/66.3/44.8 Sp. = 34.8/66.3/89.9 for criterions ≥30.9/44.0/56.2, respectively [52] For prediction of GS 6–7 Se. = 89.9/60.0/37.4 Sp. = 26.0/61.6/90.4 for criterions ≥ 28.0/42.2/55.5, respectively [52] Se = 80% for PCa/ csPCa and biopsy GS Sp. = 46.1/45.5/46.4, respectively [53] Se. = 95% of PHI Sp. = 14.1/16.3/27.4, respectively [53] AUC = 0.78 for PCa detection and for HG PCa (75% men were European) In Asian men group AUC = 0.76 for PCa detection and AUC = 0.77 for HG PCa Se. = 99–53% with corresponding Sp. = 10–72% in European group for cut-off = 25–55 [33] Se. = 96–27% with corresponding Sp. = 36–96% in Asian group for cut-off = 25–55 [33] AUC =0.704 for any cancer AUC = 0.711 for Gleason ≥ 7 [54] AUC = 0.76 for PCa detection Se. = 89% Sp. = 34% [55] AUC = 0.82 for HG PCa detection, Se. = 93% Sp. = 34% [55]	[33,49–55]

Table 3. Cont.

Commercial Product	Biomarkers	Purpose	Indication	Cohort	Avoid Biopsies	Specimen	FDA Approved	Method	Predictive Capacity	Ref.
4Kscore	tPSA, fPSA, iPSA, hK2		Diagnosis: initial biopsy, repeat biopsy. Prognosis.	n = 531 [54] n = 16,762 [55] n = 1012 [56] n = 11,134 [57] n = 611 [58] n = 718 [59] n = 2872 [60]	Avoided 29% of biopsies, delayed diagnosis 10% of HG PCa [54]. 43% avoided and delayed diagnosis of 2.4%, Gleason ≥7 for 9% 4Kscore cutoff [56]. 58% avoided and delayed diagnosis of 4.7% Gleason ≥7 for 15% 4Kscore cutoff [56]. reduction of 94.9%, 47.1% and 9.3% biopsies in men with low-risk, intermediate-risk and high-risk aggressive PCa, respectively [58] For different threshold 4%, 5%, 7.5%, 10% 58%, 66%, 75% and 80% of reduced biopsies while missed diagnose of HG PCa 1%, 2%, 7%, 2%, respectively [60]	Blood plasma	No	Blood was collected into tubes containing K2EDTA, inverted, centrifuged at 1600× g and frozen at 70 °C within 4 hours from collection. Frozen plasma was stored until shipped in dry ice to OPKO Labs (Nashville, TN, USA) for analysis. The analytical laboratory was blinded to all sample information and clinical data. Samples were thawed immediately before analysis. Then, tPSA, fPSA, iPSA and hK2 were measured. Statistical analysis was conducted by using R, version 3.1.1 (http://www.r-project.org/ accessed on 11 November 2016) [59].	AUC = 0.69 for any cancer AUC = 0.718 for Gleason ≥7 [54] AUC = 0.72 for PCa detection Se. = 74% Sp. = 60% [55] AUC = 0.81 for HG PCa detection Se. = 87% Sp. = 61% [55] AUC = 0.82 [56] AUC = 0.81 [57] AUC = 0.75 [59] AUC = 0.876 for 4Kscore AUC = 0.888 for 4Kscore with RPCRP Se. = 87% Sp. = 71% [60]	[54–60]
Mi-Prostate Score	PCA3, T2:ERG mRNA, tPSA		Diagnosis: Initial Biopsy, repeat biopsy. Prognosis *	n = 497 [42] n = 1225 [43] n = 48 [44] n = 1077 [45] n = 1525 [46]	Total of 35% of biopsies and missing 13% of ≥GG2 PCa [42] avoided 35–47% of biopsies while delaying the diagnosis of 1.0–2.3% of ≥GG2 [43]. Avoided 67% of biopsies at the risk of a false-negative rate of 20% [44]. Avoided 33% of unnecessary biopsies, missing 7% of HG PCa [45] for threshold ≤10; avoided 32% of unnecessary biopsies, missing 3.7% of GG ≥ 2 cancers [46]	Post-DRE first void urine	No	Urine samples were obtained immediately after DRE, refrigerated and processed within 4 h by mixing with an equal volume of urine transport medium and stored below −70 °C until analysis. The amount of T2:ERG and PSA mRNA was determined by TMA. Statistical analyses was performed using R version 2.10.1 (R Foundation for Statistical Computing, http://www.R-project.org accessed on 16 May 2015 [43]	AUC = 0.842 ** Se. = 88.1% Sp. = 49.6% [42] AUC = 0.772 [43] AUC = 0.88 for detection of PCa AUC = 0.772 for csPCa Se. = 80% Sp. = 90% [44] In the developmental cohort(n = 516) Se. = 95% Sp. = 39% In the validation cohort (n = 561) Se. = 93% Sp. = 33% [45] NPV = 97% Se. = 96% for threshold ≤10 [46]	[42–46]

163

Table 3. *Cont.*

Commercial Product	Biomarkers	Purpose	Indication	Cohort	Avoid Biopsies	Specimen	FDA Approved	Method	Predictive Capacity	Ref.
ExoDx Prostate IntelliScore	Exosomal mRNA ERG, PCA3 and SPDEF		Diagnosis: initial biopsy, repeat biopsy. Prognosis.*	Validation cohort $n = 519$ (training cohort $n = 255$) [48] $n = 503$ [61] $n = 229$ [62]	Total of 20% of all biopsies, 26% of unnecessary biopsies, and missing 7% of ≥GG2 PCa [61]. 26% of all biopsies, 27% of unnecessary biopsies and 2.1% delayed detection of ≥ GG2 [62].	Urine	No [1]	Urine samples were collected in a 15–20 mL container and stored at 4 °C. for up to 5 days before shipment to a central laboratory (Exosome Diagnostics, Inc. Waltham, MA, USA) for EPI assay analysis [61]. R software version 3.6.1 (R Core Team, 2019, Vienna, Austria) was used for reporting and data analysis. Two-tailed p values ≤ 0.05 were considered statistically significant. [63]	AUC = 0.73 (combined with SOC[2]) AUC = 0.71 NPV = 91% PPV = 36% Se. = 92% Sp. = 34% [48] AUC = 0.70 NPV = 89% Se. = 93% [61] AUC = 0.66 NPV = 92% Se. = 82% [62]	[48,61,62]
SelectMDx	HOXC6, KLK3, DLX1 mRNA and PSAd		Diagnosis: Initial Biopsy.	First cohort $n = 519$ Second $n = 386$ [64] $n = 1955$ [65] $n = 172$ [66]	Total of 42% of all, 53% of unnecessary biopsies [64].	Post-DRE first void urine	No	Approximately 30 mL of the first urine passed was collected into a collection cup after the DRE was performed. The urine was immediately transferred to a urine sample transport tube (Hologic San Diego, CA, USA) and samples were shipped at room temperature to a central laboratory and stored at −80 °C. Statistical analysis was performed using SPSS v.20.0 (IBM Corp., Armonk, NY, USA) and R v.3.2.1 (R Foundation for Statistical Computing, Vienna, Austria) [64].	AUC = 0.86 AUC = 0.90 (+ clinical parameters) NPV = 94% PPV = 27% Se. = 91% Sp. = 36% [64] AUC = 0.82–0.85 NPV = 95% Se. = 89–93% [65] Sp. = 47–53% [65] AUC = 0.83 NPV = 98% Se. = 96% [66]	[64–66]

Table 3. Cont.

Commercial Product	Biomarkers	Purpose	Indication	Cohort	Avoid Biopsies	Specimen	FDA Approved	Method	Predictive Capacity	Ref.
Stockholm3 Model [3]	tPSA, fPSA, hK2, MIC1 and MSMB (with genetic markers based on 232–254 SNPs) ***		Diagnosis: Initial Biopsy.	Validation cohort n = 111,819 (training cohort n = 32,453) [67] n = 59,159 [68] n = 533 [69] Two cohorts: n = 56,282 and n = 47,688 [70]	S3M could reduce the number of biopsies by 32% and could avoid 44% of benign biopsies [67], reduction in total biopsies 33–52% and avoid 42–62% of benign biopsies, while missing 10–20% GS $\geq$ 7 [68]. Total of 38% of all biopsy avoided, delaying diagnosis for 6% of men with GG $\geq$ 2 cancer [69] reduction in total biopsies 53% and avoided 76% of benign biopsies [70].	Blood plasma	No	Prior to prostate biopsy, sample blood was collected for testing. Biopsy results were used to validate the Stockholm3 test results. The program R version 3.4.2(R Foundation for Statistical Computing, http://www.R-project.org accessed on 31 August 2018) was used to perform the statistical analyses [69].	AUC = 0.69 for all prostate cancers AUC = 0.74 for Gleason $\geq$ 7 [67] AUC = 0.75 for Gleason $\geq$ 7 [68] AUC = 0.89 for GG $\geq$ 2 [69]	[67–70]

[1]—ExoDx received FDA Breakthrough Designation status in June 2019. Since October 2019, Medicare Administrative Contractor (MAC) National Government Services, Inc. has issued a final Local Coverage Decision (LCD) L37733, which covers the ExoDx Prostate test before an initial prostate biopsy. [2]—SOC: prostate-specific antigen level, age, race, family history) [3]—Predictors in S3m include prostate examination (DRE and prostate volume), clinical variables (first-degree family history of PCa, and a previous biopsy, age) *—Similar to PCA3, the utility for prognosis remains controversial. **—AUC for ERSPC risk calculator plus PCA3 plus TMPRSS2-ERG ***—The initial S3M version had also included intact PSA, but due to interference between kallikreins in the immunosorbent assay with the allergen chip, it was removed from the S3M. Recently, a novel biomarker, HOXB13 SNP, a rare germline mutation of the HOXB13 gene with a high impact on prostate cancer risk, was included. Abbreviations: AUC—area under the curve, csPCa—clinically significant prostate cancer, DLX1—distal-less homeobox 1, DRE—digital rectal exam, ERG—estrogen-regulated gene, fPSA—free PSA, GG—grade group, GS—Gleason score, HG—high grade, hK2—human kallikrein-related peptidase 2, HOXC6—homeobox C6, iPSA—intact PSA, KLK3—kallikrein-related peptidase 3, LG—low grade, n—number of patients participating in study, mRNA—messenger RNA, NPV—negative predictive value, PCa—prostate cancer, PHI—Prostate Health Index, PPV—positive predictive value, PSA—prostate-specific antigen, PSAd—PSA density, p2PSA—(−2) proPSA, Ref—references, SNPs—single-nucleotide polymorphisms, Se—sensitivity, Sp—specificity, STHLM3—Stockholm-3, SPDEF-SAM pointed domain-containing ETS transcription factor, T2-ERG—transmembrane protease serine 2-ERG, tPSA: total PSA, 4K—four-kallikrein panel.

A validation study [48] conducted in 2016 on a (training) cohort of 255 patients initially and a separate validation cohort of 519 patients tested the ability of the ExoDx test in combination with SOC (PSA level, age, race and family history) to identify PCa GS (Gleason score) ≥ 7 in men aged ≥ 50 awaiting their first biopsy (PSA 2–20 ng/mL and/or suspicious DRE). With a cut-off value of 15.6, ExoDx alone demonstrated a sensitivity of =91.89%, NPV = 91.3% and AUC of 0.71 for distinguishing GS $\geq$ 7 from GS 6 and benign PCa. When the test was combined with SOC (AUC = 0.73), the ExoDx test outperformed SOC alone (AUC = 0.63) and the PCPTRC risk calculator (AUC = 0.62) in differentiating PCa GS $\geq$ 7 od GS 6 and benign PCa.

In a study published 2 years later [61], McKiernan et al. evaluate the clinical utility of ExoDx in comparison with standard clinical parameters for distinguishing grade (GG) $\geq$ 2 PCa from GG1 PCa and benign disease (in men eligible for the first biopsy) and conducted a prospective study of 503 patients aged ≥ 50 years with PSA "grey zone" (2–10 ng/mL). The results obtained were similar to previous studies. The combined model of ExoDx and SOC achieved the highest value (AUC = 0.71), and ExoDx alone (AUC = 0.70) was better at predicting GG2 PCa at initial biopsy than SOC (AUC = 0.62). A value of 15.6 was confirmed as the recommended cut-off point to distinguish patients at high risk of GG2 PCa at their initial biopsy. At a cut-off point of 15.6, a high negative predictive value of NPV = 89% was achieved (Table 3), preventing 26% of unnecessary biopsies and 20% of all biopsies (with only 7% of $\geq$GS7 PCa missed).

These two prospective studies validating over 1000 patients [48,61] showed that ExoDx (AUC 0.71 and AUC 0.70, respectively) was better at predicting clinically significant PCa at first biopsy than existing risk calculators and PCPT-RC (AUC 0.63), ERSPC-RC (AUC 0.58) and PSA alone (AUC 0.58). Both studies show that this test is useful for risk stratification of $\geq$GG2 due to GG1 cancer and benign disease and improves identification of patients at higher risk of advanced prostate cancer and helps avoid unnecessary biopsies.

Other work [48,61] confirmed the utility of ExoDx for primary biopsy but lacked confirmation for use on repeat biopsy. McKiernan et al. conducted a study [62] in 229 patients qualified for repeat biopsy; an AUC of 0.66 and an NPV of 92% (irrespective of other clinical features) were achieved at a previously validated cut-off point of 15.6, which would avoid 26% of unnecessary biopsies, omitting only 2.1% of patients with HG PCa (Table 3). Furthermore, in this study, AUC curves and net health benefits analyses showed better performance of ExoDx than the ERSPC and PSA risk calculator in predicting HG-PCa in men with a prior negative prostate biopsy. A total of 71.6% of patients were Caucasian, 14.4% African American and the study was completed on the most ethnically diverse group. The vast majority of publications are from the USA, and no studies have been completed on Asian or African populations or, more widely, on African Americans. It is not known what the cut-off values should be and what the diagnostic and prognostic accuracy is for a multiethnic population.

The clinical utility of ExoDx Prostate was recently evaluated in 1094 patients scheduled for their first biopsy (with PSA 2–10 ng/mL). This first study [63] of PCa biomarkers with a blinded control arm showed that ExoDx helped avoid unnecessary biopsies when the test was negative and increased the detection of HG PCa by 30% compared with a control arm without ExoDx (SOC alone). Compared to SOC, the test missed 49% fewer HG PCa. The study showed that ExoDx improved patient stratification and influenced the decisions made by (68%) urologists about biopsy (with rising PSA being the main reason for not following ExoDx results).

3.4. SelectMDx

SelectMDx (MDxHealth, Inc., Irvine, CA, USA) is a urine-based test after DRE that measures three biomarkers: DLX1 (progression gene), HOXC6 (cell proliferation gene), KLK3 (reference gene) and clinical risk factors (age, DRE, PSA and prostate volume, which can be calculated from the TRUS measurements substituted into the formula:

height × width × length × 0.523). HOXC6 and DLX1 mRNA levels are assessed to estimate the risk of PCa on biopsy and the presence of high-risk cancer.

Men with elevated PSA levels in the "grey zone" (4–10 ng/mL) and/or an abnormal DRE result are subjects for whom an initial biopsy is considered. The result determines whether the patient is at high or low risk of PCa. It supports clinical decision-making and stratifies patients into those who may benefit from biopsy and early cancer detection and others for whom it is better to avoid this invasive procedure and continue with routine screening or active surveillance.

In a study [64], 386 men with an elevated PSA (≥ 3 ng/mL), abnormal DRE or family history of PCa, awaiting initial or repeat biopsy were studied. The predictive model (which included DRE as an additional risk factor) achieved an AUC = 0.86 in predicting high-grade cancer (after biopsy). Moreover, it was shown that with a cut-off point of -2.8, a 98% NPV with a sensitivity of 96%, the risk of GS ≥ 7 PCa was very low. For GS = 7 PCa, a 53% reduction in unnecessary biopsies was achieved while missing only 2% of cases with csPCa.

A study [65] was performed on a multinational (Netherlands, France, Germany) group of 715 patients with PSA < 10 ng/mL, before the initial prostate biopsy. SelectMDx achieved very high predictive values (AUC = 0.82 with 89% sensitivity, 53% specificity and NPV = 95% (Table 3), outperforming the PCPTRC 2.0. risk calculator (AUC = 0.70). This supports the use of the SelectMDx (MDxHealth, Inc., Irvine, CA, USA) test for the detection of HG PCa prior to the initial prostate biopsy.

To evaluate the clinical utility of SelectMDx, 418 patients who had an initial biopsy were studied. A total of 165 of them were positive. The number of biopsies performed within 3 months of the test was reviewed. For patients with a positive result, 71 patients (43%) were biopsied—27 of these patients were identified as having cancer, including 10 with a grade > 2. During this time, 9 patients with negative SelectMDx test results (3.6%) were biopsied—4 were identified as having cancer—all with a grade ≤ 2. SelectMDx has been shown to have a significant impact on decisions about the frequency and timing of biopsies. When the test was positive, the time period was shorter (median: 2 months) and the number of biopsies was five times higher than when SelectMDx was negative (median: 5 months). The test assisted urologists in their decision-making and is, therefore, a useful tool in daily urological practice [71].

A typical dilemma for the urologist deciding on a repeat prostate biopsy was presented in a case report of two men [72]. Both patients had already had their first negative biopsy, with normal DRE results, serum PSA levels of 3–10 ng/mL, no family history of PCa (and a negative ERSPC RC4 risk score). In these considerations, the European Association of Urology (EAU) recommendations [4] suggest the inclusion of mpMRI, RC and/or liquid biopsy tests. The mpMRI is the most accurate tool for localisation of PCa, but this imaging modality performed in the second patient did not show the presence of a tumour. The SelectMDx test showed the presence of PCa and therefore played a key role in individualising the need for repeat biopsy. In the mentioned report, NPV = 98%, and the risk score correlates with the mpMRI results, but it describes only two cases; therefore it suggests and indicates the need for further studies in risk stratification for repeat biopsy using the SelectMDx test.

It is unknown what the cut-off values should be and what the diagnostic and prognostic accuracy is for a multi-ethnic population. There is a lack of studies on Asian or African American populations.

4. Serum Biomarkers

Serum biomarkers, determined from blood samples, are produced by healthy and abnormal cells. PSA is undoubtedly the most widely studied cancer biomarker, but its clinical utility due to low specificity and specificity raises the need to find a test with better diagnostic values. Three tests that may have a positive impact on clinical practice are described below.

4.1. Prostate Health Index (PHI)

The Prostate Health Index (PHI; Beckman Coulter Inc., Brea, CA, USA), determined from a serum test, includes total PSA (tPSA), free, non-protein-bound PSA (fPSA) and (–2)proPSA (the fPSA isoform resulting from incomplete processing of the PSA precursor).

Determination of PHI values is indicated in men with PSA levels in the "grey zone" (4–10 ng /mL) and an unsuspected digital rectal examination (DRE) result [53], at age $\geq$50 years. The PHI score is calculated from the formula: $((-2)proPSA/fPSA) \times \sqrt{PSA}$.

The above formula indicates that men with lower fPSA, higher total PSA and (–2)proPSA are at an increased risk of development of clinically significant PCa [73].

This was confirmed by a study [74] in which the authors demonstrated that a low fPSA with a high total PSA indicates a risk of more aggressive PCa.

High PHI values indicate an increased likelihood of detecting prostate cancer, so when a biopsy (initial or repeat) is recommended, consideration should be given to using this less invasive method.

The PHI score has a greater diagnostic value than considering each of the indices (tPSA, fPSA) separately [49,51], improves the detection of PCa [75], improves clinical decision-making and predicts PCa aggressiveness [49,76,77]. Although the PHI score mainly provides information on overall PCa risk, studies [49,53] show an association between PHI value and prediction of PCa GS $\geq$ 7. A study [49] reported AUC = 0.72 to distinguish PCa GS $\geq$ 4 + 3 from GS $\leq$ 3 + 4 or no PCa.

Teams of researchers Lepor et al. [78] and Loeb et al. [53] showed that PHI is more specific in detecting csPCa than tPSA and/or fPSA. Furthermore, they concluded that this test might be useful in active surveillance and prediction of adverse outcomes after prostatectomy. Guazzoni et al. [52] showed that this was due to (–2)proPSA, as at GS $\geq$7, both PHI and (–2)proPSA were significantly elevated.

De la Calle et al., based on a multicentre study [50], showed that PHI is a predictor of PCa GS $\geq$ 7 (AUC = 0.78–0.82). When the PHI cut-off value of 24 is taken, 36% of unnecessary biopsies are avoided, while only 2.5% of high-grade cancers are missed. With a PHI cut-off point of 25, 40% of biopsies would be missed, and detection of lower grade PCa cases (GS = 6) would be reduced by 25%. However, this is associated with an underdetection of approximately 5% of clinically significant cancer cases.

A study [79] also found a significant effect of PHI on biopsy decisions. The study included 506 men diagnosed using PHI score and 683 without PHI determination, who were the control group. In both groups, men had PSA in the range of 4–10 ng/mL and unsuspecting DRE results. PHI score influenced medical management in 73% of patients; when the score was low, biopsy was postponed, and when it was high or moderate (PHI $\geq$ 36), biopsy was performed. Men who had a PHI test had fewer biopsies than the control group: 36.4% vs. 60.3%, respectively.

In response to these publications, Ehdaie and Carlsson [80] expressed concern about excluding men from a biopsy on the basis of PHI values and the risk of overlooking aggressive cancer, pointing out that the rate of an omitted PCa was 30%.

The authors of the paper [79], in response [81] to [80], maintain that the biopsy was safely postponed. They cite NCCN and AUA recommendations that men without biopsy who are in the diagnostic grey area will be monitored more closely or with additional methods. In a second response [82], it was shown that due to the small number of high-grade cancers, the study would not allow drawing firm conclusions.

In a study [33] (Table 2) involving a European (n = 503) and an Asian (n = 1652) population, the use of PHI established the recommended cut-off points for the above ethnic groups. More biopsies were avoided in the Asian group (56% vs. 40%). This study also identified the need to establish differential cut-off points for different ethnic groups. The authors of the publication recommended cut-off points for csPCa: PHI > 40 for European men and PHI > 30 for men of Asian origin. This result is not surprising, as Asians have a four times lower risk of prostate cancer than Europeans.

To assess the ability of PHI to detect Gleason grade 2–5 (GGG) PCa in Black men, 158 patients with elevated PSA levels and 135 controls were recruited [32] (Table 2). With PSA $\geq$ 4.0 and PHI $\geq$ 35.0, 33.0% of unnecessary biopsies were avoided, but 17.3% of GGG 2–5 PCa were missed. With PSA $\geq$ 4.0 and PHI $\geq$ 28.0, 17.9% were avoided, and the sensitivity of detecting GGG 2–5 PCa was 90.4%. These results indicate that PHI $\geq$ 28.0 can be safely used to avoid unnecessary biopsies in Black men, although it is associated with a risk of missed detection of GGG 2–5 PCa.

Currently, some researchers are considering the use of the PHI density index (PHID—calculated as PHI divided by prostate volume) in diagnosis to identify csPCa. Tosoian et al. [83] showed that the prevalence of csPCa is associated with higher PHID and has a higher diagnostic value compared to PHI (AUC= 0.84 vs. 0.76). Their study indicates that PHID can prevent 38% of unnecessary biopsies while failing to detect only 2% of csPCa.

Another article [84] examined the diagnostic efficacy of PHI and PHID in terms of avoiding unnecessary biopsies. The results indicate that PHI (AUC = 0.722) and PHID (AUC = 0.739) have a higher diagnostic value than PSA, f-PSA% and PSAD (AUC = 0.595, 0.612 and 0.698, respectively). The combined sensitivity of PHI and PHID was 98%, avoiding 20% of biopsies in the non-diagnosis of only one patient with csPCa. Therefore, the use of the PHID density index may be a promising tool in the evaluation of PCa.

4.2. 4Kscore

The 4Kscore (Opko Health Inc., Miami, FL, USA) is a test developed to identify HG-PCa in patients with a suspicious DRE or elevated serum PSA. The test measures the levels of four kallikreins (4K): total PSA (tPSA), free PSA (fPSA), intact PSA (iPSA) and serum levels of human kallikrein 2 (hK2). It then compares the values obtained in the algorithm with the patient's age, DRE and results of previous prostate biopsies. Based on this information, the algorithm generates a percentage probability score to predict HG PCa even years in advance. This assessment allows further management to be determined depending on the outcome of the test and a decision to perform an initial or subsequent biopsy. This test is recommended primarily for men with a genetic family history. However, it can be performed by any man over 35 years of age who wants to assess his personalised risk of disease in the future.

The 4K test, although not designed to assess the predicted course of already diagnosed prostate cancer, has also been used in patients with csPCa to identify candidates for more intensive therapy. It has also been used to improve treatment selection and thus increase the chance of cure in patients suspected of having an underestimated malignancy. The 4Kscore provides an estimate of a patient's risk of developing distant metastases within 10 years.

Parekh and colleagues [56] on a validation cohort of 1012 indicated that the 4Kscore was better at predicting clinically significant PCa than the Prostate Cancer Prevention Trial Risk Calculator 2.0 (PCPT RC) (AUC 0.82 vs. 0.74) (Table 3). This study also indicates that, depending on the cut-off point, 30–58% of biopsies were reducible, while missing only 1.3–4.7% of HG PCa. A threshold of 1–7.5% is considered low risk, allowing safe delay of biopsy and continued follow-up with PSA. A cut-off of 9% reduces the number of biopsies to 43%, with 2.4% of csPCa cases missed [56,58]. At a cut-off of 15%, this test avoids prostate biopsies performed for indolent cancer by up to 58% and misses 4.7% [56]. A cut-off score of $\geq$20% indicates the need for biopsy due to the high risk of csPCa.

A comprehensive systematic review [57] including 12 studies (11,134 patients) showed, almost identically to the above study, an AUC = 0.81 for the 4Kscore in detecting csPCa (Table 3).

In a study [85], 43,692 asymptomatic men (unscreened, PCa-free, with low PSA values) were followed for 20 years, and the 4Kscore was evaluated for early detection of malignant prostate cancer. This work aimed to estimate the risk of prostate cancer metastasis or death by analysing the 4Kscore and PSA. It turned out that already at the time of blood collection, the 4Kscore indicated in whom an aggressive form of prostate cancer would appear. The

4Kscore significantly improved the detection of HG PCa in men with moderately elevated PSA. The authors concluded that men with an elevated PSA but a low 4Kscore could be safely observed by performing blood marker tests instead of direct biopsy. They indicated that men with a low 4Kscore have a very low long-term risk of death from prostate cancer or metastasis.

In a study [58] involving 611 patients, the 4Kscore test was ordered to assess the risk of aggressive prostate cancer in men with abnormal PSA and/or DRE results. Patients were divided into three risk groups: low, medium and high. The test results influenced biopsy decisions in 88.7% of men, where the biopsy avoidance rates were: 94.0%, 52.9% and 19.0% for the low, intermediate and high-risk groups, respectively. The risk category assessed by the 4Kscore was closely related to biopsy outcome, confirming the usefulness of the test in clinical practice.

A case–control study [35] (Table 2) evaluated the 4Kscore in 1667 prostate cancer cases and 691 control men with PSA ≥ 2 ng/mL. The men were from a variety of ethnic groups, including African American, Hispanic, Japanese, Native Hawaiian and Caucasian men. Results showed that across all ethnic groups, the 4Kscore was better at detecting both general and aggressive prostate cancer than tPSA or tPSA + fPSA. Therefore, the 4Kscore has broad clinical applicability and can be used for prostate cancer screening in a multiethnic population.

A study [56] conducted in the USA, evaluating the efficacy of the 4Kscore, examined 1012 men scheduled for prostate biopsy. The diagnostic performance in detecting HG-PCa was evaluated, showing an AUC of 0.82. African American (AA) patients comprised only 8.1% of the study group, which meant that the results were not representative of AA. For this reason, a validation study [34] (Table 2) was conducted on a population with a higher proportion of AA patients. The study included 366 men, 205 of whom were African American. The results of the study showed no significant difference in predicting tumour aggressiveness in this population, showing AUC = 0.81; therefore, the 4Kscore can be used to make biopsy decisions in both African Americans and non-African Americans.

A study [60] aimed at reducing unnecessary biopsies and overdiagnosis of benign PCa used the 4Kscore and the RPCRP risk calculator to predict csPCa at biopsy. A study of 2873 men showed that RPCRP and 4Kscore had very similar performance (AUC = 0.868 vs. AUC = 0.876), and their combination gave even better results (AUC = 0.888). This indicates that adding further predictors is a compromise between clinical utility, cost and patient burden.

4.3. Stockholm3 Model

Stockholm 3 Model (S3M) combines serum biomarkers (total PSA, free PSA, free/total PSA ratio, hK2, MIC1 and MSMB with genetic markers (254 single nucleotide polymorphisms [SNPs] and an unclassified variable for SNP HOXB13). The test also takes into account clinical data (age, previous prostate biopsy—family history, use of 5-alpha reductase inhibitors) and prostate examination (DRE, prostate volume). The S3M available in Sweden, Norway, Denmark and Finland is in clinical use for predicting the risk of aggressive prostate cancer and assessing the need for biopsy. The S3M research team at Karolinska Institute is currently working with two major laboratories in Europe, as well as laboratories in the US and Canada, to introduce the test in additional countries around the world. Additional validation studies have been conducted in Germany, the Netherlands and the UK. Studies on non-Caucasian populations (e.g., Hispanics, African Americans, Asians) are also planned. If the S3M is negative, the man has a low or normal risk of prostate cancer and is recommended to be followed up in 6 years. If the test is positive, it is recommended that the man is referred to a urologist. The urologist measures the volume of the prostate gland and carries out a DRE. If the prostate volume and/or the DRE test is abnormal, a biopsy is recommended. Otherwise, a Stockholm3 test in 2 years is recommended.

A study involving 59,159 men [67] compared S3M with PSA ≥ 3 ng/mL, a screening test for prostate cancer. The study was designed so that both tests detected the same number of Gleason score (GS) ≥ 7 tumours, and the tests were graded on the number of biopsies needed to achieve this. The results showed that S3M, in detecting tumours with a Gleason score of at least 7, has significantly higher specificity, sensitivity and AUC (0.74 versus 0.56 for PSA) for csPCa. Patients with a S3M score ≥ 11% were recommended to be referred to a urologist for further diagnosis. With a retained GS sensitivity ≥ 7, S3M avoided 32% of prostate biopsies (Table 3). In benign tumours, the level of biopsies avoided was 44%. In addition, the authors indicated that S3M could detect aggressive cancer even in men with PSA levels of 1.5–3 ng/mL, and the number of tumours with a Gleason score ≤ 6 was reduced by 17%, reducing overdiagnosis.

A study [68] described how, after fitting S3M to more data, the updated S3M slightly improved the AUC in predicting prostate cancer GS ≥ 7 compared with previously published results [67] (0.75 vs. 0.74). Each additional predictor (including DRE, previous biopsies and prostate volume) increased the AUC by up to one unit. The combination of predictors helps to increase the accuracy of diagnosis while reducing the number of unnecessary biopsies.

Studies [67–70,86] prove that S3M reduces overdiagnosis and the number of prostate biopsies while maintaining sensitivity for clinically significant prostate cancer.

In a short report [86], the authors evaluated how the S3M threshold affects the number of cancers detected and the number of biopsies performed. They collected data from a validation cohort of 47,688 men (with PSA ≥ 1 ng/mL) and then calculated the percentage of biopsies avoided and the percentage of cancer detections for different cut-off points of the S3M test. They noted that as the cut-off point increased, the number of cancers detected and biopsies performed decreased. They considered it reasonable to use S3M test values between 7% and 14% for the cut-off point for biopsy decisions, where cut-off values below 10% would increase sensitivity for Gleason score tumours ≥ 7 compared with PSA ≥ 3 ng/mL. They noted that the threshold could be selected to fit different health systems and even individual men.

Long-term follow-up of the replacement of PSA (as part of the standard prostate cancer diagnostic procedure) with Stockholm3 in prostate cancer detection in primary care in the Stavanger region of Norway showed that the implementation was beneficial. Compared with PSA, S3M reduced the proportion of clinically insignificant PCa (from 58% to 35%) and the number of biopsies performed (from 29.0% to 20.8%). In addition, it increased the proportion of biopsies positive for csPCa from 42% to 65%. This management may also lead to a reduction in healthcare costs. It has been estimated that direct healthcare costs decreased by 23–28% per male studied [87].

S3M is not suitable for men who have previously been diagnosed with or treated for prostate cancer or who are under follow-up after prostate cancer. It has no proven value for men diagnosed with prostate cancer or who have undergone a biopsy or other examination by a urologist within the last 6 months. It does not replace biopsy in men under active monitoring. This test was not evaluated on men younger than 50 years or older than 70 years and was restricted to an ethnically homogeneous population (Stockholm County, Sweden). The S3M was shown to be superior to prostate-specific antigen (PSA) as a screening tool for prostate cancer in all men aged 50–70 years. Furthermore, the S3M test can be performed in cases where the PSA value is > 1.5. The S3M has been shown to be superior in detecting, now often overlooked, aggressive cancer in men with PSA levels of 1.5–3 ng/mL. The S3M may reduce unnecessary biopsies without compromising the ability to diagnose prostate cancer with a Gleason score of at least 7.

5. Tissue Biomarkers

Tissue biomarkers analyse changes in nucleic acid expression and composition of tissue collected during needle core biopsy of the prostate. The main concept is to detect changes in the histologically normal field neighbouring prostate cancer. This helps to verify whether the patient requires an additional biopsy.

ConfirmMDx (MDxHealth)

ConfirmMDx (MDxHealth, Inc., Irvine, CA, USA) is a tissue-based epigenetic assay that uses methylation-specific PCR (MSP) to analyse three prostate-cancer-related changes in DNA methylation patterns of suppressor genes (GSTP1, APC and RASSF1) in biopsy tissue (formalin-fixed and paraffin-embedded). All these biomarkers were isolated from biopsy-positive tissue.

ConfirmMDx is a molecular test clinically validated for predicting prostate cancer risk in men who have had a traditional thick-needle prostate biopsy that did not reveal the presence of cancer cells in collected histopathological material. In many cases, prostate biopsy results are falsely negative. The biopsy specimen may not be cancerous, and the histopathological result will not reveal the presence of cancer. However, due to the "halo effect", tissue with a normal morphological appearance will show epigenetic changes, indicating the presence of cancer. Using the test, histopathological material that has already been taken—during a prostate biopsy—can be re-examined in a detailed epigenetic analysis quantifying the level of methylation of promoter regions of three genes in benign prostate tissue, assessing with high accuracy the presence of cancer cells in neighbouring areas.

ConfirmMDx offers the opportunity to avoid unnecessary repeat biopsies. It allows a decision to be made on whether to include (rule-in) or exclude (rule-out) therapy. High-risk men with a previously negative biopsy may have undetected cancer after the test. Such patients with a previous "false negative" biopsy result should be included for repeat biopsy and appropriate treatment.

It also allows low-risk men to be excluded from repeat biopsies, which protects the patient from unnecessary stress and possible complications and reduces healthcare costs. This test increases the negative predictive value.

In the MATLOC study [28] involving 498 men with histopathologically negative prostate biopsies who had repeat biopsies within 30 months, positive results (cases) and negative results (controls) were reported. The clinical impact of a panel of epigenetic markers was assessed, showing for all cancers: NPV, sensitivity and specificity of 90%, 68% and 64%, respectively (Table 1). The results showed that in a multivariate model including patient age, PSA, DRE and histopathological features of the first biopsy, the epigenetic test was a significant independent predictor. At the same time, it was shown that the addition of this test could improve the diagnostic process for prostate cancer and reduce the number of unnecessary biopsies.

This was confirmed in the multicentre DOCUMENT study [29], which validated the clinical ability of ConfirmMDx to predict negative histopathological results in repeat prostate biopsies. For this purpose, archived core tissue samples from prostate biopsies with negative prostate cancer from 350 patients were evaluated. All patients had repeat biopsies after 24 months with negative (control) or positive (cases) histopathological results. The epigenetic test was shown to be a significant independent predictor of PCa detection after repeat biopsy and showed an NPV of 88%, with a sensitivity = 62% and specificity= 64% (Table 1).

Van Neste et al. [30] conducted a study on a cohort of 803 men, stratified according to their general methylation status (positive or negative) as defined in MATLOC [28] and DOCUMENT [29]. This study demonstrated an NPV of 96% for csPCa, and that methylation intensity was strongly correlated with the cancer stage. In assessing the prediction of GS $\geq$ 7 PCa after repeat biopsy, ConfirmMDx reached an AUC = 0.76 (Table 1). The decision curve analysis indicated the high clinical utility of the risk score as a decision tool in repeat biopsy. This indicates that ConfirmMDx is a much better predictor compared to currently used indicators such as PSA and risk calculator (PCPT-RC).

A population of 211 African American men undergoing repeat biopsy was studied to compare the accuracy of predicting repeat biopsy outcomes with previous studies conducted in predominantly Caucasian populations [31] (Table 2). The specificity of this epigenetic assay was 60.0% and the sensitivity was 74.1% for detecting PCa at repeat biopsy. For detection of all PCa and GS ≥ 7 PCa, the NPV was 78.8% and 94.2%, respectively (Table 1). This study showed no significant differences in sensitivity and specificity between this test and previously described validation studies involving predominantly Caucasian populations and indicates usefulness for African Americans in risk stratification after an initially negative biopsy.

Wojno et al. [88] in 2014 already noted a reduced number of biopsies in clinical practice in centres using ConfirmMDx. They studied 138 patients with a median PSA level of 4.7 ng/mL and previous negative biopsies. They indicated a 4.4% repeat biopsy rate in ConfirmMDx-negative men, compared with a 43% prior repeat biopsy rate, indicating a potential 10-fold reduction.

A later study [89] in 2019 confirmed the impact of ConfirmMDx on biopsy decision-making. A total of 605 men with a median PSA level of 6.8 ng/mL and previous negative biopsies were studied. There was a six times higher repeat biopsy rate in ConfirmMDx positive men than in men with a negative test result.

ConfirmMDx enables a higher degree of accuracy (previously unattainable by prostate biopsy procedures alone) and has clinical, financial [90] and health benefits by reducing the number of medically unnecessary and expensive repeat biopsies that are part of the current standard of care.

6. The Financial Aspect

A prostate MRI costs between 500 and 2500 USD in the United States, depending on whether the patient is insured. Approximately 1 million American men are currently referred for prostate biopsy each year. If all of these men underwent an MRI instead, costs could reach 3 billion USD per year.

In a paper [91] addressing the costs associated with prostate biopsy and its potential complications, the authors analysed charges for the procedure and related claims for all Medicare Fee-for-Service patients over a 2-year period (January 2014–December 2015). The study included 234,819 prostate biopsy cases and associated costs.

Uncomplicated biopsies cost about 1750 USD, those with one complication were already more expensive at 4060 USD, and for patients requiring hospital admission, the cost was as high as 13,840 USD (average cost was 2020 USD). The most common complication of biopsy is bleeding and infection, which can be prevented using biomarker tests from urine or blood. The cost of tests based on these is higher than the commonly used PSA but lower than biopsy, which makes it a cost-effective option.

In a paper [92], Santhianathen et al. conducted a cost-effectiveness analysis of biomarkers for 2018. Costs were obtained directly from pharmaceutical companies (these were as reported by Prostate Cancer Markers): PHI 499 USD, 4Kscore 1185 USD, SelectMDx 500 USD and ExoDx 760 USD the cost of ExoDx was estimated using data from the CMS (Centers for Medicare and Medicaid Services) Clinical Laboratory Fee Schedule). Discounted QALYs and costs were estimated; for example, a 50-year-old male with an elevated PSA level (3 ng/mL or greater). The cost of the current SOC strategy of ultrasound-guided transrectal biopsy was 3863 USD and the discounted QALY (an indicator of an individual's or group's health status expressing quality-adjusted life expectancy) was 18.0853. Each of the biomarkers tested improved the QALY compared with SOC. The ExoDx index provided the highest QALY with an incremental cost-effectiveness ratio of 58,404 USD per QALY. The study showed that before biopsy in men with elevated PSA levels, the use of SelectMDx (MDxHealth, Inc., Irvine, CA, USA) or EPI (Exosome Diagnostics, Waltham, MA, USA) assesses were cost-effective, PHI (Beckman Coulter Inc., Brea, CA, USA), was found to be more expensive and less efficient.

In an economic evaluation, Nicholson et al. [93], comparing diagnostic value for money, found that the PHI test and PCA3 were no more cost-effective than clinical evaluation, which also generates more QALYs.

Reports [94] on SelectMDx support data from four European countries [95], which showed that SelectMDx in the initial diagnosis of prostate cancer saves healthcare costs and increases QALYs compared with the current standard of care based on prostate biopsy for elevated prostate-specific antigen [95].

This was confirmed in a study by Govers et al. [96] on a population of US men with elevated PSA. Results were related to QALYs and cost of care from a payer (Medicare) perspective. Routine use of SelectMDx to guide biopsy decisions was shown to be beneficial—gaining an average of 0.045 QALYs and saving 1694 USD per patient.

Based on studies [92,95,96], it can be concluded that a SelectMDx-based strategy improves health outcomes and reduces costs.

The publication [90] focused on determining the impact of ConfirmMDx on the healthcare budget. It examined whether costs are recovered by avoiding unnecessary biopsies.

The implementation of ConfirmMDx created a hypothetical commercial health plan in which direct costs were calculated over a 1-year horizon using 2013 Medicare fee-for-service rates. The study concluded that the net cost of a commercial health plan with 1 million members would be reduced by approximately 500,000 USD if patients with histopathologically negative biopsies were screened using an epigenetic test to distinguish between patients who should undergo repeat biopsies and those who should not. The use of this genetic test may reduce healthcare costs and improve clinical management.

STHLM3 is not a commercially available test, for which reason its price is unknown. However, it is expected to be similar to other biomarkers currently available (>224 USD). These tests are more expensive than the common PSA, but are more reliable and can be performed less frequently due to their better diagnostic value. It also avoids biopsies, reduces overdiagnosis and allows a treatment plan to be customised to the patient and thus also reduces costs.

7. Guidelines

The National Comprehensive Cancer Network (NCCN) guidelines—version 2.2020—recommend considering the use of biomarkers for the early detection of prostate cancer, indicating that the specificity of screening can be improved in assessing the indication for biopsy (Grade C recommendation). They indicate the possibility of using the Prostate Health Index (PHI), SelectMDx, 4Kscore and ExoDx to assess the likelihood of high-grade cancer (Gleason score $\geq 3 + 4$, GG ≥ 2).

The NCCN guidelines also address post-biopsy management. They indicate the possibility of using tests to improve specificity in high-risk patients despite a negative prostate biopsy result: 4Kscore, PHI, percentage free PSA, PCA3 and ConfirmMDx (included/added from 2020). The recommendations for the management of benign biopsy results themselves have changed "PSA and DRE 6–24 months apart and consideration of per cent free PSA, 4Kscore, PHI, PCA3 or ConfirmMDx and/or mpMRI and/or improved prostate biopsy techniques. Repeat prostate biopsy, depending on risk". However, the guidelines note that the extent to which tests are validated in different populations varies and that it is unclear what the optimal combination of tests with MRI would be. In the current NCCN guidelines, MPS is listed as a biomarker requiring additional testing.

The EAU gives a strong recommendation for the use of risk-calculators and imaging in asymptomatic men with PSA levels of 2–10 ng/mL, while giving a weak recommendation (strength rating—weak) for urine and blood biomarkers to avoid biopsy [4].

The FDA has approved PCA3 and PHI. ExoDx received FDA Breakthrough Design recognition in June 2019. SelectMDx has not been reviewed by the FDA due to the agency determining that such approval is not necessary but includes CAP (College of American Pathologists) and CLIA (Clinical Laboratory Improvement Amendments) accreditations.

ConfirmMDx and 4Kscore do not have FDA recommendations but are accredited by CAP and CLIA.

8. Biomarkers and mpMRI

Multiparametric magnetic resonance imaging (mpMRI) is a promising new tool for the diagnosis, prognosis and monitoring of PCa.

The European Society of Urology guidelines on PCa recommends the use of mpMRI before prostate biopsy in previously untreated patients with suspected PCa [4].

In addition to the high sensitivity of mpMRI in detecting hg-PCa, mpMRI also has disadvantages, i.e., low specificity, high cost, the need for expensive, specialised equipment, low sensitivity in predicting the presence of extra-urethral expansion and the requirement for an expert review. Current research focuses on comparing biomarker tests with mpMRI and also on the extent to which they can complement each other.

A study [66] on 172 men showed promising correlation results between SelectMDx and mpMRI. There was a statistically significant difference in the SelectMDx score between PI-RADS 3 and 4 ($p < 0.01$) and between PI-RADS 4 and 5. The SelectMDx score was better than the PCA3 score in predicting outcome for suspected PCa on mpMRI (AUC = 0.83 for SelectMDx versus 0.65 for PCA3), suggesting the possibility of using the SelectMDx test to stratify patients for mpMRI.

The combination of 4KScore (AUC = 0.70) with mpMRI (AUC = 0.74) resulted in a prognostic improvement (AUC = 0.82 for 4KScore and mpMRI combined) in detecting aggressive PCa [97,98].

In a recent article [94], the authors demonstrated that the 4KScore, used in addition to mpMRI, can reduce unnecessary SBx (without worsening the diagnosis of csPCa) and identify patients who would benefit from undergoing TBx alone. An evaluation of 408 men showed a reduction (39.5%) in unnecessary biopsies and a reduction in detection (33.9%) of GG1 disease, with 5.2% (diagnosed with SBx) and 1.1% (diagnosed with SBx combined with TBx) missing.

In another study [99], 266 men who were not biopsied underwent three strategies using 4Kscore, mpMRI and combination PSA density (PSAD) to determine the safest method to skip biopsy. The first strategy starts by assessing the 4Kscore value. If it was >7.5, indicating an intermediate or high risk of csPCa, mpMRI was performed. If it was negative and the 4Kscore value was above 7.5 but below 18 (intermediate risk), the patient remained under clinical observation, but in case of a positive mpMRI result, a biopsy was performed. The second strategy started with mpMRI and was similar thereafter. In the third strategy, PSAD was calculated in case of a positive mpMRI result. The results confirmed that 4Kscore combined with mpMRI gave a better AUC = 0.82 than each method alone: 4Kscore (AUC = 0.70), mpMRI (AUC = 0.74). The best strategy seems to be an initial biopsy if the 4Kscore was >7.5%, followed by mpMRI and another biopsy for those with positive mpMRI (PIRADS $\geq$ 3) or 4Kscore >18%. This would avoid 34.2% of prostate biopsies while missing 2.7% of clinically significant PCa. However, this model is more expensive and requires external validation in a multicentre study, but it gives us an idea of how we can improve the selection of men for biopsy using biomarkers and mpMRI.

PHI, total PSA, PSAD and the ability of mpMRI to identify csPCa were compared in a group of 395 men [100]. In detecting csPCa for PSA, PSAD, Pi-RADS and PHI, the AUCs were as follows, respectively: 59.5, 64.9, 62.5 and 68.9 in patients undergoing biopsy, and for patients with a previous negative biopsy: 55.4, 69.3, 64.4 and 71.2. This indicates that PHI had comparable results to mpMRI and outperformed other indices.

Adding PHI to mpMRI leads to increased predictive accuracy of csPCa and a reduction of up to 50% in unnecessary biopsies (for men with PI-RADS 3–5 and PHI $\geq$ 30). Moreover, combination AUC outperforms PHI and mpMRI alone (AUC were 0.87, 0.73 and 0.83, respectively [101].

A study [102] performed prostate cancer diagnosis using a combination of Stockholm3 and mpMRI. Targeted biopsies or mpMRI were performed only in men at higher risk as assessed by S3M. When maintaining the number of detected FG cancers $\geq$2, there was a 42% saving of biopsies and a 46% reduction in FG1 detection. Using a combination of S3M and MRI TBx, the detection of GG 1 tumours and the number of biopsies needed were almost halved, with no reduction in sensitivity in detecting GG 2 cancers, compared with using SBx.

9. Discussion

PSA is a highly sensitive screening test. However, it lacks specificity, resulting in a high rate of unnecessary prostate biopsies. The liquid biopsy tests are more expensive than the commonly used PSA, but because of their better diagnostic value, they can be performed less frequently and avoid other more costly procedures such as biopsy or mpMRI.

It seems important to differentiate the tests in terms of their advantages and disadvantages and to demonstrate which biomarker may be most useful in a given clinical situation.

PHI, 4Kscore, SelectMDx and ExoDx offer better specificity than PSA and can help identify men with $GS \geq 7$ PCa. MPS also outperforms PSA and each of its components in HG PCa detection, and its performance in men with suspicious PSA levels helps to validate the need for initial or repeat biopsy.

STHLM3 is also significantly superior to PSA and can detect HG PCa even in men with PSA levels ≥ 1.5 ng/mL. However, this test has not yet been validated on multiethnic groups, nor have tests comparing it with other liquid biopsy tests been developed.

In men at increased risk of PCa with a previous negative biopsy, additional information can be obtained with the Progensa-PCA3 urine test, MPS, ExoDx and the 4Kscore, PHI and STHLM3 serum tests or the tissue-based epigenetic test (ConfirmMDx).

PCA3 reduces prostate biopsy rates in men undergoing repeat biopsy, but there is still no consensus on the cut-off value.

As PCA3 increases with cancer aggressiveness, tests based on it—Progensa PCA3, MiPS and ExoDx—show the ability to distinguish between cancers with high and low Gleason scores, indicating high utility in therapeutic decision-making.

As ExoDx uses an algorithm independent of PSA and its derivatives, clinical factors (features) and standard of care (SOC), it is feasible (in the US) to perform at home. The patient takes a sample, hands it over to a courier and then discusses the result with the doctor via telehealth. This novelty (ExoDx Prostate At-Home Collection) seems particularly useful in times of coronavirus pandemics and for people living far from medical care.

PHI is significantly better than SelectMDx in diagnosing any PCa, while SelectMDx is significantly better than PHI in diagnosing csPCa.

The 4Kscore assesses the risk of detecting HG PCa if a biopsy is performed. It has been shown to have a better detection rate for HG PCa than the modified PCPTRC and SOC. In addition, 4Kscore can predict HG PCa even years in advance and assess the risk of distant metastasis, e.g., in genetically burdened men. It thus helps to non-invasively avoid prostate biopsy for men in whom it is not necessary and identifies men at higher risk for whom an early intervention is beneficial.

Hendriks [103] and colleagues undertook a comparison of the diagnostic values of two FDA-approved tests, PHI and PCA3, for primary and repeated biopsy. Unfortunately, after compiling all studies published before 2017, they were unable to draw clear conclusions due to the conflicting results of the articles analysed. Study [104] notes that although in a double-blind study of PCA3 vs. PHI, PCA3 is superior to PHI in cancer prediction accuracy, when considering only significant PCa, PHI remains the most accurate predictor. For this reason, the authors recommend using PHI instead of PCA3 in population-based screening.

In a study [54] on 531 men (PSA 3–15 ng/mL) who underwent an initial biopsy, 4Kscore and PHI had similar AUCs in predicting PCa (AUC = 0.69 and 0.74, respectively) and csPCa (0.72 vs. 0.71, respectively).

Russo et al. indicated in their systematic review [55] the high diagnostic accuracy of PHI and 4Kscore. Both tests were tested on multiethnic groups and showed high diagnostic value in them. Although both biomarkers provide similar diagnostic accuracy in the detection of general and high-grade PCa and reduce the number of unnecessary biopsies, it should be borne in mind that there are disturbing reports on PHI [80–82].

Furthermore, PHI should not be interpreted as absolute proof of the presence or absence of prostate cancer. Elevated PSA and PHI can be observed not only in patients with prostate cancer but also with benign diseases. PHI results should be interpreted taking into account clinical factors or family history, and individual clinical decisions should be made based on them.

Vedder et al. [105] added PCA3 and 4Kscore to the ERSCPC risk calculator and compared performance. They showed that 4Kscore was better than PCA3 in predicting PCA in men (with PSA ≥ 3.0 ng/mL) (AUC 0.78 and 0.62, respectively). However, when no PSA limit was set, PCA3 performed better than 4Kscore (AUC 0.63 vs. 0.56). When added to ERSPC, both biomarkers slightly improved the prediction of PCa, with no significant differences (in performance) between them.

Additionally, the previously mentioned study [60] confirmed that adding ERSPC to the 4Kscore improves diagnostic value. However, it is worth recalling that the 4Kscore is the most expensive of the tests compiled in our review.

In addition, it is important to remember that drugs such as 5-alpha-reductase inhibitors: finasteride, dutasteride and anti-androgen therapy can affect the levels of PSA and other biomarkers. Such medications should be discontinued for at least 6 months prior to the study. Samples for the test should be taken when the clinician is satisfied that the prostate tissue has recovered, normally no less than 6 months after the date of the last biopsy or any other prostate procedure. The impact of these procedures on the performance of the test has not yet been assessed.

10. Conclusions

Recently, molecular characterisation of PCa has become increasingly important, and a wide range of biomarker-based liquid biopsy tests are commercially available to assist urologists in clinical decision-making. The prostate cancer liquid biopsy biomarkers listed above have a high NPV and therefore help prevent unnecessary biopsies. As mentioned earlier, numerous publications [16–18] have not shown a correlation between PCA3 values and prostate cancer aggressiveness (Gleason score). Given this fact and reports of unexplained PCA3 well above the cut-off [15] without cancer on biopsy, it is reasonable to use newer, more sensitive and specific diagnostic tools to detect patients requiring prompt and radical treatment. For example, PCA3 in combination with other biomarkers such as TMPRSS2: ERG fusion [36] in Mi-Prostate Score [41–46] or ERG and SPDEF in ExoDx Prostate IntelliScore [47,48,61–63], where it shows better diagnostic and prognostic potential.

From a clinical point of view, it is critical to identify assays for the early detection of aggressive PCa subtype when it can still be treated effectively. Recent years have led to the development of totally non-invasive tests i.e., (ExoDx Prostate At-Home Collection) where first catch, nondigital rectal examination urine specimens appeared helpful in identifying aggressive (Gleason score 7–10) PCa in a racially diverse patient cohort. Similarly, the four-kallikrein panel showed effectiveness in identifying aggressive PCa in a multiethnic population.

It seems that in the near future, molecular biomarkers, clinical and histopathological features and diagnostic imaging will have to be used in a complementary rather than a competitive manner to ensure the best possible selection of patients for mpMRI and eventual biopsy.

Author Contributions: All authors made substantial contributions to this work; writing—original draft preparation: M.M.; supervision and review of the paper: M.S. and J.A.S. All authors have read and agreed to the published version of the manuscript.

Funding: This research was funded by Collegium Medicum University of Zielona Góra.

Conflicts of Interest: J.A.S. is a consultant for MDxHealth. M.M. and M.S declare no conflict of interest.

Abbreviations

AUC	Area under the curve
BMI	Body Mass Index
csPCa	Clinically significant prostate cancer
DCA	Decision curve analysis
DRE	Digital Rectal Examination
EAU	European Association of Urology
EPI	ExoDx Prostate IntelliScore
ERSPC RC	European Randomised study of Screening for Prostate Cancer risk calculator
fPSA	Free non-protein-bound PSA
GG	Grade Group
GS	Gleason score
HG	High grade
hK2	Human kallikrein 2
iPSA	Intact PSA
ISUP	International Society of Urological Pathology
LG	Low grade
lncRNA	Long non-coding RNA
MiPS	Michigan Prostate Score
mpMRI	Multiparametric Magnetic Resonance Imaging
MRI	Magnetic Resonance Imaging
NPV	Negative predictive value
NCCN	National Comprehensive Cancer Network
PCa	Prostate cancer
PCA3	Prostate Cancer Antigen 3
PCPT-RC	Prostate Cancer Prevention Trial Risk Calculator
PHI	Prostate Health Index
PHID	Prostate Health Index density
PI-RADS	Prostate Imaging-Reporting and Data System
PPV	Positive predictive value
PSA	Prostate-specific antigen
PSAD	Prostate-specific antigen density
QALY	Quality-adjusted life year
RP	Radical prostatectomy
SBx	Systematic Biopsy
SNPs	Single-nucleotide polymorphisms
SOC	Standard of care
STHLM3	Stockholm 3
S3M	Stockholm 3 Model
TBx	Targeted biopsy
TNM	Tumor-Node-Metastasis (Staging System)
tPSA	Total PSA
TRUS	Transrectal ultrasound
TRUS-Bx	Transrectal ultrasound (TRUS) guided biopsy
4Kscore	Four-kallikrein score

References

1. Siegel, R.L.; Miller, K.D.; Jemal, A. Cancer statistics. *CA Cancer J. Clin.* **2020**, *70*, 7–30. [CrossRef] [PubMed]
2. Bray, F.; Ferlay, J.; Soerjomataram, I.; Siegel, R.L.; Torre, L.A.; Jemal, A. Global cancer statistics 2018: GLOBOCAN estimates of incidence and mortality worldwide for 36 cancers in 185 countries. *CA Cancer J. Clin.* **2018**, *68*, 394–424. [CrossRef] [PubMed]
3. Fernández-Lázaro, D.; Hernández, J.L.G.; García, A.C.; Martínez, A.C.; Mielgo-Ayuso, J. Liquid Biopsy as Novel Tool in Precision Medicine: Origins, Properties, Identification and Clinical Perspective of Cancer's Biomarkers. *Diagnostics* **2020**, *10*, 215. [CrossRef]
4. Mottet, N.; van den Bergh, R.C.N.; Briers, E.; van den Broeck, T.; Cumberbatch, M.G.; De Santis, M.; Cornford, P. EAU-EANM-ESTRO-ESUR-SIOG Guidelines on Prostate Cancer—2020 Update. Part 1: Screening, Diagnosis, and Local Treatment with Curative Intent. *Eur. Urol.* **2020**, *79*, 243–262. [CrossRef]
5. Lilja, H.; Ulmert, D.; Vickers, A.J. Prostate-specific antigen and prostate cancer: Prediction, detection and monitoring. *Nature reviews. Cancer* **2008**, *8*, 268–278. [PubMed]

6. Graif, T.; Loeb, S.; Roehl, K.A.; Gashti, S.N.; Griffin, C.; Yu, X.; Catalona, W.J. Under Diagnosis and Over Diagnosis of Prostate Cancer. *J. Urol.* **2007**, *178*, 88–92. [CrossRef] [PubMed]
7. Lundström, K.J.; Drevin, L.; Carlsson, S.; Garmo, H.; Loeb, S.; Stattin, P.; Bill-Axelson, A. Nationwide population based study of infections after transrectal ultrasound guided prostate biopsy. *J. Urol.* **2014**, *192*, 1116–1122. [CrossRef] [PubMed]
8. Bruyère, F.; Malavaud, S.; Bertrand, P.; Decock, A.; Cariou, G.; Doublet, J.D.; Bernard, L.; Bugel, H.; Conquy, S.; Sotto, A.; et al. Prosbiotate: A Multicenter, Prospective Analysis of Infectious Complications after Prostate Biopsy. *J. Urol.* **2015**, *193*, 145–150. [CrossRef] [PubMed]
9. Hessels, D.; Klein Gunnewiek, J.M.T.; van Oort, I.; Karthaus, H.F.; van Leenders, G.J.; van Balken, B.; Kiemeney, L.; Witjes, J.; Schalken, J.A. DD3PCA3-based Molecular Urine Analysis for the Diagnosis of Prostate Cancer. *Eur. Urol.* **2003**, *44*, 8–16. [CrossRef]
10. Fenner, A. PCA3 as a Grade Reclassification Predictor. *Nat. Rev. Urol.* **2017**, *14*, 390. [CrossRef]
11. Pal, R.P.; Maitra, N.U.; Mellon, J.K.; Khan, M.A. Defining prostate cancer risk before prostate biopsy. *Urol. Oncol. Semin. Orig. Investig.* **2013**, *31*, 1408–1418. [CrossRef]
12. Marks, L.S.; Fradet, Y.; Deras, I.L.; Blase, A.; Mathis, J.; Aubin, S.M.; Cancio, A.T.; Desaulniers, M.; Ellis, W.J.; Rittenhouse, H.; et al. PCA3 Molecular Urine Assay for Prostate Cancer in Men Undergoing Repeat Biopsy. *Urology* **2007**, *69*, 532–535. [CrossRef] [PubMed]
13. Rodríguez, S.V.M.; García-Perdomo, H.A. Diagnostic accuracy of prostate cancer antigen 3 (PCA3) prior to first prostate biopsy: A systematic review and meta-analysis. *Can. Urol. Assoc. J.* **2019**, *14*, E214–E219. [CrossRef] [PubMed]
14. Wei, J.T.; Feng, Z.; Partin, A.W.; Brown, E.; Thompson, I.; Sokoll, L.; Chan, D.W.; Lotan, Y.; Kibel, A.S.; Busby, J.E.; et al. Can Urinary PCA3 Supplement PSA in the Early Detection of Prostate Cancer? *J. Clin. Oncol.* **2014**, *32*, 4066–4072. [CrossRef] [PubMed]
15. Roobol, M.J.; Schröder, F.H.; Van Leenders, G.L.; Hessels, D.; van den Bergh, R.C.; Wolters, T.; Van Leeuwen, P.J. Performance of Prostate Cancer Antigen 3 (PCA3) and Prostate-Specific Antigen in Prescreened Men: Reproducibility and Detection Characteristics for Prostate Cancer Patients with High PCA3 Scores ($\geq$100). *Eur. Urol.* **2010**, *58*, 893–899. [CrossRef] [PubMed]
16. Auprich, M.; Chun, F.K.; Ward, J.F.; Pummer, K.; Babaian, R.; Augustin, H.; Luger, F.; Gutschi, S.; Budäus, L.; Fisch, M.; et al. Critical Assessment of Preoperative Urinary Prostate Cancer Antigen 3 on the Accuracy of Prostate Cancer Staging. *Eur. Urol.* **2011**, *59*, 96–105. [CrossRef]
17. Leyten, G.H.J.; Wierenga, E.A.; Sedelaar, J.P.; Van Oort, I.M.; Futterer, J.J.; Barentsz, J.O.; Schalken, J.A.; Mulders, P.F.A. Value of PCA3 to Predict Biopsy Outcome and Its Potential Role in Selecting Patients for Multiparametric MRI. *Int. J. Mol. Sci.* **2013**, *14*, 11347–11355. [CrossRef]
18. Vlaeminck-Guillem, V.; Devonec, M.; Colombel, M.; Rodriguez-Lafrasse, C.; Decaussin-Petrucci, M.; Ruffion, A. Urinary PCA3 Score Predicts Prostate Cancer Multifocality. *J. Urol.* **2011**, *185*, 1234–1239. [CrossRef]
19. Aubin, S.M.; Reid, J.; Sarno, M.J.; Blase, A.; Aussie, J.; Rittenhouse, H.; Rittmaster, R.; Andriole, G.L.; Groskopf, J. PCA3 Molecular Urine Test for Predicting Repeat Prostate Biopsy Outcome in Populations at Risk: Validation in the Placebo Arm of the Dutasteride REDUCE Trial. *J. Urol.* **2010**, *184*, 1947–1952. [CrossRef]
20. Progensa PCA3 Assay. Available online: https://www.hologic.com/sites/default/files/2019-05/502083-IFU-PI_003_01.pdf (accessed on 20 December 2020).
21. Feibus, A.H.; Sartor, O.; Moparty, K.; Chagin, K.; Kattan, M.W.; Ledet, E.; Levy, J.; Lee, B.; Thomas, R.; Silberstein, J.L. Clinical Use of PCA3 and TMPRSS2:ERG Urinary Biomarkers in African-American Men Undergoing Prostate Biopsy. *J. Urol.* **2016**, *196*, 1053–1060. [CrossRef]
22. Ochiai, A.; Okihara, K.; Kamoi, K.; Oikawa, T.; Shimazui, T.; Murayama, S.; Tomita, K.; Umekawa, T.; Uemura, H.; Miki, T. Clinical utility of the prostate cancer gene 3 (PCA3) urine assay in Japanese men undergoing prostate biopsy. *BJU Int.* **2013**, *111*, 928–933. [CrossRef] [PubMed]
23. Shen, M.; Chen, W.; Yu, K.; Chen, Z.; Zhou, W.; Lin, X.; Weng, Z.; Li, C.; Wu, X.; Tao, Z. The diagnostic value of PCA3 gene-based analysis of urine sediments after digital rectal examination for prostate cancer in a Chinese population. *Exp. Mol. Pathol.* **2011**, *90*, 97–100. [CrossRef] [PubMed]
24. Wang, F.-B.; Chen, R.; Ren, S.-C.; Shi, X.; Zhu, Y.-S.; Zhang, W.; Jing, T.-L.; Zhang, C.; Gao, X.; Hou, J.G.; et al. Prostate cancer antigen 3 moderately improves diagnostic accuracy in Chinese patients undergoing first prostate biopsy. *Asian J. Androl.* **2017**, *19*, 238–243. [CrossRef] [PubMed]
25. Ramos, C.G.; Valdevenito, R.; Vergara, I.; Anabalon, P.; Sanchez, C.; Fulla, J. PCA3 sensitivity and specificity for prostate cancer detection in patients with abnormal PSA and/or suspicious digital rectal examination. First Latin American experience. *Urol. Oncol.* **2013**, *31*, 1522–1526. [CrossRef] [PubMed]
26. Hessels, D.; van Gils, M.P.M.Q.; van Hooij, O.; Jannink, S.A.; Witjes, J.A.; Verhaegh, G.W.; Schalken, J.A. Predictive value of PCA3 in urinary sediments in determining clinico-pathological characteristics of prostate cancer. *Prostate* **2010**, *70*, 10–16. [CrossRef]
27. Chevli, K.K.; Duff, M.; Walter, P.; Yu, C.; Capuder, B.; Elshafei, A.; Jones, J.S.; Malczewski, S.; Kattan, M.W. Urinary PCA3 as a Predictor of Prostate Cancer in a Cohort of 3,073 Men Undergoing Initial Prostate Biopsy. *J. Urol.* **2014**, *191*, 1743–1748. [CrossRef]
28. Stewart, G.D.; Van Neste, L.; Delvenne, P.; Delrée, P.; Delga, A.; McNeill, S.A.; O'Donnell, M.; Clark, J.; Van Criekinge, W.; Bigley, J.; et al. Clinical Utility of an Epigenetic Assay to Detect Occult Prostate Cancer in Histopathologically Negative Biopsies: Results of the MATLOC Study. *J. Urol.* **2013**, *189*, 1110–1116. [CrossRef]

29. Partin, A.W.; Van Neste, L.; Klein, E.A.; Marks, L.S.; Gee, J.R.; Troyer, D.A.; Rieger-Christ, K.; Jones, J.S.; Magi-Galluzzi, C.; Mangold, L.A.; et al. Clinical Validation of an Epigenetic Assay to Predict Negative Histopathological Results in Repeat Prostate Biopsies. *J. Urol.* **2014**, *192*, 1081–1087. [CrossRef]
30. Van Neste, L.; Partin, A.W.; Stewart, G.D.; Epstein, J.I.; Harrison, D.J.; Van Criekinge, W. Risk score predicts high-grade prostate cancer in DNA-methylation positive, histopathologically negative biopsies. *Prostate* **2016**, *76*, 1078–1087. [CrossRef]
31. Waterhouse, R.L.; Van Neste, L.; Moses, K.A.; Barnswell, C.; Silberstein, J.L.; Jalkut, M.; Tutrone, R.; Sylora, J.; Anglade, R.; Murdock, M.; et al. Evaluation of an Epigenetic Assay for Predicting Repeat Prostate Biopsy Outcome in African American Men. *Urology* **2019**, *128*, 62–65. [CrossRef] [PubMed]
32. Babajide, R.; Carbunaru, S.; Nettey, O.S.; Watson, K.S.; Holloway-Beth, A.; McDowell, T.; Ben Levi, J.; Murray, M.; Stinson, J.; Hollowell, C.M.P.; et al. Performance of Prostate Health Index in Biopsy Naïve Black Men. *J. Urol.* **2021**, *205*, 718–724. [CrossRef] [PubMed]
33. Chiu, P.K.-F.; Ng, C.-F.; Semjonow, A.; Zhu, Y.; Vincendeau, S.; Houlgatte, R.; Lazzeri, M.; Guazzoni, G.; Stephan, C.; Haese, A.; et al. A Multicentre Evaluation of the Role of the Prostate Health Index (PHI) in Regions with Differing Prevalence of Prostate Cancer: Adjustment of PHI Reference Ranges is Needed for European and Asian Settings. *Eur. Urol.* **2019**, *75*, 558–561. [CrossRef]
34. Punnen, S.; Freedland, S.J.; Polascik, T.J.; Loeb, S.; Risk, M.C.; Savage, S.; Mathur, S.C.; Uchio, E.; Dong, Y.; Silberstein, J.L. A Multi-Institutional Prospective Trial Confirms Noninvasive Blood Test Maintains Predictive Value in African American Men. *J. Urol.* **2018**, *199*, 1459–1463. [CrossRef]
35. Darst, B.F.; Chou, A.; Wan, P.; Pooler, L.; Sheng, X.; Vertosick, E.A.; Conti, D.V.; Wilkens, L.R.; Le Marchand, L.; Vickers, A.J.; et al. The Four-Kallikrein Panel Is Effective in Identifying Aggressive Prostate Cancer in a Multiethnic Population. *Cancer Epidemiol. Biomark. Prev.* **2020**, *29*, 1381–1388. [CrossRef]
36. Merdan, S.; Tomlins, S.A.; Barnett, C.L.; Morgan, T.M.; Montie, J.E.; Wei, J.T.; Denton, B.T. Assessment of long-term outcomes associated with urinary prostate cancer antigen 3 and TMPRSS2:ERG gene fusion at repeat biopsy. *Cancer* **2015**, *121*, 4071–4079. [CrossRef] [PubMed]
37. Mao, Z.; Ji, A.; Yang, K.; He, W.; Hu, Y.; Zhang, Q.; Zhang, D.; Xie, L. Diagnostic performance of PCA3 and hK2 in combination with serum PSA for prostate cancer. *Medicine* **2018**, *97*, e12806. [CrossRef]
38. Alshalalfa, M.; Verhaegh, G.W.; Gibb, E.A.; Santiago-Jiménez, M.; Erho, N.; Jordan, J.; Yousefi, K.; Lam, L.L.; Kolisnik, T.; Chelissery, J.; et al. Low PCA3 expression is a marker of poor differentiation in localized prostate tumors: Exploratory analysis from 12,076 patients. *Oncotarget* **2017**, *8*, 50804–50813. [CrossRef] [PubMed]
39. Soares, J.C.; Soares, A.C.; Rodrigues, V.C.; Melendez, M.E.; Santos, A.C.; Faria, E.F.; Oliveira, O.N. Detection of the Prostate Cancer Biomarker PCA3 with Electrochemical and Impedance-Based Biosensors. *ACS Appl. Mater. Interfaces* **2019**, *11*, 46645–46650. [CrossRef] [PubMed]
40. Yamkamon, V.; Htoo, K.; Yainoy, S.; Suksrichavalit, T.; Tangchaikeeree, T.; Eiamphungporn, W. Urinary PCA3 detection in prostate cancer by magnetic nanoparticles coupled with col-orimetric enzyme-linked oligonucleotide assay. *EXCLI J.* **2020**, *19*, 501–513. [CrossRef] [PubMed]
41. Tomlins, S.A.; Rhodes, D.R.; Perner, S.; Dhanasekaran, S.M.; Mehra, R.; Sun, X.-W.; Varambally, S.; Cao, X.; Tchinda, J.; Kuefer, R.; et al. Recurrent Fusion of TMPRSS2 and ETS Transcription Factor Genes in Prostate Cancer. *Science* **2005**, *310*, 644–648. [CrossRef] [PubMed]
42. Leyten, G.H.J.M.; Hessels, D.; Jannink, S.A.; Smit, F.P.; de Jong, H.; Cornel, E.B.; de Reijke, T.M.; Vergunst, H.; Kil, P.; Knipscheer, B.C.; et al. Prospective Multicentre Evaluation of PCA3 and TMPRSS2-ERG Gene Fusions as Diagnostic and Prognostic Urinary Biomarkers for Prostate Cancer. *Eur. Urol.* **2014**, *65*, 534–542. [CrossRef]
43. Tomlins, S.A.; Day, J.R.; Lonigro, R.J.; Hovelson, D.; Siddiqui, J.; Kunju, L.P.; Dunn, R.L.; Meyer, S.; Hodge, P.; Groskopf, J.; et al. Urine TMPRSS2:ERG Plus PCA3 for Individualized Prostate Cancer Risk Assessment. *Eur. Urol.* **2016**, *70*, 45–53. [CrossRef]
44. Salami, S.S.; Schmidt, F.; Laxman, B.; Regan, M.M.; Rickman, D.S.; Scherr, D.; Bueti, G.; Siddiqui, J.; Tomlins, S.A.; Wei, J.T.; et al. Combining urinary detection of TMPRSS2:ERG and PCA3 with serum PSA to predict diagnosis of prostate cancer. *Urol. Oncol. Semin. Orig. Investig.* **2013**, *31*, 566–571. [CrossRef]
45. Sanda, M.G.; Feng, Z.; Howard, D.H.; Tomlins, S.A.; Sokoll, L.J.; Chan, D.W.; Regan, M.M.; Groskopf, J.; Chipman, J.; Patil, D.H.; et al. Association Between Combined TMPRSS2:ERG and PCA3 RNA Urinary Testing and Detection of Aggressive Prostate Cancer. *JAMA Oncol.* **2017**, *3*, 1085. [CrossRef]
46. Tosoian, J.J.; Trock, B.J.; Morgan, T.M.; Salami, S.S.; Tomlins, S.A.; Spratt, D.E.; Siddiqui, J.; Kunju, L.P.; Botbyl, R.; Chopra, Z.; et al. Use of the MyProstateScore Test to Rule Out Clinically Significant Cancer: Validation of a Straightforward Clinical Testing Approach. *J. Urol.* **2021**, *205*, 732–739. [CrossRef] [PubMed]
47. Denzer, K.; Kleijmeer, M.J.; Heijnen, H.F.; Stoorvogel, W.; Geuze, H.J. Exosome: From internal vesicle of the multivesicular body to intercellular signaling device. *J. Cell Sci.* **2000**, *113*, 3365–3374. [CrossRef] [PubMed]
48. McKiernan, J.; Donovan, M.J.; O'Neill, V.; Bentink, S.; Noerholm, M.; Belzer, S.; Skog, J.; Kattan, M.W.; Partin, A.; Andriole, G.; et al. A Novel Urine Exosome Gene Expression Assay to Predict High-grade Prostate Cancer at Initial Biopsy. *JAMA Oncol.* **2016**, *2*, 882–889. [CrossRef]
49. Catalona, W.J.; Partin, A.W.; Sanda, M.G.; Wei, J.T.; Klee, G.G.; Bangma, C.H.; Slawin, K.M.; Marks, L.S.; Loeb, S.; Broyles, D.L.; et al. A Multicenter Study of (−2)Pro-Prostate Specific Antigen Combined with Prostate Specific Antigen and Free Prostate Specific

Antigen for Prostate Cancer Detection in the 2.0 to 10.0 ng/ml Prostate Specific Antigen Range. *J. Urol.* **2011**, *185*, 1650–1655. [CrossRef]
50. De La Calle, C.; Patil, D.; Wei, J.T.; Scherr, D.; Sokoll, L.; Chan, D.W.; Siddiqui, J.; Mosquera, J.M.; Rubin, M.A.; Sanda, M.G. Multicenter Evaluation of the Prostate Health Index to Detect Aggressive Prostate Cancer in Biopsy Naïve Men. *J. Urol.* **2015**, *194*, 65–72. [CrossRef]
51. Boegemann, M.; Stephan, C.; Cammann, H.; Vincendeau, S.; Houlgatte, A.; Jung, K.; Blanchet, J.-S.; Semjonow, A. The percentage of prostate-specific antigen (PSA) isoform (–2)proPSA and the Prostate Health Index improve the diagnostic accuracy for clinically relevant prostate cancer at initial and repeat biopsy compared with total PSA and percentage free PSA in men. *BJU Int.* **2015**, *117*, 72–79. [CrossRef]
52. Guazzoni, G.F.; Lazzeri, M.; Nava, L.; Lughezzani, G.; Larcher, A.; Scattoni, V.; Gadda, G.M.; Bini, V.; Cestari, A.; Buffi, N.M.; et al. Preoperative Prostate-Specific Antigen Isoform p2PSA and Its Derivatives, %p2PSA and Prostate Health Index, Predict Pathologic Outcomes in Patients Undergoing Radical Prostatectomy for Prostate Cancer. *Eur. Urol.* **2012**, *61*, 455–466. [CrossRef]
53. Loeb, S.; Sanda, M.G.; Broyles, D.L.; Shin, S.S.; Bangma, C.H.; Wei, J.T.; Partin, A.; Klee, G.G.; Slawin, K.M.; Catalona, W.J.; et al. The Prostate Health Index Selectively Identifies Clinically Significant Prostate Cancer. *J. Urol.* **2015**, *193*, 1163–1169. [CrossRef]
54. Nordström, T.; Vickers, A.; Assel, M.; Lilja, H.; Grönberg, H.; Eklund, M. Comparison Between the Four-kallikrein Panel and Prostate Health Index for Predicting Prostate Cancer. *Eur. Urol.* **2015**, *68*, 139–146. [CrossRef]
55. Russo, G.I.; Regis, F.; Castelli, T.; Favilla, V.; Privitera, S.; Giardina, R.; Cimino, S.; Morgia, G. A Systematic Review and Meta-analysis of the Diagnostic Accuracy of Prostate Health Index and 4-Kallikrein Panel Score in Predicting Overall and High-grade Prostate Cancer. *Clin. Genitourin. Cancer* **2017**, *15*, 429–439.e1. [CrossRef]
56. Parekh, D.J.; Punnen, S.; Sjoberg, D.D.; Asroff, S.W.; Bailen, J.L.; Cochran, J.S.; Concepcion, R.; David, R.D.; Deck, K.B.; Dumbadze, I.; et al. A Multi-institutional Prospective Trial in the USA Confirms that the 4Kscore Accurately Identifies Men with High-grade Prostate Cancer. *Eur. Urol.* **2015**, *68*, 464–470. [CrossRef] [PubMed]
57. Zappala, S.M.; Scardino, P.T.; Okrongly, D.; Linder, V.; Dong, Y. Clinical performance of the 4Kscore Test to predict high-grade prostate cancer at biopsy: A meta-analysis of us and European clinical validation study results. *Rev. Urol.* **2017**, *19*, 149–155. [CrossRef]
58. Konety, B.; Zappala, S.M.; Parekh, D.J.; Osterhout, D.; Schock, J.; Chudler, R.M.; Oldford, G.M.; Kernen, K.M.; Hafron, J. The 4Kscore®Test Reduces Prostate Biopsy Rates in Community and Academic Urology Practices. *Rev. Urol.* **2015**, *17*, 231–240.
59. Lin, D.W.; Newcomb, L.F.; Brown, M.D.; Sjoberg, D.D.; Dong, Y.; Brooks, J.D.; Carroll, P.R.; Cooperberg, M.; Dash, A.; Ellis, W.J.; et al. Evaluating the Four Kallikrein Panel of the 4Kscore for Prediction of High-grade Prostate Cancer in Men in the Canary Prostate Active Surveillance Study. *Eur. Urol.* **2017**, *72*, 448–454. [CrossRef] [PubMed]
60. Verbeek, J.F.M.; Bangma, C.H.; Kweldam, C.F.; van der Kwast, T.H.; Kümmerlin, I.P.; van Leenders, G.J.L.H.; Roobol, M.J. Reducing unnecessary biopsies while detecting clini-cally significant prostate cancer including cribriform growth with the ERSPC Rotterdam risk cal-culator and 4Kscore. *Urol. Oncol. Semin. Orig. Investig.* **2019**, *37*, 138–144. [CrossRef]
61. McKiernan, J.; Donovan, M.J.; Margolis, E.; Partin, A.; Carter, B.; Brown, G.; Torkler, P.; Noerholm, M.; Skog, J.; Shore, N.; et al. A Prospective Adaptive Utility Trial to Validate Performance of a Novel Urine Exosome Gene Expression Assay to Predict High-grade Prostate Cancer in Patients with Prostate-specific Antigen 2–10 ng/ml at Initial Biopsy. *Eur. Urol.* **2018**, *74*, 731–738. [CrossRef] [PubMed]
62. McKiernan, J.; Noerholm, M.; Tadigotla, V.; Kumar, S.; Torkler, P.; Sant, G.; Alter, J.; Donovan, M.J.; Skog, J. A urine-based Exosomal gene expression test stratifies risk of high-grade prostate Cancer in men with prior negative prostate biopsy undergoing repeat biopsy. *BMC Urol.* **2020**, *20*, 138. [CrossRef]
63. Tutrone, R.; Donovan, M.J.; Torkler, P.; Tadigotla, V.; McLain, T.; Noerholm, M.; Skog, J.; McKiernan, J. Clinical utility of the exosome based ExoDx Prostate(IntelliScore) EPI test in men presenting for initial Biopsy with a PSA 2–10 ng/mL. *Prostate Cancer Prostatic Dis.* **2020**, *23*, 607–614. [CrossRef]
64. Van Neste, L.; Hendriks, R.J.; Dijkstra, S.; Trooskens, G.; Cornel, E.B.; Jannink, S.A.; de Jong, H.; Hessels, D.; Smit, F.P.; Melchers, W.J.G.; et al. Detection of High-grade Prostate Cancer Using a Urinary Molecular Biomarker–Based Risk Score. *Eur. Urol.* **2016**, *70*, 740–748. [CrossRef]
65. Haese, A.; Trooskens, G.; Steyaert, S.; Hessels, D.; Brawer, M.; Vlaeminck-Guillem, V.; Ruffion, A.; Tilki, D.; Schalken, J.; Groskopf, J.; et al. Multicenter Optimization and Validation of a 2-Gene mRNA Urine Test for Detection of Clinically Significant Prostate Cancer before Initial Prostate Biopsy. *J. Urol.* **2019**, *202*, 256–263. [CrossRef]
66. Hendriks, R.J.; van der Leest, M.; Dijkstra, S.; Barentsz, J.O.; van Criekinge, W.; van de Kaa, C.A.H.; Schalken, J.A.; Mulders, P.; van Oort, I.M. A urinary biomarker-based risk score correlates with multiparametric MRI for prostate cancer detection. *Prostate* **2017**, *77*, 1401–1407. [CrossRef]
67. Grönberg, H.; Adolfsson, J.; Aly, M.; Nordström, T.; Wiklund, P.; Brandberg, Y.; Thompson, J.; Wiklund, F.; Lindberg, J.; Clements, M.; et al. Prostate cancer screening in men aged 50–69 years (STHLM3): A prospective population-based diagnostic study. *Lancet Oncol.* **2015**, *16*, 1667–1676. [CrossRef]
68. Ström, P.; Nordström, T.; Aly, M.; Egevad, L.; Grönberg, H.; Eklund, M. The Stockholm-3 Model for Prostate Cancer Detection: Algorithm Update, Biomarker Contribution, and Reflex Test Potential. *Eur. Urol.* **2018**, *74*, 204–210. [CrossRef]

69. Möller, A.; Olsson, H.; Grönberg, H.; Eklund, M.; Aly, M.; Nordström, T. The Stock-holm3 blood-test predicts clinically-significant cancer on biopsy: Independent validation in a multi-center community cohort. *Prostate Cancer Prostatic Dis.* **2018**, *22*, 137–142. [CrossRef]
70. Eklund, M.; Nordström, T.; Aly, M.; Adolfsson, J.; Wiklund, P.; Brandberg, Y.; Thompson, J.; Wiklund, F.; Lindberg, J.; Presti, J.C.; et al. The Stockholm-3 (STHLM3) Model can Improve Prostate Cancer Diagnostics in Men Aged 50–69 yr Compared with Current Prostate Cancer Testing. *Eur. Urol. Focus* **2018**, *4*, 707–710. [CrossRef]
71. Shore, N.; Hafron, J.; Langford, T.; Stein, M.; Dehart, J.; Brawer, M.; Hessels, D.; Schalken, J.; Van Criekinge, W.; Groskopf, J.; et al. Urinary Molecular Biomarker Test Impacts Prostate Biopsy Decision Making in Clinical Practice. *Urol. Pr.* **2019**, *6*, 256–261. [CrossRef]
72. Minnee, P.; Hessels, D.; Schalken, J.A.; Van Criekinge, W. Clinically significant Prostate Cancer diagnosed using a urinary molecular biomarker-based risk score: Two case reports. *BMC Urol.* **2019**, *19*, 1–6. [CrossRef] [PubMed]
73. Loeb, S.; Catalona, W.J. The Prostate Health Index: A new test for the detection of prostate cancer. *Ther. Adv. Urol.* **2013**, *6*, 74–77. [CrossRef]
74. Catalona, W.J.; Southwick, P.; Slawin, K.M.; Partin, A.W.; Brawer, M.K.; Flanigan, R.C.; Patel, A.; Richie, J.P.; Walsh, P.C.; Scardino, P.T.; et al. Comparison of percent free PSA, PSA density, and age-specific PSA cutoffs for prostate cancer detection and staging. *Urology* **2000**, *56*, 255–260. [CrossRef]
75. Stephan, C.; Kahrs, A.-M.; Cammann, H.; Lein, M.; Schrader, M.; Deger, S.; Miller, K.; Jung, K. A (−2)proPSA-based artificial neural network significantly improves differentiation between prostate cancer and benign prostatic diseases. *Prostate* **2009**, *69*, 198–207. [CrossRef]
76. Jansen, F.H.; van Schaik, R.H.; Kurstjens, J.; Horninger, W.; Klocker, H.; Bektic, J.; Wildhagen, M.F.; Roobol, M.J.; Bangma, C.H.; Bartsch, G. Prostate-Specific Antigen (PSA) Isoform p2PSA in Combination with Total PSA and Free PSA Improves Diagnostic Accuracy in Prostate Cancer Detection. *Eur. Urol.* **2010**, *57*, 921–927. [CrossRef] [PubMed]
77. Guazzoni, G.; Nava, L.; Lazzeri, M.; Scattoni, V.; Lughezzani, G.; Maccagnano, C.; Dorigatti, F.; Ceriotti, F.; Pontillo, M.; Bini, V.; et al. Prostate-Specific Antigen (PSA) Isoform p2PSA Significantly Improves the Prediction of Prostate Cancer at Initial Extended Prostate Biopsies in Patients with Total PSA Between 2.0 and 10 ng/ml: Results of a Prospective Study in a Clinical Setting. *Eur. Urol.* **2011**, *60*, 214–222. [CrossRef]
78. Lepor, A.; Catalona, W.J.; Loeb, S. The Prostate Health Index. *Urol. Clin. North Am.* **2016**, *43*, 1–6. [CrossRef]
79. White, J.; Shenoy, B.V.; Tutrone, R.F.; Karsh, L.I.; Saltzstein, D.R.; Harmon, W.J.; Broyles, D.L.; Roddy, T.E.; Lofaro, L.R.; Paoli, C.J.; et al. Clinical utility of the Prostate Health Index (phi) for biopsy decision management in a large group urology practice setting. *Prostate Cancer Prostatic Dis.* **2017**, *21*, 78–84. [CrossRef] [PubMed]
80. Ehdaie, B.; Carlsson, S. Reply to 'Clinical utility of the Prostate Health Index (phi) for biopsy decision management in a large group urology practice setting'. *Prostate Cancer Prostatic Dis.* **2018**, *21*, 446–447. [CrossRef] [PubMed]
81. White, J.; Tutrone, R.F. Reply to Letter to the Editor re: 'Clinical utility of the Prostate Health Index (phi) for biopsy decision management in a large group urology practice setting'. *Prostate Cancer Prostatic Dis.* **2018**, *21*, 604. [CrossRef]
82. White, J.; Tutrone, R.F.; Reynolds, M.A. Second Reply to Letter to the Editor re: "Clinical utility of the Prostate Health Index (phi) for biopsy decision management in a large group urology practice setting". *Prostate Cancer Prostatic Dis.* **2019**, *22*, 639–640. [CrossRef]
83. Tosoian, J.J.; Druskin, S.C.; Andreas, D.; Mullane, P.; Chappidi, M.; Joo, S.; Ghabili, K.; Mamawala, M.; Agostino, J.; Carter, H.B.; et al. Prostate Health Index density improves detection of clinically significant prostate cancer. *BJU Int.* **2017**, *120*, 793–798. [CrossRef] [PubMed]
84. Schulze, A.; Christoph, F.; Sachs, M.; Schroeder, J.; Stephan, C.; Schostak, M.; Koenig, F. Use of the Prostate Health Index and Density in 3 Outpatient Centers to Avoid Unnecessary Prostate Biopsies. *Urol. Int.* **2020**, *104*, 181–186. [CrossRef] [PubMed]
85. Vertosick, E.A.; Häggström, C.; Sjoberg, D.D.; Hallmans, G.; Johansson, R.; Vickers, A.J.; Stattin, P.; Lilja, H. Prespecified 4-Kallikrein Marker Model at Age 50 or 60 for Early Detection of Lethal Prostate Cancer in a Large Population Based Cohort of Asymptomatic Men Followed for 20 Years. *J. Urol.* **2020**, *204*, 281–288. [CrossRef]
86. Nordström, T.; Grönberg, H.; Adolfsson, J.; Egevad, L.; Aly, M.; Eklund, M. Balancing Overdiagnosis and Early Detection of Prostate Cancer using the Stockholm-3 Model. *Eur. Urol. Focus* **2018**, *4*, 385–387. [CrossRef]
87. Viste, E.; Vinje, C.A.; Lid, T.G.; Skeie, S.; Evjen-Olsen, Ø.; Nordström, T.; Thorsen, O.; Gilje, B.; Janssen, E.A.M.; Kjosavik, S.R. Effects of replacing PSA with Stockholm3 for diagnosis of clinically significant prostate cancer in a healthcare system—The Stavanger experience. *Scand. J. Prim. Heal. Care* **2020**, *38*, 315–322. [CrossRef]
88. Wojno, K.J.; Costa, F.J.; Cornell, R.J.; Small, J.D.; Pasin, E.; Van Criekinge, W.; Bigley, J.W.; Van Neste, L. Reduced Rate of Repeated Prostate Biopsies Observed in ConfirmMDx Clinical Utility Field Study. *Am. Health Drug Benefits* **2014**, *7*, 129–134.
89. Yonover, P.; Steyaert, S.; Cohen, J.; Ruiz, C.; Grafczynska, K.; Garcia, E.; Dehart, J.; Brawer, M.; Groskopf, J.; Van Criekinge, W. Mp24-02 clinical utility of confirmmdx for prostate cancer in a community urology practice. *J. Urol.* **2019**, *201* (Suppl. 4), e334. [CrossRef]
90. Aubry, W.; Lieberthal, R.; Willis, A.; Bagley, G.; Willis, S.M., 3rd; Layton, A. Budget impact model: Epigenetic assay can help avoid unnecessary repeated prostate biopsies and re-duce healthcare spending. *Am. Health Drug Benefits* **2013**, *6*, 15–24.

91. Weiner, A.; Manjunath, A.; Kirsh, G.M.; Scott, J.A.; Concepcion, R.D.; Verniero, J.; Kapoor, D.A.; Shore, N.D.; Schaeffer, E.M. The Cost of Prostate Biopsies and their Complications: A Summary of Data on All Medicare Fee-for-Service Patients over 2 Years. *Urol. Pr.* **2020**, *7*, 145–151. [CrossRef]
92. Sathianathen, N.J.; Kuntz, K.M.; Alarid-Escudero, F.; Lawrentschuk, N.L.; Bolton, D.M.; Murphy, D.G.; Weight, C.; Konety, B.R. Incorporating Biomarkers into the Primary Prostate Biopsy Setting: A Cost-Effectiveness Analysis. *J. Urol.* **2018**, *200*, 1215–1220. [CrossRef]
93. Nicholson, A.; Mahon, J.; Boland, A.; Beale, S.; Dwan, K.; Fleeman, N.; Hockenhull, J.; Dundar, Y. The clinical effectiveness and cost-effectiveness of the PROGENSA®prostate cancer antigen 3 assay and the Prostate Health Index in the diagnosis of prostate cancer: A systematic review and economic evaluation. *Heal. Technol. Assess.* **2015**, *19*, 1–192. [CrossRef]
94. Falagario, U.G.; Lantz, A.; Jambor, I.; Martini, A.; Ratnani, P.; Wagaskar, V.; Treacy, P.; Veccia, A.; Bravi, C.A.; Bashorun, H.O.; et al. Using biomarkers in patients with positive multiparametric magnetic resonance imaging: 4Kscore predicts the presence of cancer outside the index lesion. *Int. J. Urol.* **2020**, *28*, 47–52. [CrossRef] [PubMed]
95. Govers, T.M.; Hessels, D.; Vlaeminck-Guillem, V.; Schmitz-Dräger, B.J.; Stief, C.G.; Martinez-Ballesteros, C.; Ferro, M.; Borque-Fernando, A.; Rubio-Briones, J.; Sedelaar, J.P.M.; et al. Cost-effectiveness of SelectMDx for prostate cancer in four European countries: A comparative modeling study. *Prostate Cancer Prostatic Dis.* **2018**, *22*, 101–109. [CrossRef]
96. Govers, T.; Caba, L.; Resnick, M.J. Cost-Effectiveness of Urinary Biomarker Panel in Prostate Cancer Risk Assessment. *J. Urol.* **2018**, *200*, 1221–1226. [CrossRef] [PubMed]
97. Margel, D.; Ber, Y.; Baniel, J. Re: Optimizing Patient's Selection for Prostate Biopsy: A Single Institution Experience with Multi-parametric MRI and the 4Kscore Test for the Detection of Aggressive Prostate Cancer. *Eur. Urol.* **2019**, *76*, 535–536. [CrossRef] [PubMed]
98. Punnen, S.; Nahar, B.; Soodana-Prakash, N.; Koru-Sengul, T.; Stoyanova, R.; Pollack, A.; Kava, B.; Gonzalgo, M.L.; Ritch, C.R.; Parekh, D.J. Optimizing patient's selection for prostate biopsy: A single institution experience with multi-parametric MRI and the 4Kscore test for the detection of aggressive prostate cancer. *PLoS ONE* **2018**, *13*, e0201384. [CrossRef]
99. Falagario, U.G.; Martini, A.; Wajswol, E.; Treacy, P.-J.; Ratnani, P.; Jambor, I.; Anastos, H.; Lewis, S.; Haines, K.; Cormio, L.; et al. Avoiding Unnecessary Magnetic Resonance Imaging (MRI) and Biopsies: Negative and Positive Predictive Value of MRI According to Prostate-specific Antigen Density, 4Kscore and Risk Calculators. *Eur. Urol. Oncol.* **2020**, *3*, 700–704. [CrossRef] [PubMed]
100. Stejskal, J.; Adamcová, V.; Záleský, M.; Novák, V.; Čapoun, O.; Fiala, V.; Dolejšová, O.; Sedláčková, H.; Veselý, Š.; Zachoval, R. The predictive value of the prostate health index vs. multiparametric magnetic resonance imaging for prostate cancer diagnosis in prostate biopsy. *World J. Urol.* **2020**, *39*, 1889–1895. [CrossRef]
101. Hsieh, P.-F.; Li, W.-J.; Lin, W.-C.; Chang, H.; Chang, C.-H.; Huang, C.-P.; Yang, C.-R.; Chen, W.-C.; Chang, Y.-H.; Wu, H.-C. Combining prostate health index and multiparametric magnetic resonance imaging in the diagnosis of clinically significant prostate cancer in an Asian population. *World J. Urol.* **2010**, *38*, 1207–1214. [CrossRef] [PubMed]
102. Grönberg, H.; Eklund, M.; Picker, W.; Aly, M.; Jäderling, F.; Adolfsson, J.; Landquist, M.; Haug, E.S.; Ström, P.; Carlsson, S.; et al. Prostate Cancer Diagnostics Using a Combination of the Stockholm3 Blood Test and Multiparametric Magnetic Resonance Imaging. *Eur. Urol.* **2018**, *74*, 722–728. [CrossRef] [PubMed]
103. Hendriks, R.J.; Van Oort, I.M.; Schalken, J.A. Blood-based and urinary prostate cancer biomarkers: A review and comparison of novel biomarkers for detection and treatment decisions. *Prostate Cancer Prostatic Dis.* **2017**, *20*, 12–19. [CrossRef] [PubMed]
104. Seisen, T.; Rouprêt, M.; Brault, D.; Léon, P.; Cancel-Tassin, G.; Compérat, E.; Renard-Penna, R.; Mozer, P.; Guechot, J.; Cussenot, O. Accuracy of the prostate health index versus the urinary prostate cancer antigen 3 score to predict overall and significant prostate cancer at initial biopsy. *Prostate* **2015**, *75*, 103–111. [CrossRef] [PubMed]
105. Vedder, M.M.; de Bekker-Grob, E.W.; Lilja, H.G.; Vickers, A.; van Leenders, G.J.; Steyerberg, E.W.; Roobol, M.J. The Added Value of Percentage of Free to Total Prostate-specific Antigen, PCA3, and a Kallikrein Panel to the ERSPC Risk Calculator for Prostate Cancer in Prescreened Men. *Eur. Urol.* **2014**, *66*, 1109–1115. [CrossRef]

Review

The Roadmap of Colorectal Cancer Screening

Enea Ferlizza [1,*], Rossella Solmi [1], Michela Sgarzi [1], Luigi Ricciardiello [2] and Mattia Lauriola [1]

[1] Department of Experimental, Diagnostic and Specialty Medicine, University of Bologna, 40138 Bologna, Italy; rossella.solmi@unibo.it (R.S.); michela.sgarzi2@unibo.it (M.S); mattia.lauriola2@unibo.it (M.L.)
[2] Gastroenterology Unit, Department of Medical and Surgical Sciences, University of Bologna, 40138 Bologna, Italy; luigi.ricciardiello@unibo.it
* Correspondence: enea.ferlizza2@unibo.it; Tel.: +39-051-209-4115

Simple Summary: Colorectal cancer (CRC) is the third most common form of cancer in terms of incidence and the second in terms of mortality worldwide. CRC develops over several years, thus highlighting the importance of early diagnosis. Fecal occult blood test screening reduces incidence and mortality. However, the participation rate remains low and the tests present a high number of false positive results. This review provides an overview of CRC screening globally and the most recent approaches aimed at improving accuracy and participation in CRC screening, while also considering the need for gender and age differentiation. New fecal tests and markers such as DNA methylation, mutation or integrity, proteins and microRNAs are explored, including recent investigations into fecal microbiota. Liquid biopsy approaches, involving novel markers, such as circulating mRNA, micro-RNA, DNA, proteins and extracellular vesicles are discussed. The approaches reported are based on quantitative PCR methods or arrays and sequencing assays that identify candidate biomarkers in blood samples.

Abstract: Colorectal cancer (CRC) is the third most common form of cancer in terms of incidence and the second in terms of mortality worldwide. CRC develops over several years, thus highlighting the importance of early diagnosis. National screening programs based on fecal occult blood tests and subsequent colonoscopy have reduced the incidence and mortality, however improvements are needed since the participation rate remains low and the tests present a high number of false positive results. This review provides an overview of the CRC screening globally and the state of the art in approaches aimed at improving accuracy and participation in CRC screening, also considering the need for gender and age differentiation. New fecal tests and biomarkers such as DNA methylation, mutation or integrity, proteins and microRNAs are explored, including recent investigations into fecal microbiota. Liquid biopsy approaches, involving novel biomarkers and panels, such as circulating mRNA, micro- and long-non-coding RNA, DNA, proteins and extracellular vesicles are discussed. The approaches reported are based on quantitative PCR methods that could be easily applied to routine screening, or arrays and sequencing assays that should be better exploited to describe and identify candidate biomarkers in blood samples.

Keywords: fecal immunochemical test (FIT); colonoscopy; flexible sigmoidoscopy; liquid biopsy; mRNA; microRNA; ctDNA; proteins; extracellular vesicles

1. Introduction

Colorectal cancer (CRC) develops over time from modifications of the normal intestinal mucosa to benign precancerous adenomas, carcinoma, and eventually aggressive metastatic cancer [1]. This transition is a complex, multifactorial process that has been characterized over the years. Adenomatous polyposis coli (APC) gene mutations or deletions leading to chromosomal instability represents one of the pathways that drives the development of CRC [2–4]. Activating mutations of the KRAS oncogene and inactivating mutations of the TP53 tumor suppressor gene further promote adenoma–carcinoma progression. Microsatellite instability (MSI), aberrant CpG island methylation phenotype (CIMP), chromosomal instability (CIN) and BRAF mutations are also associated with the transition and development of CRC. Colorectal cancer is also linked to different risk factors such as older age, male sex, adverse lifestyle habits (smoking, increased consumption of red meat and alcohol), chronic intestinal diseases, clinical history of polyps, genes, and heredity [5].

The slow growth of this cancer makes the identification of precancerous lesions and early detection of cancer fundamental for defeating the disease. Screening is, thus, essential to reduce the incidence and mortality of CRC. In fact, CRC mortality is gradually decreasing in industrialized countries due to the widespread adoption of screening programs [5]. Today, the implementation of screening opportunities is crucial, and the research in this field is prolific globally.

This review is divided in two parts. The first focuses on the CRC status, screening methodologies, national screening programs and their application worldwide. The second part firstly examines the main drawbacks of the fecal occult blood test (FOBT), which is the golden standard screening test worldwide, and then focuses on the strategies aimed at improving CRC screening using liquid biopsy approaches and suitable candidate biomarkers (mRNA, miRNA, ctDNA, microvesicles).

2. Colorectal Cancer Status in Europe and in the World

CRC is the third most common form of cancer in terms of incidence and the second in terms of mortality worldwide, with 1.9 million new cases and 930,000 deaths reported in 2020 [6]. There are important geographical discrepancies regarding the incidence and mortality of CRC (Figure 1, Tables S1 and S2). Australia and New Zealand show the highest incidence, followed by Europe and North America [7–10]. The highest reported mortality rates are in central Eastern Europe. The lowest CRC incidence is registered in South Asia and in Africa, where also the lowest mortality rates are recorded, although in these areas the highest mortality to incidence ratio is recorded.

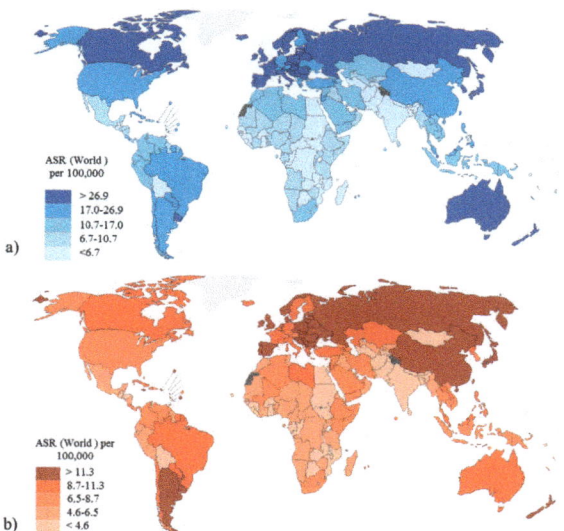

Figure 1. World and colorectal cancer in 2020. (**a**) Estimated age standardized incidence rate (100,000) for world countries; (**b**) Estimated age standardized mortality rate (100,000) for the world countries. Modified from Global Cancer Observatory (GBO) 2020, International Agency for Research on Cancer, World Health Organization [6].

In Europe, CRC is the second most common oncological disease in terms of incidence and mortality, with 519,820 new cases and 244,824 deaths registered in 2020 [6]. The highest incidence was observed in central Eastern Europe and there are substantial differences between European countries (Table S3; Figure 2), with respect to both risk factors, linked to different lifestyles, and screening policies [5,9,11].

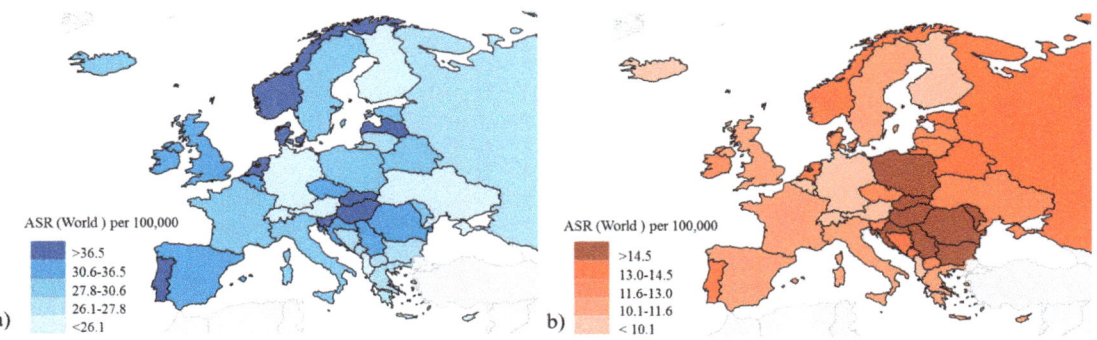

Figure 2. Colorectal cancer in Europe in 2020. (**a**) Estimated age standardized incidence rate (100,000) for European countries; (**b**) Estimated age standardized mortality rate (100,000) for the European countries. Modified from Global Cancer Observatory 2020, International Agency for Research on Cancer, World Health Organization [6].

In Italy, CRC is the third most common oncological disease in terms of incidence and the second in terms of mortality, with 49,327 new cases and 19,258 deaths reported in 2018 (Table S4). The 5-year survival is 66%, with no differences between men and women [12].

3. Colorectal Cancer Screening

3.1. Advantages of Screening in Terms of Incidence and Mortality

CRC screening programs have been shown to reduce incidence and mortality [7,13–16]. Soon after the activation of screening programs, the incidence showed an increase in the short-term, which tended to decrease over the subsequent years [13,17] and the cancers that were detected were more often diagnosed at earlier stages [16]. Notably, populations with active screening programs have shown an impressive reduction in mortality from 22 to 68% [16–21].

3.2. Screening Tests

3.2.1. Stool-Based Tests

There are currently three types of screening for detecting CRC: stool-based, imaging, and endoscopic tests [5]. Stool-based tests (fecal occult based test, FOBT) shows the presence of hem (gFOBT) or human globin (FIT) of hemoglobin in stool samples. The gFOBT has a long history and consists of a colorimetric assay which uses the guaiac reaction [22]. FIT is an immunochemical test, which exploits a specific antibody. It has replaced gFOBT because it is more sensitive and accurate at detecting CRC (sensitivity 69–95% vs. 25–38%) and does not require dietary restrictions [23–25].

Patients with positive FIT tests are referred for a colonoscopy for further investigations. In order to respond to the best cost–benefit strategy, numerous studies have tried to fix the optimal cut-off of FIT. The most commonly used value currently appears to be 100 ng/mL, corresponding to 20 µg of Hb per g of stool [24]. A high variability has been recorded in FIT screening between different centers and kits, with the analytical performance depending on antibody characteristics (mono or polyclonal), buffer volume or composition of collection vials [7,26]. Other drawbacks of FIT are related to the less than optimal level of enrollment in screening programs, the high number of false positive results (15–30%), the poor ability to detect serrated polyps, and the low sensitivity for adenomas [5,7,27].

3.2.2. Imaging Tests

Imaging tests include the double-contrast barium enema (DCBE), computed tomographic colonography (CTC), and colon capsule endoscopy (CCE). Today, novel imaging methods are often used rather than DCBE [5]. CTC was introduced about 20 years ago and provides endoluminal images of the air-distended colon, reconstructed by computed tomography or magnetic resonance [5]. CCE is recognized by the European Society of Gastrointestinal Endoscopy as an acceptable screening method for CRC (with a sensitivity of 84% and specificity of 93%) [28]. However, these methods require intensive bowel preparation and are more expensive than colonoscopy and biopsies cannot be performed [5,28].

3.2.3. Endoscopic Tests

Endoscopic tests consist of flexible sigmoidoscopy (FS) and colonoscopy (CS). FS visualizes only the distal gastrointestinal tract, but does not detect lesions in the proximal colon. The advantages of FS include the fact that no dietary restrictions are required and it involves minimal bowel preparation [5,19,20]. Colonoscopy represents the gold standard for diagnosis, with a high sensitivity and specificity for detecting cancerous and precancerous lesions (97–98%) in the entire large bowel and the distal part of the small bowel [18]. During the procedure, it is also possible to perform biopsies for histological evaluation. However, colonoscopy is an expensive and risky method, since complications such as bleeding or bowel perforation occur in approximately 0.1–0.2% of patients [5,7,19].

3.3. Screening Status in Europe and the World

In 2012, the European Union drew up guidelines on CRC screening and diagnosis, recommending the use of national screening programs based on FIT, FS or CS [29].

In Italy, the national screening program is FIT-based, which is recommended every two years and carried out for the population deemed at risk (50–69 years) [7,12,30–32]. According to the most recent data, screening coverage was between 90 and 96%, depending on the geographic area; however, the overall participation was still low, ranging from 60% in the north to 23% in the south (Tables S5 and S6).

Several European countries, including the extra-EU, have developed an ongoing or planned national or regional screening program, with an invitation system for the population considered at risk (Table 1). The majority use the fecal occult blood test (gFOBT or FIT), with wide differences in participation rates.

Table 1. Colorectal cancer (CRC) screening in Europe [15,33–39].

Country	Program	Test [1]	Cut-Off [2]	Target Age [3]	Interval [3] (years)	Invited [4] (%)	Participation [5] (%)
Albania	NA	NA	NA	NA	NA	NA	NA
Austria	Regional/Opportunistic	FIT/gFOBT/CS	NA	40–80	1	NA	NA
Belarus	NA	NA	NA	NA	NA	NA	NA
Belgium	Regional	FIT/CS	15	50–74	2	99.2	27.7
Bosnia and Herzegovina	Regional/Opportunistic	gFOBT	NA	>50	NA	NA	NA
Bulgaria	No/Opportunistic	gFOBT	NA	NA	NA	NA	NA
Croatia	National	gFOBT	NA	50–74	2	100	15.3
Cyprus	Pilot/Planned	FIT	NA	50–69	2	NA	NA
Czech Republic	Regional/Opportunistic	FIT/gFOBT/CS	15	50–79	2	NA	22.7
Denmark	National	FIT	20	50–74	2	25	64
Estonia	Pilot/Planned	FIT	NA	60–69	2		
Finland	Pilot/Planned	gFOBT	NA	60–69	2	23.9	66.4
France	Regional	FIT/gFOBT	30	50–74	2	99.1	26.5

Table 1. Cont.

Country	Program	Test [1]	Cut-Off [2]	Target Age [3]	Interval [3] (years)	Invited [4] (%)	Participation [5] (%)
Germany	Opportunistic/Pilot/planned	FIT/gFOBT/CS	NA	50–74	2–10	NA	NA
Greece	No/Opportunistic	gFOBT/CS	NA	50–74	NA	NA	NA
Hungary	Pilot/Planned	FIT	20	50–70	2	21.1	36.7
Iceland	Opportunistic/Planned	gFOBT/CS	NA	55–75	2–10	NA	30
Ireland	National	FIT	20	60–69	2	10.9	39.6
Italy	National	FIT/FS [6]	20	50–74 [6]	2	75	42
Latvia	No/Opportunistic	gFOBT	NA	50–74	NA	NA	11.1
Lithuania	Opportunistic/Pilot/Planned	FIT	NA	50–74	2	NA	53.1
Luxembourg	Opportunistic/Planned	FIT/gFOBT/CS	NA	55–74	2	NA	NA
Macedonia	No/Opportunistic	FIT	NA	NA	NA	NA	NA
Malta	National	FIT	16–20	55–66	2	100	45.4
Montenegro	Regional	FIT	NA	50–74	2	NA	33.3
The Netherlands	National	FIT	47	55–75	2	38.5	71.2
Norway	Regional/Pilot	FIT	NA	55–64	2	NA	64.8
Poland	National	CS	NA	55–64	10	12.5	16.7
Portugal	Regional	FIT/gFOBT	20	50–70	2	1.6	62
Romania	No/Opportunistic	NA	NA	NA	NA	NA	NA
Russian Federation	Opportunistic/Pilot	FIT/CS	NA	48–75	NA	NA	NA
Serbia	National	FIT	NA	50–74	2	NA	58.4
Slovakian Republic	No/Opportunistic	FIT/gFOBT/CS	NA	NA	NA	NA	NA
Slovenia	National	FIT	20	50–69	2	93	47.1
Spain	National	FIT/gFOBT	20	50–69	2	14.2	50.2
Sweden	Regional	gFOBT	NA	60–69	2	100	62.7
Switzerland	No/Opportunistic	FIT/CS	NA	50–69	2–10	NA	22
Ukraine	NA	NA	NA	NA	NA	NA	NA
United Kingdom	National	FIT/gFOBT/FS	NA	50–74 [6]	2	100	56.1

[1] Guaiac fecal occult blood test (gFOBT), fecal immunochemical test (FIT), colonoscopy (CS), flexible sigmoidoscopy (FS), not available (NA). [2] Cut-off for FIT in μg Hb/g feces [3]. Target age and interval screening according to the national programs. [4] Percentage of people of the target age invited to participate in the screening. [5] Percentage of invited people that participated in the screening. [6] Regional or national differences.

National and regional organized screening programs using the fecal test (gFOBT or FIT) have also been reported for Canada, Brazil, Argentina, Chile and Uruguay, which obtained high participation rates in the pilot studies (90.1–79.7%) [10,36].

On the other hand, in the USA, the U.S. Preventive Service Task Force recommends that asymptomatic adults aged from 50–75 have a screening test on a voluntary basis, and a national screening program is still not available. The choice for the average risk population in the USA is between stool-based tests (gFOBT, FIT, FIT-DNA), direct visualization tests (FS, CS, CTC) or the serological DNA test (SEPT9). As of 2018, 68.8% of people aged 50–75 with health insurance are reported to be up-to-date with colorectal cancer screening [40].

In 2015, recommendations for CRC screening in the Asia Pacific region were updated [41] and the Asia Cohort Consortium focused also on health outcomes in Asian populations [42]. Few countries (Australia, China, Japan, New Zealand, South Korea, Thailand and Taiwan) have national or regional screening programs and these are mainly based on the fecal test (FIT or gFOBT). Additionally, in these regions the national participation rates were rather low (13–41.3%) [10,36].

Among the countries in the eastern Mediterranean, only Israel reported an organized screening program based on FIT designed for people aged 50–74 years, while in other countries, only opportunistic screening has been adopted (Jordan, Qatar and the United Arabic Emirates). Finally, regarding Africa, the adoption of organized screening may have a limited impact due to the relatively low incidence of CRC and the limited economic resources [10,36].

4. Disadvantages of Fecal Tests (FIT/gFOBT), Room for Improvement

In order to fully benefit from screening programs, the participation rate should be higher than 80%, thus, low take-up is one of the main drawbacks of all the screening programs, with large differences among and within countries [10,16,17,19,31,36,43,44]. Numerous studies have addressed the pitfalls of FIT, including the high rate of false positives and negatives. A false positive FIT can create unnecessary psychological distress and superfluous requests for colonoscopies, with associated healthcare costs. Between 8 and 32% of FIT positive participants do not have significant lesions [19,45,46]. On the other hand, false-negative FIT results can delay CRC diagnosis and dissuade participants from subsequent evaluations [47–49].

The FIT sensitivity for CRC ranges from 91 to 71%, according to the Hb cut-off, with specificity ranging from 90 to 95%. For advanced adenoma (AA), the sensitivity falls from 40 to 25%, with specificity ranging from 90 to 95% [50,51]. There are several risk factors for CRC: gender, age, obesity, alcohol consumption, current or former smoking and the use of drugs, such as non-steroidal anti-inflammatory drugs (NSAIDs) or anticoagulants [24,27,47]. Differences in FIT performance by sex and age have been described. The pooled sensitivity of FIT for advanced neoplasia (AN) was higher in women than in men, with pooled specificities of 92 and 94%, respectively. Accordingly, De Klerk et al., found that the highest risk of a false positive was found for females and the use of NSAIDs [47]. Interestingly, low-dose aspirin was associated with a higher risk of false positives (FPs), suggesting a possible effect on bleeding of early lesions. Several factors appear to be associated with an increased risk of false-positive FIT: male sex, older age (>65), obesity, and current smoking [27,52].

The importance of improving FIT screening has led to various strategies aimed at finding a balance between resources, participation rates and large populations. Some countries have decided to increase the FIT positivity threshold, with the hope of reducing the false positive results and optimizing colonoscopy performance. However, a high threshold leads to a decreased sensitivity and an increased specificity only for advanced neoplasia [53]. This solution reduces the number of false positives, but also increases the false negatives. The decision to increase the FIT cut-off value "simply" to reduce the number of colonoscopies seems more "cost-effective" than "patient-effective", and the adjustment of the FIT cut-off value cannot be the only viable solution.

A novel approach to FIT screening was developed by Senore et al., who evaluated the sum of quantitative FIT results during consecutive negative screening rounds [54]. Subjects with a cumulative fecal Hb level ≥ 20 µg/g showed an 18-fold increase in their cumulative AN (CRC and AA) risk over the subsequent two rounds. This is an interesting approach; however, the number of false positive FITs would still be very high. Another option is to select gender-specific or age-specific cut-offs in FIT screening [53,55,56]. In a stratification model, patients could be assigned to different risk levels of finding AN, by combining different risk factors (such as sex, age or Hb value), or considering FIT separately from the prediction model. The risk-stratification based on prediction models might be better at predicting neoplastic outcomes, including all FIT results, and might enlarge the eligible population including younger subjects (<50 years) and/or people with a history of familiarity for CRC.

5. New Tests

5.1. Fecal Tests

In recent years, new tests have been developed to optimize CRC screening and diagnosis. Since the first studies on *RAS* oncogene mutations [57], DNA alterations in stools have been investigated, as well as proteins and microRNAs.

A recent systematic review evaluated the performance of FIT combined with other stool markers, including DNA methylation, mutation or integrity markers (PHACTR3, APC, p53, KRAS and BRAF), proteins (transferrin, calprotectin and calgranulin) and microRNA (miR-106a) [58]. Notably, the largest increase in sensitivity for CRC was found with long DNA as a measure of DNA integrity in the APC gene and p53, with a specificity of 98%. Among the protein markers, combining transferrin or calgranulin C tests to FIT yielded a slight increase in sensitivity [58,59]. However, calprotectin led to a significant increase in sensitivity for all adenomas, from 53 to 86%, but the specificity decreased from 68 to 26% [58,60].

Following the first studies on DNA and the feasibility of the prototype of a multitarget panel assay in stools [61], Imperiale et al. further developed and tested the multitarget stool DNA ColoGuard (MT-sDNA; Exact Science) in a large screening setting. This test is currently an alternative stool test, which was approved by the Food and Drug Administration (FDA) and is currently employed in the USA [62]. The MT-sDNA test, ColoGuard, combining stool DNA markers (methylated BMP3 and NDRG4 promoter regions, mutant KRAS) with the results of FIT, undoubtedly represented a milestone of stool-based molecular testing in CRC screening. It was tested in nearly 10,000 people, showing a significantly better sensitivity than FIT in predicting any stage of CRC (92.3 vs. 73.8%, respectively) and AA (42.2 vs. 23.8%), and retaining a high specificity (89.9 vs. 96.4%). Similar results were obtained by Bosch et al., in a cohort of 1014 people [63]. By comparing the single-application performance of the MT-sDNA test with FIT, MT-sDNA showed a greater sensitivity for AA than FIT at the lowest cut-off tested (10 μg Hb/g of feces), with a slight decrease in specificity (94 vs. 98%). No significant difference was highlighted to distinguish between proximal and distal AA.

A similar FIT-DNA test kit, ColoClear, manufactured by New Horizon Health Technology Corporation Limited in Hangzhou, China, calculates a risk prediction by combining the FIT test with the detection of the KRAS gene mutation, NDRG4 and BMP3 methylation. ColoClear was tested in 839 subjects, obtaining a sensitivity for CRC and AA of 97.5 and 53.1%, respectively, with a combined sensitivity for predicting AN (CRC and AA) of 88.9% and a specificity of 89.1% [64]. Moreover, no significant difference was highlighted for proximal or distal colon CRC, while sensitivity for distal AA was higher than for proximal AA (61 vs. 30%).

Another stool test, developed in Italy, with the collaboration of Diatech Pharmacogenetics, is based on the evaluation of stool DNA integrity [65,66]. The authors carried out a quantitative evaluation based upon fluorescence amplification of different genomic DNA targets called fluorescence long DNA (FL-DNA). FL-DNA showed a 70% sensitivity and 87% specificity in detecting CRC in stool samples from subjects recruited by a regional screening program based on FIT positivity.

Microbiome-based tests could represent a new frontier in the CRC detection. Grobbee et al. measured fecal microbiota in FIT positive subjects. An overall increase in total bacterial content (16S) was associated with patients affected by high grade dysplasia and CRC [67]. Similarly, a new non-invasive CRC screening test based on microbiome data was employed to reduce the false positive rate of FIT [68]. The authors targeted specific genomic DNA bacterial sequences: Eubacteria (EUB) as the total bacterial load, *Faecalibacterium prausnitzii* (B10), *Subdoligranulum variabile* (B46), *Ruminococcus, Roseburia* and *Coprococcus* (B48), *Roseburia intestinalis* (RSBI), *Gemella morbillorum* (GMLL), *Peptostreptococcus stomatis* (PTST), *Bacteroides fragilis* (BCTF), *Collinsella intestinalis* (CINT), and *Bacteroides thetaiotaomicron* (BCTT). GMLL, PTST and BCTF correlated significantly with AN. Although the sensitivity values for bacterial markers alone were much lower than FIT performance, a final algorithm consisting of the combination of FIT with three ratios between bacterial markers (PTST/EUB, BCTF/EUB, BCTT/ EUB) decreased the number of false positive results by 50%, obtaining a sensitivity of 80% and a specificity of 90%. Finally, panels of proteins were tested in stool samples to identify AA or AN subjects, however, the levels of sensitivity and specificity were quite low (54 vs. 13%) (Hp, LRG1, RBP4, and FN1; 62 vs. 40%) [69].

Nevertheless, despite these efforts to improve fecal-based screening, the main drawbacks remain the low participation rate and the costs [70,71].

5.2. A New Alternative: Liquid Biopsy

The analysis of tumor-derived biomarkers in biological fluids has the potential to increase the participation rate [72]. Peripheral blood is one of the most studied biological fluids, and an accurate blood test could be an attractive alternative for asymptomatic, average-risk individuals who are reluctant to undergo screening by a stool test or endoscopy. Two independent surveys [73,74] showed that a blood sample would be preferred to a stool sample in a screening setting. In a clinical trial, 12% of people who refused to enroll in a stool-based screening, agreed to perform the blood-based test [71]. Thus, blood CRC biomarkers remain very attractive and are under investigation, including several molecules from nucleic acids such as DNA and various types of RNA (messenger, mRNA; micro, miRNA; long non-coding, lncRNA) to proteins, from circulating tumor cells to microvesicles. There is also growing interest in biomarker combination, which could obtain a higher sensitivity than single biomarker-based tests [72,75,76].

Figure 3 summarizes the blood-based tests discussed in this review. The next part of this review focuses, above all, on mRNA, including microRNAs and lncRNA. DNA, proteins and microvesicles are also briefly discussed.

			Sensitivity[1]	Specificity[1]	Advantages	Drawbacks
Routine tests	Stool-based	gFOBT	25-38	91-98	Non-invasive Economical	High Variability Low participation High number of FPs
		FIT 10µg/g	77-88	80-95		
		FIT 20µg/g	69-93	93-96		
		FIT >20µg/g	66-86	94-98		
	Imaging	DCBE			Good sensitivity	Drastic bowel preparation Expensive
		CTC	84	93		
		CCE				
	Endoscopic	FS	90-92[2]	NA	Simultaneous biopsies excision Diagnostic gold standard	Drastic bowel preparation Expensive Potential complications Invasive
		CS	97-98	97-98		
New tests	Stool-based	ColoGuard	92.3	89.9	Improvement of FIT performances Non-invasive	Low participation Further validation needed
		ColoClear+FIT	88.9[3]	89.1[3]		
		FL-DNA+FIT	70	87		
		Bacterial DNA+FIT	80	90		
	Blood-based mRNA	CELTiC	83-90[3]	76-81[3]	High participation Non-invasive	Further validation needed
		ColonSentry	72	70		
		COLOX	79.5	90		
	miRNA	Mir-21	72	85		
		Mir-29a+Mir92a	69-89	70-89.1		
	DNA	EpiproColon 2.0	61.2-82.2	83.6-95.1		
		BCAT1+IKZF1 methylation	66	94		
		CancerSEEK	70[4]	99[4]		
	Proteins	IFNg, EMMPRIN, ERBB4, PSA, CD69, AREG, HGF, CEA	44-65	80-90		
		AFP, CA19-9, CEA, hs-CRP, CyFra21-1, Ferritin, Galectin-3, TIMP-1	80	50		

Figure 3. Routine and new tests used or proposed for colorectal cancer screening. Guaiac fecal occult blood test (gFOBT), fecal immunochemical test (FIT), double-contrast barium enema (DCBE), computed tomographic colonography (CTC), colon capsule endoscopy (CCE), colonoscopy (CS), flexible sigmoidoscopy (FS), fluorescent long DNA (FL-DNA), not available (NA), false positive (FP). [1] The reported percentages of sensitivity and specificity refer to colorectal cancer. [2] Refers only to distal colorectal cancer. [3] Refers to advanced neoplasia. [4] Refers to the average value among 8 cancer types.

5.2.1. mRNA

The emergence of RNA sequencing (RNA-seq) technologies with the evolution of next generation sequencing (NGS) is promising for the diagnosis, prognosis and therapy of cancers including CRC. However, this method is very expensive and, paradoxically, provides too much information that is not yet fully exploitable for the purpose of early diagnosis [77].

In 2016, Rodia et al. [78] used a novel bioinformatic approach to search for specific RNAs with high differential gene expression between CRC and normal blood. The genes showing the highest significant difference were analyzed by qRT-PCR in blood samples of healthy and CRC patients. The authors reported that *CEACAM6, LGALS4, TSPAN8* and *COL1A2* (known as CELTiC) discriminated between the two groups with a sensitivity and specificity of 92 and 67%, respectively. The CELTiC panel was subsequently analyzed in a population of FIT positive subjects, confirming its ability to identify patients with high-risk lesions (CRC and AA), and appeared able to discriminate false positive FIT and low risk patients (non-advanced adenoma and polyps) [79]. In 2020, the CELTiC panel was measured in blood samples from healthy FIT negative subjects, reporting significant gender differences for *CEACAM6* and *COL1A2*, thus highlighting the importance of gender as a potential factor in the comparison between healthy and FIT false positive subjects. The CELTiC panel obtained high AUCs when comparing healthy to AN, low risk, FIT false positive subjects or a combination of these groups, with good sensitivities and specificities ranging from 83 to 90% and from 76 to 81%, respectively. These results confirmed the need for additional studies to better define gender- and age-specific reference intervals for the early diagnosis of CRC [80].

A similar approach was applied using ColonSentry, a panel of seven mRNAs [81,82]. Six out of seven genes (*ANXA3, CLEC4D, LMNB1, PRRG4, TNFAIP6* and *VNN1*) were overexpressed in the blood of CRC patients, and one (*IL2RB*) was under expressed with a blinded validation test set resulting in 72% sensitivity and 70% specificity, with similar predictive values for left- and right-sided CRC.

A similar test is COLOX [83,84], a panel of 29 mRNAs (BCL3, IL1B, PTGS2, MAP2K3, PTGES, PPARG, MMP11, CCR1, EGR1, CACNB4, CES1, IL8, S100A8, CXCL11, ITGA2, NME1, JUN, TNFSF13B, CXCR3, MAPK6, CD63, ITGB5, GATA2, LTF, MMP9, CXCL10, MSL1, RHOC, FXYD5) measured in the peripheral blood mononuclear cells. Few individual genes showed significant differences among age classes, but the whole panel was not affected by the age of the patient. Twelve mRNAs (BCL3, IL1B, PTGS2, PTGES, PPARG, MMP11, CCR1, EGR1, CACNB4, CES1, IL8, S100A8) were able to differentiate between the control group and CRC, and five mRNAs (CES1, CXCL11, IL1B, ITGA2, NME1) identified large adenomas. The authors also found seven markers specifically able to differentiate between large adenomas and CRC (BCL3, PTGES, PPARG, MMP11, IL8, TNFSF13B, CXCR3).

Other mRNAs have also been investigated as blood markers of CRC [85–87]. In 2018, Alamro et al. reported significantly higher mRNA expression of inflammatory genes (COX-2, TNF-α, NF-κB, IL-6) in blood samples of 20 CRC compared to 15 healthy controls without significant association with gender, age or tumor localization [85]. A case–control study performed on 83 CRC patients and 11 healthy donors resulted in significantly higher levels of circulating HMGA2 mRNA in CRC patients with an AUC of 0.932 and a sensitivity of 86.8%. The authors highlighted also a significant association with tumor localization, reporting a greater expression in patients with colon cancer and right-sided CRC, but not with age or gender of the patients [86]. Hamm et al. performed a genome-wide expression analysis on RNA obtained from peripheral blood monocytes collected from 329 subjects (128 healthy, 160 CRC, 41 other gastric diseases) divided in different cohorts [87]. Twenty-three genes showed differential expression between healthy and CRC. By testing different statistical models, the authors reported sensitivity values from 80 to 100%, specificity from 92.3 to 93.3%, and AUCs from 0.86 to 0.99. However, the panel was not tested for the evaluation of preneoplastic lesions (i.e., polyps) or tumor localization [87].

5.2.2. miRNA

MicroRNAs (miRNAs) are small non-coding RNAs (~20–22 nucleotides) that regulate gene expression through repression or degradation of mRNAs. miRNAs seem to be promising plasma biomarkers associated with the onset of CRC, and several studies have searched for specific panels of miRNA capable of increasing both the sensitivity and specificity of screening [4,7,88–91]. miR-7, miR-17-3p, miR-18, miR-21, miR-29a, miR-31, miR-92a, miR-93, miR-155, miR-181b, miR-200c, miR-221, miR-409-3p, let-7g are some of the miRNAs tested, either individually or as a panel in plasma or serum of patients affected by CRC. However, only a few have been confirmed as diagnostic CRC biomarkers by more than one study [89–93]. Among the most studied, mir-21 and mir-29 family (mir-29a, mir-29b and mir-29c) are overexpressed in CRC and associated with CRC progression and metastasis [7,91]. miR-21 is one of the most investigated diagnostic markers of CRC, identified in several biological fluids (plasma, serum, whole blood) [72,94].

Clusters of miRNA have been taken into consideration and the mir-17–92 cluster, also called oncomiR-1, is one of the most studied clusters in relation to CRC. This cluster contains different members such as miR-17, miR-18a, miR19a, miR-19b, miR-20a and miR-92a, and evidence suggests that miR-29a and miR-92a may have a good sensitivity (69 to 89%) and specificity (70 to 89.1%) in CRC detection. In addition, the combination of miR-92a and miR-29a appears to increase the performance of single miRNAs for detecting AN [75,91,95]. Other members, such as miR-19a and miR-19b, were upregulated in plasma from CRC patients compared to healthy individuals, and their combination obtained an AUC of 0.82. The combination with another four plasma miRNAs (miR-18a, miR-29a, miR-15b and miR-335) showed promising results in differentiating between controls and CRC (AUC 0.95) or AA (AUC 0.91) and with similar performances for proximal (AUC 0.97) and distal (AUC 0.95) CRC [75,96]. These results highlight the importance of combinatorial approaches involving specific panels of miRNAs. Other panels of miRNA have recently been reported [97–99]. A panel of seven miRNAs (miR-103a-3p, miR-127-3p, miR-151a-5p, miR-17-5p, miR-181a-5p, miR-18a-5p and miR-18b-5p) was identified and evaluated in a four-stage experiment (screening, training, testing and external validation) involving a total of 139 CRC patients and 132 controls. The performances of the panel obtained an AUC of 0.895 with a sensitivity and specificity of 76.9 and 86.7%, respectively, without significant associations between serum levels of the analyzed miRNAs and age, gender or location [99].

However, the use of mRNA and miRNA is still limited due to the lack of extensive clinical validations.

5.2.3. DNA

In addition to RNA, DNA has also been widely studied in liquid biopsies searching for CRC biomarkers. Cell-free DNA (cfDNA) and the tumor-derived fraction termed as circulating tumor DNA (ctDNA) are of great interest. cfDNA mutations in genes frequently associated with tumorigenesis have been assessed for the early detection of the most common tumor types, including CRC [72,100,101]. KRAS mutations were detected in plasma from CRC patients, however it was also reported that 0.45–20% of healthy individuals may carry genomic alterations in cfDNA, with particular regard to TP53 and KRAS variants [72]. The low abundance of tumor-derived DNA is one of the main challenges for early detection, as well as cancer-associated mutations accumulated with age. Aberrant DNA methylation is a feature of most solid cancers and is a promising biomarker for early diagnosis [75,100,102]. Septin 9 (SEPT9), a GTP-binding protein belonging to the Septin family, is one of the most widely studied DNA markers in blood in relation to CRC. In fact, CRCs show an atypical methylation status of SEPT9 gene. The EpiProcolon assay, which detects circulating methylated SEPT9 (mSEPT9), was recently approved by the FDA [102,103]. The values of sensitivity obtained from independent studies for mSEPT9 ranged from 48.2 to 95.6%, with specificity ranging from 79.1 to 99.1%. In the most recent studies, the sensitivity of the EpiproColon test 2.0 ranged from 61.2 to 82.2% and the specificity from 83.6 to 95.1%, showing a better performance than carcinoembryonic antigen (CEA) and/or FIT tests in the screening of asymptomatic populations [75,102–104]. However, mSEPT9 is not able to distinguish between CRC and polyps or adenomas and seems not affected by tumor localization, but may be affected by age or sex, suggesting that age- and sex-specific cut-offs are required to better optimize the screening and diagnostic procedures [75,102,104]. The combination of mSEPT9 with the FIT test seems to improve the sensitivity for CRC and AA detection obtaining 94 and 43%, respectively, but at the cost of losing the specificity [76].

Other ctDNA markers, such as BCAT1 and IKZF1, have been studied. BCAT1 and IKZF1 methylation obtained a sensitivity of 66% and a specificity of 94% for CRC detection in a prospective study analyzing more than 2000 individuals, including 129 people with CRC [75,105]. A different approach was recently applied by Cohen et al. [106] who described a multi-analyte test (CancerSEEK) to identify eight common cancers, including CRC, by determining the levels of circulating proteins and mutations in ctDNA. The median sensitivity of CancerSEEK was 73 and 78% for stage II and III cancers, respectively, and 43% for stage I cancers. In particular, 14 out of the 16 genes tested (AKT1, APC, BRAF, CDKN2A, CTNNB1, FBXW7, FGFR2, GNAS, KRAS, NRAS, PIK3CA, PPP2R1A, PTEN, TP53) were detected in plasma samples of CRC patients. CRC was also the type of cancer detected with the highest prediction accuracy.

Another approach [107] analyzed cfDNA and ctDNA by applying targeted error correction sequencing (TEC-Seq) for the sensitive and specific detection of low-abundance sequence alterations using NGS in commonly altered cancer genes. In plasma samples from 44 healthy individuals and 194 patients affected by CRC (n = 42), lung (n = 65), ovarian (n = 42) or breast (n = 45) cancers, the authors analyzed a panel of 55 cancer driver genes. cfDNA was significantly higher in cancer patients than in healthy individuals and, within CRC patients, stage IV showed significantly higher cfDNA than stages I to III. In addition, 83% of CRC patients had detectable alterations in driver genes (ctDNA). These detection rates were higher in patients with stages II, III and IV, (89, 90 and 94%, respectively) and were also detected in half of the patients with stage I cancer, suggesting that larger panels of ctDNA may improve the ability to detect small tumors and pre-neoplastic lesions.

cfDNA in the blood samples of CRC patients has also been studied using NGS, machine-learning approaches, genome sequencing and digital sequencing technologies [108–111]. The various models applied by Wan et al. [108] obtained a variable sensitivity (71–85%) with 85% specificity, showing promising preliminary results.

5.2.4. Proteins

Carcinoembryonic antigen (CEA) and carbohydrate antigen (CA19-9) are two of the most studied gastrointestinal tumor-associated proteins in blood (or plasma/serum) [90,112–114]. Serum CEA and/or CA19-9 levels are significantly higher in CRC patients compared to healthy subjects and are well-known cancer markers. However, CEA and CA19-9 concentrations may also be high in other conditions or tumors and their usefulness as CRC screening biomarkers is still an open issue. However, today CEA and CA19-9 are used and approved in clinical practice to detect metastatic disease, recurrence, or to monitor response to treatments [114–120].

Proteomic approaches have recently been applied to blood samples (or plasma/serum) of CRC patients to search for new biomarkers in screening or diagnosis [121–124]. Chen and colleagues performed protein profiling quantifying tumor-associated protein biomarkers in CRC and healthy control plasma samples. Seventeen proteins showed significantly different concentrations between CRC and controls, nine were overexpressed (CEA, GDF-15, AREG, IL-6, CXCL10, CXCL9, PSA, TNFα, cathepsin-D), and eight were downregulated (HGF receptor, CXCL5, ERBB4, FLT3L, CD69, EMMPRIN, VEGFR-2, Caspase-3). Carcinoembryonic antigen (CEA), growth differentiation factor 15 (GDF-15), and amphiregulin (AREG) were the most significant. In addition, applying a logistic regression model, the authors constructed a multi-marker prediction algorithm including eight markers (IFNg,

EMMPRIN, ERBB4, PSA, CD69, AREG, HGF receptor and CEA) reporting moderate sensitivities (44–65%) at high specificities (80–90%) [117].

Finally, an additional panel of eight plasma proteins including AFP, CA19-9, CEA, hs-CRP, CyFra21-1, Ferritin, Galectin-3 and TIMP-1 was tested in 4698 subjects including CRC, AA, non-advanced adenomas and extracolonic cancers [115,125]. All the individual biomarkers significantly identified AN (CRC + AA) and the multivariable model including all the biomarkers and age and gender obtained an AUC of 0.76, with 80% sensitivity and 50% specificity. However, the various models tested, including a combination of the eight proteins, showed moderate performances (90% specificity, 19% sensitivity) in discriminating AA from other conditions (non-advanced adenomas, non-colonic tumors, healthy).

5.2.5. Extracellular Vesicles

Extracellular vesicles (EVs), such as exosomes (EXOs), microvesicles (MVs) and large oncosomes, may contain promising biomarkers. Three main categories divide EVs on the basis of biogenesis and approximate size: EXOs (~40–100 nm) derive from multivesicular bodies within the cells; MVs (~100 nm–1 μm) are formed from the outward budding of the plasma membrane; apoptotic bodies (APs) (~1–5 μm) arise from dying cells undergoing apoptosis [72]. In addition to these classes, some cancer-specific subtypes of EVs have been identified: oncosomes (~100–400 nm) produced by non-transformed cells, whose contents can determine oncogenic effects, and large oncosomes (~1–10 μm) derived from malignant cells [126]. EVs contain proteins, RNA, DNA and lipids, which reflect in part the composition of the cell of origin. By protecting nucleic acids from degradation, EVs could also be considered a better source for tumor molecular profiling compared with cell-free nucleic acids [126,127]. EVs and EXOs are also secreted by cancer cells and in a greater amount than normal cells, therefore, increasing the transfer of RNAs, growth factors and chemokines participating in cancer progression [128–130]. Examples of molecules identified in EVs at increased levels include surface proteins detected by flow cytometry, such as the epithelial cell adhesion molecule (EpCAM), CD9, CD81, CD63 and CD147 in the bloodstream of CRC patients [126,131–133].

One of the drawbacks of studying EVs and exosomes is the lack of a standardized protocol to isolate them from blood and to extract their content or surface material [133–137]. Despite the differences in EV isolation and although most of the studies are case–controls, miRNA is one of the classes most studied as a biomarker in EVs.

Ogata-Kawata et al. [138] evaluated 88 CRC patients and 11 controls to assess the ability of serum EV-miRNAs. In particular, miR-21, miR-23a, and miR-1246 differentiated CRC patients (all stages) from controls [138]. By comparing serum EV-miRNA from CRC patients to healthy controls, Yan et al., found that miR-486 was significantly upregulated, while miR-548c was significantly downregulated [139]. Liu et al. also reported an increase in miR-486 levels in the serum EVs of CRC patients compared to healthy subjects [140]. Peng et al. further confirmed the downregulation of serum EV miR-548c in CRC patients, also finding an association with shorter survival and liver metastases [141].

Other authors have evaluated panels of EV-miRNA. Min et al. [142] analyzed EV-miRNA from blood samples of early-stage CRC patients and non-cancerous controls. The authors found 38 miRNAs upregulated and 57 downregulated in CRC patients compared to healthy controls, some of which, such as Let-7b-3p, miR-150-3p, miR-145-3p, miR-139-3p, had already been reported in the plasma of CRC patients. ROC curve analysis of the single miRNAs reached AUCs of 0.792, 0.686, 0.692, and 0.679, respectively. On the other hand, a logistic model including let-7b-3p, miR-139-3p, and miR-145-3p, confirmed the increased potential of panels of EV-miRNAs compared to individuals ones with an AUC of 0.927 [142].

Cha et al. [143] evaluated eight mRNA markers (MYC, VEGF, CDX2, CD133, CEA, CK19, EpCAM, and CD24) extracted from plasma EVs. Of the eight mRNAs, the combination of VEGF and CD133 showed statistically significant differences between healthy and CRC, and obtained an AUC of 0.96 with 100% sensitivity and 80% specificity in discriminating between the two groups.

EVs also contain other types of RNA, such as long non-coding RNA (lncRNA) or mRNA which are dysregulated in CRC. The presence and performance of lnc-RNA and mRNA has been assessed in APs, MVs and EXOs [144]. In a first screening, in sera of CRC patients and healthy subjects, 21 lncRNAs and 16 mRNAs showed significant differences between EXOs of healthy and CRC samples. In the subsequent validation phase, tested in 30 CRC, 20 adenoma and 30 healthy subjects, the combination of lncRNA breast cancer anti-estrogen resistance 4 (BCAR4) with two mRNAs (keratin-associated protein 5–4, KRTAP5-4, and melanoma antigen family A3, MAGEA3) provided the greatest predictive ability, with an AUC of 0.877.

6. Conclusions

There is a long history of CRC screening tests and several studies have attempted to discover cancer biomarkers in stool or blood samples. However, most of the identified biomarkers, (mRNAs, miRNAs, ctDNA, EVs) have only been evaluated in preliminary case–control studies.

In order to improve the screening and the diagnosis of CRC, large-scale randomized studies are needed to confirm the clinical benefits and the usefulness of these tests. In particular, RNA-seq and NGS, could be used to describe and characterize the evolution and development of CRC, in order to discover new and earlier biomarkers, thus improving outcomes. On the other hand, qRT-PCR may be simpler and cheaper when applied to panels of biomarkers aimed at higher levels of performance in terms of sensitivity, specificity, accuracy and speed of execution.

In addition to the differences in sensitivity and specificity between tests, and sometimes the lack of extensive investigating trials, the main drawback remains the low participation rate. However, the use of blood samples may change this trend. Liquid biopsy could be also used to assess the prognosis, response to therapies and during follow-up.

Finally, the development of algorithms, including those derived with artificial intelligence, which associate outcome-influencing parameters such as gender and age with candidate markers, will be a further tool to improve the current efficacy of CRC screening.

Supplementary Materials: The following are available online at https://www.mdpi.com/2072-6694/13/5/1101/s1, Table S1: World and colorectal cancer in 2020. New cases, incidence and mortality rates estimation in continents. Obtained from Global Cancer Observatory (GBO) 2020, International Agency for Research on Cancer (http://gco.iarc.fr/ (accessed on 21 January 2021) [6]), World Health Organization, Table S2: World and colorectal cancer in 2020. New cases, incidence and mortality rates estimation in world areas. Obtained from Global Cancer Observatory (GBO) 2020, International Agency for Research on Cancer (http://gco.iarc.fr/ (accessed on 21 January 2021) [6]), World Health Organization, Table S3: Europe and colorectal cancer in 2020. New cases estimation, incidence and mortality rates in European countries for males and females. Obtained from Global Cancer Observatory (GBO) 2020, International Agency for Research on Cancer (http://gco.iarc.fr/ (accessed on 21 January 2021) [6]), World Health Organization, Table S4: Italy and colorectal cancer. New cases estimation, incidence and mortality for regions of Italy for males and females, Table S5: Colorectal cancer screening in Italy. Invitation rates of the target population and participation rates of the invited people in northern central and southern of Italy, Table S6: Colorectal cancer screening in Italy. Population, percentage of people older than 65 years and participation in FIT screening of the invited target population for each region of Italy.

Author Contributions: Writing—original draft preparation, E.F. and R.S.; figures and editing, M.S.; writing—review, L.R. and M.L. All authors have read and agreed to the published version of the manuscript.

Funding: This work was supported by Spin-Off Projects (J34I19003100002 Finanziamento regione Emilia Romagna per Assegni Call for Business Plan 2019) and POC (Proof of Concept) grants from the University of Bologna and the Emilia-Romagna regional administration (co-funded with EU resources FSE 2014/2020).

Data Availability Statement: Data are contained within the article or supplementary material. The data presented in this study are available in Supplementary Tables S1–S6.

Conflicts of Interest: R.S. and M.L. have intellectual property rights on an international patent pending (WO2016/185451: method and kit for diagnosis of colorectal cancer; USA patent number 10900085; European patent office number EP3298165). R.S., L.R. and M.L. have intellectual property rights on an international patent pending (WO/2019/138303: new prognostic method). The remaining authors (E.F. and M.S.) declared no conflict of interest. The funders had no role in the design of the study; in the collection, analysis, or interpretation of data; in the writing of the manuscript, or in the decision to publish the results.

References

1. Vogelstein, B.; Fearon, E.R.; Hamilton, S.R.; Kern, S.E.; Preisinger, A.C.; Leppert, M.; Smits, A.M.M.; Bos, J.L. Genetic alterations during colorectal-tumor development. *N. Engl. J. Med.* **1988**, *319*, 525–532. [CrossRef] [PubMed]
2. Puccini, A.; Berger, M.D.; Naseem, M.; Tokunaga, R.; Battaglin, F.; Cao, S.; Hanna, D.L.; McSkane, M.; Soni, S.; Zhang, W.; et al. Colorectal cancer: Epigenetic alterations and their clinical implications. *Biochim. Biophys. Acta Rev. Cancer* **2017**, *1868*, 439–448. [CrossRef]
3. Brenner, H.; Kloor, M.; Pox, C.P. Colorectal cancer. *Lancet* **2014**, *383*, 1490–1502. [CrossRef]
4. Siskova, A.; Cervena, K.; Kral, J.; Hucl, T.; Vodicka, P.; Vymetalkova, V. Colorectal adenomas—genetics and searching for new molecular screening biomarkers. *Int. J. Mol. Sci.* **2020**, *21*, 3260. [CrossRef]
5. Hadjipetrou, A.; Anyfantakis, D.; Galanakis, C.G.; Kastanakis, M.; Kastanakis, S. Colorectal cancer, screening and primary care: A mini literature review. *World J. Gastroenterol.* **2017**, *23*, 6049–6058. [CrossRef]
6. Global Cancer Observatory. Available online: https://gco.iarc.fr/ (accessed on 21 January 2021).
7. Rosso, C.; Cabianca, L.; Gili, F.M. Non-invasive markers to detect colorectal cancer in asymptomatic population. *Minerva Biotecnol.* **2019**, *31*, 23–29. [CrossRef]
8. Bray, F.; Ferlay, J.; Soerjomataram, I.; Siegel, R.L.; Torre, L.A.; Jemal, A. Global cancer statistics 2018: GLOBOCAN estimates of incidence and mortality worldwide for 36 cancers in 185 countries. *CA Cancer J. Clin.* **2018**, *68*, 394–424. [CrossRef]
9. Ferlay, J.; Colombet, M.; Soerjomataram, I.; Mathers, C.; Parkin, D.M.; Piñeros, M.; Znaor, A.; Bray, F. Estimating the global cancer incidence and mortality in 2018: GLOBOCAN sources and methods. *Int. J. Cancer* **2019**, *144*, 1941–1953. [CrossRef] [PubMed]
10. Schreuders, E.H.; Ruco, A.; Rabeneck, L.; Schoen, R.E.; Sung, J.J.Y.; Young, G.P.; Kuipers, E.J. Colorectal cancer screening: A global overview of existing programmes. *Gut* **2015**, *64*, 1637–1649. [CrossRef] [PubMed]
11. Gini, A.; Jansen, E.E.L.; Zielonke, N.; Meester, R.G.S.; Senore, C.; Anttila, A.; Segnan, N.; Mlakar, D.N.; de Koning, H.J.; Lansdorp-Vogelaar, I.; et al. Impact of colorectal cancer screening on cancer-specific mortality in Europe: A systematic review. *Eur. J. Cancer* **2020**, *127*, 224–235. [CrossRef]
12. Associazione Italiana di Oncologia Medica; Associazione Italiana dei Registri Tumori; Progressi delle Aziende Sanitarie per la Salute in Italia; Società Italiana di Anatomia Patologica e Citodiagnostica. *I numeri del Cancro in Italia*, 9th ed.; Intermedia Editore: Brescia, Italy, 2019.
13. Arnold, M.; Sierra, M.S.; Laversanne, M.; Soerjomataram, I.; Jemal, A.; Bray, F. Global patterns and trends in colorectal cancer incidence and mortality. *Gut* **2017**, *66*, 683–691. [CrossRef]
14. Kaminski, M.F.; Robertson, D.J.; Senore, C.; Rex, D.K. Optimizing the quality of colorectal cancer screening worldwide. *Gastroenterology* **2020**, *158*, 404–417. [CrossRef] [PubMed]
15. Senore, C.; Basu, P.; Anttila, A.; Ponti, A.; Tomatis, M.; Vale, D.B.; Ronco, G.; Soerjomataram, I.; Primic-Žakelj, M.; Riggi, E.; et al. Performance of colorectal cancer screening in the European Union Member States: Data from the second European screening report. *Gut* **2019**, *68*, 1232–1244. [CrossRef] [PubMed]
16. Vicentini, M.; Zorzi, M.; Bovo, E.; Mancuso, P.; Zappa, M.; Manneschi, G.; Mangone, L.; Giorgi Rossi, P. Impact of screening programme using the faecal immunochemical test on stage of colorectal cancer: Results from the IMPATTO study. *Int. J. Cancer* **2019**, *145*, 110–121. [CrossRef] [PubMed]
17. Rossi, P.G.; Vicentini, M.; Sacchettini, C.; Di Felice, E.; Caroli, S.; Ferrari, F.; Mangone, L.; Pezzarossi, A.; Roncaglia, F.; Campari, C.; et al. Impact of screening program on incidence of colorectal cancer: A cohort study in Italy. *Am. J. Gastroenterol.* **2015**, *110*, 1359–1366. [CrossRef] [PubMed]
18. Brenner, H.; Stock, C.; Hoffmeister, M. Effect of screening sigmoidoscopy and screening colonoscopy on colorectal cancer incidence and mortality: Systematic review and meta-analysis of randomised controlled trials and observational studies. *BMJ* **2014**, *348*, g2467. [CrossRef]
19. Grobbee, E.J.; van der Vlugt, M.; van Vuuren, A.J.; Stroobants, A.K.; Mallant-Hent, R.C.; Lansdorp-Vogelaar, I.; Bossuyt, P.M.M.; Kuipers, E.J.; Dekker, E.; Spaander, M.C.W. Diagnostic yield of one-time colonoscopy vs one-time flexible sigmoidoscopy vs multiple rounds of mailed fecal immunohistochemical tests in colorectal cancer screening. *Clin. Gastroenterol. Hepatol.* **2020**, *18*, 667–675. [CrossRef] [PubMed]

20. Holme, Ø.; Løberg, M.; Kalager, M.; Bretthauer, M.; Hernán, M.A.; Aas, E.; Eide, T.J.; Skovlund, E.; Lekven, J.; Schneede, J.; et al. Long-term effectiveness of sigmoidoscopy screening on colorectal cancer incidence and mortality in women and men. *Ann. Intern. Med.* **2018**, *168*, 775–782. [CrossRef] [PubMed]
21. Jodal, H.C.; Helsingen, L.M.; Anderson, J.C.; Lytvyn, L.; Vandvik, P.O.; Emilsson, L. Colorectal cancer screening with faecal testing, sigmoidoscopy or colonoscopy: A systematic review and network meta-analysis. *BMJ Open* **2019**, *9*, e032773. [CrossRef] [PubMed]
22. Greegor, D.H. Diagnosis of large-bowel cancer in the asymptomatic patient. *J. Am. Med. Assoc.* **1967**, *201*, 943–945. [CrossRef]
23. Lee, J.K.; Liles, E.G.; Bent, S.; Levin, T.R.; Corley, D.A. Accuracy of fecal immunochemical tests for colorectal cancer. *Ann. Intern. Med.* **2014**, *160*, 171–181. [CrossRef]
24. Selby, K.; Levine, E.H.; Doan, C.; Gies, A.; Brenner, H.; Quesenberry, C.; Lee, J.K.; Corley, D.A. Effect of sex, age, and positivity threshold on fecal immunochemical test accuracy: A systematic review and meta-analysis. *Gastroenterology* **2019**, *157*, 1494–1505. [CrossRef] [PubMed]
25. Stracci, F.; Zorzi, M.; Grazzini, G. Colorectal cancer screening: Tests, strategies, and perspectives. *Front. Public Health* **2014**, *2*, 210. [CrossRef]
26. Cusumano, V.T.; May, F.P. Making FIT count: Maximizing appropriate use of the fecal immunochemical test for colorectal cancer screening programs. *J. Gen. Intern. Med.* **2020**, *35*, 1870–1874. [CrossRef] [PubMed]
27. Amitay, E.L.; Cuk, K.; Niedermaier, T.; Weigl, K.; Brenner, H. Factors associated with false-positive fecal immunochemical tests in a large German colorectal cancer screening study. *Int. J. Cancer* **2019**, *144*, 2419–2427. [CrossRef]
28. Spada, C.; Hassan, C.; Bellini, D.; Burling, D.; Cappello, G.; Carretero, C.; Dekker, E.; Eliakim, R.; de Haan, M.; Kaminski, M.F.; et al. Imaging alternatives to colonoscopy: CT colonography and colon capsule. European Society of Gastrointestinal Endoscopy (ESGE) and European Society of Gastrointestinal and Abdominal Radiology (ESGAR) Guideline—Update 2020. *Endoscopy* **2020**, *52*, 1127–1141.
29. Malila, N.; Senore, C.; Armaroli, P. European guidelines for quality assurance in colorectal cancer screening and diagnosis. *Endoscopy* **2012**, *44*, SE31–SE48. [CrossRef] [PubMed]
30. ONS (Osservatorio Nazionale Screening). *Rapporto 2019*; ONS: Firenze, Italy, 2019; pp. 1–45.
31. Venturelli, F.; Sampaolo, L.; Carrozzi, G.; Zappa, M.; Giorgi Rossi, P. Associations between cervical, breast and colorectal cancer screening uptake, chronic diseases and health-related behaviours: Data from the Italian PASSI nationwide surveillance. *Prev. Med.* **2019**, *120*, 60–70. [CrossRef]
32. Mancini, S.; Bucchi, L.; Giuliani, O.; Ravaioli, A.; Vattiato, R.; Baldacchini, F.; Ferretti, S.; Sassoli de Bianchi, P.; Mezzetti, F.; Triossi, O.; et al. Proportional incidence of interval colorectal cancer in a large population-based faecal immunochemical test screening programme. *Dig. Liver Dis.* **2020**, *52*, 452–456. [CrossRef]
33. Basu, P.; Ponti, A.; Anttila, A.; Ronco, G.; Senore, C.; Vale, D.B.; Segnan, N.; Tomatis, M.; Soerjomataram, I.; Primic Žakelj, M.; et al. Status of implementation and organization of cancer screening in The European Union Member States—Summary results from the second European screening report. *Int. J. Cancer* **2018**, *142*, 44–56. [CrossRef] [PubMed]
34. Vale, D.B.; Anttila, A.; Ponti, A.; Senore, C.; Sankaranaryanan, R.; Ronco, G.; Segnan, N.; Tomatis, M.; Žakelj, M.P.; Elfström, K.M.; et al. Invitation strategies and coverage in the population-based cancer screening programmes in the European Union. *Eur. J. Cancer Prev.* **2019**, *28*, 131–140. [CrossRef]
35. Ponti, A.; Anttila, A.; Ronco, G.; Senore, C.; Basu, P.; Segnan, N.; Tomatis, M.; Primic-Žakelj, M.; Dillner, J.; Fernan, M.; et al. *Cancer Screening in the European Union. Report on the Implementation of the Council Recommendation on Cancer Screening*; International Agency for Research on Cancer: Lyon, France, 2017; pp. 1–313.
36. Navarro, M.; Nicolas, A.; Ferrandez, A.; Lanas, A. Colorectal cancer population screening programs worldwide in 2016: An update. *World J. Gastroenterol.* **2017**, *23*, 3632–3642. [CrossRef] [PubMed]
37. Cardoso, R.; Guo, F.; Heisser, T.; Hoffmeister, M.; Brenner, H. Utilisation of colorectal cancer screening tests in european countries by type of screening offer: Results from the european health interview survey. *Cancers* **2020**, *12*, 1409. [CrossRef] [PubMed]
38. Altobelli, E.; D'Aloisio, F.; Angeletti, P.M. Colorectal cancer screening in countries of European Council outside of the EU-28. *World J. Gastroenterol.* **2016**, *22*, 4946–4957. [CrossRef]
39. Altobelli, E.; Rapacchietta, L.; Marziliano, C.; Campagna, G.; Profeta, V.; Fagnano, R. Differences in colorectal cancer surveillance epidemiology and screening in the WHO European Region. *Oncol. Lett.* **2019**, *17*, 2531–2542. [CrossRef] [PubMed]
40. United States Preventive Services Taskforce Recommendation: Colorectal Cancer Screening. Available online: https://www.uspreventiveservicestaskforce.org/uspstf/draft-update-summary/colorectal-cancer-screening3 (accessed on 21 January 2021).
41. Sung, J.J.Y.; Ng, S.C.; Chan, F.K.L.; Chiu, H.M.; Kim, H.S.; Matsuda, T.; Ng, S.S.M.; Lau, J.Y.W.; Zheng, S.; Adler, S.; et al. An updated Asia Pacific Consensus Recommendations on colorectal cancer screening. *Gut* **2015**, *64*, 121–132. [CrossRef]
42. The Asia Cohort Consortium. Available online: https://www.asiacohort.org/index.html (accessed on 22 February 2021).
43. Kooyker, A.I.; Toes-Zoutendijk, E.; Opstal-van Winden, A.W.J.; Spaander, M.C.W.; Buskermolen, M.; Vuuren, H.J.; Kuipers, E.J.; Kemenade, F.J.; Ramakers, C.; Thomeer, M.G.J.; et al. The second round of the Dutch colorectal cancer screening program: Impact of an increased fecal immunochemical test cut-off level on yield of screening. *Int. J. Cancer* **2020**, *147*, 1098–1106. [CrossRef]
44. Rutka, M.; Bor, R.; Molnár, T.; Farkas, K.; Pigniczki, D.; Fábián, A.; Győrffy, M.; Bálint, A.; Milassin, Á.; Szűcs, M.; et al. Efficacy of the population-based pilot colorectal cancer screening, Csongrád county, Hungary, 2015. *Turk. J. Med. Sci.* **2020**, *50*, 756–763. [CrossRef]

45. Nielson, C.M.; Petrik, A.F.; Jacob, L.; Vollmer, W.M.; Keast, E.M.; Schneider, J.L.; Rivelli, J.S.; Kapka, T.J.; Meenan, R.T.; Mummadi, R.R.; et al. Positive predictive values of fecal immunochemical tests used in the STOP CRC pragmatic trial. *Cancer Med.* **2018**, *7*, 4781–4790. [CrossRef] [PubMed]
46. Berry, E.; Miller, S.; Koch, M.; Balasubramanian, B.; Argenbright, K.; Gupta, S. Lower abnormal fecal immunochemical test cut-off values improve detection of colorectal cancer in system-level screens. *Clin. Gastroenterol. Hepatol.* **2020**, *18*, 647–653. [CrossRef]
47. de Klerk, C.M.; Vendrig, L.M.; Bossuyt, P.M.; Dekker, E. Participant-related risk factors for false-positive and false-negative fecal immunochemical tests in colorectal cancer screening: Systematic review and meta-analysis. *Am. J. Gastroenterol.* **2018**, *113*, 1778–1787. [CrossRef]
48. Stegeman, I.; De Wijkerslooth, T.R.; Stoop, E.M.; Van Leerdam, M.; Van Ballegooijen, M.; Kraaijenhagen, R.A.; Fockens, P.; Kuipers, E.J.; Dekker, E.; Bossuyt, P.M. Risk factors for false positive and for false negative test results in screening with fecal occult blood testing. *Int. J. Cancer* **2013**, *133*, 2408–2414. [CrossRef]
49. Wong, M.C.S.; Ching, J.Y.L.; Chan, V.C.W.; Lam, T.Y.T.; Luk, A.K.C.; Ng, S.S.M.; Sung, J.J.Y. Factors associated with false-positive and false-negative fecal immunochemical test results for colorectal cancer screening. *Gastrointest. Endosc.* **2015**, *81*, 596–607. [CrossRef]
50. Gies, A.; Cuk, K.; Schrotz-King, P.; Brenner, H. Direct comparison of diagnostic performance of 9 quantitative fecal immunochemical tests for colorectal cancer screening. *Gastroenterology* **2018**, *154*, 93–104. [CrossRef]
51. Imperiale, T.F.; Gruber, R.N.; Stump, T.E.; Emmett, T.W.; Monahan, P.O. Performance characteristics of fecal immunochemical tests for colorectal cancer and advanced adenomatous polyps: A systematic review and meta-analysis. *Ann. Intern. Med.* **2019**, *170*, 319–329. [CrossRef] [PubMed]
52. Ibañez-Sanz, G.; Garcia, M.; Mila, N.; Hubbard, R.A.; Vidal, C.; Binefa, G.; Benito, L.; Moreno, V. False-positive results in a population-based colorectal screening program: Cumulative risk from 2000 to 2017 with biennial screening. *Cancer Epidemiol. Biomarkers Prev.* **2019**, *28*, 1909–1916. [CrossRef]
53. van de Veerdonk, W.; Hoeck, S.; Peeters, M.; Van Hal, G. Towards risk-stratified colorectal cancer screening. Adding risk factors to the fecal immunochemical test: Evidence, evolution and expectations. *Prev. Med.* **2019**, *126*, 105746. [CrossRef]
54. Senore, C.; Zappa, M.; Campari, C.; Crotta, S.; Armaroli, P.; Arrigoni, A.; Cassoni, P.; Colla, R.; Fracchia, M.; Gili, F.; et al. Faecal haemoglobin concentration among subjects with negative FIT results is associated with the detection rate of neoplasia at subsequent rounds: A prospective study in the context of population based screening programmes in Italy. *Gut* **2020**, *69*, 523–530. [CrossRef] [PubMed]
55. Blom, J.; Löwbeer, C.; Elfström, K.M.; Sventelius, M.; Öhman, D.; Saraste, D.; Törnberg, S. Gender-specific cut-offs in colorectal cancer screening with FIT: Increased compliance and equal positivity rate. *J. Med. Screen.* **2019**, *26*, 92–97. [CrossRef]
56. Mannucci, A.; Zuppardo, R.A.; Rosati, R.; Di Leo, M.; Perea, J.; Cavestro, G.M. Colorectal cancer screening from 45 years of age: Thesis, antithesis and synthesis. *World J. Gastroenterol.* **2019**, *25*, 2565–2580. [CrossRef]
57. Sidransky, D.; Tokino, T.; Hamilton, S.R.; Kinzler, K.W.; Levin, B.; Frost, P.; Vogelstein, B. Identification of *ras* oncogene mutations in the stool of patients with curable colorectal tumors. *Science* **1992**, *256*, 102–105. [CrossRef]
58. Niedermaier, T.; Weigl, K.; Hoffmeister, M.; Brenner, H. Fecal immunochemical tests combined with other stool tests for colorectal cancer and advanced adenoma detection: A systematic review. *Clin. Transl. Gastroenterol.* **2016**, *7*, e175. [CrossRef] [PubMed]
59. Gies, A.; Cuk, K.; Schrotz-King, P.; Brenner, H. Fecal immunochemical test for hemoglobin in combination with fecal transferrin in colorectal cancer screening. *United Eur. Gastroenterol. J.* **2018**, *6*, 1223–1231. [CrossRef]
60. Widlak, M.M.; Neal, M.; Daulton, E.; Thomas, C.L.; Tomkins, C.; Singh, B.; Harmston, C.; Wicaksono, A.; Evans, C.; Smith, S.; et al. Risk stratification of symptomatic patients suspected of colorectal cancer using faecal and urinary markers. *Color. Dis.* **2018**, *20*, 335–342. [CrossRef]
61. Ahlquist, D.A.; Skoletsky, J.E.; Boynton, K.A.; Harrington, J.J.; Mahoney, D.W.; Pierceall, W.E.; Thibodeau, S.N.; Shuber, A.P. Colorectal cancer screening by detection of altered human DNA in stool: Feasibility of a multitarget assay panel. *Gastroenterology* **2000**, *119*, 1219–1227. [CrossRef] [PubMed]
62. Imperiale, T.F.; Levin, T.R.; Lidgard, G.P.; Itzkowitz, S.H.; Ahlquist, D.A.; Ransohoff, D.F.; Lavin, P.; Berger, B.M. Multitarget stool DNA testing for colorectal-cancer screening. *N. Engl. J. Med.* **2014**, *370*, 1287–1297. [CrossRef]
63. Bosch, L.J.W.; Melotte, V.; Mongera, S.; Daenen, K.L.J.; Coupé, V.M.H.; Van Turenhout, S.T.; Stoop, E.M.; De Wijkerslooth, T.R.; Mulder, C.J.J.; Rausch, C.; et al. Multitarget stool DNA test performance in an average-risk colorectal cancer screening population. *Am. J. Gastroenterol.* **2019**, *114*, 1909–1918. [CrossRef]
64. Mu, J.; Huang, Y.; Cai, S.; Li, Q.; Song, Y.; Yuan, Y.; Zhang, S.; Zheng, S. Plausibility of an extensive use of stool DNA test for screening advanced colorectal neoplasia. *Clin. Chim. Acta* **2020**, *501*, 42–47. [CrossRef]
65. Rengucci, C.; De Maio, G.; Menghi, M.; Scarpi, E.; Guglielmo, S.; Fusaroli, P.; Caletti, G.; Saragoni, L.; Casadei Gardini, A.; Zoli, W.; et al. Improved stool DNA integrity method for early colorectal cancer diagnosis. *Cancer Epidemiol. Biomarkers Prev.* **2014**, *23*, 2553–2560. [CrossRef]
66. Calistri, D.; Rengucci, C.; Casadei Gardini, A.; Frassineti, G.L.; Scarpi, E.; Zoli, W.; Falcini, F.; Silvestrini, R.; Amadori, D. Fecal DNA for noninvasive diagnosis of colorectal cancer in immunochemical fecal occult blood test-positive individuals. *Cancer Epidemiol. Biomarkers Prev.* **2010**, *19*, 2647–2654. [CrossRef]

67. Grobbee, E.J.; Lam, S.Y.; Fuhler, G.M.; Blakaj, B.; Konstantinov, S.R.; Bruno, M.J.; Peppelenbosch, M.P.; Kuipers, E.J.; Spaander, M.C.W. First steps towards combining faecal immunochemical testing with the gut microbiome in colorectal cancer screening. *United Eur. Gastroenterol. J.* **2020**, *8*, 293–302. [CrossRef] [PubMed]
68. Malagón, M.; Ramió-Pujol, S.; Serrano, M.; Serra-Pagès, M.; Amoedo, J.; Oliver, L.; Bahí, A.; Mas–de–Xaxars, T.; Torrealba, L.; Gilabert, P.; et al. Reduction of faecal immunochemical test false-positive results using a signature based on faecal bacterial markers. *Aliment. Pharmacol. Ther.* **2019**, *49*, 1410–1420. [CrossRef]
69. Komor, M.A.; Bosch, L.J.W.; Coupé, V.M.H.; Rausch, C.; Pham, T.V.; Piersma, S.R.; Mongera, S.; Mulder, C.J.J.; Dekker, E.; Kuipers, E.J.; et al. Proteins in stool as biomarkers for non-invasive detection of colorectal adenomas with high risk of progression. *J. Pathol.* **2020**, *250*, 288–298. [CrossRef]
70. Redwood, D.G.; Blake, I.D.; Provost, E.M.; Kisiel, J.B.; Sacco, F.D.; Ahlquist, D.A. Alaska native patient and provider perspectives on the multitarget stool DNA test compared with colonoscopy for colorectal cancer screening. *J. Prim. Care Community Health* **2019**, *10*, 215013271988429. [CrossRef]
71. Symonds, E.L.; Cock, C.; Meng, R.; Cole, S.R.; Fraser, R.J.L.; Young, G.P. Uptake of a colorectal cancer screening blood test in people with elevated risk for cancer who cannot or will not complete a faecal occult blood test. *Eur. J. Cancer Prev.* **2018**, *27*, 425–432. [CrossRef]
72. Normanno, N.; Cervantes, A.; Ciardiello, F.; De Luca, A.; Pinto, C. The liquid biopsy in the management of colorectal cancer patients: Current applications and future scenarios. *Cancer Treat. Rev.* **2018**, *70*, 1–8. [CrossRef] [PubMed]
73. Benning, T.M.; Dellaert, B.G.C.; Dirksen, C.D.; Severens, J.L. Preferences for potential innovations in non-invasive colorectal cancer screening: A labeled discrete choice experiment for a Dutch screening campaign. *Acta Oncol.* **2014**, *53*, 898–908. [CrossRef]
74. Osborne, J.M.; Flight, I.; Wilson, C.J.; Chen, G.; Ratcliffe, J.; Young, G.P. The impact of sample type and procedural attributes on relative acceptability of different colorectal cancer screening regimens. *Patient Prefer. Adherence* **2018**, *12*, 1825–1836. [CrossRef] [PubMed]
75. Marcuello, M.; Vymetalkova, V.; Neves, R.P.L.; Duran-Sanchon, S.; Vedeld, H.M.; Tham, E.; van Dalum, G.; Flügen, G.; Garcia-Barberan, V.; Fijneman, R.J.; et al. Circulating biomarkers for early detection and clinical management of colorectal cancer. *Mol. Asp. Med.* **2019**, *69*, 107–122. [CrossRef]
76. Niedermaier, T.; Weigl, K.; Hoffmeister, M.; Brenner, H. Fecal immunochemical tests in combination with blood tests for colorectal cancer and advanced adenoma detection—Systematic review. *United Eur. Gastroenterol. J.* **2018**, *6*, 13–21. [CrossRef]
77. Byron, S.A.; Van Keuren-Jensen, K.R.; Engelthaler, D.M.; Carpten, J.D.; Craig, D.W. Translating RNA sequencing into clinical diagnostics: Opportunities and challenges. *Nat. Rev. Genet.* **2016**, *17*, 257–271. [CrossRef] [PubMed]
78. Rodia, M.T.; Ugolini, G.; Mattei, G.; Montroni, I.; Zattoni, D.; Ghignone, F.; Veronese, G.; Marisi, G.; Lauriola, M.; Strippoli, P.; et al. Systematic large-scale meta-analysis identifies a panel of two mRNAs as blood biomarkers for colorectal cancer detection. *Oncotarget* **2016**, *7*, 30295–30306. [CrossRef]
79. Rodia, M.T.; Solmi, R.; Pasini, F.; Nardi, E.; Mattei, G.; Ugolini, G.; Ricciardiello, L.; Strippoli, P.; Miglio, R.; Lauriola, M. *LGALS4*, *CEACAM6*, *TSPAN8*, and *COL1A2*: Blood markers for colorectal cancer—validation in a cohort of subjects with positive fecal immunochemical test result. *Clin. Colorectal Cancer* **2018**, *17*, e217–e228. [CrossRef]
80. Ferlizza, E.; Solmi, R.; Miglio, R.; Nardi, E.; Mattei, G.; Sgarzi, M.; Lauriola, M. Colorectal cancer screening: Assessment of *CEACAM6*, *LGALS4*, *TSPAN8* and *COL1A2* as blood markers in faecal immunochemical test negative subjects. *J. Adv. Res.* **2020**, *24*, 99–107. [CrossRef] [PubMed]
81. Chao, S.; Ying, J.; Liew, G.; Marshall, W.; Liew, C.-C.; Burakoff, R. Blood RNA biomarker panel detects both left- and right-sided colorectal neoplasms: A case-control study. *J. Exp. Clin. Cancer Res.* **2013**, *32*, 44. [CrossRef] [PubMed]
82. Marshall, K.W.; Mohr, S.; El Khettabi, F.; Nossova, N.; Chao, S.; Bao, W.; Ma, J.; Li, X.J.; Liew, C.C. A blood-based biomarker panel for stratifying current risk for colorectal cancer. *Int. J. Cancer* **2010**, *126*, 1177–1186. [CrossRef] [PubMed]
83. Ciarloni, L.; Ehrensberger, S.H.; Imaizumi, N.; Monnier-Benoit, S.; Nichita, C.; Myung, S.J.; Kim, J.S.; Song, S.Y.; Kim, T.; Van Der Weg, B.; et al. Development and clinical validation of a blood test based on 29-gene expression for early detection of colorectal cancer. *Clin. Cancer Res.* **2016**, *22*, 4604–4611. [CrossRef]
84. Ciarloni, L.; Hosseinian, S.; Monnier-Benoit, S.; Imaizumi, N.; Dorta, G.; Ruegg, C. Discovery of a 29-gene panel in peripheral blood mononuclear cells for the detection of colorectal cancer and adenomas using high throughput real-time PCR. *PLoS ONE* **2015**, *10*, e0123904. [CrossRef]
85. Alamro, R.; Mustafa, M.; Al-Asmari, A. Inflammatory gene mRNA expression in human peripheral blood and its association with colorectal cancer. *J. Inflamm. Res.* **2018**, *11*, 351–357. [CrossRef]
86. Sahengbieke, S.; Wang, J.; Li, X.; Wang, Y.; Lai, M.; Wu, J. Circulating cell-free high mobility group AT-hook 2 mRNA as a detection marker in the serum of colorectal cancer patients. *J. Clin. Lab. Anal.* **2018**, *32*, e22332. [CrossRef]
87. Hamm, A.; Prenen, H.; Van Delm, W.; Di Matteo, M.; Wenes, M.; Delamarre, E.; Schmidt, T.; Weitz, J.; Sarmiento, R.; Dezi, A.; et al. Tumour-educated circulating monocytes are powerful candidate biomarkers for diagnosis and disease follow-up of colorectal cancer. *Gut* **2016**, *65*, 990–1000. [CrossRef] [PubMed]
88. Bustin, S.A.; Murphy, J. RNA biomarkers in colorectal cancer. *Methods* **2013**, *59*, 116–125. [CrossRef]
89. Moridikia, A.; Mirzaei, H.; Sahebkar, A.; Salimian, J. MicroRNAs: Potential candidates for diagnosis and treatment of colorectal cancer. *J. Cell. Physiol.* **2018**, *233*, 901–913. [CrossRef]

90. Nikolaou, S.; Qiu, S.; Fiorentino, F.; Rasheed, S.; Tekkis, P.; Kontovounisios, C. Systematic review of blood diagnostic markers in colorectal cancer. *Tech. Coloproctol.* **2018**, *22*, 481–498. [CrossRef]
91. Thomas, J.; Ohtsuka, M.; Pichler, M.; Ling, H. MicroRNAs: Clinical relevance in colorectal cancer. *Int. J. Mol. Sci.* **2015**, *16*, 28063–28076. [CrossRef] [PubMed]
92. Ganepola, G.A.; Nizin, J.; Rutledge, J.R.; Chang, D.H. Use of blood-based biomarkers for early diagnosis and surveillance of colorectal cancer. *World J. Gastrointest. Oncol.* **2014**, *6*, 83–97. [CrossRef] [PubMed]
93. Nikolouzakis, T.K.; Vassilopoulou, L.; Fragkiadaki, P.; Sapsakos, T.M.; Papadakis, G.Z.; Spandidos, D.A.; Tsatsakis, A.M.; Tsiaoussis, J. Improving diagnosis, prognosis and prediction by using biomarkers in CRC patients. *Oncol. Rep.* **2018**, *39*, 2455–2472. [CrossRef]
94. Yu, W.; Wang, Z.; Shen, L.; Wei, Q. Circulating microRNA-21 as a potential diagnostic marker for colorectal cancer: A meta-analysis. *Mol. Clin. Oncol.* **2016**, *4*, 237–244. [CrossRef]
95. Huang, Z.; Huang, D.; Ni, S.; Peng, Z.; Sheng, W.; Du, X. Plasma microRNAs are promising novel biomarkers for early detection of colorectal cancer. *Int. J. Cancer* **2010**, *127*, 118–126. [CrossRef]
96. Herreros-Villanueva, M.; Duran-Sanchon, S.; Martín, A.C.; Pérez-Palacios, R.; Vila-Navarro, E.; Marcuello, M.; Diaz-Centeno, M.; Cubiella, J.; Diez, M.S.; Bujanda, L.; et al. Plasma microRNA signature validation for early detection of colorectal cancer. *Clin. Transl. Gastroenterol.* **2019**, *10*, e00003. [CrossRef]
97. Yamada, A.; Horimatsu, T.; Okugawa, Y.; Nishida, N.; Honjo, H.; Ida, H.; Kou, T.; Kusaka, T.; Sasaki, Y.; Yagi, M.; et al. Serum MIR-21, MIR-29a, and MIR-125b are promising biomarkers for the early detection of colorectal neoplasia. *Clin. Cancer Res.* **2015**, *21*, 4234–4242. [CrossRef]
98. Zanutto, S.; Ciniselli, C.M.; Belfiore, A.; Lecchi, M.; Masci, E.; Delconte, G.; Primignani, M.; Tosetti, G.; Dal Fante, M.; Fazzini, L.; et al. Plasma miRNA-based signatures in CRC screening programs. *Int. J. Cancer* **2020**, *146*, 1164–1173. [CrossRef]
99. Zhang, H.; Zhu, M.; Shan, X.; Zhou, X.; Wang, T.; Zhang, J.; Tao, J.; Cheng, W.; Chen, G.; Li, J.; et al. A panel of seven-miRNA signature in plasma as potential biomarker for colorectal cancer diagnosis. *Gene* **2019**, *687*, 246–254. [CrossRef] [PubMed]
100. Khakoo, S.; Georgiou, A.; Gerlinger, M.; Cunningham, D.; Starling, N. Circulating tumour DNA, a promising biomarker for the management of colorectal cancer. *Crit. Rev. Oncol. Hematol.* **2018**, *122*, 72–82. [CrossRef]
101. The Cancer Genome Atlas. Available online: https://portal.gdc.cancer.gov/ (accessed on 22 February 2021).
102. Song, L.; Jia, J.; Peng, X.; Xiao, W.; Li, Y. The performance of the SEPT9 gene methylation assay and a comparison with other CRC screening tests: A meta-analysis. *Sci. Rep.* **2017**, *7*, 3032. [CrossRef] [PubMed]
103. Hu, J.; Hu, B.; Gui, Y.C.; Tan, Z.B.; Xu, J.W. Diagnostic value and clinical significance of methylated SEPT9 for colorectal cancer: A meta-analysis. *Med. Sci. Monit.* **2019**, *25*, 5813–5822. [CrossRef]
104. He, N.; Song, L.; Kang, Q.; Jin, P.; Cai, G.; Zhou, J.; Zhou, G.; Sheng, J.; Cai, S.; Wang, J.; et al. The pathological features of colorectal cancer determine the detection performance on blood ctDNA. *Technol. Cancer Res. Treat.* **2018**, *17*, 1–9. [CrossRef] [PubMed]
105. Pedersen, S.K.; Symonds, E.L.; Baker, R.T.; Murray, D.H.; McEvoy, A.; Van Doorn, S.C.; Mundt, M.W.; Cole, S.R.; Gopalsamy, G.; Mangira, D.; et al. Evaluation of an assay for methylated BCAT1 and IKZF1 in plasma for detection of colorectal neoplasia. *BMC Cancer* **2015**, *15*, 654. [CrossRef]
106. Cohen, J.D.; Li, L.; Wang, Y.; Thoburn, C.; Afsari, B.; Danilova, L.; Douville, C.; Javed, A.A.; Wong, F.; Mattox, A.; et al. Detection and localization of surgically resectable cancers with a multi-analyte blood test. *Science* **2018**, *359*, 926–930. [CrossRef]
107. Phallen, J.; Sausen, M.; Adleff, V.; Leal, A.; Hruban, C.; White, J.; Anagnostou, V.; Fiksel, J.; Cristiano, S.; Papp, E.; et al. Direct detection of early-stage cancers using circulating tumor DNA. *Sci. Transl. Med.* **2017**, *9*, eaan2415. [CrossRef]
108. Wan, N.; Weinberg, D.; Liu, T.-Y.; Niehaus, K.; Ariazi, E.A.; Delubac, D.; Kannan, A.; White, B.; Bailey, M.; Bertin, M.; et al. Machine learning enables detection of early-stage colorectal cancer by whole-genome sequencing of plasma cell-free DNA. *BMC Cancer* **2019**, *19*, 832. [CrossRef] [PubMed]
109. Choi, I.S.; Kato, S.; Fanta, P.T.; Leichman, L.; Okamura, R.; Raymond, V.M.; Lanman, R.B.; Lippman, S.M.; Kurzrock, R. Genomic profiling of blood-derived circulating tumor DNA from patients with colorectal cancer: Implications for response and resistance to targeted therapeutics. *Mol. Cancer Ther.* **2019**, *18*, 1852–1862. [CrossRef] [PubMed]
110. Odegaard, J.I.; Vincent, J.J.; Mortimer, S.; Vowles, J.V.; Ulrich, B.C.; Banks, K.C.; Fairclough, S.R.; Zill, O.A.; Sikora, M.; Mokhtari, R.; et al. Validation of a plasma-based comprehensive cancer genotyping assay utilizing orthogonal tissue- and plasma-based methodologies. *Clin. Cancer Res.* **2018**, *24*, 3539–3549. [CrossRef] [PubMed]
111. Zill, O.A.; Banks, K.C.; Fairclough, S.R.; Mortimer, S.A.; Vowles, J.V.; Mokhtari, R.; Gandara, D.R.; Mack, P.C.; Odegaard, J.I.; Nagy, R.J.; et al. The landscape of actionable genomic alterations in cell-free circulating tumor DNA from 21,807 advanced cancer patients. *Clin. Cancer Res.* **2018**, *24*, 3528–3538. [CrossRef]
112. Kuespert, K.; Pils, S.; Hauck, C.R. CEACAMs: Their role in physiology and pathophysiology. *Curr. Opin. Cell Biol.* **2006**, *18*, 565–571. [CrossRef]
113. Beauchemin, N.; Arabzadeh, A. Carcinoembryonic antigen-related cell adhesion molecules (CEACAMs) in cancer progression and metastasis. *Cancer Metastasis Rev.* **2013**, *32*, 643–671. [CrossRef]
114. Jelski, W.; Mroczko, B. Biochemical markers of colorectal cancer—present and future. *Cancer Manag. Res.* **2020**, *12*, 4789–4797. [CrossRef] [PubMed]

115. Rasmussen, L.; Nielsen, H.J.; Christensen, I.J. Early detection and recurrence of colorectal adenomas by combination of eight cancer-associated biomarkers in plasma. *Clin. Exp. Gastroenterol.* **2020**, *13*, 273–284. [CrossRef]
116. Hall, C.; Clarke, L.; Pal, A.; Buchwald, P.; Eglinton, T.; Wakeman, C.; Frizelle, F. A review of the role of carcinoembryonic antigen in clinical practice. *Ann. Coloproctol.* **2019**, *35*, 294–305. [CrossRef]
117. Chen, H.; Zucknick, M.; Werner, S.; Knebel, P.; Brenner, H. Head-to-head comparison and evaluation of 92 plasma protein biomarkers for early detection of colorectal cancer in a true screening setting. *Clin. Cancer Res.* **2015**, *21*, 3318–3326. [CrossRef]
118. Halilovic, E.; Rasic, I.; Sofic, A.; Mujic, A.; Rovcanin, A.; Hodzic, E.; Kulovic, E. The importance of determining preoperative serum concentration of carbohydrate antigen 19-9 and carcinoembryonic antigen in assessing the progression of colorectal cancer. *Med. Arch.* **2020**, *74*, 346–349. [CrossRef]
119. Baqar, A.R.; Wilkins, S.; Staples, M.; Angus Lee, C.H.; Oliva, K.; McMurrick, P. The role of preoperative CEA in the management of colorectal cancer: A cohort study from two cancer centres. *Int. J. Surg.* **2019**, *64*, 10–15. [CrossRef]
120. Koulis, C.; Yap, R.; Engel, R.; Jardé, T.; Wilkins, S.; Solon, G.; Shapiro, J.D.; Abud, H.; McMurrick, P. Personalized medicine—current and emerging predictive and prognostic biomarkers in colorectal cancer. *Cancers* **2020**, *12*, 812. [CrossRef] [PubMed]
121. Ushigome, M.; Nabeya, Y.; Soda, H.; Takiguchi, N.; Kuwajima, A.; Tagawa, M.; Matsushita, K.; Koike, J.; Funahashi, K.; Shimada, H. Multi-panel assay of serum autoantibodies in colorectal cancer. *Int. J. Clin. Oncol.* **2018**, *23*, 917–923. [CrossRef]
122. Chauvin, A.; Boisvert, F.-M. Clinical proteomics in colorectal cancer, a promising tool for improving personalised medicine. *Proteomes* **2018**, *6*, 49. [CrossRef] [PubMed]
123. Ahn, S.B.; Sharma, S.; Mohamedali, A.; Mahboob, S.; Redmond, W.J.; Pascovici, D.; Wu, J.X.; Zaw, T.; Adhikari, S.; Vaibhav, V.; et al. Potential early clinical stage colorectal cancer diagnosis using a proteomics blood test panel. *Clin. Proteomics* **2019**, *16*, 34. [CrossRef]
124. Ivancic, M.M.; Megna, B.W.; Sverchkov, Y.; Craven, M.; Reichelderfer, M.; Pickhardt, P.J.; Sussman, M.R.; Kennedy, G.D. Noninvasive detection of colorectal carcinomas using serum protein biomarkers. *J. Surg. Res.* **2020**, *246*, 160–169. [PubMed]
125. Wilhelmsen, M.; Christensen, I.J.; Rasmussen, L.; Jørgensen, L.N.; Madsen, M.R.; Vilandt, J.; Hillig, T.; Klærke, M.; Nielsen, K.T.; Laurberg, S.; et al. Detection of colorectal neoplasia: Combination of eight blood-based, cancer-associated protein biomarkers. *Int. J. Cancer* **2017**, *140*, 1436–1446. [CrossRef]
126. Desmond, B.J.; Dennett, E.R.; Danielson, K.M. Circulating extracellular vesicle microRNA as diagnostic biomarkers in early colorectal cancer—a review. *Cancers* **2019**, *12*, 52. [CrossRef]
127. González-Masiá, J.A.; García-Olmo, D.; García-Olmo, D.C. Circulating nucleic acids in plasma and serum (CNAPS): Applications in oncology. *Onco. Targets. Ther.* **2013**, *6*, 819–832.
128. Mousavi, S.; Moallem, R.; Hassanian, S.M.; Sadeghzade, M.; Mardani, R.; Ferns, G.A.; Khazaei, M.; Avan, A. Tumor-derived exosomes: Potential biomarkers and therapeutic target in the treatment of colorectal cancer. *J. Cell. Physiol.* **2019**, *234*, 12422–12432. [CrossRef] [PubMed]
129. Rashed, M.H.; Bayraktar, E.; Helal, G.K.; Abd-Ellah, M.; Amero, P.; Chavez-Reyes, A.; Rodriguez-Aguayo, C. Exosomes: From garbage bins to promising therapeutic targets. *Int. J. Mol. Sci.* **2017**, *18*, 538. [CrossRef]
130. Shao, Y.; Shen, Y.; Chen, T.; Xu, F.; Chen, X.; Zheng, S. The functions and clinical applications of tumor-derived exosomes. *Oncotarget* **2016**, *7*, 60736–60751. [CrossRef] [PubMed]
131. Menck, K.; Bleckmann, A.; Wachter, A.; Hennies, B.; Ries, L.; Schulz, M.; Balkenhol, M.; Pukrop, T.; Schatlo, B.; Rost, U.; et al. Characterisation of tumour-derived microvesicles in cancer patients' blood and correlation with clinical outcome. *J. Extracell. Vesicles* **2017**, *6*, 1340745. [CrossRef] [PubMed]
132. Tugutova, E.A.; Tamkovich, S.N.; Patysheva, M.R.; Afanas'ev, S.G.; Tsydenova, A.A.; Grigor'eva, A.E.; Kolegova, E.S.; Kondakova, I.V.; Yunusova, N.V. Relation between tetraspanin-associated and tetraspanin-non-associated exosomal proteases and metabolic syndrome in colorectal cancer patients. *Asian Pac. J. Cancer Prev.* **2019**, *20*, 809–815. [CrossRef]
133. Tamkovich, S.N.; Yunusova, N.V.; Stakheeva, M.N.; Somov, A.K.; Frolova, A.E.; Kiryushina, N.A.; Afanasyev, S.G.; Grigor'eva, A.E.; Laktionov, P.P.; Kondakova, I.V. Isolation and characterization of exosomes from blood plasma of breast cancer and colorectal cancer patients. *Biochem. Suppl. Ser. B Biomed. Chem.* **2017**, *11*, 291–295.
134. Menck, K.; Bleckmann, A.; Schulz, M.; Ries, L.; Binder, C. Isolation and characterization of microvesicles from peripheral blood. *J. Vis. Exp.* **2017**, *119*, e55057. [CrossRef]
135. Coumans, F.A.W.; Brisson, A.R.; Buzas, E.I.; Dignat-George, F.; Drees, E.E.E.; El-Andaloussi, S.; Emanueli, C.; Gasecka, A.; Hendrix, A.; Hill, A.F.; et al. Methodological guidelines to study extracellular vesicles. *Circ. Res.* **2017**, *120*, 1632–1648. [CrossRef]
136. Mateescu, B.; Kowal, E.J.K.; van Balkom, B.W.M.; Bartel, S.; Bhattacharyya, S.N.; Buzás, E.I.; Buck, A.H.; de Candia, P.; Chow, F.W.N.; Das, S.; et al. Obstacles and opportunities in the functional analysis of extracellular vesicle RNA—An ISEV position paper. *J. Extracell. Vesicles* **2017**, *6*, 1286095. [CrossRef]
137. Belov, L.; Matic, K.J.; Hallal, S.; Best, O.G.; Mulligan, S.P.; Christopherson, R.I. Extensive surface protein profiles of extracellular vesicles from cancer cells may provide diagnostic signatures from blood samples. *J. Extracell. Vesicles* **2016**, *5*, 25355. [CrossRef]
138. Ogata-Kawata, H.; Izumiya, M.; Kurioka, D.; Honma, Y.; Yamada, Y.; Furuta, K.; Gunji, T.; Ohta, H.; Okamoto, H.; Sonoda, H.; et al. Circulating exosomal microRNAs as biomarkers of colon cancer. *PLoS ONE* **2014**, *9*, e92921. [CrossRef]
139. Yan, S.; Han, B.; Gao, S.; Wang, X.; Wang, Z.; Wang, F.; Zhang, J.; Xu, D.; Sun, B. Exosome-encapsulated microRNAs as circulating biomarkers for colorectal cancer. *Oncotarget* **2017**, *8*, 60149–60158. [CrossRef]

140. Liu, C.; Eng, C.; Shen, J.; Lu, Y.; Yoko, T.; Mehdizadeh, A.; Chang, G.J.; Rodriguez-Bigas, M.A.; Li, Y.; Chang, P.; et al. Serum exosomal miR-4772-3p is a predictor of tumor recurrence in stage II and III colon cancer. *Oncotarget* **2016**, *7*, 76250–76260. [CrossRef] [PubMed]
141. Peng, Z.Y.; Gu, R.H.; Yan, B. Downregulation of exosome-encapsulated miR-548c-5p is associated with poor prognosis in colorectal cancer. *J. Cell. Biochem.* **2019**, *120*, 1457–1463. [CrossRef] [PubMed]
142. Min, L.; Zhu, S.; Chen, L.; Liu, X.; Wei, R.; Zhao, L.; Yang, Y.; Zhang, Z.; Kong, G.; Li, P.; et al. Evaluation of circulating small extracellular vesicles derived miRNAs as biomarkers of early colon cancer: A comparison with plasma total miRNAs. *J. Extracell. Vesicles* **2019**, *8*, 1643670. [CrossRef]
143. Cha, B.S.; Park, K.S.; Park, J.S. Signature mRNA markers in extracellular vesicles for the accurate diagnosis of colorectal cancer. *J. Biol. Eng.* **2020**, *14*, 4. [CrossRef]
144. Dong, L.; Lin, W.; Qi, P.; Xu, M.D.; Wu, X.; Ni, S.; Huang, D.; Weng, W.W.; Tan, C.; Sheng, W.; et al. Circulating long RNAs in serum extracellular vesicles: Their characterization and potential application as biomarkers for diagnosis of colorectal cancer. *Cancer Epidemiol. Biomarkers Prev.* **2016**, *25*, 1158–1166. [CrossRef] [PubMed]

Review

Position of Circulating Tumor Cells in the Clinical Routine in Prostate Cancer and Breast Cancer Patients

Gerit Theil *,†, Paolo Fornara and Joanna Bialek †

Medical Faculty, Martin Luther University Halle-Wittenberg, Clinic of Urology, 06120 Halle (Saale), Germany; paolo.fornara@uk-halle.de (P.F.); joanna.bialek@uk-halle.de (J.B.)
* Correspondence: gerit.theil@uk-halle.de; Tel.: +49-3455573122
† These authors contributed equally to this work.

Received: 5 November 2020; Accepted: 10 December 2020; Published: 15 December 2020

Simple Summary: Many different therapies are applied to fight tumor disease. Blood-based biosources, like circulating tumor cells (CTCs), offer the opportunity to monitor the healing progression and the real-time response to the therapy. In this review, we analyze the outcomes of the clinical trials and scientific studies of prostate and breast cancer performed in the decade between April 2010 and April 2020. Additionally, we describe the abstracts from the 4th Advances in Circulating Tumor Cells (ACTC) meeting in 2019. We discuss the potential therapeutic opportunities related to the CTCs and the challenges ahead in the routine treatment of cancer.

Abstract: Prostate cancer and breast cancer are the most common cancers worldwide. Anti-tumor therapies are long and exhaustive for the patients. The real-time monitoring of the healing progression could be a useful tool to evaluate therapeutic response. Blood-based biosources like circulating tumor cells (CTCs) may offer this opportunity. Application of CTCs for the clinical diagnostics could improve the sequenced screening, provide additional valuable information of tumor dynamics, and help personalized management for the patients. In the past decade, CTCs as liquid biopsy (LB) has received tremendous attention. Many different isolation and characterization platforms are developed but the clinical validation is still missing. In this review, we focus on the clinical trials of circulating tumor cells that have the potential to monitor and stratify patients and lead to implementation into clinical practice.

Keywords: prostate cancer; breast cancer; circulating tumor cells; treatment decisions

1. Introduction

Cancer is caused by multiple molecular alterations in normal host cells that act together to drive uncontrolled cell self-renewal, growth and invasion, and lead to malignant transformation and progression. The majority of cancer-associated deaths (approximately 90%) are induced by metastatic disease rather than the primary cancer [1]. The early detection of cancer and subsequent noninvasive tumor profiling and monitoring should be enabled in every cancer patient. Thus, there is an unmet clinical need for biomarkers to fulfill the claim in precision oncology.

Blood-based biosources such as circulating tumor cells (CTCs), cell-free DNA (cfDNA), tumor-educated platelets and cell-free nucleic acids (circulating tumor DNA, long non-coding RNA, messenger RNA and microRNA) offer this opportunity. These biomarkers, summarized as liquid biopsy (LB), could provide information on urgent cancer characteristics [2]. CTCs detach from primary or metastatic tumors to enter the bloodstream, from which a small CTC population has the ability to metastasize to multiple organs [3]. In addition, CTCs are genetically unstable, evading immune defenses and modification metabolism [3,4]. These characteristics reflect the dynamic and heterogeneous phenotype of CTCs. Furthermore, at present, we know that cancer cells survive after infiltrating distant organs and can be present for years in the bone marrow as disseminated tumor cells (DTCs), which are correlated with an increased risk of eventual clinical recurrence [5].

The concentration of CTCs in the blood is very low, and a single CTC is in the background of millions of blood cells. Nonetheless, CTCs could serve as a comprehensive window into metastatic disease for the real-time monitoring of therapy responses.

LB in the form of CTCs received tremendous attention following approval of the automated CellSearch® system (Menarini Silicon Biosystems Inc, Huntington Valley, USA) by the Food and Drug Administration (FDA). Thus, the importance of CTC enumeration as a surrogate marker for survival benefits in breast and prostate cancer patients was commenced [6,7].

The clinical utility and reliable information of CTCs as useful biomarkers must still be demonstrated in the standard care of cancer therapy. In the latest guidelines (version 3) the Prostate Cancer Clinical Trials Working Group (PCWG) determined that for the outcome assessment of patients enrolled in clinical trials, the incorporation of CTC enumeration (using CellSearch platform) must be the endpoint [8]. This decision illustrates and emphasizes the importance of the serial biological profiling of cancer. Moreover, it promotes CTCs in the field of personalized cancer treatment, supplying unique information on individual cancer-associated variations in tumor burden.

In this review, we analyze current clinical studies and focus on the clinical application of CTCs in prostate and breast cancer patients. We outline important results of clinical trials that may be translated into clinical practice.

2. Evidence Acquisition

A literature review was performed via PubMed/Medline and the Cochrane Library (January 2010–April 2020). In addition, abstracts from the 4th Advances in Circulating Tumor Cells (ACTC) meeting in 2019 were searched for relevant abstracts. The search terms included CTCs, CTC, circulating tumor cell, circulating tumor cells, prostate cancer, breast cancer and clinical trial. All clinical trials reporting fewer than 20 patients were excluded (Figure 1).

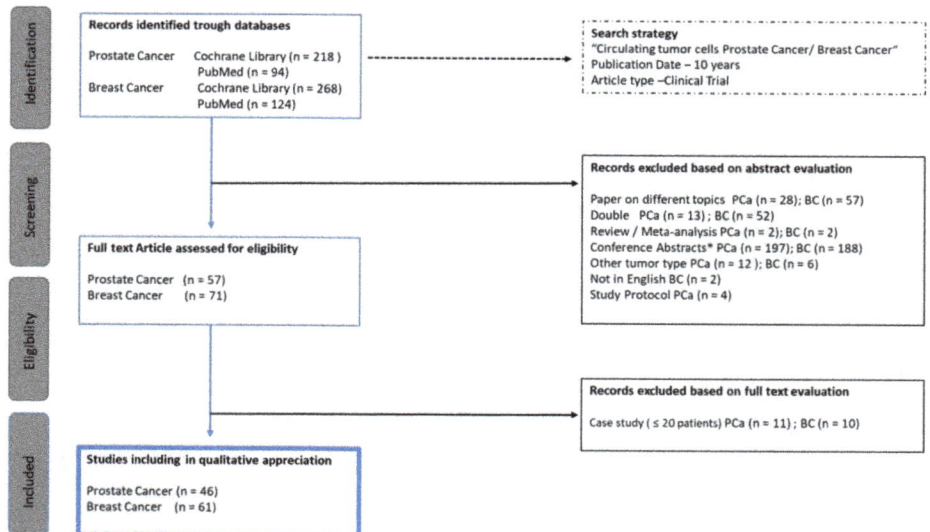

Figure 1. PRISMA flow diagram. BC = breast cancer; PCa = prostate Cancer; n = number; * other abstracts than ACTC (4th Advances in Circulating Tumor Cells) meeting 2019.

3. Isolation Methods

The efficient capture of CTCs remains a technical challenge. Over the last decade, several platforms have been developed to detect these rare cells. CTC enrichment can be achieved based on physical and biological properties [9,10].

Currently, the most commonly used standardized method for detection and enrichment is the CellSearch® system, based on affinity to epithelial cell adhesion molecule (EpCAM). The CellSearch® system detects and enumerates cells expressing the epithelial marker EpCAM and cytokeratins (CK) 8/18+ and/or 19+, and the additional treatment of the blood samples with a CD45 antibody allows the elimination of leukocytes. A semiautomated fluorescence microscope captures images that are manually reviewed for the following CTC criteria: round to oval morphology, visible nucleus (DAPI positive), size > 4 µm, positive staining for cytokeratin, and negative staining for CD45 [11]. Recently, cells have been labeled with other markers, such as human epidermal growth factor receptor 2 (HER2), insulin-like growth factor 1 receptor (IGF-1R), epidermal growth factor receptor (EGFR), androgen receptor (AR), Ki-67 (a marker of proliferation) or vascular endothelial growth factor receptor 2 (VEGFR-2) [12].

Not all CTCs express EpCAM. It has been demonstrated that CTCs can undergo epithelial-mesenchymal transition (EMT) and can lose EpCAM expression. Therefore, analysis with the CellSearch® system must be excluded, and other techniques based on physical properties must be employed. Mononuclear cells can be isolated by density gradient centrifugation using Ficoll or a more effective porous barrier. Cancer cells differ from normal cells in their electromagnetic charge. This feature is used during separation with dielectrophoretic field-flow fractionation (depFFF). Finally, most CTCs are larger than leukocytes and erythrocytes, and, as previously described, prepared buffy coat can be passed through porous membranes by selecting larger cells [13]. The Epic Sciences CTC platform uses only red blood cell lysates, and approximately 3×10^6 nucleated cells are dispensed onto 10–12 glass micro slides and frozen at −80 °C until examination, allowing the unbiased detection and molecular characterization of CTCs [14]. The CTC-iChip also offers the opportunity for antigen-independent CTC isolation with negative depletion of normal blood cells. CTCs can be individually selected and analyzed as single cells suited for a detailed transcriptome analysis [15].

Numerous CTC isolation platforms were presented at the 4th ACTC meeting in 2019. It is not surprising that the CellSearch® system was the most common platform used in the investigations. Interestingly, applications of the CellSearch® system together with antibody-independent CTC isolation platforms in one blood sample were also common. An outstanding technology of a new in vivo approach, an indwelling intravascular aphaeretic CTC isolation system based on an EpCAM functionalized chip, was presented. Currently, the system has only been proven in a canine model [16]. Furthermore, many investigators utilized the physical properties of CTCs by label-independent isolation CTC methods (Isoflux—Fluxion Biosciences, Inc., Alameda, USA; Screencells—Sarcelles, France; Parsortix—ANGLE, Guildford, UK; ISET—Rarecells Diagnostics, Paris, France, and RosetteSep—STEMCELL Technologies, Vancouver, BC, Canada). The Vortex Biosciences VTX-1 instrument (Vortex Biosciences, Pleasanton, CA, USA) is a label-free CTC isolation system that enables the detection of gene expression in both CTCs and exosomal cfRNA from the same blood sample. An important issue is single-CTC isolation and analysis. For this purpose, different CTC isolation methods were combined. Workflows for single-cell selection using the QIAscout single-cell isolation platform were described. Pereira-Veiga et al. [17] used VyCAP (VyCAP BV, Enschede, Netherlands), DEPArry (Menarini Silicon Biosystems S.p.A, Bologna Italy) and manual micromanipulation for single-cell isolation. The authors concluded that the VyCAP Puncher system yielded a higher recovery rate and that the analysis at the single cell level provided better insights into CTCs' heterogeneity [17].

4. Evidence Synthesis Prostate Cancer

It was estimated that over 1.3 million new cases of prostate cancer and 359,000 associated deaths worldwide occurred in 2018, accounting for the second most frequent cancer and the fifth leading cause of cancer-related death in men [18]. The disease phenotypes varied from indolent to aggressive. One challenge for clinicians is to determine the optimal sequencing therapies for patients who present intermediate, high-risk localized, locally advanced or metastatic prostate cancer (mPCa) to minimize overtreatment and improve outcomes. The treatment of mPCa is becoming increasingly complex [19].

In total, we screened 94 titles, and reviewed 46 full-text papers (Figure 1). From these 57 publications, 11 describe less than 20 and 46 describe 20 or more prostate cancer patients. Over 40 clinical trials evaluated the value of CTCs in metastatic castration-resistant prostate cancer (mCRPC) patients. There are only a few studies analyzing the value of CTC in new metastatic hormone-sensitive PCa ($n = 1$), neuroendocrine PCa ($n = 1$) and localized PCa ($n = 3$). In 10 phase III, 17 phase II and 10 phase I/II trials integrated measuring outcomes related to CTC enumeration or characterization.

The studies like SWOG S0421, MAINSAIL, COU-AA-301, analyzed the first-line docetaxel-base treatment with or without additional agents. The efficacy of therapy among the patients was determined enumeration of baseline CTC count and CTC count after defined cycle of chemotherapy or after treatment. The combination of CTC count with other serum markers were analyzed in the trials COU-AA-301, ELM-PC4 and IMMC38 as prognosis or surrogate biomarker for survival in mCRPC patients. Nine clinical trials evaluated androgen receptor (AR) splice variants (AR-Vs) in CTCs as marker responsible for castration-resistant prostate cancer progression. The baseline CTC-derived AR-V7 status as a biomarker of the response or resistance to therapies was analyzed in PROPHECY, TAXNEGY trial and three further monocentric trials.

In 18 studies, CTCs additional markers were analyzed: (1) standard genes (EpCAM, CK 8, 18, 19, and CD45-), (2) prostate-specific membrane antigen (PSMA)—protein, (3) osteoblast regulators—mRNA, (4) telomerase activity—mRNA, (5) TMPRESS2—mRNA, (6) AR-V7—mRNA/DNA. (Table S1).

4.1. CTC Enumeration—Prostate Cancer

Most clinical trials describing evaluations of the CTC count were published in the early 2000s. Multiple groups confirmed the prognostic value of the CTC count [20–24]. The SWOG S0421 trial validated the CTC count as a prognostic factor in mCRPC patients who received first-line docetaxel-based therapy. The CTC counts of 263 blood samples at baseline (day 0) and of 231 blood samples at day 21 were evaluated. It has been repeatedly acknowledged that a CTC count ≥5 per 7.5 mL at baseline determined with the CellSearch® system was associated with a high tumor burden and poor disease outcomes. Additionally, a higher CTC count is correlated with worse bone pain, higher prostate-specific antigen (PSA) levels, more liver disease, lower hemoglobin levels and higher alkaline phosphatase levels [21].

A subgroup analysis from the MAINSAIL trial demonstrated that in mCRPC patients ($n = 208$), an increased CTC count from <5 cells/7.5 mL to ≥5 cells/7.5 mL after three cycles of docetaxel chemotherapy predicted significantly shorter overall survival (OS) (HR: 5.24, $p = 0.025$). A reduction in the CTC count from ≥5 cells/7.5 mL to <5 cells/7.5 mL was associated with the best prognosis ($p = 0.003$). Interestingly, there was no correlation of the baseline CTC count with the PSA level [25].

Heller et al. [26] determined the CTC count in combination with common prognostic laboratory measures (lactate dehydrogenase, LDH; PSA; hemoglobin, and alkaline phosphatase, ALK = ALPHA) in patients with CRPC. Their objective was to quantify a risk model to predict short-term versus long-term survival. For this purpose, data from patients enrolled in the phase III registration trial of abiraterone acetate (AA) plus prednisone (COU-AA-301; NCT00638690) and the registration trial of a similar design evaluating orteronel plus prednisone (ELM-PC4; NCT01193244) were used. The results suggested that the incorporation of CTC measurement into ALPHA as a prediction error of survival was 3.75 months (SE, 0.22) versus 3.95 months (SE, 0.28) for ALPHA alone [26]. In the COU-AA301 study (AA plus prednisone versus prednisone alone), the Prentice criteria [27] were also applied to test CTC counts and LDH as surrogates for OS at the individual mCRPC patient level. The combination of CTC count and LDH level satisfied the Prentice criteria and highlighted its clinical utility [28].

Furthermore, a baseline CTC count <5 cells was analyzed in mCRPC patients who were enrolled in the COU-AA301 and IMMC38 trials. Among 259 (50.7%) patients in the COU-AA301 trial and 212 (50.4%) in the IMMC38 trial, zero CTCs were detected at baseline. Patients were treated with an AR-targeting drug (COU-AA301) and chemotherapy (IMMC38) [29]. CTC progression was defined as any increase in the CTC count and conversion of a CTC count from <5 to ≥5 CTCs during the first 12 weeks of treatment. These data revealed that CTC progression is positively associated with poor OS. Interestingly, patients with CTC progression also had a significantly lower PSA response rate than those without progression [29].

The SWOG 0925 trial, demonstrated in metastatic hormone-sensitive prostate cancer patients treated with androgen deprivation therapy (ADT) ± cixutumumab (n = 105), that a low CTC count (0 versus 1 to 4 versus ≥5/7.5 mL blood) correlated with the PSA level [30].

A pooled analysis of five randomized trials (n = 4196 patients) investigated the CTC count as a predictor of prolonged survival in mCRPC patients. The CTC counts were determined by using the CellSearch® system. The response measure endpoints were CTC0, CTC conversion and PSA levels at baseline and week 13. Patients who had a CTC count ≥1 at baseline and zero CTCs at week 13 were defined as CTC0. CTC conversion was defined as patients with a CTC count ≥5 at baseline and ≤4 at week 13. The results revealed that the use of the CTC0 count improved the ability to assess the treatment response [31].

Moreover, these results revealed that CTC conversion data from trials on treatment efficacy are highly reliable and can be obtained in a short time [26,31].

There are a marginal number of studies which investigate the role of CTC in localized prostate cancer patient in our search results. Murray et al. [32,33] examine the role of prostate cancer-specific CTC in patients after radical prostatectomy (RP) [32] and in patients with biochemical failure after RP [33]. The CTC were enriched by using density gradient isolation and detected by PSA immunocytochemistry. The author mentioned these CTCs as circulating prostate cells (CPCs) and postulated an identical phenotype to DTCs. They summarized that more CTC were detected in patients with positive margins, extracapsular extension, and vascular and lymphatic infiltration, which associated with biochemical failure [32]. The second publication of Murray and colleagues demonstrated a significant correlation between CPC detection and clinical variables such as progression-free survival (PFS) after a long follow-up period of 15 years. Additionally, CPC must express PSA and α-Methylacyl CoA racemase (P504S) and CD82 (tumor suppressor gene). Patients with CD82-positive CPC have a better biochemical failure-free survival at five years similar to CPC negative patients [33]. It is important to note from these results that blood samples (EDTA) were stored at 4 °C and analyzed within 48 h, which could be relevant for half-life of CPC, and also, no CPC number was provided.

4.2. Functional Characterization of CTCs—Prostate Cancer

The enumeration of CTCs has prognostic value in metastatic hormone-sensitive and metastatic castration-resistant prostate cancers. Nevertheless, CTCs still must demonstrate incremental value in predictive accuracy relative to known biomarkers. Therefore, the enumeration of CTCs has not yet become a standard in the clinic. Molecular phenotyping could aid in the investigation of prostate cancer-specific markers and be routinely incorporated into the care of patients with prostate cancer. CTC profiling can demonstrate the DNA, RNA or protein characteristics of pooled or single cells and reflect the real-time phenotype of the primary tumor or metastasis [34].

A potential predictive marker of sensitivity to the androgen biosynthesis inhibitor AA, the androgen-driven transmembrane protease serine 2 (TMPRSS2)–v-ets erythroblastosis virus E26 oncogene homolog (ERG), is the focus of several studies [35,36].

A proof of principle study described an analytically validated polymerase chain reaction (PCR)-based assay to detect TMPRSS2-ERG fusions in CTCs. However, the TMPRSS2-ERG fusion status in CTCs has a limited role as a predictive biomarker of sensitivity to AA in post chemotherapy-treated CRPC patients [37].

Additional effort has been directed at the molecular characterization of telomerase activity in CTCs. In men with mCRPC treated with first-line docetaxel ± atrasentan (SWOG 0421 trial), telomerase activity (TA) was investigated in live-captured CTCs in parallel to baseline CTC counts. The CTC TA measurement was performed with a slot microfilter. The authors analyzed TA in CTC lysates by qPCR-telomeric repeat amplification (TRAP) and concluded that CTC TA was an independent predictive marker for OS in men with a CTC count ≥5. Limitations of the TA assay were its low sensitivity; only 47% of patients with ≥5 CTCs captured by the CellSearch® system showed prognostic TA [38].

The prostate-specific membrane antigen (PSMA) is overexpressed in prostate cancer including advanced stage [39]. Interestingly, the PSMA overexpression is correlated with higher tumor grade, androgen deprivation and increased in mCRPC patient [40]. A phase II clinical trial suggested the application of CTCs as a selection tool for the safety and efficacy of newly developed drugs. Patients with progressing mCRPC (chemotherapy-naïve; n = 42) were treated with docetaxel-encapsulating nanoparticles functionalized with PSMA molecules (BIND-014). The structure of this particle allows binding of PSMA-expressing tumor tissue or cells. CTC enumerations were performed with the CellSearch® system and revealed that 39 of the 42 chemotherapy-naïve mCRPC patients were positive for CTCs. CTC conversion from an unfavorable count (≥5) to a favorable count (<5) was noted in 13 of 26 patients after treatment. The Epic Sciences platform allows the detection of nucleated cells after red blood lysis, and PSMA staining on CTCs and CTC clusters was subsequently used. In 16 (89%) of 18 patients with PSMA-positive CTCs detected, the number of CTCs was reduced after treatment. The PSMA expression levels can help select patients who are likely to benefit from PSMA-directed treatment. The authors concluded that PSMA expression on CTCs could serve as a patient selection biomarker for an early response in further clinical trials [41]. Another phase II trial investigated a PSMA antibody-drug conjugate (PSMA ADC) coupled to monomethyl auristatin E, which binds to PSMA-positive cells and induces cytotoxicity. The study demonstrated a decline in the CTC count of ≥50% in 78% (60/77) of mCRPC patients and conversion (from ≥5 cells/7.5 mL blood to less than 5 cells/7.5 mL blood) in 47% (36/77) of mCRPC patients. The highest CTC response (≥50%) was documented in patients with combined high PSMA expression on CTCs and low levels of neuroendocrine markers (94%) [42].

Armstrong and colleagues investigated treatment with radium-223 in men with bone metastases and progressive mCRPC in a small phase II trial. Evidence of prostate cancer osteomimicry biomarkers in CTCs was examined. These osteoblast regulators, such as bone alkaline phosphatase (B-ALP, gene ALPL), osteopontin, osteocalcin, osteoblast cadherin, runt-related transcription factor 2, bone gamma carboxyglutamate protein, tumor necrosis factor ligand superfamily 11, activator of NF-kappa-B ligand and secreted protein acidic and cysteine rich, and osteonectin, were examined at the RNA level. The authors concluded that genomic and phenotypic evidence supports osteomimicry in CTCs and tumor biopsies of men with mCRPC [43]. In multiple phase I/II studies, the CTC genotype or phenotype was used as a tool to determine possible direct drug targets or downstream drug targets [44–47].

AR splice variants (AR-Vs) represent a crucial mechanism responsible for castration-resistant prostate cancer progression. All variants have common structural features resulting from deletion of the ligand-binding domain (LBD) as a consequence of alternative splicing. Interestingly, the absence of the LBD can confer constitutive, androgen-independent activity [48]. AR-V7 is the most discussed splice variant in humans. A current study focusing on the mRNA analysis of CTCs revealed information on the clinical efficacy of anti-androgen treatment—CTCs which harbor the androgen receptor splice variant (AR-V7) and, despite the absence of the LBD, confer constitutive androgen-independent activity [48]. In 2014, Antonarakis et al. [49] analyzed the baseline CTC-derived AR-V7 status as a biomarker of the response or resistance to therapies in a small cohort of 62 men with enzalutamide- or abiraterone acetate (AA)-pretreated mCRPC. The results demonstrated that the outcomes of treatment with both drugs were significantly worse in patients who harbored the AR-V7 splice variant in CTCs than in those who did not harbor AR-V7. This study led to increased interest in the prospective utility of CTCs for the serial monitoring of second-generation AR antagonists as a mechanism of treatment resistance. Thus, CTCs might help to predict failure to treatment with enzalutamide and AA but not docetaxel or cabazitaxel [49–51].

The PROPHECY trial (ClinicalTrials.gov identifier: NCT02269982) compared two CTC platforms: the Johns Hopkins University modified-AdnaTest CTC AR-V7 mRNA assay and the Epic Sciences CTC nuclear-specific AR-V7 protein assay. The AdnaTest uses antibodies against EpCAM and HER2 for CTC capture. The Epic Sciences CTC platform uses only red blood cell lysates, and approximately 3×10^6 nucleated cells are dispensed onto 10–16 glass micro slides. The prognostic significance of baseline CTC AR-V7 evaluated based on radiographic or clinical progression-free survival (PFS) in 118 men with high-risk mCRPC was determined. Both assays demonstrated significantly different PFS rates in AR-V7-positive men with mCRPC compared with AR-V7-negative men. Interestingly, the percentage agreement between the AR-V7 CTC assays was 82% (86/105). The authors concluded by two blood-based assays that the AR-V7 splice variant in CTCs may optimize treatment selection beyond the clinical assessment of prognosis [52].

In the phase II TAXYNERGY trial (ClinicalTrials.gov identifier: NCT01718353), Antonarakis et al. [53] analyzed an early taxane switch in men with chemotherapy-naïve, metastatic castration-resistant prostate cancer. The assumption was that the clinical response was associated with taxane drug-target engagement (DTE), which results in microtubule bundling (MTB) and nuclear AR localization (ARNL) in CTCs [53]. CTCs were isolated from PSMA-based microfluidic devices [54]. This study demonstrated that the combination of real-time CTC-based %ARNL and MTB mRNA detection at an early time point could be used to indicate a benefit in men treated with taxanes [53].

Furthermore, the nuclear localization of the AR-V7 protein in CTCs from 161 men with mCRPC was analyzed as a treatment-specific biomarker. The authors verified that nuclear localization of the AR-V7 protein was associated with superior survival with taxane therapy over androgen receptor signaling (ARS)-directed therapy (HR, 0.24; 95% CI, 0.10–0.57: $p = 0.035$) in a clinical setting [55]. The same group analyzed the phenotypic heterogeneity of CTCs (179 unique patients) to obtain supporting information for the treatment choice of androgen receptor signaling inhibitors (ARSIs) and taxanes in mCRPC patients. To quantify the heterogeneity of CTCs, the Shannon index was used. The phenotypic features were subdivided into 15 subtypes: "A"–"O". For example, the following characteristics were associated with cell type A: low cytokeratin, no AR expression, and large cell size. The following characteristics were associated with cell type F: frequently in the histological cluster of 2 CTCs, AR expression and high cytokeratin expression. CTCs were processed with the Epic system. The authors concluded that low CTC phenotypic heterogeneity was associated with prolonged OS in patients treated with an ARSI. A higher Shannon index was associated with better OS for patients treated with taxanes [56]. The advantage of CTC analysis is to allow single-cell determination of nuclear AR-V7 protein localization in different CTC subtypes. In this context, Miyamato et al. demonstrated the RNA-based digital CTC quantification of prostate-derived transcripts as predictive of the AA response in men with mCRPC. Single-cell RNA analysis of CTCs was performed on the androgen-responsive transcripts PSA, KLK2, and TMRESS2, and on anterior gradient protein 2 homolog, PSMA, homeobox B13 and the androgen-independent transcripts FAT1 and STEAP2. The authors concluded that the digital scoring of CTC mRNA of prostate-derived transcripts can be used in high-throughput analyses in clinical practice [57].

In a second study, in a population of 34 localized prostate cancer patients prior to radical prostatectomy, Miyamoto et al. analyzed digital CTC quantification by whole transcriptome amplification and multiplex droplet digital PCR of a panel of 8 genes (AGR2, FAT1, FOLH1, HOXB13, KLK2, KLK3, STEAP2, TMPRESS2). Efficient risk stratification of localized PCa patients to guide optimal treatment by digital CTC quantification was only recently achieved by using the differential weighting of 6 genes from the panel, therefore, the authors could predict early presence or absence of prostate cancer dissemination in localized disease [57].

Among 108 posters, 16 reported the analysis of prostate cancer patients. Five abstracts from the ACTC meeting in 2019 discussed the possible roles of AR and splice variants in mCRPC patients. In this context, new quantification platforms for the proteins or mRNAs of CTCs were presented. Hofmann et al. [17] analyzed an mRNA in situ padlock probe to detect AR and AR-V7 on CellCollector. AR-V7 was detectable in 91% (10/11) of advanced prostate cancer patients. An analysis by Markou et al. [17] demonstrated an RT-qPCR assay for determination of the proto-oncogene PIM-1 mRNA in the EpCAM-positive fraction of mCRPC patients (n = 50). They concluded that PIM-1 mRNA expression should be further assessed to profile CTCs. CTC clusters (two or more CTCs) were evaluated in metastatic breast cancer (MBC) patients (n = 57) and mPCa patients (n = 57) in relation to PFS and OS. CTCs were captured with the CellSearch® system. The authors confirmed the further prognostic value of the CTC cluster compared with the CTC count alone [17].

5. Evidence Synthesis—Breast Cancer

Breast cancer is one of the most common cancers among women worldwide, with an incidence of approximately 17,000,000 cases per year and the highest mortality among women [58,59]. In the early status, breast cancer is relatively treatable, whereas metastasis oft finishes with death of the patient. Even though decision on the proper therapy is not an easy step, one must consider the molecular subtype of the tumor.

Over 70 clinical trials analyzing CTCs in breast cancer patients have been performed. From these 71 publications, 10 describe less than 20 and 61 describe 20 and more patients bearing diverse breast carcinomas of various stages: 17 with primary breast cancer (PBC) and 24 with MBC. Forty-one studies were described as a part of clinical trials, or at least some of the patients were included in clinical studies (trial phase I, 2; II, 18; III, 15). The most frequently used system for the CTC isolation in breast cancer patients is the CellSearch® system (n = 47 trails). Patients bearing primary breast cancer were analyzed in several studies like SUCCESS, NeoALLTO, and BEVERLY-2. The efficacy of the given therapy among the patients was examined by enumeration of CTCs before and after applied study-specific therapies (patients included in SUCCESS, BEVERLY-2) as well as by evaluating and comparing the rate of pathological complete response (pCR) in HER2/ErbB2-overexpressing and/or HER2/ErbB2-amplified PBC (patients included in NeoALLTO). In two studies, analysis of additional staining of CTCs (Barriere et al.) or additional factors in blood (patients included in SUCCESS I) were performed. Changes of CTC number and/or phenotype by metastatic breast cancer (MBC) patients were analyzed in other trials. Six of them (OnSITE, CirCe01 and some multi- or mono-centric studies) focused only on analysis of the enumeration, when the others (CAMELLIA, TBCRC, NEOZOTAC side-study, LANDSCAPE, BCA2001, AVALUZ) investigated phenotypical changes among CTCs after applied therapies.

In 27 studies, CTCs' additional markers were analyzed: (1) tumor markers (MUC1 and HER2), (2) subsidiary markers of stemness (CD44 and BMI1), (3) EMT markers (TWIST, AKT2, PI3KA, ALDH1, and vimentin), (4) apoptotic genes expression, (5) study-related genes (AGTR1), and (6) 55-CTC-specific genes (Table S2).

5.1. CTC Enumeration—Breast Cancer

The basic analysis of CTCs is their enumeration. Most investigators have noted the prognostic significance of the CTC counts for OS [60,61] as well as chemotherapy [62–69]. In analysis performed by Trapp et al. [70], describing the early-stage high risk PBC patients of the SUCCESS A trial, an association between the presence of CTCs two years after chemotherapy with zoledronate and shortened OS and disease-free survival (DFS) was observed. Symonds et al. [71] revealed that a decrease in CTC numbers from baseline to the first assessment after employing nab-paclitaxel and bevacizumab therapy followed by maintenance therapy with bevacizumab and erlotinib by metastatic TNBC patients correlated with prolonged PFS and OS. Similar results were noticed in the OnSITE study analyzing the CTC number before and after the second cycle of treatment of the HER2-negative advanced BC patients pre-treated with anthracyclines and taxanes, who received eribulin as third-line chemotherapy show similar results [72]. Liang et al. [73] analyzed the number of CTCs as one of the efficacy parameters for the comparison of three therapy strategies for tumor cryoablation with natural killer (NK) cells therapy (I, cryoablation; II, cryoablation + NK cell therapy; III, NK cell therapy-Herceptin) and Herceptin for patients with HER2-overexpressing recurrent BC. The reduced number of CTCs after combined therapies correlated partially with prolonged PFS [73]. According to the prospective study with newly diagnosed metastatic BC patients treated with the systemic therapy, the detection of ≥5 CTCs or CTC clusters can predict a growing hazard ratio and worse PFS and OS during therapy [74]. Identification of one or more CTCs before PBC resection correlates with poor relapse-free survival and OS [75]. Similarly, Goodman et al. [76] observed longer OS, local recurrence-free survival (LRFS), and disease-free survival (DFS) in CTC-positive patients enrolled in the phase III SUCCESS study and from the National Cancer Database (NCDB), who underwent radiotherapy than in patients who did not. On the other hand, after comparing two therapies (anthracycline-containing chemotherapy and anthracycline-free chemotherapy), Schramm et al. [77] did not notice any differences in the number of CTCs between the therapies in the HER2-negative patients with early BC included in SUCCESS C study. Similarly, none or only the weak prognostic relevance of CTC count was demonstrated by Jueckstock et al. [78] and Hepp et al. [79] The authors employed node-positive or high risk node-negative BC patients of the SUCCESS A study before adjuvant taxane-based chemotherapy or patients of the SUCCESS study receiving the therapy based on fluorouracil, epirubicin and cyclophosphamide (FEC) followed either by docetaxel vs. by docetaxel supplemented with gemcitabine, respectively [78,79].

Among studies enrolling 50 (Lelievre et al. [80]) or 28 (Tokudome et al. [81]) patients, CTCs were detected in only 8 or 9 at baseline and in only 1 or 5 at the final analysis, respectively. Similarly, during an analysis of the influence of therapy on HER2-negative breast cancer patients, Gonzalez-Angulo et al. [82] found CTCs only in 9 (28.1%) patients at baseline and in 3 (13.6%) at the end of the 18-week study, whereas Agelaki et al. [83] found a decrease in the percentage of HER2-positive CTCs from 93.45% (total number of patients = 21) to 66.6% after the first cycle. Paoletti et al. [84] described a study with 45 estrogen receptor-positive (ER)/HER2-negative metastatic or locoregionally recurrent disease BC patients enrolled in the CTC analysis, but only 11 presented ≥5 CTCs. Similarly, Pierga et al. [85] analyzed 41 HER2-positive MBC patients displaying brain metastasis, but the number of patients with CTCs decreased from 20 (≥1 CTC) and 9 (≥5 CTC) at baseline to 11 and 3, respectively, at day 21. However, elevated levels of CTCs were noted at baseline in both groups as strong prognostic factors [84,85]. In contrast, an analysis of ER expression on CTCs in patients with positive primary tumors ($n = 16$) showed intrapatient heterogeneity, though the small number of included patients had the authors conclude the link to the mechanism of the escape from the endocrine therapy [86]. On the other hand, after an analysis of ER/HER2 expression on CTCs in small MBC and LABC/IBC patient cohorts ($n = 36$), Somlo et al. [87] suggested that the pilot trial may help to validate CTC-based targeted therapy.

5.2. Functional Characterization of CTCs—Breast Cancer

Molecular characterization of a tumor allows division into different subtypes: ER/PR-positive and HER2-positive. Each subtype responds to different therapies [88–90]. The other subtype does not express any of the receptors (triple-negative, TNBC) or respond to any of the hormonal target therapies [91]. However, many reports indicate that independent of tumor size, histopathological grade, ER/PR status or axillary lymph node involvement, the HER2 status of a patient can change [92]. First, a negative primary tumor may create positive metastases, which can influence the therapy decision [93]. Therefore, the classification of a tumor using biopsy is crucial. However, due to the location of the metastasis, biopsy is not always possible. LB is less invasive and can be used to monitor disease development. Even if HER2 expression among the CTCs within each patient is heterogeneous, revealing the strongly positive cells in blood samples allows us to suggest a positive HER2 tumor status. This was demonstrated with the study investigating the efficacy of neoadjuvant treatment on inflammatory BC (HER2-negative—BEVERLY 01 or HER2-positive—BEVERLY 02) with non-metastatic patients (M0) or prospective study with M1 BC [94]. The detection rate of CTCs and the determination of their HER2 status could be a good clinical strategy during treatment [93,95,96].

CTC investigations can also support the prediction of survival in combination with other factors. In their studies with inflammatory BC patients of the BEVERLY-2 survival data, Pierga et al. [97] revealed that the analysis of CTCs and pathologic complete response (pCR) is a good combination of parameters for creating a subgroup with a very good survival prognosis. Based on a comparison of two patient groups with early-stage breast cancer, CTC-positive and CTC-negative, included in phase I SUCCESS study (FEC therapy described previously), König et al. [98] demonstrated that the cytokine profile could also serve as a marker for CTC involvement in disease progression. Among the T-helper cell 2 cytokine (Th2) levels, interleukins 8 and 13 (IL-8, IL-13) were highly secreted by the CTC-negative group of patients negative for progesterone receptor. This correlation was not observed in CTC-positive patients. Similarly, there was an association between the IL-4 level and survival (patients who died had a high IL-4 level) in hormone receptor-negative patients in the CTC-negative but not CTC-positive group. Vilsmaier et al. [99] speculated that Th1 cytokines (Il-1α and Il-1β) are also involved in the release of CTCs in breast cancer patients. Conversely, in a similar patient cohort, (phase I SUCCESS), the expression of two vascular markers, soluble fms-like tyrosine kinase-1 (sFlt1) and placental growth factor (PlGF), in correlation to the CTC status, found their increased expression in CTC-negative patients was analyzed [99]. This suggests that the overexpression of both markers in tumor cells inhibits invasion of the decanted tumor cells into blood vessels [100].

Several trials analyzed apoptosis in CTCs. Negative breast cancer responds to apoptosis, triggering tigatuzumab treatment and Paoletti et al. [101] hypothesized the induction of apoptosis in CTCs among metastatic TNBC patients treated with tigatuzumab in the included studies. However, no prognostic effect was observed [101]. Other studies describing the influence of zoledronic acid (ZA, an inhibitor of tumor growth and inductor of apoptosis) infusion on the CTC value in patients with locally advanced BC (group 1) and those with bone metastasis only (group 2) revealed a decrease in the CTC number after 48 h; however, the difference was minimalized after 14 days. Additionally, an analysis of the apoptotic marker M30 in CTCs after 14 days showed increased levels of apoptotic CTCs [102]. Similarly, in their analysis with patients bearing progressive metastatic BC shared in the cohorts, depending on the treatment (endocrine therapy, texane-based or non-texane based chemotherapy). Smerage et al. [103] hypothesized that the presence of M30-positive CTCs was associated with a good prognosis, but the results of their studies after any of the therapies produced the opposite outcomes. In patients with high numbers of CTCs, high levels of M30-CTC correlated with a poor prognosis, while high CTC-Bcl-2 (B-cell lymphoma 2) levels were associated with a good prognosis [103].

Another mechanism that is strongly correlated with cell invasion is EMT. Analyses of this process are crucial since it leads to downregulation of the expression of epithelial markers such as EpCAM. Such a situation can result in false-negative findings considering EpCAM-based CTC isolation. Guan et al. [104] examined CTCs of the HER2-negative metastatic BC women participating in CAMELLIA trial. This phase III study analyzed the metronomic capecitabine chemotherapy vs. intermittent capecitabine maintenance therapy following the capecitabine complemented with docetaxel first-line chemotherapy. Examined CTCs were isolated with a method based on cell size, enumerated and expression of epithelial markers (EpCAM, CK8/18/19), and mesenchymal markers was analyzed (Twist and Vimentin). The investigation of Guan et al. [104] revealed that the group of patients who secreted EMT-CTCs experienced shorter PFS than the group of non-EMT-CTC patients. Additionally, HER2-negative patients demonstrated almost two times higher EMT-CTC counts than HER2-positive patients [104]. Barriere et al. [105] analyzed CTCs in patients with early breast cancer. They detected cells with dedifferentiated characteristics (mesenchymal phenotype, stem cell phenotype or both) in 37.6% of the analyzed patients, and the predominant epithelial phenotype was present in 8.75% of probes. They concluded that the molecular analysis of CTCs is more relevant than enumeration only [105]. An analysis of both epithelial and mesenchymal CTCs is also suggested as a good tool to predict the responsiveness to eribulin [106].

Among 108 posters, 25 described the analysis of breast cancer patients. Many presentations addressed the relation of CTCs with therapy. An analysis of blood samples from 36 BC patients showed a positive association between the level of EMT-CTCs before treatment and the effect of neoadjuvant therapy. Kallergi et al. [17] examined the expression of cytokeratin/C-X-C chemokine receptor type 4/transcription factor jun-B (CK/CXCR4/JUNB) in DTCs isolated from the bone marrow (BM) of 39 HR-positive, HER2-negative breast cancer patients and compared them to that in breast cancer cell lines. They concluded that DTCs expressing these proteins appear to create a subgroup of BC patients at high risk of relapse. After an analysis of epithelial-mesenchymal plasticity (EMP) of primary tumors and their CTCs, Hassan et al. [17] concluded that CTCs may provide important information regarding the progression of cancer. Strati et al. [17] performed an analysis on 100 patients with early breast cancer and 19 healthy donors, and confirmed the importance of CTCs as prognostic factors after detecting the overexpression of TWIST1 and stem cell transcripts. Considering the treatment selection in metastatic TNBC, Zang et al. [17] examined the association between the levels of CTCs and programmed cell death 1 ligand 1 (PD-L1) expression. They suggested this as a potential predictive therapy marker in metastatic triple-negative breast cancer. In addition, the copy number alteration (CNA) profiles of CTCs may be linked to OS. An analysis of the ERα gene (ESR1) (Tzanikou et al. [17]) and PIK3CA gene (Stergioupoulou et al. [17]) revealed that hot-spot mutations in single CTCs is also possible.

6. Perspectives of Real-Time Monitoring

Anti-tumor therapies are long and exhaustive for patients. Real-time monitoring of the healing process could be a useful tool to evaluate therapeutic responses. The application of LB for clinical diagnostics could improve sequence screening, provide additional valuable information on tumor dynamics (early response, mutation-based resistance to target therapy) and help personalize management for patients (Table 1).

Table 1. Specification feature of CTCs (circulating tumor cells) for potential clinical application.

Cancer Type	Characterization	Clinical Utility of CTCs Validated in Trials	Implementation in Clinical Practice	Reference
Prostate Cancer	AR/AR-Splice Variants	Prognosis Treatment Selection Therapy Monitoring Drug resistance	requires further evaluation	[49–53,55]
	PSMA	Therapy monitoring	basis for future evaluation	[41,42]
	Enumeration	Prognosis	potential clinical application	[20–31]
Breast Cancer	HER2	Therapy monitoring Prognosis	basis for future evaluation basis for future evaluation	[87,94–96,105] [105]
	EMT	Prognosis Therapy monitoring	basis for future evaluation requires future evaluation	[104] [105]
	Apoptosis	Therapy monitoring	basis for future evaluation	[101–103]
	Enumeration	Prognosis Therapy monitoring	potential clinical application requires future evaluation basis for future evaluation requires future evaluation	[60,61] [74,97,102] [87,93,101–106]

AR: androgen receptor, PSAM: prostate-specific membrane antigen, HER2: human epidermal growth factor receptor 2, EMT: epithelial-mesenchymal transition, CTC: circulating tumor cells.

This less invasive method allows the analysis of CTCs, among other factors. Furthermore, CTCs may offer the opportunity for the real-time monitoring of cancer progression. Additionally, CTCs are sources of DNA, RNA and protein, which provide information on tumor heterogeneity. A good opportunity for this approach is CTC single-cell analysis [34]. At the 4th ACTC meeting, some new approaches for single-CTC isolation were presented.

The first step of involvement of CTCs as prognostic and predictive biomarkers in the clinic was the approval of the CellSearch® system for CTC enumeration. Within the scope of validation, it was calculated that a cutoff of ≥5 CTCs/7.5 mL blood of breast or prostate cancer patients is associated with an unfavorable prognosis [6,7]. The CellSearch® system was the best validated CTC platform, as demonstrated by the large number of publications and the large number of analyzed patient samples. In our review, 8659 prostate cancer and 12,994 breast cancer patients were examined with the CellSearch® system. In summary, most studies used the only FDA-approved technique to detect CTCs; however, this system also has limitations (EpCAM-based detection, sample size).

For a long time, the main applications of CTCs in clinical trials were based only on enumeration. Many investigators suggest that enumeration could be an independent prognostic factor of DFS or OS, or at least in combination with other agents [21,26,28,62,69,72,77]. The statement of the PCWG published by Scher et al. [8] confirmed the importance of CTC enumeration.

However, in breast cancer patients, many reports indicate no relation between the CTC number and therapy response [77–79]. In prostate cancer patients, the detection rate of CTCs is approximately 80%, but unfortunately, not all active progressive cancers have measurable CTCs in the blood [107].

The clinical trials demonstrated the possibility of real-time monitoring in metastatic cancer patients. Future studies must develop a gold standard in combination of CTC isolation/characterization technology for the ability to identify the best treatment. This is the main requirement for patients to benefit from early therapeutic intervention.

7. Discussion

Cancer cells can enter and are motile in blood circulation long before the tumor diagnosed. They offer a possibility for an early cancer diagnosis than standard diagnostics tools of imaging or biomarkers. The CTC analysis can provide insight into personalized cancer characteristics.

CTCs are relatively rare cells with a heterogenetic phenotype and are difficult to capture. At present, there are several platforms that can be used to capture or characterize CTCs. All of these platforms have advantages and disadvantages. The user must be able to decide which is the best and most suitable method for individual cancer patients. However, it is possible that in metastatic or late stage patients, no CTCs are detected. One reason for such conditions, could be the necrotic changes, or reduced tumor vascularization or extreme heterogenic CTC population. Another reason is the variation of the CTC frequency in a single blood sample, which also includes the short half-life of CTC and possible circadian rhythms of CTCs. Furthermore, it must be considered that phenotypic changes in CTC, in terms of epithelial-to-mesenchymal transition (EMT), reflects features of and results in downregulation of cell surface marker EpCAM.

The EMT process is also implicated in the generation of cancer stem cells (CSC), which are cells with abilities to self-renewal [108]. Such changes increase aggressiveness of the tumor cells, provoke their dissemination from the tumor and induce metastases. Detection of CSC as subpopulation CTCs and EMT-CTCs by patients would be crucial information for the treatment and potential therapy resistance.

A more precise analysis of CTCs would be applying the markers like Vimentin (EMT) or CD44 (CSC) at any level (DNA, RNA, protein) in regard to their clinical utility. The broader knowledge on intratumor heterogeneity and dynamic genetic and physiological changes in CTCs undergoing EMT or stem cell CTCs enforces continuous widening of the spectrum of specific markers to be analyzed.

The tumor-released cells can circulate cell-clusters composed of different CTC subpopulation. Only a few of the reviewed trials analyzed CTC-clusters additional to CTCs. The presence of CTC-clusters supported the previous diagnosis obtained with CTCs [41,74].

In our opinion, the crucial point of the efficient CTC analysis, enabling to include them into diagnostics, is their isolation. The combination of the EpCAM-based CellSearch® system with antibody-independent CTC isolation platforms in one blood sample extends the best suitable CTC isolation method. This enables the isolation of all subpopulation of CTCs including CTC, EMT-CTC, CSC and can provide a precise outcome. Approbation of such a system or platform as an additional or supporting diagnostic tool would make the therapy more precise.

Such an analysis could be helpful to avoid the low response rate to therapy. A good example is the detection of the HER2 status in circulating breast cancer cells, which could be opposite to that in the primary tumor [93]. A combined analysis of additional factors, as described by Paoletti et al. [101] (multiparameter CTC-Endocrine Therapy Index (CTC-ETI)), supplies more information and may predict resistance to endocrine therapy. In preclinical studies, researchers analyzed the combined results of the enumeration and expression of ER, HER2 and Ki-67 in MBC patients and demonstrated strong analytical validity of the technique through intrapatient heterogeneity [84].

The molecular characterization of AR mRNA in CTCs and the detection of AR-V7 in CTCs can be used as a tool to guide treatment decisions for men with advanced prostate cancer [51,52]. The benefits for patients are that they are protected against the unnecessary side effects of ineffective treatments.

During their investigations, Smerage et al. [103] revealed that in patients with high numbers of CTCs after treatment, the number of cells did not change; thus, the number of apoptotic cells increased. High levels of M30-CTC (apoptotic cells) correlate with a poor prognosis, whereas high levels of CTC-Bcl-2 (anti-apoptotic cells) are associated with a good prognosis [103]. This is a very important observation, as the initiation of apoptosis in cells as a reaction to environmental stress leads to many morphological and biochemical changes, and probably to the production and secretion of substances that can be spread via the bloodstream to other organs, causing damage [109].

The signals from the tumor microenvironment (TME) and the microenvironment of CTC population can modify the protein pattern of disseminated cells, preparing it to create metastasis. One of the proteins induced by TME but also by chemotherapy is prostaglandin (PG)-endoperoxide synthase 2 (COX-2), which finally promotes the carcinogenesis and the rate of cancer recurrence, reducing the survival rate. COX-2 is implicated in the suppression of the apoptosis causing the resistance of tumor cells. The downstream signaling protein of the COX-2 is, among others, the Bcl-2, the anti-apoptotic marker increased in CTCs analyzed by Smerage et al. [107] (reviewed in [110]). Since both the positive and the negative signals may be initiated by the therapy, the monitoring of the treatment is crucial. LB, as the CTCs, is an excellent tool to analyze the changes during the long healing process. Although this hypothesis needs further investigation, if it is correct, it could bear significant consequences (Table 1). The critical point of preanalytical variables on LB in prostate cancer was also discussed in the plenary session of the ACTC 2019 meeting [17]; however, this issue is also applicable for other cancer types, such as breast cancer. Howerd Scher pointed out, in this session, the importance of introducing robust quality controls in all steps, such as pre-analytical procedures, and sample collection and processing, as well as analytical steps like molecular assay specification and bioinformatics algorithm, and data reporting. This standardization is absolutely necessary for its application into clinical routine. Alix-Panabieres [111] summarized in her research, the need for more intervention studies on the implementation of CTCs in the clinic. All abstracts at the 4th ACTC meeting described diverse proofs of concept in CTC isolation and characterization, and confirmed the heterogeneity in this field.

Cabel et al. [112] summarized an analysis of the utility of CTCs in clinical trials. They noted three main concepts of the studies: (1) CTCs serve as surrogate tumor material, (2) CTC enumeration can be used to monitor therapy, and (3) specific biological features of CTCs and their relation to metastatic spread. This conclusion is still relevant. The clinical validity of CTC enumeration by the CellSearch® system is very high; however, its utility is still not standardized in the clinic and requires further investigations. Moreover, the clinical relevance of CTC characterization indicates the current therapeutic target HER2/ER in breast cancer patients and AR-V targets elucidates the resistance mechanism for prostate cancer patients. Even if there is a great need to identify predictive markers for therapy, it remains a great challenge to implement CTCs as a robust standard tool in the clinic.

8. Conclusions and Future Perspectives

LB, in the form of CTCs, is an excellent tool to monitor the disease development and the progress of the therapy. One must emphasize that different isolation methods may give disparate results; therefore, in the clinic, it is crucial to use systems of comparable validation. There is still a great need to develop and standardize the platform adequate to each tumor type. This crucial step can be achieved by a combination of already existing methods based on EpCAM detection and physical properties of the cells. A milestone would be an acceptance of LB in clinical diagnostic routine as a basic or at least as a supporting factor detecting disease progress. Application of additional staining markers on CTCs, specific for each type of the tumor, could help to monitor the changes of the tumor-derived cells during early-stage therapy and help to make a decision on continuing or stopping the therapy. However, as we know with the example of PSA, to become the useful tumor biomarker, the protein or factor must reach the high trust of the clinicians based on long positive experience. It was demonstrated in mCRPC that the CTC count is independent of the PSA level. In conclusion, the CTC count, in contrast to the PSA level, is not directly affected by ADT. Another clinical application of the cell count is the kinetic of CTC number, which reveal a higher discriminatory power for overall survival [26,31].

Supplementary Materials: The following are available online at http://www.mdpi.com/2072-6694/12/12/3782/s1, Table S1: Main clinical Trials for Prostate Cancer, Table S2: Main clinical Trials for Breast Cancer.

Author Contributions: G.T. and J.B. conceived and designed the manuscript. G.T., J.B. and P.F. all aided in the original draft preparation, review, and editing. All authors have read and agreed to the published version of the manuscript.

Funding: This research received no external funding.

Conflicts of Interest: The authors declare no conflict of interest.

References

1. Lambert, A.W.; Pattabiraman, D.R.; Weinberg, R.A. Emerging biological principles of metastasis. *Cell* **2017**, *168*, 670–691. [CrossRef] [PubMed]
2. Alix-Panabieres, C.; Pantel, K. Circulating tumor cells: Liquid biopsy of cancer. *Clin. Chem.* **2013**, *59*, 110–118. [CrossRef] [PubMed]
3. Massague, J.; Obenauf, A.C. Metastatic colonization by circulating tumour cells. *Nature* **2016**, *529*, 298–306. [CrossRef] [PubMed]
4. Welch, D.R.; Hurst, D.R. Defining the hallmarks of metastasis. *Cancer Res.* **2019**, *79*, 3011–3027. [CrossRef]
5. Braun, S.; Vogl, F.D.; Naume, B.; Janni, W.; Osborne, M.P.; Coombes, R.C.; Schlimok, G.; Diel, I.J.; Gerber, B.; Gebauer, G.; et al. A pooled analysis of bone marrow micrometastasis in breast cancer. *N. Engl. J. Med.* **2005**, *353*, 793–802. [CrossRef] [PubMed]
6. De Bono, J.S.; Scher, H.I.; Montgomery, R.B.; Parker, C.; Miller, M.C.; Tissing, H.; Doyle, G.V.; Terstappen, L.W.; Pienta, K.J.; Raghavan, D. Circulating tumor cells predict survival benefit from treatment in metastatic castration-resistant prostate cancer. *Clin. Cancer Res.* **2008**, *14*, 6302–6309. [CrossRef]
7. Cristofanilli, M.; Budd, G.T.; Ellis, M.J.; Stopeck, A.; Matera, J.; Miller, M.C.; Reuben, J.M.; Doyle, G.V.; Allard, W.J.; Terstappen, L.W.; et al. Circulating tumor cells, disease progression, and survival in metastatic breast cancer. *N. Engl. J. Med.* **2004**, *351*, 781–791. [CrossRef]
8. Scher, H.I.; Morris, M.J.; Stadler, W.M.; Higano, C.; Basch, E.; Fizazi, K.; Antonarakis, E.S.; Beer, T.M.; Carducci, M.A.; Chi, K.N.; et al. Trial design and objectives for castration-resistant prostate cancer: Updated recommendations from the prostate cancer clinical trials working group 3. *J. Clin. Oncol.* **2016**, *34*, 1402–1418. [CrossRef]
9. Habli, Z.; AlChamaa, W.; Saab, R.; Kadara, H.; Khraiche, M.L. Circulating tumor cell detection technologies and clinical utility: Challenges and opportunities. *Cancers* **2020**, *12*, 1930. [CrossRef]
10. Ferreira, M.M.; Ramani, V.C.; Jeffrey, S.S. Circulating tumor cell technologies. *Mol. Oncol.* **2016**, *10*, 374–394. [CrossRef]

11. Allard, W.J.; Matera, J.; Miller, M.C.; Repollet, M.; Connelly, M.C.; Rao, C.; Tibbe, A.G.; Uhr, J.W.; Terstappen, L.W. Tumor cells circulate in the peripheral blood of all major carcinomas but not in healthy subjects or patients with nonmalignant diseases. *Clin. Cancer Res.* **2004**, *10*, 6897–6904. [CrossRef] [PubMed]
12. Swennenhuis, J.F.; van Dalum, G.; Zeune, L.L.; Terstappen, L.W. Improving the CellSearch(R) system. *Expert Rev. Mol. Diagn.* **2016**, *16*, 1291–1305. [CrossRef] [PubMed]
13. Paoletti, C.; Hayes, D.F. Circulating tumor cells. *Adv. Exp. Med. Biol.* **2016**, *882*, 235–258. [PubMed]
14. Werner, S.L.; Graf, R.P.; Landers, M.; Valenta, D.T.; Schroeder, M.; Greene, S.B.; Bales, N.; Dittamore, R.; Marrinucci, D. Analytical validation and capabilities of the epic CTC platform: Enrichment-free circulating tumour cell detection and characterization. *J. Circ. Biomark.* **2015**, *4*, 3. [CrossRef]
15. Fachin, F.; Spuhler, P.; Martel-Foley, J.M.; Edd, J.F.; Barber, T.A.; Walsh, J.; Karabacak, M.; Pai, V.; Yu, M.; Smith, K.; et al. Monolithic chip for high-throughput blood cell depletion to sort rare circulating tumor cells. *Sci. Rep.* **2017**, *7*, 10936. [CrossRef]
16. Kim, T.H.; Wang, Y.; Oliver, C.R.; Thamm, D.H.; Cooling, L.; Paoletti, C.; Smith, K.J.; Nagrath, S.; Hayes, D.F. A temporary indwelling intravascular aphaeretic system for in vivo enrichment of circulating tumor cells. *Nat. Commun.* **2019**, *10*, 1478. [CrossRef]
17. Lianidou, E.; Pantel, K. Hellenic society of liquid biopsy. In Proceedings of the 4th ACTC | Advances in Circulating Tumor Cells, Corfu, Greece, 2–5 October 2019.
18. Bray, F.; Ferlay, J.; Soerjomataram, I.; Siegel, R.L.; Torre, L.A.; Jemal, A. Global cancer statistics 2018: GLOBOCAN estimates of incidence and mortality worldwide for 36 cancers in 185 countries. *CA Cancer J. Clin.* **2018**, *68*, 394–424. [CrossRef]
19. Nuhn, P.; de Bono, J.S.; Fizazi, K.; Freedland, S.J.; Grilli, M.; Kantoff, P.W.; Sonpavde, G.; Sternberg, C.N.; Yegnasubramanian, S.; Antonarakis, E.S. Update on systemic prostate cancer therapies: Management of metastatic castration-resistant prostate cancer in the era of precision oncology. *Eur. Urol.* **2019**, *75*, 88–99. [CrossRef]
20. Antonarakis, E.S.; Heath, E.I.; Posadas, E.M.; Yu, E.Y.; Harrison, M.R.; Bruce, J.Y.; Cho, S.Y.; Wilding, G.E.; Fetterly, G.J.; Hangauer, D.G.; et al. A phase 2 study of KX2-391, an oral inhibitor of Src kinase and tubulin polymerization, in men with bone-metastatic castration-resistant prostate cancer. *Cancer Chemother. Pharmacol.* **2013**, *71*, 883–892. [CrossRef]
21. Goldkorn, A.; Ely, B.; Quinn, D.I.; Tangen, C.M.; Fink, L.M.; Xu, T.; Twardowski, P.; van Veldhuizen, P.J.; Agarwal, N.; Carducci, M.A.; et al. Circulating tumor cell counts are prognostic of overall survival in SWOG S0421: A phase III trial of docetaxel with or without atrasentan for metastatic castration-resistant prostate cancer. *J. Clin. Oncol.* **2014**, *32*, 1136–1142. [CrossRef]
22. Smith, M.R.; Sweeney, C.J.; Corn, P.G.; Rathkopf, D.E.; Smith, D.C.; Hussain, M.; George, D.J.; Higano, C.S.; Harzstark, A.L.; Sartor, A.O.; et al. Cabozantinib in chemotherapy-pretreated metastatic castration-resistant prostate cancer: Results of a phase II nonrandomized expansion study. *J. Clin. Oncol.* **2014**, *32*, 3391–3399. [CrossRef] [PubMed]
23. Smith, M.; de Bono, J.; Sternberg, C.; le Moulec, S.; Oudard, S.; de Giorgi, U.; Krainer, M.; Bergman, A.; Hoelzer, W.; de Wit, R.; et al. Phase III study of cabozantinib in previously treated metastatic castration-resistant prostate cancer: COMET-1. *J. Clin. Oncol.* **2016**, *34*, 3005–3013. [CrossRef] [PubMed]
24. Thalgott, M.; Heck, M.M.; Eiber, M.; Souvatzoglou, M.; Hatzichristodoulou, G.; Kehl, V.; Krause, B.J.; Rack, B.; Retz, M.; Gschwend, J.E.; et al. Circulating tumor cells versus objective response assessment predicting survival in metastatic castration-resistant prostate cancer patients treated with docetaxel chemotherapy. *J. Cancer Res. Clin. Oncol.* **2015**, *141*, 1457–1464. [CrossRef] [PubMed]
25. Vogelzang, N.J.; Fizazi, K.; Burke, J.M.; de Wit, R.; Bellmunt, J.; Hutson, T.E.; Crane, E.; Berry, W.R.; Doner, K.; Hainsworth, J.D.; et al. Circulating tumor cells in a phase 3 study of docetaxel and prednisone with or without lenalidomide in metastatic castration-resistant prostate cancer. *Eur. Urol.* **2017**, *71*, 168–171. [CrossRef]
26. Heller, G.; Fizazi, K.; McCormack, R.; Molina, A.; MacLean, D.; Webb, I.J.; Saad, F.; de Bono, J.S.; Scher, H.I. The added value of circulating tumor cell enumeration to standard markers in assessing prognosis in a metastatic castration-resistant prostate cancer population. *Clin. Cancer Res.* **2017**, *23*, 1967–1973. [CrossRef]
27. Prentice, R.L. Surrogate endpoints in clinical trials: Definition and operational criteria. *Stat. Med.* **1989**, *8*, 431–440. [CrossRef]

28. Scher, H.I.; Heller, G.; Molina, A.; Attard, G.; Danila, D.C.; Jia, X.; Peng, W.; Sandhu, S.K.; Olmos, D.; Riisnaes, R.; et al. Circulating tumor cell biomarker panel as an individual-level surrogate for survival in metastatic castration-resistant prostate cancer. *J. Clin. Oncol.* **2015**, *33*, 1348–1355. [CrossRef]
29. Lorente, D.; Olmos, D.; Mateo, J.; Dolling, D.; Bianchini, D.; Seed, G.; Flohr, P.; Crespo, M.; Figueiredo, I.; Miranda, S.; et al. Circulating tumour cell increase as a biomarker of disease progression in metastatic castration-resistant prostate cancer patients with low baseline CTC counts. *Ann. Oncol.* **2018**, *29*, 1554–1560. [CrossRef]
30. Yu, E.Y.; Li, H.; Higano, C.S.; Agarwal, N.; Pal, S.K.; Alva, A.; Heath, E.I.; Lam, E.T.; Gupta, S.; Lilly, M.B.; et al. SWOG S0925: A randomized phase II study of androgen deprivation combined with cixutumumab versus androgen deprivation alone in patients with new metastatic hormone-sensitive prostate cancer. *J. Clin. Oncol.* **2015**, *33*, 1601–1608. [CrossRef]
31. Heller, G.; McCormack, R.; Kheoh, T.; Molina, A.; Smith, M.R.; Dreicer, R.; Saad, F.; de Wit, R.; Aftab, D.T.; Hirmand, M.; et al. Circulating tumor cell number as a response measure of prolonged survival for metastatic castration-resistant prostate cancer: A comparison with prostate-specific antigen across five randomized phase III clinical trials. *J. Clin. Oncol.* **2018**, *36*, 572–580. [CrossRef]
32. Murray, N.P.; Reyes, E.; Orellana, N.; Fuentealba, C.; Badinez, L.; Olivares, R.; Porcell, J.; Duenas, R. Secondary circulating prostate cells predict biochemical failure in prostate cancer patients after radical prostatectomy and without evidence of disease. *Sci. World J.* **2013**, *2013*, 762064. [CrossRef] [PubMed]
33. Murray, N.P.; Aedo, S.; Fuentealba, C.; Reyes, E. 10 Year Biochemical Failure Free Survival of Men with CD82 Positive Primary Circulating Prostate Cells Treated by Radical Prostatectomy. *Asian Pac. J. Cancer Prev.* **2018**, *19*, 1577–1583. [PubMed]
34. Keller, L.; Pantel, K. Unravelling tumour heterogeneity by single-cell profiling of circulating tumour cells. *Nat. Rev. Cancer* **2019**, *19*, 553–567. [CrossRef] [PubMed]
35. Gopalan, A.; Leversha, M.A.; Satagopan, J.M.; Zhou, Q.; Al-Ahmadie, H.A.; Fine, S.W.; Eastham, J.A.; Scardino, P.T.; Scher, H.I.; Tickoo, S.K.; et al. TMPRSS2-ERG gene fusion is not associated with outcome in patients treated by prostatectomy. *Cancer Res.* **2009**, *69*, 1400–1406. [CrossRef] [PubMed]
36. Song, C.; Chen, H. Predictive significance of TMRPSS2-ERG fusion in prostate cancer: A meta-analysis. *Cancer Cell Int.* **2018**, *18*, 177. [CrossRef]
37. Danila, D.C.; Anand, A.; Sung, C.C.; Heller, G.; Leversha, M.A.; Cao, L.; Lilja, H.; Molina, A.; Sawyers, C.L.; Fleisher, M.; et al. TMPRSS2-ERG status in circulating tumor cells as a predictive biomarker of sensitivity in castration-resistant prostate cancer patients treated with abiraterone acetate. *Eur. Urol.* **2011**, *60*, 897–904. [CrossRef]
38. Goldkorn, A.; Ely, B.; Tangen, C.M.; Tai, Y.C.; Xu, T.; Li, H.; Twardowski, P.; Veldhuizen, P.J.; Agarwal, N.; Carducci, M.A.; et al. Circulating tumor cell telomerase activity as a prognostic marker for overall survival in SWOG 0421: A phase III metastatic castration resistant prostate cancer trial. *Int. J. Cancer* **2015**, *136*, 1856–1862. [CrossRef]
39. Silver, D.A.; Pellicer, I.; Fair, W.R.; Heston, W.D.; Cordon-Cardo, C. Prostate-specific membrane antigen expression in normal and malignant human tissues. *Clin. Cancer Res.* **1997**, *3*, 81–85.
40. Sweat, S.D.; Pacelli, A.; Murphy, G.P.; Bostwick, D.G. Prostate-specific membrane antigen expression is greatest in prostate adenocarcinoma and lymph node metastases. *Urology* **1998**, *52*, 637–640. [CrossRef]
41. Autio, K.A.; Dreicer, R.; Anderson, J.; Garcia, J.A.; Alva, A.; Hart, L.L.; Milowsky, M.I.; Posadas, E.M.; Ryan, C.J.; Graf, R.P.; et al. Safety and Efficacy of BIND-014, a Docetaxel Nanoparticle Targeting Prostate-Specific Membrane Antigen for Patients with Metastatic Castration-Resistant Prostate Cancer: A Phase 2 Clinical Trial. *JAMA Oncol.* **2018**, *4*, 1344–1351. [CrossRef]
42. Petrylak, D.P.; Vogelzang, N.J.; Chatta, K.; Fleming, M.T.; Smith, D.C.; Appleman, L.J.; Hussain, A.; Modiano, M.; Singh, P.; Tagawa, S.T.; et al. PSMA ADC monotherapy in patients with progressive metastatic castration-resistant prostate cancer following abiraterone and/or enzalutamide: Efficacy and safety in open-label single-arm phase 2 study. *Prostate* **2020**, *80*, 99–108. [CrossRef] [PubMed]
43. Armstrong, A.J.; Gupta, S.; Healy, P.; Kemeny, G.; Leith, B.; Zalutsky, M.R.; Spritzer, C.; Davies, C.; Rothwell, C.; Ware, K.; et al. Pharmacodynamic study of radium-223 in men with bone metastatic castration resistant prostate cancer. *PLoS ONE* **2019**, *14*, e0216934. [CrossRef] [PubMed]

44. Bradley, D.A.; Daignault, S.; Ryan, C.J.; Dipaola, R.S.; Cooney, K.A.; Smith, D.C.; Small, E.; Mathew, P.; Gross, M.E.; Stein, M.N.; et al. Cilengitide (EMD 121974, NSC 707544) in asymptomatic metastatic castration resistant prostate cancer patients: A randomized phase II trial by the prostate cancer clinical trials consortium. *Investig. New Drugs* **2011**, *29*, 1432–1440. [CrossRef] [PubMed]
45. Chi, K.N.; Yu, E.Y.; Jacobs, C.; Bazov, J.; Kollmannsberger, C.; Higano, C.S.; Mukherjee, S.D.; Gleave, M.E.; Stewart, P.S.; Hotte, S.J. A phase I dose-escalation study of apatorsen (OGX-427), an antisense inhibitor targeting heat shock protein 27 (Hsp27), in patients with castration-resistant prostate cancer and other advanced cancers. *Ann. Oncol.* **2016**, *27*, 1116–1122. [CrossRef] [PubMed]
46. McHugh, D.; Eisenberger, M.; Heath, E.I.; Bruce, J.; Danila, D.C.; Rathkopf, D.E.; Feldman, J.; Slovin, S.F.; Anand, B.; Chu, R.; et al. A phase I study of the antibody drug conjugate ASG-5ME, an SLC44A4-targeting antibody carrying auristatin E, in metastatic castration-resistant prostate cancer. *Investig. New Drugs* **2019**, *37*, 1052–1060. [CrossRef]
47. Schweizer, M.T.; Haugk, K.; McKiernan, J.S.; Gulati, R.; Cheng, H.H.; Maes, J.L.; Dumpit, R.F.; Nelson, P.S.; Montgomery, B.; McCune, J.S.; et al. A phase I study of niclosamide in combination with enzalutamide in men with castration-resistant prostate cancer. *PLoS ONE* **2018**, *13*, e0198389. [CrossRef] [PubMed]
48. Hu, R.; Dunn, T.A.; Wei, S.; Isharwal, S.; Veltri, R.W.; Humphreys, E.; Han, M.; Partin, A.W.; Vessella, R.L.; Isaacs, W.B.; et al. Ligand-independent androgen receptor variants derived from splicing of cryptic exons signify hormone-refractory prostate cancer. *Cancer Res.* **2009**, *69*, 16–22. [CrossRef]
49. Antonarakis, E.S.; Lu, C.; Wang, H.; Luber, B.; Nakazawa, M.; Roeser, J.C.; Chen, Y.; Mohammad, T.A.; Chen, Y.; Fedor, H.L.; et al. AR-V7 and resistance to enzalutamide and abiraterone in prostate cancer. *N. Engl. J. Med.* **2014**, *371*, 1028–1038. [CrossRef]
50. Onstenk, W.; Sieuwerts, A.M.; Kraan, J.; Van, M.; Nieuweboer, A.J.; Mathijssen, R.H.; Hamberg, P.; Meulenbeld, H.J.; de Laere, B.; Dirix, L.Y.; et al. Efficacy of Cabazitaxel in Castration-resistant Prostate Cancer Is Independent of the Presence of AR-V7 in Circulating Tumor Cells. *Eur. Urol.* **2015**, *68*, 939–945. [CrossRef]
51. Antonarakis, E.S.; Lu, C.; Luber, B.; Wang, H.; Chen, Y.; Nakazawa, M.; Nadal, R.; Paller, C.J.; Denmeade, S.R.; Carducci, M.A.; et al. Androgen Receptor Splice Variant 7 and Efficacy of Taxane Chemotherapy in Patients with Metastatic Castration-Resistant Prostate Cancer. *JAMA Oncol.* **2015**, *1*, 582–591. [CrossRef]
52. Armstrong, A.J.; Halabi, S.; Luo, J.; Nanus, D.M.; Giannakakou, P.; Szmulewitz, R.Z.; Danila, D.C.; Healy, P.; Anand, M.; Rothwell, C.J.; et al. Prospective Multicenter Validation of Androgen Receptor Splice Variant 7 and Hormone Therapy Resistance in High-Risk Castration-Resistant Prostate Cancer: The PROPHECY Study. *J. Clin. Oncol.* **2019**, *37*, 1120–1129. [CrossRef] [PubMed]
53. Antonarakis, E.S.; Tagawa, S.T.; Galletti, G.; Worroll, D.; Ballman, K.; Vanhuyse, M.; Sonpavde, G.; North, S.; Albany, C.; Tsao, C.K.; et al. Randomized, noncomparative, phase II trial of early switch from docetaxel to cabazitaxel or vice versa, with integrated biomarker analysis, in men with chemotherapy-naive, metastatic, castration-resistant prostate cancer. *J. Clin. Oncol.* **2017**, *35*, 3181–3188. [CrossRef] [PubMed]
54. Kirby, B.J.; Jodari, M.; Loftus, M.S.; Gakhar, G.; Pratt, E.D.; Chanel-Vos, C.; Gleghorn, J.P.; Santana, S.M.; Liu, H.; Smith, J.P.; et al. Functional characterization of circulating tumor cells with a prostate-cancer-specific microfluidic device. *PLoS ONE* **2012**, *7*, e35976. [CrossRef]
55. Scher, H.I.; Lu, D.; Schreiber, N.A.; Louw, J.; Graf, R.P.; Vargas, H.A.; Johnson, A.; Jendrisak, A.; Bambury, R.; Danila, D.; et al. Association of AR-V7 on circulating tumor cells as a treatment-specific biomarker with outcomes and survival in castration-resistant prostate cancer. *JAMA Oncol.* **2016**, *2*, 1441–1449. [CrossRef] [PubMed]
56. Scher, H.I.; Graf, R.P.; Schreiber, N.A.; McLaughlin, B.; Jendrisak, A.; Wang, Y.; Lee, J.; Greene, S.; Krupa, R.; Lu, D.; et al. Phenotypic heterogeneity of circulating tumor cells informs clinical decisions between AR signaling inhibitors and taxanes in metastatic prostate cancer. *Cancer Res.* **2017**, *77*, 5687–5698. [CrossRef] [PubMed]
57. Miyamoto, D.T.; Lee, R.J.; Kalinich, M.; LiCausi, J.A.; Zheng, Y.; Chen, T.; Milner, J.D.; Emmons, E.; Ho, U.; Broderick, K.; et al. An RNA-based digital circulating tumor cell signature is predictive of drug response and early dissemination in prostate cancer. *Cancer Discov.* **2018**, *8*, 288–303. [CrossRef] [PubMed]
58. Ferlay, J.; Soerjomataram, I.; Dikshit, R.; Eser, S.; Mathers, C.; Rebelo, M.; Parkin, D.M.; Forman, D.; Bray, F. Cancer incidence and mortality worldwide: Sources, methods and major patterns in GLOBOCAN 2012. *Int. J. Cancer* **2015**, *136*, E359–E386. [CrossRef]

59. Ghoncheh, M.; Pournamdar, Z.; Salehiniya, H. Incidence and Mortality and Epidemiology of Breast Cancer in the World. *Asian Pac. J. Cancer Prev.* **2016**, *17*, 43–46. [CrossRef]
60. Tarhan, M.O.; Gonel, A.; Kucukzeybek, Y.; Erten, C.; Cuhadar, S.; Yigit, S.C.; Atay, A.; Somali, I.; Dirican, A.; Demir, L.; et al. Prognostic significance of circulating tumor cells and serum CA15-3 levels in metastatic breast cancer, single center experience, preliminary results. *Asian Pac. J. Cancer Prev.* **2013**, *14*, 1725–1729. [CrossRef]
61. Wallwiener, M.; Hartkopf, A.D.; Baccelli, I.; Riethdorf, S.; Schott, S.; Pantel, K.; Marmé, F.; Sohn, C.; Trumpp, A.; Rack, B.; et al. The prognostic impact of circulating tumor cells in subtypes of metastatic breast cancer. *Breast Cancer Res. Treat.* **2013**, *137*, 503–510. [CrossRef]
62. Rack, B.; Schindlbeck, C.; Jückstock, J.; Andergassen, U.; Hepp, P.; Zwingers, T.; Friedl, T.W.; Lorenz, R.; Tesch, H.; Fasching, P.A.; et al. Circulating tumor cells predict survival in early average-to-high risk breast cancer patients. *J. Natl. Cancer Inst.* **2014**, *106*, dju066. [CrossRef] [PubMed]
63. Georgoulias, V.; Apostolaki, S.; Bozionelou, V.; Politaki, E.; Perraki, M.; Georgoulia, N.; Kalbakis, K.; Kotsakis, A.; Xyrafas, A.; Agelaki, S.; et al. Effect of front-line chemotherapy on circulating CK-19 mRNA-positive cells in patients with metastatic breast cancer. *Cancer Chemother. Pharmacol.* **2014**, *74*, 1217–1225. [CrossRef] [PubMed]
64. Bian, L.; Liu, Y.; Wang, T.; Zhang, S.; Shao, Z.; Tong, Z.; Song, E.; Wang, X.; Liao, N.; Jiang, Z. Prediction value for dynamic changes of circulating tumor cell in therapeutic response and prognosis of Chinese metastatic breast cancer patients. *Zhonghua Yi Xue Za Zhi* **2014**, *94*, 265–268. [PubMed]
65. Rugo, H.S.; Chien, A.J.; Franco, S.X.; Stopeck, A.T.; Glencer, A.; Lahiri, S.; Arbushites, M.C.; Scott, J.; Park, J.W.; Hudis, C.; et al. A phase II study of lapatinib and bevacizumab as treatment for HER2-overexpressing metastatic breast cancer. *Breast Cancer Res. Treat.* **2012**, *134*, 13–20. [CrossRef]
66. Pierga, J.Y.; Hajage, D.; Bachelot, T.; Delaloge, S.; Brain, E.; Campone, M.; Diéras, V.; Rolland, E.; Mignot, L.; Mathiot, C.; et al. High independent prognostic and predictive value of circulating tumor cells compared with serum tumor markers in a large prospective trial in first-line chemotherapy for metastatic breast cancer patients. *Ann. Oncol.* **2012**, *23*, 618–624. [CrossRef]
67. Helissey, C.; Berger, F.; Cottu, P.; Diéras, V.; Mignot, L.; Servois, V.; Bouleuc, C.; Asselain, B.; Pelissier, S.; Vaucher, I.; et al. Circulating tumor cell thresholds and survival scores in advanced metastatic breast cancer: The observational step of the CirCe01 phase III trial. *Cancer Lett.* **2015**, *360*, 213–218. [CrossRef]
68. Smerage, J.B.; Barlow, W.E.; Hortobagyi, G.N.; Winer, E.P.; Leyland-Jones, B.; Srkalovic, G.; Tejwani, S.; Schott, A.F.; O'Rourke, M.A.; Lew, D.L.; et al. Circulating tumor cells and response to chemotherapy in metastatic breast cancer: SWOG S0500. *J. Clin. Oncol.* **2014**, *32*, 3483–3489. [CrossRef]
69. Jiang, Z.F.; Cristofanilli, M.; Shao, Z.M.; Tong, Z.S.; Song, E.W.; Wang, X.J.; Liao, N.; Hu, X.C.; Liu, Y.; Wang, Y.; et al. Circulating tumor cells predict progression-free and overall survival in Chinese patients with metastatic breast cancer, HER2-positive or triple-negative (CBCSG004): A multicenter, double-blind, prospective trial. *Ann. Oncol.* **2013**, *24*, 2766–2772. [CrossRef]
70. Trapp, E.; Janni, W.; Schindlbeck, C.; Jückstock, J.; Andergassen, U.; de Gregorio, A.; Alunni-Fabbroni, M.; Tzschaschel, M.; Polasik, A.; Koch, J.G.; et al. Presence of circulating tumor cells in high-risk early breast cancer during follow-up and prognosis. *J. Natl. Cancer Inst.* **2019**, *111*, 380–387. [CrossRef]
71. Symonds, L.; Linden, H.; Gadi, V.; Korde, L.; Rodler, E.; Gralow, J.; Redman, M.; Baker, K.; Wu, Q.V.; Jenkins, I.; et al. Combined Targeted Therapies for First-line Treatment of Metastatic Triple Negative Breast Cancer-A Phase II Trial of Weekly Nab-Paclitaxel and Bevacizumab Followed by Maintenance Targeted Therapy with Bevacizumab and Erlotinib. *Clin. Breast Cancer* **2019**, *19*, e283–e296. [CrossRef]
72. Manso, L.; Antón, F.M.; Perón, Y.I.; Mingorance, J.I.D.; García, P.B.; González, M.J.E.; Martínez-Jañez, N.; López-González, A.; Garate, C.O.; García, A.B.; et al. Safety of eribulin as third-line chemotherapy in HER2-negative, advanced breast cancer pre-treated with taxanes and anthracycline: OnSITE study. *Breast J.* **2019**, *25*, 219–225. [CrossRef] [PubMed]
73. Liang, S.; Niu, L.; Xu, K.; Wang, X.; Liang, Y.; Zhang, M.; Chen, J.; Lin, M. Tumor cryoablation in combination with natural killer cells therapy and Herceptin in patients with HER2-overexpressing recurrent breast cancer. *Mol. Immunol.* **2017**, *92*, 45–53. [CrossRef] [PubMed]

74. Larsson, A.M.; Jansson, S.; Bendahl, P.O.; Jörgensen, C.L.T.; Loman, N.; Graffman, C.; Lundgren, L.; Aaltonen, K.; Rydén, L. Longitudinal enumeration and cluster evaluation of circulating tumor cells improve prognostication for patients with newly diagnosed metastatic breast cancer in a prospective observational trial. *Breast Cancer Res.* **2018**, *20*, 48. [CrossRef] [PubMed]
75. Hall, C.S.; Karhade, M.G.; Bauldry, J.B.B.; Valad, L.M.; Kuerer, H.M.; DeSnyder, S.M.; Lucci, A. Prognostic value of circulating tumor cells identified before surgical resection in nonmetastatic breast cancer patients. *J. Am. Coll. Surg.* **2016**, *223*, 20–29. [CrossRef] [PubMed]
76. Goodman, C.R.; Seagle, B.L.; Friedl, T.W.P.; Rack, B.; Lato, K.; Fink, V.; Cristofanilli, M.; Donnelly, E.D.; Janni, W.; Shahabi, S.; et al. Association of circulating tumor cell status with benefit of radiotherapy and survival in early-stage breast cancer. *JAMA Oncol.* **2018**, *4*, e180163. [CrossRef] [PubMed]
77. Schramm, A.; Schochter, F.; Friedl, T.W.P.; de Gregorio, N.; Andergassen, U.; Alunni-Fabbroni, M.; Trapp, E.; Jaeger, B.; Heinrich, G.; Camara, O.; et al. Prevalence of circulating tumor cells after adjuvant chemotherapy with or without anthracyclines in patients with HER2-negative, hormone receptor-positive early breast cancer. *Clin. Breast Cancer* **2017**, *17*, 279–285. [CrossRef] [PubMed]
78. Jueckstock, J.; Rack, B.; Friedl, T.W.; Scholz, C.; Steidl, J.; Trapp, E.; Tesch, H.; Forstbauer, H.; Lorenz, R.; Rezai, M.; et al. Detection of circulating tumor cells using manually performed immunocytochemistry (MICC) does not correlate with outcome in patients with early breast cancer-Results of the German SUCCESS-A-trial. *BMC Cancer* **2016**, *16*, 401. [CrossRef]
79. Hepp, P.; Andergassen, U.; Jäger, B.; Trapp, E.; Alunni-Fabbroni, M.; Friedl, T.W.; Hecker, N.; Lorenz, R.; Fasching, P.; Schneeweiss, A.; et al. Association of CA27.29 and Circulating Tumor Cells Before and at Different Times After Adjuvant Chemotherapy in Patients with Early-stage Breast Cancer-The SUCCESS Trial. *Anticancer Res.* **2016**, *36*, 4771–4776. [CrossRef]
80. Lelièvre, L.; Clézardin, P.; Magaud, L.; Roche, L.; Tubiana-Mathieu, N.; Tigaud, J.D.; Topart, D.; Raban, N.; Mouret-Reynier, M.A.; Mathevet, P. Comparative study of neoadjuvant chemotherapy with and without zometa for management of locally advanced breast cancer with serum VEGF as primary endpoint: The NEOZOL study. *Clin. Breast Cancer* **2018**, *18*, e1311–e1321. [CrossRef]
81. Tokudome, N.; Ito, Y.; Takahashi, S.; Kobayashi, K.; Taira, S.; Tsutsumi, C.; Oto, M.; Oba, M.; Inoue, K.; Kuwayama, A.; et al. Detection of circulating tumor cells in peripheral blood of heavily treated metastatic breast cancer patients. *Breast Cancer* **2011**, *18*, 195–202. [CrossRef]
82. Gonzalez-Angulo, A.M.; Lei, X.; Alvarez, R.H.; Green, M.C.; Murray, J.L.; Valero, V.; Koenig, K.B.; Ibrahim, N.K.; Litton, J.K.; Nair, L.; et al. Phase II randomized study of ixabepilone versus observation in patients with significant residual disease after neoadjuvant systemic therapy for HER2-negative breast cancer. *Clin. Breast Cancer* **2015**, *15*, 325–331. [CrossRef] [PubMed]
83. Agelaki, S.; Kalykaki, A.; Markomanolaki, H.; Papadaki, M.A.; Kallergi, G.; Hatzidaki, D.; Kalbakis, K.; Mavroudis, D.; Georgoulias, V. Efficacy of lapatinib in therapy-resistant HER2-positive circulating tumor cells in metastatic breast cancer. *PLoS ONE* **2015**, *10*, e0123683. [CrossRef] [PubMed]
84. Paoletti, C.; Schiavon, G.; Dolce, E.M.; Darga, E.P.; Carr, T.H.; Geradts, J.; Hoch, M.; Klinowska, T.; Lindemann, J.; Marshall, G.; et al. Circulating biomarkers and resistance to endocrine therapy in metastatic breast cancers: Correlative results from AZD9496 oral SERD phase I trial. *Clin. Cancer Res.* **2018**, *24*, 5860–5872. [CrossRef] [PubMed]
85. Pierga, J.Y.; Bidard, F.C.; Cropet, C.; Tresca, P.; Dalenc, F.; Romieu, G.; Campone, M.; Aït-Oukhatar, C.M.; le Rhun, E.; Gonçalves, A.; et al. Circulating tumor cells and brain metastasis outcome in patients with HER2-positive breast cancer: The LANDSCAPE trial. *Ann. Oncol.* **2013**, *24*, 2999–3004. [CrossRef]
86. Babayan, A.; Hannemann, J.; Spötter, J.; Müller, V.; Pantel, K.; Joosse, S.A. Heterogeneity of estrogen receptor expression in circulating tumor cells from metastatic breast cancer patients. *PLoS ONE* **2013**, *8*, e75038. [CrossRef]
87. Somlo, G.; Lau, S.K.; Frankel, P.; Hsieh, H.B.; Liu, X.; Yang, L.; Krivacic, R.; Bruce, R.H. Multiple biomarker expression on circulating tumor cells in comparison to tumor tissues from primary and metastatic sites in patients with locally advanced/inflammatory, and stage IV breast cancer, using a novel detection technology. *Breast Cancer Res. Treat.* **2011**, *128*, 155–163. [CrossRef]
88. Reis-Filho, J.S.; Pusztai, L. Gene expression profiling in breast cancer: Classification, prognostication, and prediction. *Lancet* **2011**, *378*, 1812–1823. [CrossRef]

89. Eroles, P.; Bosch, A.; Perez-Fidalgo, J.A.; Lluch, A. Molecular biology in breast cancer: Intrinsic subtypes and signaling pathways. *Cancer Treat. Rev.* **2012**, *38*, 698–707. [CrossRef]
90. Kapp, A.V.; Jeffrey, S.S.; Langerod, A.; Borresen-Dale, A.L.; Han, W.; Noh, D.Y.; Bukholm, I.R.; Nicolau, M.; Brown, P.O.; Tibshirani, R. Discovery and validation of breast cancer subtypes. *BMC Genom.* **2006**, *7*, 231. [CrossRef]
91. Crown, J.; O'Shaughnessy, J.; Gullo, G. Emerging targeted therapies in triple-negative breast cancer. *Ann. Oncol.* **2012**, *23* (Suppl. 6), vi56–vi65. [CrossRef]
92. Jaeger, B.A.S.; Neugebauer, J.; Andergassen, U.; Melcher, C.; Schochter, F.; Mouarrawy, D.; Ziemendorff, G.; Clemens, M.; Abel, E.V.; Heinrich, G.; et al. The HER2 phenotype of circulating tumor cells in HER2-positive early breast cancer: A translational research project of a prospective randomized phase III trial. *PLoS ONE* **2017**, *12*, e0173593. [CrossRef] [PubMed]
93. Fehm, T.; Müller, V.; Aktas, B.; Janni, W.; Schneeweiss, A.; Stickeler, E.; Lattrich, C.; Löhberg, C.R.; Solomayer, E.; Rack, B.; et al. HER2 status of circulating tumor cells in patients with metastatic breast cancer: A prospective, multicenter trial. *Breast Cancer Res. Treat.* **2010**, *124*, 403–412. [CrossRef] [PubMed]
94. Ligthart, S.T.; Bidard, F.C.; Decraene, C.; Bachelot, T.; Delaloge, S.; Brain, E.; Campone, M.; Viens, P.; Pierga, J.Y.; Terstappen, L.W. Unbiased quantitative assessment of Her-2 expression of circulating tumor cells in patients with metastatic and non-metastatic breast cancer. *Ann. Oncol.* **2013**, *24*, 1231–1238. [CrossRef] [PubMed]
95. Ignatiadis, M.; Litière, S.; Rothe, F.; Riethdorf, S.; Proudhon, C.; Fehm, T.; Aalders, K.; Forstbauer, H.; Fasching, P.A.; Brain, E.; et al. Trastuzumab versus observation for HER2 nonamplified early breast cancer with circulating tumor cells (EORTC 90091-10093, BIG 1-12, Treat CTC): A randomized phase II trial. *Ann. Oncol.* **2018**, *29*, 1777–1783. [CrossRef]
96. Azim, H.A.; Rothé, F., Jr.; Aura, C.M.; Bavington, M.; Maetens, M.; Rouas, G.; Gebhart, G.; Gamez, C.; Eidtmann, H.; Baselga, J.; et al. Circulating tumor cells and response to neoadjuvant paclitaxel and HER2-targeted therapy: A sub-study from the NeoALTTO phase III trial. *Breast* **2013**, *22*, 1060–1065. [CrossRef]
97. Pierga, J.Y.; Petit, T.; Lévy, C.; Ferrero, J.M.; Campone, M.; Gligorov, J.; Lerebours, F.; Roché, H.; Bachelot, T.; Charafe-Jauffret, E.; et al. Pathological response and circulating tumor cell count identifies treated HER2+ inflammatory breast cancer patients with excellent prognosis: BEVERLY-2 survival data. *Clin. Cancer Res.* **2015**, *21*, 1298–1304. [CrossRef]
98. Konig, A.; Vilsmaier, T.; Rack, B.; Friese, K.; Janni, W.; Jeschke, U.; Andergassen, U.; Trapp, E.; Juckstock, J.; Jager, B.; et al. Determination of interleukin-4, -5, -6, -8 and -13 in serum of patients with breast cancer before treatment and its correlation to circulating tumor cells. *Anticancer Res.* **2016**, *36*, 3123–3130.
99. Vilsmaier, T.; Rack, B.; König, A.; Friese, K.; Janni, W.; Jeschke, U.; Weissenbacher, T. Influence of Circulating Tumour Cells on Production of IL-1α, IL-1β and IL-12 in Sera of Patients with Primary Diagnosis of Breast Cancer Before Treatment. *Anticancer Res.* **2016**, *36*, 5227–5236. [CrossRef]
100. Vilsmaier, T.; Rack, B.; Janni, W.; Jeschke, U.; Weissenbacher, T. Angiogenic cytokines and their influence on circulating tumour cells in sera of patients with the primary diagnosis of breast cancer before treatment. *BMC Cancer* **2016**, *16*, 547. [CrossRef]
101. Paoletti, C.; Li, Y.; Muñiz, M.C.; Kidwell, K.M.; Aung, K.; Thomas, D.G.; Brown, M.E.; Abramson, V.G.; Irvin, W.J., Jr.; Lin, N.U.; et al. Significance of circulating tumor cells in metastatic triple-negative breast cancer patients within a randomized, phase II trial: TBCRC 019. *Clin. Cancer Res.* **2015**, *21*, 2771–2779. [CrossRef]
102. Foroni, C.; Milan, M.; Strina, C.; Cappelletti, M.; Fumarola, C.; Bonelli, M.; Bertoni, R.; Ferrero, G.; Maldotti, M.; Takano, E.; et al. Pure anti-tumor effect of zoledronic acid in naïve bone-only metastatic and locally advanced breast cancer: Proof from the "biological window therapy". *Breast Cancer Res. Treat.* **2014**, *144*, 113–121. [CrossRef] [PubMed]
103. Smerage, J.B.; Budd, G.T.; Doyle, G.V.; Brown, M.; Paoletti, C.; Muniz, M.; Miller, M.C.; Repollet, M.I.; Chianese, D.A.; Connelly, M.C.; et al. Monitoring apoptosis and Bcl-2 on circulating tumor cells in patients with metastatic breast cancer. *Mol. Oncol.* **2013**, *7*, 680–692. [CrossRef] [PubMed]

104. Guan, X.; Ma, F.; Li, C.; Wu, S.; Hu, S.; Huang, J.; Sun, X.; Wang, J.; Luo, Y.; Cai, R.; et al. The prognostic and therapeutic implications of circulating tumor cell phenotype detection based on epithelial-mesenchymal transition markers in the first-line chemotherapy of HER2-negative metastatic breast cancer. *Cancer Commun.* **2019**, *39*, 1–10. [CrossRef] [PubMed]
105. Barriere, G.; Riouallon, A.; Renaudie, J.; Tartary, M.; Rigaud, M. Mesenchymal characterization: Alternative to simple CTC detection in two clinical trials. *Anticancer Res.* **2012**, *32*, 3363–3369.
106. Horimoto, Y.; Tokuda, E.; Murakami, F.; Uomori, T.; Himuro, T.; Nakai, K.; Orihata, G.; Iijima, K.; Togo, S.; Shimizu, H.; et al. Analysis of circulating tumour cell and the epithelial mesenchymal transition (EMT) status during eribulin-based treatment in 22 patients with metastatic breast cancer: A pilot study. *J. Transl. Med.* **2018**, *16*, 287. [CrossRef]
107. Onstenk, W.; de Klaver, W.; de Wit, R.; Lolkema, M.; Foekens, J.; Sleijfer, S. The use of circulating tumor cells in guiding treatment decisions for patients with metastatic castration-resistant prostate cancer. *Cancer Treat. Rev.* **2016**, *46*, 42–50. [CrossRef]
108. Scheel, C.; Weinberg, R.A. Cancer stem cells and epithelial-mesenchymal transition: Concepts and molecular links. *Semin. Cancer Biol.* **2012**, *22*, 396–403. [CrossRef]
109. Paoletti, C.; Muniz, M.C.; Thomas, D.G.; Griffith, K.A.; Kidwell, K.M.; Tokudome, N.; Brown, M.E.; Aung, K.; Miller, M.C.; Blossom, D.L.; et al. Development of circulating tumor cell-endocrine therapy index in patients with hormone receptor-positive breast cancer. *Clin. Cancer Res.* **2015**, *21*, 2487–2498. [CrossRef]
110. Hashemi Goradel, N.; Najafi, M.; Salehi, E.; Farhood, B.; Mortezaee, K. Cyclooxygenase-2 in cancer: A review. *J. Cell Physiol.* **2019**, *234*, 5683–5699. [CrossRef]
111. Alix-Panabieres, C. The future of liquid biopsy. *Nature* **2020**, *579*, S9. [CrossRef]
112. Cabel, L.; Proudhon, C.; Gortais, H.; Loirat, D.; Coussy, F.; Pierga, J.Y.; Bidard, F.C. Circulating tumor cells: Clinical validity and utility. *Int. J. Clin. Oncol.* **2017**, *22*, 421–430. [CrossRef] [PubMed]

Publisher's Note: MDPI stays neutral with regard to jurisdictional claims in published maps and institutional affiliations.

© 2020 by the authors. Licensee MDPI, Basel, Switzerland. This article is an open access article distributed under the terms and conditions of the Creative Commons Attribution (CC BY) license (http://creativecommons.org/licenses/by/4.0/).

Article

Liquid Biopsy-Based Exo-oncomiRNAs Can Predict Prostate Cancer Aggressiveness

Xavier Ruiz-Plazas [1,2,†], Antonio Altuna-Coy [1,†], Marta Alves-Santiago [1,2], José Vila-Barja [2,‡], Joan Francesc García-Fontgivell [1,3], Salomé Martínez-González [3], José Segarra-Tomás [1,2,*,§] and Matilde R. Chacón [1,*,§]

[1] Disease Biomarkers and Molecular Mechanisms Group, IISPV, Joan XXIII University Hospital, Universitat Rovira i Virgili, 43007 Tarragona, Spain; xarupl@gmail.com (X.R.-P.); antonio.altuna@iispv.cat (A.A.-C.); martalves@hotmail.es (M.A.-S.); fontgi@yahoo.es (J.F.G.-F.)
[2] Urology Unit, Joan XXIII University Hospital, 43007 Tarragona, Spain; jose.vila@urv.cat
[3] Pathology Unit, Joan XXIII University Hospital, 43007 Tarragona, Spain; mgonzalez.hj23.ics@gencat.cat
* Correspondence: jsegarra.hj23.ics@gencat.cat (J.S.-T.); mrch2424@gmail.com (M.R.C.); Tel.: +34-977295500 (ext. 3406) (J.S.-T. & M.R.C.)
† These authors contributed equally to this work.
‡ This study is dedicated to the memory of José Vila-Barja, who passed away on 20 February 2020.
§ These authors are joint senior authors on this work.

Simple Summary: The main problem encountered in the management of prostate cancer (PCa) is the inability to distinguish slow-growing indolent tumors from aggressive tumors. It is therefore important to explore non-invasive assays for the early detection of this aggressive subtype, when it can still be treated effectively. The presence of the TWEAK cytokine in biofluids of the PCa microenvironment might drive the secretion of extracellular vesicles (EVs) containing exo-oncomicroRNAs capable of modifying the tumor microenvironment. These exo-oncomicroRNAs are potentially useful as PCa biomarkers. We identified 2 exo-oncomiRNAs isolated from semen EVs by the action of TWEAK in the tumor microenvironment and, we determined their usefulness as biomarkers of PCa prognostic. We also established, for the first time, that TWEAK modulates potential exo-oncomiRNA targets, both tightly linked to cancer progression. In conclusion, our study shows that semen detection of TWEAK-regulated exo-oncomiRNAs can improve PCa prognosis, opening new avenues for diagnosis and treatment.

Abstract: Liquid biopsy-based biomarkers, including microRNAs packaged within extracellular vesicles, are promising tools for patient management. The cytokine tumor necrosis factor-like weak inducer of apoptosis (TWEAK) is related to PCa progression and is found in the semen of patients with PCa. TWEAK can induce the transfer of exo-oncomiRNAs from tumor cells to body fluids, and this process might have utility in non-invasive PCa prognosis. We investigated TWEAK-regulated exo-microRNAs in semen and in post-digital rectal examination urine from patients with different degrees of PCa aggressiveness. We first identified 14 exo-oncomiRNAs regulated by TWEAK in PCa cells in vitro, and subsequently validated those using liquid biopsies from 97 patients with PCa. Exo-oncomiR-221-3p, -222-3p and -31-5p were significantly higher in the semen of high-risk patients than in low-risk peers, whereas exo-oncomiR-193-3p and -423-5p were significantly lower in paired samples of post-digital rectal examination urine. A panel of semen biomarkers comprising exo-oncomiR-221-3p, -222-3p and TWEAK was designed that could correctly classify 87.5% of patients with aggressive PCa, with 85.7% specificity and 76.9% sensitivity with an area under the curve of 0.857. We additionally found that TWEAK modulated two exo-oncomiR-221-3p targets, *TCF12* and *NLK*. Overall, we show that liquid biopsy detection of TWEAK-regulated exo-oncomiRNAs can improve PCa prognosis prediction.

Keywords: exosomes; prostate cancer; exo-oncomiRNAS; TWEAK; semen

Citation: Ruiz-Plazas, X.; Altuna-Coy, A.; Alves-Santiago, M.; Vila-Barja, J.; García-Fontgivell, J.F.; Martínez-González, S.; Segarra-Tomás, J.; Chacón, M.R. Liquid Biopsy-Based Exo-oncomiRNAs Can Predict Prostate Cancer Aggressiveness. *Cancers* **2021**, *13*, 250. https://doi.org/10.3390/cancers 13020250

Received: 18 November 2020
Accepted: 7 January 2021
Published: 11 January 2021

Publisher's Note: MDPI stays neutral with regard to jurisdictional claims in published maps and institutional affiliations.

Copyright: © 2021 by the authors. Licensee MDPI, Basel, Switzerland. This article is an open access article distributed under the terms and conditions of the Creative Commons Attribution (CC BY) license (https://creativecommons.org/licenses/by/4.0/).

1. Introduction

Prostate cancer (PCa) is the most commonly diagnosed cancer and the fifth leading cause of cancer-related death in men in the developed world [1]. The incidence and morbidity of PCa continues to increase, likely due to changes in eating habits and the aging of the population [2]. A major challenge in the management of PCa is the inability to distinguish slow-growing and indolent tumors from aggressive tumors, which can lead to under-treatment of patients with aggressive tumors and over-treatment of those with indolent tumors. The prostate-specific antigen (PSA) test together with the tumor-nodes-metastasis (TNM) stage and the Gleason score of prostate biopsy [3] are considered indisputable prognostic factors to guide treatment decision-making. Among them, only PSA is objective, making it the most extensively studied biomarker in PCa [4]. However, its lack of specificity for clinically significant tumors has led to a rise in the number of prostate biopsies performed, with a consequent increase in the diagnosis of insignificant tumors and over-treatment of patients. Accordingly, the establishment of predictive biomarkers that can distinguish between aggressive and indolent PCa would be highly valuable in clinical practice, and could reduce the risk of over-diagnosis/over-treatment. In the context of biomarker discovery, liquid biopsy has proved to be a promising non-invasive modality for cancer diagnosis and prognosis that enables the assessment of circulating molecules in biological fluids, including serum, urine and semen [5].

Inflammation predisposes to the development of cancer and promotes all stages of tumorigenesis [6]. Inflammatory molecules—including cytokines and growth factors—released by immune cells of the inflammatory tumor microenvironment can have a direct effect on pre-malignant and cancer cells by enhancing their proliferation and resistance to cell death and environmental stress, thereby directly promoting tumor growth and progression [6]. Tumor necrosis factor-like weak inducer of apoptosis (TWEAK) is an inflammatory cytokine that governs tumor growth by promoting inflammation and inducing angiogenesis [7], and is produced by several cells of the immune system (natural killer cells and macrophages, among others) [8]. TWEAK can typically be found as a membrane-anchored (mTWEAK) protein on the surface of cells, but it can also be released as a soluble form (sTWEAK) by proteolytic processing. Both forms function through binding to their bona fide receptor Fn14 [7], forming a receptor–ligand pair. The role of the TWEAK/Fn14 axis has been established in some solid cancers, including breast and brain cancer [7]. We have demonstrated that low serum levels of sTWEAK in head and neck cancer are related to low survival rates, a finding that we later confirmed in a large cohort of patients, overall pointing to sTWEAK as a robust non-invasive biomarker of this disease [9,10]. We have also established a non-invasive biomarker panel with high negative predictive value to classify PCa aggressiveness that included sTWEAK levels and Fn14 mRNA expression [11].

The release of extracellular vesicles (EVs) from cells is an active process and has been shown to be a mechanism of cell-to-cell communication [12]. Exosomes are small (nanometer-size) extracellular cargo vesicles that are secreted after the fusion of endosomes with a plasma membrane, and are released by all cell types including cancer cells [13,14]. Exosomes can induce functional changes to receiving cells in the premetastatic niche—a specialized tumor microenvironment—for instance, aiding PCa cells to overcome the low-androgen conditions in distant metastatic organs [15]. Exosome secretion has long been linked to inflammation [16] and several experimental models have been employed to characterize the role of EVs in the development and progression of inflammatory diseases. The presence of sTWEAK in PCa tumors can not only contribute to fuel tumor progression [17,18], but might also promote the secretion of EVs, which will likely have an impact on the premetastatic niche, favoring the process of migration and proliferation. This is the case for exosomes derived from TWEAK-stimulated macrophages in epithelial ovarian cancer, which have been demonstrated to be internalized by the cancer cells and inhibit cell metastasis [19].

Oncogenic shuttle miRNAs (exo-oncomiRNAs), which show long-term stability in circulation and other body fluids, have been identified in exosomes [20]. Liquid biopsy exo-oncomiRNAs are thus potentially informative diagnostic and/or prognostic biomarkers and might also be helpful in understanding how tumor cells transfer their oncogenic potential to the environment [21]. Several studies have demonstrated that exomiRNAs isolated from liquid biopsy might be useful for the diagnostic and risk classification of PCa [22,23]. In this context, the most consistently reported deregulated exomiRNAs identified as promising PCa diagnostic biomarkers in both urine and blood are miR-141, miR-375, miR-21 and Let-7 [24–27]. While miR-141 is also frequently identified to be useful for risk classification in serum [27], other exomiRNAs have been proposed as having prognostic potential such as the combination of miR-1290 and miR-375 in plasma [24] and miR-2909 in urine [28]. The literature is more scarce surrounding semen, and only miRNA-342-3p and miRNA-374b-5p have been proposed as candidates for prognosis, and miRNA-142-3p and miRNA-142-5p were described as having diagnostic potential [29]. Overall, more extensive cohort studies are needed (especially using semen) to validate the identified exomiRNAs.

Exo-oncomiRNAs can be useful tools for non-invasive diagnosis and therapy monitoring in cancer; therefore, in the present study we sought to investigate whether exo-oncomiRNAs are shuttled into biofluids by the action of sTWEAK in the tumor microenvironment, and to determine their usefulness as prognostic PCa biomarkers in two different liquid biopsies: semen and post-digital rectal examination urine. We also aimed to examine the downstream targets of exo-oncomiRNAs, which might be important for the control of PCa.

2. Results

2.1. Extracellular Vesicle-Derived Exo-oncomiRNAs Are Differentially Expressed in Liquid Biopsies from Patients with Prostate Cancer Based on the Degree of Cancer Aggressiveness

We sought to search for a useful and practicable biomarker panel capable of differentiating aggressive from non-aggressive forms of PCa in liquid biopsy-based exo-oncomiRNAs isolated from EVs and secreted under sTWEAK stimulating conditions. The search was divided into two phases: the initial phase was established to isolate the EV-cargo (exo-oncomiRNAs) secreted into cell culture medium of two PCa cell lines—PC-3 and LNCaP—treated or not with sTWEAK; in the second phase, we assayed for expression using a real-time PCR array of 752 miRNA target onco-miRNAs. We specifically chose an androgen-independent line (PC-3) and an androgen-dependent line (LNCaP). Although the two cell lines do not cover the entire spectrum of PCa, they allowed us to implement a first approach to identify possible exo-miRNAs expressed through the influence of TWEAK [30].

Isolated EVs were confirmed by transmission electron microscopy (TEM) and by Western blot analysis of selected EV markers in order to comply with the guidelines of the International Society of Extracellular Vesicles [31]. Results confirmed the presence of EVs within the expected range (30–100 nm), which were enriched for CD9, CD63 and CD81 markers (Figure 1). The detailed results of immunoblotting are shown in Figure S1.

By screening a 752-miRNA panel, the following 14 exo-oncomiRNAs were selected from the first phase study that were significantly altered after sTWEAK treatment, comparing either PC-3 or/and LNCaP cell lines, which accomplished the following criteria: cycle threshold (Ct) < 33 and at least >1.8-fold-over-expression when comparing both sTWEAK-stimulated cell lines (Table S1): miR-125b-1-3p, miR-193b-3p, miR-221-3p, miR-222-3p, miR-23a-3p, miR-27a-3p, miR-29a-3p, miR-31-5p, miR-497-5p, miR-643, miR-663b, miR-940, miR-9-5p and miR-99a-3p.

In the second phase of the experimental approach, we evaluated the expression levels of the 14 selected exo-oncomiRNAs in EVs isolated from liquid biopsy (semen and post-digital rectal examination urine) from 97 patients with low- or high-risk PCa. Pathological and clinical characteristics of patients are listed in Table 1. Gleason grade (GG) criteria and TNM classification was determined in accordance with the International Society of Urological Pathology (ISUP). Complementary examinations included prostate volume, measured by transrectal ultrasound, and PSA, as in standard clinical practice.

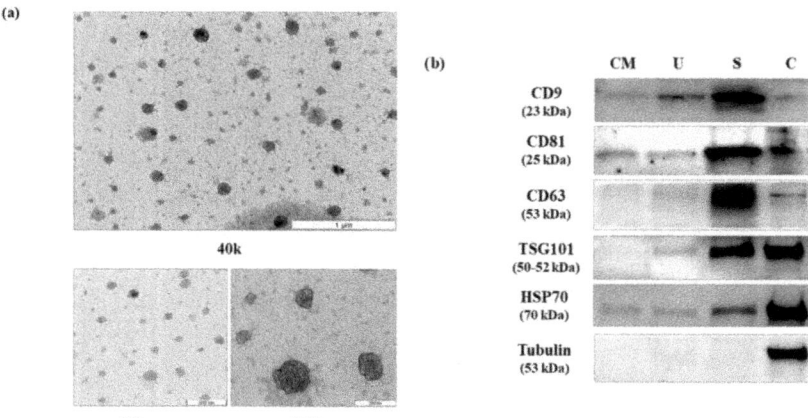

Figure 1. Characterization of isolated extracellular vesicles. (**a**) Analysis of extracellular vesicles (EVs) by electron microscopy at different magnification. (**b**) Western blot image of protein extracts prepared from EVs isolated from PC-3 culture media (CM), post-digital rectal examination urine (U), semen (S) and total cell extract from PC-3 (C), and tested with the following antibodies: CD9, CD81, CD63, TSG101, HSP70 and tubulin. Uncropped Western Blot image is available in Figure S1.

Analysis of the expression pattern of the 14 selected exo-oncomiRNAs in liquid biopsy of semen and post-digital rectal examination urine from patients with high-risk (ISUP Group III, IV and V) and low-risk (ISUP Group I and II) PCa revealed significant differences in the following five exo-oncomiRNAs: exo-oncomiR-221-3p, exo-oncomiR-222-3p, exo-oncomiR-31-5p, which were up-regulated in semen of high-risk patients versus low-risk patients; and exo-oncomiR-193-3p and exo-oncomiR-423-5p, which were down-regulated in post-digital rectal examination urine samples of high-risk patients (Figure 2). There were no significant differences between the studied groups for the remaining nine exo-oncomiRNAs (Table S2).

Table 1. Clinical and pathological characteristics of the studied cohort.

Patient's Characteristics	Mean ± SD	N
Age (years)	63.5 ± 6.35	97
Prostatic Volume (c.c)	47.49 ± 23.09	97
Testosterone (nmol/L)	14.37 ± 5.07	97
Total PSA (ng/mL)	9.57 ± 7.92	97
		N (%)
BMI (kg/m^2)	<25	25 (25.8)
	$25 \leq x \leq 29.99$	50 (51.5)
	≥ 30	19 (19.6)
Total PSA (ng/mL)		
	<4	8 (8.2)
	$4 \leq x < 10$	60 (61.9)
	≥ 10	29 (29.9)

Table 1. *Cont.*

Patient's Characteristics		Mean ± SD	N
ISUP-GG			
	Low Risk	Group I	32 (33.0)
		Group II	25 (25.8)
	High Risk	Group III	23 (23.7)
		Group IV	10 (10.3)
		Group V	7 (7.2)
T pathological stage			
		≤T2a	68 (70.1)
		T3,T4	29 (29.9)
N pathological stage			
		NX	57 (58.8)
		N0	34 (35.1)
		N1	6 (6.2)

Abbreviations: BMI, body mass index; ISUP-GG, International Society of Urological Pathology Gleason Grade groups based on the Gleason score as follows: (Gleason score ≤ 6—group I; 3 + 4 = 7 group II; 4 + 3 = 7 group III; 4 + 4 = 8—group IV; and 9–10—group V); PSA, prostate-specific antigen; T stage, Tumor category; N node, category. The bolded words differentiate the clinical and pathological characteristics from the rest of the table.

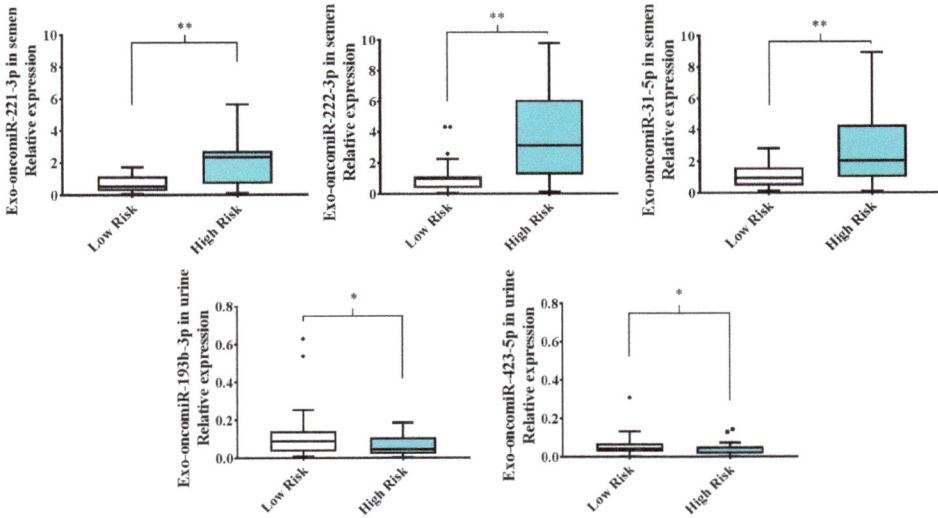

Figure 2. Exo-oncomiRNAs are differentially expressed in liquid biopsy from patients with prostate cancer. Box plots of relative expression of the 5 discriminatory exo-oncomiRNAs analyzed in semen and post-digital rectal examination urine liquid biopsies from patients with low- and high-risk PCa. Results are expressed as mean values ± SD. Statistical differences between groups are indicated: * $p < 0.05$; ** $p < 0.01$.

We also examined for clinical and metabolic differences between the high-risk and low-risk groups (Table 2). Univariate analysis showed that only total PSA was significantly higher in the high-risk group than in the low-risk group ($p = 0.007$), whereas sTWEAK semen levels were significantly lower in the high-risk group than in the low-risk group ($p = 0.009$) (Table 2), as has been reported [11]. We then tested for correlations between the five differentially expressed exo-oncomiRNAs and clinical and metabolic parameters using Spearman's bivariate correlation coefficient test. The most relevant associations observed were the significant negative correlations between semen sTWEAK levels and the expression levels of exo-oncomiR-221-3p, exo-oncomiR-222-3p, and exo-oncomiR31-5p (r = −0.375 p = 0.017, r = −0.387 p = 0.013 and r = −0.364 p = 0.021, respectively) (Figure S2).

Table 2. Anthropometric and analytical characteristics according to ISUP-GG criteria.

Patient's Stratification	ISUP GG Classification		
	Low-Risk	High-Risk	
	(Group I and II)	(Group III, IV and V)	
	N = 57	N = 40	
	Mean ± SD	Mean ± SD	p-Value
Anthropometric parameters			
Age (years)	62.46 ± 6.74	64.96 ± 5.52	0.066
BMI (kg/m^2)	27.97 ± 4.07	27.64 ± 3.46	0.718
Prostatic volume (c.c)	48.68 ± 24.56	45.81 ± 21	0.687
Glycemic profile			
Glucose (mmol/L)	5.82 ± 1.1	6.29 ± 2.26	0.388
Insulin (pmol/L)	89.36 ± 58.39	87.42 ± 47.09	0.841
HOMA-IR	3.46 ± 2.58	3.67 ± 2.71	0.841
HbA1c (%)	5.74 ± 0.64	5.92 ± 0.84	0.364
Lipid profile			
Cholesterol (mmol/L)	5.03 ± 1.06	5.03 ± 1.1	0.957
HDL cholesterol (mmol/L)	1.49 ± 0.73	1.42 ± 0.39	0.672
LDL cholesterol (mmol/L)	3.28 ± 1.3	3 ± 0.88	0.503
Triglycerides (mmol/L)	1.36 ± 0.74	1.55 ± 0.96	0.711
Hepatic profile			
AST (μkat/L)	0.39 ± 0.19	0.33 ± 0.07	0.171
ALT (μkat/L)	0.42 ± 0.22	0.36 ± 0.11	0.402
GGT (μkat/L)	0.7 ± 0.85	0.65 ± 0.48	0.887
Renal profile			
Uric acid (μmol/L)	368.05 ± 83.1	456.2 ± 529.55	0.376
Urea (mmol/L)	14.26 ± 3.23	14.8 ± 5.12	0.808
Creatinine (μmol/L)	85.83 ± 18.44	80.05 ± 13.97	0.072
Hormonal profile			
SHBG (nmol/L)	46.13 ± 52.46	40.02 ± 16.36	0.814
Testosterone (nmol/L)	14.93 ± 4.62	13.55 ± 5.63	0.101
Tumoral marker			
Total PSA (μg/L)	7.71 ± 4.8	12.24 ± 10.43	0.007
Biofluid Biomarker profile			
Semen cytokines (pg/mg of total protein)			
sTWEAK	989.62 ± 685.75	617.25 ± 447.57	0.009
Exo-oncomiRNAs in semen—Relative expression levels			
miR-221-3p	0.75 ± 0.6	2.17 ± 1.7	0.002
miR-222-3p	2.01 ± 2.79	3.79 ± 2.92	0.006
miR-31-5p	1.05 ± 0.73	2.75 ± 2.27	0.004
Exo-oncomiRNAs in urine—Relative expression levels			
miR-193-3p	0.12 ± 0.12	0.06 ± 0.05	0.037
miR-423-5p	0.05 ± 0.05	0.04 ± 0.03	0.034

BMI, body mass index; HOMA-IR, homeostatic model assessment of insulin resistance; HbA1c, Hemoglobin A1c; HDL, high-density lipoprotein; LDL, low-density lipoprotein; AST, aspartate aminotransferase; ALT, alanine aminotransferase; GGT, γ-Glutamyltransferase; SHBG, sex hormone-binding globulin; PSA, prostate-specific antigen; sTWEAK, soluble tumor necrosis factor-like weak inducer of apoptosis.

2.2. Semen Levels of Exo-oncomiR-221-3p May Help Identify an Aggressive Prostate Cancer Phenotype

We developed a partial least square-discriminant analysis (PLS-DA) model to evaluate the potential of the five selected exo-oncomiRNAs plus PSA in serum, sTWEAK in semen, age, prostatic volume, and testosterone, for the stratification of patients. Cross-validation analyses showed that a one-component model had an accuracy of 72.02% (R^2 = 0.2073 and Q^2 = 0.1493) (Figure S3a) indicating that is a good predictive model [32]. With regards to the importance of individual components, variable importance in projection (VIP) scores highlighted age, exo-oncomiR-222-3p in semen, exo-oncomiR-31-5p in semen, PSA in serum, sTWEAK in semen and exo-oncomiR-221-3p in semen as the most important variables (Figure 3). The VIP model estimated that exo-oncomiR-221-3p in semen and sTWEAK in semen had more influence than total PSA (Figure 3).

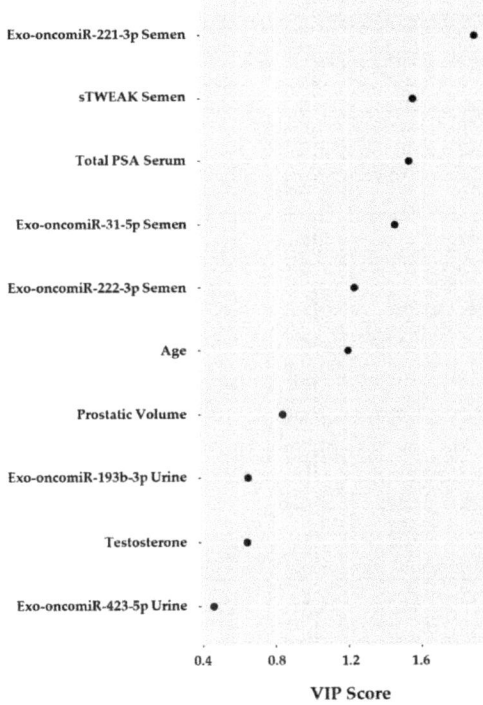

Figure 3. Variable importance in projection (VIP) scores. Selected variables: total PSA, testosterone, prostatic volume, age, sTWEAK in semen, exo-oncomiR-221-3p, exo-oncomiR222-3p, exo-oncomiR31-5p, exo-oncomiR-193-3p and exo-oncomiR-423-5p are shown in the model. Variables with scores close to or greater than 1 were considered important in the model.

Variables with VIP score $\geq$ 1 were considered important in the model for determining PCa aggressiveness. To evaluate the usefulness of exo-oncomiRNAs as potential prognosis biomarkers of PCa aggressiveness in liquid biopsy, we performed logistic regression and receiver operating characteristic (ROC) curve analysis combining the following variables: exo-oncomiR-221-3p, exo-oncomiR-222-3p, exo-oncomiR-31-5p, total PSA, sTWEAK levels and age; Table 3 lists the different combinations. Results showed that the area under the curve (AUC) of each individual variable was below 0.8. Thus, we used a multivariate regression model combining each potential biomarker to test which combination was more suitable for correct diagnosis. Notably, we observed that the presence of exo-oncomiR-221-

3p outperformed the other individual variables alone or in combination. The best panel in our study to distinguish PCa aggressiveness was that composed by exo-oncomiR-221-3p, exo-oncomiR-222-3p and semen sTWEAK, which could correctly classify 87.5% of patients, with an AUC of 0.857 and with 85.7% specificity and 76.9% sensitivity (Table 3) (Figure S3).

Table 3. Exo-oncomiRNAs-based models as diagnostic classifiers.

ROC Model	AUC	Error	p-Value	95% CI Lower	95% CI Upper	Sensivity (%)	Specificity (%)	% Correct Diagnosis
Age	0.610	0.058	0.066	0.496	0.724	85	75.4	62.9
Total PSA	0.662	0.056	0.007	0.552	0.772	85	31.6	63.9
sTWEAK	0.708	0.072	0.009	0.567	0.848	85.7	52.8	71.9
exo-oncomiR-221-3p	0.79	0.078	0.002	0.638	0.943	86.7	55.6	78.6
exo-oncomiR-222-3p	0.758	0.08	0.006	0.601	0.915	86.7	74.1	66.7
exo-oncomiR-31-5p	0.768	0.082	0.004	0.607	0.929	86.7	48.1	76.2
Total PSA + Age	0.704	0.054	0.001	0.597	0.810	85	70.2	67
Total PSA + sTWEAK	0.738	0.072	0.003	0.597	0.879	85.7	47.2	71.9
Total PSA + exo-oncomiR-221-3p	0.864	0.063	<0.001	0.74	0.998	86.7	55.6	83.3
Total PSA + exo-oncomiR-222-3p	0.78	0.071	0.003	0.641	0.919	86.7	55.6	73.8
Total PSA + exo-oncomiR-31-5p	0.832	0.07	<0.001	0.695	0.969	86.7	51.9	81
sTWEAK + Age	0.709	0.069	0.009	0.574	0.844	85.7	50	66.7
sTWEAK + exo-oncomiR-221-3p	0.841	0.073	<0.001	0.698	0.983	85.7	69.2	82.5
sTWEAK + exo-oncomiR-222-3p	0.745	0.086	0.012	0.576	0.913	85.7	42.3	70
sTWEAK + exo-oncomiR-31-5p	0.808	0.077	0.001	0.657	0.958	85.7	61.5	77.5
exo-oncomiR-221-3p + Age	0.802	0.077	0.001	0.651	0.954	86.7	33.3	76.2
exo-oncomiR-221-3p + exo-oncomiR-222-3p	0.802	0.078	0.001	0.65	0.955	86.7	63	76.2
exo-oncomiR-221-3p + exo-oncomiR-31-5p	0.8	0.079	0.001	0.646	0.954	86.7	55.6	81
exo-oncomiR-222-3p + Age	0.751	0.081	0.008	0.592	0.909	86.7	66.7	73.8
exo-oncomiR-222-3p + exo-oncomiR-31-5p	0.8	0.077	0.001	0.649	0.951	86.7	55.6	81
exo-oncomiR-31-5p + Age	0.778	0.078	0.003	0.625	0.930	86.7	44.4	73.8
Total PSA + sTWEAK + Age	0.746	0.067	0.002	0.614	0.878	85.7	44.4	73.7
Total PSA + sTWEAK + exo-oncomiR-221-3p	0.863	0.068	<0.001	0.73	0.996	85.7	69.2	85
Total PSA + sTWEAK + exo-oncomiR-222-3p	0.758	0.086	0.008	0.59	0.926	85.7	46.2	75
Total PSA + sTWEAK + exo-oncomiR-31-5p	0.824	0.076	0.001	0.675	0.974	85.7	73.1	77.5
Total PSA + exo-oncomiR-221-3p + Age	0.889	0.056	<0.001	0.780	0.998	85.7	37	83.3
Total PSA + exo-oncomiR-221-3p + exo-oncomiR-222-3p	0.872	0.06	<0.001	0.755	0.988	86.7	59.3	83.3
Total PSA + exo-oncomiR-221-3p + exo-oncomiR-31-5p	0.854	0.067	<0.001	0.724	0.985	86.7	51.9	83.3
Total PSA + exo-oncomiR-222-3p + Age	0.840	0.064	<0.001	0.714	0.965	86.7	37	83.3
Total PSA + exo-oncomiR-222-3p + exo-oncomiR-31-5p	0.849	0.069	<0.001	0.713	0.985	86.7	59.3	83.3
Total PSA + exo-oncomiR-31-5p + Age	0.862	0.061	<0.001	0.743	0.981	86.7	37	83.3
sTWEAK + exo-oncomiR-221-3p + Age	0.854	0.067	<0.001	0.723	0.986	85.7	23.7	77.5
sTWEAK + exo-oncomiR-221-3p + exo-oncomiR-222-3p	0.857	0.069	<0.001	0.721	0.993	85.7	76.9	87.5

Table 3. Cont.

ROC Model	AUC	Error	p-Value	95% CI Lower	95% CI Upper	Sensivity (%)	Specificity (%)	% Correct Diagnosis
sTWEAK + exo-oncomiR-221-3p + exo-oncomiR-31-5p	0.841	0.073	<0.001	0.698	0.983	85.7	69.2	82.5
sTWEAK + exo-oncomiR-222-3p + Age	0.764	0.078	0.006	0.611	0.917	85.7	50	72.5
sTWEAK + exo-oncomiR-222-3p + exo-oncomiR-31-5p	0.83	0.073	0.001	0.687	0.972	85.7	53.8	82.5
sTWEAK + exo-oncomiR-31-5p + Age	0.821	0.074	0.001	0.677	0.966	85.7	34.6	75
exo-oncomiR-221-3p + exo-oncomiR-222-3p + exo-oncomiR-31-5p	0.807	0.076	0.001	0.658	0.956	86.7	51.9	83.3
exo-oncomiR-221-3p + exo-oncomiR-222-3p + Age	0.820	0.074	0.001	0.675	0.965	86.7	25.9	76.2
exo-oncomiR-221-3p + exo-oncomiR-31-5p + Age	0.812	0.074	0.001	0.668	0.957	86.7	37	78.6
exo-oncomiR-222-3p + exo-oncomiR-31-5p + Age	0.802	0.075	0.001	0.655	0.950	86.7	44	78.6
Total PSA + sTWEAK + exo-oncomiR-221-3p + exo-oncomiR-222-3p	0.86	0.071	<0.001	0.721	0.999	85.7	69.2	85
Total PSA + sTWEAK + exo-oncomiR-221-3p + exo-oncomiR-31-5p	0.86	0.069	<0.001	0.724	0.995	85.7	69.2	85
Total PSA + sTWEAK + exo-oncomiR-222-3p + exo-oncomiR-31-5p	0.83	0.076	0.001	0.682	0.978	85.7	65.4	82.5
Age + Total PSA + sTWEAK + exo-oncomiR-221-3p	0.879	0.62	<0.001	0.757	1	85.7	23.1	85
Age + Total PSA + sTWEAK + exo-oncomiR-222-3p	0.808	0.074	0.001	0.662	0.953	85.7	50	82.5
Age + Total PSA + sTWEAK + exo-oncomiR-31-5p	0.849	0.069	<0.001	0.715	0.983	85.7	53.2	82.5
Age + Total PSA + exo-oncomiR-221-3p + exo-oncomiR-222-3p	0.894	0.053	<0.001	0.789	0.998	86.7	56.7	83.3
Age + Total PSA + exo-oncomiR-221-3p + exo-oncomiR-31-5p	0.879	0.059	<0.001	0.764	0.994	86.7	54.3	83.3
Age + Total PSA + exo-oncomiR-222-3p + exo-oncomiR-31-5p	0.867	0.059	<0.001	0.752	0.982	86.7	49.2	81
Age + sTWEAK + exo-oncomiR-221-3p + exo-oncomiR-222-3p	0.868	0.061	<0.001	0.748	0.988	86.7	46.5	80
Age + sTWEAK + exo-onxomiR-221-3p + exo-oncomiR-31-5p	0.857	0.067	<0.001	0.726	0.988	85.7	76.9	80
Age + sTWEAK + exo-oncomiR-222-3p + exo-oncomiR-31-5p	0.832	0.070	0.001	0.695	0.969	86.7	46.5	80
Age + exo-oncomiR-221-3p + exo-oncomiR-222-3p + exo-oncomiR-31-5p	0.820	0.072	0.001	0.678	0.962	86.7	48.1	81

Table 3. Cont.

ROC Model	AUC	Error	p-Value	95% CI		Sensivity (%)	Specificity (%)	% Correct Diagnosis
				Lower	Upper			
Total PSA + exo-oncomiR-221-3p + exo-oncomiR-222-3p + exo-oncomiR-31-5p	0.874	0.061	<0.001	0.754	0.995	86.7	55.6	83.3
sTWEAK + exo-oncomiR-221-3p + exo-oncomiR-222-3p + exo-oncomiR-31-5p	0.86	0.068	<0.001	0.726	0.994	85.7	73.1	85
Total PSA + sTWEAK + exo-oncomiR-221-3p + exo-oncomiR-222-3p + exo-oncomiR-31-5p	0.865	0.069	<0.001	0.73	1	85.7	69.2	85
Age + Total PSA + sTWEAK + exo-oncomiR-221-3p + exo-oncomiR-222-3p	0.879	0.062	<0.001	0.757	1	86.7	70.4	87.5
Age + Total PSA + sTWEAK + exo-oncomiR-221-3p + exo-oncomiR-31-5p	0.879	0.062	<0.001	0.758	1	86.7	57.9	85
Age + Total PSA + sTWEAK + exo-oncomiR-222-3p + exo-oncomiR-31-5p	0.857	0.065	<0.001	0.729	0.985	85.7	58.3	85
Age + Total PSA + exo-oncomiR-221-3p + exo-oncomiR-222-3p + exo-oncomiR-31-5p	0.896	0.053	<0.001	0.793	0.999	86.7	57.9	83.3
Age + sTWEAK + exo-oncomiR-221-3p + exo-oncomiR-222-3p + exo-oncomiR-31-5p	0.874	0.061	<0.001	0.755	0.992	85.7	63.8	82.5
Age + Total PSA + sTWEAK + exo-oncomiR-221-3p + exo-oncomiR-222-3p + exo-oncomiR-31-5p	0.879	0.062	<0.001	0.757	1	85.7	68.9	87.5

Receiver operating characteristic (ROC) curve values showing the predictive efficiency for distinguishing PCa aggressiveness. Percentage of correct diagnostic values was obtained by multivariate models (backward stepwise, conditional method). AUC, area under the curve; 95% CI (confidence interval). The bolded words represent the ROC Models.

2.3. TWEAK Modulates Potential Predicted Targets for oncomiR-221-3p

Several studies have shown that oncomiR-221 and oncomiR-222 are dysregulated in many cancers [33], including PCa [34,35], which is in line with our findings showing deregulated exo-oncomiR-221-3p and exo-oncomiR-222-3p in semen liquid biopsy of PCa. In vitro analysis showed that oncomiR-221-3p expression was found significantly up-regulated by sTWEAK only in PC-3 cells, both internally and in secreted EVs, and not in LNCaP cells, indicating that sTWEAK can potentially modulate oncomiR221-3p downstream targets (Figure 4a).

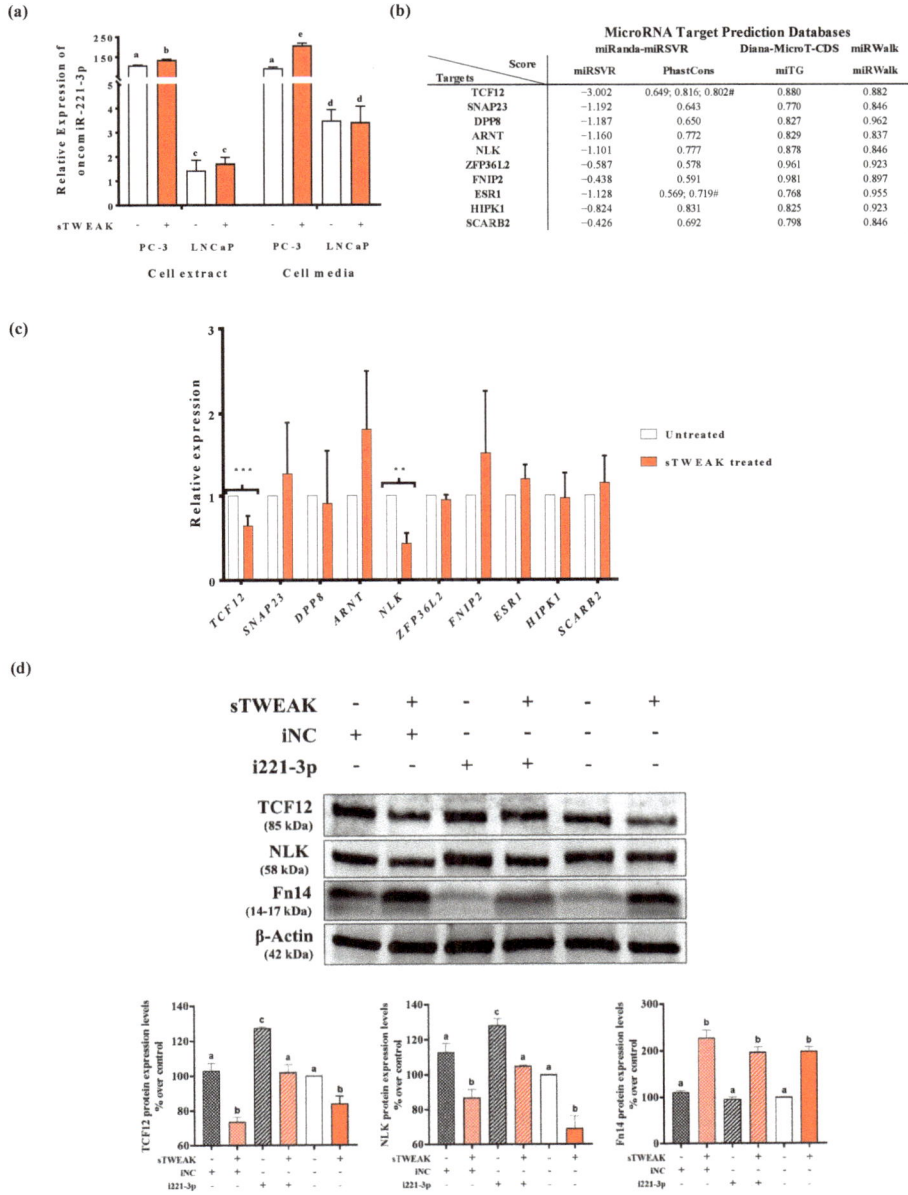

Figure 4. sTWEAK regulates oncomiR-221-3p expression and down-regulates NLK and TCF12 targets. (**a**) oncomiR-221-3p expression in PC-3 and LNCaP cell extracts and in extracellular vesicles (EVs) isolated from cell media. Different lettering over boxes indicates statistical differences. Significant differences are established at $p < 0.05$. Data are expressed as mean ± SEM (n = 4 experiments). (**b**) Selected targets for oncomiR-221-3p by 3 different target prediction algorithms. # conserved elements in multiply-aligned sequences. (**c**) qRT-PCR mRNA expression of selected oncomiR-221-3p targets in PC-3 cells before and after treatment with sTWEAK. Significant differences: ** $p < 0.01$; *** $p < 0.001$. Data are expressed as mean ± SEM (n = 6 experiments).

Transcription factor 12 (TCF12); synaptosome associated protein 23 (SNAP23); dipeptidyl peptidase 8 (DPP8); aryl hydrocarbon receptor nuclear translocator (ARNT); Nemo-like kinase (NLK); ZFP36 ring finger protein-like 2 (ZFP36L2); folliculin interacting protein 2 (FNIP2); estrogen receptor 1 (ESR1); homeodomain interacting protein kinase 1 (HIPK1); and scavenger receptor class B-ember 2 (SCARB2). (**d**) Expression of TCF12, NLK and Fn14 protein in PC-3 cells transfected with the oncomiR-221-3p inhibitor and further treated with sTWEAK. Representative Western blots are presented (top). The membranes were tested with the corresponding antibody. iNC: inhibitor negative control, i221-3p: inhibitor miR-221-3p; Nemo-like kinase (NLK); transcription factor 12 (TCF12); fibroblast growth factor 14 (Fn14). Relative protein expression levels are shown (bottom), which were normalized to the corresponding control β-actin. Different lettering over boxes indicates statistical differences. Significant differences are established at $p < 0.05$. Data are expressed as mean ± SEM ($n = 3$ experiments). Full-length blots and gels are presented in Figure S4.

To demonstrate a direct effect of sTWEAK on oncomiR-221-3p targets, we first searched for possible oncomiR-221-3p targets and selected only those shared by the miRanda, Diana-MicroT-CDS and miRWalk databases. We obtained 69 genes with putative target sites for oncomiR-221-3p in their 3′untranslated regions. We then ranked the candidate genes by the miRSVR score (the lower the score, the stronger the match to the seed region); if two or more targets had a similar miRSVR score we considered the higher score from the Diana-microT-CDS algorithm, miTG. With these criteria, we selected 10 possible targets implicated in cancer and/or inflammation (Figure 4b, Table S3), shared also by onco-miR222-3p, because both miRNAS are encoded in tandem and contain identical seed sequences separated by 727 bases [33] (Table S4). Of the 10 targets only NLK (Nemo-like kinase) and TCF12 (transcription factor 12) expression levels were found to be reduced in PC-3 cells after sTWEAK treatment for 24 h by real-time PCR (Figure 4c) and Western blotting (Figure 4d). The stimulatory effect of sTWEAK was accompanied by the increased expression of its receptor Fn14 (Figure 4d).

Finally, we performed in vitro experiments using PC-3 cells and an oncomiR-221-3p inhibitor, which consistently influenced the expression of its target genes as demonstrated by the reduced expression of TCF12 and NLK proteins when compared with non-treated counterparts (Figure 4d). As anticipated, combined sTWEAK stimulation and oncomiR-221-3p inhibition resulted in a significant down-regulation of NLK and TCF12 protein levels (Figure 4d). The detailed results of immunoblotting are shown in Figure S4.

3. Discussion

Histopathological biopsy analysis is a common method for the diagnosis of PCa. This procedure, however, only enables the analysis of part of the prostatic gland and, because of the typical multifocal nature of PCa, information from a single biopsy is often insufficient and does not reflect the dynamics of the tumor in the prostate.

Diagnosis of cancer through the use of liquid biopsy has proven to be particularly useful as a non-invasive method of diagnosis and disease progression monitoring [36]. In a similar line, exosomal miRNAs isolated in the context of cancer, termed exo-oncomiRNAs, are promising biomarkers in part due to their stability in body fluids and ease of detection and quantification at low cost. Additionally, exo-oncomiRNAs have a very important role in modulating several critical cancer processes, including proliferation, migration, and angiogenesis, through their regulation of important target genes within the tumor environment [37].

Some exo-oncomiRNAs (e.g., miR-375, miR -21 and miR-141 [24,38]) from biofluids including blood and urine are known to have diagnostic and prognostic capacity in PCa. However, inconsistencies in identified, dysregulated exo-oncomiRNA profiles have been reported, likely due to a lack of standardized exosomal isolation and miRNA quantification techniques [23]. Despite these challenges, exo-oncomiRNAs remain highly promising biomarker candidates to aid in PCa diagnosis and prognosis.

We previously established a non-invasive biomarker panel with high negative predictive value to classify PCa aggressiveness. Specifically, this biofluid signature comprised the following biomarkers: total PSA serum levels, semen levels of sTWEAK, fasting serum glycemia, and mRNA expression levels of *Fn14*, *KLK2* (a gene that encodes a protease that activates pre-PSA) and two chemokine receptors (*CXCR2* and *CCR3*) in semen cell sediment. This panel can identify PCa aggressiveness with 90.9% success [11]. Although this panel could aid the clinical prognosis of PCa by outperforming the classical clinical biomarkers (age, T-classification, and total PSA serum levels), it requires the measurement of seven different biomarkers and uses two different biological samples—serum and semen.

In the aforementioned study, we observed that in patients with high-risk PCa, the decrease in sTWEAK levels in semen was accompanied by an increase in *Fn14* mRNA levels in seminal cell sediment, pointing to an active process of ligand–receptor interaction that may favor cell proliferation and migration, as described in PCa cell models [17,18]. Accordingly, the presence of TWEAK in PCa tumors could not only fuel tumor progression, but might also promote the secretion of exo-oncomiRNAs contained within EVs, which will likely have an impact on the tumor microenvironment.

In the search for an improved prognostic panel for PCa focusing on TWEAK-induced exo-oncomiRNAs, we show here that five exo-oncomiRNAs (exo-oncomiR-221-3p, exo-oncomiR-222-3p, exo-oncomiR-31-5p, exo-oncomiR-193b-3p, exo-oncomiR-423-5p) are significantly dysregulated between low- and high-risk PCa. VIP analysis of selected variables (including age, exo-oncomiRNA levels in semen and urine and, several analytical parameters) showed that variables with VIP scores greater than 1, considered of importance in the model for determining PCa aggressiveness, included only the three exo-oncomiRNAs expressed in semen (exo-oncomiR-221-3p, exo-oncomiR-222-3p and exo-oncomiR-31-5p). This finding may not be causal. Because 25% of semen is derived from prostatic tissue [39], its contents are more likely to contain prostate disease-specific exo-oncomiRNAs [29] than post-digital rectal examination urine samples [40].

After testing several logistic regression models followed by ROC analysis including the 3 selected biomarkers (exo-oncomiR-221-3p, exo-oncomiR-222-3p and sTWEAK), the measurements in semen liquid biopsy had the best prognostic accuracy (AUC = 0.857, $p = 0.001$) when compared with the ROC curve analysis using only serum PSA levels (AUC = 0.662, $p < 0.007$). This new model can outperform the classical PSA biomarker by 23.6% for a correct diagnosis, improving the classification efficacy up to 87.5%. If we include the two selected exo-oncomiRNAs (exo-oncomiR-221-3p, exo-oncomiR-222-3p) plus PSA levels in serum, the model can predict PCa severity better than is commonly reported by PSA screening alone; however, the model composed of sTWEAK, exo-oncomiR-221-3p and exo-oncomiR-222-3p—all measured in semen—improves not only the percentage of positively diagnosed patients by 2.25%, but increases the specificity by 8%.

MiR-221 and miR-222 are encoded tandemly in chromosome Xp11.3, and are highly homologous miRNAs sharing the same "seed sequences" [33,41]. In vivo studies have demonstrated that miR-221/222 down-regulation impairs the growth of PCa xenografts, pointing to miR-221-3p as an oncogenic miRNA in PCa [42]. In the present study, we observed that the addition of exo-oncomiR-221-3p expression levels in semen improves all prognostic model panel combinations. miR-221 is overexpressed in a variety of epithelial cancers including breast, liver, bladder, pancreas, gastric, colorectal cancer, melanoma, papillary thyroid carcinoma and glioblastoma [33]. Additionally, miR-221 has been found to be related to cancer progression in cervical squamous cell carcinoma [43], confers adriamycin resistance in breast cancer [44], and is a biomarker in hepatocellular carcinoma [45], diffuse large B cell lymphomas [46] and lung adenocarcinoma [47].

Studies on the expression of miR-221 in PCa (which is referred to as mir-221-3p in MirBase), have used only PCa tissue [48,49] and have found the levels to be up-regulated. Here we show, for the first time to our knowledge that the expression levels of miR-221-3p in PCa biofluids are higher in high-risk patients than in low-risk peers, and we additionally show that miR-221-3p is up-regulated in PC-3 secreted EVs and cell extracts.

Mechanistically, in vitro studies have determined that miR-221-3p promotes proliferation of PCa cells [50].

MiR-221 directly targets NLK in neuroblastoma cells [51]. Accumulating evidence demonstrates that NLK has a pivotal role in cell proliferation, migration, invasion, and apoptosis by regulating a variety of transcriptional molecules [52]. NLK expression in PCa metastases is decreased in comparison with normal prostate epithelium and primary PCa [53]. Our findings show that oncomiR-221-3p inhibits NLK protein expression in PC-3 cells and that expression is further reduced by sTWEAK. An additional predicted and experimentally-demonstrated target for miR-221-3p is TCF12, a transcription factor member of the helix–loop–helix protein family found to be extensively expressed in many tissues [54]. As a target of miR-221, TCF12 has been related to survival after diagnosis of colon cancer [55], and there is evidence to suggest that TCF12 is involved in cell migration and differentiation [56]. Interestingly, the status of TCF12 has been found to be an independent predictor of biochemical recurrence-free survival in PCa [57]. We show here that miR-221-3p likely regulates TCF12 in PC-3 cells and its expression is, in turn, regulated by sTWEAK. While our findings point to the possibility that regulation of NLK or TCF12 might be a therapeutic approach against PCa tumors, further research and validation either in preclinical models or other established PCa cell lines will be needed to test their functional relevance in cell proliferation, invasion and chemosensitivity to cytotoxic agents.

Overall, our results reveal that TWEAK inflammation-induced exo-oncomiRNAs are components of an improved PCa prognostic panel based only on information obtained from a unique liquid biopsy, semen. Additionally, we reveal that a TWEAK inflammatory challenge in PCa cells can potentiate oncomiR-221-3p action.

4. Materials and Methods

4.1. Cell Culture

The PC-3 and LNCaP cell lines were purchased from Sigma-Aldrich (Barcelona, Spain). PC-3 cells were cultured in Ham's F-12K (Kaighn's) medium (1:1 mixture) with L-glutamate (Gibco, Fisher Scientific SL, Madrid, Spain), and LNCaP cells were cultured in RPMI 1640 medium supplemented with 1 mM sodium pyruvate (Gibco). Cultures were also supplemented with 10% fetal bovine serum, 1× antibiotic-antimycotic solution (Gibco), and 5 µg/mL plasmocin, and cultured in a humidified 5% CO_2 atmosphere at 37 °C. Cells were grown in exosome-deprived serum overnight before stimulation for 24 h with 100 ng/mL human recombinant (hr) TWEAK (PeproTech, BioNova Cientifica, Barcelona, Spain).

4.2. Extracellular Vesicle Isolation from Cell Culture Media and Exo-oncomiRNA Expression Profile Using TaqMan Low-Density Arrays

Exosomes and other extracellular vesicles from cell culture media (PC-3 and LNCaP) were isolated and exo-oncomiRNAs were extracted using the exoRNeasy Serum/Plasma Maxi Kit (Qiagen, BioNova Cientifica, Madrid, Spain). For exo-oncomiRNA screening, the miRCURY LNA Universal RT microRNA PCR, Polyadenylation and cDNA Synthesis Kit (Exiqon, BioNova Cientifica, s.l. Madrid, Spain) was used for reverse transcription. cDNA was diluted and assayed by qRT-PCR according to the protocol in a 7900HT Fast Real-Time PCR System (Applied Biosystems, Thermo Fisher Scientic, Waltham, MA, USA). Each exo-oncomiRNA was assayed using ExiLENT SYBR Green Master Mix on the Human panel I+II, V5, miRCURY LNA miRNA miRNome PCR Panel (Qiagen) that included 752 mature human cancer-related miRNAs. Fluorescence readings and expression records of the microRNAs during the qRT-PCR were performed with the SDS 2.3 program (Applied Biosystems, Foster City, CA, USA). From the quantitative analysis by qRT-PCR of all miRNAs analyzed, we only considered those miRNAs that showed expression levels with a Ct < 33. Then, using the GeneGlobe program (Qiagen) [58], C_T values for each sample were normalized to the arithmetic mean of the following reference miRNAs, hsa-miR-423-5p, SNORD38B, SNORD49A, hsa-miR-191-5p, hsa-miR-103a-3p and U6 small nuclear

RNA. The fold change expression of each exo-oncomiRNA was calculated with the formula $2^{-\Delta\Delta Ct}$ where each miRNA, regardless of the condition, was first normalized to the C_T of an endogenous control and then we calculated the $\Delta\Delta Ct = \Delta Ct$ sample treated sTWEAK $-\Delta Ct$ untreated controls. The exo-oncomiRNAs with $p \leq 0.05$ when comparing cell type and condition and with an increase ≥ 1.8-fold were considered for further analysis.

4.3. Extracellular Vesicle Analysis

Extracellular vesicles from culture media, post-digital rectal examination urine and semen were obtained using exoRNeasy Serum/Plasma Maxi Kit just before miRNA isolation by the addition of 500 µL of elution buffer XE. The isolated EVs were further concentrated using a 100,000 Da cut-off concentrator (Amicon Ultra-0.5 mL Centrifugal Filters, Millipore). Samples were then ultrasonicated 3 times during 1 min bouts. Total protein was quantified using the BCA method (Pierce). A total amount of 10 µg EV protein and 10 µg total PC-3 cell extract were loaded on 4–15% SDS-PAGE gels and immunoblotted with polyclonal rabbit antibodies against: EXOAB-CD9A1, EXOAB-CD81A-1, EXOAB-CD63A-1, EXOABHsp70A-1, EXOAB-TSG101-1 (System Biology, Palo Alto, CA, USA), and the mouse monoclonal antibody for tubulin (Thermo Fisher Scientific, Waltham, MA, USA). HRP-conjugated goat anti-mouse or anti-rabbit (both from SBI) were used as secondary antibodies. All Western blots were developed with SuperSignal West Femto chemiluminescen substrate (Pierce Biotechnology, Boston, MA, USA) and visualized with the VersaDoc imaging system and Quantity One software (Bio-Rad, Barcelona, Spain) (Supplementary Materials).

4.4. Transmission Electron Microscopy Analysis

EVs were placed on carbon-coated copper grids (200 mesh), allowed to dry, and incubated in osmium tetroxide vapors for 15–30 min. TEM images were collected using a JEOL 1011 transmission electron microscope operating at 80 kV with a megaview III camera.

4.5. Patients

Our studied patient cohort comprised 97 consecutive patients with PCa who had undergone radical prostatectomy by open surgery at the University Hospital Joan XXIII, Tarragona, between 2015 and 2019—laparoscopic or robotic surgery (intraperitoneal or extraperitoneal—with or without bilateral ilio-obturator lymphadenectomy, according to the estimated risk of lymphadenopathy based on the Briganti nomogram [56]. Patients were stratified according to the 2014 ISUP-GG and TNM classification [57,58]. Patients were stratified into two categories: low-risk (ISUP Group I and II) and high-risk (ISUP Groups III, IV and V). Written informed consent prior to their inclusion was provided by all patients. The study was approved by our local ethics committee and performed according to the provisions of the Declaration of Helsinki (Biomedical Research Law 14/2007, Royal Decree of Biobanks 1716/2011, Organic Law15/1999 of September 13 Protection of Personal Data) [11]. Clinical parameters, tumor aggressiveness, and metabolic status of all patients were documented. All methods were approved and performed in accordance with guidelines and regulations of the Ethical Committee for Clinical Research (CEIm) from Pere Virgili Research Institute (Ref. CEim171/2017) (http://www.iispv.cat/plataformes_de_suport/en_comite-iispv.html). Patient's inclusion criteria were as follows: older than 18 years, diagnosed with PCa by prostate biopsy in our center or any other, and treated by radical prostatectomy in our center. Exclusion criteria were patients with a previous history of cancer, patients older than 75 years, and those who had received any prior treatment before radical prostatectomy for PCa, as described [11].

4.6. Analytical Methods

Plasma glucose, cholesterol, triglyceride, high-density lipoprotein cholesterol, insulin levels and hepatic profile and renal profile was performed as described [59]. Levels of sTWEAK in semen were determined in duplicate using commercially available human enzyme-linked immunosorbent assay (ELISA) DuoSet Kits (R&D Systems Europe, Abingdon, UK).

4.7. Sample Processing

Serum/plasma: blood samples were collected after a fast of at least 12 h, or 2 h after an oral glucose tolerance test. Samples were centrifuged at 4 °C and stored at −80 °C.

Post-digital rectal examination urine: urine samples were collected prior to prostate biopsy or surgical intervention. Samples were centrifuged (2000× *g*, 10 min, 4 °C), and stored at −80 °C.

Semen: semen samples were centrifuged at 2000× *g* for 15 min at 22 °C to separate spermatozoa from semen plasma, and the supernatant (semen plasma) was stored at −80 °C.

All samples were processed and stored at the Institut d'Investigació Sanitària Pere Virgili (IISPV) BioBanc (B.0000853 + B.0000854) integrated in the Spanish National Biobanks Platform (PT13/0010/0029 and PT13/0010/0062) for its collaboration.

4.8. Extracellular Vesicles Extraction from Liquid Biopsy and Exo-onocomiRNA Quantitative Real-Time PCR Profiling

Extracellular vesicles and exo-miRNAs were isolated and extracted from urine and semen samples using the exoRNeasy Serum/Plasma Maxi Kit or Midi Kit (Qiagen) [12]. The miRCURY LNA Universal RT microRNA PCR, Polyadenylation and cDNA Synthesis Kit (Exiqon, BioNova Cientifica, s.l. Madrid, Spain) was used for reverse transcription. The expression profile of the 14 selected exo-oncomiRNAs was further analyzed in urine and semen samples in duplicate, using individual primers on a 7900HT Fast Real-Time PCR System (Applied Biosystems). Data were analyzed by SDS 2.3 and RQ Manager 1.2 (Applied Biosystems) using the $2^{-\Delta\Delta Ct}$ method. All values of Ct > 35 were excluded for further analysis.

4.9. Target Search by Bioinformatic Analysis

The targets of the selected exo-oncomiRNAs were searched using three target prediction software packages: (1) The miRanda algorithm (www.microRNA.org) was used to find potential target sites for miRNAs in the genomic sequence. From the miRanda algorithm results, we used the mirSVR score and PhastCons score to decipher which targets were potentially predicted. The mirSVR score is an estimate of the miRNA effect on the mRNA expression level; the more negative the score, the greater the inhibitory effect. PhastCons scores measure the conservation of nucleotide positions across vertebrates of any possible interaction; the higher the PhastCons value, the more conservative across vertebrates and the more important is the complementarity of the miRNA and the target [60]. (2) The Diana-MicroT-CDS predicts targets through the microT-CDS algorithm giving a miTG score, which is a general score for the predicted interaction. The closer the score is to 1, the greater the prediction confidence [61]. (3) Finally, we used the miRWalk platform. The calculated score is generated by executing the TarPmiR algorithm for miRNA target site prediction. The closer the score is to 1, the greater is the confidence prediction, in the same way as Diana-MicroT-CDS [62].

Candidate targets with an miR-SVR score equal or to less than −0.1; a PhastCons value equal or greater to 0.56 and, an miTG and miRWalk score equal or greater to 0.8 were considered as potential targets of exo-oncomiR-221-3p.

4.10. Functional Studies

Functional studies were performed in PC-3 cells cultured in 6-well plates and grown at 90% confluence. Cells were transfected using Lipofectamine 3000 (Invitrogen, Thermo Fisher Scientific, Waltham, MA, USA) and OPTI-MEM medium (Gibco) with a locked nucleic acid probe containing a specific sequence antisense oligonucleotide targeting exo-oncomiR-221-3p, miRCURY LNA exo-oncomiR-221-3p Power Inhibitor (Qiagen). A scrambled miRNA sequence, miRCURY LNA Power Inhibitor Control A (Qiagen), served as a negative control. Each condition was incubated for 24 h in a humidified 5% CO_2 atmosphere at 37 °C before stimulation for 24 h with 100 ng/mL TWEAK (PeproTech) in serum-free media. After stimulation, cells were harvested for protein and RNA analysis. Expression analyses of target genes were performed using commercial individual primers on a 7900HT Fast Real-Time PCR System (Applied Biosystems). Protein analysis and Western blotting were performed using standard protocols. Nitrocellulose membranes were probed with the following primary antibodies that were purchased from Cell Signaling Technology (Danvers, MA, USA): TCF12/HEB (#11825) and NLK (#94350) and NF-κB2 p100 (#4882). An anti-β-actin (A11126) antibody was purchased from Sigma-Aldrich. The standard molecular weight marker used was purchased from New England Biolabs Inc. (Herts, UK). Western blots were developed with SuperSignal West Femto chemiluminescen substrate (Pierce Biotechnology, Boston, MA, USA) and visualized with VersaDoc imaging system and Quantity One software (Bio-Rad) (Supplementary Materials).

4.11. Statistical Analysis

For in vitro assays, experimental results are presented as mean ± standard error of the mean (SEM) of 3–4 experiments. Statistical significance was assessed with Student's *t*-test. Results with $p < 0.05$ were considered statistically significant.

For human samples studies, the sample size was calculated to determine differences between exo-oncomiRNA expression levels in liquid biopsy with respect to the degree of aggressiveness of the tumor (low-risk/high-risk) in those patients diagnosed with PCa. We assumed a two-fold change difference between groups and identical standard deviation (SD) between the groups; therefore, a minimum of 35 patients was needed in each group (bilateral alpha error 0.05, power 90%). Statistical analysis were performed as described [11]. Briefly, for anthropometric and clinical variables, data are expressed as mean ± SD. Before statistical analysis, normal distribution was evaluated using Levene's test. The non-parametric Mann–Whitney U-test was used to analyze the differences in anthropometric and clinical data and absolute expression levels of the exomiR candidates between patients according to ISUP-GG—low-risk (Group I and II) and high-risk (Group III, Group IV, and Group V). A *p*-value less than 0.05 was considered statistically significant. Spearman's Rho test was used as a correlation analysis between anthropometric, clinical, and exo-miRNAs data. Partial least square discriminant analysis (PLS-DA) and VIP analysis models and binary logistic regression analysis were developed for selected variables. ROC curve analysis was performed to evaluate the best predictive model. The statistical software SPSS Statistics 24.0 (IBM, Madrid, Spain) package and R software (http://cran.r-project.org) were used for analysis.

5. Conclusions

TWEAK, exo-oncomiR -221-3p and exo-oncomiR-222-3p are proposed as an improved PCa prognostic panel based on information obtained from a unique biofluid, semen. Further studies in larger cohorts of PCa will be needed as a next step to confirm/validate our panel before it can be adopted in clinical practice.

Supplementary Materials: The following are available online at https://www.mdpi.com/2072-6694/13/2/250/s1. Table S1: 14 identified exo-oncomiRNAS, Table S2: Exo-oncomiRNAS non-significantly changed when comparing urine or semen biofluids from patients with PCa stratified by risk (low or high), Table S3: Role of elected exo-miR-221-3p targets in cancer, Table S4: List of the selected exo-oncomiR-221-3p and exo-oncomiR-222-3p target's scores, Figure S1: Complete Western blot (WB) results referring to Figure 1b, Figure S2: Spearman's correlation, Figure S3a,b: PLS-DA analysis, Figure S4: Complete WB results referring to Figure 4d.

Author Contributions: Conceptualization, M.R.C., X.R.-P., J.S.-T.; methodology, A.A.-C.; sample collection, review of clinical data and patient recruitment, M.A.-S., J.F.G.-F., J.V.-B. and S.M.-G.; writing—review and editing, M.R.C., X.R.-P., J.S.-T. and A.A.-C.; funding acquisition, M.R.C. and X.R.-P. All authors have read and agreed to the published version of the manuscript.

Funding: This study was founded by Instituto de Salud Carlos III through projects PI17/00877, PI20/00418 (co-founded by the European Regional Development Fund/European Social Found; "A way to make future"/"Investing in your future") and a grant awarded by Fundació Vallformosa "IV Premi Martí Via". No payment has been received to write this article by a pharmaceutical company or other agency.

Institutional Review Board Statement: The study was conducted according to the guidelines of the Declaration of Helsinki, and approved by the Institutional Review Board (or Ethics Committee) of Institut d'Investigació Sanitària Pere Virgili (Ref.CEim 171/2017).

Informed Consent Statement: Informed consent was obtained from all subjects involved in the study.

Data Availability Statement: The data presented in this study is contained within the article and the supplementary material.

Acknowledgments: The authors wish to particularly acknowledge the patients enrolled in this study for their participation and to the IISPV BioBanc (B.0000853 + B.0000854) integrated in the Spanish National Biobanks Platform (PT13/0010/0029 and PT13/0010/0062) for its collaboration.

Conflicts of Interest: The authors declare no conflict of interest.

References

1. Bray, F.; Ferlay, J.; Soerjomataram, I.; Siegel, R.L.; Torre, L.A.; Jemal, A. Global cancer statistics 2018: GLOBOCAN estimates of incidence and mortality worldwide for 36 cancers in 185 countries. CA. *Cancer J. Clin.* **2018**, *68*, 394–424. [CrossRef]
2. Markozannes, G.; Tzoulaki, I.; Karli, D.; Evangelou, E.; Ntzani, E.; Gunter, M.J.; Norat, T.; Ioannidis, J.P.; Tsilidis, K.K. Diet, body size, physical activity and risk of prostate cancer: An umbrella review of the evidence. *Eur. J. Cancer* **2016**, *69*, 61–69. [CrossRef] [PubMed]
3. Epstein, J.I.; Egevad, L.; Amin, M.B.; Delahunt, B.; Srigley, J.R.; Humphrey, P.A. The 2014 International Society of Urological Pathology (ISUP) Consensus Conference on Gleason Grading of Prostatic Carcinoma. *Am. J. Surg. Pathol.* **2015**, *40*, 1. [CrossRef] [PubMed]
4. Saini, S. PSA and beyond: Alternative prostate cancer biomarkers. *Cell. Oncol.* **2016**, *39*, 97–106. [CrossRef]
5. Di Meo, A.; Bartlett, J.; Cheng, Y.; Pasic, M.D.; Yousef, G.M. Liquid biopsy: A step forward towards precision medicine in urologic malignancies. *Mol. Cancer* **2017**, *16*, 1–14. [CrossRef]
6. Greten, F.R.; Grivennikov, S.I. Inflammation and Cancer: Triggers, Mechanisms, and Consequences. *Immunity* **2019**, *51*, 27–41. [CrossRef]
7. Winkles, J.A. The TWEAK-Fn14 cytokine-receptor axis: Discovery, biology and therapeutic targeting. *Nat. Rev. Drug Discov.* **2008**, *7*, 411–425. [CrossRef]
8. Maecker, H.; Varfolomeev, E.; Kischkel, F.; Lawrence, D.; LeBlanc, H.; Lee, W.; Hurst, S.; Danilenko, D.; Li, J.; Filvaroff, E.; et al. TWEAK attenuates the transition from innate to adaptive immunity. *Cell* **2005**, *123*, 931–944. [CrossRef]
9. Terra, X.; Gómez, D. External validation of sTWEAK as a prognostic noninvasive biomarker for head and neck squamous cell carcinoma. *Head Neck* **2014**, *36*, 1391. [CrossRef]
10. Avilés-Jurado, F.X.; Terra, X.; Gómez, D.; Flores, J.C.; Raventós, A.; Maymó-Masip, E.; León, X.; Serrano-Gonzalvo, V.; Vendrell, J.; Figuerola, E.; et al. Low blood levels of sTWEAK are related to locoregional failure in head and neck cancer. *Eur. Arch. Oto-Rhino-Laryngol.* **2015**, *272*, 1733–1741. [CrossRef]
11. Ruiz-Plazas, X.; Rodríguez-Gallego, E.; Alves, M.; Altuna-Coy, A.; Lozano-Bartolomé, J.; Portero-Otin, M.; García-Fontgivell, J.F.; Martínez-González, S.; Segarra, J.; Chacón, M.R. Biofluid quantification of TWEAK/Fn14 axis in combination with a selected biomarker panel improves assessment of prostate cancer aggressiveness. *J. Transl. Med.* **2019**, *17*, 1–13. [CrossRef] [PubMed]
12. Enderle, D.; Spiel, A.; Coticchia, C.M.; Berghoff, E.; Mueller, R.; Schlumpberger, M.; Sprenger-Haussels, M.; Shaffer, J.M.; Lader, E.; Skog, J.; et al. Characterization of RNA from exosomes and other extracellular vesicles isolated by a novel spin column-based method. *Wiley Interdiscip. Rev. RNA* **2012**, *3*, 286–293. [CrossRef] [PubMed]
13. Ruivo, C.F.; Adem, B.; Silva, M.; Melo, S.A. The biology of cancer exosomes: Insights and new perspectives. *Cancer Res.* **2017**, *77*, 6480–6488. [CrossRef] [PubMed]
14. Tkach, M.; Théry, C. Communication by Extracellular Vesicles: Where We Are and Where We Need to Go. *Cell* **2016**, *164*, 1226–1232. [CrossRef] [PubMed]
15. Pan, J.; Ding, M.; Xu, K.; Yang, C.; Mao, L.J. Exosomes in diagnosis and therapy of prostate cancer. *Oncotarget* **2017**, *8*, 97693–97700. [CrossRef]
16. Huang, C.; Luo, W.-F.; Ye, Y.-F.; Lin, L.; Wang, Z.; Luo, M.-H.; Song, Q.-D.; He, X.-P.; Chen, H.-W.; Kong, Y.; et al. Characterization of Inflammatory Factor-Induced Changes in Mesenchymal Stem Cell Exosomes and Sequencing Analysis of Exosomal microRNAs. *World J. Stem Cells* **2019**, *11*, 859–890. [CrossRef] [PubMed]

17. Huang, M.; Narita, S.; Tsuchiya, N.; Ma, Z.; Numakura, K.; Obara, T.; Tsuruta, H.; Saito, M.; Inoue, T.; Horikawa, Y.; et al. Overexpression of Fn14 promotes androgen-independent prostate cancer progression through MMP-9 and correlates with poor treatment outcome. *Carcinogenesis* **2011**, *32*, 1589–1596. [CrossRef]
18. Yin, J.; Morrissey, C.; Barrett, B.; Corey, E.; Ylaya, K.; Hewitt, S.; Fang, L.; Tillman, H.; Lake, R.; Vessella, R.; et al. AR-Regulated TWEAK-FN14 Pathway Promotes Prostate Cancer Bone Metastasis. *Cancer Res.* **2014**, *74*, 4306–4317. [CrossRef]
19. Hu, Y.; Li, D.; Wu, A.; Qiu, X.; Di, W.; Huang, L.; Qiu, L. TWEAK-stimulated macrophages inhibit metastasis of epithelial ovarian cancer via exosomal shuttling of microRNA. *Cancer Lett.* **2017**, *393*, 60–67. [CrossRef]
20. Valadi, H.; Ekström, K.; Bossios, A.; Sjöstrand, M.; Lee, J.J.; Lötvall, J.O. Exosome-mediated transfer of mRNAs and microRNAs is a novel mechanism of genetic exchange between cells. *Nat. Cell Biol.* **2007**, *9*, 654–659. [CrossRef]
21. Svoronos, A.A.; Engelman, D.M.; Slack, F.J. OncomiR or tumor suppressor? The duplicity of MicroRNAs in cancer. *Cancer Res.* **2016**, *76*, 3666–3670. [CrossRef] [PubMed]
22. Tang, Z.; Li, D.; Hou, S.; Zhu, X. The cancer exosomes: Clinical implications, applications and challenges. *Int. J. Cancer* **2020**, *146*, 2946–2959. [CrossRef] [PubMed]
23. Wang, J.; Ni, J.; Beretov, J.; Thompson, J.; Graham, P.; Li, Y. Exosomal microRNAs as liquid biopsy biomarkers in prostate cancer. *Crit. Rev. Oncol. Hematol.* **2020**, *145*, 4–10. [CrossRef] [PubMed]
24. Huang, X.; Yuan, T.; Liang, M.; Du, M.; Xia, S.; Dittmar, R.; Wang, D.; See, W.; Costello, B.A.; Quevedo, F.; et al. Exosomal miR-1290 and miR-375 as Prognostic Markers in Castration-resistant Prostate Cancer. *Eur. Urol.* **2015**, *67*, 33–41. [CrossRef] [PubMed]
25. Endzeliņš, E.; Berger, A.; Melne, V.; Bajo-Santos, C.; Soboļevska, K.; Ābols, A.; Rodriguez, M.; Šantare, D.; Rudnickiha, A.; Lietuvietis, V.; et al. Detection of circulating miRNAs: Comparative analysis of extracellular vesicle-incorporated miRNAs and cell-free miRNAs in whole plasma of prostate cancer patients. *BMC Cancer* **2017**, *17*, 1–13. [CrossRef]
26. Foj, L.; Ferrer, F.; Serra, M.; Arévalo, A.; Gavagnach, M.; Giménez, N.; Filella, X. Exosomal and Non-Exosomal Urinary miRNAs in Prostate Cancer Detection and Prognosis. *Prostate* **2017**, *77*, 573–583. [CrossRef] [PubMed]
27. Li, Z.; Ma, Y.Y.; Wang, J.; Zeng, X.F.; Li, R.; Kang, W.; Hao, X.K. Exosomal microRNA-141 is upregulated in the serum of prostate cancer patients. *OncoTargets Ther.* **2015**, *9*, 139–148. [CrossRef]
28. Wani, S.; Kaul, D.; Mavuduru, R.S.; Kakkar, N.; Bhatia, A. Urinary-exosomal miR-2909: A novel pathognomonic trait of prostate cancer severity. *J. Biotechnol.* **2017**, *259*, 135–139. [CrossRef]
29. Barceló, M.; Castells, M.; Bassas, L.; Vigués, F.; Larriba, S. Semen miRNAs Contained in Exosomes as Non-Invasive Biomarkers for Prostate Cancer Diagnosis. *Sci. Rep.* **2019**, *9*, 1–16. [CrossRef]
30. Lee, J.M.; Im, G.-I. PTHrP isoforms have differing effect on chondrogenic differentiation and hypertrophy of mesenchymal stem cells. *Biochem. Biophys. Res. Commun.* **2012**, *421*, 819–824. [CrossRef]
31. Konoshenko, M.Y.; Lekchnov, E.A.; Vlassov, A.V.; Laktionov, P.P. Isolation of Extracellular Vesicles: General Methodologies and Latest Trends. *Biomed. Res. Int.* **2018**, *2018*, 8545347. [CrossRef] [PubMed]
32. Fay, M.P.; Shih, J.H. Permutation tests using estimated distribution functions. *J. Am. Stat. Assoc.* **1998**, *93*, 387–396. [CrossRef]
33. Song, F.; Ouyang, Y.; Che, J.; Li, X.; Zhao, Y.; Yang, K.; Zhao, X.; Chen, Y.; Fan, C.; Yuan, W. Potential value of miR-221/222 as diagnostic, prognostic, and therapeutic biomarkers for diseases. *Front. Immunol.* **2017**, *8*, 1–9. [CrossRef]
34. Goto, Y.; Kojima, S.; Nishikawa, R.; Kurozumi, A.; Kato, M.; Enokida, H.; Matsushita, R.; Yamazaki, K.; Ishida, Y.; Nakagawa, M.; et al. MicroRNA expression signature of castration-resistant prostate cancer: The microRNA-221/222 cluster functions as a tumour suppressor and disease progression marker. *Br. J. Cancer* **2015**, *113*, 1055–1065. [CrossRef]
35. Coarfa, C.; Fiskus, W.; Eedunuri, V.K.; Rajapakshe, K.; Foley, C.; Chew, S.A.; Shah, S.S.; Geng, C.; Shou, J.; Mohamed, J.S.; et al. Comprehensive proteomic profiling identifies the androgen receptor axis and other signaling pathways as targets of microRNAs suppressed in metastatic prostate cancer. *Oncogene* **2016**, *35*, 2345–2356. [CrossRef]
36. Lobo, J.; Gillis, A.J.; van den Berg, A.; Dorssers, L.C.; Belge, G.; Dieckmann, K.P.; Roest, H.P.; van der Laan, L.J.W.; Gietema, J.; Robert, J. Hamilton Looijenga Identification and Validation Model for Informative Liquid Biopsy-Based microRNA Biomarkers: Insights from Germ Cell Tumor In Vitro, in Vivo and Patient-Derived Data João. *Cells* **2019**, *8*, 1637. [CrossRef]
37. Peng, Y.; Croce, C.M. The role of microRNAs in human cancer. *Signal Transduct. Target. Ther.* **2016**, *1*, 15004. [CrossRef]
38. Agaoglu, F.Y.; Kovancilar, M.; Dizdar, Y.; Darendeliler, E.; Holdenrieder, S.; Dalay, N.; Gezer, U. Investigation of miR-21, miR-141, and miR-221 in blood circulation of patients with prostate cancer. *Tumor Biol.* **2011**, *32*, 583–588. [CrossRef]
39. Drabovich, A.P.; Saraon, P.; Jarvi, K.; Diamandis, E.P. Seminal plasma as a diagnostic fluid for male reproductive system disorders. *Nat. Rev. Urol.* **2014**, *11*, 278–288. [CrossRef]
40. Eskra, J.N.; Rabizadeh, D.; Pavlovich, C.P.; Catalona, W.J.; Luo, J. Approaches to urinary detection of prostate cancer. *Prostate Cancer Prostatic Dis.* **2019**, *22*, 362–381. [CrossRef]
41. Song, Q.; An, Q.; Niu, B.; Lu, X.; Zhang, N.; Cao, X. Role of miR-221/222 in Tumor Development and the Underlying Mechanism. *J. Oncol.* **2019**, *2*, 7252013. [CrossRef]
42. Mercatelli, N.; Coppola, V.; Bonci, D.; Miele, F.; Costantini, A.; Guadagnoli, M.; Bonanno, E.; Muto, G.; Frajese, G.V.; De Maria, R.; et al. The inhibition of the highly expressed mir-221 and mir-222 impairs the growth of prostate carcinoma xenografts in mice. *PLoS ONE* **2008**, *3*, e4029. [CrossRef] [PubMed]

43. Zhou, C.F.; Ma, J.; Huang, L.; Yi, H.Y.; Zhang, Y.M.; Wu, X.G.; Yan, R.M.; Liang, L.; Zhong, M.; Yu, Y.H.; et al. Cervical squamous cell carcinoma-secreted exosomal miR-221-3p promotes lymphangiogenesis and lymphatic metastasis by targeting VASH1. *Oncogene* **2019**, *38*, 1256–1268. [CrossRef] [PubMed]
44. Pan, X.; Hong, X.; Lai, J.; Cheng, L.; Cheng, Y.; Yao, M.; Wang, R.; Hu, N. Exosomal MicroRNA-221-3p Confers Adriamycin Resistance in Breast Cancer Cells by Targeting PIK3R1. *Front. Oncol.* **2020**, *10*, 441. [CrossRef] [PubMed]
45. Ghosh, S.; Bhowmik, S.; Majumdar, S.; Goswami, A.; Chakraborty, J.; Gupta, S.; Aggarwal, S.; Ray, S.; Chatterjee, R.; Bhattacharyya, S.; et al. The exosome encapsulated microRNAs as circulating diagnostic marker for hepatocellular carcinoma with low alpha-fetoprotein. *Int. J. Cancer* **2020**, *147*, 2934–2947. [CrossRef]
46. Larrabeiti-Etxebarria, A.; Lopez-Santillan, M.; Santos-Zorrozua, B.; Lopez-Lopez, E.; Garcia-Orad, A. Systematic review of the potential of MicroRNAs in diffuse large B cell lymphoma. *Cancers* **2019**, *11*, 144. [CrossRef]
47. Wu, Y.; Wei, J.; Zhang, W.; Xie, M.; Wang, X.; Xu, J. Serum exosomal miR-1290 is a potential biomarker for lung adenocarcinoma. *OncoTargets Ther.* **2020**, *13*, 7809–7818. [CrossRef]
48. Kristensen, H.; Thomsen, A.R.; Haldrup, C.; Dyrskjøt, L.; Høyer, S.; Borre, M.; Mouritzen, P.; Ørntoft, T.F.; Sørensen, K.D. Novel diagnostic and prognostic classifiers for prostate cancer identified by genome-wide microRNA profiling. *Oncotarget* **2016**, *7*, 30760–30771. [CrossRef]
49. Kurul, N.O.; Ates, F.; Yilmaz, I.; Narli, G.; Yesildal, C.; Senkul, T. The association of let-7c, miR-21, miR-145, miR-182, and miR-221 with clinicopathologic parameters of prostate cancer in patients diagnosed with low-risk disease. *Prostate* **2019**, *79*, 1125–1132. [CrossRef]
50. Galardi, S.; Mercatelli, N.; Giorda, E.; Massalini, S.; Frajese, G.V.; Ciafrè, S.A.; Farace, M.G. miR-221 and miR-222 expression affects the proliferation potential of human prostate carcinoma cell lines by targeting p27Kip1. *J. Biol. Chem.* **2007**, *282*, 23716–23724. [CrossRef]
51. He, X.Y.; Tan, Z.L.; Mou, Q.; Liu, F.J.; Liu, S.; Yu, C.W.; Zhu, J.; Lv, L.Y.; Zhang, J.; Wang, S.; et al. microRNA-221 enhances MYCN via targeting nemo-like kinase and functions as an oncogene related to poor prognosis in neuroblastoma. *Clin. Cancer Res.* **2017**, *23*, 2905–2918. [CrossRef] [PubMed]
52. Huang, Y.; Yang, Y.; He, Y.; Li, J. The emerging role of Nemo-like kinase (NLK) in the regulation of cancers. *Tumor Biol.* **2015**, *36*, 9147–9152. [CrossRef] [PubMed]
53. Emami, K.H.; Brown, L.G.; Pitts, T.E.M.; Sun, X.; Vessella, R.L.; Corey, E. Nemo-like kinase induces apoptosis and inhibits androgen receptor signaling in prostate cancer cells. *Prostate* **2009**, *69*, 1481–1492. [CrossRef] [PubMed]
54. Kishore, S.; Jaskiewicz, L.; Burger, L.; Hausser, J.; Khorshid, M.; Zavolan, M. A quantitative analysis of CLIP methods for identifying binding sites of RNA-binding proteins. *Nat. Methods* **2011**, *8*, 559–567. [CrossRef]
55. Mullany, L.E.; Herrick, J.S.; Wolff, R.K.; Stevens, J.R.; Samowitz, W.; Slattery, M.L. Transcription factor-microRNA associations and their impact on colorectal cancer survival. *Mol. Carcinog.* **2017**, *56*, 2512–2526. [CrossRef]
56. Chen, W.S.; Chen, C.C.; Chen, L.L.; Lee, C.C.; Huang, T.S. Secreted heat shock protein 90α (HSP90α) induces nuclear factor-κB-mediated TCF12 protein expression to down-regulate E-cadherin and to enhance colorectal cancer cell migration and invasion. *J. Biol. Chem.* **2013**, *288*, 9001–9010. [CrossRef]
57. Chen, Q.B.; Liang, Y.K.; Zhang, Y.Q.; Jiang, M.Y.; Han, Z.D.; Liang, Y.X.; Wan, Y.P.; Yin, J.; He, H.C.; Zhong, W. De Decreased expression of TCF12 contributes to progression and predicts biochemical recurrence in patients with prostate cancer. *Tumor Biol.* **2017**, *39*. [CrossRef]
58. Zhang, M.; Xiao, X.; Xiong, D.; Liu, Q. Accurate normalization of real-time quantitative RT-PCR data by geometric averaging of multiple internal control genes. *J. Artif. Intell. Res.* **2014**, *50*, 1–30. [CrossRef]
59. Vendrell, J.; Maymó-Masip, E.; Tinahones, F.; García-España, A.; Megia, A.; Caubet, E.; García-Fuentes, E.; Chacón, M.R. Tumor necrosis-like weak inducer of apoptosis as a proinflammatory cytokine in human adipocyte cells: Up-regulation in severe obesity is mediated by inflammation but not hypoxia. *J. Clin. Endocrinol. Metab.* **2010**, *95*, 2983–2992. [CrossRef]
60. Riffo-Campos, Á.L.; Riquelme, I.; Brebi-Mieville, P. Tools for sequence-based miRNA target prediction: What to choose? *Int. J. Mol. Sci.* **2016**, *17*, 1987. [CrossRef]
61. Willkomm, S.; Zander, A.; Grohmann, D. Drug Target miRNA. *Methods Mol. Biol.* **2017**, *1517*, 291–304. [CrossRef] [PubMed]
62. Sticht, C.; De La Torre, C.; Parveen, A.; Gretz, N. Mirwalk: An online resource for prediction of microrna binding sites. *PLoS ONE* **2018**, *13*, e0206239. [CrossRef] [PubMed]

Article

Molecular Signature of Extracellular Vesicular Small Non-Coding RNAs Derived from Cerebrospinal Fluid of Leptomeningeal Metastasis Patients: Functional Implication of miR-21 and Other Small RNAs in Cancer Malignancy

Kyue-Yim Lee [1,2], Yoona Seo [3,4], Ji Hye Im [1,2], Jiho Rhim [3,4], Woosun Baek [3,4], Sewon Kim [5], Ji-Woong Kwon [2], Byong Chul Yoo [3], Sang Hoon Shin [2], Heon Yoo [2,3], Jong Bae Park [3], Ho-Shin Gwak [1,2,4,*] and Jong Heon Kim [3,4,*]

[1] Department of Cancer Control, National Cancer Center Graduate School of Cancer Science and Policy, Goyang 10408, Korea; 70564@ncc.re.kr (K.-Y.L.); 75262@ncc-gcsp.ac.kr (J.H.I.)
[2] Neuro-Oncology Clinic, National Cancer Center, Goyang 10408, Korea; jwkwon@ncc.re.kr (J.-W.K.); nsshin@ncc.re.kr (S.H.S.); heonyoo@ncc.re.kr (H.Y.)
[3] Department of Cancer Biomedical Science, National Cancer Center Graduate School of Cancer Science and Policy, Goyang 10408, Korea; yoona.seo@ncc.re.kr (Y.S.); jhrhim@ncc.re.kr (J.R.); bws@ncc.re.kr (W.B.); yoo_akh@ncc.re.kr (B.C.Y.); jbp@ncc.re.kr (J.B.P.)
[4] Division of Cancer Biology, Research Institute, National Cancer Center, Goyang 10408, Korea
[5] Macrogen Inc., Seoul 08511, Korea; raphaelfriend30@gmail.com
* Correspondence: nsghs@ncc.re.kr (H.-S.G.); jhkim@ncc.re.kr (J.H.K.)

Simple Summary: Leptomeningeal metastasis (LM) is a lethal complication in which cancer metastasizes to the meninges. Currently, there are neither definitive treatments nor diagnosis methods for LM patients. In this study, we suggest the examination of small non-coding RNA (smRNA) populations of extracellular vesicles (EVs) derived from the cerebrospinal fluid (CSF) as a potential vehicle for diagnosis and treatment strategies. Systemic and quantitative analysis of smRNA subpopulations from LM CSF EVs showed unique expression patterns between LM patients and healthy donors. In addition, LM CSF EVs smRNAs appeared to be associated with LM pathogenesis suggesting they may be viable targets for novel diagnostic and treatment strategies.

Abstract: Leptomeningeal metastasis (LM) is a fatal and rare complication of cancer in which the cancer spreads via the cerebrospinal fluid (CSF). At present, there is no definitive treatment or diagnosis for this deleterious disease. In this study, we systemically and quantitatively investigated biased expression of key small non-coding RNA (smRNA) subpopulations from LM CSF extracellular vesicles (EVs) via a unique smRNA sequencing method. The analyzed subpopulations included microRNA (miRNA), Piwi-interacting RNA (piRNA), Y RNA, small nuclear RNA (snRNA), small nucleolar RNAs (snoRNA), vault RNA (vtRNA), novel miRNA, etc. Here, among identified miRNAs, miR-21, which was already known to play an essential oncogenic role in tumorigenesis, was thoroughly investigated via systemic biochemical, miR-21 sensor, and physiological cell-based approaches, with the goal of confirming its functionality and potential as a biomarker for the pathogenesis and diagnosis of LM. We herein uncovered LM CSF extravesicular smRNAs that may be associated with LM-related complications and elucidated plausible pathways that may mechanistically contribute to LM progression. In sum, the analyzed smRNA subpopulations will be useful as targets for the development of therapeutic and diagnostic strategies for LM and LM-related complications.

Keywords: leptomeningeal metastasis; cerebrospinal fluid; extracellular vesicle; biomarker; RNA sequencing; small non-coding RNA; microRNA

1. Introduction

Leptomeningeal metastasis (LM) is a fatal and rare complication of cancer in which the cancer spreads to the meninges surrounding the brain and spinal cord via the cerebrospinal fluid (CSF) [1,2]. LM occurs in approximately 5% of people with cancer and is usually a terminal-stage cancer [1–3]. The median survival of LM patients is approximately 4–8 weeks [4]. At present, there is no definitive treatment or diagnosis of this deleterious disease. CSF is a clear, colorless body fluid found in the brain and spinal cord. It is produced by specialized ependymal cells in the choroid plexuses of the ventricles of the brain [1]. CSF acts as a cushion or buffer, providing basic mechanical and immunological protection to the brain inside the skull [5]. As a potential vehicle for the spread of LM, CSF can be considered a good resource for the identification of new biological markers for the diagnosis derived from LM cells [6–8].

Extracellular vesicles (EVs) mostly consist of exosomes and microvesicles (MVs) of different origin; however, there are limitations in current techniques to completely separate the MV from the exosome since they are overlapped in size and share surface biomarkers [9,10]. EVs are thought to function in intercellular communication. They contain not only essential macromolecules (e.g., proteins and lipids) from their cell of origin, but also functional RNA molecules that can be delivered to recipient cells to undergo translation or perform other functions [11].

Most of the RNAs found in EVs are small non-coding RNAs (smRNAs) of less than 200-nucleotides (nt) in length [12]. Recent studies have indicated that the smRNAs contained within EVs are generally enriched for functional species, such as the well-studied microRNAs (miRNAs) [13]. This finding suggests that EVs are likely to have a direct influence on gene expression of recipient cells upon internalization [14]. In addition to miRNAs, the advent of next-generation sequencing (NGS) revealed the presence of a broad spectrum of additional smRNAs in cells, most of which may be incorporated into EVs [13]. These additional smRNAs include Piwi-interacting RNA (piRNA), Y RNA, small nuclear RNA (snRNA), small nucleolar RNA (snoRNA), vault RNA (vtRNA), tRNA-derived small RNA (tsRNA), ribosomal RNA (rRNA), and small interfering RNA (siRNA) [15]. So far, only the miRNAs have been confirmed to sustain gene-regulatory functions upon cell-to-cell transfer [16]. Thus, researchers have focused on EV-related miRNAs for potential therapeutic exploitation.

miRNAs are considered to be strong prognostic markers and key therapeutic targets in various human diseases, especially cancer [17,18]. More than 600 different miRNAs encoded in the human genome negatively regulate gene expression at the posttranscriptional level by inducing translational repression and/or destabilizing specific mRNAs by targeting their 3' untranslated regions (UTRs) [19]. Thousands of different genes can be subject to regulation by a single miRNA or miRNA family [20]. Although the action mechanisms and oncogenic roles of miRNAs are relatively well understood in cancer, the roles of EV-related miRNAs and their potential applications in monitoring tools and therapeutic approaches are still under investigation.

In this report, we describe for the first time a comprehensive smRNA profile from the EVs of LM patient CSF, as obtained via an unbiased polyadenylation-based smRNA library construction procedure and subsequent NGS analysis. We performed deep sequencing on a subpopulation of relatively well-characterized smRNAs found in CSF EVs from LM patients and healthy control donors (HCs). The significance of biased expression of smRNA subpopulations was extensively validated using various biochemical methods. Moreover, the functionality of miR-21, which was found to be the most essential among the LM CSF EV-relevant smRNAs, was verified using a newly developed multipurpose lentivirus-based miR-21 monitoring system and physiological cell-based approaches. Finally, we discuss the potential roles of miR-21 and other essential smRNAs in the progression of LM and their usefulness for diagnosing LM.

2. Results

2.1. Biochemical and Molecular Characterization of EVs from CSF of LM Patients

Characteristics of the 19 LM patients and 16 HCs are summarized in Table 1 and Table S2.

Table 1. Clinical characteristics of CSF samples and their applications in this study ($n = 24$).

Patients No.	Gender	Age	Patient Group	Primary Disease	Sample Site	Applications
LM1	Female	67	LM	NSCLC	Intraventricular	NGS, ddPCR, Luc, WB, SL
LM2	Female	67	LM	NSCLC	Lumbar	NGS, ddPCR, ExoView, WB, SL
LM3	Female	63	LM	NSCLC	Lumbar	NGS, ddPCR, Luc, SL
LM4	Male	44	LM	NSCLC	Lumbar	NGS, ddPCR, Luc
LM5	Female	54	LM	NSCLC	Lumbar	NGS, ddPCR, M.A., Luc
LM6	Male	54	LM	NSCLC	Lumbar	NGS, ddPCR, Luc
LM7	Male	69	LM	NSCLC	Lumbar	NGS, ddPCR, M.A., Luc
LM8	Female	36	LM	Breast cancer	Lumbar	NGS, ddPCR, ExoView qRT-PCR, Luc,
LM9	Female	55	LM	NSCLC	Intraventricular	ddPCR (E.V.), qRT-PCR, Luc
LM10	Male	62	LM	NSCLC	Intraventricular	ddPCR (E.V.), Luc
LM11	Male	65	LM	NSCLC	Intraventricular	Luc
LM12	Female	63	LM	NSCLC	Intraventricular	ddPCR (E.V.), Luc
LM13	Male	68	LM	NSCLC	Lumbar	Luc
LM14	Male	56	LM	NSCLC	Intraventricular	Luc
HC1	Female	61	Healthy control	Unruptured an	Cisternal	NGS
HC2	Female	60	Healthy control	Unruptured an	Cisternal	NGS, ddPCR
HC3	Male	50	Healthy control	Unruptured an	Cisternal	ddPCR
HC4	Female	73	Healthy control	Unruptured an	Cisternal	ddPCR
HC5	Female	45	Healthy control	Unruptured an	Lumbar	ExoView, WB, ddPCR (E.V.)
HC6	Male	69	Healthy control	Unruptured an	Lumbar	ExoView, M.A., Luc, ddPCR (E.V.)
HC7	Female	55	Healthy control	Unruptured an	Lumbar	qRT-PCR, M.A., Luc, ddPCR (E.V.)
HC8	Male	61	Healthy control	Unruptured an	Lumbar	qRT-PCR
HC9	Male	59	Healthy control	Unruptured an	Intraventricular	ddPCR (E.V.)
HC10	Male	40	Healthy control	Unruptured an	Intraventricular	ddPCR (E.V.)

LM, leptomeningeal metastasis; HC, healthy control; NSCLC, non-small cell lung cancer; Unruptured an, unruptured aneurysms; NGS, next generation sequencing; ddPCR, droplet digital PCR; qRT-PCR, real-time reverse transcription polymerase chain reaction; M.A., migration assay; Luc, luciferase assay; WB, western blot; SL, splinted ligation; E.V., external validation.

The mean age of all patients was 59.97 years (range, 36–73 years). Overall, 19 patients were female and 16 were male. Non-small cell lung cancer (NSCLC) was the most frequent primary cancer type among LM patients except LM8, 17, and 19 (breast cancer). To identify potential diagnostic small RNA (smRNA) biomarkers in LM patient CSF EVs, we first optimized a procedure for isolating EVs from a minimal amount of CSF (Figure 1A). All sample preparations and analyses were performed in accordance with our recently published work [8]. EVs were isolated from 2 mL (LM patients) or 4 mL (HCs) of CSF, and then the total RNAs were purified and further analyzed via NGS. As shown in Figure 1B, the presence and characteristics of the EVs were first verified using ExoView Tetraspanin Chip on the LM or HC CSF samples [8]. Beads bearing CD9/CD63/CD81 antibody-captured EVs were measured for their mean fluorescence intensity. Similar fluorescence patterns were obtained between LM patients and HCs, indicating that EVs were present in both LM and HC CSFs (Figure 1B, bottom).

After removing cellular components and fragments, we measured the concentrations and sizes of the presumed EVs from LM CSF by nanoparticle tracking analysis (NTA) using a NanoSight NS300 (for details, see Supplementary Materials and Methods; Figure 1C and Table S1). Nineteen LM patients exhibited a relatively high mean EV concentrations compared with 16 HCs ($p = 0.004$) which showed a similar pattern as we previously reported [8].

The EVs were isolated from LM CSF using a miRCURY exosome isolation kit (Figure 1D), and the purified EV pellets were used for western blotting (WB) to identify well-known EV markers [11,21–24]. As shown in Figure 1E and Figure S1, the representative EV markers, flotillin-1, CD63, CD81, and CD9, were clearly detected in HC and LM CSFs, and their expression patterns were similar between the groups. This confirmed that our EVs possessed biochemical characteristics similar to those reported previously. The absence of GM-130 (a Golgi marker) and cytochrome C (Cyto. C, a mitochondrial marker) excluded potential contamination with cellular vesicular structures, such as those from the Golgi and mitochondria [8].

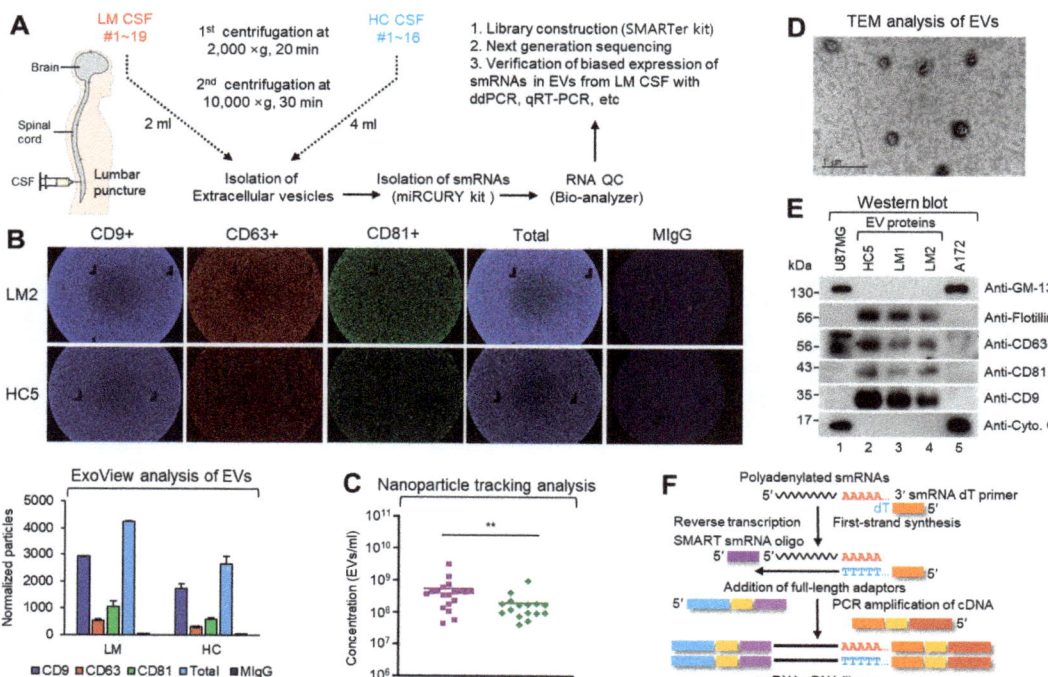

Figure 1. Biochemical and molecular characterization of EVs from CSF of LM patients. (**A**) Brief schematic flowchart and analytical procedure for patient sample collection, smRNA preparation, library construction, NGS, and biochemical analysis of EVs from CSF of LMs and HCs. CSF samples were obtained from patients harboring leptomeningeal metastasis, from intraventricular puncture ($n = 12$), cisternal puncture ($n = 6$) or lumbar puncture ($n = 17$). (**B**) The relative expression of EV specific markers was measured using ExoView R100 platform and Tetraspanin kit. CSF samples were incubated with the mixture of CD9-, CD63-, and CD81-capture antibodies on the Tetraspanin chip. After washing with washing buffer, the fluorescence images of chips were obtained with the confocal microscopy-based camera. Each antibody conjugated particles in 50-200 nm size were measured with the ExoView built-in software (bottom; bar graphs). Mouse IgG (MIgG) was used for negative control. The data represent the mean values of three independent experiments ($n = 3$) and error bars in the graph represent ± standard deviation. (**C**) Nanoparticle tracking analysis (NTA) revealed the difference in the EV concentrations between LM (purple rectangle, $n = 19$) and HC (green rhombus, $n = 16$). NTA was repeated 3 times per sample and an average value was provided ($n = 3$, two-tailed t-test, ** $p < 0.01$). (**D**) Transmission electron microscopy (TEM) shows bi-membranous vesicles of purified CSF EVs. scale bar: 1 µm. (**E**) Western blot analysis of EV markers (Flotillin-1, CD63, CD81 and CD9) and cellular proteins (GM-130 and cytochrome C in U87MG and A172 glioma cells) in HC and LM CSF EVs. (**F**) Schematic representation cDNA synthesis procedure of SMARTer smRNA-Seq method. Priming of a polyadenylated smRNA by a 3' smRNA dT primer is followed by the synthesis of the first-strand cDNA by a MMLV-derived reverse transcriptase (RT). Extension and template switching then allows for a complete single stranded cDNA. PCR amplification allows for the addition of library adaptors and the synthesis of second-strand cDNA.

Next, we sought to isolate total RNA from the EVs, as this would allow us to use NGS to analyze smRNA profiles. Total RNAs were isolated with a miRCURY RNA isolation kit and the criterion was evaluated based on the RNA size in the Bioanalyzer and the total amount of RNA in RiboGreen (Figure S1) [13].

For quality control during smRNA library construction, more than 10 commercially available RNA isolation kits were tested in the initial step. However, only the miRCURY RNA Isolation Kit-Cell & Plant (#300110, Exiqon; Qiagen, Hilden, Germany) was effective for RNA sequencing quality control and further library construction. As shown in Figure S1, smRNAs derived from HC and LM CSF-isolated EVs were analyzed using an Agilent Bioanalyzer with smRNA chips and showed relatively fair smRNA profiles for both HC and LM EVs. When we set out to construct the smRNA library, we tested numerous kits but obtained successful results only using the SMARTer smRNA-Seq Kit for Illumina (Takara Bio Inc., Shiga, Japan), which involves polyadenylation-mediated cDNA amplification of smRNA (Figure 1F). We then subjected the EV smRNA obtained from two healthy controls (HC1 and HC2) and eight LM patients (LM1-8) to smRNA NGS, which yielded averages of 6.4 million reads and 13.2 million reads, respectively. Uniquely clustered reads were then sequentially aligned to reference genome, miRBase v21, and the non-coding RNA database, Rfam 9.1, to identify known miRNAs and other types of smRNA subpopulations [24–26].

2.2. Essential Subpopulation of smRNAs Show Biased Expression Patterns in LM CSF EVs

From among the annotated smRNA populations, we identified and focused on well-known classes that were present and asymmetrically distributed in HC versus LM CSF EVs. The most abundant housekeeping RNAs and contaminating mRNA fragments were excluded from the comparison and further analysis. We determined the relative distributions of the ten most abundant classes between HC and LM EVs, which corresponded to 33.1% of all aligned reads. These RNA classes included miRNA, piRNA, Y RNA, snoRNA, snRNA, vtRNA, novel miRNA, and scRNA (Figure 2).

The remaining aligned reads of EVs represented rRNA, tRNA, etc. All of the sorted and identified RNA classes exhibited differential distribution between HC and LM. Interestingly, as shown in Figure 2B, all of these smRNA subpopulations were relatively enriched in EVs from LM CSF compared to HC.

2.3. Analysis of Relative Expression Profiles and Related Cellular Pathways of miRNA in EVs from LM Patient CSF

To profile the EV miRNAs, we obtained approximately 3.7 million to 24.3 million raw reads using Illumina HiSeq. The raw reads of EV miRNAs from LM patient CSF were preprocessed, analyzed with miRDeep2, and then trimmed for adapter sequences [27,28]. Differentially expressed miRNAs between HC and LM CSF EVs were determined by selecting those with |fold change| ≥ 2 and p-value < 0.05. A total of 46 significantly differentially expressed miRNAs were identified, and hierarchical clustering showed that the miRNA expression profile of LM EVs was markedly distinct from that of HC (Figure 3 and Figure S2).

Next, we validated the nine differentially expressed miRNAs using droplet digital polymerase chain reaction (ddPCR) [29]. We selected hsa-miR-21-5p, hsa-miR-19b-3p, hsa-miR-25-3p, hsa-miR-200c-3p, hsa-miR-19a-3p, and hsa-miR-34b-3p for validation of the biased expression between LM and HC. As expected, these miRNA molecules were significantly enriched in LM CSF EVs compared with HC (Figure 4A). In contrast, hsa-miR-423-5p, hsa-miR-1273g-3p, and hsa-miR-4271, which were found to be significantly downregulated in LM CSF EVs, were validated as being significantly downregulated in our ddPCR analysis (Figure 4B).

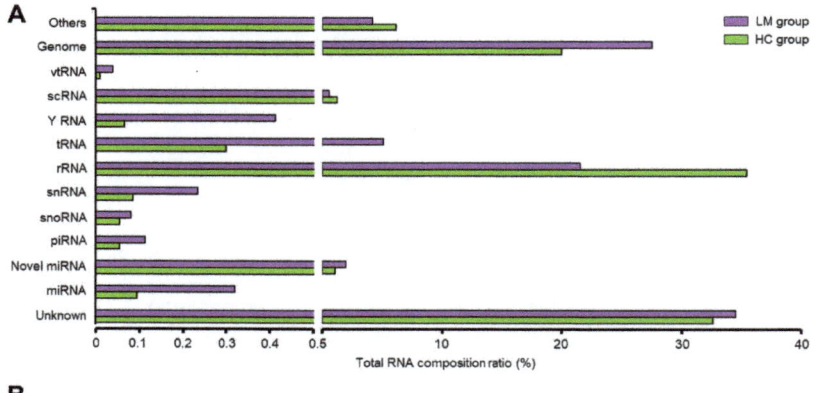
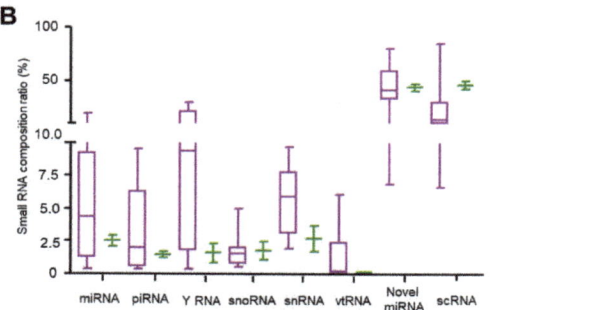

Figure 2. The proportion of smRNA subtypes showed biased expression between CSF EVs from LM and healthy control (HC). (**A**) smRNA sequencing data were processed for 10 types of general sm RNAs subclasses including genome and unknown population in RNA central database. Bar graph showed the average percentage of each smRNA subtype. (**B**) After removing reads mapped to rRNA, tRNA, and others, eight subtypes (miRNA, piRNA, Y RNA, snoRNA, snRNA, vtRNA, novel miRNA, and scRNA) of biologically intriguing smRNAs were processed. Boxplot showed the median percentage value and interquartile range of each smRNA subtypes. (LM, purple box, $n = 8$; HC, green bar, $n = 2$).

As shown in Figure 3E, hsa-miR-21-5p was ranked first in relative expression. We thus selected miR-21 for further analysis. This expression was validated using and conventional TaqMan probe-based real-time reverse transcription PCR (qRT-PCR). hsa-miR-200c-3p, which was the fifth-ranked miRNA in LM CSF EVs, was also selected for further analysis since it is known to play important roles in the metastasis and mobilization of cancer cells. As shown in Figure 4C, qRT-PCR result showed more than hundred folds elevation of both miRNAs in EVs from LM CSF which was already analyzed as significantly upregulated by NGS. The biased expression of miR-21 and miR-200c in LM CSF EVs was particularly notable since the expression levels were found to be relatively lower in CSF EVs from glioblastoma multiforme (GBM) patients than those from LM CSF. These data suggest that some yet-unknown driving mechanism governing the biased expression of both miRNAs is more potent in LM CSF EVs than in GBM CSF EVs. This may explain the poorer outcome of LM.

Out of the 46 miRNAs classified herein, additional essential miRNAs, such as hsa-miR-191-5p and hsa-miR-93-5p, also showed higher expression in LM CSF EVs than HC and GBM EVs (Figure 4D). In contrast, hsa-miR-204-5p and hsa-let-7b-5p, which are well-known to play tumor-suppressive roles [30], showed severe downregulation in LM CSF EVs compared with GBM CSF EVs (Figure S3).

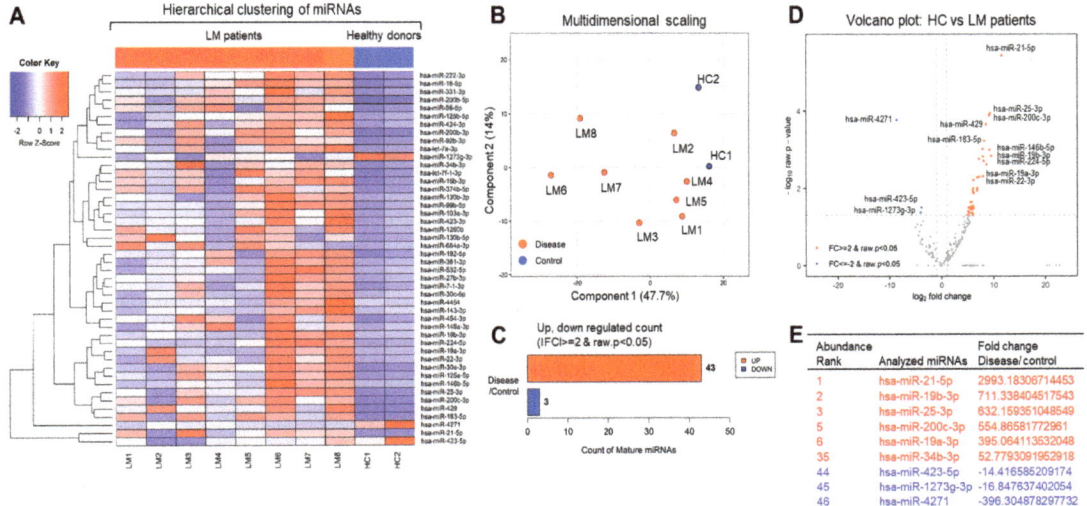

Figure 3. Distinct expression of miRNAs in EVs from LM patient CSF. Analysis of relative miRNA expression profile in EVs extracted from CSF of LMs and HCs. Hierarchical clustering analysis of significantly expressed miRNA was visualized via (**A**) heatmap showing z score of extravesicular miRNA from HC (*n* = 2) and patients with LM (*n* = 8) with 43 upregulated (red) and 3 downregulated (blue) miRNAs. (**B**) Multidimensional scaling (MDS) map of HC and LM was generated with proximity (calculated in Euclidean distance) indicating similarity, and (**C**) count of up- and downregulated mature miRNAs were found by fold change and *p*-value. (**D**) Volcano plot shows differentially expressed miRNA in HC and LM patients with the x-axis showing log2 fold-change and y-axis showing −log10 of the *p*-value from LM versus HC miRNA expression counts. (**E**) Table displays the fold change of 6 upregulated and 3 downregulated miRNAs in LM compared to HC ranked in the order of abundance.

Furthermore, we visualized the expression of miR-21-5p using a well-established conventional biochemical approach. Splinted ligation demonstrated that our ddPCR and qRT-PCR results were not caused by non-specific amplification of unwanted RNA species. As shown in Figure 4E and Figure S16, ligated bands of LM EV-derived and cellular miR-21-5p (from K562 cells) migrated to the same point as the synthetic miR-21-5p control. Moreover, similarly high and biased expression of miR-21-5p was observed in additional LM patient CSF EVs (Figure S3A), as assessed by ddPCR analysis. Our NGS (comprehensive high-throughput) and standard biochemical approaches clearly demonstrated that miR-21 is present in the EVs from CSF of LM patients. Collectively, these results suggested that the biased existence of specific population of miRNAs may contribute to LM progression.

To investigate the predicted functions of the miRNAs that were differentially expressed in LM EVs and HC EVs, we used the Database for Annotation, Visualization and Integrated Discovery (DAVID) functional annotation tool to perform Gene Ontology (GO) analysis. The significantly enriched GO terms included biological processes and molecular functions (Figure 5).

The 43 miRNAs upregulated in LM CSF EVs represented processes such as cellular response to hypoxia, positive regulation of cell proliferation, positive regulation of transcription, and positive regulation of gene expression in the biological process category. The selected miRNA, miR-21-5p, is involved in wound healing, cellular response to hypoxia, positive regulation of transcription, and negative regulation of apoptotic process.

We also used a Kyoto Encyclopedia of Genes and Genomes (KEGG) pathway enrichment analysis to examine the signaling pathways associated with each of the 43 miRNAs found to be differentially expressed in LM CSF EVs versus HC. Our KEGG analysis revealed that the pathways of glioma, small cell lung cancer, pathways in cancer, and miRNAs in cancer were particularly relevant. For miR-21-5p, the pathways of HIF-1 signaling, MAPK signaling, cancer and miRNAs in cancer were found to be significant.

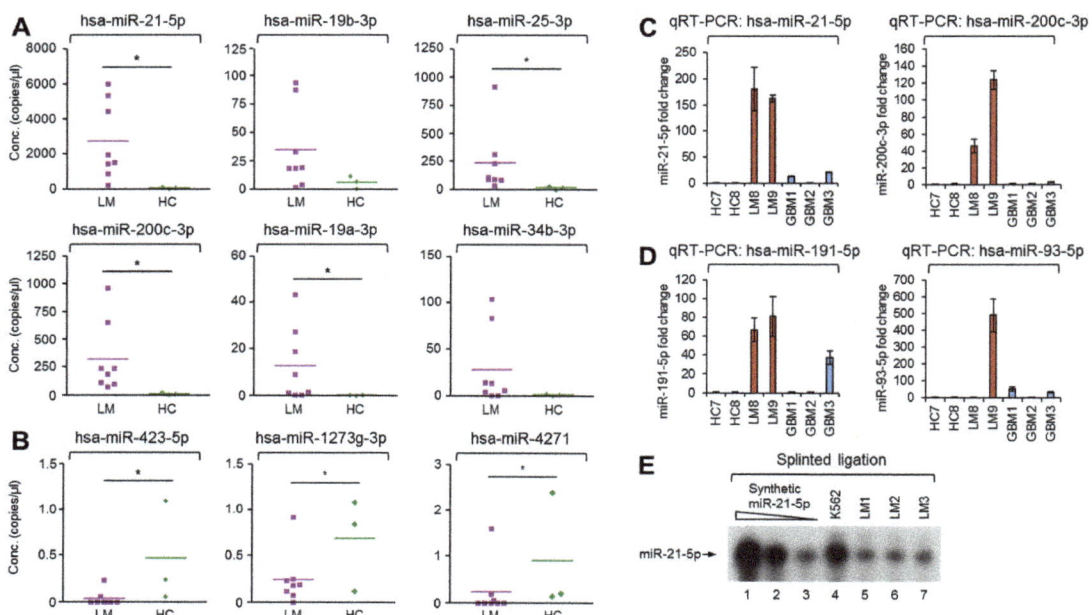

Figure 4. Biochemical verification of biased miRNA expression in EVs derived from LM and HC by ddPCR. (**A**) 6 upregulated miRNAs were analyzed by ddPCR (ddPCR Supermix for Probes; hsa-miR-21-5p, hsa-miR-19b-3p, hsa-miR-25-3p, hsa-miR-200c-3p, hsa-miR-19a-3p, and hsa-miR-34b-3p) in EVs from LM (purple rectangle, $n = 8$) vs HC (green rhombus, $n = 3$). (**B**) 3 downregulated miRNAs were analyzed by ddPCR (QX200 ddPCR EvaGreen Supermix; hsa-miR-423-5p, hsa-miR-1273g-3p, and hsa-miR-4271) in EVs from LM (purple rectangle, $n = 8$) vs HC (green rhombus, $n = 3$) (Mann-Whitney U test, * $p < 0.05$). (**C**) 2 upregulated miRNAs were analyzed by qRT-PCR (TaqMan Advanced miRNA Assay probes; miR-21-5p and miR-200c-3p) in CSF EVs from 2 LMs (LM8 and LM9), 2 HCs (HC7 and HC8) and 3 glioblastoma multiforme patients (GBM1-3). (**D**) 2 upregulated miRNAs were analyzed by qRT-PCR (TaqMan Advanced miRNA Assay probes; hsa-miR-191-5p and hsa-miR-93-5p) in CSF EVs from 2 LMs (LM8 and LM9), 2 HCs (HC7 and HC8), and 3 glioblastoma multiforme patients (GBM1-3). The data represent the mean values of three independent experiments ($n = 3$) and error bars in the graph represent ± standard deviation. (**E**) Splinted ligation was performed with [^{32}P] 5′-end-labeled oligonucleotide probe specific for mature miR-21-5p as described in Supplementary Materials and Methods. The experiments were repeated at least three times with similar results. The data shown in panel (**E**) is a representative image.

2.4. miRNA Sensor-Based Investigation of miR-21 Functionality in EVs from LM Patient CSF

To demonstrate whether the miRNA in the EVs from LM CSF could be physically incorporated into the miRNA-induced silencing complex (miRISC) and be functional within the miRNA machinery, we devised a lentivirus-based miR-21-sensing reporter. In this system, firefly luciferase is transcribed under the control of the cytomegalovirus (CMV) promoter and the translation of its mRNA is governed by five consecutive miR-21-targeting sites located in the 3′UTR. The devised system offers the ability to easily monitor the positive role of miR-21-containing LM CSF. The expression of firefly luciferase can be easily normalized using human phosphoglycerate kinase 1 (hPGK) promoter-driven *Renilla* luciferase activity in a dual luciferase assay (see Figure 6A). More details on this and other miRNA-monitoring sensors were described in our recent report [31]. In the present study, the binding of miR-21 to its target sites in the 3′UTR of the firefly luciferase mRNA repressed its translation, such that firefly luciferase activity was inversely correlated with the cellular level of miR-21 [31]. When we used the developed system to examine cellular expression in various cell lines, we found that miR-21 was highly expressed in NSCLC A549 cells and was barely detectable in 293T [31].

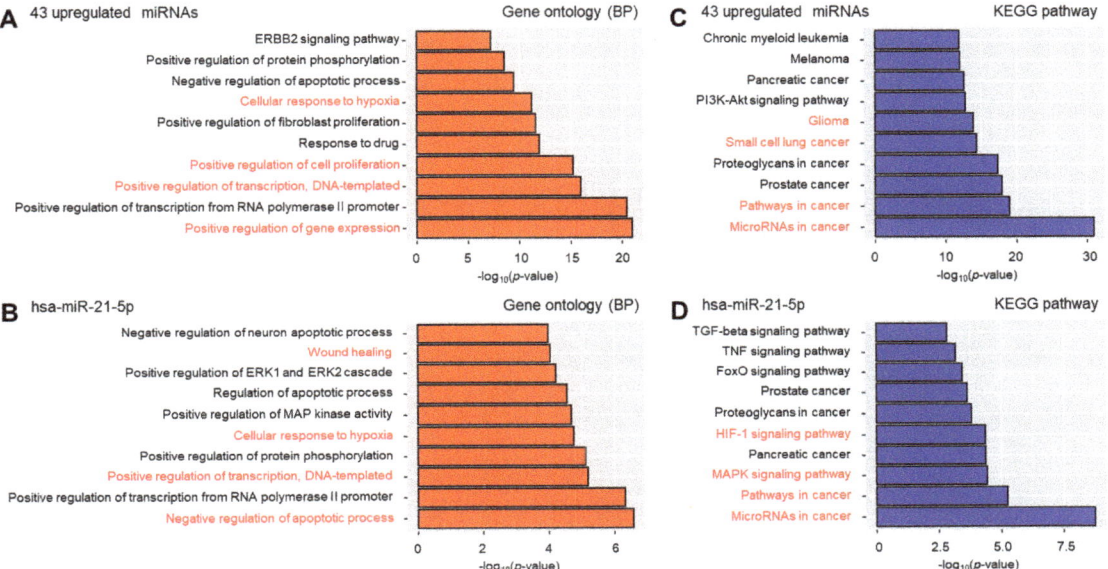

Figure 5. Analysis of the miRNA target gene enriched pathway. The experimentally validated target genes of 43 upregulated miRNAs (**A**,**C**) or hsa-miR-21-5p (**B**,**D**) were searched from miRTargetLink Human database. The target gene enriched pathways were computed from the Database for Annotation, Visualization and Integrated Discovery (DAVID). Bar graph depicted the top 10 highly enriched and biologically relevant pathways on Gene Ontology (GO) and Kyoto Encyclopedia of Genes and Genomes (KEGG) pathway. BP, biological processes.

Next, to test the specific reactivity of luc-miR-21 sensor against LM CSF containing potentially functional miR-21s, we treated LM CSF directly onto luc-miR-21 sensor-bearing 293T cells. As shown in Figure 6B, in most cases, the LM CSF markedly inhibited the translation of firefly luciferase in luc-miR-21 sensor-bearing 293T cells. The physiological activity of LM CSF harboring miR-21 was further confirmed by directly applying isolated LM and HC CSF EVs to luc-miR-21 sensor-bearing A549 cells. As shown in Figure 6C, LM CSF EV inhibited the translation of firefly luciferase in luc-miR-21 sensor-bearing A549 cells, whereas HC CSF EVs did not.

These data collectively demonstrated that LM CSF and LM CSF EVs contain functional miR-21, indicating that this miRNA is physically incorporated into the cellular miRISC. This event can potentiate downstream cellular oncogenic signal cascades triggered by miR-21, which may induce LM malignancy among targeted recipient cells in patients.

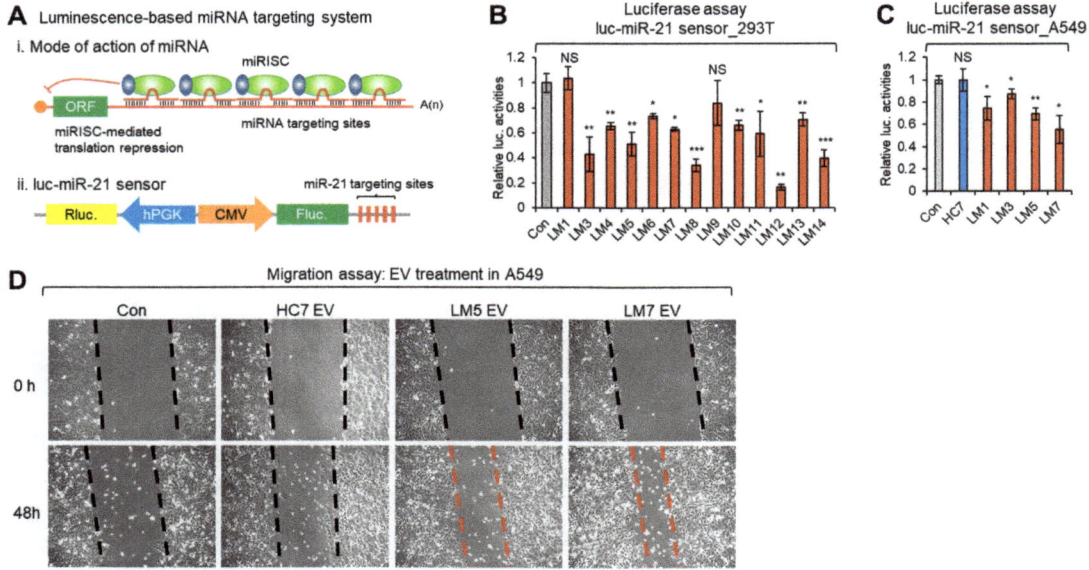

Figure 6. miRNA sensor and cellular migration-based investigation of extravesicular miR-21 functionality of LM patient CSF. (**A**) Mode of action of miRNA targeting (i), and schematic depiction of luminescence-based miR-21 sensor system (ii). miRISC; miRNA-induced silencing complex. (**B,C**) Relative firefly luciferase activities showed inverse correlation with functional miR-21 expression level in CSFs (**B**) and EVs (**C**), which binds to the miRNA targeting sites in 3′UTR of firefly luciferase in each luc-miR-21 sensor-bearing 293T (**B**) and NSCLC A549 (**C**) cells. *Renilla* luciferase activities were used in normalization of firefly luciferase activities. Data represent the mean values of three independent experiments (n = 3). Error bars in the graph represent ± standard deviation, and the *p*-values compare each LM CSF group to medium control (Con). * $p < 0.05$, ** $p < 0.01$, *** $p < 0.001$; NS, not significant. (**D**) Representative images of A549 migration assays after 48 h treatment with EVs from LM CSF (LM5 and LM7) and HC (HC7). Phase contrast microscopy images were taken with Axio Observer (100×, Zeiss).

2.5. Effects of miR-21-Containing LM CSF EVs on the Migratory Phenotype of NSCLC A549 Cells

To test whether miR-21 from LM CSF EVs can potentiate downstream cellular oncogenic signal cascades, we used a migration (wound-healing) assay. First, to elucidate the effect of miR-21 on the migration of A549 cells directly in vitro, we transfected the cells with synthetic miRNA mimics or negative control nucleotides and performed a migration assay. As shown in the Figure S4, the migratory phenotype of A549 cells was markedly enhanced at the edge of the scratch at 48 h in cells transfected with synthetic miR-21-5p, but not in those subjected to mock transfection or transfected with negative control nucleotides. Based on this observation, we next tested the effect of LM CSF EVs on NSCLC A549 cell motility. As shown in Figure 6D, Figures S13 and S17, the migratory phenotype of A549 cells was markedly enhanced at the edge of the scratch at 48 h in cells treated with LM CSF EVs, compared to untreated control cells or those treated with HC EVs. These results collectively suggested that EVs from LM CSF can promote the motility of NSCLC A549 cells in vitro, and these EVs contain a key regulator(s) of migration; thus, may be mediated by miR-21.

2.6. Biased Expression of Piwi-Interacting RNAs and Y RNAs in EVs from LM Patient CSF

Interestingly, NGS analysis of the LM CSF EVs revealed the biased expression of Piwi-interacting RNAs (piRNAs). piRNAs form RNA-protein complexes through interactions with Piwi-subfamily proteins. The piRNA complexes are mostly involved in the epigenetic and posttranscriptional silencing of transposable elements [32]. As far as we know, this is the first report of piRNA expression in LM CSF EVs. As shown in Figure 7, hierarchical clustering revealed that piRNA expression could be clearly distinguished between LM CSF EVs and HC. Our analysis showed that 35 piRNAs were significantly upregulated in LM CSF EVs, while 19 were significantly downregulated. Our results suggested that piRNA could be a pathogenic index that could be used to discriminate LM. The analyzed LM samples showed similar clustering groups, whereas HC samples showed a very distinctive pattern (Figure S5). We further used ddPCR to validate two highly ranked piRNAs among the differentially expressed molecules. As shown in Figure 7E, hsa-piR-36340 and hsa-piR-33415 were significantly enriched in LM CSF EVs compared with HC.

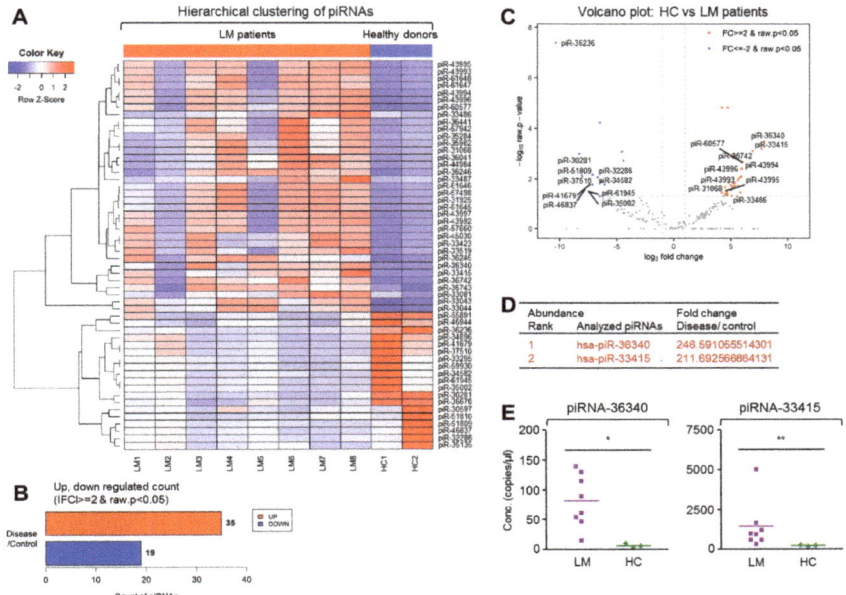

Figure 7. Distinct expression of Piwi-interacting RNAs in EVs from LM patient CSF. Analysis of relative Piwi-interacting RNA (piRNA) expression profile in EVs extracted from LMs and HCs. Hierarchical clustering analysis of significantly expressed piRNA was visualized via (**A**) heatmap showing z score of extravesicular piRNA from HC ($n = 2$) and patients with LM ($n = 8$) with 54 piRNA satisfying FC2 and raw p-value. (**B**) Count of up- and downregulated piRNA was found by fold change and raw p-value. (**C**) Volcano plot shows differentially expressed piRNA in HC and LM patients with the x-axis showing log2 fold-change and y-axis showing -log10 of the raw p-value from LM versus HC piRNA expression counts. (**D**) Table displays the fold change of two piRNA in LM compared to HC ranked in the order of abundance. (**E**) Level of the two piRNAs was confirmed in HC ($n = 3$) and LM EVs ($n = 8$) using ddPCR (piRNA-36340: Mann-Whitney U test, * $p < 0.05$; piRNA-33415: Unpaired t-test with Welch's correction, ** $p < 0.01$).

Another interesting smRNA species, Y RNAs, also showed markedly biased expression patterns between LM CSF EVs and HC. The members of this well-characterized smRNA subpopulation act as components of the Ro60 ribonucleoprotein particle, which is a target of autoimmune antibodies in patients with systemic lupus erythematosus [33]. Y RNAs are also necessary for DNA replication through their interactions with chromatin and initiation proteins [34]. In the LM CSF EVs, the number of upregulated Y RNAs (156) was greater than that seen for miRNA or piRNA, whereas only seven Y RNAs were downregulated. Our data suggested that Y RNA could be very useful as diagnostic biomarkers for LM. We used ddPCR to validate Y RNA-Y69, which was highly ranked as being differentially expressed in LM CSF EVs. As shown in Figure 8E, this confirmed that Y RNA-Y69 was significantly enriched in LM CSF EVs compared with HC CSF EVs.

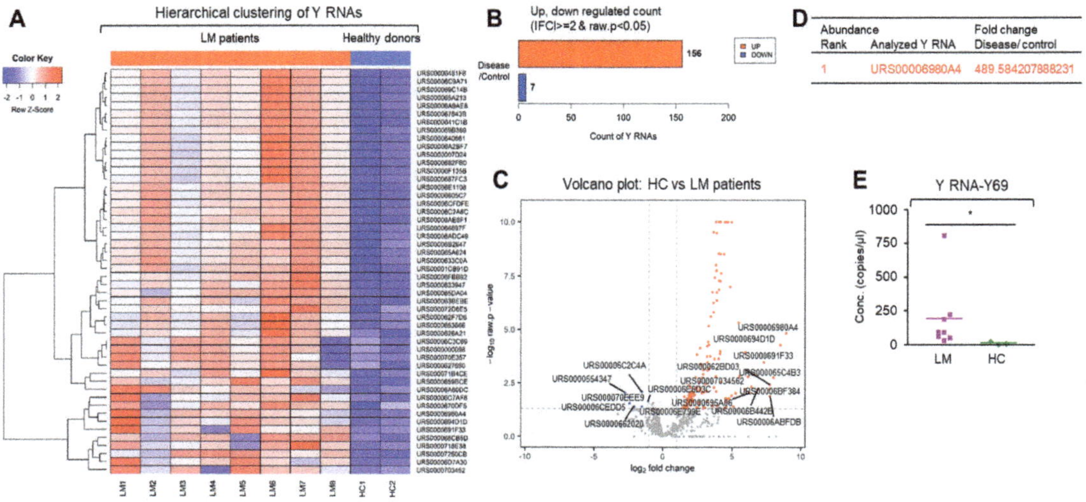

Figure 8. Distinct expression of Y RNAs in EVs from LM patient CSF. Analysis of relative Y RNA expression profile in EVs extracted from LMs and HCs. Hierarchical clustering analysis of significantly expressed Y RNA was visualized via (**A**) heatmap showing z score of extravesicular Y RNA from HC ($n = 2$) and patients with LM ($n = 8$) with 50 Y RNA that best satisfied FC2 value and adjusted p-value. (**B**) Count of up- and downregulated Y RNA was found by fold change and raw p-value. (**C**) Volcano plot shows differentially expressed Y RNA in HC and LM patients with the x-axis showing log2 fold-change and y-axis showing -log10 of the raw p-value from LM versus HC Y RNA expression counts. (**D**) Table displays the fold change of a Y RNA in LM compared to HC ranked in the order of abundance. (**E**) The upregulated Y RNA-Y69 expression was confirmed in HC ($n = 3$) and LM EVs ($n = 8$) using ddPCR (Mann-Whitney U test, * $p < 0.05$).

2.7. Biased Expression of Essential Small RNA Subpopulations in EVs from LM Patient CSF: snRNA, snoRNA, vtRNA, Novel miRNA, and scRNA

We additionally identified meaningful and essential smRNA subpopulations that showed markedly biased expression in our NGS analysis of LM CSF EVs versus HC. As shown in Figures S10 and S11, the analyzed reads were annotated for identification of novel miRNAs, which were predicted from mature, star, and loop sequences according to the RNAfold algorithm miRDeep2 [28]. In contrast to the well-identified miRNA, the potential novel miRNA identified based on this analysis were largely downregulated (48, |fold change| ≥ 2 and p-value < 0.05), with only nine exhibiting upregulation. This suggested that, except for the nine cases, most of the predicted novel miRNAs may play a role in LM progression, opposing the function of known miRNAs. Future work is warranted to examine whether these predicted novel miRNAs are expressed and functional. We used ddPCR to validate the highly ranked upregulated novel miRNA-973. As shown in Figure S10, novel miRNA-973 was significantly enriched in LM EVs compared with HC. The data indicated that at least the upregulated novel miRNAs are processed in the LM CSF EVs and thus may exert positive functions in LM pathogenesis.

Finally, we examined essential smRNAs, such as snRNA, snoRNA, vtRNA, and scRNA (Figures S6–S9, S11, and S12). We found that snRNA, snoRNA, and vtRNA were highly enriched in LM CSF EVs. Without exception, all of the identified snRNA, snoRNA, and vtRNA were detected to a much greater extent in LM CSF EVs compared to HC. We could speculate that the skewed distribution of snoRNAs in LM CSF EVs is likely a reflection of their biological function(s) in the cell, whereas the observed enrichment of several snoRNA fragments may be attributed to a miRNA-like cytoplasmic function.

snRNA-200C70, vtRNA-6C57F3, and scRNA-F3729, which were found to be differentially expressed in LM CSF EVs versus HC, were checked by ddPCR. As shown in Figures S6, S9, and S12, they were significantly enriched in LM EVs compared with HC. This suggests that these smRNAs may contribute to LM pathogenesis.

3. Discussion

3.1. Comprehensive and Quantitative Analysis of Essential smRNA Subpopulation in EVs from LM patient CSF

LM is generally considered to be a late complication of solid tumors [3]. LM causes fatal cancer complications; thus its early diagnosis and monitoring are essential for effective treatment and improved disease prognosis [2]. To date, we lack any clear biomarkers that reflect the disease progression or molecular mechanisms of LM; this can be a major obstacle to finding effective treatments. Therefore, efforts to identify biomarkers for monitoring and investigating the mechanisms of the disease are critical to improve the prognosis of LM and explore new therapeutic options. In several studies on LM, CSF cytology and magnetic resonance imaging methods were used for diagnosis; however, these methods are highly dependent on the examiner and are limited by reader-to-reader variability – entailing a high risk for non-specific results [2,6]. Moreover, repeated lumbar punctures for CSF cytology are not favorable for patients [35]. The data from this study strongly indicate that analysis of smRNAs in CSF could be a promising approach for developing minimally invasive assays for LM detection.

In this study, we examined CSF smRNA levels in individual LM patients as potential markers for monitoring disease diagnosis. This is the first comprehensive and systemic analysis of the smRNA molecular profiling in a standard set of CSF samples from individual LM and non-LM patients. The identification of specific CSF smRNAs in this study suggests new ways to diagnose and monitor LM and may contribute to efforts to explore the mechanisms of LM.

The use of smRNA as LM biomarkers could offer several benefits. First, smRNAs, essentially miRNAs, are functional in the biological system, and thus could be easily applied for biochemical and physiological assays [13,20]. Here, we demonstrated that LM CSF EV miRNA are physiologically functional using a specific miR-21 biosensor and various cell-based assays. Second, smRNAs have sequence specificity and may be amplified with various biochemical assays, making them suitable for disease-specific diagnostics [20]. Third, smRNAs are useful because they are typically less than 200-bp in size, and thus do not need to be fragmented prior to library preparation [36,37]. Finally, storage conditions do not influence the quality or distribution of the recovered smRNAs [38]. Therefore, smRNA could be better biomarkers for the diagnosis of LM compared to macromolecules such as large RNA, DNA, proteins, etc.

Our approach and quantitative-qualitative analyses of the smRNAs from LM CSF EVs are superior to those of previous reports, as indicated by the following lines of evidence. First, in this study, we set up a small-scale (2–4 mL of CSF) pipeline for analyzing the entire small RNA transcriptome of CSF EVs. We successfully conducted RNA-sequencing of extravesicular smRNAs using 3′-end polyadenylation-based SMARTer smRNA-Seq commercial kit. We tested around 10 other kits for total RNA isolation that did not pass library quality control and/or were not successful for library construction. In addition, this protocol, which does not require laborious ultracentrifugation or large volumes of CSF, may be practical in the clinical setting.

Previous reports showed that conventional adaptor-based approaches yielded biased results for smRNA populations [37]. Compared to the conventional method, the polyadenylated method offers easier construction of the smRNA library, an improved success rate for library construction, and a decreased risk for bias due to the 5′ and 3′ end nucleotide composition. To our knowledge, this is the first report of the global identification of smRNAs from the LM CSF EVs of individual patients. In addition, our study utilized small-scale volume of CSF, unlike previous studies in which samples were pooled or required large volumes of CSF [13]. Yagi et al. previously performed NGS-based miRNA profiling of CSF exosomes. However, the authors used CSF from several healthy volunteers, did not analyze disease-related smRNAs, and examined only miRNAs. Here, we systemically analyzed the extravesicular smRNA profile at an individual level. In this way, we may achieve our goal of efficient personalized and diseased-oriented analysis of smRNA populations. In addition to relevant miRNAs, we first analyzed and identified other essential smRNA subpopulations, including piRNA, Y RNA, snRNA, snoRNA, and vtRNA, in the LM CSF EV context.

Second, the biased expression of LM CSF extravesicular smRNA was systemically, biochemically, and functionally validated through various biochemical and cellular approaches. Interestingly, we found that the potential LM pathogenesis-related miRNAs were highly upregulated in LM CSF EVs. Our study also revealed that the ratios of different types of extravesicular smRNAs differed between LM patients and HC. Furthermore, identification of new smRNA population will increase our knowledge of LM.

Finally, the presence of smRNAs separated and identified on a large scale was quantitatively confirmed using ddPCR (also see Figure S14 for normalized expression) as well as qRT-PCR. These validated smRNA species, which may be analyzed using simple and precise biochemistry, could potentially be used as cost-saving and efficient biomarkers. We urgently need new strategies for the molecular diagnosis and liquid biopsy of LM using simple and effective biochemical approaches.

3.2. Implication of miRNAs for LM Pathogenesis and as Essential Biomarkers for LM Diagnosis

Through the analysis of smRNA populations via a polyadenylation-mediated library construction method, we identified 46 miRNAs that showed meaningful differences between LM CSF EVs and HC. We biochemically validated the differential expression of hsa-miR-21-5p, hsa-miR-19b-3p, hsa-miR-25-3p, hsa-miR-200c-3p, hsa-miR-19a-3p, and hsa-miR-34b-3p, which were significantly higher in LM patients than in HCs. The top-scoring miRNA, miR-21, is intriguing since it has been suggested as a critical biomarker of various cancers and is generally considered to act as an oncogene by negatively regulating various tumor-suppressive target mRNAs [39]. For example, miR-21 plays significant roles in central nervous system (CNS)-related tumors [40]. Our group and others previously showed that miR-21 expression is tightly correlated with the malignancy of glioma [31,41], and we have recently shown that miR-21 exhibited significant biased expression after chemotherapeutic treatment [8]. Thus, the previous and present findings suggest that miR-21 could be an essential biomarker for the treatment response of LM. Moreover, our experimental results suggest that miRNAs, especially miR-21, are key components of LM pathogenesis and characterization.

We systemically validated the function of extravesicular miR-21 in LM CSF via our newly developed miR-21 sensor [31]. Using this sensitive system, we were confident that LM CSF contains functional miRNAs, not just meaningless smRNA fragments. This was also confirmed with isolated EVs. Our data indicated that LM CSF EVs, which contain higher level of functional miR-21, positively affected the migratory phenotype of NSCLC A549 cells, whereas HC CSF EV did not.

Another top-scoring miRNA, miR-19b, is known to enhance proliferation and apoptosis resistance via the EGFR signaling pathway by targeting PP2A and BIM in non-small cell lung cancer [42]. miR-200c, which is a member of the miR-200 family, regulates epithelial-mesenchymal transition by downregulating ZEB1/2 and upregulating E-cadherin [43]. miR-200c also has important functions in proliferation, invasion, and metastasis [44]. Furthermore, miR-200c is a well-established prognostic and diagnostic marker in different cancer types. Another study investigating serum miRNAs as cancer biomarkers showed that miR-200c is associated with NSCLC, suggesting that it could potentially be useful for diagnosis [45].

Our GO analyses showed that the target genes of 43 miRNAs, including top-scored miR-21, were involved in regulating cell migration and cell differentiation. Based on the existing literature and our experimental results, we speculate that miR-21 containing 43 miRNAs may have a critical function for LM pathogenesis [8,17]. The increase of miRNA expression may activate proliferation, invasion, and migration, which are closely related to cancer metastasis mechanisms; thus promoting the occurrence of LM. We also performed KEGG pathway analysis for the target genes of the 43 miRNAs and found that they were involved in the pathways related to glioma, small cell lung cancer, pathways in cancer, and miRNAs in cancer. Many of these pathways have close connections to CNS metastasis, and abnormal signaling is likely to play a crucial role in the development of LM. Accumulating evidence has demonstrated that the activation of MAPK, PI3K-Akt, and HIF-1 signaling pathways are important for cancer progression. Further molecular investigations are needed and such work will likely provide new perspectives on the mechanisms of LM and suggest new intervention methods in LM treatment.

3.3. Implication of smRNA Subpopulations for LM Pathogenesis and as Potential Biomarkers for LM Diagnosis: piRNA, Y RNA, and other smRNAs

One of the smRNAs that showed significantly biased expression in our NGS analysis was piRNA. The piRNAs comprise the largest class of smRNA molecules expressed in animal cells. They are around 26–31-nt long, and form RNA-protein complexes through interactions with Piwi-subfamily proteins [46]. The formed piRNA complexes are mostly involved in the epigenetic and posttranscriptional silencing of transposable elements, but can also contribute to regulating other genetic elements in germline cells [32]. Consistent with our data, a growing number of studies have shown that piRNA and Piwi proteins, which are abnormally expressed in various cancers, may serve as novel biomarkers and therapeutic targets. However, the functions of piRNAs in cancer and their underlying mechanisms are not fully understood [47]. As shown in Figure 7, hierarchical clustering of piRNA expression yielded clearly different patterns in LM CSF EVs compared with HC. Thirty-five piRNAs were significantly upregulated and 19 piRNAs were significantly downregulated in LM CSF EVs versus HC. Notably, our hierarchical clustering suggested that piRNAs could be used as an index to discriminate LM. To our knowledge, this is the first study to show that biased expression piRNA in LM CSF EVs. Therefore, the essential role of piRNA for the LM complication should be pursued in the near future.

Another intriguing smRNA subpopulation that was highly upregulated in EVs from LM CSF is Y RNAs. Y RNAs are highly conserved non-coding RNAs of ~100-nt in size. These smRNAs are components of the Ro60 ribonucleoprotein particle, which is a target of autoimmune antibodies in patients with systemic lupus erythematosus. These RNAs are involved in various basic intracellular processes, such as DNA replication and RNA quality control [33]. Several studies have reported the abundant presence of Y RNAs in cell culture EV and body fluids. However, relatively little work has focused on the function and biomarker potential of extracellular Y RNAs. Notably, Y RNAs are overexpressed in some human tumors and required for cell proliferation, and miRNA-sized breakdown products of Y RNAs may be involved in other unknown pathological conditions [34]. As shown in Figure 8, hierarchical clustering showed that Y RNA expression is also distinctive in LM CSF EVs compared with HC. Our results revealed that 156 Y RNAs were significantly upregulated in LM CSF EVs, but only seven Y RNAs were significantly downregulated. The essential role of Y RNA for the LM complication also should be pursued in near future.

snoRNAs are a class of smRNA molecules that primarily guide chemical modifications of other RNAs, mainly rRNAs, tRNAs, and snRNAs. snoRNAs can function as miRNAs. It has been shown that human ACA45 snoRNA can be processed into a 21-nt long mature miRNA by Dicer [48]. Bioinformatic analyses have revealed that putatively snoRNA-derived, miRNA-like fragments occur in different organisms [49]. Mutations and aberrant expression of snoRNAs have been reported in cell transformation, tumorigenesis, and metastasis, indicating that snoRNAs may serve as biomarkers and/or therapeutic targets for cancer [50]. During LM development, aberrant expression of snoRNA may contribute to LM-related complications. Alternatively, the miRNA-like activity of snoRNA may affect recipient cells upon LM CSF EV delivery.

snRNAs represent a class of small RNA molecules that are found within the splicing speckles and Cajal bodies of the cell nucleus in eukaryotic cells. snRNA average ~150-nt in length, and their primary function is in the processing of pre-messenger RNA (hnRNA) in the nucleus and the U1 spliceosomal RNA is recurrently mutated in multiple cancers [51,52].

A major obstacle in cancer treatment is the development of chemoresistance, and vtRNAs are known to play a role in this phenomenon. vtRNAs may facilitate the export of certain chemotherapeutic drugs through binding-site specific interactions. In addition, recent studies suggest that vtRNAs may inhibit drug activity through interfering with drug target sites; thus conferring drug resistance during chemotherapy of LM [53].

With the exception of miRNAs, the smRNA subclasses have limited biochemical analysis tools and functional assays, making it difficult to functionally verify their cellular roles in LM. Furthermore, the lack of an animal model for LM further complicates the research. In the near future, we can hope that new proper model systems will be developed to facilitate the study of LM.

4. Materials and Methods

4.1. Collection of Clinical Samples and Preparatory Process

CSF samples were collected after approval of Institutional Review Board in National Cancer Center, Korea (NCC2014-0135) in accordance with the ethical guidelines outlined in the Declaration of Helsinki. The informed consent was obtained from all patients. CSF samples were obtained from each patient before the lumbar, intraventricular, and cisternal puncture. Obtained CSF was centrifuged within 1 h at $2000 \times g$ for 20 min for removing cells and cellular debris at room temperature. After first centrifugation, CSFs were further centrifuged at $10,000 \times g$ for 30 min and kept frozen at $-80\ °C$.

4.2. ExoView Analysis of EVs in CSFs

The physical and biological properties of EVs in CSF samples were characterized by using ExoView R-100 (NanoView Biosciences, Boston, MA, USA) and ExoView Tetraspanin kits (NanoView Biosciences) including anti-CD81, anti-CD63 and anti-CD9 immobilized chips, labeling agents, washing solutions (solution A and B) and blocking agent (NanoView Biosciences). The three-capture antibody spots in each tetraspanin case were arrayed in one chip thus, average and standard deviation could be measured in one chip. Briefly, the 35 µL of diluted sample with solution A was dropped on the ExoView Tetraspanin chip and incubated overnight (16 h) at room temperature (RT). After the incubation process, the sample loaded chip was washed by 1 mL of solution A for 3 min and this was repeated by three times. Subsequently, the EVs on the chip were labeled by using 250 µL of a mixture of anti-CD81/Alexa Fluor 555 (green), anti-CD63/Alexa Fluor 647 (red), and anti-CD9/Alexa Fluor 488 (blue) and incubated for 1 h at RT to measure the colocalization of tetraspanin on the surface of EVs. In this case, the fluorescein-labeled antibody was diluted in mixture of solution A and blocking solution with 1:600. Finally, the chip was rinsed by 1 ml of solution A and B and dried at RT. The EV captured chip was scanned by ExoView R-100 via nScan software (NanoView Biosciences) and data were analyzed through NanoViewer 2.9 software (NanoView Biosciences).

4.3. Isolation of EVs from LM Patient CSFs

EVs were isolated from CSF of LM and HC individuals. The CSF samples were centrifuged twice and the cleared samples were used for the isolation of EVs with miR-CURY Exosome Cell/Urine/CSF Kit (#300102, Exiqon) according to the manufacturer's instructions. In brief, CSF sample gently mixed Precipitation Buffer B, then the mixtures were vortexed and incubated for 60 min at 2–8 °C for precipitating exosome pellet. After centrifugation at $10,000 \times g$ for 30 min at 20 °C, then the supernatant was removed.

4.4. EV RNA Extraction and Measurement

EV RNA extraction was performed using the miRCURY RNA Isolation Kit - Cell & Plant (#300110, Exiqon) following the manufacturer's instructions. In brief, exosome pellet was lysed with the provided lysis solution supplemented with 96–100% ethanol, and the mixture was loaded to the column. The EV RNA was washed and eluted with 15–50 μL of RNase-free water. The extracted RNA concentration was calculated by Quant-IT RiboGreen (Invitrogen; Thermo Fisher Scientific, Waltham, MA, USA). RNA size was confirmed using Agilent RNA 6000 Pico Kit and Small RNA Kit on Agilent 2100 Bioanalyzer (Agilent Technologies, Waldbronn, Germany).

4.5. Small RNA Library Construction and Sequencing

The 10 ng of RNA isolated from each sample was used to construct sequencing libraries with the SMARTer smRNA-Seq Kit for Illumina (Takara Bio Inc.), following the manufacturer's protocol. Input RNA was first polyadenylated in order to provide a priming sequence for an oligo(dT) primer. cDNA synthesis was primed by the 3′ smRNA dT Primer, which incorporates an adapter sequence at the 5′ end of each first-strand cDNA molecule. In the template-switching step, PrimeScript Reverse Transcriptase utilized the SMART smRNA Oligo as a template for the addition of a second adapter sequence to the 3′ end of each first-strand cDNA molecule. In the next step, full-length Illumina adapters (including index sequences for sample multiplexing) were added during PCR amplification. The forward PCR primer was then bound to the sequence added by the SMART smRNA Oligo, while the reverse PCR primer was bound to the sequence added by the 3′ smRNA dT Primer. The amplified libraries were purified from 6% Novex TBE-PAGE gels (Thermo-Fisher Scientific) to excise the fraction over 138-bp (over than 18-bp of cDNA plus 120-bp of adaptor). The resulting library cDNA included sequences required for clustering on an Illumina flow cell. The libraries were gel purified, and validated by checking the size, purity, and concentration on the Agilent 2100 Bioanalyzer. The libraries were then quantified using qPCR according to the qPCR quantification protocol guide (KAPA Library Quantification kits for Illumina Sequencing platforms; Roche, Basel, Switzerland) and qualified using the TapeStation D1000 ScreenTape (Agilent Technologies). The libraries were pooled in equimolar amounts, and sequenced on an Illumina HiSeq 2500 (Illumina, San Diego, CA, USA) instrument to generate 101 base reads. Image decomposition and quality values calculation were performed using the modules of the Illumina pipeline. Adapter trimming was assessed using the Cutadapt program.

4.6. Data Analysis of smRNA Sequencing

Uniquely clustered reads were then sequentially aligned to reference genome, miRBase v21 and non-coding RNA database Rfam 9.1 to identify known miRNAs and other type of smRNAs. Hierarchical cluster heatmaps, dendrograms, volcano plots, and multidimensional scaling (MDS) of extracted EV smRNA were performed by Macrogen (Seoul, Korea). Various hierarchical clustering analysis was performed using complete linkage and Euclidean distance as measures of similarity in differentially expressed patterns of the smRNA with the criteria of |fold change| ≥ 2 and p-value < 0.05. Distance matrices were processed by MDS to obtain a dimensionally reduced map of gene coordinates with distance between plots as a measure of similarity. All data analysis and visualization were performed using R 3.6.3 (www.r-project.org) (Table S4).

4.7. Synthesis of smRNA cDNA and Droplet Digital PCR (ddPCR)

Approximately 2 ng of purified total EV RNA was reverse transcribed to generate cDNA with TaqMan Advanced miRNA cDNA Synthesis Kit (A28007, Applied Biosystems, Foster City, CA, USA) according to the manufacturer's instruction. Four μL of cDNA was mixed ddPCR Supermix for Probes (No dUTP, Bio-Rad Laboratories, Inc., Hercules, CA, USA) and TaqMan Advanced miRNA Assay probes (Applied Biosystems; hsa-miR-21-5p, hsa-miR-19-3p, hsa-miR-25-3p, hsa-miR-200c-3p, hsa-miR-19a-3p, and hsa-miR-34b-

3p). Four µL of cDNA was mixed with the QX200 ddPCR EvaGreen Supermix (Bio-Rad Laboratories, Inc.) and oligomers for the following smRNAs; hsa-miR-423-5p, hsa-miR-1273g-3p, hsa-miR-4271, piRNA-33415, piRNA-36340, Y RNA-Y69, snRNA-200C70, vtRNA-6C57F3, novel miRNA-973, and scRNA-F3729 (Table S3). Each 20 µL of reaction mixture was mixed with 70 µL of droplet generation oil and partitioned into up to 20,000 nL-sized droplets by a QX299 droplet generator (Bio-Rad Laboratories, Inc.). Final 40 µL droplet mixture was used for the PCR reaction following cycling protocol. ddPCR Supermix for probes; 95 °C for 5 min (DNA polymerase activation), followed by 40 cycles of 95 °C for 30 s (denaturation) and 55 °C for 1 min (annealing) followed by post-cycling steps of 98 °C for 10 min (enzyme inactivation) and an infinite 4 °C hold by Applied Biosystems 7900HT Sequence Detection System. QX200 ddPCR EvaGreen Supermix; 95 °C for 5 min (DNA polymerase activation), followed by 39 cycles of 95 °C for 30s (denaturation) and 60 °C for 1 min (annealing) followed by post-cycling steps of 1 cycle of 4 °C for 5 min, and 1 cycle of 90 °C for 5 min and an infinite 4 °C hold. Cycling between the temperatures was set to a ramp rate of 2.5 °C/s. Annealing temperature of each oligomer set was optimized in QX200 ddPCR EvaGreen Supermix-based PCR reaction. The amplified PCR product of the nucleic acid target in the droplets were quantified in the FAM channels using QC200 Droplet Reader (Bio-Rad Laboratories, Inc.) and analyzed using QuantaSoft v.1.7.4.0917 software (Bio-Rad Laboratories, Inc.). The concentration (smRNA copies/µL) value generated by QuantaSoft was converted to smRNA copies/nL of CSF. The Mann-Whitney U test and unpaired t-test with Welch's correction were used to compare significant differences in each smRNA expression between two groups according to the result of normality test (Shapiro-Wilk test). * p-value < 0.05, ** p-value < 0.01.

4.8. The Gene Ontology of Biological Processes and the Kyoto Encyclopedia of Genes and Genomes Pathways

The putative target genes of miRNA were searched from miRTargetLink [54] providing a network algorithm based on the miRTarBase and experimentally validated interactions. The Gene Ontology (GO) of biological processes and the Kyoto Encyclopedia of Genes and Genomes (KEGG) pathways were computed using DAVID Bioinformatics [55]. 455 target genes of 43 miRNAs were identified from miRTargetLink database and gene set enrichment analysis was assessed.

4.9. Luciferase Assay of miR-21 Sensor-Bearing Cell Lines

Two luminescence-based miR-21 sensor-bearing cell lines (293T and A549) were established by lentiviral infection, and then harvested 16–20 h after treatment with designated CSFs or CSF EV concentrates. Cells were lysed with Passive Lysis Buffer (Promega, Madison, WI, USA), and the aliquot of lysates was analyzed by measuring luminescence signals with Dual-Luciferase Reporter Assay System (Promega) in a reader (SpectraMax L, Molecular Devices, San Jose, CA, USA). The miR-21 sensor signal from firefly luciferase was normalized with that from *Renilla* luciferase. The normalized quantification data was used in comparing the relative luciferase activities. Data are presented as the mean ± standard deviation determined from at least three independent experiments. Differences were assessed by two-tailed Student's t-test using Excel software (Microsoft, Redmond, WA, USA). * p-value < 0.05, ** p-value < 0.01, *** p-value < 0.001; NS, not significant.

4.10. Migration Assay

Human NSCLC A549 cells were plated on 6-well plates at a density of 6×10^5 per well and cultured up to 80–90% confluence for 2 days before scratching. Then 4–6 lines were scratched in each confluent monolayer with a sterile 200 µL tip. Dislodged cells were removed by washing with warm Dulbecco's phosphate-buffered saline, and the remaining cells were replenished with fresh medium. A representative line with similar width were selected and photographed in each group at starting point, and then treated with concentrated EV dissolved medium or transfected with miR-21 mimics. After 24 h and 48 h, snapshots of the scratch were taken with the microscope (100×) (Axio Observer, Zeiss, Oberkochen, Germany).

5. Conclusions

We reveal herein the molecular profile of CSF extravesicular smRNAs that are potentially associated with LM pathogenesis. We successfully identified various LM-associated smRNA populations that showed significantly biased expression patterns in LM. Our extensive NGS analysis and relevant biochemical and cell-based validations demonstrated that miRNAs and the other analyzed smRNA subpopulations may be useful targets for the development of therapeutic and diagnostic strategies for LM. Further investigation is critically needed to address the potential of extravesicular smRNAs as a novel pharmacological target for LM and LM-related complications. Additional experiments and bioinformatic analyses may reveal novel means to explore the underlying mechanisms of LM and provide additional promising targets for treating LM patients.

Supplementary Materials: The following are available online at https://www.mdpi.com/2072-6694/13/2/209/s1, Figure S1: Bioanalyzer analysis of the size distribution of RNA from HC and LM CSF EVs; Figure S2: Hierarchical clustering dendrogram of miRNAs among LM patients; Figure S3: Biochemical verification of biased miRNA expression in EVs derived from LM and HC by ddPCR and qRT-PCR; Figure S4: Investigation of synthetic miR-21 functionality on cellular migratory phenotype of NSCLC A549 cells; Figure S5: Hierarchical clustering analysis of piRNAs and Y RNAs in EVs from LM patient CSF; Figure S6: Distinct expression of snRNAs in EVs from LM patient CSF; Figure S7: Distinct expression of snoRNAs in EVs from LM patient CSF; Figure S8: Hierarchical clustering analysis of snRNAs and snoRNAs in EVs from LM patient CSF; Figure S9: Distinct expression of vtRNAs in EVs from LM patient CSF; Figure S10: Distinct expression of novel miRNAs in EVs from LM patient CSF; Figure S11: Hierarchical clustering analysis of vtRNAs and novel miRNAs in EVs from LM patient CSF; Figure S12: Distinct expression of scRNAs in EVs from LM patient CSF; Figure S13: Western blot analysis of the validated target genes of miR-21 in LM CSF EV-treated A549 migration assay samples; Figure S14: Biochemical verification of normalized smRNA expression in EVs derived from LM and HC by ddPCR; Figure S15: Original images of western blot related with Figure 1E; Figure S16: Original image of splinted ligation related with Figure 4E; Figure S17: Original images of western blot related with Figure S13B; Table S1: Experimental raw data of the CSF samples in this study; Table S2: Clinical characteristics of additional CSF samples and their applications in this study (n = 11); Table S3: Oligonucleotides used in ddPCR analysis; Table S4: List of smRNA sequencing expression raw data by descending fold change and p-value.

Author Contributions: Conceptualization, J.H.K., H.-S.G., J.B.P., and B.C.Y. Methodology and sample collection, H.-S.G., J.-W.K., S.H.S., and H.Y. Data acquisition and curation, K.-Y.L., Y.S., J.H.I., J.R., W.B., and S.K. Writing of original draft, J.H.K. Review and editing, all authors. All authors have read and agreed to the published version of the manuscript.

Funding: This work was supported by research grants from the National Cancer Center, Korea (NCC-1910090, NCC-2010273, NCC-2010320, NCC-2110530, NCC-2110531, and NCC-2110532) and the National Research Foundation of Korea (NRF) grant funded by the Korea government (MSIT) (NRF-2017R1A2B4008257 and NRF-2019R1A2C1089145).

Institutional Review Board Statement: The study was conducted according to the guidelines of the Declaration of Helsinki, and approved by the Institutional Review Board of National Cancer Center, Korea (NCC2014-0135; date of approval: August 23, 2014).

Informed Consent Statement: Informed consent was obtained from all subjects involved in the study.

Data Availability Statement: The raw data that support the findings of this study are available from the corresponding authors upon reasonable request.

Acknowledgments: We specially thank Jeyoung Woo (Macrogen, Seoul, Korea) for helpful assistance for smRNA data analysis and interpretation. Quantum Design Korea (Seoul, Korea) is also acknowledged for helpful assistance of the ExoView analysis.

Conflicts of Interest: The authors declare no conflict of interest.

Abbreviations

EVs	Extracellular vesicles
LM	Leptomeningeal metastasis
NTA	Nanoparticle tracking analysis
CSF	Cerebrospinal fluid
ddPCR	Droplet digital polymerase chain reaction
NSCLC	Non-small cell lung cancer
smRNA	Small non-coding RNA
qRT-PCR	Real time reverse transcription PCR
GO	Gene Ontology
DAVID	Database for Annotation, Visualization and Integrated Discovery
KEGG	Kyoto Encyclopedia of Genes and Genomes
miRISC	miRNA-induced silencing complex
MDS	Multidimensional scaling
FC	Fold change

References

1. Boire, A.; Zou, Y.; Shieh, J.; Macalinao, D.G.; Pentsova, E.; Massagué, J. Complement component 3 adapts the cerebrospinal fluid for leptomeningeal metastasis. *Cell* **2017**, *168*, 1101–1113. [CrossRef] [PubMed]
2. Cheng, H.; Perez-Soler, R. Leptomeningeal metastases in non-small-cell lung cancer. *Lancet Oncol.* **2018**, *19*, e43–e55. [CrossRef]
3. Remon, J.; Le Rhun, E.; Besse, B. Leptomeningeal carcinomatosis in non-small cell lung cancer patients: A continuing challenge in the personalized treatment era. *Cancer Treat. Rev.* **2017**, *53*, 128–137. [CrossRef] [PubMed]
4. Pace, A.; Fabi, A. Chemotherapy in neoplastic meningitis. *Crit. Rev. Oncol. Hematol.* **2006**, *60*, 194–200. [CrossRef] [PubMed]
5. Segal, M.B. The choroid plexuses and the barriers between the blood and the cerebrospinal fluid. *Cell. Mol. Neurobiol.* **2000**, *20*, 183–196. [CrossRef] [PubMed]
6. Im, J.H.; Yoo, B.C.; Lee, J.H.; Kim, K.H.; Kim, T.H.; Lee, K.Y.; Kim, J.H.; Park, J.B.; Kwon, J.W.; Shin, S.H.; et al. Comparative cerebrospinal fluid metabolites profiling in glioma patients to predict malignant transformation and leptomeningeal metastasis with a potential for preventive personalized medicine. *EPMA J.* **2020**, *11*, 469–484. [CrossRef]
7. Cordone, I.; Masi, S.; Summa, V.; Carosi, M.; Vidiri, A.; Fabi, A.; Pasquale, A.; Conti, L.; Rosito, I.; Carapella, C.; et al. Overexpression of syndecan-1, MUC-1, and putative stem cell markers in breast cancer leptomeningeal metastasis: A cerebrospinal fluid flow cytometry study. *Breast Cancer Res.* **2017**, *19*, 46. [CrossRef]
8. Lee, K.-Y.; Im, J.H.; Lin, W.; Gwak, H.-S.; Kim, J.H.; Yoo, B.C.; Kim, T.H.; Park, J.B.; Park, H.J.; Kim, H.J.; et al. Nanoparticles in 472 human cerebrospinal fluid: Changes in extracellular vesicle concentration and miR-21 expression as a biomarker for leptomeningeal metastasis. *Cancers* **2020**, *12*, 2745. [CrossRef]
9. Akers, J.C.; Ramakrishnan, V.; Nolan, J.P.; Duggan, E.; Fu, C.-C.; Hochberg, F.H.; Chen, C.C.; Carter, B.S. Comparative analysis of technologies for quantifying extracellular vesicles (EVs) in clinical cerebrospinal fluids (CSF). *PLoS ONE* **2016**, *11*, e0149866. [CrossRef]
10. Santiago-Dieppa, D.R.; Steinberg, J.; Gonda, D.; Cheung, V.J.; Carter, B.S.; Chen, C.C. Extracellular vesicles as a platform for 'liquid biopsy' in glioblastoma patients. *Expert Rev. Mol. Diagn.* **2014**, *14*, 819–825. [CrossRef]
11. Raposo, G.; Stoorvogel, W. Extracellular vesicles: Exosomes, microvesicles, and friends. *J. Cell Biol.* **2013**, *200*, 373–383. [CrossRef] [PubMed]
12. Kim, K.M.; Abdelmohsen, K.; Mustapic, M.; Kapogiannis, D.; Gorospe, M. RNA in extracellular vesicles. *Wiley Interdiscip. Rev. RNA* **2017**, *8*, 1413. [CrossRef] [PubMed]
13. Yagi, Y.; Ohkubo, T.; Kawaji, H.; Machida, A.; Miyata, H.; Goda, S.; Roy, S.; Hayashizaki, Y.; Suzuki, H.; Yokota, T. Next-generation sequencing-based small RNA profiling of cerebrospinal fluid exosomes. *Neurosci. Lett.* **2017**, *636*, 48–57. [CrossRef] [PubMed]
14. Tetta, C.; Ghigo, E.; Silengo, L.; Deregibus, M.C.; Camussi, G. Extracellular vesicles as an emerging mechanism of cell-to-cell communication. *Endocrine* **2013**, *44*, 11–19. [CrossRef] [PubMed]
15. Li, C.; Qin, F.; Hu, F.; Xu, H.; Sun, G.; Han, G.; Wang, T.; Guo, M. Characterization and selective incorporation of small non-coding RNAs in non-small cell lung cancer extracellular vesicles. *Cell Biosci.* **2018**, *8*, 2. [CrossRef]
16. Schwarzenbach, H.; Gahan, P.B. MicroRNA shuttle from cell-to-cell by exosomes and its impact in cancer. *Noncoding RNA* **2019**, *5*, 28. [CrossRef]
17. Kopkova, A.; Sana, J.; Machackova, T.; Vecera, M.; Radova, L.; Trachtova, K.; Vybihal, V.; Smrcka, M.; Kazda, T.; Slaby, O.; et al. Cerebrospinal fluid microRNA signatures as diagnostic biomarkers in brain tumors. *Cancers* **2019**, *11*, 1546. [CrossRef]
18. Sato-Kuwabara, Y.; Melo, S.A.; Soares, F.A.; Calin, G.A. The fusion of two worlds: Non-coding RNAs and extracellular vesicles—diagnostic and therapeutic implications (review). *Int. J. Oncol.* **2015**, *46*, 17–27. [CrossRef]
19. Bartel, D.P. MicroRNAs: Target recognition and regulatory functions. *Cell* **2009**, *136*, 215–233. [CrossRef]
20. Flynt, A.S.; Lai, E.C. Biological principles of microRNA-mediated regulation: Shared themes amid diversity. *Nat. Rev. Genet.* **2008**, *9*, 831–842. [CrossRef]

21. Min, L.; Zhu, S.; Chen, L.; Liu, X.; Wei, R.; Zhao, L.; Yang, Y.; Zhang, Z.; Kong, G.; Li, P.; et al. Evaluation of circulating small extracellular vesicles derived miRNAs as biomarkers of early colon cancer: A comparison with plasma total miRNAs. *J. Extracell. Vesicles* **2019**, *8*, 1643670. [CrossRef] [PubMed]
22. Yekula, A.; Minciacchi, V.R.; Morello, M.; Shao, H.; Park, Y.; Zhang, X.; Muralidharan, K.; Freeman, M.R.; Weissleder, R.; Lee, H.; et al. Large and small extracellular vesicles released by glioma cells in vitro and in vivo. *J. Extracell. Vesicles* **2020**, *9*, 1689784. [CrossRef] [PubMed]
23. Street, J.M.; Barran, P.E.; Mackay, C.L.; Weidt, S.; Balmforth, C.; Walsh, T.; Chalmers, R.T.; Webb, D.J.; Dear, J.W. Identification and proteomic profiling of exosomes in human cerebrospinal fluid. *J. Transl. Med.* **2012**, *10*, 5. [CrossRef] [PubMed]
24. Cheng, L.; Sharples, R.A.; Scicluna, B.J.; Hill, A.F. Exosomes provide a protective and enriched source of miRNA for biomarker profiling compared to intracellular and cell-free blood. *J. Extracell. Vesicles* **2014**, *3*, 1–14. [CrossRef]
25. Griffiths-Jones, S.; Moxon, S.; Marshall, M.; Khanna, A.; Eddy, S.R.; Bateman, A. Rfam: Annotating non-coding RNAs in complete genomes. *Nucleic Acids Res.* **2005**, *33*, 121–124. [CrossRef]
26. Griffiths-Jones, S.; Grocock, R.J.; Van Dongen, S.; Bateman, A.; Enright, A.J. miRBase: microRNA sequences, targets and gene nomenclature. *Nucleic Acids Res.* **2006**, *34*, D140–D144. [CrossRef]
27. Burgos, K.L.; Javaherian, A.; Bomprezzi, R.; Ghaffari, L.; Rhodes, S.; Courtright, A.; Tembe, W.; Kim, S.; Metpally, R.; Van Keuren-Jensen, K. Identification of extracellular miRNA in human cerebrospinal fluid by next-generation sequencing. *RNA* **2013**, *19*, 712–722. [CrossRef]
28. Friedländer, M.R.; Mackowiak, S.D.; Li, N.; Chen, W.; Rajewsky, N. miRDeep2 accurately identifies known and hundreds of novel microRNA genes in seven animal clades. *Nucleic Acids Res.* **2011**, *40*, 37–52. [CrossRef]
29. Chen, W.W.; Balaj, L.; Liau, L.M.; Samuels, M.L.; Kotsopoulos, S.K.; Maguire, C.A.; Loguidice, L.; Soto, H.; Garrett, M.C.; Zhu, L.D.; et al. BEAMing and droplet digital PCR analysis of mutant IDH1 mRNA in glioma patient serum and cerebrospinal fluid extracellular vesicles. *Mol. Ther. Nucleic Acids* **2013**, *2*, e109. [CrossRef]
30. Hong, B.S.; Ryu, H.S.; Kim, N.; Kim, J.; Lee, E.; Moon, H.; Kim, K.H.; Jin, M.-S.; Kwon, N.H.; Kim, S.; et al. Tumor suppressor microRNA-204-5p regulates growth, metastasis, and immune microenvironment remodeling in breast cancer. *Cancer Res.* **2019**, *79*, 1520–1534. [CrossRef]
31. Seo, Y.; Kim, S.S.; Kim, N.; Cho, S.; Park, J.B.; Kim, J.H. Development of a miRNA-controlled dual-sensing system and its application for targeting miR-21 signaling in tumorigenesis. *Exp. Mol. Med.* **2020**, *52*, 1989–2004. [CrossRef] [PubMed]
32. Siomi, M.C.; Sato, K.; Pezic, D.; Aravin, A.A. PIWI-interacting small RNAs: The vanguard of genome defence. *Nat. Rev. Mol. Cell Biol.* **2011**, *12*, 246–258. [CrossRef] [PubMed]
33. Gulia, C.; Signore, F.; Gaffi, M.; Gigli, S.; Votino, R.; Nucciotti, R.; Bertacca, L.; Zaami, S.; Baffa, A.; Santini, E.; et al. Y RNA: An overview of their role as potential biomarkers and molecular targets in human cancers. *Cancers* **2020**, *12*, 1238. [CrossRef]
34. Christov, C.P.; Trivier, E.; Krude, T. Noncoding human Y RNAs are overexpressed in tumours and required for cell proliferation. *Br. J. Cancer* **2008**, *98*, 981–988. [CrossRef]
35. van Bussel, M.T.J.; Pluim, D.; Kerklaan, B.M.; Bol, M.; Sikorska, K.; Linders, D.T.C.; van den Broek, D.; Beijnen, J.H.; Schellens, J.H.M.; Brandsma, D. Circulating epithelial tumor cell analysis in CSF in patients with leptomeningeal metastases. *Neurology* **2020**, *94*, e521–e528. [CrossRef] [PubMed]
36. Head, S.R.; Komori, H.K.; LaMere, S.A.; Whisenant, T.; Van Nieuwerburgh, F.; Salomon, D.R.; Ordoukhanian, P. Library construction for next-generation sequencing: Overviews and challenges. *Biotechniques* **2014**, *56*, 61–64, 66, 68. [CrossRef]
37. Buschmann, D.; Haberberger, A.; Kirchner, B.; Spornraft, M.; Riedmaier, I.; Schelling, G.; Pfaffl, M.W. Toward reliable bi-omarker signatures in the age of liquid biopsies—how to standardize the small RNA-Seq workflow. *Nucleic Acids Res.* **2016**, *44*, 5995–6018. [CrossRef]
38. Chen, X.; Ba, Y.; Ma, L.; Cai, X.; Yin, Y.; Wang, K.; Guo, J.; Zhang, Y.; Chen, J.; Guo, X.; et al. Characterization of microRNAs in serum: A novel class of biomarkers for diagnosis of cancer and other diseases. *Cell Res.* **2008**, *18*, 997–1006. [CrossRef]
39. Si, M.-L.; Zhu, S.; Wu, H.; Lu, Z.; Wu, F.; Mo, Y.-Y. miR-21-mediated tumor growth. *Oncogene* **2007**, *26*, 2799–2803. [CrossRef]
40. Gabriely, G.; Wurdinger, T.; Kesari, S.; Esau, C.C.; Burchard, J.; Linsley, P.S.; Krichevsky, A.M. MicroRNA 21 promotes glioma invasion by targeting matrix metalloproteinase regulators. *Mol. Cell. Biol.* **2008**, *28*, 5369–5380. [CrossRef]
41. Yin, J.; Park, G.; Lee, J.E.; Choi, E.Y.; Park, J.Y.; Kim, T.H.; Park, N.; Jin, X.; Jung, J.E.; Shin, D.; et al. DEAD-box RNA helicase DDX23 modulates glioma malignancy via elevating miR-21 biogenesis. *Brain* **2015**, *138*, 2553–2570. [CrossRef] [PubMed]
42. Baumgartner, U.; Berger, F.; Hashemi Gheinani, A.; Burgener, S.S.; Monastyrskaya, K.; Vassella, E. miR-19b enhances proliferation and apoptosis resistance via the EGFR signaling pathway by targeting PP2A and BIM in non-small cell lung cancer. *Mol. Cancer* **2018**, *17*, 44. [CrossRef]
43. Zhou, X.; Wang, Y.; Shan, B.; Han, J.; Zhu, H.; Lv, Y.; Fan, X.; Sang, M.; Liu, X.-D.; Liu, W. The downregulation of miR-200c/141 promotes ZEB1/2 expression and gastric cancer progression. *Med. Oncol.* **2014**, *32*, 428. [CrossRef] [PubMed]
44. Sulaiman, S.A.; Ab Mutalib, N.-S.; Jamal, R. miR-200c Regulation of metastases in ovarian cancer: Potential role in epithelial and mesenchymal transition. *Front. Pharmacol.* **2016**, *7*, 271. [CrossRef]
45. Ceppi, P.; Mudduluru, G.; Kumarswamy, R.; Rapa, I.; Scagliotti, G.V.; Papotti, M.; Allgayer, H. Loss of miR-200c expression induces an aggressive, invasive, and chemoresistant phenotype in non-small cell lung cancer. *Mol. Cancer Res.* **2010**, *8*, 1207–1216. [CrossRef] [PubMed]
46. Seto, A.G.; Kingston, R.E.; Lau, N.C. The coming of age for Piwi proteins. *Mol. Cell* **2007**, *26*, 603–609. [CrossRef] [PubMed]

47. Chalbatani, G.M.; Dana, H.; Memari, F.; Gharagozlou, E.; Ashjaei, S.; Kheirandish, P.; Marmari, V.; Mahmoudzadeh, H.; Mozayani, F.; Maleki, A.R.; et al. Biological function and molecular mechanism of piRNA in cancer. *Pr. Lab. Med.* **2019**, *13*, e00113. [CrossRef]
48. Ender, C.; Krek, A.; Friedländer, M.R.; Beitzinger, M.; Weinmann, L.; Chen, W.; Pfeffer, S.; Rajewsky, N.; Meister, G. A human snoRNA with microRNA-like functions. *Mol. Cell* **2008**, *32*, 519–528. [CrossRef]
49. Taft, R.J.; Glazov, E.A.; Lassmann, T.; Hayashizaki, Y.; Carninci, P.; Mattick, J.S. Small RNAs derived from snoRNAs. *RNA* **2009**, *15*, 1233–1240. [CrossRef]
50. Liang, J.; Wen, J.; Huang, Z.; Chen, X.-P.; Zhang, B.-X.; Chu, L. Small nucleolar RNAs: Insight into their function in cancer. *Front. Oncol.* **2019**, *9*, 587. [CrossRef]
51. Matera, A.G.; Terns, R.M.; Terns, M.P. Non-coding RNAs: Lessons from the small nuclear and small nucleolar RNAs. *Nat. Rev. Mol. Cell. Biol.* **2007**, *8*, 209–220. [CrossRef] [PubMed]
52. Shuai, S.; Suzuki, H.; Diaz-Navarro, A.; Nadeu, F.; Kumar, S.A.; Gutierrez-Fernandez, A.; Delgado, J.; Pinyol, M.; López-Otín, C.; Puente, X.S.; et al. The U1 spliceosomal RNA is recurrently mutated in multiple cancers. *Nature* **2019**, *574*, 712–716. [CrossRef] [PubMed]
53. Gopinath, S.C.; Wadhwa, R.; Kumar, P.K. Expression of noncoding vault RNA in human malignant cells and its importance in mitoxantrone resistance. *Mol. Cancer Res.* **2010**, *8*, 1536–1546. [CrossRef] [PubMed]
54. Hamberg, M.; Backes, C.; Fehlmann, T.; Hart, M.; Meder, B.; Meese, E.; Keller, A. MiRTargetLink—miRNAs, genes and interaction networks. *Int. J. Mol. Sci.* **2016**, *17*, 564. [CrossRef]
55. Huang, D.W.; Sherman, B.T.; Lempicki, R.A. Systematic and integrative analysis of large gene lists using DAVID bioinformatics resources. *Nat. Protoc.* **2009**, *4*, 44–57. [CrossRef] [PubMed]

Article

Subpopulations of Circulating Cells with Morphological Features of Malignancy Are Preoperatively Detected and Have Differential Prognostic Significance in Non-Small Cell Lung Cancer

Emanuela Fina [1,*], Davide Federico [2,3], Pierluigi Novellis [4,†], Elisa Dieci [4,†], Simona Monterisi [1], Federica Cioffi [4], Giuseppe Mangiameli [3,4], Giovanna Finocchiaro [5], Marco Alloisio [3,4] and Giulia Veronesi [4,*,†]

1. Humanitas Research Center, IRCCS Humanitas Research Hospital, Manzoni 56, 20089 Rozzano, MI, Italy; simonterisi@gmail.com
2. Division of Pathology, IRCCS Humanitas Research Hospital, Manzoni 56, 20089 Rozzano, MI, Italy; davide.federico@humanitas.it
3. Department of Biomedical Sciences, Humanitas University, Rita Levi Montalcini 4, 20072 Pieve Emanuele, MI, Italy; giuseppe.mangiameli@cancercenter.humanitas.it (G.M.); marco.alloisio@cancercenter.humanitas.it (M.A.)
4. Division of Thoracic Surgery, Humanitas Cancer Center, IRCCS Humanitas Research Hospital, Manzoni 56, 20089 Rozzano, MI, Italy; pierluigi.novellis84@gmail.com (P.N.); dieci.elisa@hsr.it (E.D.); federica.cioffi@humanitas.it (F.C.)
5. Division of Oncology and Hematology, IRCCS Humanitas Research Hospital, Manzoni 56, 20089 Rozzano, MI, Italy; giovanna.finocchiaro@cancercenter.humanitas.it
* Correspondence: emanuela.fina@humanitasresearch.it or emanuela1.fina@gmail.com (E.F.); veronesi.giulia@hsr.it (G.V.)
† Present address: Department of Thoracic Surgery, IRCCS San Raffaele Scientific Institute, Olgettina 60, 20132 Milano, MI, Italy.

Simple Summary: Lung cancer is by far the main cause of cancer-related deaths among both men and women. Early detection of malignant nodules and non-invasive monitoring of disease status is essential to increase the chance of cure. In this study, we analyzed the frequency and the biological features of circulating tumor cells, i.e., cells released from the tumor and in transit in the bloodstream, in patients with a diagnosis of non-small cell lung cancer undergoing surgical resection, with the aim to develop a blood-based diagnostic test and to promptly identify patients at risk of post-operative disease recurrence.

Abstract: Background: Non-small cell lung cancer (NSCLC) frequently presents when surgical intervention is no longer feasible. Despite local treatment with curative intent, patients might experience disease recurrence. In this context, accurate non-invasive biomarkers are urgently needed. We report the results of a pilot study on the diagnostic and prognostic role of circulating tumor cells (CTCs) in operable NSCLC. Methods: Blood samples collected from healthy volunteers ($n = 10$), nodule-negative high-risk individuals enrolled in a screening program ($n = 7$), and NSCLC patients ($n = 74$) before surgery were analyzed (4 mL) for the presence of cells with morphological features of malignancy enriched through the ISET® technology. Results: CTC detection was 60% in patients, while no target cells were found in lung cancer-free donors. We identified single CTCs (sCTC, 46%) and clusters of CTCs and leukocytes (heterotypic clusters, hetCLU, 31%). The prevalence of sCTC (sCTC/4 mL ≥ 2) or the presence of hetCLU predicted the risk of disease recurrence within the cohort of early-stage (I–II, $n = 52$) or advanced stage cases (III–IVA, $n = 22$), respectively, while other tumor-related factors did not inform prognosis. Conclusions: Cancer cell hematogenous dissemination occurs frequently in patients with NSCLC without clinical evidence of distant metastases, laying the foundation for the application of cell-based tests in screening programs. CTC subpopulations are fine prognostic classifiers whose clinical validity should be further investigated in larger studies.

Keywords: circulating tumor cells; lung cancer; early diagnosis; cancer biomarkers

---

Citation: Fina, E.; Federico, D.; Novellis, P.; Dieci, E.; Monterisi, S.; Cioffi, F.; Mangiameli, G.; Finocchiaro, G.; Alloisio, M.; Veronesi, G. Subpopulations of Circulating Cells with Morphological Features of Malignancy Are Preoperatively Detected and Have Differential Prognostic Significance in Non-Small Cell Lung Cancer. *Cancers* 2021, 13, 4488. https://doi.org/10.3390/cancers13174488

Academic Editor: David Wong

Received: 31 July 2021
Accepted: 26 August 2021
Published: 6 September 2021

Publisher's Note: MDPI stays neutral with regard to jurisdictional claims in published maps and institutional affiliations.

Copyright: © 2021 by the authors. Licensee MDPI, Basel, Switzerland. This article is an open access article distributed under the terms and conditions of the Creative Commons Attribution (CC BY) license (https://creativecommons.org/licenses/by/4.0/).

1. Introduction

Lung cancer is the tumor with the highest fatality rate worldwide, both in males and females [1], due to its aggressiveness and biological heterogeneity, with non-small cell lung cancer (NSCLC) representing the most frequent histological subtype [2,3]. Importantly, since approximately 75% of patients present symptoms when the disease has advanced locally or disseminated at distant sites [4], lung cancer mortality is also a large consequence of late diagnoses. Therefore, early detection is a key for improving patient's survival [5], and the US National Lung Screening Trial (NLST), the Nelson study and other non-randomized trials have actually demonstrated a significant reduction in mortality (20–30%) and morbidity upon screening programs based on a thoracic scan by low-dose computed tomography (LDCT) [6,7]. At present, uncertainties on high costs, risk of radiation exposure and false positives are obstacles to the large-scale implementation of such screening in Europe [8–10], and controversies exist in the management of subjects with indeterminate or premalignant nodules, which have to be monitored for a long time, and in some cases biopsied, increasing the risk of subsequent morbidities [11]. Moreover, notwithstanding the considerable advantages for patients who are diagnosed with the early-stage disease compared to those with unresectable tumors [12,13], unfortunately, 30% to 50% of cases who receive an indication for local treatment experience disease recurrence and die despite surgery with curative intent [14,15]. Lung cancer still lacks accurate biomarkers, and staging is no longer considered an accurate prognostic factor since patients with the disease at the same stage may undergo recurrence with variable incidence [16,17]. In this context, both non-invasive diagnostic tests and novel prognostic and predictive biomarkers are urgently needed to better stratify patients at risk of recurrence upon surgery and adjuvant therapies [18–20].

The measurement of blood biomarkers is an attractive approach to monitor cancer appearance and evolution: (i) suitable to be repeated, (ii) minimally invasive and (iii) believed to represent the systemically diffused expression of tumor heterogeneity [21–23]. Among all possible non-invasive biomarkers, circulating tumor cells (CTCs) are largely informative as they represent cancer cells in transit in blood, with the expected ability to re-seed the site of origin and/or to colonize distant organs [24], and which can be enumerated and characterized at DNA, RNA, protein and morphological level, thus providing access to a considerable amount of information. Importantly, hematogenous dissemination is now considered an early event in tumor progression [25], and CTCs were actually observed in the blood of patients without clinically detectable metastases in several solid tumors [26]. In lung cancer, a seminal work demonstrated that CTC analysis with the Isolation by Size of Epithelial/Tumor cells (ISET®) technology, which enables the vast majority of hematopoietic cells to be excluded by blood sample filtration through a porous membrane, can anticipate the detection of malignant nodules by computed tomography scan in patients with chronic obstructive pulmonary disease (COPD) who had eventually developed stage IA tumor [27], fostering the introduction of cell-based tests in diagnostic trials.

CTCs can be enriched and detected by several techniques based on the physical and biological differences between cancer cells and blood cell types; however, accurate detection is hampered by their rarity and phenotypical heterogeneity [28]. Size-based enrichment of CTCs coupled to cytomorphological analysis are unbiased with respect to the expression of protein markers and may reach a sensitivity of one cell per blood volume [29,30]. On the basis of these considerations, we have analyzed blood samples of lung cancer-free individuals and operable NSCLC patients before surgery, using the ISET® technology. Previous works in early-stage NSCLCs demonstrated that CTCs can be observed by ISET® in about 50% of patients and that they have prognostic significance in the preoperative setting [31,32], while in a multicentric screening trial, CTC analysis was able to identify only 26% of lung cancers detected at first LDCT scan [33].

Here, we have described atypical cells and searched for cells with morphological features of malignancy upon staining with standard cytological colorants, with the aim to assess the prognostic significance of CTC count and possible CTC subpopulations in operable NSCLC and to further explore the applicability of a CTC-based test in screening programs by comparison with a cohort of lung cancer-free individuals.

2. Materials and Methods

2.1. Study Design

We designed and carried out a prospective observational study, following the STROBE guidelines and the approval by the Ethics Committee of Humanitas Research Hospital in Rozzano (Milan, Italy). Signed informed consent was obtained from all patients. Consecutive patients with a confirmed diagnosis of NSCLC, aged 18 years or older, not pregnant, treatment naïve, without prior cancer detected within the previous 5 years or a second malignancy if there was evidence of active disease and candidate to surgical resection were enrolled from January 2017 to September 2018 and admitted in the Thoracic Surgery Division of Humanitas Cancer Center. During this phase of the study, blood samples were collected from lung cancer patients detected outside the smoker's health multiple action (SMAC) screening program, the same day or the day before surgical intervention. Blood samples for CTC analysis were also collected from high-risk individuals enrolled in the SMAC screening program before the LDCT scan, starting from May 2019. Inclusion criteria were: heavy smokers, i.e., ≥30 packs per year, for more than 30 years, or former-smokers aged 55 years or older, who had ceased smoking within the 15 years prior to enrollment in the study, absence of symptoms of lung cancer, such as worsening of cough, hoarseness, hemoptysis and weight loss. Exclusion criteria were: previous diagnosis of lung cancer, extrapulmonary cancer history in the last 5 years (excluding in situ tumors or skin epidermoid tumor), chest CT scan performed in the last 18 months, severe lung or extrapulmonary diseases that may preclude or invalidate appropriate therapy in case of diagnosis of malignant pulmonary neoplasia. CTC analysis was performed on individuals without LDCT-detected pulmonary nodules. Blood samples from a group of healthy volunteers aged 30–50 years were analyzed as a negative control to optimize the target cell identification. The data were collected by a review of electronic medical records. The TNM staging manual of the American Joint Committee on Cancer (AJCC) 8th edition was applied. Cases considered for survival analyses include patients whose disease recurrence or death was clearly documented and attributed to NSCLC, while patients whose status was not available were excluded.

2.2. Blood Collection and CTC Enrichment

Samples of peripheral venous blood were drawn from patients or lung cancer-free donors using a 21G needle, collected in K_2EDTA BD Vacutainer® tubes (Becton Dickinson Italia, Milan, Italy), preserved at room temperature under gentle agitation and processed within 1.5 h using the Isolation by Size of Epithelial/Tumor cells (ISET®) technology (Rarecells® Diagnostics, Paris, France). Briefly, 10-mL blood samples were diluted 1:10 with a proprietary Rarecells® Buffer, which lysates red blood cells, and fixed with 37% formaldehyde solution (Sigma-Aldrich, St. Louis, MO, USA) at a final concentration of about 0.7% for 10 minutes and under gentle agitation. The blood was filtered through a filtration block containing a polycarbonate membrane, which hosts ten porous filter spots (8-μm-diameter cylindrical pores), each spot representing the filtration product of 1 mL of blood. The membranes were stored at 4 °C until cytological staining.

2.3. CTC Detection Method and Identification Criteria

Four ISET membrane spots per sample, which are the equivalent of 4 mL of blood, were stained using May–Grünwald and Giemsa colorants following these steps: incubation with a pure May–Grünwald solution (Sigma, St. Louis, MO, USA) for 5 min, then with May–Grünwald 50% diluted in distilled water for 5 min and finally with a Giemsa solution (Sigma) 10% diluted in distilled water for 20 min. Stained membranes were mounted with Organo/Limonene Mount™ mounting medium (Sigma) and examined under a bright-field microscope (Olympus BX51) using a 10× objective. Areas of interest were subsequently digitalized at a 40× magnification for cytomorphological analysis. All images were analyzed by a referral cytopathologist, without knowledge of disease status and outcome, on the basis of the classical morphological criteria of malignancy, also described by Hofman and colleagues [34], and other morphological criteria to define populations of non-malignant circulating cells, as reported by Wechsler [35]. Images of atypical cells recorded during the analysis of a training set of healthy volunteers were also taken into account to exclude cells with uncertain malignancy. CTCs were defined as cells presenting with a nucleus-to-cytoplasm ratio >0.75 and nucleus diameter >20 μm, and with at least one of the following characteristics: nuclear border irregularities or nuclear hyperchromasia. Clusters were defined as groups of at least two CTCs or groups of at least one CTC and at least another cell type without features of malignancy, juxtaposed or in direct contact at the cell membrane level. Clusters of cells with homogenous chromatin staining and nucleus-to cytoplasm ratio <0.70, generally oval-shaped, were defined as clusters of epithelial-like cells. Large macrophage-like cells were defined as cells with a longer diameter >30 μm, with low nucleus-to-cytoplasm ratio, abundant pale basophilic cytoplasm and showing several cell shapes, such as fusiform, tadpole-like, round or oblong. Naked nuclei were defined as hyperchromatic and irregularly shaped nuclei with a longer diameter >20 μm and without apparent cytoplasm. Samples were called CTC-positive when at least one CTC was observed in a total of four filter spots, which corresponds to 4 mL of blood. All samples were considered evaluable for cytomorphological analysis according to cellularity, the prevalence of damaged cells and staining quality.

2.4. Spike-in Experiments

A549 ATCC® CCL-185™and NCI-H460 [H460] ATCC® HTB-177™ lung cancer cell lines were kindly provided by the European Institute of Oncology in Milan. Cells were propagated according to the instructions provided by the American Type Culture Collection (Manassas, VA, USA). Before performing spike-in experiments, the cells were detached with Trypsin-Versene® solution (Lonza, Basel, Switzerland), counted using the Trypan Blue 0.4% solution as a vital dye exclusion assay (viability >99% in all tests) and injected in 10 mL of blood collected from healthy volunteers at a dilution of 1000 cells per milliliter of blood (i.e., 1000 cells per membrane spot). The spiked-in samples were incubated for 30 min at room temperature under gentle agitation until chemical fixation and filtration as described before. The membranes were stained with Hematoxylin (Histo-line Laboratories Srl, Milan, Italy) for 3 min and Eosin Y aqueous solution (Histo-line Laboratories Srl) for 1 min, or with May–Grünwald and Giemsa solutions (Sigma) as described before and observed at the Olympus BX51 under a 20× magnification objective.

2.5. Statistical Analysis

Associations between categorical variables were tested by Fisher's exact test using contingency tables. Differences in discrete variables were tested using the Mann–Whitney and Kruskal–Wallis test. Linear regression analysis and Pearson's correlation coefficient r were used to estimate the correlation between cell numbers and age. Cox proportional-hazards regression was used to investigate the prognostic role of CTC status or number on recurrence-free survival, with relative hazard ratios (HR) and 95% confidence interval (CI). Significance in the probability of time-to-event was tested by log-rank test. Each selected factor was investigated in univariable analysis. All tests for the comparison of experimental groups were performed using GraphPad Prism, version 7.04. Survival and Cox regression analyses were performed, and Kaplan–Meier plots were constructed using MedCalc, version 12.7, and SPSS, version 26.0, respectively. All tests were two-sided, and significance statements refer to a p-value < 0.05.

3. Results

3.1. Cancer Cell Hematogenous Dissemination Is a Frequent Event in NSCLC Patients without Clinical Evidence of Metastasis

We performed the prospective collection of blood samples before surgical intervention ($n = 74$), and we analyzed blood samples of young volunteers ($n = 10$) and high-risk lung cancer-free individuals ($n = 7$), as assessed by LDCT as negative controls (Table 1). The number of cases locally treated with curative intent was 32, 20 and 17 for stages I, II and III, respectively. Five patients had stage IV disease at baseline: three underwent only diagnostic surgical procedure at pleural or lung nodules, while two underwent surgery with radical intent, one case presenting with a big, excavated lung lesion with the paraneoplastic syndrome and no possibility to undergo chemotherapy, the other case underwent segmental resection plus resection of a local pleural lesion with partial decortication for intraoperative diagnosis of a small pleural metastasis. In the control cases, cells with features of malignancy were not observed, while some nuclei larger than 20 μm, hyperchromatic or with heterogeneous chromatin stain, without apparent cytoplasm, sometimes overlapping, were found in 7 out of the 17 (41%) cases (Supplementary Figure S1). Images were digitalized and taken into account during the morphological analysis of patients' blood samples in order to exclude indeterminate atypical figures and avoid misleading interpretations. Of the total population of NSCLC patients, 44 cases (59.5%) were called CTC-positive, as they had at least one cell with clear features of malignancy detected in 4 mL of blood, 57.5% with stage I–III and 80% with stage IVA tumor (*p*-value = 0.4922). The CTC status did not correlate with patient demographics nor with smoking habits, as also with pathologic tumor stage, histology and invasiveness features, except for a statistically significant higher CTC positivity rate observed in cases without peritumoral neoplastic angioinvasion. Interestingly, the proportion of CTC-positive cases with stage I or II NSCLC was considerable (57.7%) with respect to the limited amount of blood analyzed in this study (Table 1).

Table 1. Circulating tumor cell (CTC) detection rate and cohorts' characteristics.

	N (%)	N CTC+ve (%)	*p*-Value
Patients with NSCLC	74 (100)	44 (59.5)	
Median (range) Age (years)	71 (43–86)		
Sex			
Female	33 (44.6)	18 (54.5)	
Male	41 (55.4)	26 (63.4)	0.4820
Smoking habits			
Current smoker	25 (33.8)	17 (68.0)	
Former smoker	32 (43.2)	18 (56.3)	
Never smoker	16 (21.6)	9 (56.3)	
Missing	1 (1.4)	0	0.7764 [a]
Tumor stage			
IA	27 (36.5)	15 (55.6)	
IB	5 (6.8)	2 (40.0)	
IIA	4 (5.4)	4 (100)	
IIB	16 (21.6)	9 (56.3)	
IIIA	11 (14.9)	7 (63.6)	
IIIB	6 (8.1)	3 (50.0)	
IIIC	0	0	
IVA	5 (6.8)	4 (80.0)	0.7964 [b]
Histology			
Adenocarcinoma	55 (74.3)	33 (60.0)	
Squamous cell carcinoma	16 (21.6)	10 (62.5)	
Other	2 (2.7)	1 (50.0)	
Missing	1 (1.4)	0	>0.9999 [c]

Table 1. Cont.

		N (%)	N CTC+ve (%)	p-Value
Patients with NSCLC		74 (100)	44 (59.5)	
Grading				
	G1	4 (5.4)	3 (75.0)	
	G2	44 (59.5)	23 (52.3)	
	G3	24 (32.4)	17 (70.8)	
	G4	0	0	
	Missing	2 (2.7)	1 (50.0)	0.2092 [d]
Visceral pleura invasion				
	PL0	56 (75.7)	31 (55.4)	
	PL1	8 (10.8)	6 (75.0)	
	PL2	5 (6.8)	3 (60.0)	
	PL3	2 (2.7)	2 (100)	
	Missing	3 (4.1)	2 (66.7)	0.2494 [e]
Peritumoral neoplastic angioinvasion				
	Absent	66 (89.2)	42 (63.6)	
	Present	6 (8.1)	1 (16.7)	
	Missing	2 (2.7)	1 (50.0)	0.0357
Lymph-node status				
	Negative	48 (64.9)	29 (60.4)	
	Positive	25 (33.8)	14 (56.0)	
	Missing	1 (1.4)	0	0.8039
Healthy volunteers		10		
	Median (range) Age (years)	35 (33–47)		
Sex				
	Female	7 (70.0)	0	-
	Male	3 (30.0)	0	-
Smoking habits				
	Smoker	2 (20.0)	0	-
	Never smoker	8 (80.0)	0	-
High-risk subjects		7		
	Median (range) Age (years)	63 (53–73)		
Sex				
	Female	3 (42.9)	0	-
	Male	4 (57.1)	0	-

[a] current/former versus never smokers. [b] stage I/II versus III/IV. [c] adenocarcinoma versus squamous cell carcinoma. [d] G1/G2 versus G3/G4. [e] PL0 vs. PL1/PL2/PL3.

3.2. Subpopulations of Circulating Atypical Cells Differentiate Operable NSCLC Patients from Lung Cancer-Free Individuals

The morphological analysis of cytological samples prepared on ISET membranes revealed the presence of a heterogeneous population of circulating atypical cells, which includes different subsets observed at a variable frequency within the two cohorts of patients and controls. We identified three subpopulations of cells with features of malignancy, hereafter referred to as (i) single CTCs ($n = 71$ cells), i.e., not in direct contact with other cells (sCTC, Figure 1a–c), (ii) homotypic CTC clusters (homCLU, $n = 2$ clusters), i.e., groups of slightly overlapping CTCs, and (iii) heterotypic clusters (hetCLU, $n = 40$ clusters), i.e., CTCs in direct contact with leukocytes (Figure 1d–f), mainly monocytes (62.5%) and neutrophil granulocytes (12.5%); we also observed (iv) a subpopulation of large cells, hereafter referred to as atypical macrophage-like cells (Figure 1g–i), (v) a subpopulation of clusters of epithelial-like cells (Figure 1j–l) and (vi) a considerable number of nuclei with a longer diameter >20 μm, in some cases with irregular membrane border and/or hyperchromasia, each of them apparently without the classical cytoplasmic rim observed in CTCs, in some cases similar to those observed in healthy donors (Supplementary Figure S1), hereafter referred to as naked nuclei (Supplementary Figure S2). In order to exclude the possibility that clusters were a result of technical artifacts due to the CTC enrichment procedure, we performed spike-in experiments with A549 and NCI-H460 cell lines at a dilution of about 1000 cells per milliliter in a total of 10 mL of blood from three healthy donors. Spiked-in, filtered and stained tumor cells were typically round-shaped and showed nucleus diameters larger than 20 μm, a nucleus-to-cytoplasm ratio around 90% and nuclear hyperchromatism. We did not observe the formation of homotypic clusters after filtration and staining, except for some doublets of NCI-H460 cells, and leukocytes were randomly interspersed among cancer cells.

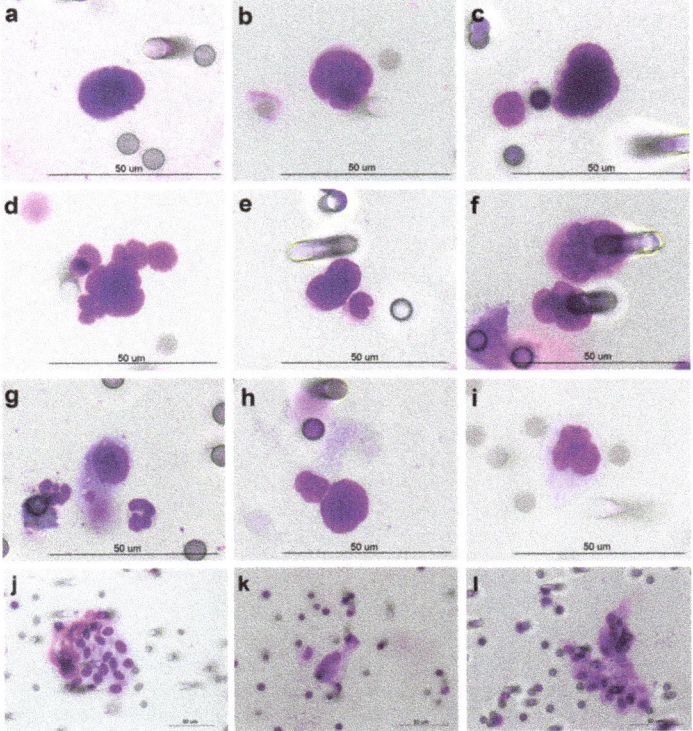

Figure 1. Subpopulations of atypical circulating cells differentiate patients with operable non-small cell lung cancer from lung cancer-free individuals. Images depict (**a**–**c**) single cells with morphological features of malignancy and (**d**–**f**) heterotypic clusters of cells with features of malignancy physically interacting with normal cells, such as (**d**) neutrophil granulocytes, (**e**) monocytes or (**f**) other indeterminate leukocytes, detected in patients, (**g**–**i**) atypical macrophage-like cells, (**g**) oblong or (**h**) tadpole-like, both detected in patients, and (**i**) irregularly shaped, detected in healthy donors, and (**j**–**l**) clusters of epithelial-like cells, detected in patients, on porous membranes stained with May–Grünwald and Giemsa. Objective magnification 40×.

In the cohort of 74 NSCLC patients, sCTC and hetCLU were detected in 34 (45.9%) and 23 (31.1%) cases, respectively, while both CTC subsets were observed in 14 (18.9%) cases. Homotypic clusters were observed in two cases only (2.7%). Neither the presence nor the prevalence or number of sCTC correlated with patients' demographics, smoking habits and tumor features, except for males and smokers, where at least two sCTC were detected at a significantly higher frequency compared to females and never smokers (Supplementary Table S1 and Figure 2a). Neither the presence nor the number of hetCLU correlated with the clinicopathological features (Supplementary Table S2 and Figure 2b).

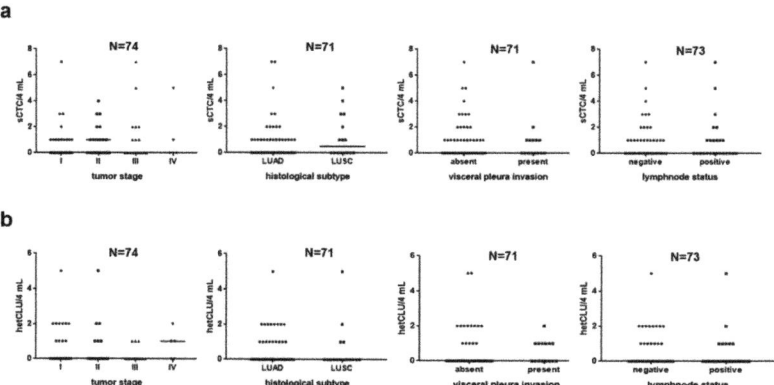

Figure 2. The number of single CTCs (sCTC) and heterotypic CTC clusters (hetCLU) is not associated with the clinicopathological features of stage I–IVA non-small cell lung cancers. Dot plots represent the distribution of (**a**) sCTC or (**b**) hetCLU count (median number, line) in 4 mL of peripheral blood according to the tumor stage, the histological subtype (LUAD, lung adenocarcinoma; LUSC, lung squamous cell carcinoma), the visceral pleura invasion and the lymph-node involvement. Differences were not significant (p-value > 0.05) by Kruskal–Wallis (tumor stage) or Mann–Whitney test.

Cells with features of atypical large macrophages were found in 12 out of 74 (16.2%) patients, and 2 out of 17 (12%) controls, whereas naked nuclei were observed in both cohorts, although with a two-fold increased frequency in patients (61/74, 82.4%) compared to controls (7/17, 41%; p-value = 0.0011). Interestingly, clusters of epithelial-like cells without apparent features of malignancy were detected in 9 out of 74 patients (12.2%) and none of the control cases. The number of patients called CTC-positive, who also had at least one atypical large macrophage, naked nucleus or cluster of epithelial-like cells were 11 (25%), 37 (84.1%) and 4 (9.1%), respectively.

3.3. The Prevalence of Single Circulating Tumor Cells Predicts the Risk of Recurrence in Patients with Surgically Treated Stage I–II NSCLCs

We then explored the utility of CTCs in serving as prognostic biomarkers to identify patients with an early-stage tumor at higher risk of disease recurrence. All stage I and II cases underwent surgical resection with curative intent. Six out of the 52 patients received post-operative adjuvant platinum-based with either vinorelbine or gemcitabine chemotherapy and/or radiotherapy. The disease status of all patients was monitored by clinical and radiological exams, except for eight cases (two stage I and six stage II) who were lost at follow-up (n = 3) or had died for unknown causes (n = 5). The median (range) observation time was 28 (1–47) months, and the total number of recurrence events was 13 (seven at intrathoracic level, five at distant sites and missing information in one case). The risk of disease recurrence in stages III and IV compared to stages I and II patients were HR 95%CI 3.45 (1.37–8.67), p-value = 0.0006, with an equal number of events (13) per group. In the stages I and II cohort, neither the tumor stage nor the grading or the lymph-node status were able to discriminate patients at higher risk of early disease recurrence. The numbers of stages I and II cases out of 44 evaluable for disease recurrence and with at least one, two, three or five sCTC, or at least one hetCLU, or CTC-positive irrespective of the subset type, were 21 (47.7%), 8 (18.2%), 5 (11.4%), 1 (2.3%), 15 (34.1%) and 26 (59.1%), respectively. No differences were found when grouping patients according to the overall CTC status or the presence of hetCLU, while patients with a prevalence of at least two or three CTCs in 4 mL of blood had a statistically significant shorter recurrence-free survival probability (HR 95% CI, cut-off two CTC/4 mL: 5.15 (1.10–24.33), p-value = 0.0009; cut-off three CTC/4 mL: 3.99 (0.47–33.57), p-value = 0.0216) compared to the counterpart with a more favorable CTC count (Figure 3 and Supplementary Table S3), demonstrating that the number of sCTC was the most relevant predictor of prognosis in early-stage operable NSCLCs.

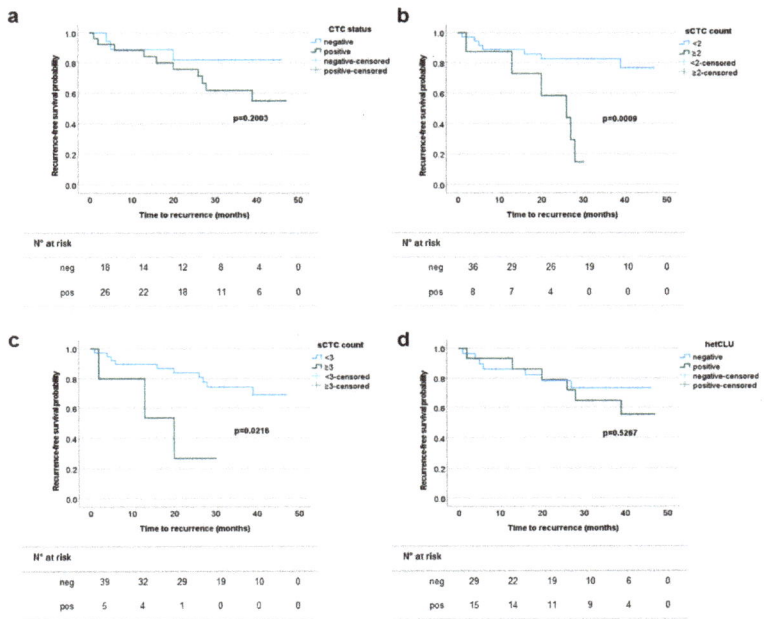

Figure 3. The prevalence of single circulating tumor cells (sCTC) correlates with reduced recurrence-free survival in stages I and II non-small cell lung cancers. Kaplan–Meier plots showing the time-to-event probability of disease recurrence according to (**a**) the presence of CTCs irrespective of the cell subset, (**b**,**c**) the prevalence of sCTC or (**d**) the presence of heterotypic circulating tumor cell clusters (hetCLU).

3.4. The Presence of Heterotypic Clusters of CTCs Predicts the Risk of Recurrence in Patients with Surgically Treated Stage III–IVA NSCLCs

We assessed the clinical significance of CTCs in the group of patients presenting with operable NSCLC at advanced stages (III and IV). Seventeen out of the 22 patients received post-operative adjuvant platinum-based with either vinorelbine or gemcitabine chemotherapy and/or radiotherapy or targeted therapy with gefitinib or afatinib for stage IV EGFR mutated tumors. The disease outcome of all patients was monitored by clinical and radiological exams, except for four stage III cases that were lost at follow-up ($n = 1$) or had died for unknown causes ($n = 3$). The median (range) observation time was 17; (1–33) months, and the total number of recurrence events was 13 (5 at intrathoracic level, 6 at distant sites and missing information in 2 cases).

In advanced NSCLC patients, neither the tumor stage nor the grading were able to accurately identify cases at higher risk of early disease recurrence, although a slight trend toward statistical significance was obtained when grouping according to the tumor stage (Supplementary Table S4). The numbers of stages III and IV cases out of 18 evaluable for disease recurrence and with at least one, two, three or five sCTCs, or at least one hetCLU, or CTC-positive irrespective of the subset type, were 8 (44.4%), 6 (33.3%), 3 (16.7%), 3 (16.7%), 5 (27.8%) and 12 (66.7%), respectively. According to the survival analysis based on the CTC status, a slight trend toward statistical significance was observed when considering the overall CTC population, while no difference was found when grouping patients according to the prevalence of sCTC. Interestingly, cases presenting with at least one hetCLU in 4 mL of blood had a statistically significant shorter recurrence-free survival probability (HR 95%CI: 3.44 (0.76–15.50), *p*-value = 0.0129) compared to patients without hetCLU (Figure 4 and Supplementary Table S4), providing evidence for the first time of the prognostic significance of hetCLU in advanced stage operable NSCLCs.

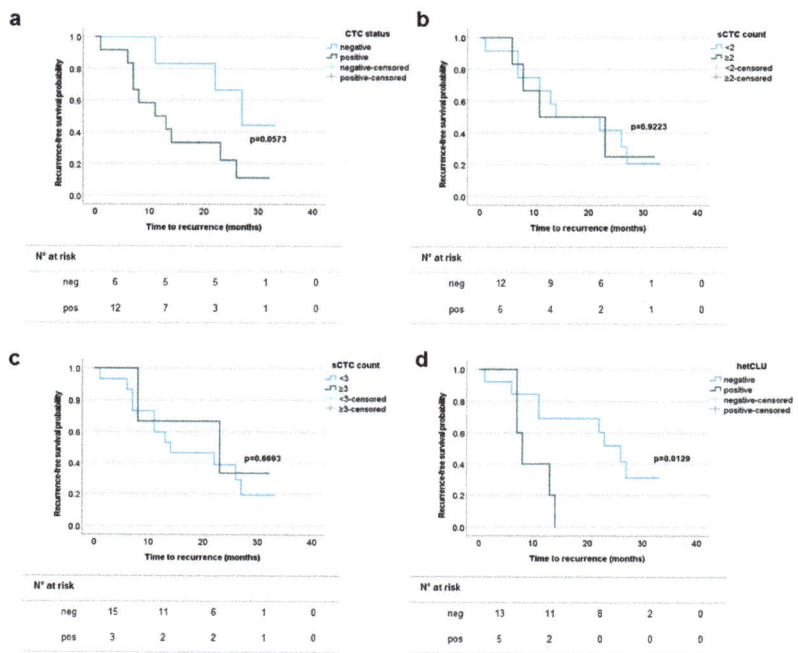

Figure 4. The presence of heterotypic circulating tumor cell clusters (hetCLU) correlates with reduced recurrence-free survival in stages III and IVA non-small cell lung cancers. Kaplan–Meier plots showing the time-to-event probability of disease recurrence according to (**a**) the presence of CTCs irrespective of the cell subset, (**b**,**c**) the prevalence of single CTCs (sCTC) or (**d**) the presence of hetCLU.

4. Discussion

In this study, we have observed that cells with morphological features of malignancy can be detected in 4 mL of peripheral blood in 60% of patients with operable NSCLC and that the CTC frequency is not dependent on the tumor stage. We have found that in NSCLC, cells with features of malignancy can circulate in physical contact with leukocytes, mainly monocytes, forming aggregates of two cells in the majority of cases, here called heterotypic clusters, with an overall frequency of 31%, without significant differences according to the disease stage. We have also documented the presence and frequency of other subpopulations of cells in patients, such as clusters of epithelial-like cells, large macrophage-like cells and naked nuclei >20 μm, which were found both in patients and lung cancer-free controls. Importantly, our study revealed different clinical messages hidden in CTCs based on the subset type as, compared to other tumor-related factors, only the CTC number at baseline was able to inform early-stage patients' risk and time of recurrence, whereas heterotypic clusters represented the most informative subpopulation of CTCs with respect to disease outcome in the advanced stage setting.

Contrarily to the traditional view that cancer cells disseminate via blood vessels within a time window closer to the clinical manifestation of secondary lesions rather than to the initiation of a tumor, CTCs are now widely recognized as events that can be detected at a considerable frequency even in patients presenting with early-stage or locally advanced disease. Since CTCs are known to be heterogeneous within the same patient, and a gold standard approach for accurate detection has not been developed yet [36], it is clear that the operational definition for the measurement of CTCs may change from study to study along with the technical approach. One of the first reports on CTC analysis in stage I–III NSCLC demonstrated that the CTC detection rate obtained by the capture of cells expressing EpCAM and cytokeratins 8/18/19 was lower compared to parallel samples analyzed by a size-based isolation approach coupled to cytomorphological analysis [32]. However, two main biological and technical aspects should be taken into account: first, some CTCs might have shaped their makeup while undergoing the epithelial-to-mesenchymal transition (EMT), thus downregulating the expression of surface adhesion proteins [37] and remaining undetectable to EpCAM-based capture; second, neutrophils have a low-to-absent expression of CD45 and, importantly, may be non-specifically bound by antibodies for some intracellular proteins, including cytokeratins [38], thus leading to CTC misidentification. Following the first demonstration that the early detection of CTCs was able to anticipate stage I lung cancer diagnosis by radiological scan in COPD patients [27], the majority of successive studies was performed using CTC enrichment methods based on the physical properties of tumor cells. The CTC positivity data obtained in our study population was 59.5%, which is slightly higher compared to the study by Hofman and colleagues in a case series of 210 NSCLC patients undergoing radical surgery [32], although the difference is not negligible if considering that we have analyzed 4 mL rather than 10 mL of blood. In other studies, the CTC detection rate was 80% by morphological analysis on 10 mL of blood from 40 chemonaïve stage IIIA and IV cases [39] and 69.5% by immunostaining for EpCAM and CD45 upon peripheral blood mononuclear cell collection and subsequent filtration in 82 cases with any stage [40].

In addition to the population of single CTCs, other subsets of cells are emerging as possible diagnostic biomarkers, as they showed high specificity in distinguishing between patients from lung cancer-free individuals: CTC clusters [41–43], which are aggregates of at least two CTCs in physical contact, held together through intercellular junctions [44], also known as circulating tumor microemboli, which can appear with infiltrated or surrounding platelets [35], or aggregates of cancer cells with non-malignant stromal or immune cells [44], as also large macrophage-like cells [35] and tumor-macrophage hybrid cells [45,46], which instead express both epithelial and leukocyte/macrophage-specific markers. The frequency of CTC clusters may vary depending on the technical approach and the disease stage [41–43,47]; therefore, further studies are needed to confirm the origin of this subset of CTCs. However, it has been becoming increasingly clear that CTC clusters may act as predictors or players in therapy resistance [48,49], and experimental evidence showed that polymorphonuclear/myeloid-derived suppressor cells interact with CTCs and promote their metastatic potential [50]. Remarkably, a non-conventional approach for CTC isolation recently showed that heterotypic clusters could be detected at a relevant frequency in many non-metastatic and metastatic solid tumors [51]. Therefore, heterotypic clusters can represent a promising biomarker and therapeutic target. Furthermore, the role of clusters of epithelial-like cells that we have observed in our cohort of patients is worth being clarified.

Studies with other technical approaches provided interesting results on the clinical role of CTCs in NSCLC in the operative setting. It was reported that cytokeratin- and EGFR-positive cells enriched by an immunomagnetic approach could be observed in stage I–III NSCLC patients at different frequencies before and 1 month after surgery and that post- but not pre-surgery detected cytokeratin-positive CTCs were associated with disease-free survival [52]. In 2019, a work with the CellSearch^{TM} system showed that tumor cells collected from the pulmonary vein during surgical procedures could be observed in 48% of cases, that using a cut-off of at least 18 CTCs in 7.5 mL of blood it was possible to predict disease relapse and that mutations in CTCs largely overlapped with those found in metastases detected 10 months later [53]. Other authors challenged the effect of surgery on CTC kinetics and found a significant decline in EpCAM-enriched CTCs a few days after surgery in all patients, and that early rebound of CTC counts was associated with disease recurrence [54]. Our data provide further evidence of the role of CTCs detected in the peripheral venous blood, and importantly, of the different significance that CTC subsets may have in classifying patients at risk of relapse when detected before surgery. Longitudinal studies are of interest to assess CTC kinetics in response to post-surgery administered systemic therapies.

Size-based approaches have already been shown to increase the sensitivity in detecting CTC clusters in metastatic NSCLCs compared to epitope-dependent methods. In 2011, vimentin-positive and cytokeratin-negative CTC clusters were described in three out of six metastatic NSCLC patients [55], and later it was found that circulating tumor microemboli, which were defined as clusters of at least three CTCs, could be observed in 43% of patients using ISET® but were undetectable by CellSearch™, which captures cells by anti-EpCAM antibodies [39]. Moreover, CTC clusters isolated by a biomarker-independent, size-based microfluidic method could be observed in 96% of patients with metastatic NSCLC, and 75% of them were EpCAM-negative [47]. In our study, with stage I–IVA NSCLC patients, the frequency of homotypic CTC clusters was negligible. The marker-free technical approach based on the direct evaluation of cytological samples enabled the visualization of CTCs in contact with leukocytes, as also of other atypical cells. The application of a filtration method coupled to morphological analysis brings many advantages. However, a crucial point that should be addressed is the reproducibility of the analysis based on cytomorphological criteria. Although they have been already defined and blindly validated by a team of 10 cytopathologists on 808 blood samples analyzed by the ISET method, such criteria hold the same limits as those used in routine cytology [56]. In our study, cells with features of malignancy were not found in lung cancer-free individuals, following blinded analysis. Larger cohorts of individuals at risk of lung cancer and the inter-reader variability should be evaluated in multicentric studies in order to increase the robustness of the CTC test and to foster its application in the clinical routine. The identification of a panel of biomarkers for CTC detection in NSCLC is also of crucial importance in this context. Searching for lung cancer-specific markers is a long-standing issue for pathologists, which consequently affects CTC studies. An attentive look at CTC and metastasis biological hallmarks may help to identify markers alternative to cytokeratin, and in fact, a recent paper showed that the glycolysis enzyme hexokinase 2 (HK2) increased the detection of CTCs in a cohort of 18 stage III lung adenocarcinoma patients without clinical evidence of distant metastases from 39% when considering cytokeratin-positive to 61% when considering HK2-positive cell subsets [57].

To summarize the results of this study, we have discovered a population of heterotypic CTC clusters in early-stage NSCLC and provided first evidence of the differential prognostic significance of single CTC count, using a low cut-off (two CTCs in 4 mL of blood), and of the presence of heterotypic clusters, in operable NSCLC patients with stages I or II and stages III or IVA, respectively, demonstrating that looking at CTC subsets rather than the overall CTC population can help to refine the classification of patients at risk of disease recurrence, and outperforming classical primary tumor-related markers; we have performed the analysis on cytological samples corresponding to 4 mL only compared to larger blood volumes (from 7.5 to 10 mL) used in other studies, scoring as CTC-positive about 60% of patients and none of the control individuals, and we have documented and described other subsets of circulating atypical cells occurring with different frequency in patients and lung cancer-free donors.

5. Conclusions

The ISET® technology for CTC enrichment coupled to cytomorphological analysis was proven as a promising approach for the development of non-invasive biomarkers in NSCLC. With a view to the implementation of a CTC-based test in screening programs, studies with larger case series and the introduction of molecular analyses to infer the origin of atypical cells would be desirable. CTC subsets also deserve consideration to be included in the clinical routine among the standard prognostic factors in order to early identify patients at risk of recurrence and refine the therapeutic strategies accordingly.

Supplementary Materials: The following are available online at https://www.mdpi.com/article/10.3390/cancers13174488/s1, Figure S1. Naked nuclei detected in the blood of lung cancer-free individuals; Figure S2. Naked nuclei detected in the blood of on-small cell lung cancer patients; Table S1. Single circulating tumor cell (sCTC) prevalence and non-small cell lung cancer (NSCLC) patients' characteristics; Table S2. Circulating tumor cell cluster (cCTC) status and NSCLC patients' characteristics; Table S3. Recurrence-free survival probability in patients with operable stage I–II NSCLC according to standard clinico–pathological parameters and CTC status; Table S4. Recurrence-free survival probability in patients with operable stage III–IV NSCLC according to standard clinico-pathological parameters and CTC status.

Author Contributions: Conceptualization, G.V. and E.F.; methodology, S.M., E.F. and D.F.; validation, E.F.; formal analysis, E.F.; investigation, E.F., S.M., D.F., P.N., G.M., G.F. and G.V.; resources, M.A. and G.V.; data curation, E.F., E.D., F.C. and S.M.; writing—original draft preparation, E.F.; writing—review and editing, G.V. and G.F.; visualization, E.F. and S.M.; supervision, G.V.; project administration, G.V.; funding acquisition, G.V. All authors have read and agreed to the published version of the manuscript.

Funding: This work was supported by specific grants by LILT (5X1000 program year 2013), the Umberto Veronesi Foundation (Milan, Italy), TRANSCAN-2 Joint Transnational Call 2016 (JTC 2016) project code TRANSCAN-045 CLEARLY, Italian Ministry of Health (Finalizzata Ministeriale 2016) project code PE-2016-02364336. E. Fina is the recipient of a postdoctoral research fellowship awarded by the Umberto Veronesi Foundation.

Institutional Review Board Statement: The study was conducted according to the guidelines of the Declaration of Helsinki and approved by the Ethics Committee of IRCCS Humanitas Research Hospital.

Informed Consent Statement: Informed consent was obtained from all patients involved in the study.

Data Availability Statement: Data is contained within the article or supplementary material.

Acknowledgments: The authors wish to acknowledge all donors for participating in the study. A special thanks also to Patrizia Paterlini-Bréchot and her team from the Université Paris Descartes and Rarecells Diagnostics for technical support during the starting phase of the project.

Conflicts of Interest: The study funders had no role in the design of the study, the collection, analysis, and interpretation of the data, the writing of the manuscript, and the decision to submit the manuscript for publication. The authors declare no competing non-financial interests but the following competing financial interest: G.V. received honoraria from Ab Medica SpA (Milan, Italy). All the other Authors declare no conflict of interest.

References

1. Siegel, R.L.; Miller, K.D.; Jemal, A. Cancer statistics, 2019. *CA Cancer J. Clin.* **2019**, *69*, 7–34. [CrossRef]
2. Herbst, R.S.; Morgensztern, D.; Boshoff, C. The biology and management of non-small cell lung cancer. *Nature* **2018**, *553*, 446–454. [CrossRef]
3. Campbell, J.D.; Alexandrov, A.; Kim, J.; Wala, J.; Berger, A.H.; Pedamallu, C.S.; Shukla, S.A.; Guo, G.; Brooks, A.N.; Murray, B.A.; et al. Distinct patterns of somatic genome alterations in lung adenocarcinomas and squamous cell carcinomas. *Nat. Genet.* **2016**, *48*, 607–616. [CrossRef] [PubMed]
4. Walters, S.; Maringe, C.; Coleman, M.; Peake, M.D.; Butler, J.; Young, N.; Bergström, S.; Hanna, L.; Jakobsen, E.; Kölbeck, K.; et al. Lung cancer survival and stage at diagnosis in Australia, Canada, Denmark, Norway, Sweden and the UK: A population-based study, 2004–2007. *Thorax* **2013**, *68*, 551–564. [CrossRef] [PubMed]
5. Allemani, C.; Matsuda, T.; Di Carlo, V.; Harewood, R.; Matz, M.; Nikšić, M.; Bonaventure, A.; Valkov, M.; Johnson, C.J.; Estève, J.; et al. Global surveillance of trends in cancer survival 2000–14 (CONCORD-3): Analysis of individual records for 37 513 025 patients diagnosed with one of 18 cancers from 322 population-based registries in 71 countries. *Lancet* **2018**, *391*, 1023–1075. [CrossRef]

6. The National Lung Screening Trial Research Team; Aberle, D.R.; Adams, A.M.; Berg, C.D.; Black, W.C.; Clapp, J.D.; Fagerstrom, R.M.; Gareen, I.F.; Gatsonis, C.; Marcus, P.M.; et al. Reduced Lung-Cancer Mortality with Low-Dose Computed Tomographic Screening. *N. Engl. J. Med.* **2011**, *365*, 395–409. [CrossRef]
7. Henschke, C.I.; Boffetta, P.; Gorlova, O.; Yip, R.; DeLancey, J.O.; Foy, M. Assessment of lung-cancer mortality reduction from CT Screening. *Lung Cancer* **2011**, *71*, 328–332. [CrossRef] [PubMed]
8. Puggina, A.; Broumas, A.; Ricciardi, W.; Boccia, S. Cost-effectiveness of screening for lung cancer with low-dose computed tomography: A systematic literature review. *Eur. J. Public Health* **2016**, *26*, 168–175. [CrossRef]
9. Veronesi, G.; Novellis, P.; Voulaz, E.; Alloisio, M. Early detection and early treatment of lung cancer: Risks and benefits. *J. Thorac. Dis.* **2016**, *8*, E1060–E1062. [CrossRef]
10. Veronesi, G.; Baldwin, D.; Henschke, C.; Ghislandi, S.; Iavicoli, S.; Oudkerk, M.; De Koning, H.; Shemesh, J.; Field, J.; Zulueta, J.; et al. Recommendations for Implementing Lung Cancer Screening with Low-Dose Computed Tomography in Europe. *Cancers* **2020**, *12*, 1672. [CrossRef]
11. Veronesi, G.; Bellomi, M.; Scanagatta, P.; Preda, L.; Rampinelli, C.; Guarize, J.; Pelosi, G.; Maisonneuve, P.; Leo, F.; Solli, P.; et al. Difficulties encountered managing nodules detected during a computed tomography lung cancer screening program. *J. Thorac. Cardiovasc. Surg.* **2008**, *136*, 611–617. [CrossRef]
12. Shah, R.; Sabanathan, S.; Richardson, J.; Mearns, A.J.; Goulden, C. Results of surgical treatment of stage I and II lung cancer. *J. Cardiovasc. Surg.* **1996**, *37*, 169–172.
13. Nesbitt, J.C.; Putnam, J.B.; Walsh, G.L.; Roth, J.A.; Mountain, C.F. Survival in early-stage non-small cell lung cancer. *Ann. Thorac. Surg.* **1995**, *60*, 466–472. [CrossRef]
14. Al-Kattan, K.; Sepsas, E.; Fountain, S.W.; Townsend, E.R. Disease recurrence after resection for stage I lung cancer. *Eur. J. Cardio-Thorac. Surg.* **1997**, *12*, 380–384. [CrossRef]
15. Hoffman, P.C.; Mauer, A.M.; Vokes, E.E. Lung cancer. *Lancet* **2000**, *355*, 479–485. [CrossRef]
16. Uramoto, H.; Tanaka, F. Recurrence after surgery in patients with NSCLC. *Transl. Lung Cancer Res.* **2014**, *3*, 242–249. [CrossRef]
17. Blanchon, F.; Grivaux, M.; Asselain, B.; Lebas, F.-X.; Orlando, J.-P.; Piquet, J.; Zureik, M. 4-year mortality in patients with non-small-cell lung cancer: Development and validation of a prognostic index. *Lancet Oncol.* **2006**, *7*, 829–836. [CrossRef]
18. Knight, S.B.; Crosbie, P.A.; Balata, H.; Chudziak, J.; Hussell, T.; Dive, C. Progress and prospects of early detection in lung cancer. *Open Biol.* **2017**, *7*, 170070. [CrossRef]
19. Mok, T.S.K. Personalized medicine in lung cancer: What we need to know. *Nat. Rev. Clin. Oncol.* **2011**, *8*, 661–668. [CrossRef] [PubMed]
20. Kerr, K.M.; Bubendorf, L.; Edelman, M.; Marchetti, A.; Mok, T.; Novello, S.; O'Byrne, K.; Stahel, R.; Peters, S.; Felip, E.; et al. Second ESMO consensus conference on lung cancer: Pathology and molecular biomarkers for non-small-cell lung cancer. *Ann. Oncol.* **2014**, *25*, 1681–1690. [CrossRef]
21. Hofman, P. Liquid biopsy for early detection of lung cancer. *Curr. Opin. Oncol.* **2017**, *29*, 73–78. [CrossRef] [PubMed]
22. Mathai, R.A.; Vidya, R.V.S.; Reddy, B.S.; Thomas, L.; Udupa, K.; Kolesar, J.; Rao, M. Potential Utility of Liquid Biopsy as a Diagnostic and Prognostic Tool for the Assessment of Solid Tumors: Implications in the Precision Oncology. *J. Clin. Med.* **2019**, *8*, 373. [CrossRef] [PubMed]
23. Heitzer, E.; Haque, I.S.; Roberts, C.E.S.; Speicher, M.R. Current and future perspectives of liquid biopsies in genomics-driven oncology. *Nat. Rev. Genet.* **2019**, *20*, 71–88. [CrossRef] [PubMed]
24. Massagué, J.; Obenauf, A.C. Metastatic colonization by circulating tumour cells. *Nature* **2016**, *529*, 298–306. [CrossRef]
25. Klein, C.A. Parallel progression of primary tumours and metastases. *Nat. Rev. Cancer* **2009**, *9*, 302–312. [CrossRef]
26. Hamilton, G.; Rath, B. Circulating Tumor Cells in the Parallel Invasion Model Supporting Early Metastasis. *Oncomedicine* **2018**, *3*, 15–27. [CrossRef]
27. Ilie, M.; Hofman, V.; Long, E.; Selva, E.; Vignaud, J.-M.; Padovani, B.; Mouroux, J.; Marquette, C.H.; Hofman, P. "Sentinel" Circulating Tumor Cells Allow Early Diagnosis of Lung Cancer in Patients with Chronic Obstructive Pulmonary Disease. *PLoS ONE* **2014**, *9*, e111597. [CrossRef]
28. Alix-Panabières, C.; Pantel, K. Challenges in circulating tumour cell research. *Nat. Rev. Cancer* **2014**, *14*, 623–631. [CrossRef]
29. Vona, G.; Sabile, A.; Louha, M.; Sitruk, V.; Romana, S.P.; Schütze, K.; Capron, F.; Franco, D.; Pazzagli, M.; Vekemans, M.; et al. Isolation by Size of Epithelial Tumor Cells: A New Method for the Immunomorphological and Molecular Characterization of Circulating Tumor Cells. *Am. J. Pathol.* **2000**, *156*, 57–63. [CrossRef]
30. Laget, S.; Broncy, L.; Hormigos, K.; Dhingra, D.M.; Benmohamed, F.; Capiod, T.; Osteras, M.; Farinelli, L.; Jackson, S.; Paterlini-Bréchot, P. Technical Insights into Highly Sensitive Isolation and Molecular Characterization of Fixed and Live Circulating Tumor Cells for Early Detection of Tumor Invasion. *PLoS ONE* **2017**, *12*, e0169427. [CrossRef] [PubMed]
31. Hofman, V.; Bonnetaud, C.; Ilié, M.; Vielh, P.; Vignaud, J.M.; Fléjou, J.F.; Lantuejoul, S.; Piaton, E.; Mourad, N.; Butori, C.; et al. Preoperative Circulating Tumor Cell Detection Using the Isolation by Size of Epithelial Tumor Cell Method for Patients with Lung Cancer Is a New Prognostic Biomarker. *Clin. Cancer Res.* **2011**, *17*, 827–835. [CrossRef] [PubMed]
32. Hofman, V.; Ilie, M.I.; Long, E.; Selva, E.; Bonnetaud, C.; Molina, T.; Venissac, N.; Mouroux, J.; Vielh, P.; Hofman, P. Detection of circulating tumor cells as a prognostic factor in patients undergoing radical surgery for non-small-cell lung carcinoma: Comparison of the efficacy of the CellSearch Assay™ and the isolation by size of epithelial tumor cell method. *Int. J. Cancer* **2010**, *129*, 1651–1660. [CrossRef]

33. Marquette, C.-H.; Boutros, J.; Benzaquen, J.; Ferreira, M.; Pastre, J.; Pison, C.; Padovani, B.; Bettayeb, F.; Fallet, V.; Guibert, N.; et al. Circulating tumour cells as a potential biomarker for lung cancer screening: A prospective cohort study. *Lancet Respir. Med.* **2020**, *8*, 709–716. [CrossRef]
34. Hofman, V.; Long, E.; Ilié, M.; Bonnetaud, C.; Vignaud, J.M.; Fléjou, J.F.; Lantuejoul, S.; Piaton, E.; Mourad, N.; Butori, C.; et al. Morphological analysis of circulating tumour cells in patients undergoing surgery for non-small cell lung carcinoma using the isolation by size of epithelial tumour cell (ISET) method. *Cytopathology* **2011**, *23*, 30–38. [CrossRef]
35. Wechsler, J. *Circulating Tumor Cells from Solid Cancers*; Sauramps Medical Press: Montpellier, France, 2015.
36. Alix-Panabières, C.; Pantel, K. Circulating Tumor Cells: Liquid Biopsy of Cancer. *Clin. Chem.* **2013**, *59*, 110–118. [CrossRef]
37. Thiery, J.P. Epithelial–mesenchymal transitions in tumour progression. *Nat. Rev. Cancer* **2002**, *2*, 442–454. [CrossRef]
38. Schehr, J.L.; Schultz, Z.D.; Warrick, J.W.; Guckenberger, D.J.; Pezzi, H.M.; Sperger, J.M.; Heninger, E.; Saeed, A.; Leal, T.; Mattox, K.; et al. High Specificity in Circulating Tumor Cell Identification Is Required for Accurate Evaluation of Programmed Death-Ligand 1. *PLoS ONE* **2016**, *11*, e0159397. [CrossRef]
39. Krebs, M.G.; Hou, J.-M.; Sloane, R.S.; Lancashire, L.; Priest, L.; Nonaka, D.; Ward, T.H.; Backen, A.; Clack, G.; Hughes, A.; et al. Analysis of Circulating Tumor Cells in Patients with Non-small Cell Lung Cancer Using Epithelial Marker-Dependent and -Independent Approaches. *J. Thorac. Oncol.* **2012**, *7*, 306–315. [CrossRef]
40. Sonn, C.-H.; Cho, J.H.; Kim, J.-W.; Kang, M.S.; Lee, J.; Kim, J. Detection of circulating tumor cells in patients with non-small cell lung cancer using a size-based platform. *Oncol. Lett.* **2017**, *13*, 2717–2722. [CrossRef] [PubMed]
41. Carlsson, A.; Nair, V.S.; Luttgen, M.S.; Keu, K.V.; Horng, G.; Vasanawala, M.; Kolatkar, A.; Jamali, M.; Iagaru, A.H.; Kuschner, W.; et al. Circulating Tumor Microemboli Diagnostics for Patients with Non–Small-Cell Lung Cancer. *J. Thorac. Oncol.* **2014**, *9*, 1111–1119. [CrossRef]
42. Mascalchi, M.; Maddau, C.; Sali, L.; Bertelli, E.; Salvianti, F.; Zuccherelli, S.; Matucci, M.; Borgheresi, A.; Raspanti, C.; Lanzetta, M.M.; et al. Circulating tumor cells and microemboli can differentiate malignant and benign pulmonary lesions. *J. Cancer* **2017**, *8*, 2223–2230. [CrossRef] [PubMed]
43. Manjunath, Y.; Upparahalli, S.V.; Suvilesh, K.N.; Avella, D.M.; Kimchi, E.T.; Staveley-O'Carroll, K.F.; Li, G.; Kaifi, J.T. Circulating tumor cell clusters are a potential biomarker for detection of non-small cell lung cancer. *Lung Cancer* **2019**, *134*, 147–150. [CrossRef] [PubMed]
44. Aceto, N. Bring along your friends: Homotypic and heterotypic circulating tumor cell clustering to accelerate metastasis. *Biomed. J.* **2020**, *43*, 18–23. [CrossRef] [PubMed]
45. Adams, D.; Martin, S.S.; Alpaugh, R.K.; Charpentier, M.; Tsai, S.; Bergan, R.C.; Ogden, I.M.; Catalona, W.; Chumsri, S.; Tang, C.-M.; et al. Circulating giant macrophages as a potential biomarker of solid tumors. *Proc. Natl. Acad. Sci. USA* **2014**, *111*, 3514–3519. [CrossRef]
46. Manjunath, Y.; Mitchem, J.B.; Suvilesh, K.N.; Avella, D.M.; Kimchi, E.T.; Staveley-O'Carroll, K.F.; Deroche, C.B.; Pantel, K.; Li, G.; Kaifi, J.T. Circulating Giant Tumor-Macrophage Fusion Cells Are Independent Prognosticators in Patients With NSCLC. *J. Thorac. Oncol.* **2020**, *15*, 1460–1471. [CrossRef]
47. Zeinali, M.; Lee, M.; Nadhan, A.; Mathur, A.; Hedman, C.; Lin, E.; Harouaka, R.; Wicha, M.; Zhao, L.; Palanisamy, N.; et al. High-Throughput Label-Free Isolation of Heterogeneous Circulating Tumor Cells and CTC Clusters from Non-Small-Cell Lung Cancer Patients. *Cancers* **2020**, *12*, 127. [CrossRef]
48. Lee, M.; Kim, E.J.; Cho, Y.; Kim, S.; Chung, H.H.; Park, N.H.; Song, Y.-S. Predictive value of circulating tumor cells (CTCs) captured by microfluidic device in patients with epithelial ovarian cancer. *Gynecol. Oncol.* **2017**, *145*, 361–365. [CrossRef]
49. Bithi, S.S.; Vanapalli, S.A. Microfluidic cell isolation technology for drug testing of single tumor cells and their clusters. *Sci. Rep.* **2017**, *7*, 41707. [CrossRef]
50. Sprouse, M.L.; Welte, T.; Boral, D.; Liu, H.N.; Yin, W.; Vishnoi, M.; Goswami-Sewell, D.; Li, L.; Pei, G.; Jia, P.; et al. PMN-MDSCs Enhance CTC Metastatic Properties through Reciprocal Interactions via ROS/Notch/Nodal Signaling. *Int. J. Mol. Sci.* **2019**, *20*, 1916. [CrossRef]
51. Akolkar, D.; Patil, D.; Crook, T.; Limaye, S.; Page, R.; Datta, V.; Patil, R.; Sims, C.; Ranade, A.; Fulmali, P.; et al. Circulating ensembles of tumor-associated cells: A redoubtable new systemic hallmark of cancer. *Int. J. Cancer* **2020**, *146*, 3485–3494. [CrossRef]
52. Bayarri-Lara, C.; Ortega, F.G.; De Guevara, A.C.L.; Puche, J.L.; Zafra, J.R.; De Miguel-Pérez, D.; Ramos, A.S.-P.; Giraldo-Ospina, C.F.; Gómez, J.A.N.; Delgado-Rodríguez, M.; et al. Circulating Tumor Cells Identify Early Recurrence in Patients with Non-Small Cell Lung Cancer Undergoing Radical Resection. *PLoS ONE* **2016**, *11*, e0148659. [CrossRef]
53. Chemi, F.; Rothwell, D.; McGranahan, N.; Gulati, S.; Abbosh, C.; Pearce, S.P.; Zhou, C.; Wilson, G.A.; Jamal-Hanjani, M.; Birkbak, N.; et al. Pulmonary venous circulating tumor cell dissemination before tumor resection and disease relapse. *Nat. Med.* **2019**, *25*, 1534–1539. [CrossRef]
54. Wu, C.-Y.; Lee, C.-L.; Fu, J.-Y.; Yang, C.-T.; Wen, C.-T.; Liu, Y.-H.; Liu, H.-P.; Hsieh, J.C.-H.; Wu, C.-F. Circulating Tumor Cells as a Tool of Minimal Residual Disease Can Predict Lung Cancer Recurrence: A longitudinal, Prospective Trial. *Diagnostics* **2020**, *10*, 144. [CrossRef] [PubMed]
55. Lecharpentier, A.; Vielh, P.; Perez-Moreno, P.; Planchard, D.; Soria, J.C.; Farace, F. Detection of circulating tumour cells with a hybrid (epithelial/mesenchymal) phenotype in patients with metastatic non-small cell lung cancer. *Br. J. Cancer* **2011**, *105*, 1338–1341. [CrossRef] [PubMed]

56. Hofman, V.J.; Ilié, M.; Bonnetaud, C.; Selva, E.; Long, E.; Molina, T.; Vignaud, J.M.; Fléjou, J.F.; Lantuejoul, S.; Piaton, E.; et al. Cytopathologic Detection of Circulating Tumor Cells Using the Isolation by Size of Epithelial Tumor Cell Method. *Am. J. Clin. Pathol.* **2011**, *135*, 146–156. [CrossRef]
57. Yang, L.; Yan, X.; Chen, J.; Zhan, Q.; Hua, Y.; Xu, S.; Li, Z.; Wang, Z.; Dong, Y.; Zuo, D.; et al. Hexokinase 2 discerns a novel circulating tumor cell population associated with poor prognosis in lung cancer patients. *Proc. Natl. Acad. Sci. USA* **2021**, *118*, 2012228118. [CrossRef] [PubMed]

Article

Circulating Tumor DNA Reflects Uveal Melanoma Responses to Protein Kinase C Inhibition

John J. Park [1,2,3], Russell J. Diefenbach [1,2], Natalie Byrne [3], Georgina V. Long [2,4,5], Richard A. Scolyer [2,5,6], Elin S. Gray [7], Matteo S. Carlino [2,3,5,†] and Helen Rizos [1,2,*,†]

1. Department of Biomedical Sciences, Faculty of Medicine, Health and Human Sciences, Macquarie University, Sydney, NSW 2109, Australia; john.park4@hdr.mq.edu.au (J.J.P.); russell.diefenbach@mq.edu.au (R.J.D.)
2. Melanoma Institute Australia, The University of Sydney, Sydney, NSW 2065, Australia; georgina.long@sydney.edu.au (G.V.L.); richard.scolyer@health.nsw.gov.au (R.A.S.); matteo.carlino@sydney.edu.au (M.S.C.)
3. Department of Medical Oncology, Westmead and Blacktown Hospitals, Sydney, NSW 2145, Australia; natalie.byrne@sydney.edu.au
4. Department of Medical Oncology, Royal North Shore Hospital and Mater Hospitals, Sydney, NSW 2065, Australia
5. Faculty of Medicine and Health, The University of Sydney, Sydney, NSW 2006, Australia
6. Tissue Pathology and Diagnostic Oncology, Royal Prince Alfred Hospital and NSW Health Pathology, Sydney, NSW 2050, Australia
7. Centre for Precision Health, Edith Cowan University, Joondalup, WA 6027, Australia; e.gray@ecu.edu.au
* Correspondence: helen.rizos@mq.edu.au; Tel.: +61-2-9850-2762
† Both authors contributed equally as senior authors.

Citation: Park, J.J.; Diefenbach, R.J.; Byrne, N.; Long, G.V.; Scolyer, R.A.; Gray, E.S.; Carlino, M.S.; Rizos, H. Circulating Tumor DNA Reflects Uveal Melanoma Responses to Protein Kinase C Inhibition. *Cancers* **2021**, *13*, 1740. https://doi.org/10.3390/cancers13071740

Academic Editor: Fabrizio Bianchi

Received: 3 March 2021
Accepted: 3 April 2021
Published: 6 April 2021

Publisher's Note: MDPI stays neutral with regard to jurisdictional claims in published maps and institutional affiliations.

Copyright: © 2021 by the authors. Licensee MDPI, Basel, Switzerland. This article is an open access article distributed under the terms and conditions of the Creative Commons Attribution (CC BY) license (https://creativecommons.org/licenses/by/4.0/).

Simple Summary: Uveal melanoma (UM) is a rare cancer, with no effective standard systemic therapy in the metastatic setting. Over 95% of UM harbor activating driver mutations that can be detected in the circulation. In this study, circulating tumor DNA (ctDNA) was measured in 17 metastatic UM patients treated with protein kinase C inhibitor (PKCi)-based therapy. ctDNA predicted response to targeted therapy and increasing UM ctDNA preceded radiological progression with a lead-time of 4–10 weeks. Next generation sequencing (NGS) of ctDNA also identified prognostic and treatment resistance mutations. Longitudinal ctDNA monitoring is useful for monitoring disease response and progression in metastatic UM and is a valuable addition to adaptive clinical trial design.

Abstract: The prognosis for patients with UM is poor, and recent clinical trials have failed to prolong overall survival (OS) of these patients. Over 95% of UM harbor activating driver mutations, and this allows for the investigation of ctDNA. In this study, we investigated the value of ctDNA for adaptive clinical trial design in metastatic UM. Longitudinal plasma samples were analyzed for ctDNA in 17 metastatic UM patients treated with PKCi-based therapy in a phase 1 clinical trial setting. Plasma ctDNA was assessed using digital droplet PCR (ddPCR) and a custom melanoma gene panel for targeted next generation sequencing (NGS). Baseline ctDNA strongly correlated with baseline lactate dehydrogenase (LDH) ($p < 0.001$) and baseline disease burden ($p = 0.002$). Early during treatment (EDT) ctDNA accurately predicted patients with clinical benefit to PKCi using receiver operator characteristic (ROC) curves (AUC 0.84, [95% confidence interval 0.65–1.0, $p = 0.026$]). Longitudinal ctDNA assessment was informative for establishing clinical benefit and detecting disease progression with 7/8 (88%) of patients showing a rise in ctDNA and targeted NGS of ctDNA revealed putative resistance mechanisms prior to radiological progression. The inclusion of longitudinal ctDNA monitoring in metastatic UM can advance adaptive clinical trial design.

Keywords: uveal melanoma; circulating tumor DNA; next generation sequencing; PKC inhibitor; liquid biopsy; treatment; response; melanoma

1. Introduction

Uveal melanoma (UM) is the most common primary intraocular malignancy [1]. The tumor arises from melanocytes within the uveal tract, with more than 90% of cases involving the choroid followed by iris and ciliary body [2]. UM is a rare cancer, affecting approximately 5–7 individuals per million each year [1,3–5]. Despite successful local treatment with either surgery or radiation therapy, approximately 50% of patients with UM will develop metastatic disease [6] with over 90% of metastases occurring in the liver [7]. Currently there is no effective systemic treatment in metastatic UM and the median progression free survival (PFS) and overall survival (OS) are 3.3 months and 10.2 months, respectively [8].

Nearly 95% of UM harbor mutually exclusive activating driver mutations in *GNAQ*, *GNA11*, *CYSTLR2* and *PLCβ4* genes [9–14]. Molecular profiling, cytogenetic and transcriptomic analysis of UM have provided accurate prognostic information [9,15]. Additional hot spot mutations affecting the *EIF1AX* and *SF3B1* genes are associated with better prognosis whereas loss of function *BAP1* gene alterations are correlated with the development of UM metastases and poor prognosis [9,16]. Somatic copy number alterations such as loss of chromosome 3, 6q and 8q are also associated with poor prognosis [17]. The specific and defined mutation profile of UM provides an excellent opportunity to investigate the utility of circulating tumor DNA (ctDNA) as a biomarker to detect the presence of metastatic disease and to rapidly monitor response to early-phase drug therapies.

In cutaneous melanoma (CM), baseline ctDNA is strongly correlated with tumor burden in patients with advanced stage disease [18] and is associated with overall response rate and PFS in patients treated with targeted therapies [18,19]. A decline in ctDNA within 8 weeks of treatment initiation also predicts response to both combined BRAF and MEK inhibition and immunotherapy in CM [19,20]. In metastatic UM, ctDNA levels correlate with tumor burden and the presence of liver metastases [21,22] and are also prognostic for PFS and OS [21]. The value of ctDNA in monitoring and predicting response to trial drug therapies has not, to the best of our knowledge, been previously investigated. This is particularly relevant in metastatic UM as there are currently no effective systemic treatments, but significant ongoing clinical trial activity evaluating novel therapies. In this study, we sought to assess ctDNA in metastatic UM patients treated with protein kinase C inhibitor (PKCi)-based therapy in a phase 1 clinical trial setting (NCT02601378). Using two methods, droplet digital PCR (ddPCR) and targeted Ion Torrent next generation sequencing (NGS), we evaluated the utility of plasma ctDNA in monitoring and predicting clinical outcomes including best response to therapy and PFS.

2. Materials and Methods

2.1. Patients and Treatment

Seventeen patients with metastatic UM with known mutations in GNAQ, GNA11 and CYSTLR2, treated with the novel PKCi, LXS196 ($n = 17$) at Westmead Hospital, Sydney, Australia as part of an experimental dose escalation phase 1 clinical trial between November 2016 to August 2018 were included in this study. Written consent was obtained from all patients with metastatic UM under approved Human Research ethics committee protocols from Royal Prince Alfred Hospital (Protocol X15-0454 and HREC/11/RPAH/444).

2.2. Patient and Disease Characteristics and Response Assessment

Patient demographics and clinicopathologic features including mutation status, Eastern Cooperative Oncology Group (ECOG) performance status, and baseline LDH levels (units/litre; U/L) were collected. Baseline disease burden was determined by the sum of the product of bi-dimensional diameters (SPOD) for every metastasis $\geq$5 mm in the long axis ($\geq$15 mm in the short axis for lymph nodes). Investigator-determined objective responses were assessed radiologically with computed tomography (CT) scans at two monthly intervals using Response Evaluation Criteria in Solid Tumors (RECIST) 1.1 criteria [23]. Clinical progression was defined by primary clinician's assessment of disease progression in patients without re-staging imaging and were classified as progressive disease (PD). Clinical benefit was defined by patients who had partial response (PR) or stable disease (SD) for ≥6 months.

2.3. Plasma Preparation

Plasma samples were collected at baseline (prior to therapy start), EDT (early during treatment between 14–30 days of commencing PKCi-based therapy) and at later time points during therapy (on-treatment samples). PROG samples were defined as plasma samples taken within 30 days (before or after) of disease progression confirmed by imaging or clinical progression as determined by the treating clinician. NGS analysis was performed on baseline plasma samples and on the last available on-treatment plasma sample. Blood (10 mL) was collected in EDTA tubes (Becton Dickinson, Franklin Lakes, NJ, USA) and processed within 4 h from blood draw. Tubes were spun at 800 g for 15 min for plasma collection, followed by a second centrifugation at 1600 g for 10 min to remove cellular debris. Plasma was stored in 1–2 mL aliquots at −80 °C.

2.4. Purification of Circulating Free DNA from Plasma

Plasma circulating free DNA was extracted using the QIAamp Circulating Nucleic Acid Kit (Qiagen, Hilden, Germany) according to the manufacturer's instructions. Circulating free DNA was purified from 1–4 mL of plasma and the final elution volume was 25 µL. Total circulating free DNA was quantified using a Qubit dsDNA high sensitivity assay kit and a Qubit fluorometer 3 (Life Technologies, Carlsbad, CA, USA) according to the manufacturer's instructions.

2.5. ddPCR Analysis of ctDNA from Plasma

The copy number of ctDNA per milliliter of plasma was determined using the QX200 ddPCR system (Bio-Rad, Hercules, CA, USA), as previously described [20]. Commercially available (GNAQ Q209P and GNA11 Q209L; Bio-Rad) and customized probes [22] (GNAQ R183C and CYSTLR2 L129Q) were used to analyze ctDNA by ddPCR. The DNA copy number/mL of plasma for mutant and wild-type circulating DNA species was determined with QuantaSoft software version 1.7.4 (Bio-Rad, Hercules, CA, USA) using a manual threshold setting. If analysis confirmed only 1 positive ctDNA mutant copy per 20 µL, the ddPCR amplification was repeated up to three times, and the plasma sample was considered positive if ctDNA was positive in at least two repeat experiments. ddPCR results are reported as ctDNA copies/mL.

2.6. Custom Melanoma Gene Panel for Targeted NGS of Circulating Free DNA

An Ion Ampliseq HD made-to-order melanoma gene panel was obtained from Life Technologies (Carlsbad, CA, USA). The panel, which consists of 123 amplicons and covers melanoma-associated mutations in 30 gene targets, has been described previously [24]. This melanoma gene panel does not cover the BAP1 gene. DNA target amplification, using 20 ng circulating free DNA as template, library construction and sequencing were performed as previously described [24]. Ion Torrent NGS results in our study are reported in mutant allele frequency (MAF).

2.7. Statistical Analysis

The Spearman rank correlation coefficient was used to test the correlation between the ctDNA copies, and the baseline LDH level, baseline SPOD, or longest diameter of liver metastatic lesion. Kruskal–Wallis test with Dunn's multiple comparison test was used to compare ctDNA copies in the clinical benefit group and no clinical benefit group. EDT ctDNA copies to predict clinical benefit was measured using Receiver Operating Characteristics (ROC) analysis. Statistical analyses were carried out using GraphPad Prism 9. Positive predictive value for EDT > 16.35 copies/mL was calculated using the following formula: Number of patients showing no clinical benefit with EDT ctDNA > 16.35 copies/mL divided by number of patients with EDT ctDNA > 16.35 copies/mL. Negative predictive value was determined as follows: Number of responding patients with EDT ctDNA ≤ 16.35 copies/mL divided by number of patients with EDT ctDNA ≤ 16.35 copies/mL.

3. Results

3.1. Patient Characteristics

Seventeen patients with metastatic UM were included in this study; 11 patients received PKCi alone and six patients received PKCi in combination with the human homolog of double minute 2 (HDM2) inhibitor (HDM201). Median follow-up duration was 20.1 weeks (range 6.3–66.0 weeks). Baseline demographic data are detailed in Table 1. The median age was 56 years and the majority of patients were male (10/17; 59%) with an ECOG status of 0 (13/17; 76%). All patients had an established UM driver mutation (GNAQ Q209P (35%), GNA11 Q209L (47%), GNAQ R183Q (12%) and CYSTLR2 L129Q (6%)), and metastatic disease involving the liver. On commencement of the treatment, 11 (65%) patients had elevated LDH levels and 13 (76%) had prior systemic treatment. The majority of patients (12/17; 70%) had a choroidal primary UM, 1 (6%) had an iris primary tumor and for the remaining 4 (24%) patients, the additional component of the primary tumor was unknown. Overall PFS was 3.8 months. Patients with RECIST 1.1 PR (2/17; 12%) or SD ≥ 6 months (4/17; 24%) were classified as the 'clinical benefit' group, while patients with SD < 6 months (7/17; 41%) or PD (4/17; 24%) were classified as having 'no clinical benefit' group (Table S1).

3.2. Baseline ctDNA Levels Are Associated with Tumor Volume and LDH Level

ctDNA was detected by ddPCR in 16/17 (94%) patients prior to commencing therapy. Median ctDNA was 157.7 copies/mL with a range of 0–7172 copies/mL. Baseline ctDNA was strongly correlated with baseline LDH (Spearman's rank $r = 0.7941$, $p < 0.001$) and baseline SPOD (Spearman's rank $r = 0.7206$, $p = 0.002$) (Figure 1A,B). As expected, the total lesion SPOD was significantly correlated to liver SPOD in these patients (Spearman's rank $r = 0.8676$, $p < 0.01$; Figure S1A); however, the baseline ctDNA did not correlate with the longest diameter of liver lesion (Spearman's rank $r = 0.4027$, $p = 0.110$) or liver SPOD ($p = 0.06$, Spearman's rank $r = 0.4632$) (Figure 1C,D). The discrepant correlation between ctDNA versus total lesion SPOD and ctDNA versus liver SPOD was influenced by the distribution of melanoma metastases in patient #6. This patient had multiple disease sites, a relatively high overall tumor burden (6022 mm^2), but very low liver disease (469 mm^2) (Figure S1A). When patient #6 was excluded, ctDNA levels were significantly correlated with the longest diameter of liver lesion (Spearman's rank $r = 0.6455$, $p < 0.01$) and liver SPOD (Spearman's rank $r = 0.7176$, $p < 0.01$).

We identified six patients in the clinical benefit group (PR, or SD ≥ 6 months; including 5/6 (83%) patients treated with PKCi monotherapy) and eleven patients in the no clinical benefit group (SD < 6 months or PD; including 6/11 (55%) patients treated with PKCi monotherapy). Baseline ctDNA, SPOD and LDH were compared in the clinical benefit versus no clinical benefit patient groups. Lower median baseline ctDNA was observed in the clinical benefit group (33.8 copies/mL, range 0–333 copies/mL) compared to patients in the no clinical benefit group (196.2 copies/mL, range 15–7172 copies/mL); however, this difference was not statistically significant (Figure S1B). Similarly, LDH, median total SPOD, longest diameter of liver lesion and LDH and liver SPOD were lower in the clinical benefit versus no clinical benefit subset; however, these differences were not significantly different (Figure S1B). The sample set was too small for multivariate analysis comparing baseline ctDNA, LDH and SPOD to best response.

Table 1. Baseline clinicopathologic characteristics of uveal melanoma patients.

Characteristics	Patients ($n = 17$)
Age, Median (range)	56 (45–73)
Sex, n (%)	
Male	10 (59%)
Female	7 (41%)
ECOG PS, n (%)	
0	13 (76%)
≥ 1	4 (24%)
Mutation, n (%)	
GNAQ^{Q209P}	6 (35%)
GNA11^{Q209L}	8 (47%)
GNAQ^{R183Q}	2 (12%)
CYSTLR2^{L129Q}	1 (6%)
Number of organs involved by metastatic disease, n (%)	
1	3 (17%)
>1	14 (83%)
Liver metastases, n (%)	17 (100%)
LDH, n (%)	
$\leq$ULN	6 (35%)
>ULN	11 (65%)
Prior Systemic Treatment [a]	
Yes	13 (76%)
No	4 (24%)
Primary Tumor Type	
Choroidal	12 (70%)
Iris	1 (6%)
Unknown	4 (24%)
Treatment	
PKCi alone	11 (65%)
PKCi + HDM2i	6 (35%)
Best Response [b], n (%)	
PR	2 (12%)
SD $\geq$ 6 months	4 (24%)
SD < 6 months	7 (41%)
PD	4 (23%)
PFS (months), median (range)	3.8 (1.7–13.1)
Number of liver lesions, median (range)	9 (1–49)
Liver SPOD (mm^2), median (range)	3595 (200–15,525)
SPOD (mm^2), median (range)	5986 (200–16,782)
Largest diameter of liver lesion (mm), median (range)	35 (11–110)

[a] Prior systemic treatment includes chemotherapy or immunotherapy. [b] Patients were stratified into response groups based on RECIST 1.1. Clinical benefit was defined by patients who had partial response (PR) or stable disease (SD) for $\geq$ 6 months. Patients with SD < 6 months or PD were classified as receiving no clinical benefit. Abbreviations: LDH, lactate dehydrogenase; ULN, upper limit of normal; PR, partial response; SD, stable disease; PD, progressive disease; SPOD, sum of the product of bi-dimensional diameters; PFS, progression-free survival; PKCi, protein kinase C inhibitor (LXS196); HDM2i, HDM2 inhibitor (HDM201); ECOG PS, Eastern Cooperative Oncology Group performance status.

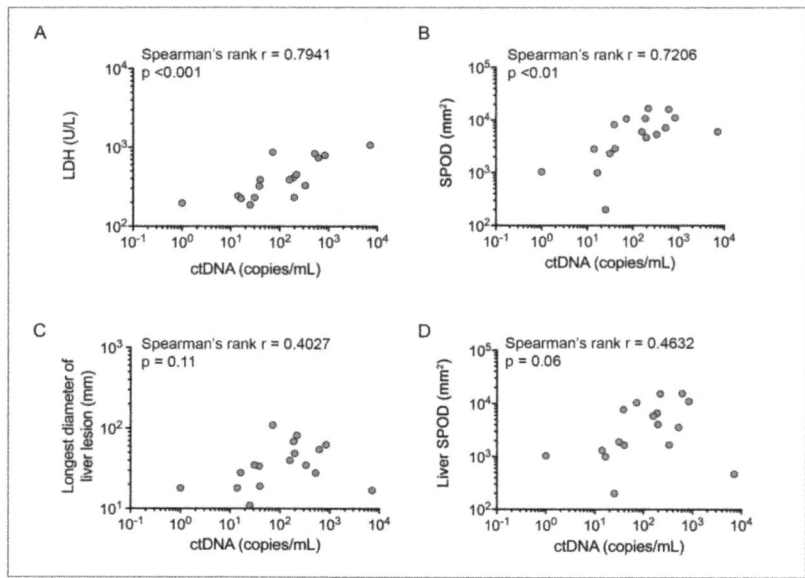

Figure 1. Relationship between uveal melanoma ctDNA (copies/mL), tumor burden and LDH. Spearman's rank correlation between ctDNA copies/mL and (**A**) LDH (U/L), $p < 0.001$, (**B**) SPOD (mm^2), $p < 0.01$, (**C**) Longest liver lesion (mm), $p = 0.11$, (**D**) Liver SPOD (mm2), $p = 0.06$. Graph shows ctDNA+1 data.

3.3. Prognostic Value of Early during Treatment (EDT) ctDNA

Paired baseline and EDT ctDNA samples were available for 16 patients (patient #15 did not have an EDT sample). Four patients (4/16; 25%) displayed undetectable ctDNA at EDT and three of these patients benefited from PKCi-based therapy (1 with PR and 2 with SD $\geq$ 6 months; SD for 9.6 and 13.1 months, respectively) and the fourth patient initially had SD but progressed at 3.7 months. Of these four patients, one (patient #5) had low disease volume, undetectable ctDNA at baseline and EDT, SD $\geq$ 6 months and a PFS of 13.1 months. Of the remaining three patients, ctDNA zero converted at EDT from baseline ctDNA levels ranging from 13–30 ctDNA copies/mL (Figure 2A). All four had a GNA11 Q209L mutation, a below median tumor burden and an LDH level below the upper limit of normal.

Another nine patients (9/12; 75%) had positive ctDNA at baseline and showed a substantial reduction but still detectable ctDNA at EDT (ctDNA reduction range 46–99%). Three of these nine patients (33%) benefited from PKCi; patient #16 achieved PR with delayed zero-conversion of ctDNA at day 57, patient #4 achieved SD $\geq$ 6 months and showed undetectable ctDNA 41 days post treatment start and patient #1 achieved SD $\geq$ 6 without zero-conversion of ctDNA. Six patients with reduced, but not undetectable ctDNA levels at EDT had no clinical benefit, with three patients showing SD < 6 months (patients #11, #13 and #14) and three patients with PD (patients #6, #8 and #10) as best response (Figure 2A, Table S1). The remaining three patients (patients #7, #12, #17) showed an increase in ctDNA from baseline to EDT and all three of these patients did not benefit from therapy (2 with SD < 6 months and 1 with, PD).

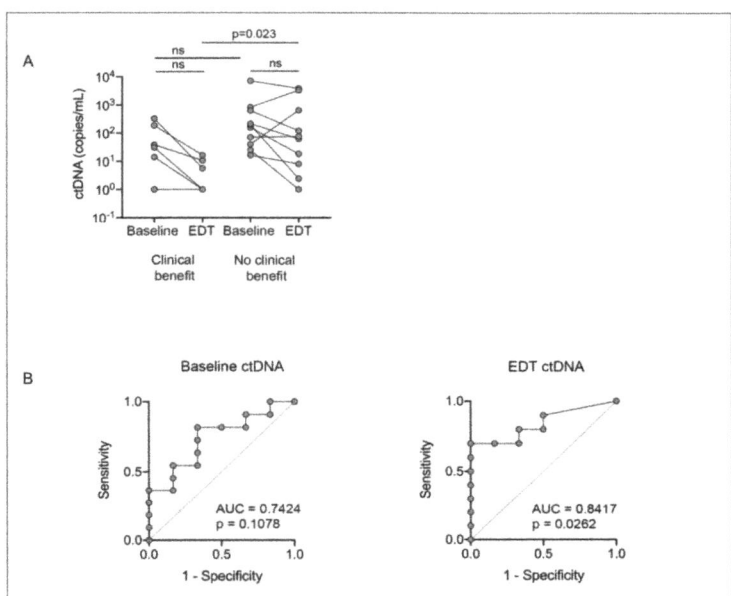

Figure 2. Predictive performance of ctDNA. (**A**) ctDNA changes from baseline to EDT in clinical benefit group ($n = 6$) and no clinical benefit group ($n = 10$) patients. Patient matched PRE-EDT ctDNA levels were compared using Wilcoxon matched-pairs signed rank test, and unpaired PRE or EDT ctDNA levels between clinical benefit and no clinical benefit patients were compared using the Mann–Whitney test. (**B**) ROC curve analysis determined a negative predictive cut-off value (i.e., value providing maximum sensitivity and specificity) for ctDNA > 16.35 copies/mL at EDT for no clinical benefit. ns, not significant; AUC, area under the curve.

EDT ctDNA levels were significantly lower in the clinical benefit patients compared to the no clinical benefit subgroup ($p = 0.023$; Figure 2A). There was no statistical difference in the baseline ctDNA level of the clinical benefit and no clinical benefit group. The changes in baseline ctDNA to EDT ctDNA in both the clinical benefit group and no clinical benefit group were also not statistically significant. The predictive accuracy of ctDNA was also examined using receiver operator characteristic classification (ROC) curves. EDT ctDNA, but not PRE ctDNA or change from PRE to EDT, accurately predicted clinical benefit to PKCi based therapy (AUC 0.84, [95% confidence interval, 0.65–1.0, $p = 0.026$]) (Figure 2B). Based on ROC curve analysis, the sensitivity and specificity for ctDNA > 16.35 copies/mL at EDT in the no clinical benefit group were 70% and 100%, respectively. The positive and negative predictive values for ctDNA > 16.35 copies/mL were 100% and 67%, respectively.

3.4. Longitudinal ctDNA Monitoring and Disease Progression

Monitoring ctDNA levels over time was also informative for establishing clinical benefit and detecting disease progression. Collectively, six patients had undetectable ctDNA during treatment (patient #2, #3, #4, #5, #9 and #16) and five of these patients (5/6; 83%) benefited from PKCi-based therapy (PR or SD $\geq$ 6 months) (Figure 3A). Patient #9 was the only patient with no clinical benefit who had an undetectable ctDNA at EDT and multiple later time points (Figure 3B, Table S1). Conversely, of the seven patients with consistently detectable ctDNA during therapy (patient #1, #7, #11, #13, #14, #15, #17) (Figure 3, Figure S2) only one patient (patient #1, PFS of 7.4 months; Figure 3C) benefited from therapy and this patient showed a 74% reduction in ctDNA level from baseline to EDT.

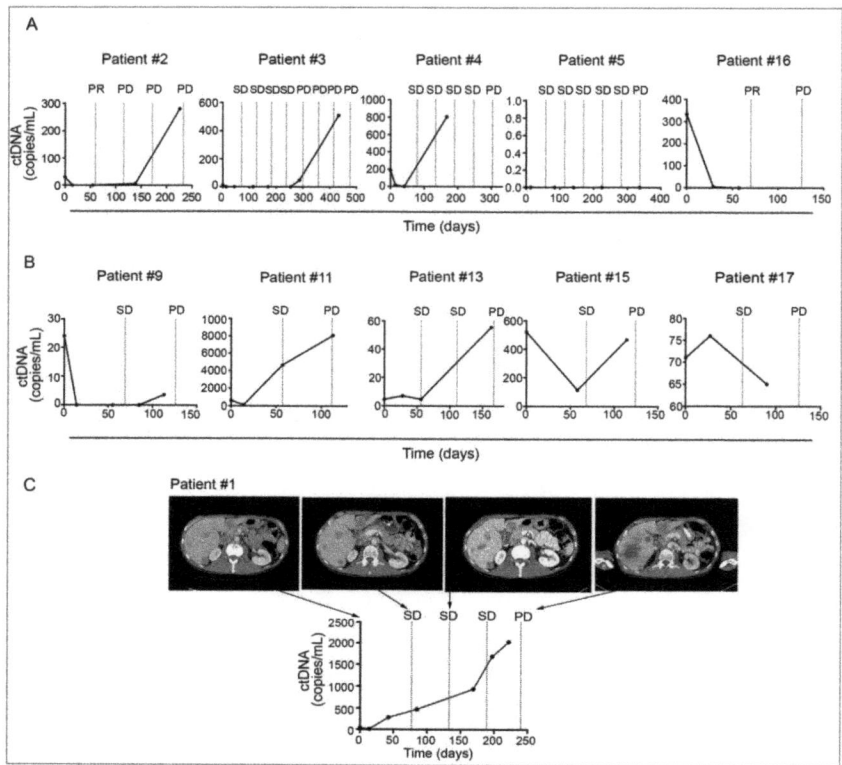

Figure 3. Monitoring of ctDNA in patients treated with PKCi in metastatic UM. ctDNA levels were collected longitudinally during treatment and correlated to CT imaging during baseline, whilst on treatment and on progression. Longitudinal ctDNA monitoring is shown for (**A**) clinical benefit patients, (**B**) no clinical benefit patients and (**C**) CT images and corresponding ctDNA data are shown for clinical benefit patient #1. Only patients #2, #3, #4, #5, #9 and #16 had undetectable ctDNA for the driver oncogene in at least one on-therapy plasma sample. SD, stable disease; PR, partial response; PD, progressive disease.

Eight out of seventeen patients also had ctDNA samples assessed within 30 days (before or after) of disease progression. In total, 7/8 patients (patient #1, #3 #7, #9, #11, #13 and #15; Figure 3, Figure S2) showed increasing ctDNA prior to radiological confirmation of disease progression with an increase in size of target lesions and new metastases as per RECIST 1.1 (Figure 4). Only patient #17 showed ctDNA levels near progression that were lower than EDT ctDNA despite CT imaging confirming disease progression (Figure 3).

3.5. Detection of Driver and Additional Mutations through Ion Torrent NGS

We next examined the ctDNA of these patients using a targeted NGS panel that included gene alterations shown to be prognostic in UM (Table 2). Paired baseline and on-treatment samples (time from baseline to on-treatment sample 0.9–11.3 months) from 17 patients were sequenced and the driver GNAQ, GNA11 and CYSLTR2 mutations identified in the tumor were confirmed by NGS in the baseline and/or on-treatment ctDNA samples in 16/17 patients. The allele frequency of tumor-associated mutations determined by ddPCR and NGS was highly correlated (Spearman's rank r = 0.968, $p < 0.001$; Figure S3A). The GNA11 Q209L driver mutation present in the UM of patient #9 was not detected in the baseline or on-treatment liquid biopsy samples using NGS, and this patient had the lowest tumor burden (SPOD = 200 mm^2), although not the lowest baseline ctDNA levels by ddPCR (24 copies/mL plasma).

Table 2. Liquid biopsy mutation analysis using Ion Torrent next generation sequencing.

Patient ID	Baseline Mutation (MAF %, LOD %)	On-Treatment Mutation (MAF %, LOD%)	Time from Baseline to on-Treatment Sample (Months)
1	GNA11^{Q209L} → (0.7, 0.6)	GNA11^{Q209L} → (23.4, 0.3) TP53^{R248Q} → (23.3, 0.3) TP53^{R342}* → (15.7, 0.3)	8.2
2	GNA11^{Q209L} → (0.5, 0.3)	GNA11^{Q209L} → (9.9, 0.6)	8.5
3	GNA11^{Q209L} → (3.0, 0.3) SF3B1^{R625H} → (1.3, 0.2)	GNA11^{Q209L} → (2.0, 0.3) SF3B1^{R625H} → (1.8, 0.3)	10.0
4	GNAQ^{R183H} → (8.5, 0.2)	GNAQ^{R183H} → (22.4, 0.2)	6.0
5	ND	GNA11^{Q209L} → (4.8, 0.6)	11.3
6	GNAQ^{Q209P} → (32.3, 0.2) SF3B1^{R625C} → (21.2, 0.1)	GNAQ^{Q209P} → (25.4, 0.2) SF3B1^{R625C} → (12.8, 0.2)	0.9
7	GNA11^{Q209L} → (22.7, 0.1) SF3B1^{R625L} → (24.0, 0.1)	GNA11^{Q209L} → (20.9, 0.2) SF3B1^{R625L} → (20.5, 0.2)	3.9
8	GNAQ^{Q209P} → (4.4, 0.1)	GNAQ^{Q209P} → (0.3, 0.2)	1.0
9	ND	ND	3.8
10	CYSLTR2^{L129Q} → (8.1, 0.3)	CYSLTR2^{L129Q} → (0.5, 0.1)	0.9
11	GNA11^{Q209L} → (13.3, 0.1)	GNA11^{Q209L} → (29.5, 0.2) TP53^{G244D} → (0.3, 0.2)	4.0
12	GNA11^{Q209L} → (3.4, 0.3) TP53^{Y220C} → (0.7, 0.3) TP53^{R248P} → (0.3, 0.3)	GNA11^{Q209L} → (14.1, 0.4)	0.9
13	GNAQ^{Q209P} → (0.8, 0.2)	GNAQ^{Q209P} → (1.0, 0.2) TP53^{R248G} → (0.3, 0.2)	5.4
14	GNAQ^{Q209P} → (6.3, 0.2) SF3B1^{R625H} → (8.6, 0.2)	GNAQ^{Q209P} → (3.1, 0.4) SF3B1^{R625H} → (6.1, 0.3)	3.8
15	GNAQ^{Q209P} → (20.1, 0.2)	GNAQ^{Q209P} → (9.7, 0.2)	3.8
16	GNAQ^{Q209P} → (5.9, 0.2)	ND	2.4
17	GNAQ^{R183Q} → (4.2, 0.2)	GNAQ^{R183Q} → (11.6, 0.3) TP53^{S215Q} → (0.4, 0.3)	3.4

ND, not detected; MAF, mutant allele frequency; LOD, limit of detection. Timing of the on-treatment plasma sample collection is also shown. *, indicates premature termination codon. All mutations shown had MAF > LOD.

In addition to the UM driver mutations, the hotspot SF3B1 R625 mutation was detected in baseline and on-treatment ctDNA samples derived from four patients with GNAQ or GNA11 driver mutations (patients #3, #6, #7 and #14; Table 2). The allele frequencies of the SF3B1 and driver GNAQ/GNA11 mutations were highly correlated in these eight ctDNA samples (Figure S3B). Of the four patients with SF3B1 mutations, one achieved SD ≥ 6 months, two had SD < 6 months and one patient had PD. The median PFS of these SF3B1 mutation-positive patients was 4.6 months, slightly longer than the median PFS of 3.8 months for the whole cohort.

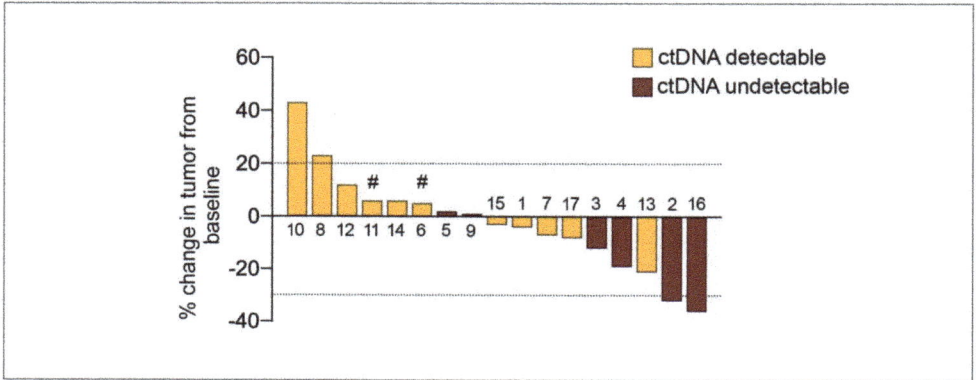

Figure 4. Treatment response in target lesions and ctDNA detectability in UM patients treated with PKCi. Percentage change in target lesions as per RECIST1.1 from 17 patients. Bars are aligned according to decreasing percentage in the sum of target lesions. Positive bars show growth in target lesions and negative bars indicate shrinkage. The dotted line corresponds to a 20% increase or 30% reduction in size of the target lesions. Patients were classified as ctDNA undetectable if at least one on-therapy plasma sample was undetectable for the driver oncogene. Patient IDs are shown above or below bars. # progression of disease with new non-target lesions.

We also identified TP53 mutations in the baseline and/or on-treatment plasma samples of 10 patients and many of these TP53 mutations are established cancer-associated mutations (e.g., S215G, Y220C, G244D, G245S, R282P, R248P/Q/G) [25] (Table S2). Of these 10 patients, five received combination PKCi+HDM2i and there was an enrichment of TP53 mutations in the on-treatment plasma samples from patients treated with PKCi+HDM2i compared to patients treated with PKCi monotherapy (Fisher exact test, $p = 0.035$). Interestingly, in four of five PKCi+HDM2i patients the TP53 variants were not identified pre-treatment, suggesting the possibility of selection during treatment. It is important to note that many TP53 mutations were detected at low allele frequencies that were below the 0.2% limit of detection of our NGS assay [24] and thus would require further validation.

4. Discussion

Currently, there is no effective systemic therapy for metastatic UM and recent clinical trials with targeted therapies and immune checkpoint inhibitors appear not to improve the OS of patients with metastatic UM [26,27]. Numerous phase 1 clinical trials are currently underway including with PKCi-based therapy. In this study, we explored the utility of ctDNA as an early marker of Phase I drug efficacy and resistance in metastatic UM [28].

We noted a strong correlation between baseline UM ctDNA levels and prognostic markers including tumor burden and LDH levels, and this is consistent with previous reports showing that elevated ctDNA reflects higher disease burden and is associated with poor prognosis in UM and CM [20,21]. It is well established that ctDNA is also associated with response, PFS and OS in metastatic CM patients treated with targeted therapies and anti-PD1-based immunotherapies [19,20]. In this study, we explored the utility of ctDNA in monitoring and predicting UM response to PKCi-based therapy. Only one other study has examined longitudinal UM ctDNA levels and treatment response. The latter was a proof-of-concept study including only three UM patients, and although it confirmed that EDT ctDNA was predictive of an anti-PD-1 inhibitor response in a cohort that included various cancer types [29], the UM patients failed to respond to PD-1 inhibitor blockade and ctDNA was only detected in two of the UM patients [29].

In this study, we show that ctDNA levels early during therapy can predict UM responses to PKCi-based targeted therapy. Interestingly, pre-treatment ctDNA did not predict response or prolonged PFS in our UM patient cohort, even though baseline ctDNA was positively correlated with disease burden and LDH. Importantly, although all responding patients with detectable baseline ctDNA showed reduced levels of ctDNA at EDT, the decrease from PRE to EDT was not significant, and ctDNA at EDT was also reduced in seven of 10 patients showing no clinical benefit to PKCi. Thus, it is the absolute level of EDT ctDNA that is indicative of treatment response in this cohort and we reported similar findings in advanced melanoma patients treated with anti-PD1-based therapy [20]. Considering the level of ctDNA is reflective of both tumor size and metabolic tumor burden [30], it is not surprising that low ctDNA early during therapy would predict treatment response. Our study also confirms that increasing UM ctDNA preceded radiological progression and did so with lead-time ranging from 4 to 10 weeks. A previous study also reported that increasing ctDNA also precedes radiologic detection of UM liver metastases [22]. Thus, the inclusion of ctDNA analysis in Phase I UM trials can provide meaningful data on patients failing to respond to novel therapies, and this may accelerate CT-based confirmation of progression or contribute to adaptive trial design, allowing for earlier access to alternate drugs.

We also utilized a custom targeted NGS panel designed for the detection of 90% of known CM gene mutations and 95% of known UM gene mutations [24]. We confirmed that the allele frequency of driver mutations identified using targeted Ion Torrent NGS was comparable to the mutant allele frequencies determined by ddPCR, although the sensitivity of NGS did not match ddPCR, and the driver GNAQ mutation was not detected by NGS in one patient, presumably due to the low volume of disease. Nevertheless, NGS was able to identify additional mutated genes (i.e., SF3B1, TP53), which have been implicated in UM prognosis and treatment response. For instance, TP53 mutations were detected in the circulation of 5/6 UM patients treated with PKCi+HDM2i. These TP53 mutations may have been selected or expanded in response to therapy as they were not detected at baseline in four patients. As these mutations were detected at low frequency they may represent clonal expansion of tumor cells or hematopoietic stem cells during the process of clonal hematopoiesis. Considering that TP53 loss confers HDM2i resistance [31] and that the TP53 mutations detected in this study are established loss-of-function alterations, ctDNA may prove valuable in the early detection of treatment resistance mechanisms. It is also worth noting that a recent study confirmed that TP53 is significantly disrupted in UM with 11/103 UM showing genomic loss or mutations affecting the TP53 gene [14].

This study was limited by the small sample size and the fact that patients were treated with two distinct treatments based on PKC inhibition. A larger patient sample in a prospective study is required to evaluate the predictive value of baseline ctDNA level and more importantly test the value of including ctDNA as a routine monitoring tool in UM clinical trials.

5. Conclusions

In summary, baseline ctDNA in metastatic UM strongly correlates with baseline LDH level and disease volume. Treatment-induced changes in ctDNA and low levels of ctDNA EDT predicted response to PKCi-based targeted therapy and the inclusion of targeted NGS yielded valuable and accurate data about driver mutation frequency and the selection of potential resistance effectors.

Despite proof of concept that ctDNA is a useful biomarker for monitoring response to therapy, in the form of evolution of resistance, and should be included in metastatic UM clinical trials, the most important challenge remains the identification of effective, durable therapies for UM.

Supplementary Materials: The following are available online at https://www.mdpi.com/article/10.3390/cancers13071740/s1, Figure S1: Relationship between PKC inhibitor-based response and ctDNA levels, LDH and disease burden, Figure S2: Longitudinal monitoring of ctDNA in patients treated with PKCi in metastatic UM, Figure S3: Significant correlation in mutant allele frequency determined by ddPCR and targeted Ion Torrent NGS, Table S1: Patient and response to therapy details, Table S2: TP53 mutations detected by Ion Torrent next generation sequencing.

Author Contributions: Conceptualization, J.J.P., R.J.D. and H.R.; methodology, J.J.P., R.J.D. and H.R.; formal analysis, J.J.P., R.J.D. and H.R.; investigation, J.J.P., N.B. and H.R.; resources, N.B., M.S.C., G.V.L., R.A.S. and E.S.G.; writing—original draft preparation, J.J.P., R.J.D. and H.R.; writing—review and editing, J.J.P., R.J.D., N.B., G.V.L., R.A.S., M.S.C. and H.R.; visualization, J.J.P., R.J.D. and H.R.; supervision, H.R.; project administration, H.R.; funding acquisition, H.R. All authors have read and agreed to the published version of the manuscript.

Funding: J.J.P. is supported by Melanoma Institute Australia Postgraduate Scholarship. This project is also supported by Sydney Vital Translational Cancer Research Centre. R.J.D. was supported in part by a donation to Melanoma Institute Australia from the Clearbridge Foundation. H.R. is supported by National Health and Medical Research Council, Research Fellowship. E.S.G. is supported by a Cancer Council of Western Australia Fellowship. G.V.L. and R.A.S. are supported by a National Health and Medical Research Council, Program Grant and Practitioner Fellowships. G.V.L. is supported by the University of Sydney Medical Foundation.

Institutional Review Board Statement: The study was conducted according to the guidelines of the Declaration of Helsinki and approved from Human Research ethics committee protocols from Royal Prince Alfred Hospital, Sydney, Australia (Protocol X15-0454 and HREC/11/RPAH/444, date of approval 4 December 2020).

Informed Consent Statement: Written informed consent was obtained from all subjects involved in this study.

Data Availability Statement: The next generation sequencing and ddPCR data presented in this study are available on request from the corresponding author. The data are not publicly available due to patient confidentiality.

Acknowledgments: We thank the patients and their families for their assistance with the study. We thank the Melanoma Institute Australia Biospecimen Bank and Westmead Hospital. The authors also gratefully acknowledge assistance from colleagues at their respective institutions.

Conflicts of Interest: The authors have no other relevant affiliations or financial involvement with any organization or entity with a financial interest in or financial conflict with the subject matter or materials discussed in the manuscript part from those disclosed. No writing assistance was utilized in the production of this manuscript. M.S.C. has served on advisory boards for Bristol-Myers Squibb, Merck Sharp and Dohme, Amgen, Novartis, Pierre Fabre, Roche, Sanofi, Merck, Ideaya, Regeneron, Nektar, Eisai and honoraria from Bristol-Myers Squibb, MSD, Novartis. R.A.S. has received fees for professional services from OBiotics, Novartis, Merck Sharp and Dohme, NeraCare, AMGEN Inc., Brsitol-Myers Squibb, Myriad Genetics, GlaxoSmithKline. G.V.L. is consultant advisor for Aduro Biotech Inc, Amgen Inc, Array Biopharma Inc, Boehringer Ingelheim International GmbH, Bristol-Myers Squibb, Highlight Therapeutics S.L., Merck Sharpe and Dohme, Novartis Pharma AG, Pierre Fabre, OBiotics Group Limited, Regeneron Pharmaceuticals Inc, SkylineDX B.V.

References

1. McLaughlin, C.C.; Wu, X.C.; Jemal, A.; Martin, H.J.; Roche, L.M.; Chen, V.W. Incidence of noncutaneous melanomas in the U.S. *Cancer* **2005**, *103*, 1000–1007. [CrossRef] [PubMed]
2. Shields, C.L.; Furuta, M.; Thangappan, A.; Nagori, S.; Mashayekhi, A.; Lally, D.R.; Kelly, C.C.; Rudich, D.S.; Nagori, A.V.; Wakade, O.A.; et al. Metastasis of uveal melanoma millimeter-by-millimeter in 8033 consecutive eyes. *Arch. Ophthalmol.* **2009**, *127*, 989–998. [CrossRef] [PubMed]
3. Vajdic, C.M.; Kricker, A.; Giblin, M.; McKenzie, J.; Aitken, J.; Giles, G.G.; Armstrong, B.K. Incidence of ocular melanoma in Australia from 1990 to 1998. *Int. J. Cancer* **2003**, *105*, 117–122. [CrossRef] [PubMed]
4. Chang, A.E.; Karnell, L.H.; Menck, H.R. The national cancer data base report on cutaneous and noncutaneous melanoma. *Cancer* **1998**, *83*, 1664–1678. [CrossRef]
5. Singh, A.D.; Turell, M.E.; Topham, A.K. Uveal melanoma: Trends in incidence, treatment, and survival. *Ophthalmology* **2011**, *118*, 1881–1885. [CrossRef]
6. Kujala, E.; Makitie, T.; Kivela, T. Very long-term prognosis of patients with malignant uveal melanoma. *Invest. Ophthalmol. Vis. Sci.* **2003**, *44*, 4651–4659. [CrossRef]
7. COMS. Assessment of metastatic disease status at death in 435 patients with large choroidal melanoma in the Collaborative Ocular Melanoma Study (COMS): COMS report no. 15. *Arch. Ophthalmol.* **2001**, *119*, 670–676. [CrossRef]

8. Khoja, L.; Atenafu, E.G.; Suciu, S.; Leyvraz, S.; Sato, T.; Marshall, E.; Keilholz, U.; Zimmer, L.; Patel, S.; Piperno-Neumann, S.; et al. Meta-analysis in metastatic uveal melanoma to determine progression free and overall survival benchmarks: An international rare cancers initiative (IRCI) ocular melanoma study. *Ann. Oncol.* **2019**, *30*, 1370–1380. [CrossRef]
9. Robertson, A.G.; Shih, J.; Yau, C.; Gibb, E.A.; Oba, J.; Mungall, K.L.; Hess, J.M.; Uzunangelov, V.; Walter, V.; Danilova, L.; et al. Integrative analysis identifies four molecular and clinical subsets in uveal melanoma. *Cancer Cell* **2017**, *32*, 204–220. [CrossRef]
10. Johansson, P.; Aoude, L.G.; Wadt, K.; Glasson, W.J.; Warrier, S.K.; Hewitt, A.W.; Hess, J.M.; Uzunangelov, V.; Walter, V.; Danilova, L.; et al. Deep sequencing of uveal melanoma identifies a recurrent mutation in PLCB4. *Oncotarget* **2016**, *7*, 4624–4631. [CrossRef]
11. Moore, A.R.; Ceraudo, E.; Sher, J.J.; Guan, Y.; Shoushtari, A.N.; Chang, M.T.; Zhang, J.Q.; Walczak, E.G.; Kazmi, M.A.; Taylor, M.T.C.B.S.; et al. Recurrent activating mutations of G-protein-coupled receptor CYSLTR2 in uveal melanoma. *Nat. Genet.* **2016**, *48*, 675–680. [CrossRef]
12. Van Raamsdonk, C.D.; Bezrookove, V.; Green, G.; Bauer, J.; Gaugler, L.; O'Brien, J.M.; Simpson, E.M.; Barsh, G.S.; Bastian, B.C. Frequent somatic mutations of GNAQ in uveal melanoma and blue naevi. *Nature* **2009**, *457*, 599–602. [CrossRef]
13. Van Raamsdonk, C.D.; Griewank, K.G.; Crosby, M.B.; Garrido, M.C.; Vemula, S.; Wiesner, T.; Obenauf, A.C.; Wackernagel, W.; Green, G.; Bouvier, N.; et al. Mutations in GNA11 in uveal melanoma. *N. Engl. J. Med.* **2010**, *363*, 2191–2199. [CrossRef]
14. Johansson, P.A.; Brooks, K.; Newell, F.; Palmer, J.M.; Wilmott, J.S.; Pritchard, A.L.; Broit, N.; Wood, S.; Carlino, M.S.; Leonard, C.; et al. Whole genome landscapes of uveal melanoma show an ultraviolet radiation signature in iris tumours. *Nat. Commun.* **2020**, *11*, 2408. [CrossRef]
15. Vichitvejpaisal, P.; Dalvin, L.A.; Mazloumi, M.; Ewens, K.G.; Ganguly, A.; Shields, C.L. Genetic analysis of uveal melanoma in 658 patients using the cancer genome atlas classification of uveal melanoma as A, B, C, and D. *Ophthalmology* **2019**, *126*, 1445–1453. [CrossRef]
16. Field, M.G.; Durante, M.A.; Anbunathan, H.; Cai, L.Z.; Decatur, C.L.; Bowcock, A.M.; Kurtenbach, S.; Harbour, J.W. Punctuated evolution of canonical genomic aberrations in uveal melanoma. *Nat. Commun.* **2018**, *9*, 116. [CrossRef]
17. Onken, M.D.; Worley, L.A.; Ehlers, J.P.; Harbour, J.W. Gene expression profiling in uveal melanoma reveals two molecular classes and predicts metastatic death. *Cancer Res.* **2004**, *64*, 7205–7209. [CrossRef]
18. Ascierto, P.A.; Minor, D.; Ribas, A.; Lebbe, C.; O'Hagan, A.; Arya, N.; Guckert, M.; Schadendorf, D.; Kefford, R.F.; Grob, J.-J.; et al. Phase II trial (BREAK-2) of the BRAF inhibitor dabrafenib (GSK2118436) in patients with metastatic melanoma. *J. Clin. Oncol.* **2013**, *31*, 3205–3211. [CrossRef]
19. Gray, E.S.; Rizos, H.; Reid, A.L.; Boyd, S.C.; Pereira, M.R.; Lo, J.; Tembe, V.; Freeman, J.; Lee, J.H.; Scolyer, R.A.; et al. Circulating tumor DNA to monitor treatment response and detect acquired resistance in patients with metastatic melanoma. *Oncotarget* **2015**, *6*, 42008–42018. [CrossRef]
20. Lee, J.; Long, G.; Boyd, S.; Lo, S.; Menzies, A.; Tembe, V.; Guminski, A.; Jakrot, V.; Scolyer, R.A.; Mann, G.J.; et al. Circulating tumour DNA predicts response to anti-PD1 antibodies in metastatic melanoma. *Ann. Oncol.* **2017**, *28*, 1130–1136. [CrossRef]
21. Bidard, F.C.; Madic, J.; Mariani, P.; Piperno-Neumann, S.; Rampanou, A.; Servois, V.; Cassoux, N.; Desjardins, L.; Milder, M.; Vaucher, I.; et al. Detection rate and prognostic value of circulating tumor cells and circulating tumor DNA in metastatic uveal melanoma. *Int. J. Cancer* **2014**, *134*, 1207–1213. [CrossRef]
22. Beasley, A.; Isaacs, T.; Khattak, M.; Freeman, J.; Allcock, R.; Chen, F.; Pereira, M.R.; Yau, K.; Bentel, J.; Vermeulen, T.; et al. Clinical application of circulating tumor cells and circulating tumor DNA in uveal melanoma. *JCO Precis. Oncol.* **2018**, *2*, 1–12. [CrossRef]
23. Eisenhauer, E.A.; Therasse, P.; Bogaerts, J.; Schwartz, L.H.; Sargent, D.; Ford, R.; Dancey, J.; Arbuck, S.; Gwyther, S.; Mooney, M.; et al. New response evaluation criteria in solid tumours: Revised RECIST guideline (version 1.1). *Eur. J. Cancer* **2009**, *45*, 228–247. [CrossRef]
24. Diefenbach, R.J.; Lee, J.H.; Menzies, A.M.; Carlino, M.S.; Long, G.V.; Saw, R.P.M.; Howle, J.R.; Spillane, A.J.; Scolyer, R.A.; Kefford, R.F.; et al. Design and testing of a custom melanoma next generation sequencing panel for analysis of circulating tumor DNA. *Cancers* **2020**, *12*, 2228. [CrossRef]
25. Petitjean, A.; Achatz, M.I.; Borresen-Dale, A.L.; Hainaut, P.; Olivier, M. TP53 mutations in human cancers: Functional selection and impact on cancer prognosis and outcomes. *Oncogene* **2007**, *26*, 2157–2165. [CrossRef] [PubMed]
26. Schank, T.E.; Hassel, J.C. Immunotherapies for the treatment of uveal melanoma-history and future. *Cancers* **2019**, *11*, 1048. [CrossRef] [PubMed]
27. Croce, M.; Ferrini, S.; Pfeffer, U.; Gangemi, R. Targeted therapy of uveal melanoma: Recent failures and new perspectives. *Cancers* **2019**, *11*, 846. [CrossRef] [PubMed]
28. Martel, A.; Baillif, S.; Nahon-Esteve, S.; Gastaud, L.; Bertolotto, C.; Romeo, B.; Mograbi, B.; Lassalle, S.; Hofman, P. Liquid biopsy for solid ophthalmic malignancies: An updated review and perspectives. *Cancers* **2020**, *12*, 3284. [CrossRef] [PubMed]
29. Cabel, L.; Riva, F.; Servois, V.; Livartowski, A.; Daniel, C.; Rampanou, A.; Lantz, O.; Romano, E.; Milder, M.; Buecher, B.; et al. Circulating tumor DNA changes for early monitoring of anti-PD1 immunotherapy: A proof-of-concept study. *Ann. Oncol.* **2017**, *28*, 1996–2001. [CrossRef]
30. McEvoy, A.C.; Warburton, L.; Al-Ogaili, Z.; Celliers, L.; Calapre, L.; Pereira, M.R.; Khattak, M.A.; Meniawy, T.M.; Millward, M.; Ziman, M.; et al. Correlation between circulating tumour DNA and metabolic tumour burden in metastatic melanoma patients. *BMC Cancer* **2018**, *18*, 726. [CrossRef]

31. Chapeau, E.A.; Gembarska, A.; Durand, E.Y.; Mandon, E.; Estadieu, C.; Romanet, V.; Wiesmann, M.; Tiedt, R.; Lehar, J.; de Weck, A.; et al. Resistance mechanisms to TP53-MDM2 inhibition identified by in vivo piggyBac transposon mutagenesis screen in an Arf(-/-) mouse model. *Proc. Natl. Acad. Sci. USA* **2017**, *114*, 3151–3156. [CrossRef]

Article

Somatic Mutation Profiling in the Liquid Biopsy and Clinical Analysis of Hereditary and Familial Pancreatic Cancer Cases Reveals *KRAS* Negativity and a Longer Overall Survival

Julie Earl [1,2,*], Emma Barreto [1,†], María E. Castillo [1,†], Raquel Fuentes [1], Mercedes Rodríguez-Garrote [1], Reyes Ferreiro [1], Pablo Reguera [1], Gloria Muñoz [3], David Garcia-Seisdedos [3], Jorge Villalón López [1], Bruno Sainz, Jr. [4,5,6], Nuria Malats [2,7] and Alfredo Carrato [1,2,8]

[1] Molecular Epidemiology and Predictive Tumor Markers Group, Ramón y Cajal Health Research Institute (IRYCIS), Carretera Colmenar Km 9100, 28034 Madrid, Spain; emma.barreto@salud.madrid.org (E.B.); marien.castillo@salud.madrid.org (M.E.C.); rfuentes@salud.madrid.org (R.F.); mercedes.rodriguez@salud.madrid.org (M.R.-G.); reyes-ferreiro@hotmail.com (R.F.); pablo.reguera@salud.madrid.org (P.R.); jorge.villalon@salud.madrid.org (J.V.L.); alfredo.carrato@salud.madrid.org (A.C.)
[2] Biomedical Research Network in Cancer (CIBERONC), C/Monforte de Lemos 3-5. Pabellón 11, 28029 Madrid, Spain; nmalats@cnio.es
[3] Translational Genomics Core Facility, Ramón y Cajal Health Research Institute (IRYCIS), 28034 Madrid, Spain; mariagloria.munoz@salud.madrid.org (G.M.); dgarcia@ebi.ac.uk (D.G.-S.)
[4] Department of Biochemistry, Universidad Autónoma de Madrid (UAM), Ramón y Cajal Health Research Institute (IRYCIS), 28034 Madrid, Spain; bsainz@iib.uam.es
[5] Instituto de Investigaciones Biomédicas "Alberto Sols" (IIBM), CSIC-UAM, C/Arzobispo Morcillo, 4, 28029 Madrid, Spain
[6] Cancer Stem Cell and Fibroinflammatory Group, Chronic Diseases and Cancer, Area 3-IRYCIS, 28029 Madrid, Spain
[7] Genetic and Molecular Epidemiology Group, Spanish National Cancer Research Centre (CNIO), 28029 Madrid, Spain
[8] Department of Medicine and Medical Specialties, Medicine Faculty, Alcala University, Plaza de San Diego, s/n, 28801 Alcalá de Henares, Spain
* Correspondence: julie.earl@live.co.uk; Tel.: +34-91-334-1307 (ext. 7877)
† These authors contributed equally to the manuscript.

Simple Summary: Pancreatic ductal adenocarcinoma (PDAC) has a poor prognosis. *KRAS* mutations occur in up to 95% of cases and render the tumor resistant to many types of therapy. Therefore, these patients are treated with traditional cytotoxic agents, according to guidelines. The familial or hereditary form of the disease accounts for up to 10–15% of cases. We hypothesized that hereditary and Familial Pancreatic Cancer cases (H/FPC) have a distinct tumor specific mutation profile due to the presence of pathogenic germline mutations and we used circulating free DNA (cfDNA) in plasma to assess this hypothesis. H/FPC cases were mainly *KRAS* mutation negative and harbored tumor specific mutations that are potential treatment targets in the clinic. Thus, we conclude that cases with a hereditary or familial background can be treated with newer and more effective agents that may ultimately improve their overall survival.

Abstract: Pancreatic ductal adenocarcinoma (PDAC) presents many challenges in the clinic and there are many areas for improvement in diagnostics and patient management. The five-year survival rate is around 7.2% as the majority of patients present with advanced disease at diagnosis that is treatment resistant. Approximately 10–15% of PDAC cases have a hereditary basis or Familial Pancreatic Cancer (FPC). Here we demonstrate the use of circulating free DNA (cfDNA) in plasma as a prognostic biomarker in PDAC. The levels of cfDNA correlated with disease status, disease stage, and overall survival. Furthermore, we show for the first time via BEAMing that the majority of hereditary or familial PDAC cases (around 84%) are negative for a *KRAS* somatic mutation. In addition, *KRAS* mutation negative cases harbor somatic mutations in potentially druggable genes such as *KIT*, *PDGFR*, *MET*, *BRAF*, and *PIK3CA* that could be exploited in the clinic. Finally, familial or hereditary cases have a longer overall survival compared to sporadic cases (10.2 vs. 21.7 months,

respectively). Currently, all patients are treated the same in the clinic with cytotoxic agents, although here we demonstrate that there are different subtypes of tumors at the genetic level that could pave the way to personalized treatment.

Keywords: liquid biopsy; cfDNA; hereditary and familial pancreatic cancer; somatic mutation profiling; potentially druggable genes

1. Introduction

The incidence and mortality rates of adenocarcinoma of the pancreas (PDAC) are almost equal [1]. Today, PDAC is the third leading cause of cancer death in the EU and by 2030 it is projected to be the second leading cause of cancer-related death [2], surpassing breast, prostate, and colorectal cancer. The overall survival at 5 years is around 7.2% due to the fact that the majority of patients present with advanced and treatment resistant disease at diagnosis. *KRAS* somatic mutations are present in around 90% of sporadic tumors and *TP53, CDKN2A,* and *SMAD4* mutations are also commonly found. However, none of these somatic changes are druggable at present time. In fact, PDAC tumors have an average of 63 somatic alterations that affect different signaling pathways [3]. The best described common risk factors for sporadic PDAC include tobacco, alcohol, diabetes, chronic pancreatitis, and obesity [4]. Family history is also an important risk factor and approximately 10% of PDAC cases have a hereditary or familial basis [5]. Hereditary pancreatic cancer are associated with a known cancer syndrome such as hereditary breast–ovarian cancer (HBOC), Peutz–Jeghers (PJ), Hereditary Pancreatitis (HP), Familial Atypical Multiple Mole Melanoma (FAMMM), and Lynch syndrome and harbor germline pathogenic mutations in genes such as *BRCA2, MLH1,* and *CDKN2A* [6]. Whereas, Familial Pancreatic Cancer (FPC) is defined as a family with at least one pair of affected first degree relatives with no identified genetic basis and account for 4–10% of PDAC patients [7,8]. The Spanish familial pancreatic cancer registry (PANGENFAM) was established in 2009 with the principal objective to characterize the phenotypic and genetic background of FPC [9].

Specific, sensitive and minimally invasive biomarkers are needed in order to accurately diagnose PDAC at a potentially curable stage and aid in patient management during treatment. CA19-9 is currently used in the clinic, although the sensitivity and specificity for the diagnosis of symptomatic PDAC is 79–81% and 82–90%, respectively [10]. Several potential biomarkers have been recently described such as a three-protein panel in urine [11], Galectin-1 (Gal-1) in serum [12], thrombospondin-2 (THBS2) in plasma [13], circulating tumor DNA (ctDNA) [14], and the IMMray™ PanCan-d 29 biomarker signature in serum [15]. The term the "liquid biopsy" was coined in 2010 [16], and is defined as the detection and analysis of molecules (e.g., protein, DNA, RNA), cells or extracellular vesicles (e.g., exosomes) that originate from the primary tumor in blood and other body fluids, such as saliva, cerebrospinal fluid, and feces. Since fresh tissue, in the form of biopsies, is scarce and prohibitive for many PDAC patients, liquid biopsies represent an attractive surrogate system to provide essential information regarding diagnosis, stage, etc. Likewise, there is an important and inherent degree of heterogeneity in primary PDAC tumors and associated metastatic lesions [17], which can only be studied in depth via liquid biopsies due to the shortage of primary and metastatic tumor tissue. Thus, research to maximize the potential information that a liquid biopsy can offer has exploded in the past decade. cfDNA, which consists of double-stranded DNA molecules of 70 to 200 bp [18,19], is released into the blood stream by apoptotic or necrotic cells or in extracellular vesicles such as exosomes. cfDNA is present in all individuals, although levels up 40 times higher are detected in patients with tumors or inflammatory disease [20]. In fact, cfDNA detection has been used previously as a prognostic and predictive marker in pancreatic cancer [21,22].

This study aimed to analyze the use of cfDNA as a prognostic and predictive marker in PDAC. Furthermore, the somatic mutation profile of Familial or Hereditary PDAC (H/FPC) and sporadic PDAC cases was also analyzed.

2. Materials and Methods

2.1. Identification and Classification of Patients

The study was approved by the local ethics committee and all patients signed the associated informed consent. PDAC patients were identified in the medical oncology and general surgery departments of the Ramon y Cajal University Hospital in Madrid, Spain. Cases that complied with the inclusion criteria of the Spanish Familial Pancreatic cancer registry (PANGENFAM) were classified as Hereditary or FPC (H/FPC cohort) [9] and included cases with and without a known genetic cause. Cases with no reported hereditary or familial pancreatic cancer syndrome were classified as sporadic cases (SP cohort). High-risk family members of H/FPC cases in the secondary screening program that were diagnosed with pancreatic cysts or intraductal papillary mucinous neoplasm (IPMN) were also included and one case that was initially included in the study as a possible PDAC, but was later confirmed as an IPMN (pancreatic cysts cohort). Blood samples in EDTA tubes were taken on entry into the study and plasma was extracted and stored until cfDNA extraction and somatic mutation analysis. All clinical and personal data was stored in a secure database Research Electronic Data Capture (REDCap: https://www.project-redcap.org/).

2.2. Isolation of cfDNA from Plasma

A total of two different methods of cfDNA isolation were compared to determine the most appropriate method for PCR and sequencing based analysis of cfDNA, the Maxwell® RSC Instrument and the QIAamp circulating nucleic acid kit (Qiagen, Venlo, Netherlands). DNA was extracted from 1 mL of plasma using the Maxwell® RSC Instrument (Promega, Madison, WI, USA) and eluted in a final volume of 50 µL from 136 patients diagnosed with PDAC and 29 individuals diagnosed with pancreatic cysts and IPMN. Plasma samples were also obtained from 40 healthy individuals with no known history of digestive disease or cancer, provided by the BioBank Hospital Ramón y Cajal-IRYCIS (PT13/0010/0002), integrated in the Spanish National Biobanks Network (ISCIII Biobank Register No. B.0000678, Spain). cfDNA was isolated as 10 plasma pools that consisted of 4 individuals in each. The concentration of cfDNA was determined in all samples using the QuantiFluor® dsDNA System (Promega, Madison, WI, USA) kit and analyzed using the Quantus Fluorometer (Promega, Madison, WI, USA).

2.3. Detection of Mutant KRAS in Plasma by BEAMing

The presence of a mutation in *KRAS* codons 12, 13, 59, 61, 117, and 146 was determined in cfDNA isolated using the Maxwell® RSC system and the OncoBEAM KRAS CRC kit (Sysmex Inostics, Hamburg, Germany) using the BEAMing technology (Sysmex Inostics, Hamburg, Germany) according to the manufacturer's instructions, which also includes positive and negative assay controls.

2.4. Sequencing of cfDNA Using the TruSight15 System (Illumina)

Cell free DNA was extracted from 1–3 mL of plasma using the QIAamp circulating nucleic acid kit (Qiagen, Venlo, Netherlands) according to the manufacturer's instructions. The DNA preparation obtained was purified using the Agentcourt AMPure XP Reagent (Beckman Coulter, Brea, CA, USA) in two successive steps in order to isolate cfDNA of approximately 160–170 bp. DNA was added to the Agentcourt AMPure XP Reagent at a ratio of $0.7\times$ and then at a ratio of $2\times$ the initial volume of DNA. The sample was then vortexed, centrifuged and incubated for 5 min at room temperature and then placed in a magnetic rack and incubated for 5 min. The supernatant containing the cfDNA was placed in a new Eppendorf and more Agentcourt AMPure XP Reagent was added for the second round of purification. The same steps were repeated and the supernatant was finally discarded. Then, 70% ethanol was added to the sample and then vortexed, centrifuged, and placed in the magnetic rack for 30 s. The supernatant was removed and the sample was washed with 70% ethanol in order to obtain a dry pellet. Finally, nuclease free water was added and the sample was vortexed, centrifuged, incubated at room temperature for 2 min and then incubated in the magnetic rack for 2 min. The purified cfDNA was analyzed with the Tape Station 2200 using the HS D1000 kit (Agilent Technologies, Santa Clara, CA, USA) to confirm the presence of a single band of approximately 160 bp. After quantification using the Qubit fluorimeter 2.0 (ThermoFisher Scientific, Waltham, MA, USA) with the HS DNA, the cfDNA was diluted to 2 ng/µL to prepare genome libraries using the TruSight Tumor 15 kit (Illumina, San Diego, CA, USA), which is optimized for low DNA input.

The sequencing panel included the genes *AKT1*, *BRAF*, *EGFR*, *ERBB2*, *FOXL2*, *GNA11*, *GNAQ*, *KIT*, *KRAS*, *MET*, *NRAS*, *PDGFRA*, *PIK3CA*, *RET*, and *TP53*. In total, 27 PDAC cases (20 H/FPC cases and 7 sporadic cases) were included in the panel sequencing analysis of cfDNA, and five of these samples were performed in duplicate. Then, two healthy individuals, one case of previous breast cancer and one previous pancreatic neuroendocrine tumor case, were also included in the study as control samples.

2.5. Identification of Pathogenic Somatic Variants

The BaseSpace Variant Interpreter (https://variantinterpreter.informatics.illumina.com/home, last accessed date: 1 July 2020) from Illumina was used for the identification of pathogenic somatic variants, which was specifically designed to analyze the sequencing data generated using the TruSight 15 kit. A detailed description of the analysis pipeline is shown in Data S1. Briefly, the variant call files (vcf) were uploaded and the "Small Variant Consequences" filter was applied that included stop gain, stop loss, splice site, missense, frameshift, deletions, insertions, and initiator codon (ATG) loss. Those variants with a deleterious, probably or possibly damaging consequence according to the SIFT and PolyPhen parameters were retained. The frequency of somatic variants was analyzed using the Catalogue of Somatic Mutations in Cancer public database (COSMIC: (https://cancer.sanger.ac.uk/cosmic, last accessed date: 1 July 2020). Samples that did not reach an average minimum amplicon coverage of 500 were excluded, and variants that passed the quality filters of genotype quality, variant frequency, and strand bias were retained.

2.6. Statistical Analysis

The R program 3.4.3 was used for statistical analysis. The Mann–Whitney test was used to analyze the differences in concentration of cfDNA between the different groups, according to disease status and stage. The Chi square test was used to study the differences in the frequency of somatic mutations in the different groups. A One-Way ANOVA test was used to determine the differences in age at diagnosis and the Fishers exact test was used to analyze the differences between sex and disease stage distributions in each group.

The "survival" package of R was used to perform survival analysis. Overall survival (OS) was determined from the date of diagnosis until the date of death (event) or the date of the last follow-up (censored). Patients were divided into 3 groups according to total cfDNA concentration for survival analysis: high ($\geq$3rd quartile), medium ($\leq$3rd quartile and $\geq$1st quartile), and low ($\leq$1st quartile). The Kaplan Meyer and Log Rank test were used to analyze overall survival among different subgroups of patient according to cfDNA levels and classification as sporadic or H/FPC cases. The analysis was adjusted for age and disease stage at diagnosis, sex and 1st line treatment and the corresponding hazard ratios were calculated.

3. Results

3.1. Patient Characteristics

In total, 184 individuals were recruited in the entire study, including 145 cases (102 sporadic and 43 familial or hereditary PDAC cases), 29 patients with pancreatic cysts, and

40 healthy controls. The demographic characteristics of the individuals included in the study are summarized Table 1 and more detailed information of the clinical characteristics of the different cohorts is provided in Table S1. The PDAC cohort included 71 males and 74 females with a median age of 69 years (29–90 years). The median age at diagnosis of sporadic cases (SP) was 70 years and 65.5 years for Hereditary or Familial PDAC cases (H/FPC), the difference was statistically significant according to the one-way ANOVA test (p = 0.000113). The Male:Female ratio for sporadic cases was 0.89 and 0.65 and this difference was not statistically significant according to the Fishers exact test. The pancreatic cyst cohort consisted of 10 males and 19 females with a median age of 53 (37–81 years) and four patients in this cohort finally underwent a surgical resection of the pancreatic lesion due to a suspicion of malignancy (Table S1). The 40 healthy individual cohort consisted of 17 females and 23 males with a median age of 39 years (18–63 years) and no known digestive diseases.

Table 1. Age and sex distribution of the individuals included in the study, according to the different groups analyzed pancreatic ductal adenocarcinoma (PDAC) cases (familial or hereditary (H/FPC) and Sporadic PDAC), pancreatic cysts and healthy controls.

Variable	All PDAC Cases	Hereditary or Familial PDAC	Sporadic	Pancreatic Cysts	Healthy Controls
Male	71	17	54	10	23
Female	74	26	58	19	17
Ratio M:F	0.96	0.65	0.89	0.53	1.35
Median age (range)	69 (29–90)	65.5 (29–84)	70 (31–90)	53 (37–81)	39 (18–63)

A total of 136 PDAC patients (94 sporadic and 42 H/FPC) were included in the cfDNA quantification analysis, 54 in the *KRAS* detection analysis in cfDNA via BEAMing (23 sporadic and 31 H/FPC) and 20 H/FPC cases and 7 sporadic cases were included in the panel sequencing analysis of cfDNA. There were no significant differences between the age, sex, and disease stage at diagnosis distribution of the cohorts used for BEAMing and sequencing analysis compared with the entire cohort.

3.2. Correlation of cfDNA Levels with Clinical Parameters

The Maxwell® RSC kit favors the isolation of small DNA fragments within the expected cfDNA size range of 150 to 200 bp with a low contamination of high-molecular weight genomic DNA (Figure S1). Thus, this method was used for the study of total cfDNA analysis in 1 mL of plasma and also for the BEAMing-based studies to avoid the amplification of non-tumor genomic DNA. The total cfDNA concentration in plasma (ng/µL) was determined for 136 patients with PDAC, 29 with pancreatic cysts and 40 healthy individuals (Figure 1a). The median cfDNA level in healthy individuals was 0.01 ng (0.005–0.09), 0.03 ng (0.005–0.7) in patients with pancreatic cysts, 0.0575 ng (0.01–4.17) in H/FPC PDAC cases, and 0.07 ng (0.005–2.055) in sporadic PDAC cases. There were significantly higher levels of cfDNA in patients with pancreatic cysts (p = 0.021), sporadic PDAC ($p \leq 0.001$) and H/FPC PDAC ($p \leq 0.001$) compared to healthy individuals. Furthermore, there was significantly higher levels of cfDNA levels in both H/FPC PDAC (p = 0.02314) and sporadic PDAC cases (p = 0.01374) compared to patients with pancreatic cysts. The median cfDNA concentration in all PDAC cases was 0.0675 ng (0.005–4.17) and there was no significant difference in cfDNA levels between H/FPC PDAC and sporadic PDAC; 0.0575 vs. 0.07 ng, respectively. The median cfDNA level in resectable cases was 0.0575 ng (0.0050–2.0000), 0.0675 ng (0.0350–2.0000) in locally advanced, and 0.07 ng (0.005–4.000) in metastatic cases (Figure S2a). Even though the median cfDNA level increased according to disease stage, the difference did not reach statistical significance. However, the levels of cfDNA were significantly higher in resectable cases ($p \leq 0.001$), locally advanced ($p \leq 0.001$), and metastatic cases ($p \leq 0.001$) compared to healthy controls. Furthermore, the cfDNA levels were also significantly higher in locally advanced (p = 0.02) and metastatic cases (p = 0.02) compared to patients with pancreatic cysts. There was no significant association between cfDNA levels and tumor size according to the Pearson Correlation (−0.1).

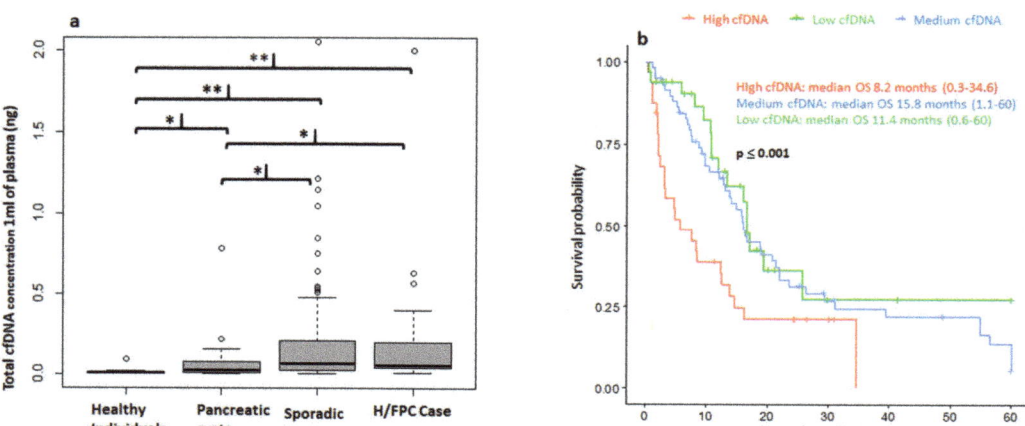

Figure 1. Correlation of circulating free DNA (cfDNA) levels with clinical characteristics. (**a**) The concentration of circulating free DNA (cfDNA) in plasma differentiates between cancer and non-cancer cases. There were significantly higher levels of total cfDNA in hereditary or familial and sporadic pancreatic ductal adenocarcinoma (PDAC) cases compared to healthy controls and patients with pancreatic cysts. ** $p \leq 0.01$; * $p \leq 0.05$. (**b**) High levels of cfDNA at diagnosis correlate with a poorer overall survival (OS). Patients were classified into 3 groups: high (>0.2037 ng), medium (≥ 0.035 ng and ≤ 0.2037), and low (<0.035 ng) cfDNA levels. Follow-up was censored at 5 years and adjusted for sex, age, and disease stage at diagnosis, sporadic or hereditary and familial, case and first line treatment. Low: $N = 33$; Medium $N = 67$ and High: $N = 31$.

Survival and cfDNA total concentration data were available for 134 PDAC cases. Cases were followed up for a median of 12 months (0.4–60 months) and survival analysis was performed based on total cfDNA levels and censored at 5 years. Patients were classified into 3 groups: high (>0.2037 ng), medium (>0.035 ng and <0.2037 ng), and low (<0.035 ng) cfDNA levels at diagnosis. The concentration of cfDNA in plasma significantly correlated with overall survival (OS) and patients with a high cfDNA concentration at diagnosis (>0.2037 ng) had a significantly shorter OS, with a median overall survival of 8.2 vs. 11.4 and 15.8 months for medium and low levels, respectively (Figure 1b). The Hazard Ratios for a low cfDNA level at diagnosis was 0.6 ($p = 0.04$), 0.5 for a medium level ($p = 0.01$) and 1.8 ($p = 0.04$) for a high level (Table S2). Furthermore, cfDNA levels were determined in plasma before and 1 month after a surgical resection of the primary tumor. There was a significant reduction in the cfDNA concentration from 0.11 ng (0.025–5.5 ng) before surgery to 0.025 ng (0.01–1.25 ng) after surgery ($p = 0.0024$) (Figure S2b), which supports the idea that cfDNA levels are indicative of tumor burden.

3.3. Analysis of Somatic Mutations in Plasma

BEAMing was performed with cfDNA isolated from 1 mL of plasma using the Maxwell® RSC kit from 54 PDAC cases, which included 31 H/FPC cases and 23 sporadic cases. The frequency of *KRAS* mutations in codon 12 and 13 was 70% in sporadic cases and 16% in familial cases (Figure 2a,b), which was statistically significant ($p \leq 0.001$), and indicated that *KRAS* somatic mutations are less frequent in H/FPC cases compared to sporadic cases. According to disease stage, KRAS positivity in sporadic and H/FPC cases was 67 vs. 17% for locally advanced cases and 75 vs. 17% in metastatic cases, respectively, which was statistically significant ($p \leq 0.001$). The same statistical analysis could not performed for the resectable cases due to the low number of cases in each in sub-group. KRAS mutation validation in primary tissue was only possible in 8 out of 54 (15%) samples due to problems associated with primary tissue availability and a high non-neoplastic cell content of primary tumors.

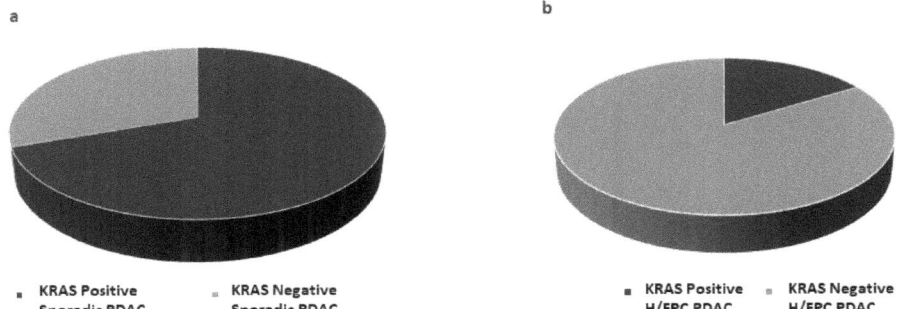

Figure 2. *KRAS* mutation status was determined in plasma from (**a**) sporadic PDAC cases (**b**) hereditary or familial PDAC (H/FPC) cases via BEAMing and mutant *KRAS* was more frequently detected in sporadic cases compared to H/FPC cases. BEAMing was performed using cfDNA isolated from 1 mL of plasma from 54 PDAC cases (31 familial cases and 23 sporadic cases). The frequency of mutant *KRAS* was 70% in sporadic cases and 16% in familial cases, which was statistically significant ($p \leq 0.001$)).

The overall DNA yield with the QIAamp circulating nucleic acid kit was higher compared to the Maxwell® RSC kit (Figure S1), although there was a higher level of genomic DNA contamination. Thus, this method was used for subsequent sequencing analyses after re-purification of fragments within the expected range of cfDNA (250 bp). Panel sequencing of 15 genes commonly mutated in primary tumors was performed in order to determine the spectrum of somatic mutations (other than *KRAS*) in H/FPC cases. A total of 3 out of 21 (15%) of H/FPC cases were positive for a *KRAS* mutation (p. Gly12Arg) (Figure 3) and BEAMing data were available for 17 of these cases. The *KRAS* status via BEAMing and sequencing was consistent in 16 cases (3 *KRAS* positive and 14 *KRAS* negative) and a *KRAS* mutation was detected by BEAMing and not by sequencing in one case. This is likely due to the lower sensitivity of sequencing for mutation detection compared to BEAMing. The overall frequency of TP53 mutations in H/FPC cases was 12 out of 21 (57%). Furthermore, mutations in *AKT*, *ERBB2*, *KIT*, and *PDGFRA* were also detected in *KRAS* negative H/FPC cases (Figure 3 and Table 2). Sequencing and BEAMing data were available for 7 sporadic cases. The presence of a *KRAS* mutation was confirmed by sequencing in 3 of 4 sporadic cases, again this is likely due to the lower sensitivity of sequencing analysis. Of the 3 *KRAS* negative cases determined by BEAMing, one was negative for mutations via sequencing, one harbored mutations in *PDGFRA* and *TP53*, and the other case had mutations in *PIK3CA*, *KIT*, *BRAF* and *ERBB2* (Table 2). The KRAS mutation frequency in these 7 sporadic cases was 43%. However, it is important to note that as 3 cases were specifically selected as they were negative for a KRAS mutation via BEAMing and 4 *KRAS* positive cases were included for comparison. Thus, there is a selection bias that is reflected in the KRAS mutation frequency.

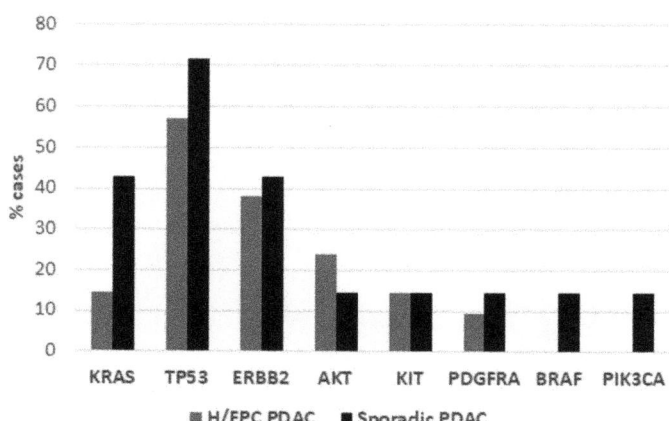

Figure 3. The frequency of somatic mutations in hereditary or familial cases and sporadic was determined by sequencing analysis.

Table 2. Summary of the somatic mutations detected in H/FPC and sporadic cases by sequencing analysis.

Case	Somatic Mutations Detected by Sequencing	KRAS Status by BEAMing
H/FPC	*KIT*: p.(Glu640Lys) and *TP53*: p.(His214Tyr) and *ERBB2*: p.(Asn850Ser)	md
	KRAS: p.(Gly12Arg) and *TP53*: p.(His193Arg)	md
	KRAS: p.(Gly12Asp)	md
	KRAS:p.(Gly12Asp) and *TP53*:p.(Tyr220Cys) and *ERBB2*:p.(Asn850Ser)	md
	AKT: p.(Lys39Asn) and *TP53*: p.(Pro250His) and *ERBB2*: p.(Arg849Gln)	nmd
	ERBB2: p.(Ala710Val)	nmd
	ERBB2: p.(Cys560Phe) and *AKT*: p.(Thr34Asn) and *TP53*: p.(Pro250His)	nmd
	PDGFRA: p.(Ala840Thr) and *AKT*: p.(Thr34Asn) and *TP53*:p.(Pro250His)	nmd
	TP53 p.(Ser215Ile) and *PDGFRA* p.(Ala840Val) and *KIT* p.(Tyr545Phe)	nmd
	TP53: p.(Glu171Gly) and *AKT*: p.(Thr34Asn)	nmd
	TP53: p.(His214Tyr)	nmd
	TP53: p.(His214Tyr)	nmd
	TP53:p.(His214Tyr) and *ERBB2*:p.(Tyr735Cys)	nmd
	TP53:p.(His214Tyr) and *KIT*:p.(Glu640Lys) and *ERBB2*:p.(Tyr735Cys) and *ERBB2*:p.(Asn850Ser)	nmd
	NEG	nmd
	NEG	nmd
	NEG	nmd
	NEG	nmd
	ERBB2:p.(Leu715Arg) and *AKT*:p.(Pro24Ser)	not tested
	NEG	not tested
	NEG	not tested
Sporadic	*KRAS*: p.(Gly12Asp) and *TP53*: p.(Gly266Val)	md
	KRAS p.(Gly12Asp) and *TP53* p.(Gly266Val)	md
	KRAS: p.(Gly12Asp) and *TP53*: p.(Gly244Ser) and *AKT*: p.(Arg23Trp) and *ERBB2*: p.(Gly732Asp)	md
	TP53: p.(Pro177Arg) and *ERBB2*: p.(Pro1130His)	md
	PDGFRA p.(Ala820Val) and *TP53* p.(Pro223Ser)	nmd
	PIK3CA: p.(Ala995Asp) and *KIT*: p.(Pro832Ser) and *BRAF*:p.(His585Tyr) and *ERBB2*: p.(Pro761Thr)	nmd
	NEG	nmd

NEG: negative for somatic mutations by sequencing analysis. md: KRAS mutation detected by BEAMing. nmd: no KRAS mutation detected by BEAMing.

Overall, 20 H/FPC and 7 sporadic cases were analyzed by sequencing analysis. Then, 3 out of 21 (14%) of H/FPC were positive for a KRAS mutation. All KRAS mutations were Gly12Asp. One case only had a KRAS mutation, another KRAS and TP53 (His193Arg), and another KRAS, TP53 (Tyr220Cys), and ERBB2 (Asn850Ser). Of the 3 sporadic cases negative for a KRAS mutation via BEAMing, one had a mutation in TP53 and PDGFR, another in PIK3CA, KIT, BRAF, and ERBB2, and the final case was mutation negative via sequencing.

According to the COSMIC database, *KRAS* and *TP53* are the most commonly mutated genes in PDAC, 64 and 47%, respectively. The *KRAS* mutations identified were known pathogenic mutations that have been previously described in COSMIC. The most frequent *KRAS* mutations found by sequencing were c.35G>A G12D and c.34G>C G12R, which are among the most frequent mutations reported in COSMIC (G12D (47%), G12V (31%), and G12R (13%)). The TP53 variant p.(His214Tyr) was identified in 5 H/FPC cases and had a high pathogenicity score (FATHMM prediction (Pathogenic (score 1.00)) but was not found in COSMIC [23].

Survival analysis was performed with hereditary or familial (H/FPC) cases and sporadic cases, which was corrected for sex, age, and stage at diagnosis and 1st line therapy. Hereditary or familial cases had a significantly longer median overall survival compared to sporadic cases; 10.2 vs. 21.7 months, respectively, ($p \leq 0.001$) (Figure 4). The Hazard Ratio for sporadic cases was 2.4 ($p \leq 0.001$) (Table S2), indicating that sporadic cases have a poorer overall survival, independently of stage at diagnosis and 1st line treatment.

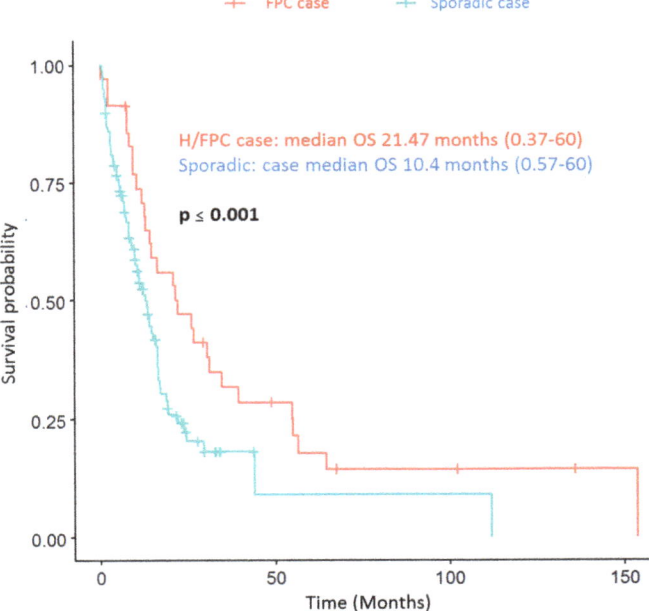

Figure 4. Hereditary or familial PDAC cases have a longer overall survival (OS) compared to sporadic cases.

Follow-up was censored at 5 years, adjusted for sex, age and disease stage at diagnosis, and first line treatment. The median OS for H/FPC cases was 21.47 months (0.37–60) and 10.4 months (0.57–60) for sporadic cases, which was statistically significant. The HR for sporadic cases was 2.4. Sporadic PDAC: N = 102 and H/FPC: N = 35.

4. Discussion

Circulating-free DNA concentration in plasma or serum has been shown to be a surrogate marker of tumor burden, survival, and recurrence in various tumor types, such as lymphoma [24], lung [25,26], prostate [27] melanoma [28], and colon [29,30]. In this study, the levels of cfDNA correlated with disease status, with higher levels found in PDAC cases vs. healthy individuals or patients with pancreatic cysts. There was a reduction in the total cfDNA levels after resection of the primary tumor, which also supports the notion that cfDNA levels may be indicative of tumor burden. Furthermore, PDAC cases with a cfDNA high concentration at diagnosis had a significantly poorer OS compared to cases with low and medium levels. The lack of primary PDAC tissue of sufficient quality and quantity is a significant problem due to the low resection rate and contamination with the stroma cells, among other reasons. Thus, the liquid biopsy is the most feasible approach for molecular studies in PDAC and we show here that molecular analysis in cfDNA is a good substitute for primary tissue. We attempted to validate *KRAS* mutations in primary tissue but were successful in only 15% of cases, which is a recurring problem in many PDAC somatic profiling studies [31,32].

Mutant *KRAS* was more frequently detected in cfDNA from sporadic cases compared to hereditary or familial cases (70 vs. 16%, respectively). This is the first time that a difference in *KRAS* mutation status frequency has been shown between sporadic and hereditary or familial cases. Previous studies have reported that the frequency of KRAS mutations is similar between familial and sporadic cases [33,34]. However, as they reply on primary tumor samples, they have some limitations, such as sample size and case selection bias. One study reported a similar frequency of *KRAS*, TP53, *CDKN2A*, and *SMAD4* mutations in sporadic and familial cases [33]. However, this study only included primary tumor-derived cell lines from 16 patients, thus there is an important sample size limitation. A second study included 39 cases and also showed a similar somatic profile between somatic and familial cases [34]. However, this study excluded cases with a known pathogenic germline mutation that predisposes them to develop PDAC. Thus, there is an important selection bias, as the cases that we hypothesize will lack KRAS somatic mutations in the primary tumor were excluded. Thus, these data cannot be directly compared with the data presented here.

Although it could be expected that the majority of sporadic cases would be positive for *KRAS* mutation, we did not detect a *KRAS* mutation in all cases. This may be due to the low cfDNA yield which appears to be dependent on disease stage. Furthermore, some patients may not actively shed cfDNA into the blood stream and, therefore, it is almost impossible to detect tumor cfDNA in these individuals. However, there was no statistically significant difference in the median cfDNA levels between sporadic and hereditary or familial cases that may account for this difference in *KRAS* detection frequency. Thus, suggesting that the somatic *KRAS* mutation frequency differs between sporadic and hereditary for familial cases. Furthermore, the *KRAS* positivity rate in the sporadic cases is consistent with previous studies based on cfDNA that report values of 70 and 62.8%. Importantly, we show for the first time that hereditary or familial cases have a longer OS survival compared to sporadic cases, even though they are treated the same regimens in the clinic. Furthermore, hereditary or familial cases were diagnosed at a significantly younger age compared with sporadic cases, 70 vs. 65.5 years, respectively. However, since the survival analysis was corrected for age and disease stage at diagnosis and 1st line adjuvant treatment, the difference in survival may be due to a distinct molecular somatic profile of H/FPC cases compared to sporadic cases.

The Maxwell system was used for BEAMing analysis as the magnetic beads-based approach has been shown to be superior to silica membrane-based methods such as the QIAamp kit [35]. In fact, we showed that Maxwell system favored the extraction of small sized fragments which correspond to cfDNA, with less genomic DNA contamination. Thus, we believe that this method is more appropriate for PCR based methods, such as BEAMing to avoid the amplification of high molecular weight genomic DNA from normal cells. The

QIAamp Kit produces the highest yield of total cfDNA compared to other commercially available kits and was therefore used for sequencing analysis [36], which is at least 10 fold less sensitive than BEAMing. The probability of finding a low number of cfDNA molecules follows the Poisson distribution and therefore this probability increases with increased sample volume. A minimum of 3.6 ng of cfDNA is needed to obtain a sensitivity of <0.1% and 36 ng are needed to obtain a sensitivity of <0.01% [35]. Therefore, the sensitivity to detect somatic mutations by sequencing analysis is much lower compared to BEAMing.

There are some important limitations with regard to the use of cfDNA as a prognostic marker that should be taken into account. Some reports have shown that cfDNA isolated from serum has a higher integrity than in plasma, although the *KRAS* allelic frequencies were much lower in serum compared to plasma, with a similar sensitivity and specificity [36]. The critical factors that influence the quality and yield of cfDNA are the time from sample extraction to processing and sample collection tube type and a second centrifugation is required to remove lysed white blood cells [36]. Plasma contains many PCR inhibitors such as heparin, hemoglobin, hormones, immunoglobulin G, and lactoferrin [37], which may be overcome for PCR applications by diluting the sample. Although, sequencing technologies are particularly sensitive to inhibitors and, thus, high quality and quantity samples are needed. Clonal hematopoiesis of indeterminate potential (CHIP) can interfere with cfDNA testing results and the parallel analysis of a paired whole blood control to exclude CHIP variants and avoid misdiagnosis has been recommended [38].

In Europe, PDAC patients are all similarly treated in the clinic. Even though it is clear that there are several PDAC sub-types at the genetic and histological level that differ in terms of prognosis and response to therapy [39]. The backbone of chemotherapy in PDAC are cytotoxic agents that target rapidly proliferating cancer cells. These include the FOLFIRINOX scheme (folinic acid, fluorouracil, irinotecan, and oxaliplatin), gemcitabine combined with nab-paclitaxel, capecitabine and gemcitabine or 5FU monotherapy according to disease stage and Eastern Cooperative Oncology Group (ECOG) status [40]. However, PDAC cases with germline and somatic mutations in DNA repair genes (*BRCA1, BRCA2, PALB2, ATM, BAP1, BARD1, BLM, BRIP1, CHEK2, FAM175A, FANCA, FANCC, NBN, RAD50, RAD51, RAD51C*), are more sensitive to platinum agents in first line and PARP inhibitors (e.g., olaparib and rucaparib) as a maintenance treatment [41,42]. Clinical guidelines recommend PDAC gene profiling and The National Comprehensive Cancer Network (NCCN) guidelines suggest clinical trial participation as first line and second line treatment options [40]. As treatment modifies the genomic composition of PDAC, cfDNA is a useful tool to identify resistance mutations, and potential new treatment targets [43].

Low prevalence focal amplifications in druggable oncogenes have been identified in PDAC such as *ERBB2, MET, FGFR1, CDK6, PIK3R3,* and *PIK3CA*, which may also be exploited in the clinic to provide alternatives to cytotoxic therapies [44]. We hypothesized that hereditary or familial PDAC cases have a distant somatic mutation spectrum due to the presence of germline pathogenic mutations in DNA repair genes. We found somatic mutations in the genes *KIT, AKT, PDGFRA, MET, PIK3CA* and *BRAF* in *KRAS* negative cases, which were mainly found in hereditary and familial cases. Thus, this subgroup of patients could be candidates for treatment with small molecule tyrosine kinase inhibitors (TKI) that inhibit *KIT* and *PDGFR* (e.g., axitinib and imatinib), new generation isoform-specific PI3K inhibitors that reduce toxicity (e.g., alpelisib), small molecule inhibitors and monoclonal antibodies against the receptor and ligands of *MET, BRAF,* and *EGFR*, among others. However, the efficacy of these treatment strategies must be confirmed in pre-clinical or clinical studies in this sub-group of patients.

We show here for the first time that hereditary or familial cases have a lower *KRAS* mutation frequency compared to sporadic cases. Although the data presented here are preliminary and should be validated in a larger cohort of patients, the observation of *KRAS* negativity in hereditary and familial cases could have an important impact in the clinic for this patient subgroup. In addition, *KRAS* negative cases harbored somatic mutations in potentially druggable genes that could potentially be exploited in the clinic.

Furthermore, they have a longer overall survival, which does not appear to be related to stage at diagnosis or 1st line treatment strategy. The main limitation of this study is the sample size. Hereditary or familial PDAC is not a common entity and thus this factor limited the number of cases that were included. Moreover, only 54 samples were included in the BEAMing analysis and 27 in the sequencing analysis due to economic constraints and 8 cases in primary tissue validation due to sample availability. However, there are some positive aspects of this study that should be highlighted. Firstly, as liquid biopsy samples were used, were able to include localized and advanced PDAC cases in the study, which provides a more representative patient cohort. Secondly, we show that the liquid biopsy is a valid alternative in many PDAC cases to primary tissue samples, due to the advent of new technologies with a high sensitivity and specificity for somatic mutation detection. Finally, this study provides preliminary data to suggest that Hereditary or Familial PDAC have a distinct somatic mutation profile compared to sporadic cases and this should be taken into account in the clinic when defining a treatment strategy.

5. Conclusions

cfDNA is a valuable source of genomic information in PDAC cases where primary tissue samples are scarce and is also useful to track the genomic changes induced by treatment and tumor dynamics, for the design of a more personalized treatment. The level of cfDNA in plasma appears to be a prognostic indicator, independently of the detection of tumor specific mutations and appears to be a valid substitute for primary tumor tissue for molecular studies in PDAC. Hereditary or familial and tend to be *KRAS* negative and harbor somatic mutations in TP53 in combination with potentially druggable mutations in genes such as *KIT, AKT, BRAF, PIK3CA,* and *PDGFR*. However, these preliminary findings must be validated in a larger cohort. Hereditary or familial PDAC cases have a greater overall survival rate, even though they are treated with the same regimens in the clinic as sporadic cases.

Supplementary Materials: The following are available online at https://www.mdpi.com/article/10.3390/cancers13071612/s1, Data S1: Analysis pipeline for the identification of pathogenic somatic variants in cfDNA using the TruSight 15 kit and The BaseSpace Variant Interpreter tool. Table S1: Demographic and clinical characteristics of the cohorts included the study. Table S2: Summary of the Cox Proportional-Hazards Model analysis with the corresponding odds ratios showing the effect size of the covariates included in the survival analysis. Figure S1: Comparative cfDNA extraction from 1 mL of plasma of 3 samples using the (a) QIAamp circulating nucleic acid kit and (b) Maxwell® RSC Instrument. Figure S2: Correlation of circulating free DNA (cfDNA) levels with clinical parameters (disease stage, disease status and surgery).

Author Contributions: Conceptualization, J.E. and A.C.; Data curation, J.E., E.B., M.E.C., R.F. (Raquel Fuentes), M.R.-G., R.F. (Reyes Ferreiro), P.R., G.M., D.G.-S., J.V.L., B.S.J., N.M. and A.C.; Funding acquisition, A.C.; Methodology, J.E., E.B., M.E.C., G.M. and D.G.-S.; Writing—original draft, J.E., E.B., M.E.C. and A.C.; Writing—review and editing, J.E., R.F. (Raquel Fuentes), M.R.-G., R.F. (Reyes Ferreiro), P.R., G.M., D.G.-S., J.V.L., B.S.J., N.M. and A.C. All authors have read and agreed to the published version of the manuscript.

Funding: This study was funded by the Instituto de Salud Carlos III (Plan Estatal de I+D+i 2013–2016): ISCIII (PI09/02221, PI12/01635, PI15/02101, and PI18/0135) and co-financed by the European Development Regional Fund "A way to achieve Europe" (ERDF), the Biomedical Research Network in Cancer: CIBERONC (CB16/12/00446), Red Temática de investigación cooperativa en cáncer: RTICC (RD12/0036/0073), La Asociación Española contra el Cáncer: AECC (Grupos Coordinados Estables 2016), Fundación Mutua Madrileña (FMM) / XVI Convocatoria de Ayudas a la Investigación en Salud 2019 and Asociación Cáncer de Páncreas (ACanPan); Asociación Española de Pancreatología (AESPANC) / IV Becas de Investigación Carmen Delgado/Miguel Pérez-Mateo 2019. The European Union's Horizon 2020 research and innovation program under grant agreement No 857381, project VISION (Strategies to strengthen scientific excellence and innovation capacity for early diagnosis of gastrointestinal cancers).

Institutional Review Board Statement: The study was conducted according to the guidelines of the Declaration of Helsinki, and approved by the Institutional Review Board of the Ramon y Cajal University Hospital (code 058/19, approved on 25th of April 2017).

Informed Consent Statement: This study was approved by the local ethics committee of the Ramón y Cajal university hospital and all patients signed an informed consent before entering into the study.

Data Availability Statement: The data presented in this study are available on request from the corresponding author. The data are not yet publicly available.

Acknowledgments: We would like to thank the collaboration of the clinical and nursing staff that have contributed to this this project, Alfonso Sanjuanbenito, Eduardo Lobo, Eduardo Lisa, Teresa Ramon y Cajal Asensio, Luis Robles Diaz, Isabel Chirivella Gonzalez, Montse Rodriguez, Eva Martínez de Castro, Alejandra Caminoa, Carmen Guillen Ponce, Celia Calcedo, María Teresa Salazar López, Andrea Santos Gil, Maria del Carmen Perez Ruiz, Manuela Hernando, Maria Jesus Casado Cespedosa, and Clara Pacheco Torres. We would also like to thank our project manager Sara Navarro for her invaluable assistance.

Conflicts of Interest: The authors have no conflicts of interest with regard to the manuscript. A.C. has received honoraria for advisory roles from Bayer, Shire and Celgene and has acted as a consultant for Bristol Myers Squibb (BMS). The funding sources of this study did not play any role in the study design, the collection, analysis and interpretation of the data, in the writing of the manuscript and in the decision to submit the paper for publication.

References

1. Ferlay, J.; Soerjomataram, I.; Dikshit, R.; Eser, S.; Mathers, C.; Rebelo, M.; Parkin, D.M.; Forman, D.; Bray, F. Cancer incidence and mortality worldwide: Sources, methods and major patterns in GLOBOCAN 2012. *Int. J. Cancer* **2015**, *136*, E359–E386. [CrossRef] [PubMed]
2. Rahib, L.; Smith, B.D.; Aizenberg, R.; Rosenzweig, A.B.; Fleshman, J.M.; Matrisian, L.M. Projecting cancer incidence and deaths to 2030: The unexpected burden of thyroid, liver, and pancreas cancers in the united states. *Cancer Res.* **2014**, *74*, 2913–2921. [CrossRef] [PubMed]
3. Jones, S.; Zhang, X.; Parsons, W.D.; Lin, J.C.; Leary, R.J.; Angenendt, P.; Mankoo, P.; Carter, H.; Kamiyama, H.; Jimeno, A.; et al. Core Signaling Pathways in Human Pancreatic Cancers Revealed by Global Genomic Analyses. *Science* **2008**, *321*, 1801–1806. [CrossRef]
4. Thomas, C. Risk factors, biomarker and imaging techniques used for pancreatic cancer screening. *Chin. Clin. Oncol.* **2017**, *6*, 61. [CrossRef]
5. Llach, J.; Carballal, S.; Moreira, L. Familial pancreatic cancer: Current perspectives. *Cancer Manag. Res.* **2020**, *12*, 743–758. [CrossRef]
6. Earl, J.; Galindo-Pumariño, C.; Encinas, J.; Barreto, E.; Castillo, M.E.; Pachón, V.; Ferreiro, R.; Rodríguez-Garrote, M.; González-Martínez, S.; Ramon y Cajal, T.; et al. A comprehensive analysis of candidate genes in familial pancreatic cancer families reveals a high frequency of potentially pathogenic germline variants. *EBioMedicine* **2020**, *53*, 102675. [CrossRef]
7. Matsubayashi, H.; Takaori, K.; Morizane, C.; Maguchi, H.; Mizuma, M.; Takahashi, H.; Wada, K.; Hosoi, H.; Yachida, S.; Suzuki, M.; et al. Familial pancreatic cancer: Concept, management and issues. *World J. Gastroenterol.* **2017**, *23*, 935–948. [CrossRef] [PubMed]
8. Petersen, G.M. Familial Pancreatic Cancer. *Semin. Oncol.* **2016**, *43*, 548–553. [CrossRef]
9. Mocci, E.; Guillen-Ponce, C.; Earl, J.; Marquez, M.; Solera, J.; Salazar-López, M.-T.; Calcedo-Arnáiz, C.; Vázquez-Sequeiros, E.; Montans, J.; Muñoz-Beltrán, M.; et al. PanGen-Fam: Spanish registry of hereditary pancreatic cancer. *Eur. J. Cancer* **2015**, *51*, 1911–1917. [CrossRef] [PubMed]
10. Ballehaninna, U.K.; Chamberlain, R.S. The clinical utility of serum CA 19-9 in the diagnosis, prognosis and management of pancreatic adenocarcinoma: An evidence based appraisal. *J. Gastrointest. Oncol.* **2012**, *3*, 105–119. [CrossRef]
11. Radon, T.P.; Massat, N.J.; Jones, R.; Alrawashdeh, W.; Dumartin, L.; Ennis, D.; Duffy, S.W.; Kocher, H.M.; Pereira, S.P.; Guarner, L.; et al. Identification of a three-biomarker panel in urine for early detection of pancreatic adenocarcinoma. *Clin. Cancer Res.* **2015**, *21*, 3512–3521. [CrossRef] [PubMed]
12. Martinez-Bosch, N.; Barranco, L.E.; Orozco, C.A.; Moreno, M.; Visa, L.; Iglesias, M.; Oldfield, L.; Neoptolemos, J.P.; Greenhalf, W.; Earl, J.; et al. Increased plasma levels of galectin-1 in pancreatic cancer: Potential use as biomarker. *Oncotarget* **2018**, *9*, 32984–32996. [CrossRef] [PubMed]
13. Kim, J.; Bamlet, W.R.; Oberg, A.L.; Chaffee, K.G.; Donahue, G.; Cao, X.-J.; Chari, S.; Garcia, B.A.; Petersen, G.M.; Zaret, K.S. Detection of early pancreatic ductal adenocarcinoma with thrombospondin-2 and CA19-9 blood markers. *Sci. Transl. Med.* **2017**, *9*, eaah5583. [CrossRef]

14. Cohen, J.D.; Javed, A.A.; Thoburn, C.; Wong, F.; Tie, J.; Gibbs, P.; Schmidt, C.M.; Yip-Schneider, M.T.; Allen, P.J.; Schattner, M.; et al. Combined circulating tumor DNA and protein biomarker-based liquid biopsy for the earlier detection of pancreatic cancers. *Proc. Natl. Acad. Sci. USA* **2017**, *114*, 201704961. [CrossRef]
15. Mellby, L.D.; Nyberg, A.P.; Johansen, J.S.; Wingren, C.; Nordestgaard, B.G.; Bojesen, S.E.; Mitchell, B.L.; Sheppard, B.C.; Sears, R.C.; Borrebaeck, C.A.K. Serum Biomarker Signature-Based Liquid Biopsy for Diagnosis of Early-Stage Pancreatic Cancer. *J. Clin. Oncol.* **2018**. [CrossRef]
16. Alix-Panabières, C.; Pantel, K. Clinical applications of circulating tumor cells and circulating tumor DNA as liquid biopsy. *Cancer Discov.* **2016**, *6*, 479–491. [CrossRef]
17. Chakraborty, J.; Langdon-Embry, L.; Cunanan, K.M.; Escalon, J.G.; Allen, P.J.; Lowery, M.A.; O'Reilly, E.M.; Gonen, M.; Do, R.G.; Simpson, A.L. Preliminary study of tumor heterogeneity in imaging predicts two year survival in pancreatic cancer patients. *PLoS ONE* **2017**, *12*, e0188022. [CrossRef] [PubMed]
18. Diaz, L.A.; Bardelli, A. Liquid biopsies: Genotyping circulating tumor DNA. *J. Clin. Oncol.* **2014**, *32*, 579–586. [CrossRef]
19. Jahr, S.; Hentze, H.; Englisch, S.; Hardt, D.; Fackelmayer, F.O.; Hesch, R.D.; Knippers, R. DNA fragments in the blood plasma of cancer patients: Quantitations and evidence for their origin from apoptotic and necrotic cells. *Cancer Res.* **2001**, *61*, 1659–1665.
20. Jung, K.; Fleischhacker, M.; Rabien, A. Cell-free DNA in the blood as a solid tumor biomarker—A critical appraisal of the literature. *Clin. Chim. Acta* **2010**, *411*, 1611–1624. [CrossRef]
21. Earl, J.; Garcia-Nieto, S.; Martinez-Avila, J.C.; Montans, J.; Sanjuanbenito, A.; Rodríguez-Garrote, M.; Lisa, E.; Mendía, E.; Lobo, E.; Malats, N.; et al. Circulating tumor cells (Ctc) and kras mutant circulating free Dna (cfdna) detection in peripheral blood as biomarkers in patients diagnosed with exocrine pancreatic cancer. *BMC Cancer* **2015**, *15*, 1–10. [CrossRef] [PubMed]
22. Lee, J.S.; Park, S.S.; Lee, Y.K.; Norton, J.A.; Jeffrey, S.S. Liquid biopsy in pancreatic ductal adenocarcinoma: Current status of circulating tumor cells and circulating tumor DNA. *Mol. Oncol.* **2019**, *13*, 1623–1650. [CrossRef] [PubMed]
23. Tate, J.G.; Bamford, S.; Jubb, H.C.; Sondka, Z.; Beare, D.M.; Bindal, N.; Boutselakis, H.; Cole, C.G.; Creatore, C.; Dawson, E.; et al. COSMIC: The Catalogue Of Somatic Mutations In Cancer. *Nucleic Acids Res.* **2019**, *47*, D941–D947. [CrossRef] [PubMed]
24. Wu, J.; Tang, W.; Huang, L.; Hou, N.; Wu, J.; Cheng, X.; Ma, D.; Qian, P.; Shen, Q.; Guo, W.; et al. The analysis of cell-free DNA concentrations and integrity in serum of initial and treated of lymphoma patients. *Clin. Biochem.* **2019**, *63*, 59–65. [CrossRef]
25. Hyun, M.H.; Sung, J.S.; Kang, E.J.; Choi, Y.J.; Park, K.H.; Shin, S.W.; Lee, S.Y.; Kim, Y.H. Quantification of circulating cell-free DNA to predict patient survival in non-small-cell lung cancer. *Oncotarget* **2017**, *8*, 94417–94430. [CrossRef]
26. Mirtavoos-Mahyari, H.; Ghafouri-Fard, S.; Khosravi, A.; Motevaseli, E.; Esfahani-Monfared, Z.; Seifi, S.; Salimi, B.; Oskooei, V.K.; Ghadami, M.; Modarressi, M.H. Circulating free DNA concentration as a marker of disease recurrence and metastatic potential in lung cancer. *Clin. Transl. Med.* **2019**, *8*, 14. [CrossRef]
27. Khani, M.; Hosseini, J.; Mirfakhraie, R.; Habibi, M.; Azargashb, E.; Pouresmaeili, F. The value of the plasma circulating cell-free DNA concentration and integrity index as a clinical tool for prostate cancer diagnosis: A prospective case-control cohort study in an Iranian population. *Cancer Manag. Res.* **2019**, *11*, 4549–4556. [CrossRef]
28. Valpione, S.; Gremel, G.; Mundra, P.; Middlehurst, P.; Galvani, E.; Girotti, M.R.; Lee, R.J.; Garner, G.; Dhomen, N.; Lorigan, P.C.; et al. Plasma total cell-free DNA (cfDNA) is a surrogate biomarker for tumour burden and a prognostic biomarker for survival in metastatic melanoma patients. *Eur. J. Cancer* **2018**, *88*, 1–9. [CrossRef] [PubMed]
29. Tarazona, N.; Gimeno-Valiente, F.; Gambardella, V.; Zuñiga, S.; Rentero-Garrido, P.; Huerta, M.; Roselló, S.; Martinez-Ciarpaglini, C.; Carbonell-Asins, J.A.; Carrasco, F.; et al. Targeted next-generation sequencing of circulating-tumor DNA for tracking minimal residual disease in localized colon cancer. *Ann. Oncol.* **2019**, *30*, 1804–1812. [CrossRef]
30. Tie, J.; Cohen, J.D.; Wang, Y.; Christie, M.; Simons, K.; Lee, M.; Wong, R.; Kosmider, S.; Ananda, S.; McKendrick, J.; et al. Circulating tumor dna analyses as markers of recurrence risk and benefit of adjuvant therapy for stage III colon cancer. *JAMA Oncol.* **2019**, *5*, 1710–1717. [CrossRef] [PubMed]
31. Qi, Z.H.; Xu, H.X.; Zhang, S.R.; Xu, J.Z.; Li, S.; Gao, H.L.; Jin, W.; Wang, W.Q.; Wu, C.T.; Ni, Q.X.; et al. The significance of liquid biopsy in pancreatic cancer. *J. Cancer* **2018**, *9*, 3417–3426. [CrossRef] [PubMed]
32. Kamyabi, N.; Bernard, V.; Maitra, A. Liquid biopsies in pancreatic cancer. *Expert Rev. Anticancer Ther.* **2019**, *19*, 869–878. [CrossRef]
33. Norris, A.L.; Roberts, N.J.; Jones, S.; Wheelan, S.J.; Papadopoulos, N.; Vogelstein, B.; Kinzler, K.W.; Hruban, R.H.; Klein, A.P.; Eshleman, J.R. Familial and sporadic pancreatic cancer share the same molecular pathogenesis. *Fam. Cancer* **2015**, *14*, 95–103. [CrossRef]
34. Roberts, N.J.; Norris, A.L.; Petersen, G.M.; Bondy, M.L.; Brand, R.; Gallinger, S.; Kurtz, R.C.; Olson, S.H.; Rustgi, A.K.; Schwartz, A.G.; et al. Whole Genome Sequencing Defines the Genetic Heterogeneity of Familial Pancreatic Cancer. *Cancer Discov.* **2016**, *6*, 166–175. [CrossRef] [PubMed]
35. Johansson, G.; Andersson, D.; Filges, S.; Li, J.; Muth, A.; Godfrey, T.E.; Ståhlberg, A. Considerations and quality controls when analyzing cell-free tumor DNA. *Biomol. Detect. Quantif.* **2019**, *17*. [CrossRef]
36. Trigg, R.M.; Martinson, L.J.; Parpart-Li, S.; Shaw, J.A. Factors that influence quality and yield of circulating-free DNA: A systematic review of the methodology literature. *Heliyon* **2018**, *4*, e00699. [CrossRef]
37. Sidstedt, M.; Hedman, J.; Romsos, E.L.; Waitara, L.; Wadsö, L.; Steffen, C.R.; Vallone, P.M.; Rådström, P. Inhibition mechanisms of hemoglobin, immunoglobulin G, and whole blood in digital and real-time PCR. *Anal. Bioanal. Chem.* **2018**, *410*, 2569–2583. [CrossRef] [PubMed]

38. Jensen, K.; Konnick, E.Q.; Schweizer, M.T.; Sokolova, A.O.; Grivas, P.; Cheng, H.H.; Klemfuss, N.M.; Beightol, M.; Yu, E.Y.; Nelson, P.S.; et al. Association of Clonal Hematopoiesis in DNA Repair Genes With Prostate Cancer Plasma Cell-free DNA Testing Interference. *JAMA Oncol.* **2020**, *7*, 107–110. [CrossRef]
39. Bailey, P.; Chang, D.K.; Nones, K.; Johns, A.L.; Patch, A.-M.; Gingras, M.-C.; Miller, D.K.; Christ, A.N.; Bruxner, T.J.C.; Quinn, M.C.; et al. Genomic analyses identify molecular subtypes of pancreatic cancer. *Nature* **2016**, *531*, 47–52. [CrossRef]
40. Ducreux, M.; Cuhna, A.S.; Caramella, C.; Hollebecque, A.; Burtin, P.; Goéré, D.; Seufferlein, T.; Haustermans, K.; Van Laethem, J.L.; Conroy, T.; et al. Cancer of the pancreas: ESMO Clinical Practice Guidelines for diagnosis, treatment and follow-up. *Ann. Oncol.* **2015**, *26*, v56–v68. [CrossRef]
41. Park, W.; Chen, J.; Chou, J.F.; Varghese, A.M.; Yu, K.H.; Wong, W.; Capanu, M.; Balachandran, V.; McIntyre, C.A.; El Dika, I.; et al. Genomic Methods Identify Homologous Recombination Deficiency in Pancreas Adenocarcinoma and Optimize Treatment Selection. *Clin. Cancer Res.* **2020**, *26*, 3239–3247. [CrossRef] [PubMed]
42. Golan, T.; Hammel, P.; Reni, M.; Van Cutsem, E.; Macarulla, T.; Hall, M.J.; Park, J.-O.; Hochhauser, D.; Arnold, D.; Oh, D.-Y.; et al. Maintenance Olaparib for Germline BRCA -Mutated Metastatic Pancreatic Cancer. *N. Engl. J. Med.* **2019**, *381*, 317–327. [CrossRef]
43. Xie, I.Y.; Gallinger, S. The genomic landscape of recurrent pancreatic cancer is modified by treatment. *Nat. Rev. Gastroenterol. Hepatol.* **2020**, *17*, 389–390. [CrossRef] [PubMed]
44. Waddell, N.; Pajic, M.; Patch, A.M.; Chang, D.K.; Kassahn, K.S.; Bailey, P.; Johns, A.L.; Miller, D.; Nones, K.; Quek, K.; et al. Whole genomes redefine the mutational landscape of pancreatic cancer. *Nature* **2015**, *518*, 495–501. [CrossRef] [PubMed]

Correction

Correction: Earl et al. Somatic Mutation Profiling in the Liquid Biopsy and Clinical Analysis of Hereditary and Familial Pancreatic Cancer Cases Reveals KRAS Negativity and a Longer Overall Survival. *Cancers* 2021, *13*, 1612

Julie Earl [1,2,*], Emma Barreto [1,†], María E. Castillo [1,†], Raquel Fuentes [1], Mercedes Rodríguez-Garrote [1], Reyes Ferreiro [1], Pablo Reguera [1], Gloria Muñoz [3], David Garcia-Seisdedos [3], Jorge Villalón López [1], Bruno Sainz, Jr. [4,5,6], Nuria Malats [2,7] and Alfredo Carrato [1,2,8]

1. Molecular Epidemiology and Predictive Tumor Markers Group, Ramón y Cajal Health Research Institute (IRYCIS), Carretera Colmenar Km 9100, 28034 Madrid, Spain; emma.barreto@salud.madrid.org (E.B.); marien.castillo@salud.madrid.org (M.E.C.); rfuentes@salud.madrid.org (R.F.); mercedes.rodriguez@salud.madrid.org (M.R.-G.); reyes-ferreiro@hotmail.com (R.F.); pablo.reguera@salud.madrid.org (P.R.); jorge.villalon@salud.madrid.org (J.V.L.); alfredo.carrato@salud.madrid.org (A.C.)
2. Biomedical Research Network in Cancer (CIBERONC), C/Monforte de Lemos 3-5. Pabellón 11, 28029 Madrid, Spain; nmalats@cnio.es
3. Translational Genomics Core Facility, Ramón y Cajal Health Research Institute (IRYCIS), 28029 Madrid, Spain; mariagloria.munoz@salud.madrid.org (G.M.); dgarcia@ebi.ac.uk (D.G.-S.)
4. Department of Biochemistry, Universidad Autónoma de Madrid (UAM), Ramón y Cajal Health Research Institute (IRYCIS), 28029 Madrid, Spain; bsainz@iib.uam.es
5. Instituto de Investigaciones Biomédicas "Alberto Sols" (IIBM), CSIC-UAM, C/Arzobispo Morcillo, 4, 28029 Madrid, Spain
6. Cancer Stem Cell and Fibroinflammatory Group, Chronic Diseases and Cancer, Area 3-IRYCIS, 28029 Madrid, Spain
7. Genetic and Molecular Epidemiology Group, Spanish National Cancer Research Centre (CNIO), 28029 Madrid, Spain
8. Department of Medicine and Medical Specialties, Medicine Faculty, Alcala University, Plaza de San Diego, s/n, 28801 Alcalá de Henares, Spain
* Correspondence: julie.earl@live.co.uk; Tel.: +34-91-334-1307 (ext. 7877)
† These authors contributed equally to the manuscript.

The authors wish to make the following corrections to this paper [1]: In the published version, Figure 4 appeared as a duplication of Figure 1b. Furthermore, the legend of Figure 2 has been corrected to accurately reflect the data shown.

The correct version of Figure 2 is as follows:

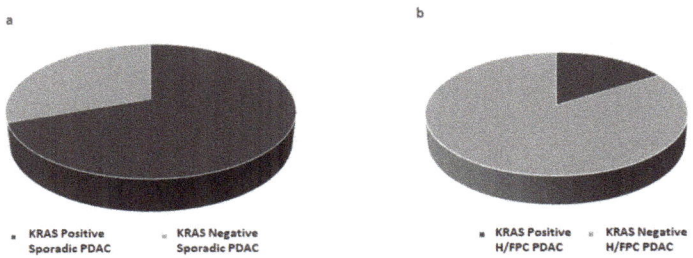

Figure 2. *KRAS* mutation status was determined in plasma from (**a**) sporadic PDAC cases (**b**) hereditary or familial PDAC (H/FPC) cases via BEAMing and mutant *KRAS* was more frequently detected in sporadic cases compared to H/FPC cases. BEAMing was performed using cfDNA isolated from 1 mL of plasma from 54 PDAC cases (31 familial cases and 23 sporadic cases). The frequency of mutant *KRAS* was 70% in sporadic cases and 16% in familial cases, which was statistically significant ($p \leq 0.001$).

The correct version of Figure 4 is as follows:

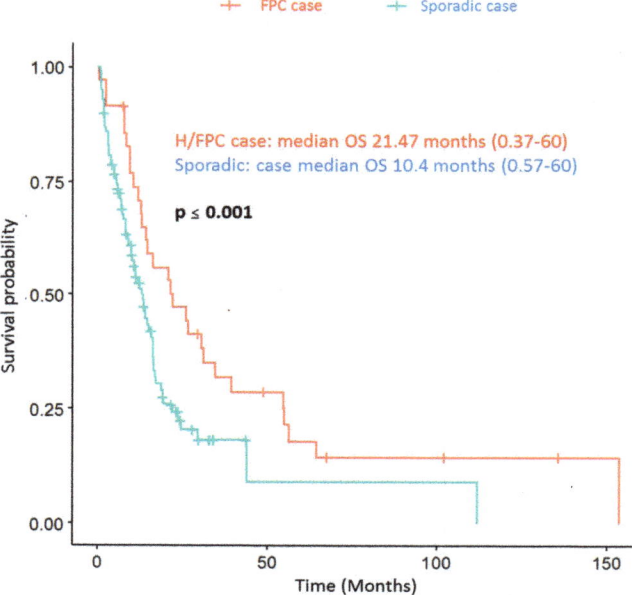

Figure 4. Hereditary or familial PDAC cases have a longer overall survival (OS) compared to sporadic cases.

We stress that these errors were purely due to human error and oversight; the corrections made do not affect or change the written portion of the figure legend, the interpretation of the results, or the final conclusions of this manuscript. The manuscript will be updated. The authors would like to apologize for any inconvenience caused. All changes have been reviewed and verified by the Academic Editors.

Author Contributions: Conceptualization, J.E. and A.C.; Data curation, J.E., E.B., M.E.C., R.F. (Raquel Fuentes), M.R.-G., R.F. (Reyes Ferreiro), P.R., G.M., D.G.-S., J.V.L., B.S.J., N.M. and A.C.; Funding acquisition, A.C.; Methodology, J.E., E.B., M.E.C., G.M. and D.G.-S.; Writing—original draft, J.E., E.B., M.E.C. and A.C.; Writing—review & editing, J.E., R.F. (Raquel Fuentes), M.R.-G., R.F. (Reyes Ferreiro), P.R., G.M., D.G.-S., J.V.L., B.S.J., N.M. and A.C. All authors have read and agreed to the published version of the manuscript.

Funding: This study was funded by the Instituto de Salud Carlos III (Plan Estatal de I+D+i 2013-2016): ISCIII (PI09/02221, PI12/01635, PI15/02101 and PI18/0135) and co-financed by the European Development Regional Fund "A way to achieve Europe" (ERDF), the Biomedical Research Network in Cancer: CIBERONC (CB16/12/00446), Red Temática de investigación cooperativa en cáncer: RTICC (RD12/0036/0073), La Asociación Española contra el Cáncer: AECC (Grupos Coordinados Estables 2016), Fundación Mutua Madrileña (FMM)/XVI Convocatoria de Ayudas a la Investigación en Salud and Asociación Cáncer de Páncreas (ACanPan); Asociación Española de Pancreatología (AESPANC)/IV Becas de Investigación Carmen Delgado/Miguel Pérez-Mateo.

Conflicts of Interest: The authors declare no conflict of interest.

Reference

1. Earl, J.; Barreto, E.; Castillo, M.E.; Fuentes, R.; Rodríguez-Garrote, M.; Ferreiro, R.; Reguera, P.; Muñoz, G.; Garcia-Seisdedos, D.; López, J.V.; et al. Somatic Mutation Profiling in the Liquid Biopsy and Clinical Analysis of Hereditary and Familial Pancreatic Cancer Cases Reveals KRAS Negativity and a Longer Overall Survival. *Cancers* **2021**, *13*, 1612. [CrossRef]

Article

Serum HPV16 E7 Oncoprotein Is a Recurrence Marker of Oropharyngeal Squamous Cell Carcinomas

Lucia Oton-Gonzalez [1], John Charles Rotondo [1], Carmen Lanzillotti [1], Elisa Mazzoni [1], Ilaria Bononi [2], Maria Rosa Iaquinta [1], Luca Cerritelli [3], Nicola Malagutti [3], Andrea Ciorba [3], Chiara Bianchini [3], Stefano Pelucchi [3], Mauro Tognon [1,*] and Fernanda Martini [1,*]

[1] Department of Medical Sciences, University of Ferrara, 44121 Ferrara, Italy; tnglcu@unife.it (L.O.-G.); rtnjnc@unife.it (J.C.R.); carmen.lanzillotti@unife.it (C.L.); elisa.mazzoni@unife.it (E.M.); mariarosa.iaquinta@unife.it (M.R.I.)

[2] Department of Translational Medicine and for Romagna, University of Ferrara, 44121 Ferrara, Italy; ilaria.bononi@unife.it

[3] ENT Unit, Department of Neuroscience and Rehabilitation, University Hospital of Ferrara, 44121 Ferrara, Italy; luca.cerritelli@unife.it (L.C.); nicola.malagutti@unife.it (N.M.); andrea.ciorba@unife.it (A.C.); chiara.bianchini@unife.it (C.B.); stefano.pelucchi@unife.it (S.P.)

* Correspondence: tgm@unife.it (M.T.); mrf@unife.it (F.M.); Tel.: +39-0532-455538 (M.T.); +39-0532-455540 (F.M.)

Simple Summary: Classical markers alone, such as HPV DNA, p16 and HPV mRNA expression, are not enough to stratify HPV-positive head and neck squamous cell carcinoma (HNSCC) patients, but when combined with serological markers, the latter are strong indicators of prognosis in oropharyngeal squamous cell carcinoma (OPSCC) patients. Specifically, HPV16 E7 oncoprotein in serum at the time of diagnosis, correlates with disease recurrence and patient overall survival. To our knowledge, this is the first study to investigate HPV E7 oncoprotein in patient serum. The E7 oncoprotein detection in serum at the time of diagnosis may be useful as a non-invasive procedure for HPV-positive OPSCC patient stratification and follow-up, helping to identify patients at risk for tumor recurrence and metastasis during follow-up, and ultimately, providing a tool for clinicians to identify patients for de-escalation treatment or those to be kept under close surveillance.

Abstract: Despite improved prognosis for many HPV-positive head and neck squamous cell carcinomas (HNSCCs), some cases are still marked by recurrence and metastasis. Our study aimed to identify novel biomarkers for patient stratification. Classical HPV markers: HPV-DNA, p16 and HPV mRNA expression were studied in HNSCC ($n = 67$) and controls ($n = 58$) by qPCR. Subsequently, ELISA tests were used for HPV16 L1 antibody and HPV16 E7 oncoprotein detection in serum at diagnosis and follow-up. All markers were correlated to relapse-free survival (RFS) and overall survival (OS). HPV-DNA was found in HNSCCs (29.85%), HPV16-DNA in 95% of cases, HPV16 E7 mRNA was revealed in 93.75%. p16 was overexpressed in 75% of HPV-positive HNSCC compared to negative samples and controls ($p < 0.001$). Classical markers correlated with improved OS ($p < 0.05$). Serological studies showed similar proportions of HPV16 L1 antibodies in all HNSCCs ($p > 0.05$). Serum E7 oncoprotein was present in 30% HPV-positive patients at diagnosis ($p > 0.05$) and correlated to HNSCC HPV16 E7 mRNA ($p < 0.01$), whereas it was associated to worse RFS and OS, especially for oropharyngeal squamous cell carcinoma (OPSCC) ($p < 0.01$). Detection of circulating HPV16 E7 oncoprotein at diagnosis may be useful for stratifying and monitoring HPV-positive HNSCC patients for worse prognosis, providing clinicians a tool for selecting patients for treatment de-escalation.

Keywords: human papillomavirus; oropharyngeal squamous cell carcinoma; treatment de-escalation; patient stratification; E7 oncoprotein; HPV DNA; HPV antibodies; ELISA

Citation: Oton-Gonzalez, L.; Rotondo, J.C.; Lanzillotti, C.; Mazzoni, E.; Bononi, I.; Iaquinta, M.R.; Cerritelli, L.; Malagutti, N.; Ciorba, A.; Bianchini, C.; et al. Serum HPV16 E7 Oncoprotein Is a Recurrence Marker of Oropharyngeal Squamous Cell Carcinomas. *Cancers* **2021**, *13*, 3370. https://doi.org/10.3390/cancers13133370

Academic Editor: Fabrizio Bianchi

Received: 18 May 2021
Accepted: 2 July 2021
Published: 5 July 2021

Publisher's Note: MDPI stays neutral with regard to jurisdictional claims in published maps and institutional affiliations.

Copyright: © 2021 by the authors. Licensee MDPI, Basel, Switzerland. This article is an open access article distributed under the terms and conditions of the Creative Commons Attribution (CC BY) license (https://creativecommons.org/licenses/by/4.0/).

1. Introduction

Human papillomavirus (HPV)-related head and neck squamous cell carcinomas (HNSCCs) are increasing worldwide [1]. Specifically, HPV-positive oropharyngeal squamous cell carcinomas (OPSCC) have increased over the past few years, with approximately 93,000 new OPSCCs diagnosed per year worldwide [2–4]. HPV-positive OPSCCs constitute a biologically distinct group of HNSCCs. Indeed, the American Joint Committee on Cancer, 8th edition, reports improved prognosis and treatment outcome for HPV-positive OPSCC patients compared to HPV-negative cases [5,6].

HPV plays an important role in OPSCC onset [7]. The main transforming activity of HPV relies on oncoproteins E6 and E7, which hamper p53 and pRb tumor suppressor protein activities, respectively. Moreover, HPV E6 and E7 oncoproteins are key players in tumor development, accounting for immune escape, angiogenesis, and the formation of a pro-proliferative microenvironment [8].

Optimizing protocols through targeted therapies and personalized treatments is paramount to increase survival rates for all patients [4]. In this context, several clinical trials, such as PATHOS (NCT02215265) or OPTIMA (NCT02258659), are now in progress [9,10] aiming to determine whether treatment de-intensification could improve quality of life for HPV-positive OPSCC patients whilst maintaining high rates of cure. Indeed, even if HPV-positive OPSCC patients usually respond to treatment de-escalation, 10–25% of HPV-positive patients present recurrences and worse prognosis [9,11–13].

Hence, correctly stratifying HPV-positive patients is necessary to select optimized treatment [14]. In an effort to improve stratification, many studies investigate HPV status, p16 overexpression which is the surrogate marker of HPV transformation [6,12], and HPV E6/E7 mRNA expression [15]. However, the current stratification system leads to several pitfalls, i.e., (i) HPV might be present as a transient infection, but not active in tumors [15]; (ii) p16 expression is not always observed in HPV-positive tumors [16]; (iii) HPV mRNA levels could be too low for detection [17,18].

Serological testing has gained interest in the past few years for HPV-positive OPSCC prognostic studies. The immune response of the host has been studied in association with both HPV-positive tumors and patient prognosis [4,19]. Serum IgG antibodies against HPV16 L1 capsid protein can be detected several years before OPSCC onset/presentation, but are also cumulative markers of viral exposure [20,21]. Antibodies to HPV16 E6 and E7 oncoproteins at the time of tumor diagnosis may be useful to predict disease-free survival in HPV-positive OPSCC patients [22,23]. However, routine testing for antibodies against HPV oncoproteins are difficult to perform due to the lack of available commercial kits.

Studies on cervical cancer have shown that detection of HPV E6 and E7 oncoproteins in cervical scrapings may constitute valuable markers for disease progression [24,25]. Moreover, the presence of HPV16 E6 and E7 oncoproteins has been demonstrated by direct ELISA in culture supernatant of HPV-positive cervical cancer cell lines SiHa and CaSki, indicating the release of viral oncoproteins from tumor cells [26].

Therefore, we have hypothesized that HPV E6 and E7 oncoproteins could be present in serum from HPV HNSCC patients, and their serum identification could be useful for prognostic purposes. The presence of HPV E6 and E7 oncoproteins in serum from HPV-associated cancer patients is yet to be investigated. Since new kits for testing serum HPV E7 protein are now commercially available new investigations can be carried out.

The aim of this study was to identify markers for HPV-positive HNSCC patient stratification. To this purpose, classical tumor markers, such as HPV DNA, p16 expression and HPV E7 mRNA were studied in different HNSCC subtypes, including OPSCC. Sera of HNSCC patients were analyzed for HPV16 L1 antibody titers and, for the first time, HPV16 E7 oncoprotein levels at the time of tumor diagnosis and during follow-up at 3, 6, 12 and 24 months. Finally, results were correlated to patient relapse-free survival (RFS) and overall survival (OS) at 24 months.

2. Materials and Methods

2.1. Patient Samples

HNSCC specimens (n = 67) from patients, mean age ± standard deviation [SD] 64.94 ± 10.88 years [y] old, were collected for analyses. Tumor-free tonsillar (TFT) samples (n = 58), from non-oncological patients who had undergone tonsillar surgery, 39 ± 15.17 y old, were used as controls. Samples were collected consecutively from 2016 to 2020 at the Ear, Nose and Throat (ENT) Clinic (University Hospital of Ferrara, Ferrara, Italy). Inclusion criteria was the histopathological detection of primary HNSCC in patients 18–95 y old, including hidden or occult SCC with lymph node cervical positive histology. Exclusion criteria were radiotherapy and/or chemotherapy treatment. The 8th edition of AJCC classification was used [5]. Tumor and TFT specimens were collected at the time of surgery. Blood specimens were also collected from HNSCC patients at the time of surgery and during patient follow-up at 3, 6, 12, and 24 months.

2.2. Nucleic Acid Extraction

DNA/RNA extractions from HNSCC (n = 67) and TFT tissues (n = 58) were carried out using the AllPrep DNA/RNA/Protein Extraction Kit (Qiagen, Milan, Italy). After quantification using the NanoDrop 2000 (Thermo Scientific, Milan, Italy), DNA and RNA samples were stored at −80 °C until analyses. DNA suitability for PCR analysis was assessed as before [27,28]. Total mRNA was retro-transcribed using the Improm II (Promega, Fitchburg, WI, USA) reverse transcription system [29].

2.3. HPV Analysis

HPV (GenBank: K02718.1) screening was performed by quantitative PCR (qPCR) using the GP5+/GP6+ primer pair (Table S1) [27]. DNA (50 ng) was analyzed in 10 µL reactions consisting of 2× SsoAdvanced Universal SYBR Green Supermix, Bio-Rad (Hercules, CA, USA) and 500 nM of each primer. QPCR analyses were performed in triplicate. Thermal cycling was: 95 °C for 5 min, 40 cycles of 95 °C for 15 s and 60 °C for 30 s. To discriminate between HPV genotypes, a final high-resolution melting (HRM) step was added from 65–95 °C, increasing 0.1 °C every 0.03 s. Recombinant plasmids containing DNA sequences from HPV types 6/11/16/18/31/33/45 were used as positive controls. An HPV-negative genomic DNA sample, and a mock sample, without DNA, were used as negative controls. HPV genotyping was done by comparing sample melting curves to the plasmid controls. Quantification of the viral DNA load was performed in comparison to the standard curve of a plasmid-HPV-type specific [29].

2.4. Gene Expression Analysis

QPCR was performed for p16 and HPV16 E7 gene expression analyses. Briefly, 50 ng of cDNA were used in 10 µL reactions using 2× SsoAdvanced Universal SYBR Green Supermix (Bio-Rad). A final concentration of 500 nM of each primer was employed (Table S1). Samples were run in triplicate, along with mock samples used as negative controls. Thermal conditions for HPV E7 and p16 were; 95 °C for 5 min and 40 cycles of 95 °C for 15 s followed by a 60 °C for 30 s. Detection of the housekeeping gene glyceraldehyde 3-phosphate dehydrogenase (GAPDH) was used for normalization of mRNA levels and the fold change was calculated using the $2^{-\Delta\Delta Ct}$ method, as done previously [30,31]. Furthermore, data was normalized against the TFT control group.

2.5. Detection of Serum HPV16 L1 Antibodies

Upon collection, blood samples were allowed to clot for 15 min at room temperature and then centrifuged at 1300 g for 15 min. Serum HPV16 L1 IgG antibodies were evaluated in HPV-positive (n = 20) and HPV-negative (n = 8) HNSCC patients at the time of diagnosis (T0) and during follow-up at 3, 6, 12 and 24 months.

HPV16 L1 IgG antibodies were analyzed with a commercial kit (HPV16 L1, Cusabio, Houston, TX, USA). The test was performed according to the manufacturer's instructions.

The signal intensity was measured as Optical Density (OD) at 450 nm (model Multiskan EX, Thermo Electron Corp., Waltham, MA, USA) [32]. The cutoff value was calculated according to the manufacturer's instructions; an OD sample/OD negative ratio, equal or greater than 2.1, was considered positive.

2.6. Serum E7 Oncoprotein Level Detection

Serum HPV16 E7 oncoprotein levels were evaluated in HPV-positive ($n = 20$) and HPV-negative ($n = 8$) HNSCC patients, using the "HPV16 E7 Oncoprotein ELISA Kit" (Cell Biolabs, San Diego, CA, USA), according to the manufacturer's instructions. The presence or absence of E7 oncoprotein was determined by considering sample absorbance above or below the cutoff value, respectively, calculated as done previously [33]. The cutoff for HPV16 E7 oncoprotein was 0.75 ng/mL. HPV16 E7 oncoprotein variation during the follow-up was assessed by the ratio between protein amount at time of relapse and at previous time point; ratios > 1 indicated increment of protein prior to relapse, while ratios < 1 indicated decrement.

2.7. Statistical Analysis

Statistical analyses were carried out using the GraphPad Prism for Windows (version 8.0, GraphPad, San Diego, CA, USA) [34]. The ANOVA test was used to compare the mean between groups for gene expression analyses. Pearson/Spearman correlation tests were used to correlate viral gene expression and HPV DNA load, and E7 oncogene and p16 expression, respectively, and to assess univariate differences of clinicopathological features according to E7 oncoprotein presence in serum. All parameters were correlated to patient's relapse-free survival (RFS) and overall survival (OS) at 24 months using the Kaplan-Meier model; statistical significance was estimated using the log-rank test. p values of less than 0.05 were considered statistically significant for all analyses.

3. Results

3.1. HPV DNA Analysis

HNSCCs and control samples were analyzed for HPV DNA sequences and genotype. HPV DNA was found in 20/67 (29.85%) of HNSCC samples, consisting of 2/20 (10%) oral squamous cell carcinoma (OSCC), 15/20 (75%) OPSCC, 2/20 (10%) hypopharyngeal cancer and 1/20 (5%) laryngeal cancer (Table 1). HPV-genotype was determined by high resolution melting (HRM) to be HPV16 in 19/20 (95%) of the HNSCC HPV-positive cases and HPV33 in 1/20 (5%) of the cases. Control DNAs were found to be HPV11-positive in 1/58 (1.7%) of the cases. Our further studies were hereafter focused on HPV type 16 due to high prevalence in HNSCC. Viral DNA load in cancer specimens ranged from 2.52×10^{-4} to 4.26×10^2 copies of HPV DNA per cell (Figure 1A).

Table 1. Clinicopathological features of HNSCC patients.

Clinicopathological Variables	HPV-Negative	HPV-Positive	p-Value	Tumor Site (HPV-Positive)					
				Oral	p-Value *	Oropharynx	p-Value	Hypopharynx	p-Value *	Larynx	p-Value *
Tumor Site											
Oral	25/47 (53.20%)	2/20 (10%)									
Oropharynx	13/47 (27.66%)	15/20 (75%)									
Hypopharynx	1/47 (2.13%)	2/20 (10%)									
Larynx	6/47 (12.77%)	1/20 (5%)									
Hidden [1]	2/47 (4.25%)	-									
Tumor Size			0.225		NA		0.031		NA		NA
T1	7/47 (14.89%)	3/20 (15%)		-		2/15 (13.33%)		1/2 (50%)		-	
T2	14/47 (29.79%)	5/20 (25%)		-		4/15 (26.67%)		1/2 (50%)		-	
T3	9/47 (19.15%)	3/20 (15%)		-		3/15 (20.00%)		-		-	
T4	17/47 (36.17%)	9/20 (45%)		2/2 (100%)		6/15 (40.00%)		-		1/1 (100%)	
Node Status			0.108		NA		0.096		NA		NA
N0	9/47 (19.15%)	3/20 (15%)		-		2/15 (13.33%)		-		1/1 (100%)	
N+	38/47 (80.85%)	17/20 (85%)		2/2 (100%)		13/15 (86.67%)		2/2 (100%)		-	
Clinical Stage			0.467		NA		0.336		NA		NA
I	1/47 (2.13%)	1/20 (5%)		-		1/15 (6.67%)		-		-	
II	7/47 (14.89%)	5/20 (25%)		-		5/15 (33.33%)		-		-	
III	7/47 (14.89%)	5/20 (25%)		-		4/15 (26.67%)		1/2 (50%)		-	
IVa	25/47 (53.19%)	9/20 (45%)		2/2 (100%)		5/15 (33.33%)		1/2 (50%)		1/1 (100%)	
IVb/c	7/47 (14.89%)	-		-		-		-		-	
Recurrence	16/47 (42.55%)	6/20 (30%)	0.0001	-	NA	4/15 (26.66%)	0.001	1/2 (50%)	NA	0/1 (0%)	NA
Persistance	4/47 (8.51%)	2/20 (10%)		1/2 (50%)		2/15 (13.33%)		-		-	
N/A	2/47 (4.25%)	2/20 (10%)		-		2/15 (13.33%)		-		-	
Tobacco consumption			0.481		NA		0.582		NA		NA
No	5/47 (10.64%)	2/20 (10%)		-		2/15 (13.33%)		-		-	
Ex Smoker	13/47 (27.66%)	7/20 (35%)		-		4/15 (26.67%)		2/2 (100%)		1/1 (100%)	
Smoker	25/47 (53.19%)	7/20 (35%)		2/2 (100%)		5/15 (33.33%)		-		-	
N/A	4/47 (8.51%)	4/20 (20%)		-		4/15 (26.67%)		-		-	

Table 1. Cont.

Clinicopathological Variables	HPV-Negative	HPV-Positive	p-Value	Tumor Site (HPV-Positive)							
				Oral	p-Value *	Oropharynx	p-Value	Hypopharynx	p-Value *	Larynx	p-Value *
Alcohol consumption			0.962		NA		0.725		NA		NA
No	10/47 (21.28%)	5/20 (25%)		-		4/15 (26.67%)		-		1/1 (100%)	
Ex consumer	3/47 (6.38%)	2/20 (10%)		1/2 (50%)		0/15 (0%)		1/2 (50%)		-	
Consumer	27/47 (57.45%)	9/20 (45%)		1/2 (50%)		7/15 (46.67%)		1/2 (50%)		-	
N/A	7/47 (14.89%)	4/20 (20%)		-		4/15 (26.67%)		-		-	
Age	64.04 ± 11.55	67.05 ± 9.05	0.709	63 ± 1.41	NA	65.73 ± 9.15	0.409	73 ± 1.41	NA	83 ± NA	NA
Gender			0.244		NA		0.289		NA		NA
Male	32/47 (68.09%)	17/20 (85%)		2/2 (100%)		12/15 (80%)		2/2 (100%)		1/1 (100%)	
Female	15/47 (31.91%)	3/20 (15%)		-		3/15 (20%)		-		-	

Clinicopathological variables in HNSCC patients both HPV-negative and HPV-positive. p-values are referred to correlation between E7 oncoprotein expression in serum and the different variables in HPV-positive HNSCC patients. [1] Hidden or occult tumors refer to SCC with lymph node cervical positive histology. p values < 0.05 were considered statistically significant. * Too few pairs were available for correlation analysis.

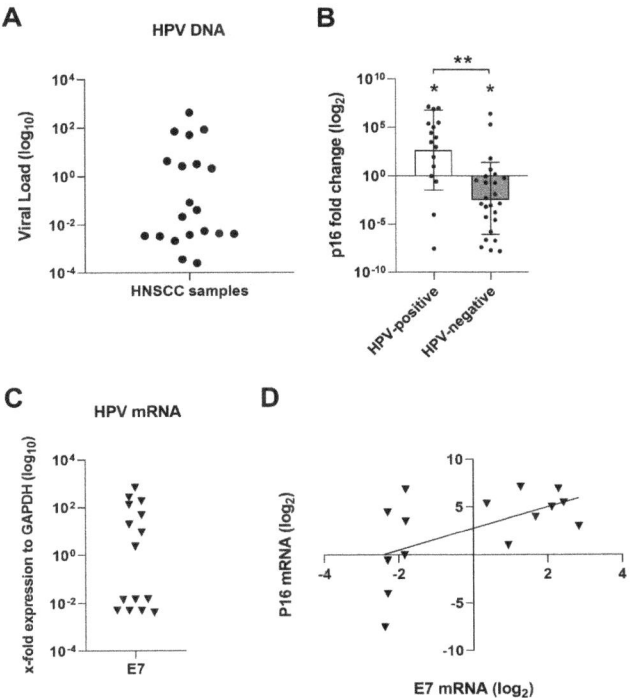

Figure 1. Analysis of classical markers for stratification of HNSCC samples. Statistical significance was indicated as * for $p < 0.05$ and ** for $p < 0.0001$; (**A**) viral load quantification of HPV-positive HNSCC samples by qPCR; (**B**) differential p16 mRNA expression in HNSCC samples analyzed by qPCR; (**C**) Viral E7 mRNA expression in HPV-positive HNSCC samples; (**D**) Spearman correlation analyses between the expression of E7 ($\log_2$) oncogene and p16 ($\log_2$) showed correlation (r = 0.59; $p < 0.05$) in HPV-positive HNSCC tumor samples.

3.2. p16 Gene Expression

Forty-one out sixty-seven HNSCC matching DNA/RNA samples were available for further analyses. P16 mRNA expression was investigated in HNSCC samples, showing upregulation in 12/16 (75%) of HPV16-positive HNSCC and in 5/25 (20%) of HPV-negative HNSCC samples compared to controls ($p < 0.001$). HPV-positive patients showed overall p16 gene upregulation compared to controls (Mean ± [SD], 2.60 ± 3.98 $\log_2$ fold, $p < 0.05$) with the exception of two samples that harboured p16 downregulated. HPV-negative samples were downregulated compared to control samples (Mean ± [SD], −2.34 ± 3.71 $\log_2$ fold, $p < 0.05$). Differences in p16 expression between HPV-positive and -negative were also significant ($p < 0.0001$) (Figure 1B).

3.3. HPV mRNA Expression

HPV-positive HNSCC samples were analyzed for HPV16 E7 gene expression by qPCR. Specifically, HPV16 E7 gene expression was analyzed in 16 HPV-positive HNSCC samples. mRNA E7 expression was detected in 15/16 (93.75%) (Figure 1C). Pearson correlation test showed no correlation between the expression levels ($\log_{10}$) of E7 and HPV DNA load (r = 0.42, $p > 0.05$). Furthermore, Spearman correlation analyses showed correlation between E7 expression and p16 up-regulation (r = 0.59; $p < 0.05$) (Figure 1D). But, HPV E7 mRNA expression did not correlate to p16 upregulation for all samples, since two samples (one OSCC and one OPSCC) presented E7 expression with p16 downregulation, and one sample presented p16 upregulation but no E7 expression; therefore, p16 is not always a good marker of HPV infection.

3.4. Serological Studies

3.4.1. HPV16 L1 Antibody Titer

Serum from all HPV-positive (n = 20) HNSCC patients (Table 1) and from HPV-negative (n = 8) HNSCC patients, consisting of 4/8 (50%) OPSCCs and 4/8 (50%) OSCCs, were tested for HPV16 L1 IgG antibodies. HPV16 L1 antibodies were found with a similar proportion in 18/20 (90%) HPV–positive HNSCC and 7/8 (87.5%) HPV-negative HNSCC patients at T0 ($p > 0.05$). HPV DNA-positive HNSCC patients presented higher Optical Density (OD) readings for antibodies anti-HPV16 L1 compared to HPV-negative (Mean ± [SD], 4.001 ± 2.11 vs. 2.29 ± 0.32; $p < 0.05$) (Figure 2A). Antibody response was further compared during follow-up at 3, 6, 12 and 24 months. Results indicated that HPV16 L1 antibody titers did not vary significantly during follow-up ($p > 0.05$) (Figure 2B).

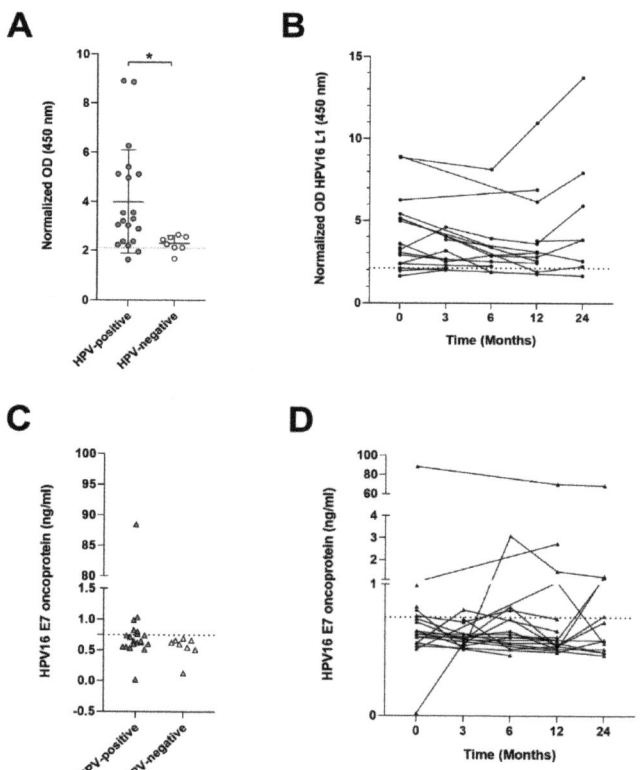

Figure 2. ELISA tests on HNSCC serum samples. Statistical significance was indicated as * for $p < 0.05$; (**A**) Serum antibody levels against HPV16 L1 in HNSCC patients. Differential OD between HPV-positive and HPV-negative patients ($p < 0.05$); (**B**) HPV16 L1 antibody variation during HPV-positive patient follow-up; (**C**) HPV16 E7 oncoprotein quantification in serum shows no difference between HPV-positive and HPV-negative patients ($p > 0.05$); (**D**) HPV16 E7 oncoprotein variation during HPV-positive patient follow-up.

3.4.2. HPV16 E7 Oncoproteins in Sera

HPV16 E7 oncoprotein (ng/mL) amounts were evaluated at the time of diagnosis and during follow-up at 3, 6, 12 and 24 months. At T0, HPV16 E7 oncoprotein was detected in 6/20 (30%) HPV-positive patient serum and no HPV-negative cases ($p > 0.05$) (Figure 2C). Variation in the amount of E7 oncoprotein during follow-up was studied. Nine out of 20 (45%) patients showed an increment in the amount of oncoprotein during follow-up; 4/9 (44.44%) patients were positive at the time of diagnosis, while 5/9 (55.55%) became positive during follow-up. Two patients out of 20 (10%) positive at the time of diagnosis, presented HPV E7 decrement over-time, and one became negative. HPV16 E7 variation in samples during follow-up resulted statistically insignificant ($p > 0.05$) (Figure 2D). Nine out 20 (45%) patients were E7 negative at T0 and remained negative during follow-up.

Finally, HPV16 E7 oncoprotein levels in serum were studied in correlation to the viral mRNA expression in the tumor samples. Results showed correlation between the amount of HPV16 E7 mRNA expressed in the tumors and E7 oncoprotein in serum ($r = 0.79$, $p < 0.01$), suggesting that circulating E7 protein may be due to release from the tumor site.

3.5. Survival Analysis

3.5.1. RFS and OS in Correlation to HPV DNA, p16 Expression and HPV mRNA

The median follow-up time of this study was 24 months. Relapse free survival (RFS) and overall survival (OS) were assessed in HPV-positive HNSCC patients compared to HPV-negative cases. Different RFS rates were observed; 72.11% and 48.77% for HPV-positive and -negative, respectively ($p > 0.05$) (Figure 3A). Furthermore, OS was improved for HPV-positive patients; 88.89% compared to 52.08% in HPV-negative OPSCCs ($p < 0.01$) (Figure 3B).

To study the effect of p16 expression on survival rate, all HNSCC samples were subdivided into p16-over or –underexpression in the tumor sample. Log_2 fold change (FC) value (with fixed interval) was used as the cut-off criteria. High and low expression were considered when FC was greater than 1 ($n = 13$) or lower than -1 ($n = 17$), respectively [35]. RFS was 73.84%, in patients carrying p16 upregulation, compared to p16 downregulation, 48.12% ($p > 0.05$) (Figure 3C). OS was 100% in patients with higher p16 expression compared to 52.94% of patients with p16 downregulation ($p < 0.01$) (Figure 3D).

RFS and OS were also assessed for HPV E7 mRNA expression in HNSCCs samples. Samples were divided into expressing E7 oncogene ($n = 15$) and non ($n = 26$). Survival proportions indicated that RFS was 64.61% in patients positive for E7 mRNA, and 48.77% in patients HPV mRNA-negative ($p > 0.05$) (Figure 3E). OS was higher in patients carrying HPV E7 mRNA, 92.85%, compared to HPV mRNA-negative, 52.08% ($p < 0.05$) (Figure 3F).

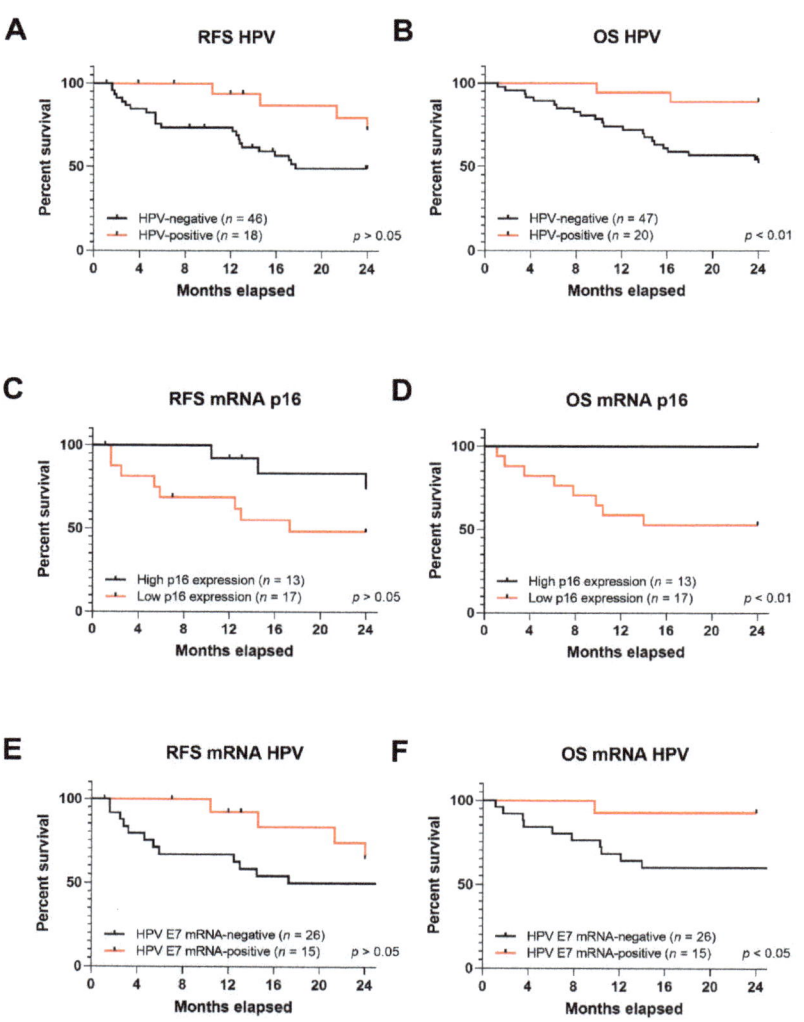

Figure 3. Kaplan-Meier (KM) curves for RFS and OS in HNSCC; KM curves for (**A**) RFS and (**B**) OS for HPV DNA presence in HNSCC tumor samples; KM curves for (**C**) RFS and (**D**) OS for p16 over- or under-expression in HNSCC samples; KM curves for (**E**) RFS and (**F**) OS for HPV E7 mRNA expression in HNSCC tumor samples. Statistical significance was indicated as $p < 0.01$ or $p < 0.05$.

3.5.2. RFS and OS in Relation to Serum HPV16 L1 Antibodies

The next step was to study the association between HPV infection serological markers, such as HPV16 L1 antibody, with patient's survival. No significant differences were observed for HPV16 L1 antibodies in RFS or OS for HPV-positive patients ($n = 20$) at the time of diagnosis and during follow up. RFS was 51.28% and 100% for HPV16 L1 antibody-positive ($n = 18$), for HPV16 L1 antibody-negative ($n = 2$) patients, respectively ($p > 0.05$) (Figure 4A). OS was also similar between HPV-positive patients and HPV16 L1 antibody positivity or negativity, at 63.64% and 100%, respectively ($p > 0.05$) (Figure 4B). Overall these results indicate that HPV16 L1 is a poor indicator of prognosis and since it is a cumulative marker of exposure, it may be used solely for epidemiological purposes.

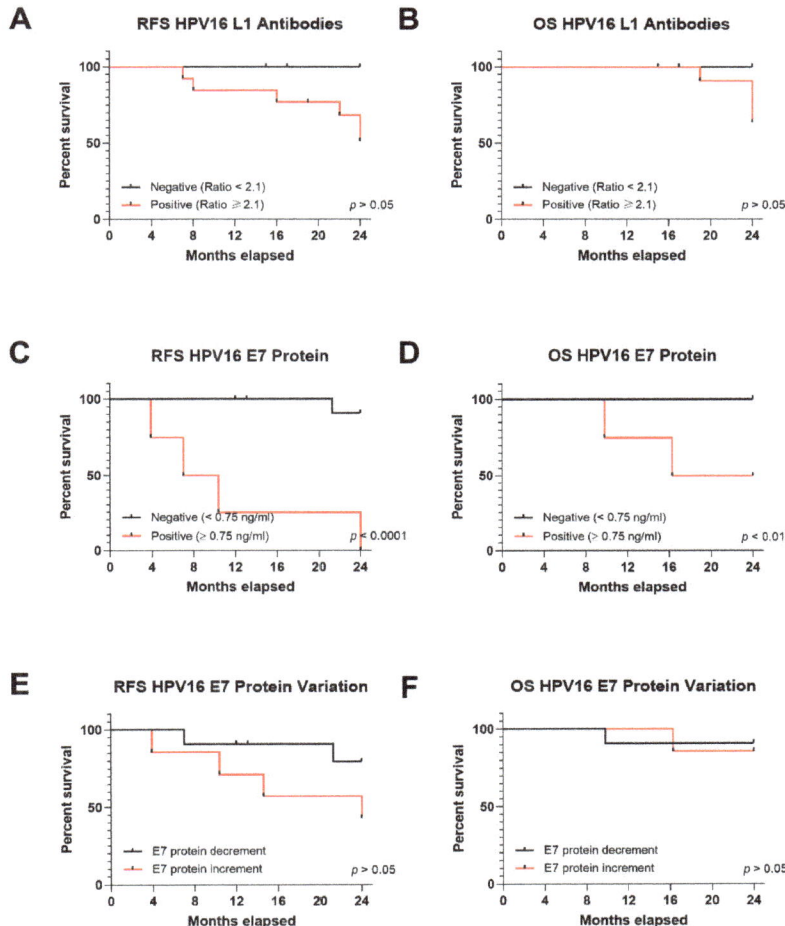

Figure 4. Kaplan-Meier (KM) curves for serological tests representing RFS and OS in HNSCC patients for HPV16 L1 and OPSCC for E7 oncoprotein; KM of (**A**) RFS and (**B**) OS for HPV16 L1 in HNSCC patients; KM of (**C**) RFS and (**D**) OS for HPV E7 oncoprotein in serum from HPV-positive OPSCC patients; KM of (**E**) RFS and (**F**) OS for increment or decrement of E7 oncoprotein in serum from OPSCC patients during follow-up. Statistical significance was indicated as $p < 0.05$.

3.5.3. RFS and OS in Relation to Serum HPV16 E7 Oncoprotein

HPV16 E7 oncoprotein in serum was correlated to patients' clinicopathological features (Table 1). Interestingly, E7 oncoprotein in serum was strongly associated to recurrence in HNSCC patients ($p < 0.0001$) and in the OPSCC subgroup ($p < 0.001$). Statistical analyses on other HNSCC subtypes were not possible due to the small sample size (Table 1). RFS was 0% for HNSCC with E7 positivity compared to 90.9% for patients testing negative for E7 protein ($p < 0.0001$) (Figure 4C). OS was 100% and 50% in patients negative and positive for E7 oncoprotein, respectively ($p < 0.01$) (Figure 4D).

The variation in serum E7 oncoprotein was also studied in correlation to patient survival. RFS was 42.85% in HNSCC patients who increased E7 oncoprotein amounts during follow-up, compared to 79.55% in those who experienced E7 decrease ($p > 0.05$) (Figure 4E). OS proportion was 85.71% for patients showing increased E7 oncoprotein, and

90.9% for those showing a decreased E7 oncoprotein ($p > 0.05$) (Figure 4F). These results highlight the importance of patient monitoring for recurrence after circulating HPV E7 oncoprotein being found at the time of diagnosis or increasing levels during follow-up.

3.6. TNM Stage in Correlation to OPSCC Patient Prognosis and E7 Oncoprotein in Serum

RFS was 72.9% and 57.14% for patients with T (1–2) and T (3–4), respectively ($p > 0.05$) (Figure 5A), while OS was 87.5% and 90%, respectively ($p > 0.05$) (Figure 5B). Similarly, no statistically significant differences were observed for RFS or OS survival rates when studied in correlation to lymph node involvement; RFS was 100% vs. 58.18% for patients without and with lymph node involvement ($p > 0.05$) (Figure 5C), while OS was similar; 100% and 86.67%, respectively ($p > 0.05$) (Figure 5D).

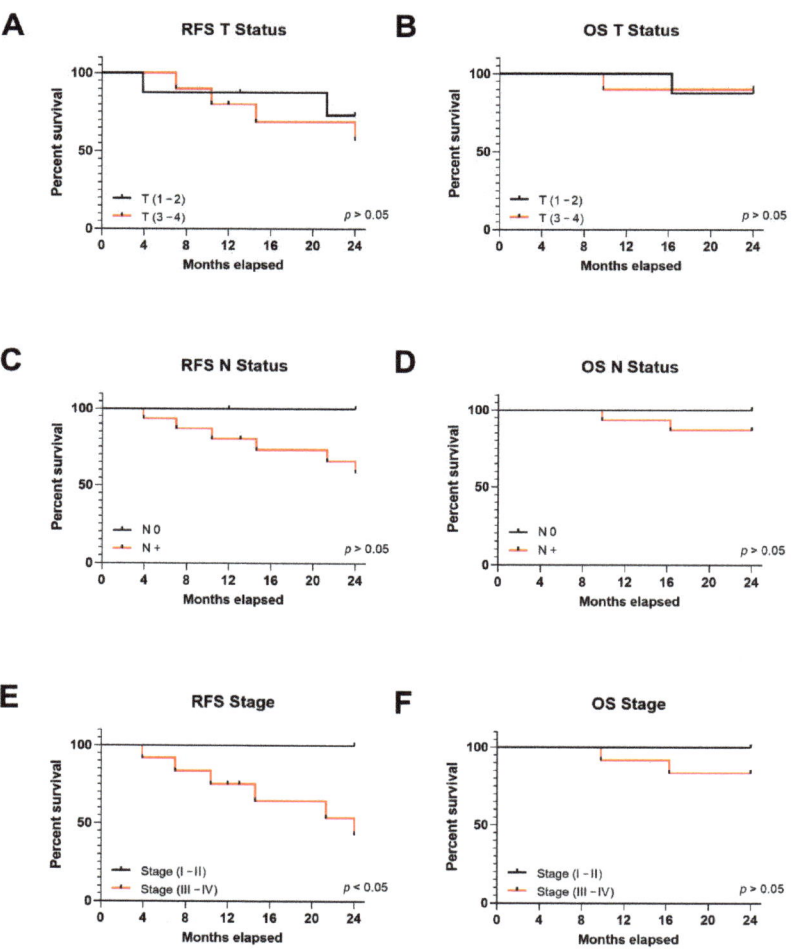

Figure 5. Kaplan-Meier (KM) curves for tumor size (T), node status (N) and stage in HPV-positive HNSCC patients representing RFS and OS; KM representing (**A**) RFS and (**B**) OS for patients divided into tumor size: T (1–2) and T (3–4); KM representing (**C**) RFS and (**D**) OS for patients divided into node status: N0 and N+; KM representing (**E**) RFS and (**F**) OS for patients divided into stages I/II and III/IV; OS for patients divided into stages I/II and III/IV. Statistical significance was indicated as $p < 0.05$.

Patients in stages III/IV are more likely to recur. Indeed, RFS for patients in stage III/IV was 42.86% compared to 100% for patients in stage I/II ($p < 0.05$) (Figure 5E), while OS was similar for both groups 83.33% and 100%, respectively ($p > 0.05$) (Figure 5F).

Lastly, we studied the correlation between serum E7 presence and tumor size, lymph node involvement and disease stage. E7 in serum correlated to tumor size ($p < 0.05$), but not to lymph node involvement in OPSCC ($p > 0.05$) (Table 1). Out of 6 HPV-positive patients with stage I/II, none presented E7 oncoprotein in serum at T0, while 5/12 (41.66%) of the patients in stage III/IV presented E7 oncoprotein in serum ($p > 0.05$).

4. Discussion

The current study aims to find markers for recurrence in HPV-positive patients. For patient stratification, we studied classical HPV markers, such as HPV DNA, p16 mRNA and viral mRNA expression. Once stratified, we studied the presence of potential serological markers, i.e., HPV16 L1 antibodies and, for the first time, the HPV E7 oncoprotein. Serological markers were then correlated to patient prognosis.

In an initial screening, we found that 29.85% of HNSCC tumor samples, including 75% OPSCC, harbored HPV-DNA, and 95% tested HPV16-positive. These findings are in accordance to other studies indicating that HPV is found in 25% of all HNSCCs, and in up to 70% of OPSCC tumors [36–39], whereas 90% of all HPV-positive tumors carried HPV type 16 [40]. P16 mRNA expression was found to be overexpressed in 75% of HPV-positive HNSCC samples. P16 is an established surrogate marker for tumors with transcriptionally active HPV, which is known to be associated with tumors that respond better to therapy and have improved outcomes [41,42]. Yet, not all HPV-positive tumors show p16 gene upregulation, as shown in herein and in previous studies [15]. Transcription of E7 viral oncogene was assessed in HPV-DNA positive patients only, showing 93.75% of HPV16 DNA positive samples expressing the HPV16 E7 oncogene, in agreement with previous studies [30].

Since HPV status has a great impact on patient prognosis for different HNSCCs, such as OPSCC and OSCC [43,44], it is important to stratify patients correctly. In our study, we found that 30% of HPV-positive patients presented recurrence within the first two years of diagnosis, similarly to other studies [9,11–13]. Classical HPV infection markers, i.e., HPV DNA, p16 and HPV mRNA showed improved patient OS in our cohort of study, but none of them correlated with recurrence.

Antibody response against L1 was studied for the prevalence of viral infection in HNSCC patients. HPV L1 antibodies are cumulative markers of past and present infection, although their presence does not imply HPV-driven tumorigenesis [45]. Indeed, in our study both HPV-positive and -negative HNSCC patients had antibodies against HPV16 L1 with a similar level of prevalence at 90% and 87.5%, respectively. Interestingly, the antibody titer in HPV-positive patients was higher compared to HPV-negative cases, which could be indicative of active infection.

We also studied the antibody response during the follow-up to monitor disease status, as proposed by Routman et al. [46], but no significant antibody titers change was observed during follow up, indicating that antibody levels against HPV L1 may not be useful in diagnosing or monitoring the disease.

Recently, the study of antibody response against HPV E6/E7 oncoproteins in OPSCC, the major subtype of HNSCC, has been proposed as a marker for disease progression. In spite of good perspectives for both diagnosis and prognosis [19,20,22,23,47], results are still under debate due to the lack of seroconversion in many patients, as was also underlined before for other diseases which are related to viral infections [48]. In a study conducted by Kreimer et al., 57.6% of OPSCC patients remained HPV E6 seronegative during follow-up [47].

To our knowledge, no previous research has been conducted to detect HPV E7 oncoprotein in HNSCC patient serum, while the availability of ELISA kits for oncoprotein detection could rapidly facilitate such study outcomes into clinical use. E7 oncoprotein in serum was specifically found with a frequency of 30% in HPV-positive samples, while the detection of circulating protein at the time of diagnosis strongly correlated to recurrence. Our data are in accordance to other studies showing higher levels of antibodies against HPV oncoproteins at the time of diagnosis in association to a significantly increased risk of recurrence [22,49]. Similar to previous studies on antibody titer variation during follow-up, our investigation found no variation, increment or decrement of E7 oncoprotein in OPSCC serum could be association with patient outcome during the two-year follow-up [20,21,50]. Nevertheless, an increase or decrease in serum E7 during follow-up was observed in patients whether experiencing recurrence or not, respectively, thus, lack of significant correlation between serum E7 level and relapse may be due to limited sample sizes.

It is to be noted that, circulating E7 protein showed correlation with high E7 mRNA expression in the tumor, suggesting that tumor sites may provide the circulating oncoprotein. Circulating E7 protein may be considered a tumorigenesis marker, representing at serological level what occurs at the tumor site. Sources of viral oncoproteins in serum have currently not been established, but some hypotheses could be proposed. Firstly, the transcriptionally active circulating tumor cells may account for the presence of serum viral oncoproteins [13]. Indeed, HPV spreading through blood cells has been previously reported [51], while HPV E6/E7 transcription in circulating tumor cells (CTCs) has been correlated to patient prognosis [13]. Secondly, invasion and the associated development of a tumor vascular bed may result in the release of E6/E7 proteins from the tumor mass, probably as a consequence of necrosis [52]. Thirdly, HPV-positive tumor cells may secrete exosomes containing viral oncoproteins, as has been reported for other DNA viruses [53]. Whatever the mechanism, HPV16 E7 oncoproteins were successfully found in HNSCC patient serum and correlated to patient prognosis. E7 oncoprotein detection in serum at the time of diagnosis displayed strong diagnostic and prognostic reliability in predicting relapses and overall survival in HPV-positive HNSCC patients, especially HPV-positive OPSCC patients. Moreover, since HPV mRNA may be present in HPV-DNA negative samples [54], the analysis of HPV mRNA should be taken into consideration in all HNSCCs to avoid HPV-driven tumors misclassification.

Moreover, E7 oncoprotein also correlated to tumor size, but not lymph node involvement or disease stage. Overall, 41.66% HNSCC patients with high disease stage III/IV presented E7 oncoprotein in serum, while none of those with low stage I/II did so, in agreement with previous serologic studies [20,55], making the detection of E7 oncoprotein in serum an excellent discriminator for HNSCC patients that may relapse, especially for OPSCC patients.

This study demonstrates for the first time, the presence of circulating E7 oncoproteins in serum from HNSCC patients using a direct ELISA assay. Our results indicate that the presence of E7 oncoprotein in OPSCC patient serum at the time of diagnosis is indicative of a higher risk of recurrence. Liquid biopsy for the detection of prognostic markers in HPV-positive OPSCC patients provides valuable information on disease progression and may help stratify and monitor patients over-time; this, can result extremely useful for patients presenting persistent or occult tumors. For future studies, in order to increase the statistical power of the study, a larger sample size for all HNSCC subtypes will be considered. This study takes medicine one step closer to correct patient stratification for therapy de-intensification. The combination of classical markers with serological markers, may be used to plan personalized treatment strategies for HPV-positive patients.

5. Conclusions

Detection of circulating HPV E7 oncoprotein at the time of diagnosis, may be used as non-invasive procedure for patient stratification and follow-up, ultimately providing a tool for clinicians to determine which patients would be good candidates for treatment de-escalation or should be kept under close surveillance.

Supplementary Materials: The following are available online at https://www.mdpi.com/article/10.3390/cancers13133370/s1, Table S1 Validated primer sets used in qPCR to detect and quantify HPV DNA and both, viral and cellular genes.

Author Contributions: Conceptualization: F.M. and S.P.; methodology, L.O.-G. and J.C.R.; software, E.M.; validation, M.T., S.P., F.M.; formal analysis, L.O.-G., J.C.R.; investigation, L.O.-G., J.C.R., I.B., M.R.I.; resources, L.C., N.M., A.C., C.B.; supervision S.P.; data curation L.O.-G., J.C.R., C.L., I.B.; writing—original draft preparation, L.O.-G.; writing—review and editing, J.C.R., C.L., M.T., S.P., F.M.; visualization, L.O.-G.; supervision, M.T., S.P., F.M.; project administration, M.T., S.P., F.M.; funding acquisition, J.C.R., M.T., F.M. All authors have read and agreed to the published version of the manuscript.

Funding: This work was partially supported by grant IG 21956 (awarded to John Charles Rotondo) and by grant IG 21617 (awarded to Mauro Tognon) from the Associazione Italiana per la Ricerca sul Cancro (AIRC), Milan, Italy.

Institutional Review Board Statement: The study was conducted according to the guidelines of the Declaration of Helsinki, and approved by the Ethics Committee of Ferrara. (Authorization n. 160986, 12 December 2016).

Informed Consent Statement: Informed written consent was obtained from all subjects involved in the study.

Data Availability Statement: The data presented in this study are available on request from the corresponding author.

Acknowledgments: The pGEM1_HPV45/pUC19_HPV52 plasmids, which are commercially available, were a kind gift from Massimo Tommasino, Infectious and Cancer Biology group, International Agency for Research on Cancer (IARC), Lyon, France.

Conflicts of Interest: The authors declare no conflict of interest. The funders had no role in the design of the study; in the collection, analyses, or interpretation of data; in the writing of the manuscript, or in the decision to publish the results.

References

1. Johnson, D.E.; Burtness, B.; Leemans, C.R.; Lui, V.W.Y.; Bauman, J.E.; Grandis, J.R. Head and neck squamous cell carcinoma. *Nat. Rev. Dis. Primers* **2020**, *6*, 1–22. [CrossRef] [PubMed]
2. Bruni, L.; Albero, G.; Serrano, B.; Mena, M.; Gómez, D.; Muñoz, J.; Bosch, F.X.; de Sanjose, S. *Human Papillomavirus and Related Diseases in Europe. Summary Report 17 June 2019*; ICO/IARC Information Centre on HPV and Cancer (HPV Information Centre): Barcelona, Spain, 2019.
3. Ferlay, J.; Colombet, M.; Soerjomataram, I.; Mathers, C.; Parkin, D.M.; Piñeros, M.; Znaor, A.; Bray, F. Estimating the global cancer incidence and mortality in 2018: GLOBOCAN sources and methods. *Int. J. Cancer* **2019**, *144*, 1941–1953. [CrossRef]
4. Wittekindt, C.; Wagner, S.; Bushnak, A.; Prigge, E.-S.; von Knebel Doeberitz, M.; Würdemann, N.; Bernhardt, K.; Pons-Kühnemann, J.; Maulbecker-Armstrong, C.; Klussmann, J.P. Increasing incidence rates of oropharyngeal squamous cell carcinoma in germany and significance of disease burden attributed to human papillomavirus. *Cancer Prev. Res.* **2019**, *12*, 375–382. [CrossRef] [PubMed]
5. Amin, M.B.; Edge, S.; Greene, F.; Byrd, D.R.; Brookland, R.K.; Washington, M.K.; Gershenwald, J.E.; Compton, C.C.; Hess, K.R.; Sullivan, D.C.; et al. (Eds.) *AJCC Cancer Staging Manual*; Springer: Chicago, IL, USA, 2017; ISBN 978-3-319-40617-6.
6. Lydiatt, W.M.; Patel, S.G.; O'Sullivan, B.; Brandwein, M.S.; Ridge, J.A.; Migliacci, J.C.; Loomis, A.M.; Shah, J.P. Head and neck cancers-major changes in the American joint committee on cancer eighth edition cancer staging Manual. *CA Cancer J. Clin.* **2017**, *67*, 122–137. [CrossRef] [PubMed]
7. De Cicco, R.; Melo Menezes, R.; Nicolau, U.R.; Pinto, C.A.L.; Villa, L.L.; Kowalski, L.P. Impact of human papillomavirus status on survival and recurrence in a geographic region with a low prevalence of HPV-related cancer: A retrospective cohort study. *Head Neck* **2020**, *42*, 93–102. [CrossRef] [PubMed]
8. Yang, R.; Klimentová, J.; Göckel-Krzikalla, E.; Ly, R.; Gmelin, N.; Hotz-Wagenblatt, A.; Řehulková, H.; Stulík, J.; Rösl, F.; Niebler, M. Combined transcriptome and proteome analysis of immortalized human keratinocytes expressing human papillomavirus 16 (HPV16) oncogenes reveals novel key factors and networks in HPV-induced carcinogenesis. *Msphere* **2019**, *4*, e00129. [CrossRef] [PubMed]
9. Hargreaves, S.; Beasley, M.; Hurt, C.; Jones, T.M.; Evans, M. Deintensification of adjuvant treatment after transoral surgery in patients with human papillomavirus-positive oropharyngeal cancer: The conception of the PATHOS study and its development. *Front. Oncol.* **2019**, *9*, 936. [CrossRef]

10. Mirghani, H.; Blanchard, P. Treatment de-escalation for HPV-driven oropharyngeal cancer: Where do we stand? *Clin. Transl. Radiat. Oncol.* **2018**, *8*, 4–11. [CrossRef] [PubMed]
11. Roberts, S.; Evans, D.; Mehanna, H.; Parish, J.L. Modelling human papillomavirus biology in oropharyngeal keratinocytes. *Philos. Trans. R. Soc. B* **2019**, *374*, 20180289. [CrossRef]
12. Cohen, E.E.W.; Bell, R.B.; Bifulco, C.B.; Burtness, B.; Gillison, M.L.; Harrington, K.J.; Le, Q.-T.; Lee, N.Y.; Leidner, R.; Lewis, R.L.; et al. The society for immunotherapy of cancer consensus statement on immunotherapy for the treatment of squamous cell carcinoma of the head and neck (HNSCC). *J. Immunother. Cancer* **2019**, *7*, 184. [CrossRef]
13. Economopoulou, P.; Koutsodontis, G.; Avgeris, M.; Strati, A.; Kroupis, C.; Pateras, I.; Kirodimos, E.; Giotakis, E.; Kotsantis, I.; Maragoudakis, P.; et al. HPV16 E6/E7 Expression in circulating tumor cells in oropharyngeal squamous cell cancers: A pilot study. *PLoS ONE* **2019**, *14*, e0215984. [CrossRef]
14. Mehanna, H.; Robinson, M.; Hartley, A.; Kong, A.; Foran, B.; Fulton-Lieuw, T.; Dalby, M.; Mistry, P.; Sen, M.; O'Toole, L.; et al. Radiotherapy plus cisplatin or cetuximab in low-risk human papillomavirus-positive oropharyngeal cancer (De-ESCALaTE HPV): An open-label randomised controlled phase 3 trial. *Lancet* **2019**, *393*, 51–60. [CrossRef]
15. Reuschenbach, M.; Tinhofer, I.; Wittekindt, C.; Wagner, S.; Klussmann, J.P. A systematic review of the HPV-attributable fraction of oropharyngeal squamous cell carcinomas in Germany. *Cancer Med.* **2019**, *8*, 1908–1918. [CrossRef] [PubMed]
16. Lechner, M.; Chakravarthy, A.R.; Walter, V.; Masterson, L.; Feber, A.; Jay, A.; Weinberger, P.M.; McIndoe, R.A.; Forde, C.T.; Chester, K.; et al. Frequent HPV-independent P16/INK4A overexpression in head and neck cancer. *Oral Oncol.* **2018**, *83*, 32–37. [CrossRef] [PubMed]
17. Kitamura, K.; Nimura, K.; Ito, R.; Saga, K.; Inohara, H.; Kaneda, Y. Evaluation of HPV16 E7 expression in head and neck carcinoma cell lines and clinical specimens. *Sci. Rep.* **2020**, *10*, 22138. [CrossRef] [PubMed]
18. Van Abel, K.M.; Moore, E.J. Focus issue: Neck dissection for oropharyngeal squamous cell carcinoma. *ISRN Surg.* **2012**, *2012*, 547017. [CrossRef] [PubMed]
19. Holzinger, D.; Wichmann, G.; Baboci, L.; Michel, A.; Höfler, D.; Wiesenfarth, M.; Schroeder, L.; Boscolo-Rizzo, P.; Herold-Mende, C.; Dyckhoff, G.; et al. Sensitivity and specificity of antibodies against HPV16 E6 and other early proteins for the detection of HPV16-driven oropharyngeal squamous cell carcinoma. *Int. J. Cancer* **2017**, *140*, 2748–2757. [CrossRef]
20. Dahlstrom, K.R.; Anderson, K.S.; Cheng, J.N.; Chowell, D.; Li, G.; Posner, M.; Sturgis, E.M. HPV serum antibodies as predictors of survival and disease progression in patients with HPV-positive squamous cell carcinoma of the oropharynx. *Clin. Cancer Res.* **2015**, *21*, 2861–2869. [CrossRef]
21. Piontek, T.; Harmel, C.; Pawlita, M.; Carow, K.; Schröter, J.; Runnebaum, I.B.; Dürst, M.; Graw, F.; Waterboer, T. Post-treatment human papillomavirus antibody kinetics in cervical cancer patients. *Phil. Trans. R. Soc. B* **2019**, *374*, 20180295. [CrossRef]
22. Spector, M.E.; Sacco, A.G.; Bellile, E.; Taylor, J.M.G.; Jones, T.; Sun, K.; Brown, W.C.; Birkeland, A.C.; Bradford, C.R.; Wolf, G.T.; et al. E6 and E7 antibody levels are potential biomarkers of recurrence in patients with advanced-stage human papillomavirus-positive oropharyngeal squamous cell carcinoma. *Clin. Cancer Res.* **2017**, *23*, 2723–2729. [CrossRef] [PubMed]
23. Lang Kuhs, K.A.; Kreimer, A.R.; Trivedi, S.; Holzinger, D.; Pawlita, M.; Pfeiffer, R.M.; Gibson, S.P.; Schmitt, N.C.; Hildesheim, A.; Waterboer, T.; et al. Human papillomavirus 16 E6 antibodies are sensitive for human papillomavirus-driven oropharyngeal cancer and are associated with recurrence. *Cancer* **2017**, *123*, 4382–4390. [CrossRef]
24. Yang, Y.-S.; Smith-McCune, K.; Darragh, T.M.; Lai, Y.; Lin, J.-H.; Chang, T.-C.; Guo, H.-Y.; Kesler, T.; Carter, A.; Castle, P.E.; et al. Direct human papillomavirus E6 whole-cell enzyme-linked immunosorbent assay for objective measurement of E6 oncoproteins in cytology samples. *Clin. Vaccine Immunol.* **2012**, *19*, 1474–1479. [CrossRef] [PubMed]
25. Kong, L.; Xiao, X.; Lou, H.; Liu, P.; Song, S.; Liu, M.; Xu, T.; Zhang, Y.; Li, C.; Guan, R.; et al. Analysis of the role of the human papillomavirus 16/18 E7 protein assay in screening for cervical intraepithelial neoplasia: A case control study. *BMC Cancer* **2020**, *20*, 999. [CrossRef] [PubMed]
26. Lee, S.J.; Cho, Y.S.; Cho, M.C.; Shim, J.H.; Lee, K.A.; Ko, K.K.; Choe, Y.K.; Park, S.N.; Hoshino, T.; Kim, S.; et al. Both E6 and E7 oncoproteins of human papillomavirus 16 inhibit IL-18-induced IFN-gamma production in human peripheral blood mononuclear and NK cells. *J. Immunol.* **2001**, *167*, 497–504. [CrossRef] [PubMed]
27. Rotondo, J.C.; Oton-Gonzalez, L.; Mazziotta, C.; Lanzillotti, C.; Iaquinta, M.R.; Tognon, M.; Martini, F. Simultaneous detection and viral DNA load quantification of different human papillomavirus types in clinical specimens by the high analytical droplet digital PCR method. *Front. Microbiol.* **2020**, *11*, 591452. [CrossRef] [PubMed]
28. Tognon, M.; Tagliapietra, A.; Magagnoli, F.; Mazziotta, C.; Oton-Gonzalez, L.; Lanzillotti, C.; Vesce, F.; Contini, C.; Rotondo, J.C.; Martini, F. Investigation on spontaneous abortion and human papillomavirus infection. *Vaccines* **2020**, *8*, 473. [CrossRef]
29. Oton-Gonzalez, L.; Rotondo, J.C.; Cerritelli, L.; Malagutti, N.; Lanzillotti, C.; Bononi, I.; Ciorba, A.; Bianchini, C.; Mazziotta, C.; De Mattei, M.; et al. Association between oncogenic human papillomavirus type 16 and Killian polyp. *Infect. Agent Cancer* **2021**, *16*, 3. [CrossRef]
30. Olthof, N.C.; Speel, E.-J.M.; Kolligs, J.; Haesevoets, A.; Henfling, M.; Ramaekers, F.C.S.; Preuss, S.F.; Drebber, U.; Wieland, U.; Silling, S.; et al. Comprehensive analysis of HPV16 integration in OSCC reveals no significant impact of physical status on viral oncogene and virally disrupted human gene expression. *PLoS ONE* **2014**, *9*, e88718. [CrossRef] [PubMed]
31. Rotondo, J.C.; Giari, L.; Guerranti, C.; Tognon, M.; Castaldelli, G.; Fano, E.A.; Martini, F. Environmental doses of perfluorooctanoic acid change the expression of genes in target tissues of common carp. *Environ. Toxicol. Chem.* **2018**, *37*, 942–948. [CrossRef]

32. Mazzoni, E.; Bononi, I.; Benassi, M.S.; Picci, P.; Torreggiani, E.; Rossini, M.; Simioli, A.; Casali, M.V.; Rizzo, P.; Tognon, M.; et al. Serum antibodies against simian virus 40 large T antigen, the viral oncoprotein, in osteosarcoma patients. *Front. Cell Dev. Biol.* **2018**, *6*, 64. [CrossRef]
33. Pietrobon, S.; Bononi, I.; Lotito, F.; Perri, P.; Violanti, S.; Mazzoni, E.; Martini, F.; Tognon, M.G. Specific detection of serum antibodies against BKPyV, a small DNA tumour virus, in patients affected by choroidal nevi. *Front. Microbiol.* **2017**, *8*, 2059. [CrossRef]
34. Contini, C.; Rotondo, J.C.; Magagnoli, F.; Maritati, M.; Seraceni, S.; Graziano, A.; Poggi, A.; Capucci, R.; Vesce, F.; Tognon, M.; et al. Investigation on silent bacterial infections in specimens from pregnant women affected by spontaneous miscarriage. *J. Cell Physiol.* **2018**, *234*, 100–107. [CrossRef]
35. Li, L.; Cai, S.; Liu, S.; Feng, H.; Zhang, J. Bioinformatics analysis to screen the key prognostic genes in ovarian cancer. *J. Ovarian Res.* **2017**, *10*, 27. [CrossRef]
36. International Agency for Research on Cancer. *Human Papillomaviruses: IARC Monographs on the Evaluation of Carcinogenic Risks to Humans*; IARC: Lyon, France, 2007; Volume 90, ISBN 978-92-832-1290-4.
37. Castellsagué, X.; Alemany, L.; Quer, M.; Halec, G.; Quirós, B.; Tous, S.; Clavero, O.; Alòs, L.; Biegner, T.; Szafarowski, T.; et al. HPV involvement in head and neck cancers: Comprehensive assessment of biomarkers in 3680 patients. *J. Natl. Cancer Inst.* **2016**, *108*, djv403. [CrossRef]
38. Betiol, J.; Villa, L.L.; Sichero, L. Impact of HPV infection on the development of head and neck cancer. *Braz. J. Med. Biol. Res.* **2013**, *46*, 217–226. [CrossRef]
39. McIlwain, W.R.; Sood, A.J.; Nguyen, S.A.; Day, T.A. Initial symptoms in patients with HPV-positive and HPV-negative oropharyngeal cancer. *JAMA Otolaryngol. Head Neck Surg.* **2014**, *140*, 441. [CrossRef]
40. Pytynia, K.B.; Dahlstrom, K.R.; Sturgis, E.M. Epidemiology of HPV-Associated Oropharyngeal Cancer. *Oral Oncol.* **2014**, *50*, 380–386. [CrossRef] [PubMed]
41. Lin, J.; Albers, A.E.; Qin, J.; Kaufmann, A.M. Prognostic significance of overexpressed P16INK4a in patients with cervical cancer: A meta-analysis. *PLoS ONE* **2014**, *9*, e106384. [CrossRef]
42. Schlecht, N.F.; Ben-Dayan, M.; Anayannis, N.; Lleras, R.A.; Thomas, C.; Wang, Y.; Smith, R.V.; Burk, R.D.; Harris, T.M.; Childs, G.; et al. Epigenetic changes in the *CDKN2A* locus are associated with differential expression of P16INK4A and P14ARF in HPV-positive oropharyngeal squamous cell carcinoma. *Cancer Med.* **2015**, *4*, 342–353. [CrossRef] [PubMed]
43. Ekanayake Weeramange, C.; Tang, K.D.; Vasani, S.; Langton-Lockton, J.; Kenny, L.; Punyadeera, C. DNA methylation changes in human papillomavirus-driven head and neck cancers. *Cells* **2020**, *9*, 1359. [CrossRef] [PubMed]
44. Mascitti, M.; Tempesta, A.; Togni, L.; Capodiferro, S.; Troiano, G.; Rubini, C.; Maiorano, E.; Santarelli, A.; Favia, G.; Limongelli, L. Histological features and survival in young patients with HPV-negative oral squamous cell carcinoma. *Oral Dis.* **2020**, *26*, 1640–1648. [CrossRef] [PubMed]
45. Pierce Campbell, C.M.; Viscidi, R.P.; Torres, B.N.; Lin, H.-Y.; Fulp, W.; Abrahamsen, M.; Lazcano-Ponce, E.; Villa, L.L.; Kreimer, A.R.; Giuliano, A.R. Human papillomavirus (HPV) L1 serum antibodies and the risk of subsequent oral HPV acquisition in men: The HIM study. *J. Infect. Dis.* **2016**, *214*, 45–48. [CrossRef] [PubMed]
46. Routman, D.M.; Jethwa, K.R.; Garda, A.E.; DeWees, T.A.; Joern, L.; Hilfrich, R.; Liu, M.C.; Price, K.A.; Moore, E.J.; Laack, N.N.; et al. HPV16 L1 capsid antibody titers and prognosis in HPV associated malignancy: Oropharyngeal, anal, cervical and vaginal cancer. *Int. J. Radiat. Oncol. Biol. Phys.* **2019**, *105*, E666. [CrossRef]
47. Kreimer, A.R.; Ferreiro-Iglesias, A.; Nygard, M.; Bender, N.; Schroeder, L.; Hildesheim, A.; Robbins, H.A.; Pawlita, M.; Langseth, H.; Schlecht, N.F.; et al. Timing of HPV16-E6 antibody seroconversion before OPSCC: Findings from the HPVC3 consortium. *Ann. Oncol.* **2019**, *30*, 1335–1343. [CrossRef] [PubMed]
48. Rizzo, R.; Pietrobon, S.; Mazzoni, E.; Bortolotti, D.; Martini, F.; Castellazzi, M.; Casetta, I.; Fainardi, E.; Di Luca, D.; Granieri, E.; et al. Serum IgG against simian virus 40 antigens are hampered by high levels of SHLA-G in patients affected by inflammatory neurological diseases, as multiple sclerosis. *J. Transl. Med.* **2016**, *14*, 216. [CrossRef]
49. Fakhry, C.; Qualliotine, J.R.; Zhang, Z.; Agrawal, N.; Gaykalova, D.A.; Bishop, J.A.; Subramaniam, R.M.; Koch, W.M.; Chung, C.H.; Eisele, D.W.; et al. Serum antibodies to HPV16 early proteins warrant investigation as potential biomarkers for risk stratification and recurrence of HPV-associated oropharyngeal cancer. *Cancer Prev. Res.* **2016**, *9*, 135–141. [CrossRef] [PubMed]
50. Schroeder, L.; Wichmann, G.; Willner, M.; Michel, A.; Wiesenfarth, M.; Flechtenmacher, C.; Gradistanac, T.; Pawlita, M.; Dietz, A.; Waterboer, T.; et al. Antibodies against human papillomaviruses as diagnostic and prognostic biomarker in patients with neck squamous cell carcinoma from unknown primary tumor: HPV antibodies as biomarker in NSCCUP patients. *Int. J. Cancer* **2018**, *142*, 1361–1368. [CrossRef]
51. Vergara, N.; Balanda, M.; Vidal, D.; Roldán, F.; Martín, H.S.; Ramírez, E. Detection and quantitation of human papillomavirus DNA in peripheral blood mononuclear cells from blood donors. *J. Med. Virol.* **2019**, *91*, 2009–2015. [CrossRef]
52. Stanley, M. Antibody reactivity to HPV E6 and E7 oncoproteins and early diagnosis of invasive cervical cancer. *Am. J. Obstet. Gynecol.* **2003**, *188*, 3–4. [CrossRef]
53. Rajagopal, C.; Harikumar, K.B. The origin and functions of exosomes in cancer. *Front. Oncol.* **2018**, *8*, 66. [CrossRef]

54. Dwedar, R.A.; Omar, N.M.; Eissa, S.A.-L.; Badawy, A.Y.A.; El-Kareem, D.A.; Ahmed Madkour, L.A.E.-F. Diagnostic and prognostic impact of E6/E7 MRNA compared to HPV DNA and P16 expression in head and neck cancers: An egyptian study. *Alex. J. Med.* **2020**, *56*, 155–165. [CrossRef]
55. Dahlstrom, K.R.; Li, G.; Hussey, C.S.; Vo, J.T.; Wei, Q.; Zhao, C.; Sturgis, E.M. Circulating human papillomavirus DNA as a marker for disease extent and recurrence among patients with oropharyngeal cancer. *Cancer* **2015**, *121*, 3455–3464. [CrossRef] [PubMed]

Review

Epigenomic and Metabolomic Integration Reveals Dynamic Metabolic Regulation in Bladder Cancer

Alba Loras [1,*,†], Cristina Segovia [2,*,†] and José Luis Ruiz-Cerdá [3,4,5]

1. Unidad Mixta de Investigación en TICs Aplicadas a la Reingeniería de Procesos Socio-Sanitarios (eRPSS), Universitat Politècnica de València-Instituto de Investigación Sanitaria La Fe, 46026 Valencia, Spain
2. Epithelial Carcinogenesis Group, Centro Nacional de Investigaciones Oncológicas (CNIO), 28029 Madrid, Spain
3. Unidad Mixta de Investigación en Nanomedicina y Sensores, Universitat Politècnica de València-Instituto de Investigación Sanitaria La Fe, 46026 Valencia, Spain; Jose.L.Ruiz@uv.es
4. Servicio de Urología, Hospital Universitario y Politécnico La Fe, 46026 Valencia, Spain
5. Departamento de Cirugía, Facultad de Medicina y Odontología, Universitat de València, 46010 Valencia, Spain
* Correspondence: alba_loras@iislafe.es (A.L.); csegovia@cnio.es (C.S.); Tel.: +34-961-245-713 (A.L.); +34-917-328-000 (C.S.)
† Both authors contributed equally in this work.

Citation: Loras, A.; Segovia, C.; Ruiz-Cerdá, J.L. Epigenomic and Metabolomic Integration Reveals Dynamic Metabolic Regulation in Bladder Cancer. *Cancers* **2021**, *13*, 2719. https://doi.org/10.3390/cancers13112719

Academic Editor: Fabrizio Bianchi

Received: 16 March 2021
Accepted: 26 May 2021
Published: 31 May 2021

Publisher's Note: MDPI stays neutral with regard to jurisdictional claims in published maps and institutional affiliations.

Copyright: © 2021 by the authors. Licensee MDPI, Basel, Switzerland. This article is an open access article distributed under the terms and conditions of the Creative Commons Attribution (CC BY) license (https://creativecommons.org/licenses/by/4.0/).

Simple Summary: Current diagnostic and follow-up methods for the clinical management of bladder cancer (BC) have limitations, and there is an urgent unmet need for non-invasive biomarkers for this highly prevalent disease. Furthermore, personalized treatments for patients could improve their quality of life and overall survival. The aim of this article is to review the literature in this area, with a primary focus on metabolic and epigenetic biomarkers of BC, as well as the targeted therapies discovered to date. We show the dynamic biological interplay established between epigenomics and metabolomics in the context of BC. These findings may be useful both for researchers and physicians in the field of BC, and could facilitate clinical decision-making regarding patients at diagnosis, prognosis, monitoring, or treatment.

Abstract: Bladder cancer (BC) represents a clinical, social, and economic challenge due to tumor-intrinsic characteristics, limitations of diagnostic techniques and a lack of personalized treatments. In the last decade, the use of liquid biopsy has grown as a non-invasive approach to characterize tumors. Moreover, the emergence of omics has increased our knowledge of cancer biology and identified critical BC biomarkers. The rewiring between epigenetics and metabolism has been closely linked to tumor phenotype. Chromatin remodelers interact with each other to control gene silencing in BC, but also with stress-inducible factors or oncogenic signaling cascades to regulate metabolic reprogramming towards glycolysis, the pentose phosphate pathway, and lipogenesis. Concurrently, one-carbon metabolism supplies methyl groups to histone and DNA methyltransferases, leading to the hypermethylation and silencing of suppressor genes in BC. Conversely, α-KG and acetyl-CoA enhance the activity of histone demethylases and acetyl transferases, increasing gene expression, while succinate and fumarate have an inhibitory role. This review is the first to analyze the interplay between epigenome, metabolome and cell signaling pathways in BC, and shows how their regulation contributes to tumor development and progression. Moreover, it summarizes non-invasive biomarkers that could be applied in clinical practice to improve diagnosis, monitoring, prognosis and the therapeutic options in BC.

Keywords: bladder cancer; metabolic pathways; metabolism; metabolomics; epigenetics; biomarkers; miRNAs

1. Introduction

BC is the second most common urological malignancy after prostate cancer, and its development has previously been shown to be strongly related to smoking, schistosomiasis infection, and occupational exposure to certain chemicals [1,2]. Worldwide, BC represents the sixth most frequent tumor with 424,082 new cases per year, and it is considered within the ten deadliest cancers [3].

According to histological criteria, 75% of newly diagnosed BCs are non-invasive (non-muscle-invasive BCs; NMIBCs) and have a 70% risk of recurrence and a 20% risk of progression, despite being treated with surgery (transurethral resection (TUR) of tumor), local chemotherapy, or non-specific immunotherapy (Bacillus Calmette–Guérin (BCG)) [1,4,5]. The remaining 25% of BCs are muscle-invasive BCs (MIBCs) and require radical cystectomy, usually followed by cisplatin-based chemotherapy. Patients with a poor performance status and/or metastatic disease have limited treatment options but may benefit from novel therapies such as immunotherapy, e.g., those that have recently been approved by the United States Food and Drug Administration (FDA) [6].

The clinical management of BC is complex. Macroscopic or microscopic urinary hematuria is one of the most prevalent symptoms in early stage BC, but alone it has low specificity (5%) since it can be present in other benign pathologies such as cystitis or urinary tract infections [7]. Therefore, urinary cytology and cystoscopy are routinely used for BC diagnosis and follow-up. Urinary cytology is a non-invasive procedure with a reasonable in-house cost, but it has poor sensitivity in low-grade BC detection. Consequently, white light cystoscopy is the gold standard for BC detection. However, this method also has drawbacks related to the omission of carcinomas in situ and preneoplasic lesions, user-dependent interpretation, invasiveness, and high costs [7–10]. In fact, the clinical management of NMIBC is one of the most expensive due to lifelong patient monitoring through cystoscopy and urinary cytology [11] to control the appearance of tumor recurrences and progression, and also due to the TUR which is carried out in recurrent BC. Due to the fact that urinary cytology and cystoscopy cannot provide prognostic information on BC disease development, the European Organization for the Research and Treatment of Cancer (EORTC) criteria are used to stratify NMIBC patients into low, intermediate, or high-risk groups, which are related to disease recurrence or progression [12]. However, although the EORTC scoring system is useful to guide the treatment of patients, it is obtained by the combination of static parameters (e.g., tumor grade and stage, number of tumors, size of tumors, presence of CIS) that do not reflect the dynamic behavior of tumors.

Taking into account the shortcomings of cystoscopy and urine cytology, research is being conducted to find specific and non-invasive BC biomarkers that provide dynamic information of tumors and improve patient management by avoiding unnecessary cystoscopies in the surveillance of them.

In the last decade, liquid biopsies have revolutionized oncology as a novel, non-invasive method to evaluate the treatment responses, assess therapy resistance, or characterize the tumor phenotype. Biomolecules such as circulating tumor cells (CTCs), circulating cell-free tumor deoxyribonucleic acid (ctDNA), messenger ribonucleic acids (mRNAs), micro-RNAs (miRNAs), long non-coding RNAs (lncRNAs), proteins and peptides, metabolites and vesicles (exosomes and endosomes) can be obtained from liquid biopsies and analyzed to provide information about the tumor [13]. Among samples used to find BC biomarkers, urine and blood have been the most frequent since the bladder releases cells and molecules into these biofluids [14]. To date, different assays based on CTCs, sediment cells, proteins, and mRNA detection have been carried out in urine or serum, and have received FDA approval for BC diagnosis and/or follow-up. Some examples are uCyt+, UroVysion, UroMark, CellSearch, CxBladder, CxBladder Monitor, Xpert BC Detection, NMP22, BTA TRAK and BTA stat [15]. However, given their low sensitivities and/or specificities, none of them have been shown to be superior to cystoscopy and have thus not yet been implemented into clinical practice. Currently, liquid biopsy is a promising non-invasive biomarker approach, and thus it may improve the management of BC.

Among molecular and analytic techniques used to identify biomarkers, metabolomics and epigenomics have developed rapidly in the past decade. Metabolic reprogramming and epigenetic modifications are two well-known hallmarks of cancer and their regulation is tightly linked to the tumor microenvironment and the eenvironment (e.g., microbiota), but is also influenced by other molecular processes (i.e., the genome, transcriptome, and proteome) [16,17]. Consequently, metabolomics and epigenomics have been found to be dynamic and closely reflect the phenotype of the tumor [18,19]. In addition, several studies have shown that the metabolome and epigenome establish bidirectional relationships in cancer cells. The metabolic reprogramming of cancer cells supports bioenergetic and biosynthetic demands of proliferation, but also alters the epigenetic landscape by modulating epigenetic metabolites. Furthermore, epigenetic mechanisms regulate metabolic gene expression to offer adaptive responses to rapidly changing environmental conditions and prolong tumor cell survival [20,21].

Understanding this dynamic relationship between metabolism and epigenetics and how they may be dysregulated in cancer is crucial to identify novel therapeutic targets and biomarkers. Additionally, it may provide a better understanding of the biological machinery underlying each tumor phenotype, thereby taking one step closer to precision medicine for the individualized treatment of patients in different types of cancer.

The review presented here is, to our knowledge, the first that specifically analyzes the interplay among epigenomics, metabolomics, and cell signaling pathways in the context of BC. It provides an overview of the complex interplay between these biochemical and molecular processes and highlights the metabolites, metabolic enzymes, miRNAs, and lncRNAs postulated as emerging clinical biomarkers and therapeutic targets in the context of BC. This review integrates the current knowledge about the pathology of BC and may help to improve clinical decision-making regarding BC patients, whether at the level of diagnosis, prognosis, monitoring, or treatment.

2. Metabolic Rewiring Controls the Epigenome in BC

It is widely known that cancer development and progression are due to genetic mutations in DNA. However, the role of metabolism and epigenetics has only been recognized in the last decade when the reprogramming of energy metabolism and epigenetic plasticity have been identified as two emerging hallmarks of cancer [22,23].

Cancer cells alter their metabolic and nutrient uptake pathways during tumor initiation, growth, and metastasis through a tightly regulated program of metabolic plasticity. This allows them to sustain the energetic and biosynthetic demands of cell proliferation and to adapt to hostile and ever-changing environments [24]. Epigenetic modifiers act on metabolic gene expression to induce changes in biochemical pathways, and many of the chemical modifications in DNA and histones derive from intermediates of cellular metabolic pathways. This indicates that fluctuations in metabolic concentrations affect the deposition and removal of chromatin modifications. Emphasizing on this last issue, several mechanisms have to be considered, such as: (i) the alteration of specific metabolites' concentrations that act as epigenetic cofactors or substrates; (ii) the generation of oncometabolites which act as inhibiting or activating epigenetic enzymes; and (iii) the translocation of metabolic enzymes and metabolites into the nucleus [20]. Below, we address these regulation processes in the context of BC.

2.1. Metabolites and DNA/Histone Methylation Processes

DNA methylation is one of the most studied epigenetic mechanisms in cancer, including BC. Methylation can be produced directly in promoter regions of cancer-related genes (CpG islands) or in residues of histones, and this can control DNA accessibility and regulate gene expression (see Section 3.1.1). In DNA or histone methylation processes, the availability of methyl groups is essential for the action of histone methyltransferases (HMTs) and DNA methyltransferases (DNMT) [20]. In this context, the methionine and folate cycles, as well as the metabolites involved in these pathways (serine, methionine, and the cofactor S-adenosyl-methionine (SAM)), have an important role in supplying one-carbon groups [20] and they are closely related to DNA methylation processes (Figure 1). High levels of these metabolites have been found in BC samples and are postulated as candidate biomarkers of BC [25,26]. SAM provides methyl groups that release S-adenosyl-homocysteine (SAH), an inhibitor of DNMTs and HMTs. Therefore, the SAM/SAH ratio is a major determinant of chromatin methylation. It is known that an increased SAM/SAH ratio correlates with hypermethylation of tumor suppressor genes and inappropriate silencing, whereas a decreased SAM/SAH ratio contributes to reduced methylation at the promoters of oncogenes (Figure 1) [27].

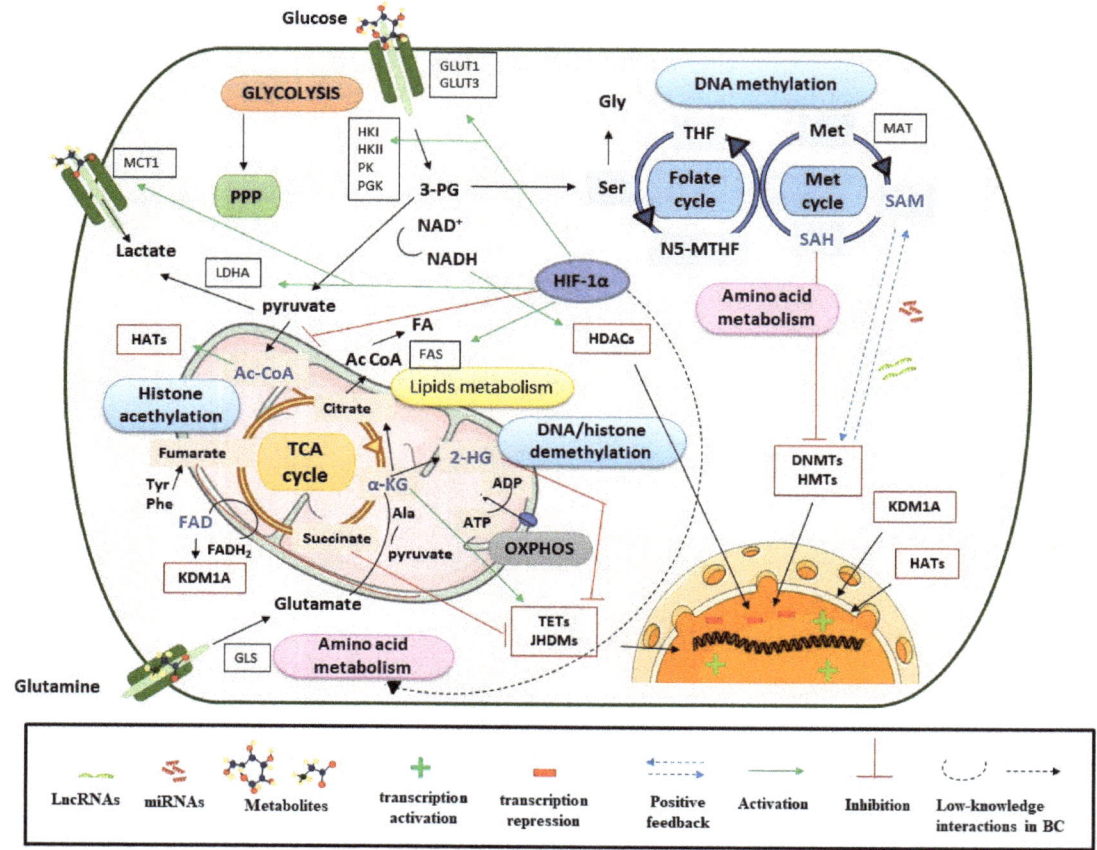

Figure 1. Metabolism controls epigenetic enzymes in BC. Ac CoA: acetyl coenzyme A; ADP: adenosine diphosphate; α-KG: alpha-ketoglutarate; ATP: adenosine triphosphate; DNMTs: DNA methyltransferases; FADH2: flavin adenine dinucleotide; GLS: glutaminase; GLUT: glucose transporter; HADAC: histone deacetylase; HATs: histone acetyltransferases; 2-HG: 2-hydroxyglutarate; HIF-1α: hypoxia-inducible factor subunit alpha; HMTs: histone methyltransferases; HK: hexokinase; JHDMs: JmjC-domain-containing histone demethylases; KDMs: lysine demethylases; LDHA: lactate dehydrogenase A; Met: methionine; NADH: nicotinamide adenine dinucleotide; N5-MTHF: 5-methyltetrahydrofolate; OXPHOX: oxidative phosphorylation; 3-PG: 3-phosphoglycerate; PGK: phosphoglycerate kinase; PI3K: phosphoinositide 3-kinases; PK: pyruvate kinase; PPP: pentose phosphate pathway; SAM: S-adenosyl methionine; Ser: serine; TCA cycle: tricarboxylic acid cycle; TETs: ten eleven translocation enzymes; THF: tetrahydrofolate.

Demethylation reactions are also susceptible to metabolic fluctuations of TCA cycle intermediates, such as α-KG, succinate, fumarate, and acetyl-CoA, which have been postulated as BC biomarkers [26,28]. They act on chromatin-modifying enzymes such as the 2-oxoglutarate-dependent dioxygenases (2-OGDO) family, which include ten eleven translocation (TET) enzymes, and the Jumonji (JHDMs) family of histone demethylases

(Figure 1) [29–31]. These enzymes catalyze the hydroxylation and demethylation of proteins and nucleic acids and play an important role in epigenetic processes. α-KG acts as a positive cofactor of 2-OGDO, thus elevated levels of α-KG from glucose and glutamine catabolism would promote demethylation processes that would influence BC epigenetic landscapes by relaxing chromatin and activating oncogene expression. Additionally, α-KG is a substrate of prolyl hydroxylase (PHDs), a type of protein that regulates hypoxia-inducible factors (HIFs) [32,33]. HIF subunit alpha (HIF-1α) regulates various processes and, under hypoxic conditions, can promote cancer cell survival. In the presence of oxygen, PHD proteins hydroxylate proline residues on HIF-1α, which leads to HIF-1α ubiquitination by Von Hippel–Lindau tumor suppressor protein (pVHL) and its proteasomal degradation [34]. In the context of cancer, most tumors have hypoxic regions and, in this case, HIF-1α is stabilized and triggers changes in glycolysis, nutrient uptake, waste handling, angiogenesis, apoptosis, and cell migration, which promote tumor survival and metastasis [34].

In BC, HIF-1α has a predictive and prognostic role; its overexpression is known to stimulate angiogenesis and can lead to a poor prognosis for patients [35]. HIF-1α is also closely linked to metabolism and regulates glycolysis, fatty acid, and amino acid pathways. HIF-1α increases glucose uptake by upregulating glucose membrane transporters (GLUT1 and GLUT3), which has been correlated with BC progression and poor overall survival [36] (Figure 1).

Furthermore, HIF-1α upregulates lactate dehydrogenase A (LDHA) to promote lactate production, regenerates nicotinamide adenine dinucleotide (NAD+), and increases the transcription of lactate transporters such as monocarboxylate transporter 1 (MCT1) [37,38]. This, in addition to high levels of hexokinase (HK), promotes a glucose flux towards pyruvate and lactate generation [39]. Previous studies have shown that lactate levels are related to BC progression and invasive or metastatic BC is known to have higher levels of this metabolite [38,39]. These metabolic processes would be in concordance with the Warburg effect [40]. Regarding lipid metabolism, several studies in cancer, including BC, have shown that HIF-1α increases the availability of fatty acids by regulating the action of fatty acid synthase (FAS), increasing fatty acid transport and reducing fatty acid oxidation [41]. In addition, HIF-1α plays an important role in amino acid metabolism, particularly in glutamine availability (Figure 1).

Cancer cells use glutamine as an energy substrate, a precursor of fatty acids, a donor of carbon and nitrogen for generating nucleotides or other amino acids, and to maintain the poll of intermediate metabolites such as acetyl-CoA or α-KG. These last metabolites are important for the anaplerotic reactions of the TCA cycle, but also for epigenetic processes by the effect that they have on HAT and HDM [34]. Due to the role that glutamine has in BC cells, several studies have investigated the inhibitory action that lncRNAs (e.g., lncRNA-p21) or miRNAs (e.g., miR-1, miR-1-3p, miR-9, miR-129) exert on GLS regulation (more details in Section 3.1.2) [42,43].

On the other hand, fumarate hydratase (FH) and succinate dehydrogenase (SDH) genes are mutated in many human cancers, including BC, which leads to the accumulation of their substrates, fumarate and succinate, respectively [30]. This is consistent with metabolomic analyses, which have shown increased levels of these metabolites in BC [28,44], but also with transcriptomic studies that have shown a strong deregulation of TCA cycle genes [14]. Among others, succinate, fumarate and α-KG are considered oncometabolites. This term refers to metabolites that are significantly elevated in tumor cells compared with control cells [45]. Succinate and fumarate, together with 2-hydroxyglutarate (2-HG), can inhibit PHDs activity under normoxic conditions [31]. Hence, succinate and fumarate could act as competitors of α-KG, inhibiting JHDMs and TET activity, and acting on bladder tumor biology through a profound impact on epigenetic effector activity [46–48] (see Figure 1). In conclusion, tumor-gene expression is regulated by epigenetic enzymes, the activity of which is dependent on metabolite availability (substrates).

2.2. Metabolites, Histone Acetylation Processes and Sirtuins

Another important metabolite, acetyl-CoA, is synthetized in several metabolic pathways (mitochondria, cytosol, and nucleus) from several sources, namely pyruvate, acetate, fatty acid β-oxidation, and amino acid catabolism. Metabolomic studies have reported elevated levels of acetyl-CoA in BC [28], and have particularly highlighted the role of glutamine as a substrate for acetyl-CoA synthesis [14]. Additionally, upregulated expression of acetyl-CoA synthase enzymes, such as ATP citrate-lyase (ACYL) or acetyl-CoA synthetase short chain family (ACSS), has been frequently found in BC cells, and some studies have reported the importance of ACSS3 for histone acetylation [49]. Acetyl-CoA acts as a cofactor which modulates kinetic and binding parameters of histone acetyltransferases (HATs). Nevertheless, CoA, the product of histone acetylation reaction, acts as an inhibitor. Therefore, the acetyl-CoA/CoA ratio has been postulated as the most important regulator of the enzymatic activity and specificity of HATs, rather than the absolute levels of acetyl-CoA [50]. In brief, high intracellular acetyl-CoA levels would trigger histone acetylation, an epigenetic marker associated with open chromatin, activating oncogenes linked with BC progression, proliferation and migration [20,30].

Another connection between metabolic processes and histone acetylation is provided by sirtuins (SIRTs), a type of NAD+-dependent histone deacetylases (HDACs) [51]. The activity of these enzymes is closely linked with the NAD+/NADH ratio, and consequently with the energy status in the cell. For example, when glycolytic activity is enhanced, the NAD+/NADH ratio decreases, thereby inhibiting SIRT catalysis [20,30]. The low NAD+/NADH ratio, together with an increase in HATs activity by elevated acetyl-CoA levels, could contribute to histone hyperacetylation and therefore an aberrant gene expression in BC [30].

2.3. Role of Metabolites in the Nucleus

Finally, the translocation or production of commonly cytosolic metabolic effectors in the nucleus can supply essential intermediates to epigenetic machinery in specific chromatin regions, which affects gene expression. Increased SAM levels in the nucleus support epigenetic methyltransferase activity at specific regions of chromatin [20]. This has been observed in cancer cells and is related to the translocation of splicing variants of MATs (S-adenosylmethionine synthetase, also known as methionine adenosyltransferase). Upregulated MAT1A levels have been reported in BC, specifically after treatment with chemotherapy, so MAT1A and possibly SAM could be related to the repression of tumor suppressor genes, (e.g., whose inhibition could confer tolerance or resistance to chemotherapy). Conversely, increased nuclear levels of acetyl-CoA can be produced by free diffusion of citrate or acetyl-CoA, but also by transient localization of the enzymes involved in its synthesis: ACSS2, ACLY, pyruvate dehydrogenase complex (PDC), and CAT (carnitine acetyltransferase). Post-translational modification of these enzymes within the nucleus or their association with lysine acetyltransferases (KATs) and transcription factors would explain their roles in chromatin regulation [48]. When there is DNA damage, ACYL is phosphorylated within the nucleus, which promotes histone H4 acetylation near sites of DNA double-strand breaks to repair them. Therefore, in response to DNA damage, ACLY phosphorylation would be enhanced, which would allow an increase in the capture of citrate or acetyl-CoA in the nucleus [48]. On the other hand, ACSS2 is recruited to specific genomic loci to supply acetyl-CoA for site-specific histone acetylation. Some studies have found that ACSS2 is translocated to the nucleus under low-glucose conditions upon phosphorylation by AMPK. Since cellular acetyl-CoA levels decrease when glucose is limited, a localized source of acetyl-CoA generated by ACSS2 could ensure the availability of this metabolite to KATs for histone acetylation [48]. PDC acts as a co-activator of signal transducers and activators of transcription 5 (STAT5) proteins. STAT5 proteins regulate specific nuclear genes in response to growth factors and cytokines which are linked to crucial cellular functions such as proliferation, differentiation, and survival. The role of STAT proteins is underscored in the field of cancer because tumors have an aberrant constitutive activation of them, which significantly contributes to tumor cell survival and malignant progression of disease [49]. Specifically in the context of BC, the findings obtained by Sun Y et al. suggested that the inhibition of STAT signaling by diindolylmethane (DIM) could decrease the invasiveness of BC, since DIM induced apoptosis in radioresistant cell lines. Therefore, DIM plus radiotherapy could be useful in overcoming such resistance [50]. Other studies performed in BC cell lines using inhibitors against STAT3/5 such as Stattic, Nifuroxazide and SH-4-54 also showed reduced survival and increased apoptosis. In a xenograft model, Static monotherapy had effects on tumors, but its combination with chemotherapy had additive effects. These findings highlight that inhibitors against STAT3/5 are promising as novel mono- and combination therapies in BC [51].

On the other hand, the regulation of NAD+/NADH levels in the nucleus is guaranteed by the activity of glycolytic enzymes (e.g., LDHA and glyceraldehyde 3-phosphate dehydrogenase (GAPDH)) since mitochondrial and nuclear membranes are impermeable to these cofactors [20]. Within the nucleus, NADH could be implicated in regulatory processes associated with histone acetylation, which in turn would influence transcriptional activity.

Last, the role of pyruvate kinase embryonic isozyme M2 (PKM2) in BC has previously been highlighted. Monomeric PKM2 translocates into the nucleus where it functions as a protein kinase that phosphorylates histones during gene transcription and chromatin remodeling [52]. Additionally, PKM2 upregulates the expression of c-Myc and cyclin D1, promoting the Warburg effect and cell cycle progression, respectively. Therefore, the role of nuclear PKM2 has been described as crucial for tumorigenesis, angiogenesis, and metastasis, and this protein has been postulated as a target for treating human cancers, including BC [53]. Numerous studies have correlated PKM2 overexpression with the development and metastasis of BC through promoting cell proliferation, migration and invasion via the mitogen-activated protein kinase (MAPK) signaling pathway [54], but also with advanced BC chemoresistance to cisplatin [55] or anticancer efficiency to pirarubicin [56]. Consequently, PKM2 could be a potential molecular prognostic marker of BC [57].

In brief, metabolic enzymes in the nucleus link metabolic flux to gene regulation, and allow nuclear membrane-impermeable metabolites to be used in epigenetic processes [58]. This metabolism–epigenetics axis would facilitate the adaptation to a changing environment around bladder tumors, providing a potential novel therapeutic target. The role that metabolites can play in modulating epigenetic enzyme action and gene expression is depicted in Figure 1.

3. Epigenetics Control Metabolic Reprogramming

Epigenetic regulation of gene expression is one of the most efficient stimulus response mechanisms. Extrinsic and intrinsic signals shape the plasticity of tumor cells to allow them to adapt to rapidly changing environmental conditions. These signals drive tumor metabolism [59], which can influence epigenetic mechanisms in several ways, such as the control of metabolite concentration necessary as cofactors or substrate for epigenetic enzymes, or the control of oncometabolites which regulate the expression of different epigenetic enzymes, among others, which have been explained above. These signals also drive metabolic reprogramming through changes in epigenetic modification patterns, achieving a rapid and coordinated response under unfavorable conditions for survival [27].

The functions of the epigenome are fundamental for the normal status of gene expression, and its alterations affect basic cellular processes such as proliferation, apoptosis or differentiation [60,61]. Epigenetics is defined as the heritable changes that occur in gene expression that do not involve alterations in the nucleotide sequence of DNA. These changes are basically divided into DNA methylation and modifications of the histone tails that allow the opening or closing of the chromatin. DNMT enzymes produce an irreversible silencing in the heterochromatin (closed chromatin state) and are divided in two groups: those involved in maintaining the methylation pattern in each cell replication (DNMT1) and those charged with *de novo* methylation (DNMT3a and DNMT3b). Chromatin regulatory elements are small molecules that regulate dynamic and reversible processes based on small post-translational modifications (PTMs) such as acetylation or methylation, among others, that make up the histone code. PTMs can be written by methyltransferases (HMTs) or acetylases (HATs), they can be deleted by demethylases (HDMs) and deacetylases (HDACs), and they can be read by different effector molecules to direct a particular transcriptional result. The main gene transcription marks are the acetylation of histones 3 and histone 4 (H3Kac, H4Kac) and methylation of histone 3 in lysine 4, 36 and 79 (H3K4me, H3K36me, H3K79me), while the best known gene repression marks are histone 3 methylation in lysines 9 and 27 and histone 4 in lysine 20 (H3K9me, H3K27me, H4K20me) [52]. The role of epigenetics is broad; it not only involves chromatin modifiers or changes in DNA methylation, but also includes non-coding RNA (ncRNA) expression (miRNA and lncRNA) that works in coordination with chromatin remodeling complexes and regulates the expression of multiple genes [62]. The metabolic reprogramming affects pivotal biological processes of the tumor cell such as survival, proliferation, or migration, among others. All these processes are the product of obtaining energy and its use in different chemical metabolic reactions. These reactions and the regulation of the use of energy to sustain demand underlie an aberrant epigenetic regulation of genes which are part of the main tumor metabolic routes and different oncogenic signaling pathways. Therefore, metabolism modulation can occur by epigenetic dysregulation of DNA methylation, histone modifications and ncRNAs.

The interplay of epigenetics and metabolic rewiring is complex. The epigenetic mechanisms, whether direct or indirect, that control metabolism are multiple and not all are well defined. However, there are two key connections between epigenetics and tumor metabolism: (i) epigenetic alterations which are directly related to the expression of metabolic enzymes; and (ii) epigenetic alterations that indirectly influence the signaling transduction cascades involved in the control of cellular metabolism.

Epigenetic alterations can also regulate major signaling transduction cascades. Many of them are well-known oncogenic pathways in cancer, especially in BC, which is considered an epigenetic disease. Chromatin remodeling gene mutations are more frequent in BC than in any other solid tumor. Importantly, they seem to be highly altered in the MIBC, where at least 89% of the alterations are in histone-modifying genes and 64% in genes associated with nucleosome positioning [63,64].

In recent studies, enhancer of zeste homolog 2 (EZH2), the main enzyme of the gene repressor complex Polycomb 2 (PRC2), has been shown to play an important role in tumor development and progression [65]. In NMIBC, EZH2 has been shown to predict recurrence and progression [66]. For example, EZH2 promotes changes in global gene expression, including aberrant expression of lncRNA HOTAIR, which acts in concert with EZH2 to mediate gene repression in high-risk tumors [67]. EZH2 also promotes the silencing of various miRNAs, such as the miR-200 family of miRNAs, which are involved in the repression of the epithelial–mesenchymal transition (EMT) and related to the increase in the probability of recurrence in the disease [68]. EZH2 is also known to cooperate with other modifiers such as DNMTs [69] and HDACs [70] to promote a permanent silencing of gene expression. A role of EZH2 in the activation of different oncogenic signaling pathways by turning off tumor suppressor genes has also been demonstrated.

DNA methylation and epigenetic enzymes that are altered in BC define different tumor subtypes and can help in the diagnosis or prognosis of patients. Below, the main epigenetic alterations that are involved in metabolic rewiring, including those are known in BC, are shown.

3.1. Epigenetic Regulation of Metabolic Enzymes and Oncogenic Pathways in Metabolism

There are different studies that show that various metabolic enzymes are altered by epigenetic events in tumor cells, as opposed to genetic mutations. These events can regulate tumor metabolic reprogramming directly or indirectly through DNA methylation, alterations in histone modification patterns, and by aberrant expression of non-coding RNAs (ncRNAs): miRNAs and lncRNAs; Figure 2), as will be shown throughout Sections 3.1.1–3.1.3.

Additionally, metabolic reprogramming is activated by oncogenic signaling cascades, such as PI3K/Akt/mTOR signaling, transcriptional factors (TFs) such as HIF, MYC or p53, and the inactivation of tumor suppressor signaling, e.g., the LKB–AMPK pathway. Many of these are present in BC, such as PIK3 or genes which upregulate c-myc glycolysis genes [71]. The PI3K/Akt/HIF-1α axis mediates glycolysis and leads to autophagy through AMPK signaling in BC cells. Additionally, the lack of AMPK signaling increases mitochondrial ROS, which enhances HIF-1α signaling [71]. These cascades are in turn connected to many other factors that lead to a dynamic network and speak to the complex behavior and biology of the tumor. These signaling pathways and activation of TFs are closely related to tumor metabolism, but in turn can be regulated by epigenetic mechanisms and as it will highlight in relation to ncRNAs (Figure 2). The implications of these signaling pathways in metabolic rewiring will help to understand their epigenetic regulation. This is briefly described below within each section.

AKT is known as the major regulator of glucose uptake and improves glucose metabolism via glycolysis and the pentose phosphate pathway [72]. MYC is usually induced by the Wnt/β-catenin, MAPK/ERK, and PI3K signaling pathways. In general, MYC induces the transcription of genes involved in glycolysis and glutaminolysis and contributes to control over redox balance [73]. MYC can also contribute to the supply of glycolysis intermediaries to the PPP for the biosynthesis of other molecules [73]. MYC can also regulate the use of glutamine by facilitating the activation of the expression of its transporters [74] or even controlling the repression of ncRNA targeting the enzyme GLS [75]. Thus, tumors that present MYC as a driver have a strong dependence on glucose or glutamine, which may indicate them as promising targets for the study of metabolic inhibitors. The classic tumor suppressor TP53 also has an emerging role in metabolic reprogramming. The activation of p53 represses the transcription of GLUT1, GLUT3 and GLUT4 transporters and activates the transcription of proteins involved in the electron transport chain for their replacement [76], and can even bypass glucose through the PPP [77]. In addition, those tumors that show a loss of p53 functionality will show a glycolytic metabolic phenotype. In hypoxic conditions, higher levels of HIF are detected. The HIFα subunits stabilize and activate the transcription of numerous glycolytic enzyme genes [78], or lactate synthesis enzyme genes, and this has been demonstrated in BC [79]. HIF can also be constitutively activated under normoxic conditions by oncogenic pathways such as PI3K/Akt/mTOR [80,81] or through the inactivation of LKB1–AMPK signaling [78].

3.1.1. DNA Methylation

DNA methylation is one of the most studied epigenetic mechanisms in cancer. Promoter hypermethylation usually occurs in tumor suppressor genes, DNA repair genes, cell cycle control, and invasiveness genes, and the silencing of their transcription causes cancer development and progression [82]. Thus, hypermethylation is correlated with the grade and stage of the tumor, with low-grade tumors being the least altered (10%) compared to high-grade (20%) and invasive tumors (30%) [83,84]. For example, the hypermethylation of GATA2, TBX2, TBX3 and ZIC4 in NIMBC is associated with progression to MIBC [84,85], but their connection with metabolic pathways is still unknown. Additionally, DNA hypomethylation in tumors leads to genomic instability and the activation of proto-oncogenes [86] and increases the risk of BC [87,88].

Numerous studies have shown that DNA methylation is related to glycolysis and glucose consumption. Indirectly, DNA methylation contributes to increased glucose uptake by silencing genes associated with glucose transporter degradation pathways such as GLUT1, allowing their overexpression [89]. Tumor suppressors, such as PTEN or LKB1, inhibit oncogenic signaling, which are central activators of glycolysis (Akt, AMPK, HIF and p53), and undergo hypermethylation of their promoters, facilitating the activation of glycolysis and the synthesis of macromolecules, thus helping to maintain the glycolytic phenotype [21,90]. Furthermore, the hypermethylation of certain metabolic enzyme genes has a direct impact on glycolysis. For example, the FBP1 gene (fructose 1,6-biphosphatase), which codes for one of the main enzymes of gluconeogenesis, shows hypermethylation of its promoter in different tumors, facilitating glycolysis [21,73,90]. In addition, the hypomethylation of promoters can activate the transcription of genes that code for glycolytic enzymes. Enzymes such as PKM2 or HK2 undergo promoter hypomethylation which allows them to be expressed and increases their availability, which leads to an accelerated glycolytic flow [91,92].

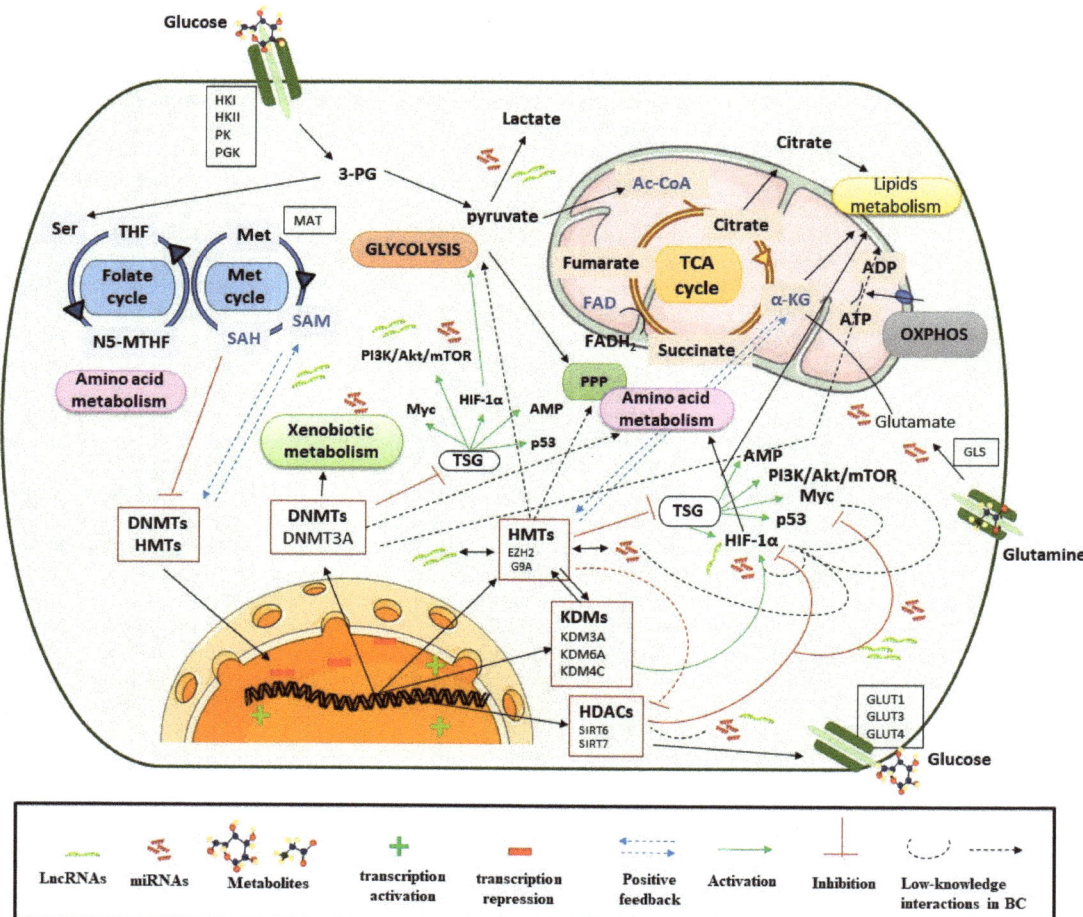

Figure 2. Epigenetic processes control metabolic pathways. Ac CoA: acetyl coenzyme A; ADP: adenosine diphosphate; Akt: alpha serine/threonine-protein kinase; α-KG: alpha-ketoglutarate; ATP: adenosine triphosphate; DNMTs: DNA methyltransferases; EZH2: enhancer of zeste homolog 2; FADH2: flavin adenine dinucleotide; G9A: euchromatic histone-lysine N-methyltransferase 2; GLS: glutaminase; GLUT: glucose transporter; HDAC: histone deacetylase; HAT: histone acetyltransferase; 2-HG: 2-hydroxyglutarate; HIF-1α: hypoxia-inducible factor subunit alpha; HMT: histone methyltransferase; HK: hexokinase; JHDMs: JmjC-domain-containing histone demethylases; KDM: lysine demethylase; LDHA: lactate dehydrogenase A; LncRNA: long non coding RNA; Met: Methionine; Myc: avian myelocytomatosis viral oncogene homolog; miRNA: microRNA; mTOR: mammalian target of rapamycin; NADH: nicotinamide adenine dinucleotide; N5-MTHF: 5-Methyltetrahydrofolate; OXPHOX: oxidative phosphorylation; 3-PG: 3-phosphoglycerate; PGK: phosphoglycerate kinase; PI3K: phosphoinositide 3-kinases; PK: pyruvate kinase; PPP: pentose phosphate pathway; SAM: S-adenosylmethionine; Ser: serine; SIRT: sirtuin; TCA cycle: tricarboxylic acid cycle; TETs: ten eleven translocation enzymes; THF: tetrahydrofolate; TSG: tumor suppressor gene.

Moreover, recent studies link DNA methylation sites to fatty acid metabolism in BC [93–97]. TIMP3 promotor hypermethylation has been described in the regulation of lipid metabolism, fatty acid oxidation and cholesterol homeostasis in response to oxidative stress [96,97]. GSTP1 (glutathione S-transferase 1) is known to be hypermethylated in BC and this enzyme is involved in xenobiotic metabolism [98,99] and can regulate glycolytic and lipidic metabolism energetics, as well as oncogenic signaling pathways in other tumors [100]. Further, human GSTM1, part of the GST superfamily, has been reported to be transcriptionally downregulated by DNA methylation in BC [101]. Some cancer cells, including BC, overexpress DNMT1, DNMT3A, and DNMT3B, which in turn leads to DNA hypermethylation of promoter regions and the silencing of tumor suppressor genes [102,103]. This DNA methylation pattern commonly negatively affects gene expression, promoting tumor growth and progression and predicting therapy outcomes [104]. In fact, the main difference between low- (LG) and high-grade (HG) BC is related to the amount of aberrant hypermethylation in specific loci, with HG BC having a greater percentage of aberrant DNA methylation patterns (over 30%) when compared to LG BC [105,106]. Elevated FHIT, CDH1, CDH13, RASSF1A and APC promoter methylation levels correlate with poor prognosis, adverse clinicopathological features, BC progression, and reduced overall survival [107].

ASS1 and SAT1 genes, which are related to amino acid metabolism, have been shown to be hypermethylated in cancer, and their association with the rewiring of cisplatin-resistant BC has also been demonstrated [108]. In patients diagnosed with BC, the ABCB1/MDR1 drug transporter exhibits a dynamic degree of methylation, changing from a hypermethylation state during carcinogenesis to a hypomethylated state during chemotherapeutic treatments. Thus, it is a possible prognostic factor for disease recurrence and treatment response in BC [109].

Finally, it should be noted that, to date, approximately 90% of identified genes involved in drug metabolism and transport that are epigenetically regulated implicate DNA methylation. Therefore, further research in this area may facilitate our understanding of the multidrug resistant response in different patients with different types of cancers [110]. Cytochrome P450 enzymes can be hypomethylated and transcriptionally activated and the metabolic response to drugs can be improved, thereby promoting resistance to treatments [110,111]. Modifications such as histone methylation or acetylation frequently work in combination to mediate the DNA methylation status of genes related to drug metabolism [112,113]. Therefore, both DNA hypomethylation and hyper methylation can contribute to the glycolytic phenotype in tumor cells and to the activation of xenobiotic metabolism.

3.1.2. Histone Modifications

Post-transcriptional modifications of histone tails regulate chromatin structure for gene activation or repression, replication, and DNA damage repair. The best studied and central modifications of epigenetic regulation include tailing amino acid methylation and acetylation marks, which are controlled by enzymes that can write them (HMTs or HATs) or can erase them (KDMs or HDACs).

Histone Methyltransferases (HMTs)

Currently, further research is still required to understand how histone modifications can control metabolic reprogramming. However, taking into account that histone methylation markers are the main regulators of gene transcription, it is highly likely that they modulate metabolism. HMTs control the methylation status of histones through repressor labels such as H3K9me, H3K27me3 and H4K20me3 [114,115], and there is evidence that some of these enzymes interfere in some way with cellular metabolic activities.

As mentioned above, EZH2 promotes global changes in gene and ncRNA expression, can cooperate with other chromatin remodelers or may co-occupy several gene loci with G9A in order to maintain gene silencing in a cooperative and coordinated manner, which has been shown in BC [116] and could explain some oncological properties of EZH2 that could favor heterogeneity in tumors [117]. As the master epigenetic regulator, EZH2 controls gene silencing through its mark H3K27me3 and it can alter the metabolic profile of tumor cells through the metabolism of glucose, lipids and amino acids, as demonstrated in recent studies [118]. EZH2 activity is affected by the availability of SAM from altered metabolic profiles, and, through its involvement in multiple metabolic pathways, EZH2 can also increase SAM gene and protein expression. Those events establish a positive feedback loop which improves the activity of EZH2 and favors tumor progression [118,119]. Furthermore, EZH2 is also influenced by the production of other metabolites involved in post-transcriptional modifications such as phosphorylation, acetylation or O-GlcNAcylation (O-linked N-acetylglucosamine modification) that can regulate the activity and stability of this enzyme [120,121].

EZH2 facilitates glucose metabolism in tumor cells in several ways. It can silence the transcription of HIF1α-directed hypoxia signaling repressors, which induces the transcription of metabolic genes such as *GLUT1*, *PDK*, or *HK2* and contributes to maintaining the Warburg effect [122]. EZH2 can also promote and regulate lipid metabolism by silencing WNT signaling pathways and overexpressing lipogenic genes such as *PPAR-γ* [123,124]. Additionally, TERT and EZH2 cooperate in the activation of PGC-1α, which is involved in the expression of FAS [125]. It has been shown that FAS is involved in the synthesis of triglycerides, and its expression is upregulated by SIRT6, a protein deacetylase known to promote adipogenesis and be repressed by EZH2 [126]. These studies demonstrate the participation of EZH2 in lipid synthesis; however, it is a field that still requires further research, especially due to the important role of EZH2 in BC. EZH2 can also regulate amino acid metabolism through multiple pathways, mainly contributing to the production of methionine for SAM which, in addition to enhancing the expression of EZH2, can affect amino acid transporters [127]. EZH2 also regulates the expression of transamination enzymes that participate in the production of α-KG to obtain glutamate, and EZH2 inactivation can upregulate glutamine metabolism [128]. New findings demonstrate that aldehyde oxidase (AOX1) is epigenetically silenced through EZH2 during the progression of advanced BC. *AOX1* silencing reconnects the tryptophan–kynurenine pathway, raising NADP levels that can increase metabolic flux through the PPP, allowing greater nucleotide synthesis [128,129]. Thus, AOX1-associated metabolites have a high predictive value for these tumors that do not have effective therapeutic opportunities.

EZH2 expression can be regulated at multiple levels. It can be transcriptionally induced by the activation of c-MYC or loss of p53 [130]. MYC regulates EZH2 directly by interacting with its transcriptional promoter or indirectly by controlling the repression of some miRNAs that silence EZH2 [131]. However, it can also be post-transcriptionally regulated by interaction with ncRNAs [132] or by the activation of signaling cascades such as PI3K–Akt [120].

Another important methyl transferase that is involved in amino acid metabolism is G9A. This enzyme can write the repression marks H3K9me1 and H3K9me2. Ribosome biogenesis and cell proliferation depend on the availability of serine, and G9A is known to increase glycolytic flux towards serine–glycine synthesis which is observed in BC [133]. G9A is overexpressed in many tumors, including BC [116], and can cooperate with TFs or with demethylases, such as KDM4C/JMJD2C, to maintain the H3K9me1 mark in the promoters of serine pathway-related genes to promote their transcriptional activation, including those for amino acid synthesis and transport [134,135]. Furthermore, like EZH2, G9A regulates transamination enzymes whose expression is activated due to H3K9 demethylation or G9A repression, and helps the production of other precursor metabolites such as NADH or α-KG, which have critical roles in the control of cellular metabolism related to cell proliferation and survival [135]. Patients with this type of epigenetic profile, as in the case of EZH2, are not suitable candidates for epigenetic inhibitor therapies as they could contribute to the metabolic reprogramming of tumors.

Histone Acetyltransferases (HATs)

The acetylation of histone lysine residues is established by HAT activity, using acetyl-CoA as an acetyl donor [136]. Therefore, tumors with high production of acetyl-CoA can destabilize acetylation levels. There are not many reports on the involvement of these enzymes in metabolic reprogramming, but it is known that the acetylation of PKM2 at the end of glycolysis decreases its activity, thus favoring the trafficking of intermediates for the biosynthesis of nucleic acids, amino acids and lipids [137], and that low levels of H4K16ac/H3K9ac are associated with a highly proliferative tumor profile, thus it is likely that these marks are related to metabolic rewiring [115,138], and would be very interesting to research.

Histone acetylation is a highly dynamic process being regulated by two enzyme family members, operating in an opposite fashion: the HATs and the HDACs. HDACs are overexpressed in various tumors, including BC, and histone acetylation levels decrease during progression to MIBC. Several studies indicate that global levels of histone acetylation are suitable biomarkers for patients with urological malignancies. For example, histone acetylation levels could be helpful to identify patients with understated pT1 tumors after TUR, identifying those who need cystectomy [139], or could help to identify how patients will respond to treatment with HDAC inhibitors because the use of these therapies results in decreased global histone acetylation levels and a poor prognosis outcome for patients [139,140].

Histone Lysine Demethylases (KDMs)

Lysine demethylases (KDMs) are often overexpressed and activated in solid tumors. These enzymes remove a methyl group from the histone tails by oxidation in a flavin adenine dinucleotide (FAD) or α-KG-dependent manner [141].

Several KDMs play an active role in metabolic rewiring in cancer, for example, KDM3A/JMJD1A, which promotes BC progression by enhancing glycolysis through the coactivation of HIF-1α [142]. KDM3A demethylates glycolytic gene promoters including GLUT1, HK2, phosphoglycerate kinase 1 (PGK1), and LDHA, among others, through H3K9me2 mark, leading to their transcriptional activation [142]. Additionally, increased H3K27ac binding on HIF-1α induces GLUT3 overexpression through KDM3A binding, further contributing to BC's glycolytic phenotype [143]. Finally, it is notable to mention that the role of KDM6A/UTX, an enzyme that catalyzes the demethylation of H3K27me2 and H3K27me3, acts as a tumor suppressor and can interact with other epigenetic elements.

Regarding glutaminolysis, the production of α-KG is reduced under hypoxic conditions because it depends on the available local levels of glutamine. Likewise, α-KG is a target for various histone demethylases [144]. The loss of KDM6A reproduces the effects of low glutamine levels, suggesting that histone demethylases may be dependent on α-KG, accentuating metabolic reprogramming [145]. It also causes important changes in the levels of H3K4me1/H3K27ac of enhancer TFs and allows methyl transferases, such as EZH2, to rewire H3K27me3 levels on the UTX-EZH2 target genes, e.g., the repressor genes of c-MYC [144] or IGFBP3, whose decreased levels are involved in glucose metabolism [146]. KDM6A is one of the most frequently mutated enzymes in BC [147]. The loss of KDM6A promotes enrichment in PRC2-regulated signaling and confers specific vulnerability to EZH2 inhibition by converting tumoral cells into inducible synthetic lethality therapeutic targets and provides a new possibility of personalized treatments in urothelial tumors [148]. Although the specific role of KDM6A in metabolic rewiring in BC is unknown, a correlation has been observed between urothelial tumors with altered KDM6A and the upregulation of DNA repair genes and mTORC1 signaling, which stimulates aerobic glycolysis and lipid and nucleotide synthesis [147,149]. Therefore, KDM6A is a fundamental part of epigenetic regulation, making it a potential candidate involved in metabolic processes.

Histone Deacetylases (HDACs)

HDACs are responsible for removing acetyl groups and are categorized into four classes. HDAC Class III, or SIRTs, has been the most extensively studied concerning roles in cell metabolism regulation [21,27,90]. SIRTs act on the activity of TFs implicated in the transcription of genes involved in glycolysis, gluconeogenesis and lipid metabolism [150]. Dysregulated expression of various HDACs has been described in urothelial tumors [151].

These enzymes could deacetylate lysines such as H3K27 to be subsequently methylated by EZH2, and even PRC2 could recruit HDACs through EED to direct cooperative gene repression [151].

SIRT1 is mainly involved in tumor suppressor in cancer, including BC [140,152,153]; it can suppress glycolysis indirectly through the deacetylation of HIF, which in turn can regulate the transcription of various glycolytic enzymes such as LDH, G6P, PFK-1, PGK-1, PGAM-1 or transporters such as GLUT1 or GLUT3 [154,155]. Moreover, SIRT1 can regulate gluconeogenesis and lipid metabolism [150].

SIRT2 is also considered a BC tumor suppressor, and can participate in metabolic dysregulation indirectly by stabilizing MYC via deacetylation of a repressor, which functions as a positive feedback loop [156]. It also contributes to gluconeogenesis by deacetylation and activation of phosphoenolpyruvate carboxykinase (PEPCK) in the absence of glucose [150]. SIRT2 silencing induces the inhibition of the HDAC6 family member, causing a significant suppression of BC cancer cell migration and invasion. This strongly supports the cooperative actions between SIRT2 and HDAC6 in urothelial malignancies [140].

SIRT3 is associated with glycolytic metabolism [157]. It can regulate glucose balance in a HIF1α-dependent manner at the mitochondrial level [158]. In contrast to SIRT3, SIRT4 is involved in the inhibition of GDH, repressing glutamine metabolism [159]. In addition, it is also able to act on the activity of pyruvate dehydrogenase that catalyzes the conversion of pyruvate to acetyl-CoA [150]. Of note, SIRT 3 and SIRT4 downregulation has been reported in BC [140,153].

SIRT6 is the most reported SIRT involved in tumor metabolic reprogramming. Its primary role is in the regulation of glucose homeostasis and lipid metabolism [150]. SIRT6 works as a suppressor by blocking the HIF-dependent glycolytic switch and MYC-dependent ribosomal biosynthesis and glutaminolysis [21]. For instance, it can directly repress the expression of glycolysis genes by the deacetylation of H3K9, such as GLUT1 [154]. Functional studies of SIRT6 in BC cell lines confirmed its role in inhibiting glycolysis [160]. It should be noted that, in addition to regulating glycolysis, SIRT6 also participates in gluconeogenesis and lipid metabolism [150]. Its over-expression can inhibit the proliferation of BC cells, while its expression decreases with the progression of BC from T2 to T4 stage [140]. Therefore, SIRT6 could be a promising druggable biomarker for BC, as would be the case of SIRT4. SIRT7 can interact directly with the MYC factor, repressing its function [161,162]. Like SIRT6 (H3K9), the repressive mark of SIRT7 (H3K18) opposes the transcription of MYC-dependent genes and can therefore regulate the metabolic alterations mediated by this factor [150,162]. SIRT7 is upregulated in many cancers, including BC [140]; however, it has been reported that in BC it may play a dual role depending on the context, suggesting that the functional importance of SIRTs may change throughout cancer progression [153]. It has been demonstrated that SIRT7 levels decrease significantly in MIBC, suggesting that SIRT7 may promote a more aggressive phenotype [152]. Although the mechanism of dysregulation of SIRT7 in BC and its putative implication in metabolic rewiring have not yet been adequately addressed, it is speculated that this could be due to epigenetic mechanisms that allow a plastic expression of SIRT7 in both carcinogenesis and tumor progression [152,153]. A lower expression of SIRT7 could be related to a positive regulation of TFs of EMT processes, such as Snail or HOTAIR, which would participate in the recruitment of EZH2 to specific genes [162]. It has been demonstrated that the expression of SIRT7 is regulated by miRNAs such as miR-125b, which in turn interacts antagonistically with the lncRNA MALAT1 in BC [163]. However, the findings on the miR-125b–MALAT1 interaction and its possible participation in metabolic regulation should be further studied and confirmed [164].

In summary of this section, this is a possible scenario of epigenetic–metabolic reprogramming in BC. On the one hand, EZH2 can interact with other chromatin remodelers such as G9a and DNMTs regulating the expression of tissue-specific gene sets in BC. EZH2 and G9a work in obtaining α-KG, in transamination reactions involved in amino acid and lipid synthesis, and promote glycolysis and the PPP. Moreover, DNMT can hypermethylate genes that enhance glycolysis, but also it plays an important role in xenobiotic metabolism. Therefore, a relationship with EZH2, and consequently with G9a, could open an exploration path in resistance to therapy and the use of combined therapies in BC. Interestingly, KDMs such as UTX or KDM3A cooperate with HIF-α and are associated with high levels of α-KG and glutamine availability, enhancing amino acid synthesis, PPP activation, glycolysis as well as lipid synthesis.

On the other hand, it has been proven that EZH2 can interact with PI3K signaling, one of the major regulators of glucose uptake that enhances glycolysis and PPP, but also with HIF, MYC and p53, either directly or indirectly. In turn, there appears to be evidence of EZH2 regulation by these potent metabolic rewires. It should be noted that EZH2 regulates the expression of HIF suppressors and that the role of sirtuins is basically to keep HIF or MYC repressed, as is the case with SIRT6 and SIRT7 in BC. There is evidence that SIRT6 may be silenced by EZH2 and that SIRT7 decreases as the disease progresses to MIBC. It can be assumed, then, that EZH2 could regulate sirtuin expression, together with ncRNAs, since EZH2 is involved in the recurrence and progression in BC. Thus, EZH2 not only controls gene silencing programs together with other chromatin remodelers, but can also regulate demethylation and deacetylation programs to orchestrate which metabolic pathways are enhanced for obtaining energy and precursors and which are maintained by the tumor cell. All this occurs through epigenetic regulation, one of the most fluid and dynamic machineries to obtain rapid responses to progress in the disease.

3.1.3. ncRNAs

Recent studies have shown that ncRNAs regulate enzymes involved in metabolic pathways such as glycolysis and the mitochondrial TCA cycle, contributing to oncogenic metabolic programming. Their aberrant gene expression contributes to the establishment of diverse mechanisms that govern the plasticity of tumor cell metabolism [21]. These small RNA molecules are non-coding, but nevertheless have a regulatory role for gene expression at the post-transcriptional level [90]. The regulatory role that miRNAs and lncRNAs can exert on metabolic enzymes and glucose, lipid and amino acid metabolic pathways in BC is discussed below [165,166].

ncRNAs can actively regulate energetic signaling by targeting key metabolic transporters and enzymes, but also by directly or indirectly controlling the expression of tumor suppressors or oncogenes in different signaling pathways.

miRNAs

There are numerous studies that link miRNAs with tumor metabolism, and they have functions in various types of cancer. miRNAs control crucial metabolic processes including glucose transport, glycolysis, the TCA cycle, glutaminolysis, altered lipid metabolism as well as amino acid biosynthesis [167], but they are also related to drug-metabolizing gene expression [106].

miRNAs can regulate the expression of numerous enzymes that participate in glucose uptake, including glycolytic enzymes such as HK2, controlled by miR-143, miR-145 and miR-155 in BC [168–170]; PKM2 regulated by miR-326 and miR133a/b [21]; or the expression of LDHA, which is controlled by miRNAs such as miR-34a or miR-200c in urothelial tumors, favoring the production of lactate [167,171], among others. However, they can also regulate the expression of glucose transporters. For instance, many miRNAs such as miR-199a, miR-138 or miR-150 can control the expression of GLUT1 [164]. Interestingly, miR-218 has been shown to regulate the expression of GLUT1, which leads to an enhancement of chemosensitivity to cisplatin in BC [172]. In addition, miR-93 and miR-133 regulate GLUT4 [173], and miR-195-5p or miR-106a improve the expression of GLUT3 in BC [174].

However, miRNAs arguably have the greatest impact on essential metabolic signaling pathways such as PI3K/Akt/mTOR and LKB1–AMPK [121,164], as well as the expression of TFs such as HIF, MYC and p53 that contribute to the metabolic phenotype of the tumor [21].

The expression of some miRNAs controls the activation of the PI3K/Akt/mTOR pathway [165]. miR-143-145 cluster and miR-133a regulate AKT expression in BC. Others regulate mTOR, such as miR-100 [175], or contribute to the inactivation of inhibitor phosphatases (PTEN), as in the case of miR-19a [176].

Regarding metabolism, miR-21 is a molecular switch of several aerobic glycolytic genes, such as GLUT1, GLUT3, LDHA, LDHB, PKM2, HK1 and HK2 in bladder tumors, and regulates glycolysis through the PI3K/Akt/mTOR pathway [177]. The AMPK pathway is one of the major sensors of cellular energy status that suppresses tumor growth during metabolic stress and is regulated by oncogenic miRNAs such as miR-451 or miR-33a/b [165]. The reactivation of LKB1–AMPK in BC cells improves apoptosis and autophagy [178,179]. However, the role of this pathway in metabolic reprogramming in BC requires further research.

These miRNAs, in turn, interact with other factors, e.g., HIF, to propagate hypoxia-induced signaling and its stabilization [180], and some miRNAs are induced by HIF1α in BC, such as miR-210, -193b, -125, or miR-145 [181]. p53 can also induce the expression of miRNAs with a partially suppressive role by inhibiting glycolytic enzymes like HK1 and HK2, such as miR-34a [165]. On the other hand, p53 is under the regulation of other oncogenic miRNAs such as miR-25, -30d, -33, -125b that can contribute to its stabilization [182].

Regarding the interaction of miRNAs with MYC, there is evidence that this factor can repress miRNAs with a suppressing role, such as miR-23a or miR-24b, or it can bind to the promoter of oncogenes of other miRNAs to promote the metabolism of glutamine [72,75].

miRNAs such as the miR-210 or miR-200 family of miRNAs can inhibit mitochondrial function by promoting glycolysis, glutaminolysis, and lactate production, which is crucial for adaptation to hypoxic microenvironments [92]. Altered expression of the miR-200 family has been reported in various cancers and is known to play an important role in BC [68]. This family includes five members located in two different clusters: miR-200a, miR-200b and miR-429 (cluster I) and miR-200c and miR-141 (cluster II). It has been shown that there is clear upregulation of cluster II in urothelial tumor tissue compared to normal tissue, and this overexpression may be mediated by the activation of specific oncogenic pathways such as MYC overexpression or p53 alterations [68]. Gene expression patterns show that miR-200 expression is involved in ncRNA metabolism and RNA splicing, and chromatin remodeling and histone modification processes are inversely related to miR-200 expression [68]. However, the expression levels of this family decrease as the disease progresses [183], suggesting a dual role of miR-200 in BC at different stages of the disease. The hypermethylation of CpG islands is known to induce high-stage miR-200 silencing in aggressive and infiltrating BC [184]. In addition, genes related to the expression of miR-200 significantly present the repressive mark H3K27me3, which indicates that the activity of EZH2 is inverse to the expression of these miRNAs and that EZH2 possibly participates in the repression of miR-200, which contributes to early recurrence, as we initially described before Section 3.1. It should be noted that studies relating to the participation of the members of this miRNA family in metabolic processes are scarce, even more so in BC, and that we have only found evidence regarding HIF-miR-200c and LDHA mechanisms [181].

miRNAs have been shown to target key enzymes involved in aerobic glycolysis, the TCA cycle and lipid metabolism [185], but also control the aberrant expression of central epigenetic enzymes, such as EZH2, and their expression can even be affected as a consequence of the hypermethylation of the gene that encodes them by both DNMT and HMT [186]. This means that in addition to metabolic dysregulation, ncRNAs further expand the intricate regulatory network in the mechanisms underlying tumorigenesis [90].

EZH2 is a direct target of miR-101 or miR-138 in BC [187]. However, these miRNAs also perform other functions. For example, in BC it has been shown that miR-101 can regulate the expression of cyclooxygenase-2, which is related to xenobiotic metabolism [188], and miR-138 in turn can control the expression of EMT factors such as ZEB2 [42].

Other miRNAs related to metabolic rewiring in BC are miR-1, miR-9 or miR-129. miR-1 acts as a tumor suppressor regulating GLS expression, which is crucial in glutamine metabolism [189]. miR-9 modulates the expression of the LASS2/CERS2 gene, which codes for a ceramide synthase, an enzyme involved in the metabolism of sphingolipids, which are part of cell membranes [190]. On the other hand, miR-129 may mediate the expression of GALNT1, an enzyme involved in protein metabolism [191].

LncRNAs

LncRNAs have been identified as important regulators of cellular metabolism. However, despite cumulative studies investigating altered expression profiles of lncRNAs during metabolic rewiring in cancer, the functional roles of these lncRNAs remain largely unexplored [192,193]. The expression pattern of an lncRNA can differ significantly depending on the metabolic process, making them crucial drivers of highly tissue-specific cancer phenotypes [194]. LncRNAs are involved in the regulation of oncogene expression, which induces metabolic reprogramming of HIF 1 α, c-MYC or p53 [195,196], but they can also interact with tumor suppressors, such as AMPK [197], oncogenic signaling pathways, metabolic enzymes or with other ncRNAs [193,194]. This network allows the maintenance of metabolic rearrangement during tumor response surrounding the microenvironment.

LncRNA-mediated glucose metabolism can occur through three different mechanisms: the alteration of the expression levels or distribution of GLUTs, the alteration of expression levels of glycolytic enzymes, or interactions with glycolytic genes and the modulation of their activity [194].

LncRNAs modulate glucose metabolism through glucose transporter regulation and glycolytic genes, and are implicated in many malignancies, including BC, such as ANRIL [198]. High tissue abundance of ANRIL in cancer is associated with aggressive clinicopathologic features, poor overall survival [199], and resistance to chemotherapy [200,201]. ANRIL interacts with signal transduction pathways in cancer such as PI3K/Akt/mTOR [202]. Furthermore, ANRIL promotes GLUT1 and LDHA expression, resulting in the upregulation of glucose uptake and the promotion of cancer progression via the Akt/mTOR pathway [194]. However, it can also interact with TFs such as c-MYC, which can transactivate ANRIL and promote tumor progression [202]. ANRIL also participates in the regulation of gene expression via mechanisms including chromatin modulation, TF binding, and miRNA regulation [203–205], and can be considered as a driver in cancer progression by increasing glucose uptake for glycolysis, and additionally, ANRIL has also been linked to fatty acid metabolism [206,207].

Other lncRNAs, such as HOTAIR and urothelial cancer-associated 1 (UCA-1), play an important role in BC. Previous data have shown that the expression of HOTAIR may be upregulated by EZH2 and that it is a predictor of disease recurrence and progression, and overall survival in MIBC [68]. Recently, its role in tumor glucose metabolism through the induction of GLUT1 gene expression has been discovered [208]. This lncRNA is also involved in OXPHO mitochondrial activity [209]. Although these metabolic roles have not been elucidated in BC, they could be part of the epigenetic regulation directed by an EZH2–lncRNAs axis that reinforces the reformulation of metabolism.

UCA-1 is the most studied lncRNA in BC. It promotes glycolysis by modulating HK2 via the activation of mTOR/STAT3 and miR-143 repression [129] and induces glutamine metabolism and redox regulation by targeting miR-16 in human BC [210]. Similarly to HOTAIR, UCA-1 has been related to mitochondrial activity [194], contributing to ARL2 induction through miR-195 inhibition in BC [192,211,212]. Therefore, UCA-1 could be related to metabolic readjustment in BC cells, although more in-depth investigations are necessary in order to prove this hypothesis. To date, it is known that UCA-1 shows association with HIF-1α in hypoxic environments [213], participates in the regulation of EMT processes though targeting miR-145–ZEB1/2 or miR-143/HMGB1 pathways [15,214,215], induces cisplatin/gemcitabine xenobiotic metabolism modulating miR-196a [216], and, curiously, it is under the regulation of miR-1 [217], one of the miRNAs involved in metabolic reprogramming in BC.

Lastly, although studies on lncRNAs and tumor metabolism in BC are scarce, it has been reported that lncRNA SLC16A1-AS1 promotes metabolic rewiring in BC disease progression [218]. This lncRNA creates an lncRNA–protein complex with E2F1, which facilitates its binding to two gene promoters: SLC16A1/MCT1, a monocarboxylate transporter in charge of lactate or pyruvate flux, and PPARα, a TF closely linked to lipid metabolism. The interaction with these metabolic effectors favors not only glycolysis, but also improves mitochondrial oxidative phosphorylation and the β-oxidation of fatty acids, allowing urothelial tumor cells to use alternative energy sources, which translates into metabolic plasticity marked by a hybrid glycolysis/OXPHOS phenotype known to facilitate BC invasiveness.

4. Non-Invasive Bladder Cancer Biomarkers

A biomarker could be defined as a characteristic which is objectively measured and capable of indicating the state of a biological process, be it normal or pathological [219]. In cancer, and specifically BC, diagnostic, prognostic, and monitoring biomarkers have been identified in biological samples, e.g., blood and urine, by high-throughput techniques [220]. In this section, we review the most important metabolomic and epigenomic studies carried out in the context of BC.

4.1. Metabolomic Studies in BC

Metabolomics identifies and quantifies endogenous and exogenous low-molecular-weight organic molecules (<1 kDa), i.e., metabolites, that are present in a biological sample [221]. Analytical platforms such as gas chromatography (GC) and liquid chromatography (LC) coupled to mass spectrometry (MS) and nuclear magnetic resonance (1H NMR) spectroscopy have been widely used in metabolomic analyses to identify potential oncological biomarkers [220,221]. Each of these techniques has advantages and limitations related to sample preparation, detection range, analysis speed, thresholds of sensitivity and specificity, so the choice of either one of these approaches depends on the aims and the requirements of the study.

In the last decade, metabolomic analyses have provided metabolomic profiles intended to be used in BC surveillance, or capable of distinguishing: (i) BC from control samples; (ii) NMIBC from MIBC samples; or (iii) low-grade from high-grade BC, with sensitivities, specificities, positive and negative predictive values (PPV, NPV) over 75 to 80%, as well as an elevated area under the curve (AUC). Although some of these studies have only provided holistic profiles [25], others have identified potential metabolites which may be used as non-invasive biomarkers. Tables 1 and 2 summarize the main metabolites identified through different metabolomic platforms in urine and serum samples, as well as the biochemical pathways in which they are involved.

Table 1. Putative identified metabolites and associated pathways in BC urine samples.

	Metabolic Biomarkers in BC		
	Urine		
Perturbed Biochemical Pathway	Levels (BC/Control)	Metabolites	Clinical Application
Glycolysis	High	Fructose [222], lactic acid [26,223]	Diagnosis
	Low	Fructose [26]	Diagnosis
TCA cycle	High	–	–
	Low	Citric acid [222–224], succinate [26]	Diagnosis
Amino acid metabolism	High	Val [20,26,222], Phe, Met, S−Adenosylmethionine, Lys [225], Leu [26,225,226], Ile, His, Ser [26], Tyr, Trp, hydroxyphenylalanine, phenylacetilglutamine, homophenylalanine, phenylglycoxylyc acid, kynurenine, hydroxyhippuric acid [221,227], hydroxytryptophan, indolacetic acid, minohippuric acid [227]	Diagnosis
	Low	Ala, PAGN, Pro, Arg [226], Asp [225,226], hippuric acid [224,227,228], creatine [26,229,230]	Diagnosis

Table 1. Cont.

Perturbed Biochemical Pathway	Levels (BC/Control)	Metabolites	Clinical Application
Metabolic Biomarkers in BC			
Urine			
GSH metabolism	High	–	–
	Low	Pyroglutamic acid [231]	Diagnosis
Taurine and hypotaurine	High	Taurine [224,230,232]	Diagnosis
	Low	–	–
Lipid metabolism	High	Carnitine [225], acetylcarnitine [26,227,230]	Diagnosis
	Low	Glycerol [222], palmitic acid [225]	Diagnosis
Nucleotide/nucleoside metabolism	High	Thymine [227], hypoxanthine, uridine [222]	Diagnosis
	Low	Adenosine [26]	Diagnosis
NAD cycle	High	–	–
	Low	Trigonelline [229]	Diagnosis

Note: This table only includes metabolites that were quantified individually in urine samples and obtained significant p_value ($p\_value < 0.05$), or metabolites that were selected as discriminants in metabolomic studies in which a validation set was used and the model's performance was good (i.e., sensitivity, specificity, PPV and NPV > 0.75). Ala: alanine; Asp: aspartate; Gly: glycine; GSH: glutathione reductase; His: histidine; Ile: isoleucine; Leu: leucine; Lys: lysine; Met: methionine; NAD: nicotinamide adenine dinucleotide; PAGN: phenylacetylglutamine; Phe: phenylalanine; Pro: proline; Ser: serine; TCA: tricarboxylic acid; Tyr: tyrosine; Val: valine; –: unknown.

Table 2. Putative identified metabolites and associated pathways in BC serum samples.

Altered Biochemical Pathway	Levels (BC/Control)	Metabolites	Clinical Application
Metabolic Biomarkers in BC			
Serum			
Glycolysis	High	Glucose [232], erythritol, D-lyxosylamine, ribonic acid [44]	Diagnosis
	Low	Lactate [232]	–
PPP	High	Ribose, gluconic acid, 2-keto-gluconic acid, xylitol, arabitol [44]	Diagnosis
	Low	–	–
Sucrose metabolism	High	Galacturonic acid, D-cellobiose, maltose [44]	Diagnosis
	Low	–	–
TCA cycle	High	Succinate, pyruvate, oxalacetate, phosphoenolpyruvate, acetyl-CoA [28], Cis-aconitic acid, fumaric acid, malic acid [44]	Diagnosis
	Low	Citrate [232]	Diagnosis
Amino acid metabolism	High	Gln, His, Malonate, Val [233], creatinine, kynurenine, norleucine [44]	Diagnosis
	Low	Tyr, Ile, Phe, Leu, Gly [232]	Diagnosis
Taurine and hypotaurine	High	Hypotaurine [44]	Diagnosis
	Low	–	–
Lipid metabolism	High	Carnitine [28]	Diagnosis
	Low	–	–
Nucleotide/nucleoside metabolism	High	Uridine, hypoxanthine [44]	Diagnosis
	Low	–	–
Organic acid	High	2-hydroxyglutaric acid, (R,R)-2,3-dihydroxybutanoic acid, 2,3,4-trihydroxybutyric acid, 2,4-dihydroxybutanoic acid, 3,4,5-trihydroxypentanoic acid, 3,4-Dihydroxybutanoic acid [44]	Diagnosis
	Low	–	–
Choline	High	Choline [232]	Diagnosis
	Low	–	–
Ketone metabolism	High	Acetoacetate [232]	Diagnosis
	Low	–	–

Note: This table only includes metabolites that were quantified individually in serum samples and obtained significant p_value ($p\_value < 0.05$), or metabolites that were selected as discriminants in metabolomic studies in which a validation set was used and the model's performance was good (i.e., sensitivity, specificity, PPV and NPV > 0.75). Gly: glycine; Gln: glutamine; His: histidine; Ile: isoleucine; Leu: leucine; Phe: phenylalanine; PPP: pentose phosphate pathway; TCA: tricarboxylic acid; Tyr: tyrosine; Val: valine; –: unknown.

Figure 3 shows the results obtained after performing an analysis of altered metabolic pathways in BC using the MetaboAnalyst 3.0 tool, and considering all the found discriminant metabolites both in urine as in serum samples. Among the set of identified metabolites in BC, a large majority are linked to pathways related to amino acid metabolism, the TCA cycle, or pyruvate metabolism. These data derived from studies performed in BC samples [14,225] share altered metabolic pathways with other types of tumors [17], highlighting that metabolic reprogramming is a common hallmark of tumors.

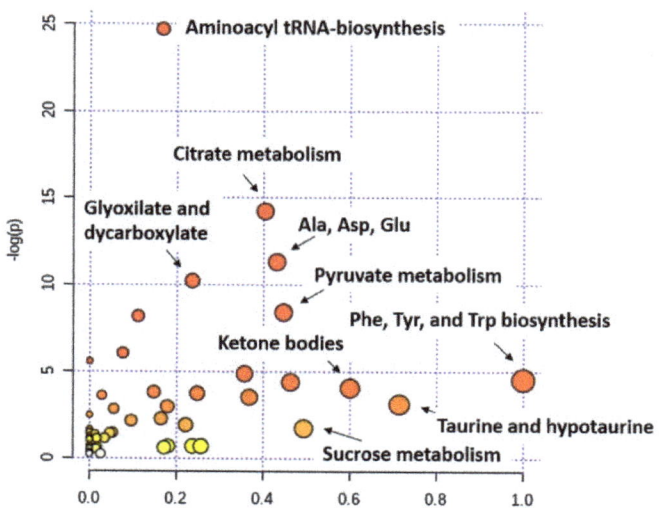

Figure 3. Analysis of altered metabolic pathways in BC when tumor and non-tumor samples (urine and serum) are compared. Note: the color and the size of each circle indicate its p-value and pathway impact value, respectively. Ala: alanine; Asp: aspartate; Glu: glutamine; Phe: phenylalanine; Tyr: tyrosine; Trp: tryptophan.

It is important to note that all these discriminant metabolites were identified in studies where tumor and non-tumor samples were compared, so clinically they could be applied as diagnostic biomarkers. Conversely, other studies identified metabolites related to BC aggressiveness using samples from LG or HG tumors [44,234–236]. For instance, Bensal et al. identified a serum metabolic profile composed of six metabolites (dimethyl amine (DMA), malonate, lactate, glutamine, histidine, and valine) able to distinguish LG and HG BC samples from control samples. All except malonate were identified as crucial to segregate LG tumors from controls; DMA, malonate, lactate, and histidine served as differentiating biomarkers of HG from controls and the combination of DMA, glutamine, and malonate was sufficient to accurately segregate LG from HG [236]. Tan et al. also performed a metabolomic study using serum samples of patients with LG and HG BC. In this case, they found a panel of serum metabolites formed by the combination of inosine, N-Acetyl-N-formyl-5-methoxykynurenamine (AFMK) and PS(O-18:0/0:0) which sufficiently discriminated not only HG BC and LG BC (AUC > 0.95), but also LG BC and healthy controls (AUC ≈ 0.99) [235]. The last notable study was carried out by Zhou et al. They observed that both HG and LG BC had distinct metabolic profiles when compared to control samples (e.g., elevated concentrations of TCA cycle metabolites or fatty acid biosynthesis metabolites). Additionally, HG tumors had higher levels of PPP intermediates, nucleotide metabolites, and amino acids than the control group. Minor differences were detected in high- and low-grade tumors, with oleic acid and serine being identified as discriminant metabolites between both groups [44]. Overall, these data suggest that differential metabolic alterations are linked to tumor aggressiveness, which allows the modulation of the processes involved in the development and progression of BC (proliferation, immune escape, differentiation, apoptosis, and invasion). Finally, other studies have identified urinary metabolic profiles as biomarkers for BC monitoring [14,228]. In these studies, samples were sequentially collected from NMIBC patients undergoing long-term disease surveillance. Metabolic profiles were able to detect recurrences in this cohort of patients by being able to observe that, after tumor removal, the metabolic profile trajectory changed towards a non-tumor phenotype in concordance with negative cystoscopy results. Metabolites linked with tryptophan, phenylalanine, arginine, proline, taurine, and hypotaurine metabolic pathways were identified as discriminant and were postulated as potential biomarkers for BC monitoring.

In conclusion, although some metabolomic studies have identified potential metabolic profiles associated with BC, their transfer to the clinic remains challenging. With further research and validation, metabolomic profiles may acquire FDA or European Medicines Agency approval as diagnostic, prognostic or monitoring biomarkers. To achieve this, the following are required: (i) the establishment of standard protocols that guarantee sample quality during collection and processing; (ii) control of the analysis quality in order to reduce preanalytical variation and batch-to-batch variability of data; (iii) a greater reproducibility among metabolites found in similar studies; and (iv) the validation of metabolic profiles in large patient populations.

4.2. Epigenetic Biomarkers in BC

The great advantage offered by epigenetic biomarkers is that they are highly dynamic, showing a reversible and measurable expression in the different stages and grades of tumors. Liquid biopsies more vastly used in the detection and surveillance of BC have used urine and serum/plasma as biofluids. These liquid biopsies present several epigenetic biomarkers, such as ctDNA for DNA alterations (mutations, CNV, methylation) or miRNAs and lncRNAs, and the detection of exosomes (microvesicles that protect small RNAs that can be found in the urine of patients).

Some epigenetic enzymes participate in metabolic reprogramming and may be considered good therapeutic targets, but they are not good biomarkers in fluids because: (i) there are other ways to measure them; (ii) there are not enough validation studies to support them as biomarkers, either for diagnosis, prognosis, recurrence or follow-up. Some of these, such as EZH2 or G9A, are good molecular markers in bladder tumor tissue, which can help to discern the nature of a tumor subtype and the prognosis of patients [116] and can also cooperate or regulate other epigenetic elements which are biomarkers in the clinic. The same situation occurs with some miRNAs and LncRNAs such as miR-143, miR-145, miR-200, miR-34a, UCA-1 or HOTAIR. They are good biomarkers in tissue expression, but they are not validated in fluids. Thanks to massive sequencing techniques, molecular panels have been achieved that mark patterns in bladder tumor subtypes [63], a fundamental tool for the classification of patients and their therapeutic opportunities [5].

DNA Methylation

Pyrosequencing techniques and comparative analysis of large databases allow us to discover DNA methylation events, also important in tumor metabolism. The hypermethylation of gene promoters occurs in 50–90% of BCs and includes a series of genes that are considered biomarkers, either in urine or serum. Some of them are *APC, ARF, BAX, BCL2, CDH1, CDKN2A, DAPK, EDNRB, EOMES, FADD, GDF15, GSTP1, LITAF, MGMT, NID2, PCDH17, POU4F2, RARβ2, RASSF1A, TCF21, TERT, TIMP3, TMS-1, TNFRSF21, TNFRSF25,* and *ZNF154* [228,237]. Many of them have been validated in large cohorts of patients and have reliable values for sensitivity, specificity and ROC curves.

Among all the alterations in the epigenetic machinery discussed previously, we must distinguish that they exist as validated biomarkers for BC, but whose implication in metabolic reprogramming has not been studied or demonstrated in BC yet. We highlight studies such as Hauser et al., who reported *TIMP3*, *APC*, *RAR-β2*, *TIG1*, *p16INK4a*, *PTGS2*, *p14ARF*, *RASSF1A* and *GSTP1* methylation promoters on the cell-free serum of BC patients for diagnosis [99,237]. The combination of methylated *CDKN2A*, *GSTP1* and *MGMT*, enzymes related to DNA repair events and drug metabolism [233,238], achieves 70% sensitivity and 100% specificity in BC detection [239]. *GSTP1* and *MGMT* in combination with *CDKN2A* and *ARF* have diagnostic power in the urine of patients with BC [239], and *GSTP1* together with *TIMP3* promoter methylation allowed the discrimination of invasive tumors [13,239]. Therefore, the identification of these genes within a population may help to better identify genetic vulnerabilities and pharmacogenetic studies and also monitoring of patients during chemotherapy. However, the cohorts of patients are not enough in some of these studies. In addition, DNA methylation patterns represent marks that can be detected throughout the development of the tumor and its progression. As the bladder tumor becomes malignant and invasive (MIBC), there are more alterations in DNA methylation, which can be used as a clinical prognostic tool. Regarding its connection with metabolism, on many occasions, we find genes associated with xenobiotic metabolism, which could give us information on resistance to previous therapies. Findings on methylation biomarkers in BC patients together with these are summarized in Table 3.

Table 3. Epigenetic marker compilation related to gene methylation status of ctDNA found in urine and serum from patients.

| Metabolic Gene/Pathway Related | Gene Status | ctDNA (Urine and Serum) | | Clinical Application |
		Urine	Serum	
Xenobiotic metabolism	Hipermet.	CDKN2A, MGMT ARF, GSPT1	–	diagnosis
Xenobiotic metabolism, lipid metabolism and β-oxidation fatty acids	Hipermet.	–	TIMP3, APC, RAR-β2, TIG1, p16INK4a, PTGS2, p14ARF, RASSF1A, GSTP1	diagnosis
Unknown	Hipermet.	GDF15, TMEFF2, VIM	–	diagnosis
Unknown	Hipermet.	PCDH10, PCDH17, APC	–	prognosis
Unknown	Hipermet.	TWIST, NID2	–	diagnosis
Unknown	Hipermet.	–	p16INK4a, p14ARF, CDH1, PCDH10, PCDH17	diagnosis

Note: Those biomarkers that have been related to tumor metabolism in BC are highlighted in red.

ncRNAs

The main epigenetic markers that can be found in liquid biopsy are miRNAs and lncRNAs. They are stable molecules that, due to their expression levels, can be related to grade, state and other characteristics of the tumor. Thus, some of them are already diagnostic (tumor initiation), prognosis (tumor in development or progression), or follow-up markers (recurrent tumor/progression/dissemination).

Most of the biomarkers described and validated for diagnosis or prognosis in patients with BC are ncRNAs, and some of them are related to the regulation of factors involved in tumor metabolism. The combination of lncRNAs offers more precise results. We underline the possible connections between epigenetics and metabolism (Tables 3–6) from BC biomarkers summarized by Lodewijk et al. [13]. Among detectable miRNAs and lncRNAs in liquid biopsy, those highlighted are or could be involved in the regulation of metabolic reprogramming in BC. We emphasize the expression of the miR-200 family, miR-21, miR-34a, miR-143 or miR-93. We can highlight that the cluster of miRNAs detected in urine for recurrence and surveillance reflects a greater number of alterations related to enzymes and metabolic pathways than the sets of miRNAs used in diagnosis (Table 4). Thus, the metabolism could be more altered as the disease progresses, and recurrence processes would represent a period of great metabolic changes for the cell.

Table 4. ncRNA markers found in urine from patient cohorts with BC.

	Urine		
Metabolic Gene/Pathway Related	Levels	Biomarker Panels	Clinical Application
GLUT1, GLUT4LDHA, HIF1	High	miR-652, miR-199a-3p, miR-140-5p, miR-93, miR-142-5p, miR-1305, miR-30a, miR-224, miR-96, miR-766, miR-223, miR-99b, miR-140-3p, let-7b, miR-141, miR-191, miR-146b-5p, miR-491-5p, miR-339-3p, miR-200c, miR-106b, miR-143, miR-429, miR-222 and miR-200a	Diagnosis
Unknown	High	miR-7-5p, miR-22-3p, miR-29a-3p, miR-126-5p, miR-200a-3p, miR-375 and miR-423-5p	Diagnosis
GLS, LHDA, HIF1,GLUT1, GLUT3, LHD, PKM2, HK2LDHA, HK1 (p53)	High	UCA1-miR-16, miR-200c, miR-205, miR-21, miR-221 and miR-34a	Recurrence and surveillance
Unknown	High/low	**NMIBC:** miR-30a-5p, let-7c-5p, miR-486-5p, miR-205-5p and let-7i-5p **NMIBC (high grade):** miR-30a-5p, let-7c-5p, miR-486-5p, miR-21-5p, miR-106b-3p, miR-151a-3p, miR-200c-3p, miR-183-5p, miR-185-5p, miR-224-5p, miR-30c-2-5p and miR-10b-5p **MIBC:** miR-30a-5p, let-7c-5p, miR-486-5p, miR-205-5p, miR-451a, miR-25-3p, miR-30a-5p and miR-7-1-5	Diagnosis/prognosis
p53, HIF1	Low	miR-125b, miR-204, miR-99a, miR-30b, and miR-532-3p.	Diagnosis
Glutamine metabolism, xenobiotic metabolism, mitochondrial activity, HIF1	High	hyal, lncRNA UCA1, microRNA-210, microRNA-96	Diagnosis (MIBC)

NOTE: In red are indicated biomarkers related to metabolism. High expression of biomarkers is highlighted in green and low expression in blue.

Table 5. miRNA markers found in serum patient cohorts with BC.

	Serum		
Metabolic Gene/Pathway Related	Level	Biomarker Panels	Clinical Application
Unknown	High	miR-422a-3p, miR-486-3p, miR-103a-3p and miR-27a-3p	Prognosis (MIBC)
Unknown	High	miR-152, miR-148b-3p, miR-3187-3p, miR-15b-5p, miR-27a-3p and miR-30a-5p	Prognosis
Unknown	High	miR-422a-3p, miR-486-3p, miR-103a-3p and miR-27a-3p	Prognosis (MIBC)
Unknown	High	miR-541, miR-200b, miR-566, miR-487 and miR-148b	Diagnosis
Unknown	Low	miR-25, miR-92a, -92b, miR-302 and miR-33b	Diagnosis
Unknown	High	miR-152	Prognosis

Table 6. lncRNA markers found in BC-derived exosomes from urine of patient cohorts.

	lncRNA-Derived Exosomes (Urine)		
Metabolic Gene/Pathway Related	Levels	Biomarker Panels	Clinical Application
GLUT1	High	HOTAIR, HOX-AS-2, MALAT1, HYMAI, LINC00477	Diagnosis (MIBC)

NOTE: biomarkers related to metabolism are in red.

On the other hand, the downregulation of miRNA expression could be a characteristic feature of invasive tumors. Some examples are miR-200, miR-1, miR-143, miR-145, miR-133a, miR-133b, and miR-125b [13], which have a tumor suppressor role and their silencing promotes reprogramming and EMT processes, and also most of them are associated with metabolic reprogramming in recurrence and xenobiotic metabolism, suggesting that their regulation program is active in more advanced stages of the disease. It is important to note that the particular role in metabolic reprogramming of ncRNAs detected in urine and in the validation sets from MIBC serum sample assays (Tables 4 and 5) has not been discovered yet.

Regarding lncRNAs, there are very few validated ones, but we can highlight a validated panel from urine samples which combines the expression of hyaluronoglucosaminidase 1 (HYAL1), miR-210, miR-96 and lncRNA UCA-1 (Table 4), thereby achieving a sensitivity of 100% and a specificity of 89.5% as a diagnosis biomarker [240]. LncRNAs such as UCA-1, MALAT1 or HOTAIR would be of great interest to carry out studies in combination of some of them with SIRTs, since there is evidence that they work in coordination in metabolic rewiring on some occasions, as we have commented previously.

In addition, miRNAs and lncRNAs are generally included in exosomes, which are crucial for the communication of cancer and stromal cells. Exosomes preserve their integrity and are very stable in liquid biopsy samples, such as serum, plasma, and urine, making them potential diagnostic and prognostic biomarkers with non-invasive methods [13]. We report a study using exosomes from the urine of patients with MIBC that carry lncRNAs (HOTAIR, HOX-AS-2, MALAT1, HYMAI, LINC00477, LOC100506688 and OTX2-AS1) as diagnostic biomarkers [241] (Table 6). The identification of exo-miRNAs associated with metastases could provide an additional tool to evaluate the follow-up of progression disease [242].

Finally, it should be mentioned that both metabolites as ncRNAs or methylation patterns in DNA are potential biomarkers that can be translated into the clinic to improve the diagnosis, prognosis and monitoring of BC. All these biomolecules provide dynamic information about tumor biology and evolution from a non-invasive and cost-effective approach.

5. Therapeutic Opportunities and Future Perspectives

Massive sequencing technologies have allowed us to advance in the management of BC, being a fundamental tool for the stratification of patients by increasingly well-known molecular subtypes. Besides, the study of omics has given a greater vision of the behavior of a tumor and its dynamics as stages progress. Thanks to studies in the field, new therapeutic options are opened. This information, together with that provided by biomarkers, makes it possible for us to talk more about precision medicine applied to patient management and clinical decision-making. Nevertheless, it is difficult to ensure that emerging therapies are implemented early in all patients, although some of them show promising results.

Clearly, a key metabolic target point is the glycolytic pathway and TCA cycle. There are potent blockers of several metabolic pathways such as GLS inhibitors, competitive G6P analogs that decrease acetyl-CoA levels [21,27], glucose transporter inhibitors [74] and even hexosamine biosynthesis pathway inhibitors that can decrease protein O-GlcNAcylation levels and reverse glucose-mediated metabolism [21]. Recent studies also suggest SAM and SAH inhibitors as potential antitumor candidates and even methionine-restricted diets [21,27,118]. SAH not only regulates intracellular levels of SAM, but the availability of SAM is critical for DNMT and HMT activity [21,27,86]. A well-known example is DzNep, an SAH inhibitor used in BC that blocks EZH2 and is closely related to H3K27me3 levels [199]. The levels of α-KG also affect the methylation status, as it is an essential cofactor for DNA demethylase (TET) and histone demethylase (JMJD) and could be affecting the levels of H3K27me3 [10]. In BC, there seems to be a positive feedback between the levels of SAM, α-KG and HMTs, so the use of inhibitors of these metabolites could lead to a promising novel strategy in this disease.

The main problem in the use of metabolic inhibitors is the heterogeneity in the metabolic profiles of the tumor cells, allowing them to escape the pharmacological effect [73]. Therefore, it is important to have information on the metabolic data from patients when selecting these inhibitors.

Regarding epigenetic therapy, two therapeutic strategies are currently used: small molecules that inhibit epigenetic enzymes and the manipulation of the expression and activity of miRNAs [243]. Epigenetic inhibitors currently approved for the treatment of cancer primarily target DNA methylation (anti-DNMTs) and histone modifications (anti-HDACs) [244].

The latest studies have focused on determining the expression pattern of HDACs in different cell lines and bladder tumors to explore their role in the development of cancer and to be able to predict the efficacy of drugs [245]. Although HDAC inhibitors are shown to be successful clinically, none of them targets class III HDACs.

Regarding SIRTs, recent studies have tried to identify SIRT6 inhibitors to understand its mechanism of action [21,22]. Other reports have shown that the inhibition of SIRT7 reduces tumor growth, and the deacetylation of its mark promotes the inhibition of miR-34a expression [150], implicated in BC. However, although there are many known compounds that regulate the activity of SIRTs, their use as therapeutic drugs is uncertain. SIRTs regulate so many cellular processes, so a change in their activity could positively affect one face of the disease but have a negative impact on the other. However, there is intense and ongoing research on its therapeutic use.

On the other hand, new epigenetic inhibitors targeting other histone remodeling enzymes (HDMTs, HMT, etc.) are being developed. These include improved DZNep analog blockers, specifically targeting EZH2 [246], G9A inhibitors [247] or agents against other HMTs [248]. Dual inhibitors are more effective, such as EZH2/EZH1 (UNC1999) [249], or against G9a/DNMT1 (CM-272) [116,250], which could represent an improved approach in cancer therapeutics, especially in BC. One of the great problems of EZH2 as a target is its relationship with acquired drug resistance. Therefore, it would be interesting to study the possible relationship between EZH2 and xenobiotic metabolism genes. EZH2 is a powerful regulator of gene expression, in addition to having other non-canonical roles, leading to the activation of different signaling pathways to maintain tumor cell viability [128]. Furthermore, EZH2 can interact with other non-histone proteins that regulate its activity by phosphorylation, such as AMPK or AKT (149, main activators of glycolytic metabolism). Therefore, it is important to develop therapeutic strategies based on the potential value of the combined intervention of EZH2 and tumor metabolic activities.

Synthetic lethality arises as a very attractive approach, based on the loss of expression of two antagonistic genes, as occurs with KDM6A (loss of expression mutation) and EZH2 (sensitization to inhibitors) in BC [148]. This strategy improves the knowledge of signaling pathways both to define with greater precision the different subtypes of tumors and to understand the drug-resistance mechanisms.

As mentioned at the beginning, there is another focus of epigenetic therapy aimed at miRNAs. Antisense oligonucleotides have been developed for silencing and synthetic miRNAs or lentiviral constructs are used to restore expression [251,252]. There are basic studies where the use of lentiviruses and antisense oligonucleotides, in combination with cisplatin or other chemotherapeutic agents, increases the sensitization of BC cells [253,254]. The use of synthetic RNAs to increase/decrease expression is also increasing. For example, the downregulation of miR-34a causes a clear inhibition of the clonogenic potential [255], restoring miR-143, and it can target HK2 [21], or miR-101 overexpression, which enhances the sensitivity to treatment [164]. On the other hand, oncogenic miRNAs that control tumor suppressive pathways (LKB1, AMPK, PTEN ...), such as miR-21, can be inhibited and allow TSG re-expression [21]. Recently, the CRISPR/Cas9 genome editing technique had been employed to knockout lncRNA UCA-1 (147). Nevertheless, the development of this type of therapy raises several challenges such as reliable administration methods, the determination of appropriate dosages, as well as the possible pleiotropic effects or resistance to therapy [165,192,194].

The ncRNAs exert an extensive and complex influence on the metabolic networks that characterize reprogramming in tumors, but currently, they mostly remain uncharacterized. Therefore, deepening the investigations of their functions and mechanisms in the regulation of metabolism is essential for the development of clinical therapies focused on patients with tumors with altered metabolism and the identification of new future biomarkers [193,194].

Apart from their possible therapeutic use in regulating the energy of cancer metabolism [165,192], the true potential application of ncRNAs lies in the growing interest in their biomarker character. Expression signatures of these ncRNAs in tumors hold great promise for the development of non-invasive diagnostic and prognostic biomarkers, which together with metabolome studies translates into a more powerful and sophisticated strategy for cancer diagnosis and treatment [192]. Additionally, exosomal-ncRNAs have several advantages over the existing approaches, such as low toxicity and target specificity [242].

The key questions to be solved in the near future are, on the one hand, that a stratification of patients must be carried out to guarantee the safety of the drugs. The metabolic and epigenetic landscape changes at various stages of tumor growth, making it difficult to design drugs that are effective [90]. Simultaneously targeting the epigenetic and metabolic pathways through combined therapies can inhibit the dynamic adaptive mechanisms of tumor cell reprogramming and obtain synergistic effects that can be evaluated in clinical studies [20].

On the other hand, resistance to therapy or a lack of pharmacological efficacy leads us to a second limitation, intertumoral and intratumoral heterogeneity—intertumoral because the pharmacodynamics of each patient tumor is different, intratumoral because the subpopulations that can be found within the tumor present different epigenetic and metabolic profiles that are regulated differently and complicate therapeutic interventions [21,73,90]. Against this, emerging and rapidly advancing tools and platforms to study epigenomic and metabolomic single cell heterogeneity are allowing the isolation and distinction of heterogeneous subpopulations of cancer cells that help us to understand the different profiles that we can find. Liquid biopsies have also been proposed as an efficient and accessible way to decipher the intratumor heterogeneity of the patient, by analyzing circulating tumor cells (CTC) and extracting circulating tumor DNA (cDNA) [21,90]. Therefore, future clinical trials should incorporate the analysis of epigenomic and metabolomic biomarkers that allow the selection of subsets of patients who may benefit from available treatments [21]. In the case of BC, although multiple genomic analyses have been carried out to study alterations in metabolic enzymes, there are no reports that integrate transcriptomic and metabolomic analyses from any type of biopsy until now [14].

Metabolites and epigenetic molecules as biomarkers are able to record and monitor the stage of the disease in real time and predict prognosis, recurrence or progression, as well as the evaluation of response to treatment and resistance. However, to demonstrate the reliability of a biomarker and the positive impact, a reliable prospective validation is unavoidably required, in addition to the need for large studies with long follow-up periods and large cohorts of patients [13,256,257]. Thus, in BC, these monitoring systems could be used as a new approach to achieve unequivocal diagnoses and minimize the invasiveness of the tests within its handling and management framework.

6. Conclusions

The metabolome and epigenome are closely intertwined in BC. EZH2 interacts with chromatin remodelers such as G9a and DNMTs, but also with HIF, MYC, and oncogenic signaling cascades, such as PI3K/Akt, to regulate metabolic reprogramming. Specifically, it appears that this regulation enhances the synthesis of α-KG from glucose and glutamine catabolism and promotes glucose flux towards glycolysis, PPP, and lipogenesis processes. Additionally, data suggest an important role of amino acids, or their metabolite derivatives, in BC, such as serine, methionine, SAM, and SAH, which are involved in one-carbon metabolism, as well as oncometabolites such as succinate, fumarate, and α-KG involved in the TCA cycle. All these metabolites have a crucial role in acting as epigenetic cofactors or substrates, but also in inhibiting or activating epigenetic enzymes that control the chromatin state and therefore gene expression. In summary, these data show the need for further research in this promising field to offer patients: (i) personalized treatments that increase their life expectancy; (ii) and non-invasive bladder cancer diagnosis and monitoring techniques that improve their quality of life, being cost-effective for health systems.

Author Contributions: Conceptualization, A.L., C.S. and J.L.R.-C.; methodology, A.L., C.S. and J.L.R.-C.; investigation, A.L. and C.S.; writing—original draft preparation, A.L. and C.S.; writing—review and editing, A.L., C.S. and J.L.R.-C.; supervision, A.L., C.S. and J.L.R.-C. All authors have read and agreed to the published version of the manuscript.

Funding: No financial support nor any kind of sponsorship was received for conducting this study.

Institutional Review Board Statement: Not applicable.

Data Availability Statement: No new data were created or analyzed in this study. Data sharing is not applicable to this article.

Acknowledgments: The authors are grateful for the professional English language editing of Camilla West, native English instructor.

Conflicts of Interest: The authors declare no conflict of interest.

References

1. Burger, M.; Catto, J.W.F.; Dalbagni, G.; Grossman, H.B.; Herr, H.; Karakiewicz, P.; Kassouf, W.; Kiemeney, L.A.; La Vecchia, C.; Shariat, S.; et al. Epidemiology and risk factors of urothelial bladder cancer. *Eur. Urol.* **2013**, *63*, 234–241. [CrossRef]
2. Kaseb, H.; Aeddula, N.R. *Bladder Cancer. En: StatPearls [Internet]*; StatPearls Publishing: Treasure Island, FL, USA. Available online: http://www.ncbi.nlm.nih.gov/books/NBK536923/ (accessed on 20 August 2020).
3. Bray, F.; Ferlay, J.; Soerjomataram, I.; Siegel, R.L.; Torre, L.A.; Jemal, A. Global cancer statistics 2018: GLOBOCAN estimates of incidence and mortality worldwide for 36 cancers in 185 countries. *Ca. Cancer J. Clin.* **2018**, *68*, 394–424. [CrossRef]
4. Anastasiadis, A.; de Reijke, T.M. Best practice in the treatment of nonmuscle invasive bladder cancer. *Adv. Urol.* **2012**, *4*, 13–32. [CrossRef] [PubMed]
5. Knowles, M.A.; Hurst, C.D. Molecular biology of bladder cancer: New insights into pathogenesis and clinical diversity. *Nat. Publ. Gr.* **2015**, *15*, 25–41. [CrossRef]
6. Ghatalia, P.; Zibelman, M.; Geynisman, D.M.; Plimack, E. Approved checkpoint inhibitors in bladder cancer: Which drug should be used when? *Adv. Med. Oncol.* **2018**, *10*. [CrossRef] [PubMed]
7. Rosser, C.J.; Urquidi, V.; Goodison, S. Urinary biomarkers of bladder cancer: An update and future perspectives. *Biomark. Med.* **2013**, *7*, 779–790. [CrossRef]
8. Cauberg Evelyne, C.; de Reijke, T.; de la Rosette, J.M.C. Emerging optical techniques in advanced cystoscopy for bladder cancer diagnosis: A review of the current literature. *Indian J. Urol.* **2011**, *27*, 245. [CrossRef] [PubMed]
9. Babjuk, M.; Böhle, A.; Burger, M.; Capoun, O.; Cohen, D.; Compérat, E.M.; Hernández, V.; Kaasinen, E.; Palou, J.; Rouprêt, M.; et al. EAU Guidelines on Non–Muscle-invasive Urothelial Carcinoma of the Bladder: Update 2016. *Eur. Urol.* **2017**, *71*, 447–461. [CrossRef]
10. Svatek, R.S.; Hollenbeck, B.K.; Holmäng, S.; Lee, R.; Kim, S.P.; Stenzl, A.; Lotan, Y. The economics of bladder cancer: Costs and considerations of caring for this disease. *Eur. Urol.* **2014**, *66*, 253–262. [CrossRef] [PubMed]
11. Leal, J.; Luengo-Fernandez, R.; Sullivan, R.; Witjes, J.A. Economic Burden of Bladder Cancer across the European Union. *Eur. Urol.* **2016**, *69*, 438–447. [CrossRef]
12. Sylvester, R.J.; Van Der Meijden, A.P.M.; Oosterlinck, W.; Witjes, J.A.; Bouffioux, C.; Denis, L.; Newling, D.W.W.; Kurth, K. Predicting recurrence and progression in individual patients with stage Ta T1 bladder cancer using EORTC risk tables: A combined analysis of 2596 patients from seven EORTC trials. *Eur. Urol.* **2006**, *49*, 466–475. [CrossRef]
13. Lodewijk, I.; Dueñas, M.; Rubio, C.; Munera-Maravilla, E.; Segovia, C.; Bernardini, A.; Teijeira, A.; Paramio, J.M.; Suárez-Cabrera, C. Liquid Biopsy Biomarkers in Bladder Cancer: A Current Need for Patient Diagnosis and Monitoring. *Int. J. Mol. Sci.* **2019**, *19*, 2514. [CrossRef] [PubMed]
14. Loras, A.; Su, C.; Mart, M.C. Integrative Metabolomic and Transcriptomic Analysis. *Cancers* **2019**, *11*, 686. [CrossRef] [PubMed]
15. De Oliveira, M.C.; Caires, H.R.; Oliveira, M.J.; Fraga, A.; Vasconcelos, M.H.; Ribeiro, R. Urinary Biomarkers in Bladder Cancer: Where Do We Stand and Potential Role of Extracellular Vesicles. *Cancers* **2020**, *12*, 1400. [CrossRef]
16. Armitage, E.G.; Ciborowski, M. Applications of Metabolomics in Cancer Studies. *Adv. Exp. Med. Biol.* **2017**, *965*, 209–234. [CrossRef]
17. Pavlova, N.N.; Thompson, C.B. The Emerging Hallmarks of Cancer Metabolism. *Cell Metab.* **2016**, *23*, 27–47. [CrossRef] [PubMed]
18. Holmes, E.; Wilson, I.D.; Nicholson, J.K. Metabolic phenotyping in health and disease. *Cell* **2008**, *134*, 714–717. [CrossRef] [PubMed]
19. Smolinska, A.; Blanchet, L.; Buydens, L.M.; Wijmenga, S.S. NMR and pattern recognition methods in metabolomics: From data acquisition to biomarker discovery: A review. *Anal. Chim. Acta* **2012**, *750*, 82–97. [CrossRef] [PubMed]
20. Crispo, F.; Condelli, V.; Lepore, S.; Notarangelo, T.; Sgambato, A.; Esposito, F.; Maddalena, F.; Landriscina, M. Metabolic Dysregulations and Epigenetics: A Bidirectional Interplay that Drives Tumor Progression. *Cells* **2019**, *8*, 798. [CrossRef]
21. Wong, C.C.; Qian, Y.; Yu, J. Interplay between epigenetics and metabolism in oncogenesis: Mechanisms and therapeutic approaches. *Oncogene* **2017**, *36*, 3359–3374. [CrossRef] [PubMed]

22. Flavahan, W.A.; Gaskell, E.; Bernstein, B.E. Epigenetic plasticity and the hallmarks of cancer. *Science* **2017**, *357*, eaal2380. [CrossRef] [PubMed]
23. Fouad, Y.A.; Aanei, C. Revisiting the hallmarks of cancer. *Am. J. Cancer Res.* **2017**, *7*, 1016–1036.
24. Frontiers-Metabolism and Transcription in Cancer: Merging Two Classic Tales-Cell and Developmental Biology [Internet]. Available online: https://www.frontiersin.org/articles/10.3389/fcell.00119/full (accessed on 20 August 2020).
25. Putluri, N.; Shojaie, A.; Vasu, V.T.; Vareed, S.K.; Nalluri, S.; Putluri, V.; Thangjam, G.S.; Panzitt, K.; Tallman, C.T.; Butler, C.; et al. Metabolomic profiling reveals potential markers and bioprocesses altered in bladder cancer progression. *Cancer Res.* **2011**, *71*, 7376–7386. [CrossRef]
26. Wittmann, B.M.; Stirdivant, S.M.; Mitchell, M.W.; Wulff, J.E.; McDunn, J.E.; Li, Z.; Dennis-Barrie, A.; Neri, B.P.; Milburn, M.V.; Lotan, Y.; et al. Bladder cancer biomarker discovery using global metabolomic profiling of urine. *PLoS ONE* **2014**, *9*, e115870. [CrossRef]
27. Miranda-Gonçalves, V.; Lameirinhas, A.; Henrique, R.; Jerónimo, C. Metabolism and epigenetic interplay in cancer: Regulation and putative therapeutic targets. *Front. Genet.* **2018**, *9*, 427. [CrossRef] [PubMed]
28. Jin, X.; Yun, S.J.; Jeong, P.; Kim, I.Y.; Kim, W.-J.; Park, S. Diagnosis of bladder cancer and prediction of survival by urinary metabolomics. *Oncotarget* **2014**, *5*, 1635–1645. [CrossRef]
29. Rasmussen, K.D.; Helin, K. Role of TET enzymes in DNA methylation, development, and cancer. *Genes Dev.* **2016**, *30*, 733–750. [CrossRef]
30. Knaap, J.A.; van der Verrijzer, C.P. Undercover: Gene control by metabolites and metabolic enzymes. *Genes Dev.* 11 Enero **2016**, *30*, 2345–2369. [CrossRef]
31. Martínez-reyes, I.; Chandel, N.S. Mitochondrial TCA cycle metabolites control. *Nat. Commun.* **2020**, 1–11. [CrossRef]
32. Majmundar, A.J.; Wong, W.J.; Simon, M.C. Hypoxia inducible factors and the response to hypoxic stress. *Mol. Cell* **2010**, *40*, 294–309. [CrossRef] [PubMed]
33. Zdzisin, B. Alpha-Ketoglutarate as a Molecule with Pleiotropic Activity: Well-Known and Novel Possibilities of Therapeutic Use. *Arch. Immunol. Exp.* **2017**, 21–36. [CrossRef] [PubMed]
34. Jun, J.C.; Rathore, A.; Younas, H.; Gilkes, D.; Polotsky, V.Y. Hypoxia-Inducible Factors and Cancer. *Curr. Sleep Med. Rep.* **2017**, *3*, 1–10. [CrossRef] [PubMed]
35. Theodoropoulos, V.E.; Lazaris, A.C.; Sofras, F.; Gerzelis, I.; Tsoukala, V.; Ghikonti, I.; Manikas, K.; Kastriotis, I. Hypoxia-inducible factor 1 alpha expression correlates with angiogenesis and unfavorable prognosis in bladder cancer. *Eur. Urol.* **2004**, *46*, 200–208. [CrossRef] [PubMed]
36. Lew, C.R.; Guin, S.; Theodorescu, D. Targeting glycogen metabolism in bladder cancer. *Nat. Rev. Urol.* **2015**, *12*, 383–391. [CrossRef] [PubMed]
37. Miao, P.; Sheng, S.; Sun, X.; Liu, J.; Huang, G. Lactate dehydrogenase A in cancer: A promising target for diagnosis and therapy. *IUBMB Life* **2013**, *65*, 904–910. [CrossRef]
38. Zhang, G.; Zhang, Y.; Dong, D.; Wang, F.; Ma, X.; Guan, F. MCT1 regulates aggressive and metabolic phenotypes in bladder cancer. *J. Cancer* **2018**, *9*. [CrossRef] [PubMed]
39. Conde, V.R.; Oliveira, P.F.; Nunes, A.R.; Rocha, C.S.; Ramalhosa, E.; Pereira, J.A.; Alves, M.G.; Silva, B.M. The progression from a lower to a higher invasive stage of bladder cancer is associated with severe alterations in glucose and pyruvate metabolism. *Exp. Cell Res.* **2015**, *335*, 91–98. [CrossRef] [PubMed]
40. DeBerardinis, R.J.; Chandel, N.S. We need to talk about the Warburg effect. *Nat. Metab.* **2020**, *2*, 127–129. [CrossRef]
41. Mylonis, I.; Simos, G.; Paraskeva, E. Hypoxia-Inducible Factors and the Regulation of Lipid Metabolism. *Cells* **2019**, *8*, 214. [CrossRef]
42. Zhang, J.; Wang, L.; Mao, S.; Liu, M.; Zhang, W.; Zhang, Z.; Guo, Y.; Huang, B.; Yan, Y.; Huang, Y.; et al. MiR-1-3p contributes to cell proliferation and invasion by targeting glutaminase in bladder cancer cells. *Cell. Physiol. Biochem.* **2018**, *51*, 513–527. [CrossRef]
43. Zhou, Q.; Zhan, H.; Lin, F.; Liu, Y.; Yang, K.; Gao, Q.; Ding, M.; Liu, Y.; Huang, W.; Cai, Z. LincRNA-p21 suppresses glutamine catabolism and bladder cancer cell growth through inhibiting glutaminase expression. *Biosci. Rep.* **2019**, *29*. [CrossRef]
44. Zhou, Y.; Song, R.; Zhang, Z.; Lu, X.; Zeng, Z.; Hu, C.; Liu, X.; Li, Y.; Hou, J.; Sun, Y.; et al. The development of plasma pseudotargeted GC-MS metabolic profiling and its application in bladder cancer. *Anal. Bioanal. Chem.* **2016**, *408*, 6741–6749. [CrossRef]
45. Khatami, F.; Aghamir, S.M.K.; Tavangar, S.M. Oncometabolites: A new insight for oncology. *Mol. Genet. Genom. Med.* **2019**, *7*. [CrossRef] [PubMed]
46. Sciacovelli, M.; Frezza, C. Oncometabolites: Unconventional triggers of oncogenic signalling cascades. *Free Radic. Biol. Med.* **2016**, *100*, 175–181. [CrossRef]
47. Xiao, M.; Yang, H.; Xu, W.; Ma, S.; Lin, H.; Zhu, H.; Liu, L.; Liu, Y.; Yang, C.; Xu, Y.; et al. Inhibition of α-KG-dependent histone and DNA demethylases by fumarate and succinate that are accumulated in mutations of FH and SDH tumor suppressors. *Genes Dev.* **2012**, *26*, 1326–1338. [CrossRef]
48. Campbell, S.L.; Wellen, K.E. Metabolic Signaling to the Nucleus in Cancer. *Mol. Cell.* **2018**, *71*, 398–408. [CrossRef] [PubMed]
49. Rani, A.; Murphy, J.J. STAT5 in Cancer and Immunity. *J. Interferon. Cytokine Res.* **2016**, *36*, 226–237. [CrossRef] [PubMed]

50. Sun, Y.; Cheng, M.K.; Griffiths, T.R.; Mellon, J.K.; Kai, B.; Kriajevska, M.; Manson, M.M. Inhibition of STAT signalling in bladder cancer by diindolylmethane: Relevance to cell adhesion, migration and proliferation. *Curr. Cancer Drug Targets* **2013**, *13*, 57–68. [CrossRef] [PubMed]
51. Hindupur, S.V.; Schmid, S.C.; Koch, J.A.; Youssef, A.; Baur, E.M.; Wang, D.; Horn, T.; Slotta-Huspenina, J.; Gschwend, J.E.; Holm, P.S.; et al. STAT3/5 Inhibitors Suppress Proliferation in Bladder Cancer and Enhance Oncolytic Adenovirus Therapy. *Int. J. Mol. Sci.* **2020**, *21*, 1106. [CrossRef] [PubMed]
52. Bernstein, B.E.; Meissner, A.; Lander, E.S. The Mammalian Epigenome. *Cell* **2007**, *128*, 669–681. [CrossRef] [PubMed]
53. Yang, W.; Lu, Z. Nuclear PKM2 regulates the Warburg effect. *Cell Cycle* **2013**, *12*, 3154–3158. [CrossRef]
54. Zhu, Q.; Hong, B.; Zhang, L.; Wang, J. Pyruvate kinase M2 inhibits the progression of bladder cancer by targeting MAKP pathway. *J. Cancer Res.* **2018**, *14*, S616–S621. [CrossRef]
55. Wang, X.; Zhang, F.; Wu, X.R. Inhibition of Pyruvate Kinase M2 Markedly Reduces Chemoresistance of Advanced Bladder Cancer to Cisplatin. *Sci. Rep.* **2017**, *7*, 1–13. [CrossRef] [PubMed]
56. Su, Q.; Tao, T.; Tang, L.; Deng, J.; Darko, K.O.; Zhou, S.; Peng, M.; He, S.; Zeng, Q.; Chen, A.F.; et al. Down-regulation of PKM2 enhances anticancer efficiency of THP on bladder cancer. *J. Cell. Mol. Med.* **2018**, 2774–2790. [CrossRef] [PubMed]
57. Huang, C.; Huang, Z.; Bai, P.; Luo, G.; Zhao, X.; Wang, X. Expression of pyruvate kinase M2 in human bladder cancer and its correlation with clinical parameters and prognosis. *Oncol. Targets* **2018**, *11*, 2075–2082. [CrossRef]
58. Boukouris, A.E.; Zervopoulos, S.D.; Michelakis, E.D. Metabolic Enzymes Moonlighting in the Nucleus: Metabolic Regulation of Gene Transcription. *Trends Biochem. Sci.* **2016**, *41*, 712–730. [CrossRef] [PubMed]
59. Vander Heiden, M.G.; DeBerardinis, R.J. Understanding the Intersections between Metabolism and Cancer Biology. *Cell* **2017**, *168*, 657–669. [CrossRef] [PubMed]
60. Liep, J.; Rabien, A.; Jung, K. Feedback networks between microRNAs and epigenetic modifications in urological tumors. *Epigenetics* **2012**, *7*, 315–325. [CrossRef]
61. Tsai, H.-C.; Baylin, S.B. Cancer epigenetics: Linking basic biology to clinical medicine. *Cell Res.* **2011**, *21*, 502–517. [CrossRef]
62. Gupta, R.A.; Shah, N.; Wang, K.C.; Kim, J.; Horlings, H.M.; Wong, D.J.; Tsai, M.-C.; Hung, T.; Argani, P.; Rinn, J.L.; et al. Long non-coding RNA HOTAIR reprograms chromatin state to promote cancer metastasis. *Nature* **2010**, *464*, 1071–1076. [CrossRef] [PubMed]
63. Robertson, A.G.; Kim, J.; Al-Ahmadie, H.; Bellmunt, J.; Guo, G.; Cherniack, A.D.; Hinoue, T.; Laird, P.W.; Hoadley, K.A.; Akbani, R.; et al. Comprehensive Molecular Characterization of Muscle-Invasive Bladder Cancer. *Cell* **2017**, *171*, 540–556.e25. [CrossRef]
64. Weinstein, J.N.; Akbani, R.; Broom, B.M.; Wang, W.; Verhaak, R.G.W.; McConkey, D.; Lerner, S.; Morgan, M.; Creighton, C.J.; Smith, C.; et al. Comprehensive molecular characterization of urothelial bladder carcinoma. *Nature* **2014**, *507*, 315–322. [CrossRef]
65. Yamaguchi, H.; Hung, M.-C. Regulation and Role of EZH2 in Cancer. *Cancer Res. Treat.* **2014**, *46*, 209–222. [CrossRef]
66. Santos, M.; Martínez-Fernández, M.; Dueñas, M.; García-Escudero, R.; Alfaya, B.; Villacampa, F.; Saiz-Ladera, C.; Costa, C.; Oteo, M.; Duarte, J.; et al. In vivo disruption of an Rb-E2F-Ezh2 signaling loop causes bladder cancer. *Cancer Res.* **2014**, *74*, 6565–6577. [CrossRef]
67. Martínez-Fernández, M.; Feber, A.; Dueñas, M.; Segovia, C.; Rubio, C.; Fernandez, M.; Villacampa, F.; Duarte, J.; López-Calderón, F.F.; Gómez-Rodriguez, M.J.; et al. Analysis of the Polycomb-related lncRNAs HOTAIR and ANRIL in bladder cancer. *Clin. Epigenetics* **2015**, *7*, 109. [CrossRef] [PubMed]
68. Martínez-Fernández, M.; Dueñas, M.; Feber, A.; Segovia, C.; García-Escudero, R.; Rubio, C.; López-Calderón, F.F.; Díaz-García, C.; Villacampa, F.; Duarte, J.; et al. A Polycomb-mir200 loop regulates clinical outcome in bladder cancer. *Oncotarget* **2015**, *6*, 42258–42275. [CrossRef]
69. Viré, E.; Brenner, C.; Deplus, R.; Blanchon, L.; Fraga, M.; Didelot, C.; Morey, L.; Van Eynde, A.; Bernard, D.; Vanderwinden, J.-M.; et al. The Polycomb group protein EZH2 directly controls DNA methylation. *Nature* **2006**, *439*, 871–874. [CrossRef]
70. Kuzmichev, A.; Nishioka, K.; Erdjument-Bromage, H.; Tempst, P.; Reinberg, D. Histone methyltransferase activity associated with a human multiprotein complex containing the Enhancer of Zeste protein. *Genes Dev.* **2002**, *16*, 2893–2905. [CrossRef]
71. Woolbright, B.L.; Ayres, M.; Taylor, J.A. Metabolic changes in bladder cancer. *Urol. Oncol. Semin. Orig. Investig.* **2018**, *36*, 327–337. [CrossRef]
72. Ward, P.S.; Thompson, C.B. Metabolic Reprogramming: A Cancer Hallmark Even Warburg Did Not Anticipate. *Cancer Cell* **2012**, *21*, 297–308. [CrossRef]
73. Thakur, C.; Chen, F. Connections between metabolism and epigenetics in cancers. *Semin. Cancer Biol.* **2019**, *57*, 52–58. [CrossRef] [PubMed]
74. Dang, C.V.; O'Donnell, K.A.; Zeller, K.I.; Nguyen, T.; Osthus, R.C.; Li, F. The c-Myc target gene network. *Semin. Cancer Biol.* **2006**, *16*, 253–264. [CrossRef]
75. Gao, P.; Tchernyshyov, I.; Chang, T.C.; Lee, Y.S.; Kita, K.; Ochi, T.; Zeller, K.I.; De Marzo, A.M.; Van Eyk, J.E.; Mendell, J.T.; et al. C-Myc suppression of miR-23a/b enhances mitochondrial glutaminase expression and glutamine metabolism. *Nature* **2009**, *458*, 762–765. [CrossRef]
76. Cairns, R.A.; Harris, I.S.; Mak, T.W. Regulation of cancer cell metabolism. *Nat. Rev. Cancer.* **2011**, *11*, 85–95. [CrossRef]
77. Trachootham, D.; Zhou, Y.; Zhang, H.; Demizu, Y.; Chen, Z.; Pelicano, H.; Chiao, P.J.; Achanta, G.; Arlinghaus, R.B.; Liu, J.; et al. Selective killing of oncogenically transformed cells through a ROS-mediated mechanism by β-phenylethyl isothiocyanate. *Cancer Cell* **2006**, *10*, 241–252. [CrossRef]

78. Semenza, G.L. HIF-1: Upstream and downstream of cancer metabolism. *Curr. Opin. Genet. Dev.* **2010**, *20*, 51–56. [CrossRef]
79. Koukourakis, M.I.; Kakouratos, C.; Kalamida, D.; Bampali, Z.; Mavropoulou, S.; Sivridis, E.; Giatromanolaki, A. Hypoxia-inducible proteins HIF1α and lactate dehydrogenase LDH5, key markers of anaerobic metabolism, relate with stem cell markers and poor post-radiotherapy outcome in bladder cancer. *Int. J. Radiat. Biol.* **2016**, *92*, 353–363. [CrossRef]
80. Plas, D.R.; Thompson, C.B. Akt-dependent transformation: There is more to growth than just surviving. *Oncogene* **2005**, *24*, 7435–7442. [CrossRef]
81. Inoki, K.; Corradetti, M.N.; Guan, K.L. Dysregulation of the TSC-mTOR pathway in human disease. *Nat. Genet.* **2005**, *37*, 19–24. [CrossRef]
82. Wolff, E.M.; Chihara, Y.; Pan, F.; Weisenberger, D.J.; Siegmund, K.D.; Sugano, K.; Kawashima, K.; Laird, P.W.; Jones, P.A.; Liang, G. Unique DNA Methylation Patterns Distinguish Noninvasive and Invasive Urothelial Cancers and Establish an Epigenetic Field Defect in Premalignant Tissue. *Cancer Res.* **2010**, *70*, 8169–8178. [CrossRef] [PubMed]
83. Catto, J.W.F.; Azzouzi, A.-R.; Rehman, I.; Feeley, K.M.; Cross, S.S.; Amira, N.; Fromont, G.; Sibony, M.; Cussenot, O.; Meuth, M.; et al. Promoter Hypermethylation Is Associated With Tumor Location, Stage, and Subsequent Progression in Transitional Cell Carcinoma. *J. Clin. Oncol.* **2005**, *23*, 2903–2910. [CrossRef]
84. Kandimalla, R.; van Tilborg, A.A.G.; Kompier, L.C.; Stumpel, D.J.P.M.; Stam, R.W.; Bangma, C.H.; Zwarthoff, E.C. Genome-wide Analysis of CpG Island Methylation in Bladder Cancer Identified TBX2, TBX3, GATA2, and ZIC4 as pTa-Specific Prognostic Markers. *Eur. Urol.* **2012**, *61*, 1245–1256. [CrossRef]
85. Lameirinhas, A.; Miranda-Gonçalves, V.; Henrique, R.; Jerónimo, C. The complex interplay between metabolic reprogramming and epigenetic alterations in renal cell carcinoma. *Genes* **2019**, *10*, 264. [CrossRef]
86. Kimura, F.; Florl, A.R.; Seifert, H.-H.; Louhelainen, J.; Maas, S.; Knowles, M.A.; Schulz, W.A. Destabilization of chromosome 9 in transitional cell carcinoma of the urinary bladder. *Br. J. Cancer* **2001**, *85*, 1887–1893. [CrossRef]
87. Nakagawa, T.; Kanai, Y.; Ushijima, S.; Kitamura, T.; Kakizoe, T.; Hirohashi, S. DNA hypomethylation on pericentromeric satellite regions significantly correlates with loss of heterozygosity on chromosome 9 in urothelial carcinomas. *J. Urol.* **2005**, *173*, 243–246. [CrossRef]
88. Lopez-Serra, P.; Marcilla, M.; Villanueva, A.; Ramos-Fernandez, A.; Palau, A.; Leal, L.; Wahi, J.E.; Setien-Baranda, F.; Szczesna, K.; Moutinho, C.; et al. A DERL3-associated defect in the degradation of SLC2A1 mediates the Warburg effect. *Nat. Commun.* **2014**, *5*, 3608. [CrossRef]
89. Leung, J.Y.; Chia, K.; Ong, D.S.T.; Taneja, R. Interweaving Tumor Heterogeneity into the Cancer Epigenetic/Metabolic Axis. *Antioxid. Redox Signal.* **2020**, *33*, 946–965. [CrossRef] [PubMed]
90. Wolf, A.; Agnihotri, S.; Munoz, D.; Guha, A. Developmental profile and regulation of the glycolytic enzyme hexokinase 2 in normal brain and glioblastoma multiforme. *Neurobiol. Dis.* **2011**, *44*, 84–91. [CrossRef]
91. Desai, S.; Ding, M.; Wang, B.; Lu, Z.; Zhao, Q.; Shaw, K.; Alfred Yung, W.K.; Weinstein, J.N.; Tan, M.; Yao, J. Tissue-specific isoform switch and DNA hypomethylation of the pyruvate kinase PKM gene in human cancers. *Oncotarget* **2014**, *5*, 8202–8210. [CrossRef]
92. Massari, F.; Ciccarese, C.; Santoni, M.; Iacovelli, R.; Mazzucchelli, R.; Piva, F.; Scarpelli, M.; Berardi, R.; Tortora, G.; Lopez-Beltran, A.; et al. Metabolic phenotype of bladder cancer. *Cancer Treat. Rev.* **2016**, *45*, 46–57. [CrossRef]
93. Cheng, S.; Wang, G.; Wang, Y.; Cai, L.; Qian, K.; Ju, L.; Liu, X.; Xiao, Y.; Wang, X. Fatty acid oxidation inhibitor etomoxir suppresses tumor progression and induces cell cycle arrest via PPARγ-mediated pathway in bladder cancer. *Clin. Sci.* **2019**, *133*, 1745–1758. [CrossRef]
94. Tian, Z.; Tian, Z.; Meng, L.; Meng, L.; Long, X.; Diao, T.; Hu, M.; Wang, M.; Liu, X.; Wang, J.; et al. DNA methylation-based classification and identification of bladder cancer prognosis-associated subgroups. *Cancer Cell Int.* **2020**, *20*, 1–11. [CrossRef] [PubMed]
95. Casagrande, V.; Mauriello, A.; Bischetti, S.; Mavilio, M.; Federici, M.; Menghini, R. Hepatocyte specific TIMP3 expression prevents diet dependent fatty liver disease and hepatocellular carcinoma. *Sci. Rep.* **2017**, *7*. [CrossRef]
96. Stöhr, R.; Kappel, B.A.; Carnevale, D.; Cavalera, M.; Mavilio, M.; Arisi, I.; Fardella, V.; Cifelli, G.; Casagrande, V.; Rizza, S.; et al. TIMP3 interplays with apelin to regulate cardiovascular metabolism in hypercholesterolemic mice. *Mol. Metab.* **2015**, *4*, 741–752. [CrossRef] [PubMed]
97. Gurioli, G.; Martignano, F.; Salvi, S.; Costantini, M.; Gunelli, R.; Casadio, V. GSTP1 methylation in cancer: A liquid biopsy biomarker? *Clin. Chem. Lab. Med.* **2018**, *56*, 705–717. [CrossRef] [PubMed]
98. Leygo, C.; Williams, M.; Jin, H.C.; Chan, M.W.Y.; Chu, W.K.; Grusch, M.; Cheng, Y.Y. DNA Methylation as a Noninvasive Epigenetic Biomarker for the Detection of Cancer. *Dis. Markers* **2017**, *2017*, 3726595. [CrossRef]
99. Louie, S.M.; Grossman, E.A.; Crawford, L.A.; Ding, L.; Camarda, R.; Huffman, T.R.; Miyamoto, D.K.; Goga, A.; Weerapana, E.; Nomura, D.K. GSTP1 Is a Driver of Triple-Negative Breast Cancer Cell Metabolism and Pathogenicity. *Cell Chem. Biol.* **2016**, *23*, 567–578. [CrossRef] [PubMed]
100. Chuang, J.-J.; Dai, Y.-C.; Lin, Y.-L.; Chen, Y.-Y.; Lin, W.-H.; Chan, H.-L.; Liu, Y.-W. Downregulation of glutathione S-transferase M1 protein in N-butyl-N-(4-hydroxybutyl)nitrosamine-induced mouse bladder carcinogenesis. *Toxicol. Appl. Pharm.* **2014**, *279*, 322–330. [CrossRef] [PubMed]
101. Li, H.-T.; Duymich, C.E.; Weisenberger, D.J.; Liang, G. Genetic and Epigenetic Alterations in Bladder Cancer. *Int. Neurourol. J.* **2016**, *20*, S84–S94. [CrossRef] [PubMed]

102. Marques-Magalhães, Â.; Graça, I.; Henrique, R.; Jerónimo, C. Targeting DNA methyltranferases in urological tumors. *Front. Pharm.* **2018**, *9*. [CrossRef] [PubMed]
103. Martinez, V.G.; Munera-Maravilla, E.; Bernardini, A.; Rubio, C.; Suarez-Cabrera, C.; Segovia, C.; Lodewijk, I.; Dueñas, M.; Martínez-Fernández, M.; Paramio, J.M. Epigenetics of Bladder Cancer: Where Biomarkers and Therapeutic Targets Meet. *Front. Genet.* **2019**, *10*, 1–27. [CrossRef]
104. Chan, M.W.Y.; Chan, L.W.; Tang, N.L.S.; Tong, J.H.M.; Lo, K.W.; Lee, T.L.; Cheung, H.Y.; Wong, W.S.; Chan, P.S.F.; Lai, F.M.M.; et al. Hypermethylation of multiple genes in tumor tissues and voided urine in urinary bladder cancer patients. *Clin. Cancer Res.* **2002**, *8*, 464–470.
105. Dulaimi, E.; Uzzo, R.G.; Greenberg, R.E.; Al-Saleem, T.; Cairns, P. Detection of bladder cancer in urine by a tumor suppressor gene hypermethylation panel. *Clin. Cancer Res.* **2004**, *10*, 1887–1893. [CrossRef] [PubMed]
106. Maruyama, R.; Toyooka, S.; Toyooka, K.O.; Harada, K.; Virmani, A.K.; Zöchbauer-Müller, S.; Farinas, A.J.; Vakar-Lopez, F.; Minna, J.D.; Sagalowsky, A.; et al. Aberrant promoter methylation profile of bladder cancer and its relationship to clinicopathological features. *Cancer Res.* **2001**, *61*, 8659–8663. [PubMed]
107. Yeon, A.; You, S.; Kim, M.; Gupta, A.; Park, M.H.; Weisenberger, D.J.; Liang, G.; Kim, J. Rewiring of cisplatin-resistant bladder cancer cells through epigenetic regulation of genes involved in amino acid metabolism. *Theranostics* **2018**, *8*, 4520–4534. [CrossRef] [PubMed]
108. Tada, Y.; Wada, M.; Kuroiwa, K.; Kinugawa, N.; Harada, T.; Nagayama, J.; Nakagawa, M.; Naito, S.; Kuwano, M. MRD1 gene overexpression and altered degree of methylation at the promoter region in bladder cancer during chemotherapeutic treatment. *Clin. Cancer Res.* **2000**, *6*, 4618–4627.
109. Peng, L.; Zhong, X. Epigenetic regulation of drug metabolism and transport. *Acta Pharm. Sin. B* **2015**, *5*, 106–112. [CrossRef] [PubMed]
110. Okino, S.T.; Pookot, D.; Li, L.C.; Zhao, H.; Urakami, S.; Shiina, H.; Igawa, M.; Dahiya, R. Epigenetic inactivation of the dioxin-responsive Cytochrome P4501A1 gene in human prostate cancer. *Cancer Res.* **2006**, *66*, 7420–7428. [CrossRef] [PubMed]
111. Ieiri, I.; Hirota, T.; Takane, H.; Higuchi, S. Epigenetic Regulation of Genes Encoding Drug-Metabolizing Enzymes and Transporters; DNA Methylation and Other Mechanisms. *Curr. Drug Metab.* **2008**, *9*, 34–38. [CrossRef] [PubMed]
112. Luo, W.; Karpf, A.R.; Deeb, K.K.; Muindi, J.R.; Morrison, C.D.; Johnson, C.S.; Trump, D.L. Epigenetic regulation of vitamin D 24-hydroxylase/CYP24A1 in human prostate cancer. *Cancer Res.* **2010**, *70*, 5953–5962. [CrossRef] [PubMed]
113. Li, B.; Carey, M.; Workman, J.L. The Role of Chromatin during Transcription. *Cell* **2007**, *128*, 707–719. [CrossRef]
114. Fraga, M.F.; Ballestar, E.; Villar-Garea, A.; Boix-Chornet, M.; Espada, J.; Schotta, G.; Bonaldi, T.; Haydon, C.; Ropero, S.; Petrie, K.; et al. Loss of acetylation at Lys16 and trimethylation at Lys20 of histone H4 is a common hallmark of human cancer. *Nat. Genet.* **2005**, *37*, 391–400. [CrossRef] [PubMed]
115. Segovia, C.; San José-Enériz, E.; Munera-Maravilla, E.; Martínez-Fernández, M.; Garate, L.; Miranda, E.; Vilas-Zornoza, A.; Lodewijk, I.; Rubio, C.; Segrelles, C.; et al. Inhibition of a G9a/DNMT network triggers immune-mediated bladder cancer regression. *Nat. Med.* **2019**, *25*, 1073–1081. [CrossRef]
116. Gupta, P.B.; Fillmore, C.M.; Jiang, G.; Shapira, S.D.; Tao, K.; Kuperwasser, C.; Lander, E.S. Stochastic State Transitions Give Rise to Phenotypic Equilibrium in Populations of Cancer Cells. *Cell* **2011**, *146*, 633–644. [CrossRef] [PubMed]
117. Zhang, T.; Gong, Y.; Meng, H.; Li, C.; Xue, L. Symphony of epigenetic and metabolic regulation—Interaction between the histone methyltransferase EZH2 and metabolism of tumor. *Clin. Epigenetics* **2020**, *12*. [CrossRef] [PubMed]
118. Papathanassiu, A.E.; Ko, J.H.; Imprialou, M.; Bagnati, M.; Srivastava, P.K.; Vu, H.A.; Cucchi, D.; McAdoo, S.P.; Ananieva, E.A.; Mauro, C.; et al. BCAT1 controls metabolic reprogramming in activated human macrophages and is associated with inflammatory diseases. *Nat. Commun.* **2017**, *8*. [CrossRef]
119. Cha, T.-L.; Zhou, B.P.; Xia, W.; Wu, Y.; Yang, C.-C.; Chen, C.-T.; Ping, B.; Otte, A.P.; Hung, M.-C. Akt-Mediated Phosphorylation of EZH2 Suppresses Methylation of Lysine 27 in Histone H3. *Science* **2005**, *310*, 306–310. [CrossRef] [PubMed]
120. Yang, M.; Vousden, K.H. Serine and one-carbon metabolism in cancer. *Nat. Rev. Cancer* **2016**, *16*, 650–662. [CrossRef]
121. Tao, T.; Chen, M.; Jiang, R.; Guan, H.; Huang, Y.; Su, H.; Hu, Q.; Han, X.; Xiao, J. Involvement of EZH2 in aerobic glycolysis of prostate cancer through miR-181b/HK2 axis. *Oncol. Rep.* **2017**, *37*, 1430–1436. [CrossRef]
122. Wan, D.; Liu, C.; Sun, Y.; Wang, W.; Huang, K.; Zheng, L. MacroH2A1.1 cooperates with EZH2 to promote adipogenesis by regulating Wnt signaling. *J. Mol. Cell Biol.* **2017**, *9*, 1–13. [CrossRef]
123. Yi, S.A.; Um, S.H.; Lee, J.; Yoo, J.H.; Bang, S.Y.; Park, E.K.; Lee, M.G.; Nam, K.H.; Jeon, Y.J.; Park, J.W.; et al. S6K1 Phosphorylation of H2B Mediates EZH2 Trimethylation of H3: A Determinant of Early Adipogenesis. *Mol. Cell* **2016**, *62*, 443–452. [CrossRef]
124. Ahmad, F.; Patrick, S.; Sheikh, T.; Sharma, V.; Pathak, P.; Malgulwar, P.B.; Kumar, A.; Joshi, S.D.; Sarkar, C.; Sen, E. Telomerase reverse transcriptase (TERT)—enhancer of zeste homolog 2 (EZH2) network regulates lipid metabolism and DNA damage responses in glioblastoma. *J. Neurochem.* **2017**, *143*, 671–683. [CrossRef]
125. Dong, Z.; Li, C.; Yin, C.; Xu, M.; Liu, S.; Gao, M. LncRNA PU.1 AS regulates arsenic-induced lipid metabolism through EZH2/Sirt6/SREBP-1c pathway. *J. Env. Sci.* **2019**, *85*, 138–146. [CrossRef] [PubMed]
126. Cormerais, Y.; Massard, P.A.; Vucetic, M.; Giuliano, S.; Tambutté, E.; Durivault, J.; Vial, V.; Endou, H.; Wempe, M.F.; Parks, S.K.; et al. The glutamine transporter ASCT2 (SLC1A5) promotes tumor growth independently of the amino acid transporter LAT1 (SLC7A5). *J. Biol. Chem.* **2018**, *293*, 2877–2887. [CrossRef] [PubMed]

127. Gu, Z.; Liu, Y.; Cai, F.; Patrick, M.; Zmajkovic, J.; Cao, H.; Zhang, Y.; Tasdogan, A.; Chen, M.; Qi, L.; et al. Loss of EZH2 reprograms BCAA metabolism to drive leukemic transformation. *Cancer Discov.* **2019**, *9*, 1228–1247. [CrossRef] [PubMed]
128. Vantaku, V.; Putluri, V.; Bader, D.A.; Maity, S.; Ma, J.; Arnold, J.M.; Rajapakshe, K.; Donepudi, S.R.; von Rundstedt, F.C.; Devarakonda, V.; et al. Epigenetic loss of AOX1 expression via EZH2 leads to metabolic deregulations and promotes bladder cancer progression. *Oncogene* **2020**, *39*, 6265–6285. [CrossRef]
129. Martínez-Fernández, M.; Rubio, C.; Segovia, C.; López-Calderón, F.F.; Dueñas, M.; Paramio, J.M. EZH2 in Bladder Cancer, a Promising Therapeutic Target. *Int. J. Mol. Sci.* **2015**, *16*, 27107–27132. [CrossRef]
130. Bracken, A.P.; Pasini, D.; Capra, M.; Prosperini, E.; Colli, E.; Helin, K. EZH2 is downstream of the pRB-E2F pathway, essential for proliferation and amplified in cancer. *EMBO J.* **2003**, *22*, 5323–5335. [CrossRef] [PubMed]
131. Benetatos, L.; Voulgaris, E.; Vartholomatos, G.; Hatzimichael, E. Non-coding RNAs and EZH2 interactions in cancer: Long and short tales from the transcriptome. *Int. J. Cancer* **2013**, *133*, 267–274. [CrossRef]
132. Ding, J.; Li, T.; Wang, X.; Zhao, E.; Choi, J.H.; Yang, L.; Zha, Y.; Dong, Z.; Huang, S.; Asara, J.M.; et al. The histone H3 methyltransferase G9A epigenetically activates the serine-glycine synthesis pathway to sustain cancer cell survival and proliferation. *Cell Metab.* **2013**, *18*, 896–907. [CrossRef] [PubMed]
133. Harding, H.P.; Zhang, Y.; Zeng, H.; Novoa, I.; Lu, P.D.; Calfon, M.; Sadri, N.; Yun, C.; Popko, B.; Paules, R.; et al. An integrated stress response regulates amino acid metabolism and resistance to oxidative stress. *Mol. Cell* **2003**, *11*, 619–633. [CrossRef]
134. Ye, J.; Mancuso, A.; Tong, X.; Ward, P.S.; Fan, J.; Rabinowitz, J.D.; Thompson, C.B. Pyruvate kinase M2 promotes de novo serine synthesis to sustain mTORC1 activity and cell proliferation. *Proc. Natl. Acad. Sci. USA* **2012**, *109*, 6904–6909. [CrossRef] [PubMed]
135. Roth, S.Y.; Denu, J.M.; Allis, C.D. Histone acetyltransferases. *Annu. Rev. Biochem.* **2001**, *70*, 81–120. [CrossRef]
136. Lv, L.; Li, D.; Zhao, D.; Lin, R.; Chu, Y.; Zhang, H.; Zha, Z.; Liu, Y.; Li, Z.; Xu, Y.; et al. Acetylation Targets the M2 Isoform of Pyruvate Kinase for Degradation through Chaperone-Mediated Autophagy and Promotes Tumor Growth. *Mol. Cell* **2011**, *42*, 719–730. [CrossRef] [PubMed]
137. Ebrahimi, A.; Schittenhelm, J.; Honegger, J.; Schluesener, H. Prognostic relevance of global histone 3 lysine 9 acetylation in ependymal tumors: Laboratory investigation. *J. Neurosurg.* **2013**, *119*, 1424–1431. [CrossRef]
138. Ellinger, J.; Schneider, A.C.; Bachmann, A.; Kristiansen, G.; Müller, S.C.; Rogenhofer, S. Evaluation of Global Histone Acetylation Levels in Bladder Cancer Patients. *Anticancer. Res.* **2016**, *36*, 3961–3964. [PubMed]
139. Giannopoulou, A.F.; Velentzas, A.D.; Konstantakou, E.G.; Avgeris, M.; Katarachia, S.A.; Papandreou, N.C.; Kalavros, N.I.; Mpakou, V.E.; Iconomidou, V.; Anastasiadou, E.; et al. Revisiting Histone Deacetylases in Human Tumorigenesis: The Paradigm of Urothelial Bladder Cancer. *Int. J. Mol. Sci.* **2019**, *20*, 1291. [CrossRef]
140. Song, Y.; Wu, F.; Wu, J. Targeting histone methylation for cancer therapy: Enzymes, inhibitors, biological activity and perspectives. *J. Hematol. Oncol.* **2016**, *9*. [CrossRef]
141. Wan, W.; Peng, K.; Li, M.; Qin, L.; Tong, Z.; Yan, J.; Shen, B.; Yu, C. Histone demethylase JMJD1A promotes urinary bladder cancer progression by enhancing glycolysis through coactivation of hypoxia inducible factor 1α. *Oncogene* **2017**, *36*, 3868–3877. [CrossRef]
142. Mimura, I.; Nangaku, M.; Kanki, Y.; Tsutsumi, S.; Inoue, T.; Kohro, T.; Yamamoto, S.; Fujita, T.; Shimamura, T.; Suehiro, J.-I.; et al. Dynamic Change of Chromatin Conformation in Response to Hypoxia Enhances the Expression of GLUT3 (SLC2A3) by Cooperative Interaction of Hypoxia-Inducible Factor 1 and KDM3A. *Mol. Cell. Biol.* **2012**, *32*, 3018–3032. [CrossRef] [PubMed]
143. Ezponda, T.; Dupéré-Richer, D.; Will, C.M.; Small, E.C.; Varghese, N.; Patel, T.; Nabet, B.; Popovic, R.; Oyer, J.; Bulic, M.; et al. UTX/KDM6A Loss Enhances the Malignant Phenotype of Multiple Myeloma and Sensitizes Cells to EZH2 inhibition. *Cell Rep.* **2017**, *21*, 628–640. [CrossRef]
144. Chang, S.; Yim, S.; Park, H. The cancer driver genes IDH1/2, JARID1C/ KDM5C, and UTX/ KDM6A: Crosstalk between histone demethylation and hypoxic reprogramming in cancer metabolism. *Exp. Mol. Med.* **2019**, *51*, 1–17. [CrossRef]
145. Miyake, H.; Nelson, C.; Rennie, P.S.; Gleave, M.E. Acquisition of chemoresistant phenotype by overexpression of the antiapoptotic gene Testosterone-repressed prostate message-2 in prostate cancer xenograft models. *Cancer Res.* **2000**, *60*, 2547–2554. [PubMed]
146. Hurst, C.D.; Alder, O.; Platt, F.M.; Droop, A.; Stead, L.F.; Burns, J.E.; Burghel, G.J.; Jain, S.; Klimczak, L.J.; Lindsay, H.; et al. Genomic Subtypes of Non-invasive Bladder Cancer with Distinct Metabolic Profile and Female Gender Bias in KDM6A Mutation Frequency. *Cancer Cell* **2017**, *32*, 701–715.e7. [CrossRef] [PubMed]
147. Ler, L.D.; Ghosh, S.; Chai, X.; Thike, A.A.; Heng, H.L.; Siew, E.Y.; Dey, S.; Koh, L.K.; Lim, J.Q.; Lim, W.K.; et al. Loss of tumor suppressor KDM6A amplifies PRC2-regulated transcriptional repression in bladder cancer and can be targeted through inhibition of EZH2. *Sci. Transl. Med.* **2017**, *9*, eaai8312. [CrossRef] [PubMed]
148. Saxton, R.A.; Sabatini, D.M. mTOR Signaling in Growth, Metabolism, and Disease. *Cell* **2017**, *168*, 960–976. [CrossRef]
149. Frydzińska, Z.; Owczarek, A.; Winiarska, K. Sirtuiny i ich rola w regulacji metabolizmu. *Postepy Biochem.* **2019**, *65*, 31–40. [CrossRef] [PubMed]
150. Niegisch, G.; Knievel, J.; Koch, A.; Hader, C.; Fischer, U.; Albers, P.; Schulz, W.A. Changes in histone deacetylase (HDAC) expression patterns and activity of HDAC inhibitors in urothelial cancers. *Urol. Oncol. Semin. Orig. Investig.* **2013**, *31*, 1770–1779. [CrossRef] [PubMed]
151. Varambally, S.; Dhanasekaran, S.M.; Zhou, M.; Barrette, T.R.; Kumar-Sinha, C.; Sanda, M.G.; Ghosh, D.; Pienta, K.J.; Sewalt, R.G.A.B.; Otte, A.P.; et al. The polycomb group protein EZH2 is involved in progression of prostate cancer. *Nature* **2002**, *419*, 624–629. [CrossRef]

152. Monteiro-Reis, S.; Lameirinhas, A.; Miranda-Gonçalves, V.; Felizardo, D.; Dias, P.C.; Oliveira, J.; Graça, I.; Gonçalves, C.S.; Costa, B.M.; Henrique, R.; et al. Sirtuins' deregulation in bladder cancer: Sirt7 is implicated in tumor progression through epithelial to mesenchymal transition promotion. *Cancers* **2020**, *12*, 1066. [CrossRef]
153. Zhong, L.; Mostoslavsky, R. SIRT6: A master epigenetic gatekeeper of glucose metabolism. *Transcription* **2010**, *1*, 17–21. [CrossRef] [PubMed]
154. Hallows, W.C.; Yu, W.; Denu, J.M. Regulation of glycolytic enzyme phosphoglycerate mutase-1 by Sirt1 protein-mediated deacetylation. *J. Biol. Chem.* **2012**, *287*, 3850–3858. [CrossRef] [PubMed]
155. Liu, P.Y.; Xu, N.; Malyukova, A.; Scarlett, C.J.; Sun, Y.T.; Zhang, X.D.; Ling, D.; Su, S.P.; Nelson, C.; Chang, D.K.; et al. The histone deacetylase SIRT2 stabilizes Myc oncoproteins. *Cell Death Differ.* **2013**, *20*, 503–514. [CrossRef]
156. Heiden, M.G.V.; Cantley, L.C.; Thompson, C.B. Understanding the warburg effect: The metabolic requirements of cell proliferation. *Science* **2009**, *324*, 1029–1033. [CrossRef] [PubMed]
157. Bell, E.L.; Emerling, B.M.; Ricoult, S.J.H.; Guarente, L. SirT3 suppresses hypoxia inducible factor 1α and tumor growth by inhibiting mitochondrial ROS production. *Oncogene* **2011**, *30*, 2986–2996. [CrossRef] [PubMed]
158. Jeong, S.M.; Xiao, C.; Finley, L.W.S.; Lahusen, T.; Souza, A.L.; Pierce, K.; Li, Y.H.; Wang, X.; Laurent, G.; German, N.J.; et al. SIRT4 has tumor-suppressive activity and regulates the cellular metabolic response to dna damage by inhibiting mitochondrial glutamine metabolism. *Cancer Cell* **2013**, *23*, 450–463. [CrossRef]
159. Wu, M.; Dickinson, S.I.; Wang, X.; Zhang, J. Expression and function of SIRT6 in muscle invasive urothelial carcinoma of the bladder. *Int. J. Clin. Exp. Pathol.* **2014**, *7*, 6504–6513.
160. Shin, J.; He, M.; Liu, Y.; Paredes, S.; Villanova, L.; Brown, K.; Qiu, X.; Nabavi, N.; Mohrin, M.; Wojnoonski, K.; et al. SIRT7 represses myc activity to suppress er stress and prevent fatty liver disease. *Cell Rep.* **2013**, *5*, 654–665. [CrossRef]
161. Barber, M.F.; Michishita-Kioi, E.; Xi, Y.; Tasselli, L.; Kioi, M.; Moqtaderi, Z.; Tennen, R.I.; Paredes, S.; Young, N.L.; Chen, K.; et al. SIRT7 links H3K18 deacetylation to maintenance of oncogenic transformation. *Nature* **2012**, *487*, 114–118. [CrossRef]
162. Battistelli, C.; Cicchini, C.; Santangelo, L.; Tramontano, A.; Grassi, L.; Gonzalez, F.J.; De Nonno, V.; Grassi, G.; Amicone, L.; Tripodi, M. The Snail repressor recruits EZH2 to specific genomic sites through the enrollment of the lncRNA HOTAIR in epithelial-to-mesenchymal transition. *Oncogene* **2017**, *36*, 942–955. [CrossRef]
163. Han, Y.; Liu, Y.; Zhang, H.; Wang, T.; Diao, R.; Jiang, Z.; Gui, Y.; Cai, Z. Hsa-miR-125b suppresses bladder cancer development by down-regulating oncogene SIRT7 and oncogenic long noncoding RNA MALAT1. *Febs Lett.* **2013**, *587*, 3875–3882. [CrossRef]
164. Xie, H.; Liao, X.; Chen, Z.; Fang, Y.; He, A.; Zhong, Y.; Gao, Q.; Xiao, H.; Li, J.; Huang, W.; et al. LncRNA MALAT1 inhibits apoptosis and promotes invasion by antagonizing miR-125b in bladder cancer cells. *J. Cancer* **2017**, *8*, 3803–3811. [CrossRef] [PubMed]
165. Beltrán-Anaya, F.O.; Cedro-Tanda, A.; Hidalgo-Miranda, A.; Romero-Cordoba, S.L. Insights into the regulatory role of non-coding RNAs in cancer metabolism. *Front. Physiol.* **2016**, *7*, 1–21. [CrossRef]
166. Pedroza-Torres, A.; Romero-Córdoba, S.L.; Justo-Garrido, M.; Salido-Guadarrama, I.; Rodríguez-Bautista, R.; Montaño, S.; Muñiz-Mendoza, R.; Arriaga-Canon, C.; Fragoso-Ontiveros, V.; Álvarez-Gómez, R.M.; et al. MicroRNAs in Tumor Cell Metabolism: Roles and Therapeutic Opportunities. *Front. Oncol.* **2019**, *9*, 1–24. [CrossRef] [PubMed]
167. Peschiaroli, A.; Giacobbe, A.; Formosa, A.; Markert, E.K.; Bongiorno-Borbone, L.; Levine, A.J.; Candi, E.; D'Alessandro, A.; Zolla, L.; Finazzi Agrò, A.; et al. MiR-143 regulates hexokinase 2 expression in cancer cells. *Oncogene* **2013**, *32*, 797–802. [CrossRef]
168. Yoshino, H.; Seki, N.; Itesako, T.; Chiyomaru, T.; Nakagawa, M.; Enokida, H. Aberrant expression of microRNAs in bladder cancer. *Nat. Rev. Urol.* **2013**, *10*, 396–404. [CrossRef] [PubMed]
169. Enokida, H.; Yoshino, H.; Matsushita, R.; Nakagawa, M. The role of microRNAs in bladder cancer. *Investig. Clin. Urol.* **2016**, *57*, S60. [CrossRef] [PubMed]
170. Woodford, M.R.; Chen, V.Z.; Backe, S.J.; Bratslavsky, G.; Mollapour, M. Structural and functional regulation of lactate dehydrogenase-A in cancer. *Future Med. Chem.* **2020**, *12*, 439–455. [CrossRef] [PubMed]
171. Li, P.; Yang, X.; Cheng, Y.; Zhang, X.; Yang, C.; Deng, X.; Li, P.; Tao, J.; Yang, H.; Wei, J.; et al. MicroRNA-218 Increases the Sensitivity of Bladder Cancer to Cisplatin by Targeting Glut1. *Cell. Physiol. Biochem.* **2017**, *41*, 921–932. [CrossRef]
172. Chen, Y.H.; Heneidi, S.; Lee, J.M.; Layman, L.C.; Stepp, D.W.; Gamboa, G.M.; Chen, B.S.; Chazenbalk, G.; Azziz, R. Mirna-93 inhibits glut4 and is overexpressed in adipose tissue of polycystic ovary syndrome patients and women with insulin resistance. *Diabetes* **2013**, *62*, 2278–2286. [CrossRef] [PubMed]
173. Fei, X.; Qi, M.; Wu, B.; Song, Y.; Wang, Y.; Li, T. MicroRNA-195-5p suppresses glucose uptake and proliferation of human bladder cancer T24 cells by regulating GLUT3 expression. *FEBS Lett.* **2012**, *586*, 392–397. [CrossRef] [PubMed]
174. Xu, C.; Zeng, Q.; Xu, W.; Jiao, L.; Chen, Y.; Zhang, Z.; Wu, C.; Jin, T.; Pan, A.; Wei, R.; et al. miRNA-100 inhibits human bladder urothelial carcinogenesis by directly targeting mTOR. *Mol. Cancer* **2013**, *12*, 207–219. [CrossRef] [PubMed]
175. Cao, Y.; Yu, S.L.; Wang, Y.; Guo, G.Y.; Ding, Q.; An, R.H. MicroRNA-dependent regulation of PTEN after arsenic trioxide treatment in bladder cancer cell line T24. *Tumor Biol.* **2011**, *32*, 179–188. [CrossRef] [PubMed]
176. Yang, X.; Cheng, Y.; Li, P.; Tao, J.; Deng, X.; Zhang, X.; Gu, M.; Lu, Q.; Yin, C. A lentiviral sponge for miRNA-21 diminishes aerobic glycolysis in bladder cancer T24 cells via the PTEN/PI3K/AKT/mTOR axis. *Tumor Biol.* **2014**, *36*, 383–391. [CrossRef] [PubMed]
177. Su, Q.; Peng, M.; Zhang, Y.; Xu, W.; Darko, K.O.; Tao, T.; Huang, Y.; Tao, X.; Yang, X. Quercetin induces bladder cancer cells apoptosis by activation of AMPK signaling pathway. *Am. J. Cancer Res.* **2016**, *6*, 498–508. [PubMed]

178. Li, F.; Yang, C.; Zhang, H.B.; Ma, J.; Jia, J.; Tang, X.; Zeng, J.; Chong, T.; Wang, X.; He, D.; et al. BET inhibitor JQ1 suppresses cell proliferation via inducing autophagy and activating LKB1/AMPK in bladder cancer cells. *Cancer Med.* **2019**, *8*, 4792–4805. [CrossRef]
179. English, S.G.; Hadj-Moussa, H.; Storey, K.B. MicroRNAs regulate survival in oxygen-deprived environments. *J. Exp. Biol.* **2018**, *221*. [CrossRef] [PubMed]
180. Blick, C.; Ramachandran, A.; Mccormick, R.; Wigfield, S.; Cranston, D.; Catto, J.; Harris, A.L. Identification of a hypoxia-regulated MIRNA signature in bladder cancer and a role for MIR-145 in hypoxia-dependent apoptosis. *Br. J. Cancer* **2015**, *113*, 634–644. [CrossRef] [PubMed]
181. Kumar, M.; Lu, Z.; Takwi, A.A.L.; Chen, W.; Callander, N.S.; Ramos, K.S.; Young, K.H.; Li, Y. Negative regulation of the tumor suppressor p53 gene by microRNAs. *Oncogene* **2011**, *30*, 843–853. [CrossRef]
182. Lee, H.; Jun, S.Y.; Lee, Y.S.; Lee, H.J.; Lee, W.S.; Park, C.S. Expression of miRNAs and ZEB1 and ZEB2 correlates with histopathological grade in papillary urothelial tumors of the urinary bladder. *Virchows Arch.* **2014**, *464*, 213–220. [CrossRef]
183. Wiklund, E.D.; Bramsen, J.B.; Hulf, T.; Dyrskjøt, L.; Ramanathan, R.; Hansen, T.B.; Villadsen, S.B.; Gao, S.; Ostenfeld, M.S.; Borre, M.; et al. Coordinated epigenetic repression of the miR-200 family and miR-205 in invasive bladder cancer. *Int. J. Cancer* **2011**, *128*, 1327–1334. [CrossRef] [PubMed]
184. Geiger, J.; Dalgaard, L.T. Interplay of mitochondrial metabolism and microRNAs. *Cell. Mol. Life Sci.* **2017**, *74*, 631–646. [CrossRef]
185. Lim, Y.Y.; Wright, J.A.; Attema, J.L.; Gregory, P.A.; Bert, A.G.; Smith, E.; Thomas, D.; Lopez, A.F.; Drew, P.A.; Khew-Goodall, Y.; et al. Epigenetic modulation of the miR-200 family is associated with transition to a breast cancer stem-celllike state. *J. Cell Sci.* **2013**, *126*, 2256–2266. [CrossRef] [PubMed]
186. Segovia, C.; Martínez-Fernández, M.; Dueñas, M.; Rubio, C.; López-Calderón, F.F.; Costa, C.; Saiz-Ladera, C.; Fernández-Grajera, M.; Duarte, J.; Muñoz, H.G.; et al. Opposing roles of PIK3CA gene alterations to EZH2 signaling in non-muscle invasive bladder cancer. *Oncotarget* **2017**, *8*, 10531–10542. [CrossRef]
187. Bu, Q.; Fang, Y.; Cao, Y.; Chen, Q.; Liu, Y. Enforced expression of miR-101 enhances cisplatin sensitivity in human bladder cancer cells by modulating the cyclooxygenase-2 pathway. *Mol. Med. Rep.* **2014**, *10*, 2203–2209. [CrossRef] [PubMed]
188. Sun, D.K.; Wang, J.M.; Zhang, P.; Wang, Y.Q. MicroRNA-138 regulates metastatic potential of bladder cancer through ZEB2. *Cell. Physiol. Biochem.* **2015**, *37*, 2366–2374. [CrossRef] [PubMed]
189. Wang, H.; Zhang, W.; Zuo, Y.; Ding, M.; Ke, C.; Yan, R.; Zhan, H.; Liu, J.; Wang, J. miR-9 promotes cell proliferation and inhibits apoptosis by targeting LASS2 in bladder cancer. *Tumor Biol.* **2015**, *36*, 9631–9640. [CrossRef] [PubMed]
190. Dyrskjot, L.; Ostenfeld, M.S.; Bramsen, J.B.; Silahtaroglu, A.N.; Lamy, P.; Ramanathan, R.; Fristrup, N.; Jensen, J.L.; Andersen, C.L.; Zieger, K.; et al. Genomic Profiling of MicroRNAs in Bladder Cancer: miR-129 Is Associated with Poor Outcome and Promotes Cell Death In vitro. *Cancer Res.* **2009**, *69*, 4851–4860. [CrossRef] [PubMed]
191. Sun, H.; Huang, Z.; Sheng, W.; Xu, M.D. Emerging roles of long non-coding RNAs in tumor metabolism. *J. Hematol. Oncol.* **2018**, *11*, 1–16. [CrossRef] [PubMed]
192. Liu, H.; Luo, J.; Luan, S.; He, C.; Li, Z. Long non-coding RNAs involved in cancer metabolic reprogramming. *Cell. Mol. Life Sci.* **2019**, *76*, 495–504. [CrossRef] [PubMed]
193. Lin, W.; Zhou, Q.; Wang, C.Q.; Zhu, L.; Bi, C.; Zhang, S.; Wang, X.; Jin, H. LncRNAs regulate metabolism in cancer. *Int. J. Biol. Sci.* **2020**, *16*, 1194–1206. [CrossRef]
194. Yang, F.; Zhang, H.; Mei, Y.; Wu, M. Reciprocal Regulation of HIF-1α and LincRNA-p21 Modulates the Warburg Effect. *Mol. Cell* **2014**, *53*, 88–100. [CrossRef] [PubMed]
195. Khan, M.R.; Xiang, S.; Song, Z.; Wu, M. The p53-inducible long noncoding RNA TRINGS protects cancer cells from necrosis under glucose starvation. *EMBO J.* **2017**, *36*, 3483–3500. [CrossRef] [PubMed]
196. Liu, X.; Xiao, Z.D.; Han, L.; Zhang, J.; Lee, S.W.; Wang, W.; Lee, H.; Zhuang, L.; Chen, J.; Lin, H.K.; et al. LncRNA NBR2 engages a metabolic checkpoint by regulating AMPK under energy stress. *Nat. Cell Biol.* **2016**, *18*, 431–442. [CrossRef] [PubMed]
197. Zhu, H.; Li, X.; Song, Y.; Zhang, P.; Xiao, Y.; Xing, Y. Long non-coding RNA ANRIL is up-regulated in bladder cancer and regulates bladder cancer cell proliferation and apoptosis through the intrinsic pathway. *Biochem. Biophys. Res. Commun.* **2015**, *467*, 223–228. [CrossRef]
198. Kong, Y.; Hsieh, C.H.; Alonso, L.C. ANRIL: A lncRNA at the CDKN2A/B locus with roles in cancer and metabolic disease. *Front. Endocrinol.* **2018**, *9*, 1–13. [CrossRef] [PubMed]
199. Xu, R.; Mao, Y.; Chen, K.; He, W.; Shi, W.; Han, Y. The long noncoding RNA ANRIL acts as an oncogene and contributes to paclitaxel resistance of lung adenocarcinoma A549 cells. *Oncotarget* **2017**, *8*, 39177–39184. [CrossRef]
200. Lan, W.G.; Xu, D.H.; Xu, C.; Ding, C.L.; Ning, F.L.; Zhou, Y.L.; Ma, L.B.; Liu, C.M.; Han, X. Silencing of long non-coding RNA ANRIL inhibits the development of multidrug resistance in gastric cancer cells. *Oncol. Rep.* **2016**, *36*, 263–270. [CrossRef]
201. Yu, G.; Liu, G.; Yuan, D.; Dai, J.; Cui, Y.; Tang, X. Long non-coding RNA ANRIL is associated with a poor prognosis of osteosarcoma and promotes tumorigenesis via PI3K/Akt pathway. *J. Bone Oncol.* **2018**, *11*, 51–55. [CrossRef]
202. Lu, Y.; Zhou, X.H.; Xu, L.; Rong, C.H.; Shen, C.; Bian, W. Long noncoding RNA ANRIL could be transactivated by c-Myc and promote tumor progression of non-small-cell lung cancer. *OncoTargets Ther.* **2016**, *9*, 3077–3084. [CrossRef]
203. Dong, X.; Jin, Z.; Chen, Y.; Xu, H.; Ma, C.; Hong, X.; Li, Y.; Zhao, G. Knockdown of long non-coding RNA ANRIL inhibits proliferation, migration, and invasion but promotes apoptosis of human glioma cells by upregulation of miR-34a. *J. Cell. Biochem.* **2018**, *119*, 2708–2718. [CrossRef]

204. Zhang, J.J.; Wang, D.D.; Du, C.X.; Wang, Y. Long noncoding RNA ANRIL promotes cervical cancer development by acting as a sponge of miR-186. *Oncol. Res.* **2018**, *26*, 345–352. [CrossRef] [PubMed]
205. Zou, Z.W.; Ma, C.; Medoro, L.; Chen, L.; Wang, B.; Gupta, R.; Liu, T.; Yang, X.Z.; Chen, T.T.; Wang, R.Z.; et al. LncRNA ANRIL is up-regulated in nasopharyngeal carcinoma and promotes the cancer progression via increasing proliferation, reprograming cell glucose metabolism and inducing sidepopulation stem-like cancer cells. *Oncotarget* **2016**, *7*, 61741–61754. [CrossRef] [PubMed]
206. Bochenek, G.; Häsler, R.; El Mokhtari, N.E.; König, I.R.; Loos, B.G.; Jepsen, S.; Schreiber, P.R.; Schaefer, A.S. The large non-coding RNA ANRIL, which is associated with atherosclerosis, periodontitis and several forms of cancer, regulates ADIPOR1, VAMP3 and C11ORF10. *Hum. Mol. Genet.* **2013**, *22*, 4516–4527. [CrossRef] [PubMed]
207. Wei, S.; Fan, Q.; Liang, Y.; Xiaodong, Z.; Ma, Y.; Zhihong, Z.; Hua, X.; Su, D.; Sun, H.; Li, H.; et al. Promotion of glycolysis by HOTAIR through GLUT1 upregulation via mTOR signaling. *Oncol. Rep.* **2017**, *38*, 1902–1908. [CrossRef]
208. Cai, B.; Song, X.Q.; Cai, J.P.; Zhang, S. HOTAIR: A cancer-related long non-coding RNA. *Neoplasma* **2014**, *61*, 379–391. [CrossRef]
209. Li, Z.; Li, X.; Wu, S.; Xue, M.; Chen, W. Long non-coding RNA UCA1 promotes glycolysis by upregulating hexokinase 2 through the mTOR-STAT3/microRNA143 pathway. *Cancer Sci.* **2014**, *105*, 951–955. [CrossRef]
210. Li, H.J.; Li, X.; Pang, H.; Pan, J.J.; Xie, X.J.; Chen, W. Long non-coding RNA UCA1 promotes glutamine metabolism by targeting miR-16 in human bladder cancer. *Jpn. J. Clin. Oncol.* **2015**, *45*, 1055–1063. [CrossRef]
211. Li, H.J.; Sun, X.M.; Li, Z.K.; Yin, Q.W.; Pang, H.; Pan, J.J.; Li, X.; Chen, W. LncRNA UCA1 Promotes Mitochondrial Function of Bladder Cancer via the MiR-195/ARL2 Signaling Pathway. *Cell. Physiol. Biochem.* **2017**, *43*, 2548–2561. [CrossRef]
212. Xue, M.; Li, X.; Li, Z.; Chen, W. Urothelial carcinoma associated 1 is a hypoxia-inducible factor-1α-targeted long noncoding RNA that enhances hypoxic bladder cancer cell proliferation, migration, and invasion. *Tumor Biol.* **2014**, *35*, 6901–6912. [CrossRef]
213. Xue, M.; Pang, H.; Li, X.; Li, H.; Pan, J.; Chen, W. Long non-coding RNA urothelial cancer-associated 1 promotes bladder cancer cell migration and invasion by way of the hsa-miR-145-ZEB1/2-FSCN1 pathway. *Cancer Sci.* **2016**, *107*, 18–27. [CrossRef] [PubMed]
214. Luo, J.; Chen, J.; Li, H.; Yang, Y.; Yun, H.; Yang, S.; Mao, X. LncRNA UCA1 promotes the invasion and EMT of bladder cancer cells by regulating the miR-143/HMGB1 pathway. *Oncol. Lett.* **2017**, *14*, 5556–5562. [CrossRef] [PubMed]
215. Pan, J.; Li, X.; Wu, W.; Xue, M.; Hou, H.; Zhai, W.; Chen, W. Long non-coding RNA UCA1 promotes cisplatin/gemcitabine resistance through CREB modulating miR-196a-5p in bladder cancer cells. *Cancer Lett.* **2016**, *382*, 64–76. [CrossRef] [PubMed]
216. Wang, T.; Yuan, J.; Feng, N.; Li, Y.; Lin, Z.; Jiang, Z.; Gui, Y. Hsa-miR-1 downregulates long non-coding RNA urothelial cancer associated 1 in bladder cancer. *Tumor Biol.* **2014**, *35*, 10075–10084. [CrossRef] [PubMed]
217. Logotheti, S.; Marquardt, S.; Gupta, S.K.; Richter, C.; Edelhäuser, B.A.H.; Engelmann, D.; Brenmoehl, J.; Söhnchen, C.; Murr, N.; Alpers, M.; et al. LncRNA-SLC16A1-AS1 induces metabolic reprogramming during Bladder Cancer progression as target and co-activator of E2F1. *Theranostics* **2020**, *10*, 9620–9643. [CrossRef]
218. Strimbu, K.; Tavel, J.A. What are biomarkers? *Curr. Opin. HIV AIDS* **2010**, *5*, 463–466. [CrossRef]
219. Patel, S.; Ahmed, S. Emerging field of metabolomics: Big promise for cancer biomarker identification and drug discovery. *J. Pharm. Biomed. Anal.* **2015**, *107*, 63–74. [CrossRef]
220. Bujak, R.; Struck-Lewicka, W.; Markuszewski, M.J.; Kaliszan, R. Metabolomics for laboratory diagnostics. *J. Pharm. Biomed. Anal.* **2015**, *113*, 108–120. [CrossRef] [PubMed]
221. Issaq, H.J.; Nativ, O.; Waybright, T.; Luke, B.; Veenstra, T.D.; Issaq, E.J.; Kravstov, A.; Mullerad, M. Detection of bladder cancer in human urine by metabolomic profiling using high performance liquid chromatography/mass spectrometry. *J. Urol.* **2008**, *179*, 2422–2426. [CrossRef]
222. Pasikanti, K.K.; Esuvaranathan, K.; Hong, Y.; Ho, P.C.; Mahendran, R.; Raman Nee Mani, L.; Chiong, E.; Chan, E.C.Y. Urinary metabotyping of bladder cancer using two-dimensional gas chromatography time-of-flight mass spectrometry. *J. Proteome Res.* **2017**, *8*, 20719–20728. [CrossRef]
223. Srivastava, S.; Roy, R.; Singh, S.; Kumar, P.; Dalela, D.; Sankhwar, S.N.; Goel, A.; Sonkar, A.A. Taurine—a possible fingerprint biomarker in non-muscle invasive bladder cancer: A pilot study by 1H NMR spectroscopy. *Cancer Biomark.* **2010**, *6*, 11–20. [CrossRef]
224. Duskova, K.; Vesely, S.; Do, J.; Silva, C.; Cernei, N. Differences in Urinary Amino Acid Patterns in Individuals with Different Types of Urological Tumor Urinary Amino Acid Patterns as Markers of Urological Tumors. *In Vivo* **2018**, *429*, 425–429. [CrossRef]
225. Alberice, J.V.; Amaral, A.F.S.; Armitage, E.G.; Lorente, J.A.; Algaba, F.; Carrilho, E.; Márquez, M.; García, A.; Malats, N.; Barbas, C. Searching for urine biomarkers of bladder cancer recurrence using a liquid chromatography-mass spectrometry and capillary electrophoresis-mass spectrometry metabolomics approach. *J. Chromatogr.* **2013**, *A 1318*, 163–170. [CrossRef]
226. Loras, A.; Trassierra, M.; Castell, J. V Bladder cancer recurrence surveillance by urine metabolomics analysis. *Sci. Rep.* **2018**, 1–10. [CrossRef]
227. Huang, Z.; Lin, L.; Gao, Y.; Chen, Y.; Yan, X.; Xing, J.; Hang, W. Bladder cancer determination via two urinary metabolites: A biomarker pattern approach. *Mol. Cell. Proteom.* **2011**, *10*, M111.007922. [CrossRef] [PubMed]
228. Reinert, T.; Modin, C.; Castano, F.M.; Lamy, P.; Wojdacz, T.K.; Hansen, L.L.; Wiuf, C.; Borre, M.; Dyrskjot, L.; Orntoft, T.F. Comprehensive Genome Methylation Analysis in Bladder Cancer: Identification and Validation of Novel Methylated Genes and Application of These as Urinary Tumor Markers. *Clin. Cancer Res.* **2011**, *17*, 5582–5592. [CrossRef] [PubMed]
229. Huang, Z.; Chen, Y.; Hang, W.; Gao, Y.; Lin, L.; Li, D.Y.; Xing, J.; Yan, X. Holistic metabonomic profiling of urine affords potential early diagnosis for bladder and kidney cancers. *Metabolomics* **2013**, *9*, 119–129. [CrossRef]

230. Kim, J.-W.; Lee, G.; Moon, S.-M.; Park, M.-J.; Hong, S.K.; Ahn, Y.-H.; Kim, K.-R.; Paik, M.-J. Metabolomic screening and star pattern recognition by urinary amino acid profile analysis from bladder cancer patients. *Metabolomics* **2010**, *6*, 202–206. [CrossRef]
231. Gamagedara, S.; Shi, H.; Ma, Y. Quantitative determination of taurine and related biomarkers in urine by liquid chromatography-tandem mass spectrometry. *Anal. Bioanal. Chem.* **2012**, *402*, 763–770. [CrossRef]
232. Cao, M.; Zhao, L.; Chen, H.; Xue, W.; Lin, D. NMR-based metabolomic analysis of human bladder cancer. *Anal. Sci.* **2012**, *28*, 451–456. [CrossRef]
233. Hoque, M.O.; Begum, S.; Topaloglu, O.; Chatterjee, A.; Rosenbaum, E.; Van Criekinge, W.; Westra, W.H.; Schoenberg, M.; Zahurak, M.; Goodman, S.N.; et al. Quantitation of Promoter Methylation of Multiple Genes in Urine DNA and Bladder Cancer Detection. *JNCI J. Natl. Cancer Inst.* **2006**, *98*, 996–1004. [CrossRef]
234. Tan, G.; Wang, H.; Yuan, J.; Qin, W.; Dong, X.; Wu, H.; Meng, P. Three serum metabolite signatures for diagnosing low-grade and high-grade bladder cancer. *Sci. Rep.* **2017**, *7*, 46176. [CrossRef] [PubMed]
235. Bansal, N.; Gupta, A.; Mitash, N.; Shakya, P.S.; Mandhani, A.; Mahdi, A.A.; Sankhwar, S.N.; Mandal, S.K. Low- and high-grade bladder cancer determination via human serum-based metabolomics approach. *J. Proteome Res.* **2013**, *12*, 5839–5850. [CrossRef]
236. Loras, A.; Martínez-Bisbal, M.C.; Quintás, G.; Gil, S.; Martínez-Máñez, R.; Ruiz-Cerdá, J.L. Urinary metabolic signatures detect recurrences in non-muscle invasive bladder cancer. *Cancers* **2019**, *11*, 914. [CrossRef]
237. Hauser, S.; Kogej, M.; Fechner, G.; Pezold Von, J.; Vorreuther, R.; Lümmen, G.; Müller, S.; Ellinger, J. Serum DNA hypermethylation in patients with bladder cancer: Results of a prospective multicenter study. *Anticancer Res.* **2013**, *33*, 779–784. [PubMed]
238. Hegi, M.E.; Liu, L.; Herman, J.G.; Stupp, R.; Wick, W.; Weller, M.; Mehta, M.P.; Gilbert, M.R. Correlation of O6-methylguanine methyltransferase (MGMT) promoter methylation with clinical outcomes in glioblastoma and clinical strategies to modulate MGMT activity. *J. Clin. Oncol.* **2008**, *26*, 4189–4199. [CrossRef]
239. Eissa, S.; Matboli, M.; Essawy, N.O.E.; Kotb, Y.M. Integrative functional genetic-epigenetic approach for selecting genes as urine biomarkers for bladder cancer diagnosis. *Tumor Biol.* **2015**, *36*, 9545–9552. [CrossRef] [PubMed]
240. Berrondo, C.; Flax, J.; Kucherov, V.; Siebert, A.; Osinski, T.; Rosenberg, A.; Fucile, C.; Richheimer, S.; Beckham, C.J. Expression of the long non-coding RNA HOTAIR correlates with disease progression in bladder cancer and is contained in bladder cancer patient urinary exosomes. *PLoS ONE* **2016**, *11*, e0147236. [CrossRef] [PubMed]
241. Tomasetti, M.; Lee, W.; Santarelli, L.; Neuzil, J. Exosome-derived microRNAs in cancer metabolism: Possible implications in cancer diagnostics and therapy. *Exp. Mol. Med.* **2017**, *49*, e285. [CrossRef]
242. Faleiro, I.; Leão, R.; Binnie, A.; Andrade De Mello, R.; Maia, A.-T.; Castelo-Branco, P. Epigenetic therapy in urologic cancers: An update on clinical trials. *Oncotarget* **2017**, *8*, 12484–12500. [CrossRef] [PubMed]
243. Arrowsmith, C.H.; Bountra, C.; Fish, P.V.; Lee, K.; Schapira, M. Epigenetic protein families: A new frontier for drug discovery. *Nat. Rev. Drug Discov.* **2012**, *11*, 384–400. [CrossRef]
244. Lavery, H.J.; Stensland, K.D.; Niegisch, G.; Albers, P.; Droller, M.J. Pathological T0 following radical cystectomy with or without neoadjuvant chemotherapy: A useful surrogate. *J. Urol.* **2014**, *191*, 898–906. [CrossRef]
245. Jiang, X.; Lim, C.Z.H.; Li, Z.; Lee, P.L.; Yatim, S.M.J.M.; Guan, P.; Li, J.; Zhou, J.; Pan, J.; Chng, W.-J.; et al. Functional Characterization of D9, a Novel Deazaneplanocin A (DZNep) Analog, in Targeting Acute Myeloid Leukemia (AML). *PLoS ONE* **2015**, *10*, e0122983. [CrossRef]
246. Sweis, R.F.; Pliushchev, M.; Brown, P.J.; Guo, J.; Li, F.; Maag, D.; Petros, A.M.; Soni, N.B.; Tse, C.; Vedadi, M.; et al. Discovery and Development of Potent and Selective Inhibitors of Histone Methyltransferase G9a. *Acs Med. Chem. Lett.* **2014**, *5*, 205–209. [CrossRef] [PubMed]
247. Tsai, C.T.; So, C.W.E. Epigenetic therapies by targeting aberrant histone methylome in AML: Molecular mechanisms, current preclinical and clinical development. *Oncogene* **2017**, *36*, 1753–1759. [CrossRef] [PubMed]
248. Konze, K.D.; Ma, A.; Li, F.; Barsyte-Lovejoy, D.; Parton, T.; MacNevin, C.J.; Liu, F.; Gao, C.; Huang, X.-P.; Kuznetsova, E.; et al. An Orally Bioavailable Chemical Probe of the Lysine Methyltransferases EZH2 and EZH1. *Acs Chem. Biol.* **2013**, *8*, 1324–1334. [CrossRef]
249. San José-Enériz, E.; Agirre, X.; Rabal, O.; Vilas-Zornoza, A.; Sanchez-Arias, J.A.; Miranda, E.; Ugarte, A.; Roa, S.; Paiva, B.; Estella-Hermoso de Mendoza, A.; et al. Discovery of first-in-class reversible dual small molecule inhibitors against G9a and DNMTs in hematological malignancies. *Nat. Commun.* **2017**, *8*, 15424. [CrossRef] [PubMed]
250. Garzon, R.; Marcucci, G.; Croce, C.M. Targeting microRNAs in cancer: Rationale, strategies and challenges. *Nat. Rev. Drug Discov.* **2010**, *9*, 775–789. [CrossRef] [PubMed]
251. Hatziapostolou, M.; Polytarchou, C.; Iliopoulos, D. MiRNAs link metabolic reprogramming to oncogenesis. *Trends Endocrinol. Metab.* **2013**, *24*, 361–373. [CrossRef] [PubMed]
252. Kunze, D.; Erdmann, K.; Froehner, M.; Wirth, M.; Fuessel, S. Enhanced Inhibition of Bladder Cancer Cell Growth by Simultaneous Knockdown of Antiapoptotic Bcl-xL and Survivin in Combination with Chemotherapy. *Int. J. Mol. Sci.* **2013**, *14*, 12297–12312. [CrossRef] [PubMed]
253. Rieger, C.; Huebner, D.; Temme, A.; Wirth, M.P.; Fuessel, S. Antisense- and siRNA-mediated inhibition of the anti-apoptotic gene Bcl-xL for chemosensitization of bladder cancer cells. *Int. J. Oncol.* **2015**, *47*, 1121–1130. [CrossRef]
254. Vinall, R.L.; Ripoll, A.Z.; Wang, S.; Pan, C.X.; Devere White, R.W. MiR-34a chemosensitizes bladder cancer cells to cisplatin treatment regardless of p53-Rb pathway status. *Int. J. Cancer* **2012**, *130*, 2526–2538. [CrossRef]

255. Costa-Pinheiro, P.; Montezuma, D. Diagnostic and prognostic epigenetic. *Epigenomics* **2015**, *7*, 1003–1015. [CrossRef] [PubMed]
256. Duquesne, I.; Weisbach, L.; Aziz, A.; Kluth, L.A.; Xylinas, E. The contemporary role and impact of urine-based biomarkers in bladder cancer. *Transl. Urol.* **2017**, *6*, 1031–1042. [CrossRef] [PubMed]
257. Pasikanti, K.K.; Esuvaranathan, K.; Ho, P.C.; Mahendran, R.; Kamaraj, R.; Wu, Q.H.; Chiong, E.; Chan, E.C.Y. Noninvasive urinary metabonomic diagnosis of human bladder cancer. *J. Proteome Res.* **2010**, *9*, 2988–2995. [CrossRef]

Article

Comprehensive Plasma Metabolomic Profile of Patients with Advanced Neuroendocrine Tumors (NETs). Diagnostic and Biological Relevance

Beatriz Soldevilla [1,2,†], Angeles López-López [3,†], Alberto Lens-Pardo [1,2], Carlos Carretero-Puche [1,2], Angeles Lopez-Gonzalvez [3], Anna La Salvia [1,4], Beatriz Gil-Calderon [1,2], Maria C. Riesco-Martinez [1,4], Paula Espinosa-Olarte [1,4], Jacinto Sarmentero [1,2], Beatriz Rubio-Cuesta [1,2], Raúl Rincón [1,2], Coral Barbas [3] and Rocio Garcia-Carbonero [1,2,4,5,*]

1. Clinical and Translational Oncology Laboratory, Gastrointestinal Unit, i+12 Research Institute Hospital 12 de Octubre, 28041 Madrid, Spain; bsoldevilla.imas12@h12o.es (B.S.); alens.imas12@h12o.es (A.L.-P.); ccarretero.imas12@h12o.es (C.C.-P.); alasalvi@ucm.es (A.L.S.); bgil.imas12@h12o.es (B.G.-C.); criesco@salud.madrid.org (M.C.R.-M.); espinosa_pau@gva.es (P.E.-O.); jacinsarmentero@yahoo.es (J.S.); beatriz230794@hotmail.com (B.R.-C.); rrincon.imas12@h12o.es (R.R.)
2. Spanish National Cancer Research Center (CNIO), 28029 Madrid, Spain
3. Centre for Metabolomics and Bioanalysis (CEMBIO), Department of Chemistry and Biochemistry, Facultad de Farmacia, Universidad San Pablo-CEU, CEU Universities Urbanización Montepríncipe, Boadilla del Monte, 28660 Madrid, Spain; ang.lopez.ce@ceindo.ceu.es (A.L.-L.); alopgon@ceu.es (A.L.-G.); cbarbas@ceu.es (C.B.)
4. Oncology Department, Hospital Universitario 12 de Octubre, 28041 Madrid, Spain
5. Department of Medicine, Faculty of Medicine, Complutense University of Madrid (UCM), 28040 Madrid, Spain
* Correspondence: rgcarbonero@gmail.com; Tel.: +34-91-3908926; Fax: +34-91-4603310
† These authors contributed equally to this study.

Simple Summary: Metabolic flexibility is one of the key hallmarks of cancer and metabolites are the final products of this adaptation, reflecting the aberrant changes of tumors. However, the metabolic plasticity of each cancer type is still unknown, and specifically to date, there are no data on metabolic profile in neuroendocrine tumors. The aim of our retrospective study was to assess the metabolomic profile of NET patients to understand metabolic deregulation in these tumors and identify novel biomarkers with clinical potential. We provided, for the first time, a comprehensive metabolic profile of NET patients and identifies a distinctive metabolic signature in plasma of potential clinical use, selecting a reduced set of metabolites of high diagnostic accuracy. We have identified 32 novel enriched metabolic pathways in NETs related with the TCA cycle, and with arginine, pyruvate or glutathione metabolism, which have distinct implications in oncogenesis and may open innovative avenues of clinical research.

Abstract: Purpose: High-throughput "-omic" technologies have enabled the detailed analysis of metabolic networks in several cancers, but NETs have not been explored to date. We aim to assess the metabolomic profile of NET patients to understand metabolic deregulation in these tumors and identify novel biomarkers with clinical potential. Methods: Plasma samples from 77 NETs and 68 controls were profiled by GC–MS, CE–MS and LC–MS untargeted metabolomics. OPLS-DA was performed to evaluate metabolomic differences. Related pathways were explored using Metaboanalyst 4.0. Finally, ROC and OPLS-DA analyses were performed to select metabolites with biomarker potential. Results: We identified 155 differential compounds between NETs and controls. We have detected an increase of bile acids, sugars, oxidized lipids and oxidized products from arachidonic acid and a decrease of carnitine levels in NETs. MPA/MSEA identified 32 enriched metabolic pathways in NETs related with the TCA cycle and amino acid metabolism. Finally, OPLS-DA and ROC analysis revealed 48 metabolites with diagnostic potential. Conclusions: This study provides, for the first time, a comprehensive metabolic profile of NET patients and identifies a distinctive metabolic signature in plasma of potential clinical use. A reduced set of metabolites of high diagnostic accuracy has been identified. Additionally, new enriched metabolic pathways annotated may open innovative avenues of clinical research.

Keywords: NETs; disease modelling; machine learning; metabolic signaling; molecular pathways; plasma metabolites; diagnostic biomarkers

1. Introduction

Reprogrammed metabolism encompasses the capacity of cells to respond or adapt their metabolic signaling to support and enable cell survival in unfavorable or hostile conditions. This ability is enhanced in cancer cells in order to improve their adaptive phenotype and maintain both viability and uncontrolled proliferation. Metabolic flexibility is therefore one of the key hallmarks of cancer [1], although the pathways involved in the metabolic plasticity of each cancer type remain to be elucidated. Metabolites are the final products of this adaptation, reflecting the aberrant changes in the genomic, transcriptomic and proteomic variability of tumors, and provide therefore useful biological and clinical information on cancer initiation and progression [2–4]. This, together with the fact that metabolomics can be easily performed in readily accessible biological samples (plasma, urine), makes metabolic profiling of cancer patients a promising tool to characterize the tumor phenotype and identify novel biomarkers of potential clinical use. Systems medicine approaches integrating high-throughput "-omic" technologies into diagnostic platforms have indeed enabled the detailed analysis of metabolic networks (known as metabolomics) in several cancers of high incidence, prevalence and mortality [5–8], but these do not include neuroendocrine neoplasms (NENs).

NENs comprise a heterogeneous family of rare tumors of increasing incidence and challenging clinical management [9]. Although they can arise in virtually any organ, the most common primary tumor sites are the lungs (25%) and the digestive tract (~65%). Well-differentiated neuroendocrine tumors (NETs) account for ~80% of all NENs, have a rather indolent clinical behavior, as compared to their exocrine counterparts, and are associated with a good to moderate prognosis depending on primary tumor site, proliferative index (ki67 or mitotic index) and tumor stage. About 20% of NETs have also the unique ability to produce and secrete amines or peptide hormones to the blood stream, the so-called "functioning tumors", that produce specific endocrine syndromes (i.e., carcinoid syndrome) that may seriously impair patients quality of life and prognosis [10]. Survival has improved over time for all NETs, likely reflecting earlier diagnosis and improvements in therapy [11,12]. However, a significant proportion of patients are still diagnosed with advanced stages of disease, highlighting the need to identify novel specific biomarkers that may contribute to an earlier detection and an increased likelihood of cure.

The study of hereditary genetic syndromes associated with an increased predisposition to develop NETs (~5%) has contributed to partially elucidate some of the mechanisms involved in their tumorigenesis [13–18]. Germline mutations in MEN1, RET, CDKN1B, VHL, NF1 and TSC1/2 are the molecular alterations most frequently detected in hereditary NENs. Although some of these mutations have a relevant representation in sporadic NETs (i.e., MEN1, TSC1/2), other molecular alterations involved in epigenetic regulation, DNA repair, telomeres regulation and chromosomal rearrangements have been also implicated [19–21]. Despite these recent advances, however, the molecular mechanisms of NET genesis and progression remain largely unraveled. In addition, few authors have explored NETs from a metabolomic perspective. A pilot prospective study analyzed urine samples from 28 gastroenteropancreatic NET patients by nuclear magnetic resonance (NMR) spectroscopy, and showed distinct metabolomic phenotypes by primary tumor site (small bowel versus pancreatic NEN) and function [22]. A second recent work described the metabolomic fingerprint of 46 small intestine NET tissues analyzed by NMR spectroscopy, suggesting the existence of complex metabolic pathways in NETs, possibly influencing tumor development and evolution, and thereby clinical outcome [23]. With the exception of these two small studies, the metabolomic profile of patients with NETs has not been studied to date.

In this context, the aim of our study was to perform a comprehensive metabolic profiling of NETs to better understand metabolic dysregulation in these tumors and identify novel biomarkers of potential clinical use. To this aim, multiplatform untargeted metabolomic analyses were performed in plasma samples of 77 patients with advanced gastrointestinal and lung NETs, and 68 non-cancer individuals (controls). The diagnostic potential and biological relevance of differential metabolites identified was assessed, and dysregulated pathways were explored to provide further insight into the molecular mechanisms involved in NET development and progression.

2. Results

2.1. Metabolomic Profiling in Plasma of Patients with Neuroendocrine Tumors

The metabolite fingerprint was assessed using a multiplatform LC-MS, GC-MS and CE-MS approach in plasma of 77 patients diagnosed with NETs and of 68 non-cancer individuals (controls). Main characteristics of the study population are summarized in Table S1. All patients had well-differentiated G1-2 NETs (33.8% G1 and 66.2% G2), the most common primary tumor site was the small intestine (58.4%) and one-third had functioning tumors (carcinoid syndrome).

Data obtained after peak alignment and filtering were used for multivariate analysis of unsupervised principal components (PCAs) to verify the distribution of QCs in each technique. System stability, performance and reproducibility of sample treatment procedures were reflected with the spontaneous grouping of these QC samples (Figure S1).

For each platform in multivariate analysis, unsupervised (Figure S2), and supervised PLS-DA and OPLS-DA models were also conducted. OPLS-DA supervised models were used to model differences between groups and were validated using permutation tests (Figure 1A–D).

All analytical techniques clearly discriminated NET patients from non-cancer individuals in the applied models. A total of 1006 metabolites were detected and univariate analysis revealed the following individually significant differential metabolites between cases and controls: 75 compounds in CE-MS, 150 in LC–MS ESI(+), 296 in LC–MS ESI(−) and 19 in GC–MS. These variables were annotated and/or identified as described in "Annotation and compound identification" in the Material and Methods section and are summarized in Table 1.

The integration of metabolic data acquired by different analytical platforms resulted in 155 identified metabolites with a differential availability in NET patients ($p < 0.05$), when compared to non-cancer individuals. Metabolite identification of some specific metabolites (arginine, glutamine, phenylalanine, among others) across more than one analytical platform significantly increases the confidence of metabolite identification (Table 1). No significant differences were found by gender, age, grade, primary tumor site and hormonal syndrome.

2.2. NETs Show a Particular Signature of Metabolites with Diagnostic Potential

The unsupervised heatmap cluster plot of the 155 metabolites identified with a differential availability in NET patients as compared to non-cancer individuals is shown in Figure S3. Given the relevant capacity to discriminate NET patients from non-cancer individuals, we applied ROC analysis and calculated the AUC of each identified metabolite to determine their individual performance as NET diagnostic biomarkers. We also calculated the variable importance in projection (VIP) score from the OPLS-DA model. VIP score estimates the importance of each metabolite in the model and therefore, their ability to discriminate NETs from non-cancer patients. Table S2 summarizes the ROC and OPLS-DA analyses of the 155 differential plasmatic metabolites identified between NETs and controls. Those with an AUC > 0.85, a VIP > 1.0 and a $|p(\text{corr})| > 0.5$ or both were considered as metabolites with biomarker potential. Twenty-seven metabolites had both an AUC > 0.85 and a VIP > 1.0 and a $|p(\text{corr})| > 0.5$; 17 metabolites had only a VIP > 1.0 and a $|p(\text{corr})| > 0.5$ and 5 metabolites had only an AUC > 0.85. Overall, thus, we identified 49 metabolites with significant diagnostic potential.

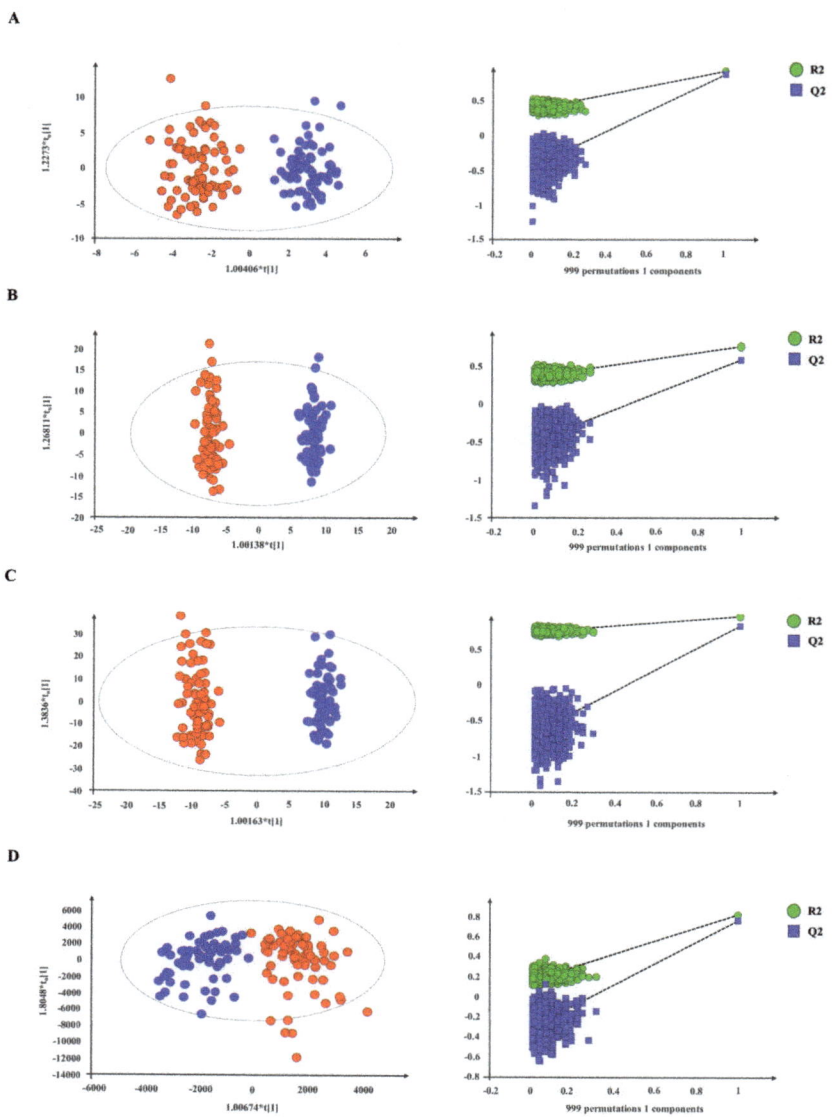

Figure 1. Supervised models show a clear separation between NET patients and non-cancer individuals. (**A**–**D**) OPLS-DA score plots and permutation tests of OPLS-DA models for each platform through 999 permutations. Panel (**A**) for CE-MS data (R2 = 0.872, Q2 = 0.843); panel (**B**), LC-MS/ESI(+) data (R2 = 0.954, Q2 = 0.871); panel (**C**), LC-MS/ESI(−) data (R2 = 0.885, Q2 = 0.788) and panel (**D**), GC-MS data (R2 = 0.781, Q2 = 0.744). Red dots, NETs (n = 77); blue dots, non-cancer individuals (n = 68).

Table 1. List of the statistically significant annotated metabolites discriminating between the plasma profiles of control ($n = 68$) and NET patients ($n = 77$) with their statistical characteristics after UVDA and MVDA (percentage of change, p-value, p(corr) and VIP) and analytical descriptors (measured mass and its deviation from the theoretical one, experimental retention time, analytical platform on which it has been detected, identification source where DB corresponds to database result, confidence level for identification according to the metabolomics standards initiative and its corresponding Human Metabolome Database (HMDB) code (http://www.hmdb.ca/, accessed on 16 December 2020)). (* = Multiple identification options).

Compound	Formula	Mass (Da)	RT (Min)	% Change	p-Value	p(corr)	VIP	Mass Error (ppm)	Analytical Platform	Identification Source	Confidence Level	HMDB Code
Triethylamine	$C_6H_{15}N$	101.1204	10.28	−53	0.0009			0	CE-MS	DB	3	HMDB0032539
					Amines							
					Amino acids, peptides and analogues							
Arginine	$C_6H_{14}N_4O_2$	174.1131	9.98	243	4.58×10^{-35}	−0.80	2.33	15	CE-MS LC-MS(+)	DB	2	HMDB0000517
Arg-Val	$C_{11}H_{23}N_5O_3$	273.1822	9.95	66	0.0004			8	CE-MS	DB	3	HMDB0028722
Aspartate	$C_4H_7NO_4$	133.0375	13.80	−32	4.98×10^{-13}			2	CE-MS	DB	2	HMDB0000191
Cys-Gly	$C_5H_{10}N_2O_3S$	178.0412	12.19	−29	8.00×10^{-9}			0	CE-MS	DB	3	HMDB0000078
Cys-Gly disulfide	$C_8H_{15}N_3O_5S_2$	297.0455	12.19	−41	1.39×10^{-8}	0.49	1.63	1	CE-MS	DB	3	HMDB0000709
Cysteineglutathione disulfide	$C_{13}H_{22}N_4O_8S_2$	426.0912	13.95	−37	7.40×10^{-6}			8	CE-MS	DB	3	HMDB0000656
Dimethyl-Arginine (symmetric)	$C_8H_{18}N_4O_2$	202.1428	10.67	25	0.0006			1	CE-MS	DB	2	HMDB0003334
GalactosylhydroxyLys	$C_{12}H_{24}N_2O_8$	324.1558	11.49	42	0.0014			8	CE-MS	DB	3	HMDB0000600
gamma-Glu-orn *	$C_{10}H_{19}N_3O_5$	261.1335	11.30	34	0.0103			4	CE-MS	DB	3	HMDB0002248
Glu-Ala *	$C_8H_{14}N_2O_5$	218.0904	14.40	135	4.38×10^{-14}			1	CE-MS	DB	3	HMDB0006248
Glu-Arg	$C_{11}H_{21}N_5O_5$	303.1554	11.49	87	6.28×10^{-14}			4	CE-MS	DB	3	HMDB0028813
Glu-Asp	$C_9H_{14}N_2O_7$	262.0801	3.78	54	0.0036	−0.52	1.55	7	LC-MS(−)	MSMS	2	HMDB0030419
Glu-hyp	$C_{10}H_{16}N_2O_6$	260.0993	12.98	37	5.12×10^{-7}			6	CE-MS	DB	3	HMDB0011161
Glu-Lys/ε-Glu-Lys	$C_{11}H_{21}N_3O_5$	275.1481	12.32	67	1.20×10^{-9}			2	CE-MS	DB	3	HMDB0029154
Glu-Lys/ε-Glu-Lys	$C_{11}H_{21}N_3O_5$	275.1481	11.37	58	1.55×10^{-7}			2	CE-MS	DB	3	HMDB0029155
Glutamine	$C_5H_{10}N_2O_3$	146.0691	1.34	69	0.0006			1	LC-MS(+) LC-MS(−)	MSMS	2	HMDB0000641
Glu-Val	$C_{10}H_{18}N_2O_5$	246.1217	14.68	48	0.0009			1	LC-MS(+)	DB	3	HMDB0028832
Glycine	$C_2H_5NO_2$	75.0320	9.75	32	0.0202			-	GC-MS	Fiehn	2	HMDB0000123
Gly-Pro	$C_7H_{12}N_2O_3$	172.0831	15.89	−20	0.0275			10	CE-MS	DB	3	HMDB0000721
Homocitrulline	$C_7H_{15}N_3O_3$	189.1118	13.52	44	0.0027			3	CE-MS	DB	3	HMDB0000679
Iminodiacetic acid	$C_4H_7NO_4$	133.0375	12.87	−37	0.0158			-	GC-MS CE-MS	Fiehn	2	HMDB0011753
Indoleacetyl glutamine	$C_{15}H_{17}N_3O_4$	303.1219	1.70	203	1.43×10^{-6}	−0.58	2.05	2	LC-MS(−) LC-MS(+)	DB	3	HMDB0013240
Leu-hyp	$C_{11}H_{20}N_2O_4$	244.1423	2.00	87	0.0018			2	LC-MS(−)	DB	3	HMDB0028867
Leu-Phe	$C_{15}H_{22}N_2O_3$	278.1630	2.14	Presented in cancer group	3.54×10^{-5}	−0.60	2.45	0	LC-MS(−)	DB	3	HMDB0013243

Table 1. Cont.

Compound	Formula	Mass (Da)	RT (Min)	% Change	p-Value	p(corr)	VIP	Mass Error (ppm)	Analytical Platform	Identification Source	Confidence Level	HMDB Code
Lys-Asp *	$C_{10}H_{19}N_3O_5$	261.1335	11.30	34	0.0103			4	CE-MS	DB	3	HMDB0028947
Methionine S-oxide	$C_5H_{11}NO_3S$	165.0478	14.06	110	3.30×10^{-9}			11	CE-MS	DB	2	HMDB0002005
N2-Methyl-lysine	$C_7H_{16}N_2O_2$	160.1208	10.67	−77	6.97×10^{-9}			2	CE-MS	DB	2	HMDB0002038
N2-Methylproline	$C_6H_{11}NO_2$	129.0791	14.59	−43	0.0031			1	CE-MS	DB	2	HMDB0059649
N6-Acetyl-hydroxy-lysine *	$C_8H_{16}N_2O_4$	204.1103	12.78	−55	0.0002			3	CE-MS	DB	3	HMDB0033891
N-acetyl-lysine	$C_8H_{16}N_2O_3$	188.1155	13.70	20	0.0002			3	CE-MS	DB	2	HMDB0000206
Ornithine	$C_5H_{12}N_2O_2$	132.0906	9.66	−39	5.58×10^{-19}	0.65	1.65	6	CE-MS	DB	2	HMDB0000214
Phenylalanine	$C_9H_{11}NO_2$	165.0770	13.88	−58	0.0050			12	CE-MS	DB	2	HMDB0000159
Pipecolic acid	$C_6H_{11}NO_2$	129.0790	10.14	49	0.0307			-	GC-MS	Fiehn	2	HMDB0000716
Pyroglutamine	$C_5H_8N_2O_2$	128.0583	11.48	68	1.41×10^{-5}			2	CE-MS	DB	3	HMDB0062558
Ser-Ala *	$C_6H_{12}N_2O_4$	176.0789	13.07	−38	1.37×10^{-8}			4	CE-MS	DB	3	HMDB0029032
Ser-hyp *	$C_8H_{14}N_2O_5$	218.0904	14.40	135	4.38×10^{-14}			1	CE-MS	DB	3	HMDB0029040
Ser-Val *	$C_8H_{16}N_2O_4$	204.1103	12.78	−55	0.0002			3	CE-MS	DB	3	HMDB0029052
Stearoyl-tyrosine *	$C_{27}H_{45}NO_4$	447.3349	5.15	−66	9.67×10^{-17}	0.75	3.74	1	LC-MS(+)	DB	3	HMDB0062343
Suberylglycine	$C_{10}H_{17}NO_5$	231.1106	1.19	Presented in cancer group	0.0001	−0.73	3.82	1	LC-MS(−)	DB	3	HMDB0000953
Thr-Ala	$C_7H_{14}N_2O_4$	190.0954	13.83	90	5.09×10^{-11}			0	CE-MS	DB	3	HMDB0029054
Thr-Gly *	$C_6H_{12}N_2O_4$	176.0789	13.07	−38	1.37×10^{-8}			4	CE-MS	DB	3	HMDB0029061
Trp-Phe	$C_{20}H_{21}N_3O_3$	351.1582	2.32	74	3.73×10^{-11}	−0.63	1.56	1	LC-MS(−)	DB	3	HMDB0029090
Val-Leu	$C_{11}H_{22}N_2O_3$	230.1616	12.83	51	0.0012			6	CE-MS	DB	3	HMDB0029131
Benzene and substituted derivatives												
Mandelic acid	$C_8H_8O_3$	152.0473	2.28	335	0.0037			3	LC-MS(−)	DB	3	HMDB0000703
3-phenylprop-2-en-1-yloxysulfonic acid	$C_9H_{10}O_4S$	214.0299	2.77	118	0.0005			2	LC-MS(−)	DB	3	HMDB0135284
Carbohydrates and carbohydrate conjugates												
Allose	$C_6H_{12}O_6$	180.0633	17.10	139	1.32×10^{-14}	0.62	1.44	-	GC-MS	Fiehn	2	HMDB0001151
Glucose	$C_6H_{12}O_6$	180.0633	17.45	174	1.98×10^{-15}	0.62	1.64	-	GC-MS	Fiehn	2	HMDB0000122
Glycerol	$C_3H_8O_3$	92.0473	9.28	45	0.0000			-	GC-MS	Fiehn	2	HMDB0000131
Mannitol	$C_6H_{14}O_6$	182.0790	17.59	196	0.0017			-	GC-MS	Fiehn	2	HMDB0000765
Phenylglucuronide PYRANOSE	$C_{12}H_{14}O_7$	270.0739	0.93	308	0.0052	−0.52	2.37	2	LC-MS(−)	DB	3	HMDB0060014
(glucose/altrose/galactose/talose)	$C_6H_{12}O_6$	180.0633	17.24	194	1.61×10^{-16}	0.63	1.71	-	GC-MS	Fiehn	2	

Table 1. Cont.

Compound	Formula	Mass (Da)	RT (Min)	% Change	p-Value	p(corr)	VIP	Mass Error (ppm)	Analytical Platform	Identification Source	Confidence Level	HMDB Code
Carboximidic acids and derivatives												
Acetylspermidine	$C_9H_{21}N_3O$	187.1671	9.15	38	6.27×10^{-6}			7	CE-MS	DB	3	HMDB0001276
Carboxylic acids and derivatives												
1-Aminocyclohexanecarboxylic acid	$C_7H_{13}NO_2$	143.0944	13.89	−32	0.0366			1	CE-MS	DB	3	HMDB0038249
di-Hydroxymelatonin *	$C_{13}H_{16}N_2O_4$	264.1110	1.34	46	0.0268			2	LC-MS(−)	DB	3	HMDB0061136
Edetic Acid	$C_{10}H_{16}N_2O_8$	292.0906	0.24	35	0.0001			1	LC-MS(+)	DB	3	HMDB0015109
Isocitric acid												HMDB0000193
Citric acid	$C_6H_8O_7$	192.0270	0.23	64	5.83×10^{-5}			0	LC-MS(−)	MSMS	2	HMDB0000094
Diazines/Pyrimidines and pyrimidine derivatives												
5,6-Dihydrothymine	$C_5H_8N_2O_2$	128.0570	12.22	56	0.0169			12	CE-MS	DB	3	HMDB0000079
Fatty acyls												
3-carboxy-4-methyl-5-propyl-2-furanpropanoic acid (CMPF)	$C_{12}H_{16}O_5$	240.0998	3.78	69	0.004			2	LC-MS(+)	MSMS	2	HMDB0061112
8-amino-7-oxo-nonanoic acid *	$C_9H_{17}NO_3$	187.1208	2.06	418	0.0220			2	LC-MS(−)	DB	3	
Arachidonic acid	$C_{20}H_{32}O_2$	304.2402	7.13	62	1.59×10^{-5}			1	LC-MS(−)	MSMS	2	HMDB0001043
beta-Phenylalanoyl-CoA *	$C_{30}H_{45}N_8O_{17}P_3S$	914.1836	2.83	73	7.07×10^{-3}			1	LC-MS(−)	DB	3	
beta-Phenylalanoyl-CoA *	$C_{30}H_{45}N_8O_{17}P_3S$	914.1836	3.62	65	8.78×10^{-3}			0	LC-MS(−)	DB	3	
DG(31:0)	$C_{34}H_{66}O_5$	554.4910	8.18	−37	0.0008			0	LC-MS(+)	DB	3	HMDB0093505
Docosapentaenoic acid	$C_{22}H_{34}O_2$	330.2558	7.25	75	6.87×10^{-9}	−0.55	1.51	1	LC-MS(−)	DB	3	HMDB0006528
Dodecenedioic acid	$C_{12}H_{20}O_4$	228.1362	3.50	191	9.87×10^{-8}	−0.60	2.24	2	LC-MS(−)	DB	3	HMDB0000933
Eicosapentaenoic acid	$C_{20}H_{30}O_2$	302.2246	6.77	65	3.88×10^{-4}			2	LC-MS(−)	DB	3	HMDB0001999
Eicosatrienoic acid	$C_{20}H_{34}O_2$	306.2558	7.39	82	1.85×10^{-10}			0	LC-MS(−)	DB	3	HMDB0010378
Eicosenoic acid	$C_{20}H_{38}O_2$	310.2871	8.34	79	2.75×10^{-10}	−0.51	1.50	1	LC-MS(−)	DB	3	HMDB0002231
Glucosylgalactosylhydroxylysine	$C_{18}H_{34}N_2O_{13}$	486.2093	12.38	46	1.73×10^{-5}			7	CE-MS	DB	3	HMDB0000585
HETE	$C_{20}H_{32}O_3$	320.2351	5.92	Presented in cancer group	1.24×10^{-5}	−0.63	2.64	0	LC-MS(−)	MSMS	2	HMDB0060101
MG(18:2)	$C_{21}H_{38}O_4$	354.2770	6.81	116	0.0007			2	LC-MS(+)	MSMS	2	HMDB0011538
MG(20:0)	$C_{23}H_{46}O_4$	386.3396	7.79	−86	1.30×10^{-12}			5	LC-MS(+)	DB	3	HMDB0072859
N-palmitoyl glutamic acid *	$C_{21}H_{39}NO_5$	385.2828	4.13	62	0.004			0	LC-MS(+)	DB	3	
Oleic acid	$C_{18}H_{34}O_2$	282.2559	20.47	67	0.0013			-	GC-MS LC-MS(+)	Fiehn	2	HMDB0000207

Table 1. Cont.

Compound	Formula	Mass (Da)	RT (Min)	% Change	p-Value	p(corr)	VIP	Mass Error (ppm)	Analytical Platform	Identification Source	Confidence Level	HMDB Code
Vaccenic acid	$C_{18}H_{34}O_2$	282.2559	20.56	23	0.0158			-	GC-MS	Fiehn	2	HMDB0041480
3-hydroxy-5-octenoylcarnitine	$C_{15}H_{27}NO_5$	301.1889	4.23	Presented in cancer group	3.57×10^{-2}	−0.72	3.96	3	LC-MS(−)	DB	3	
3-Hydroxy-5-tetradecenoylcarnitine *	$C_{21}H_{39}NO_5$	385.2828	4.13	62	0.004			0	LC-MS(+)	DB	3	HMDB0013330
9-Decenoylcarnitine	$C_{17}H_{31}NO_4$	313.2232	12.99	−15	0.0325			7	CE-MS	DB	3	HMDB0013205
Arachidonoylcarnitine *	$C_{27}H_{45}NO_4$	447.3349	5.15	−66	9.67×10^{-17}	0.75	3.74	1	LC-MS(+)	DB	3	HMDB0062343
α-Linolenyl carnitine	$C_{25}H_{43}NO_4$	421.3192	4.91	−41	1.79×10^{-8}	0.53	2.26	0	LC-MS(+)	DB	3	HMDB0006319
Linoleyl carnitine	$C_{25}H_{45}NO_4$	423.3349	5.14	−59	4.03×10^{-16}	0.73	3.40	0	LC-MS(+)	DB	3	HMDB0006469
Oleoylcarnitine												HMDB0006351
Elaidic carnitine	$C_{25}H_{47}NO_4$	425.3504	5.43	−41	1.41×10^{-8}	0.51	2.40	1	LC-MS(+)	MSMS	2	HMDB0006464
Flavonoids												
Anthraniloyl-CoA	$C_{28}H_{41}N_8O_{17}P_3S$	886.1523	0.23	186	3.47×10^{-21}	−0.69	2.39	3	LC-MS(−)	DB	3	
Glycerolipids												
11-Oxo-androsterone glucuronide	$C_{25}H_{36}O_9$	480.2359	3.43	−45	8.06×10^{-6}	0.53	2.06	1	LC-MS(−)	DB	3	HMDB0010338
Glycerophospholipids												
LPC(16:0)-OH	$C_{24}H_{50}NO_8P$	511.3274	4.12	25	0.008			0	LC-MS(+)	MSMS	2	
LPC(16:0)-OH	$C_{24}H_{50}NO_8P$	511.3274	4.21	37	0.0005			0	LC-MS(−)	MSMS	2	
LPC(18:0)-OH	$C_{26}H_{54}NO_8P$	539.3587	4.71	27	0.0100			0	LC-MS(+)	MSMS	2	
LPC(18:2)-OH	$C_{26}H_{50}NO_8P$	535.3274	4.41	Presented in cancer group	2.19×10^{-6}			0	LC-MS(−)	MSMS	2	
LPA(13:0)	$C_{16}H_{33}O_7P$	368.1963	5.23	−57	4.95×10^{-3}	0.56	2.29	7	LC-MS(−)	DB	3	HMDB0114760
LPC(22:1)	$C_{30}H_{60}NO_7P$	577.4107	6.99	66	2.14×10^{-5}			2	LC-MS(−)	MSMS	2	HMDB0010399
LPE(16:0)	$C_{21}H_{44}NO_7P$	453.2855	5.65	25	0.008			0	LC-MS(+)	MSMS	2	HMDB0011473
LPE(20:5)	$C_{25}H_{42}NO_7P$	499.2699	5.08	179	2.80×10^{-6}			0	LC-MS(+)	MSMS	2	HMDB0011489
LPE(20:5)	$C_{25}H_{42}NO_7P$	499.2699	5.16	55	0.001			0	LC-MS(+)	MSMS	2	HMDB0011489
LPE(22:6)	$C_{27}H_{44}NO_7P$	525.2855	5.37	28	0.001			0	LC-MS(+)	MSMS	2	HMDB0011526
LPE(22:6)	$C_{27}H_{44}NO_7P$	525.2855	5.45	39	0.0005			0	LC-MS(+)	MSMS	2	HMDB0011496
LPE(P-16:0)	$C_{21}H_{44}NO_6P$	437.2906	5.83	36	0.002			0	LC-MS(−)	DB	3	HMDB0011152
LPI(16:1)	$C_{25}H_{47}O_{12}P$	570.2805	5.51	97	2.63×10^{-3}			1	LC-MS(−)	MSMS	2	
LPS(18:0)	$C_{24}H_{48}NO_9P$	525.3066	6.71	104	9.62×10^{-6}			2	LC-MS(−)	MSMS	2	

Table 1. Cont.

Compound	Formula	Mass (Da)	RT (Min)	% Change	p-Value	p(corr)	VIP	Mass Error (ppm)	Analytical Platform	Identification Source	Confidence Level	HMDB Code
PC(32:0)	$C_{40}H_{80}NO_8P$	733.5622	10.47	51	3.91×10^{-7}			2	LC-MS(+)	DB	3	HMDB0007871
PC(38:2)	$C_{46}H_{88}NO_8P$	813.6247	11.94	39	0.0004			0	LC-MS(+)	MSMS	2	HMDB0007987
PC(38:5)	$C_{46}H_{82}NO_8P$	807.5778	9.82	−28	5.05×10^{-5}			2	LC-MS(+)	MSMS	2	HMDB0008156
PE(34:2) PE(O-34:3)	$C_{39}H_{74}NO_7P$	699.5203	10.25	−33	0.001			1	LC-MS(+)	MSMS	2	HMDB0011343
PE(38:6)	$C_{43}H_{74}NO_8P$	763.5151	9.31	−31	0.0087			1	LC-MS(+)	MSMS	2	HMDB0009294
PG(20:2)	$C_{26}H_{49}O_9P$	536.3114	7.41	−58	3.44×10^{-11}	0.59	2.33	4	LC-MS(−)	DB	3	
PG(28:0)	$C_{34}H_{67}O_{10}P$	666.4471	7.64	115	8.69×10^{-15}	−0.61	2.75	5	LC-MS(+)	DB	3	HMDB0116681
PS(39:5)	$C_{45}H_{78}NO_{10}P$	823.5363	10.46	25	1.60×10^{-7}			5	LC-MS(+)	DB	3	
Hydroxy acids and derivatives												
3-Hydroxydodecanedioic acid	$C_{12}H_{22}O_5$	246.1467	2.79	Presented in cancer group	2.42×10^{-6}	−0.68	2.97	0	LC-MS(−)	MSMS	2	HMDB0000413
3-Hydroxydodecanoic acid	$C_{12}H_{24}O_3$	216.1725	4.62	Presented in cancer group	2.32×10^{-11}	−0.72	2.79	2	LC-MS(−)	MSMS	2	HMDB0000387
Imidazoles												
Methylimidazole	$C_4H_6N_2$	82.0538	11.48	64	6.41×10^{-6}			9	CE-MS	DB	3	HMDB0034174
Urocanate Nicotinamide N-oxide	$C_6H_6N_2O_2$	138.0435	11.02	24	0.001			4	CE-MS	DB	3	HMDB0002730
Imidazopyrimidines/Purines and purine derivatives												
Hypoxanthine	$C_5H_4N_4O$	136.0402	14.19	−34	8.20×10^{-6}			13	CE-MS	DB	2	HMDB0000157
Indoles and derivatives												
3-Indoleacetic acid	$C_{10}H_9NO_2$	175.0633	17.94	131	2.62×10^{-7}			-	GC-MS	Fiehn	2	HMDB0000197
5-Hydroxyindole	C_8H_7NO	133.0527	0.76	37	3.82×10^{-2}			1	LC-MS(−)	DB	3	HMDB0059805
5-Hydroxyindoleacetaldehyde	$C_{10}H_9NO_2$	175.0633	2.51	150	2.63×10^{-5}	−0.51	2.55	1	LC-MS(+)	MSMS	2	HMDB0004073
5-Hydroxyindoleacetic acid	$C_{10}H_9NO_3$	191.0582	0.80	Presented in cancer group	3.28×10^{-8}			0	LC-MS(+)	MSMS	2	HMDB0000763
Lactones												
N-(4-Coumaroyl)-homoserine lactone	$C_{13}H_{13}NO_4$	247.0845	1.34	55	0.0006			2	LC-MS(+)	DB	3	
Organic acids and derivatives												
(Homo)2-aconitate *	$C_8H_{10}O_6$	202.0477	0.26	−75	7.87×10^{-59}	0.84	2.98	11	LC-MS(−)	DB	3	HMDB0001311
Lactic acid	$C_3H_6O_3$	90.0317	6.06	−77	4.72×10^{-36}	−0.92	2.58	-	GC-MS	Fiehn	2	HMDB0000243
Pyruvic acid	$C_3H_4O_3$	88.0160	5.89	−58	1.06×10^{-14}	−0.69	1.78	-	GC-MS	Fiehn	2	
Succinylacetoacetate *	$C_8H_{10}O_6$	202.0477	0.26	−75	7.87×10^{-59}	0.84	2.98	11	LC-MS(−)	DB	3	HMDB0240258

Table 1. Cont.

Compound	Formula	Mass (Da)	RT (Min)	% Change	p-Value	p(corr)	VIP	Mass Error (ppm)	Analytical Platform	Identification Source	Confidence Level	HMDB Code
Organic sulfuric acids and derivatives												
Indoxylsulfuric acid	$C_8H_7NO_4S$	213.0095	1.00	42	3.47×10^{-2}			1	LC-MS(−)	MSMS	2	HMDB0000682
p-Phenolsulfonic acid	$C_6H_6O_4S$	173.9986	0.65	174	3.45×10^{-3}			1	LC-MS(−)	MSMS	2	HMDB0060015
Organonitrogen compounds												
Phosphocholine	$C_5H_{14}NO_4P$	183.0660	5.37	32	0.003			1	LC-MS(+)	MSMS	2	
Organooxygen compounds												
4-Hydroxycyclohexylcarboxylic acid	$C_7H_{12}O_3$	144.0786	0.76	789	5.08×10^{-10}	−0.71	2.99	1	LC-MS(−)	DB	3	HMDB0001988
Acetyl-N-formyl-5-methoxykynurenamine (AFMK) *	$C_{13}H_{16}N_2O_4$	264.1110	1.34	46	0.0268			2	LC-MS(−)	DB	3	HMDB0004259
Phenols												
4-Methylcatechol	$C_7H_8O_2$	124.0524	1.23	−43	1.68×10^{-2}			2	LC-MS(−)	DB	3	HMDB0000873
Prenol lipids												
Retinol	$C_{20}H_{30}O$	286.2296	7.04	34	0.0003			2	LC-MS(+)	MSMS	2	HMDB0000305
Purine nucleosides												
1-Methyladenosine	$C_{11}H_{15}N_5O_4$	281.1132	12.49	41	3.68×10^{-16}	−0.58	1.20	3	CE-MS	DB	2	HMDB0003331
Pyridines and derivatives												
Norcotinine	$C_9H_{10}N_2O$	162.0790	11.48	34	0.0039			2	CE-MS	DB	3	HMDB0001297
Piperideine	C_5H_9N	83.0739	13.77	28	0.0239			5	CE-MS	DB	3	
Serotonine	$C_{10}H_{12}N_2O$	176.0954	11.40	228	0.0007			3	CE-MS	DB	2	HMDB0001046
Quinolines and derivatives												
8-Hydroxycarteolol	$C_{16}H_{24}N_2O_4$	308.1732	13.90	−31	0.0246			1	CE-MS	DB	3	HMDB0060990
Quinoline	C_9H_7N	129.0578	2.51	99	0.0008			1	LC-MS(+)	DB	3	HMDB0033731
Sphingolipids												
Cer(35:0)	$C_{35}H_{69}NO_3$	551.5277	11.39	58	4.24×10^{-8}			3	LC-MS(−)	DB	3	HMDB0004950
Cer(36:1)	$C_{36}H_{71}NO_3$	565.5434	11.74	42	0.0003			1	LC-MS(+)	DB	3	HMDB0012087
SM(36:0)	$C_{41}H_{85}N_2O_6P$	732.6145	10.14	41	0.0014			1	LC-MS(+)	MSMS	2	
Sphingosine-1-phosphate	$C_{18}H_{38}NO_5P$	379.2488	5.02	−30	2.17×10^{-8}	0.50	2.06	0	LC-MS(+)	MSMS	2	HMDB0000277

Table 1. Cont.

Compound	Formula	Mass (Da)	RT (Min)	% Change	p-Value	p(corr)	VIP	Mass Error (ppm)	Analytical Platform	Identification Source	Confidence Level	HMDB Code
Steroids and steroid derivatives												
12a-Hydroxy-3-oxocholadienic acid	$C_{24}H_{34}O_4$	386.2457	4.27	144	3.66×10^{-6}	−0.54	2.12	0	LC-MS(−)	MSMS	2	HMDB0000385
Biliverdin	$C_{33}H_{34}N_4O_6$	582.2478	4.73	300	1.47×10^{-16}	−0.72	2.62	0	LC-MS(−)	DB	3	HMDB0001008
Hydroxy-3-oxo-4-cholestenoate	$C_{27}H_{42}O_4$	430.3083	5.53	54	2.78×10^{-6}				LC-MS(+) LC-MS(−)	MSMS	2	HMDB0006472
Calcitroic acid	$C_{23}H_{34}O_4$	374.2457	7.39	84	6.25×10^{-11}	−0.54	1.76	6	LC-MS(−)	DB	3	
Chenodeoxycholic acid 3-glucuronide *	$C_{30}H_{48}O_{10}$	568.3248	4.33	94	2.10×10^{-3}			1	LC-MS(−)	DB	3	HMDB0000483
Cholestane-3,7,12,24,25-pentol	$C_{27}H_{48}O_5$	452.3501	7.77	−56	3.81×10^{-10}	0.56	2.03	2	LC-MS(−)	DB	3	HMDB0010355
Cholestane-3,7,12,25-tetrol-3-glucuronide	$C_{33}H_{56}O_{10}$	612.3873	4.71	97	4.89×10^{-5}			0	LC-MS(−)	DB	3	HMDB0015459
Cortisone acetate	$C_{23}H_{30}O_6$	402.2042	3.86	63	5.89×10^{-4}			9	LC-MS(−)	DB	3	HMDB0010348
Dehydroepiandrosterone 3-glucuronide	$C_{25}H_{36}O_8$	464.2410	3.91	−51	1.01×10^{-5}			2	LC-MS(−)	DB	3	HMDB0010327
Dehydroisoandrosterone 3-glucuronide *	$C_{30}H_{48}O_{10}$	568.3248	4.33	94	2.10×10^{-3}			1	LC-MS(−)	DB	3	HMDB0002596
Deoxycholic acid 3-glucuronide *	$C_{33}H_{54}O_{11}$	626.3666	4.41	88	1.02×10^{-4}			1	LC-MS(−)	DB	3	
25-O-D-glucopyranoside ecdysone	$C_{21}H_{36}O_2$	320.2715	6.35	−51	4.21×10^{-5}			0	LC-MS(−)	DB	3	HMDB0004025
Pregnanediol	$C_{21}H_{34}O_5S$	398.2127	3.64	55	5.27×10^{-3}			0	LC-MS(−)	DB	3	HMDB0240591
Pregnanolone sulfate	$C_{24}H_{40}O_4$	392.2927	4.34	188	3.93×10^{-2}	−0.51	2.35	1	LC-MS(−)	MSMS	2	HMDB0000946
Ursodeoxycholic acid	$C_{24}H_{40}O_7S$	472.2494	3.75	130	4.38×10^{-3}	−0.54	2.36	1	LC-MS(−)	DB	3	HMDB0002642
Sterol lipids												
24-Hydroxygeminivitamin D3	$C_{32}H_{54}O_5$	518.3971	6.79	−37	2.82×10^{-7}	−0.68	2.63	1	LC-MS(+)	DB	3	
Tetrapyrroles and derivatives												
Bilirubin	$C_{33}H_{36}N_4O_6$	584.2635	3.85	674	3.79×10^{-8}			4	LC-MS(−) LC-MS(+)	MSMS	2	HMDB0000054

Next, to determine which molecules were directly related to NETs independent of other clinical factors, we performed a logistic regression model (LRM) for each selected metabolite adjusted for gender, age, glycemia, creatinine levels and selected concomitant drugs as potential confounding variables before the inclusion in the model (Table S3). Only up to two significant drug associations ($p < 0.05$) were included in the LRM of each metabolite in order to avoid overfitting due to excessive explicative features. Out of the 49 metabolites assessed, only one metabolite, suberyl-glycine, was not significantly contributing to explain the classification output in its model (Table S4), indicating a poorer diagnostic ability. Thus, 48 metabolites were significantly contributing to explain the classification model (NET vs. non-cancer patients), independent of other confounding factors. The unsupervised heatmap cluster plot of these 48 metabolites clearly discriminated two clusters, one gathered all NET patients and the other one all the non-cancer individuals (Figure 2).

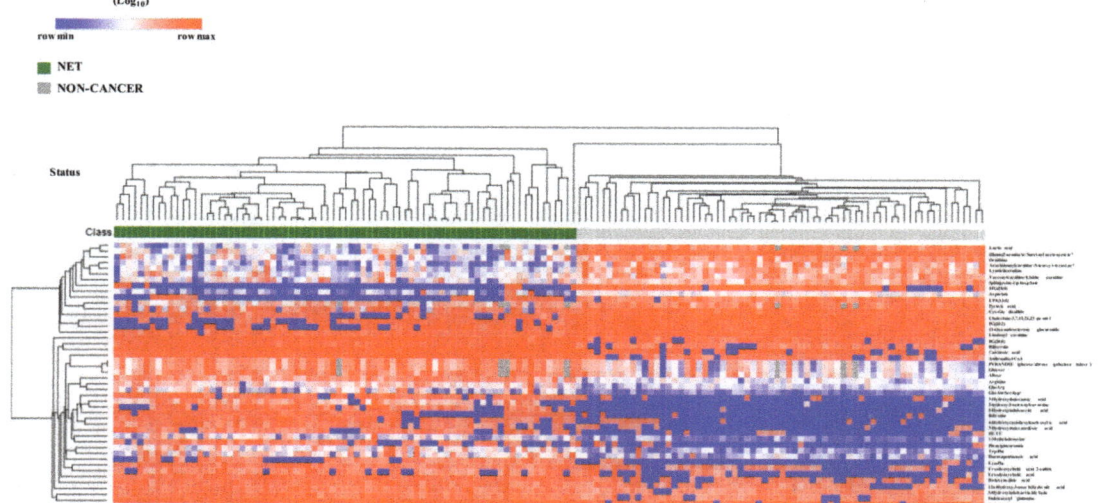

Figure 2. Metabolite biomarker candidates show high diagnostic potential of NET patients. Unsupervised hierarchical heatmap of differential plasmatic metabolites ($n = 48$) between NET ($n = 77$) and non-cancer ($n = 68$) patients. All samples are shown in columns and metabolites in rows. Hierarchical clustering was performed on rows and columns using One minus Pearson correlation metric and average as linkage method. Individual values were coded as colors, ranging from blue (row minimum) to red (row maximum). This analysis clearly discriminated two clusters, one encompassing all NET patients and the other one all non-oncologic control patients.

Finally, to validate the identity, differential expression and biomarker potential of these 48 metabolites, we performed a targeted metabolomic analysis (LC-QQQ-MS) of 13 selected metabolites based on their nature, stability and ability to be analyzed and quantified (Table S5), as described in Material and Methods.

Arginine, 1-methyladenosine, biliverdin, 5-hidroxyindolacetic acid, linoleoylcarnitine, oleoylcarnitine, sphingosine-1-phosphate, 15-hidroxyeicosatetraenoic acid, ursodeoxycholic acid and ursodeoxycholic acid 3-sulfate identities were confirmed as potential diagnostic biomarkers of NETs, while bilirubin, 3-hidroxydodecanodioic acid and 3-hidroxydodecanoic were not (Table S6), probably due to their instability. Distribution of plasma abundance and ROC curves of the validated metabolites with diagnostic potential in NETs and non-cancer patients were showed in Figure 3.

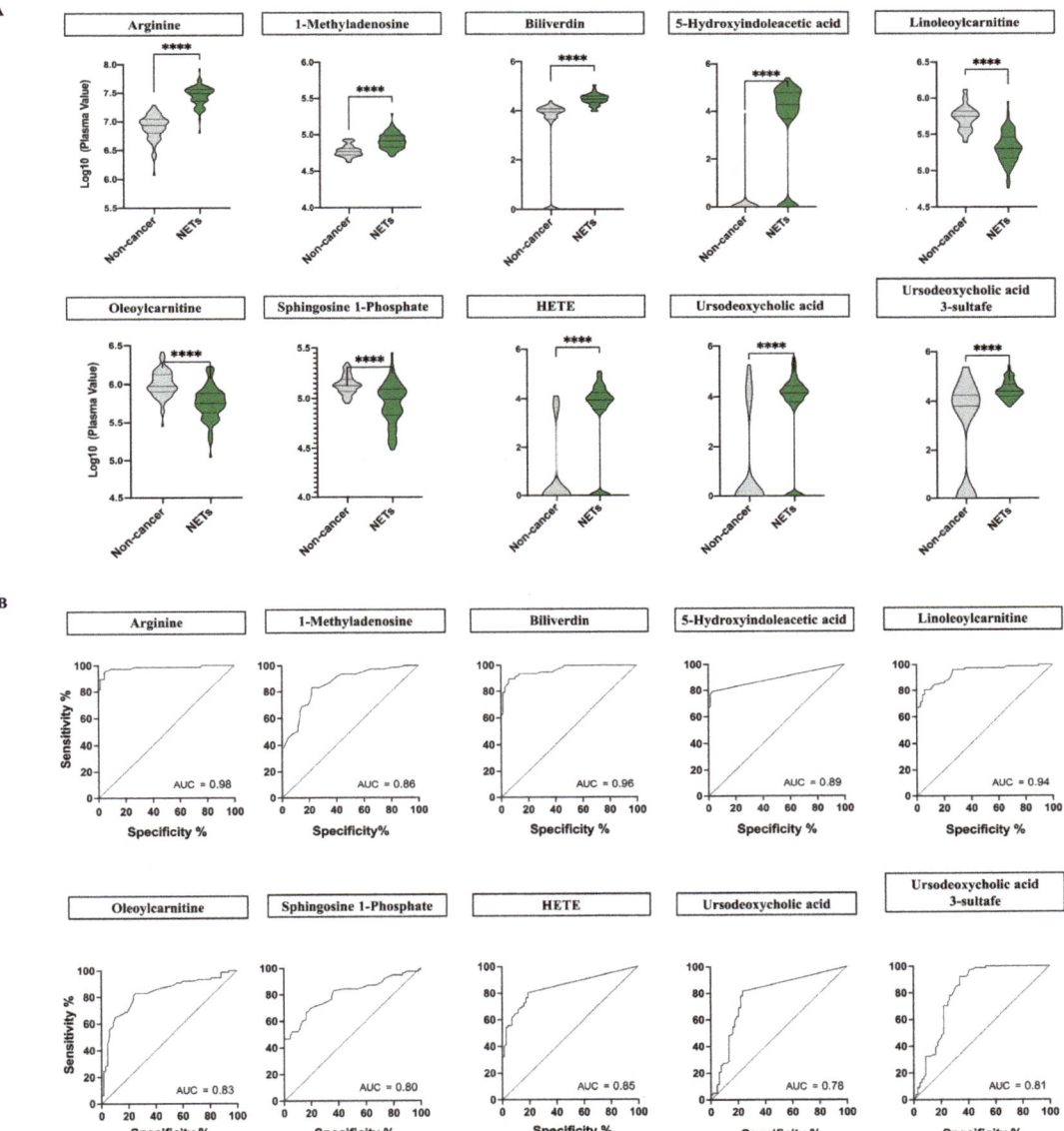

Figure 3. (**A**) Plasma abundance of the 10 validated metabolites with diagnostic potential in NETs. Violin plots and non-parametric Mann-Whitney U test were performed to assess the abundance and distribution of the 10 validated metabolites in NET patients ($n = 77$) and non-cancer individuals ($n = 68$). Log plasma values in NET patients are shown in green whereas non-cancer individuals are plotted in grey. Continuous lines correspond to the median values whereas dashed lines relate to quartiles Overall, all the validated metabolites showed significant differences (****; $p < 0.0001$) between NET and non-cancer individuals. (**B**) Receiver operating characteristic (ROC) curves of the 10 validated metabolites with biomarker potential in NETs. The curves were built based on the area under the curve (AUC) analysis of patients ($n = 77$) and non-cancer individuals ($n = 68$). The optimal cut-off points were selected according to the maximization of the Youden Index. Overall, all the validated metabolites showed high biomarker potential with AUC > 0.75.

2.3. Biochemical and Functional Nature of Identified Metabolites in NET Patients

To better understand the molecular nature of metabolites involved in NETs we classified the 155 differential metabolites according to their biochemical class. The main predominant categories identified were amino acids, peptides and their derivatives (27.7%), fatty acids (16.1%), glycerophopholipids (14.2%), steroids and derivatives (9.7%) and carbohydrates and their conjugates (3.9%) (Figure 4A). Seventy-two percent of all differential metabolites were upregulated in NET patients. The proportion of up- or downregulated metabolites by biochemical class is summarized in Figure 4B.

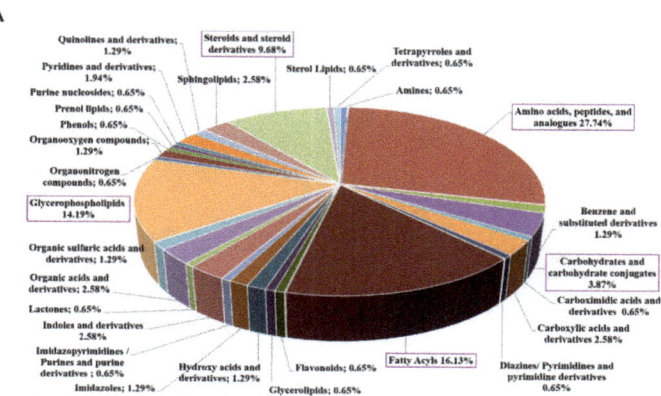

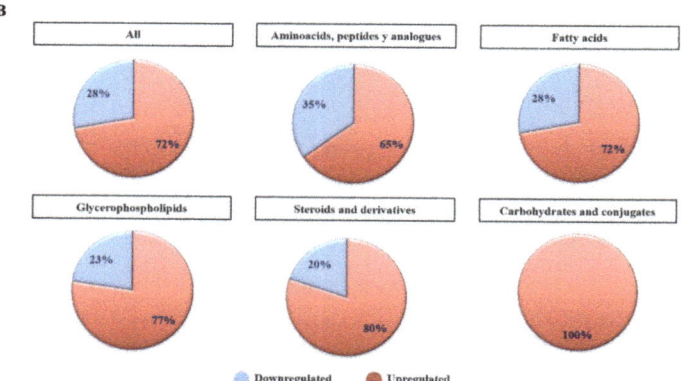

Figure 4. Biochemical classification of differential plasma metabolites in NET patients. (**A**) Pie chart showing the percentage distribution of biochemical classes of differential and annotated plasmatic metabolites (n = 155) in NETs (n = 77) vs. non-cancer (n = 68) individuals. Metabolites were classified into 28 classes according to their biochemical nature. This analysis revealed amino acids, peptides and analogues (n = 43; 27.74%), fatty acyls (n = 26; 16.13%), glycerophospholipids (n = 22; 14.19%), steroids and steroid derivatives (n = 16; 9.68%), and carbohydrates and carbohydrate conjugates (n = 6; 3.87%) as the most represented biochemical classes. (**B**) Bar chart summarizing the percentage of up- and downregulated plasmatic metabolites in the main biochemical classes found to be biologically relevant in NET (n = 77) vs. non-cancer (n = 68) individuals. Overall, upregulation (72%) prevailed over downregulation (28%) in our metabolite set (n =155). This upregulated vs. downregulated metabolite trend remained constant for all main biochemical classes: amino acids, peptides and analogues (65% vs. 35%), fatty acyls (72% vs. 28%), glycerophospholipids (77% vs. 23%), steroids and steroid derivatives (80% vs. 20%), and carbohydrates and carbohydrate conjugates (100%).

Interestingly, the main predominant categories of lipids following the official classification [24] were glycerophospholipids, fatty acids and sterols. Glycerophospholipids represent the largest class of lipids found to be altered. Lyso forms were increased in contrast to glycerophosphocolines and glycerophosphoethanolamines which were downregulated. Oxidized lysoglycerophospholipids (oxLPCs) were found with increased abundances in the NET group, and there are very few studies to date where oxLPCs have been measured. Fatty acids and their derivatives account for about 30% of the measured lipids, and it is easily noticed that there was an overall decrease in carnitine levels. In addition to that, we have been able to identify increased levels of oxidized derivatives of arachidonic acid (HETE) [25]. Other predominant group, sterols, consisted mainly of bile acids and had a high content in the NET group.

NET patients also presented higher levels of sugars and citric acid and lower levels of lactic and pyruvic acid, which are key metabolites in glycolysis and the tricarboxylic acid cycle (TCA)). A remarkable increase in serotonin and its principal metabolite, 5-hydroxyindoleacetic acid, was also detected in the NET cohort, a hormone secreted in excess in patients with carcinoid syndrome.

2.4. Differential Metabolites in NET Patients Are Related with Molecular Pathways Associated with Cancer

Considering the functional relevance of the selected metabolites documented in the literature, we found several enriched pathways related to oncogenesis, some specifically involved in NET development, such as angiogenesis, mTOR pathway, tryptophan metabolism, Warburg effect, oxidative stress, and urea cycle, among others (Figure S5A). The proportion of metabolites up- or downregulated by the pathway is summarized in Figure S5B.

Pathway analysis of the 155 differential metabolites identified several molecular dysregulated pathways in NETs. Metabolite Pathway Analysis (MPA) showed that arginine biosynthesis, arginine and proline metabolism, amimoacyl-tRNA biosynthesis, citrate cycle and pyruvate, glutathione, glyoxylate, dicarboxylate, alanine, aspartate and glutamate metabolisms were the most commonly dysregulated pathways in NET patients (Figure 5A and Table S7). Moreover, Metabolite Set Enrichment Analysis (MSEA), where we used different sets of metabolites from MPA, suggested that tryptophan metabolism and urea cycle, among others, were also dysregulated in NETs (Figure 5B and Table S8).

MPA and MSEA performed with the 48 metabolites with greater diagnostic potential confirmed arginine biosynthesis, arginine and proline metabolism, amimoacyl-tRNA biosynthesis, citrate cycle, pyruvate and glutathione metabolism, and urea cycle were the most relevant dysregulated pathways in NET patients (Figure S4, Tables S9 and S10).

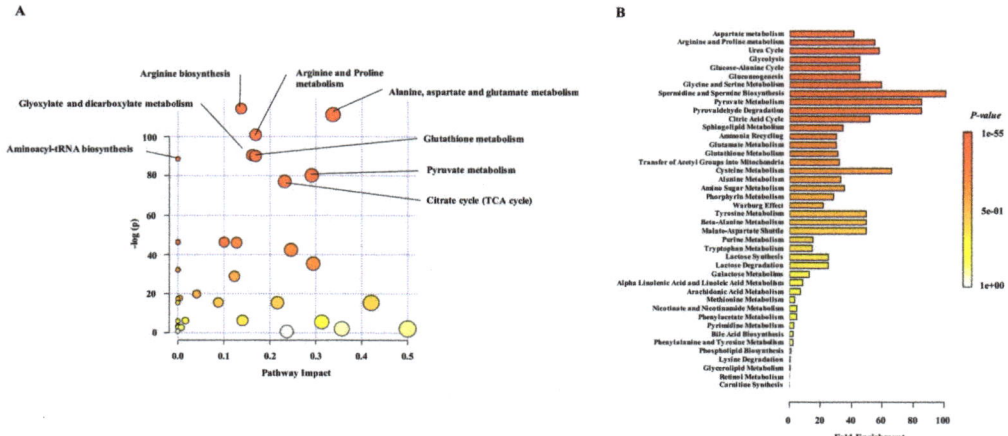

Figure 5. Pathway analysis of differential plasma metabolites in NET patients. (**A**) Metabolite Pathway Analysis (MPA) representing the significant enriched pathways (FDR < 0.05) by availability of selected metabolites (n = 155) in plasma of NET patients. The x-axis indicates the impact of matched metabolites of our dataset on the pathway from the topology analysis. The -log(pval) is plotted in the y axis and shows to which extent the pathway is enriched. Circle size represents the impact factor of matched metabolites in the pathway, and circle color indicates the pathway enrichment significance. The most enriched pathways among the 32 significant ones were: arginine biosynthesis (FDR: 1.0143×10^{-48}); alanine, aspartate and glutamate metabolism (FDR: 2.2363×10^{-47}); arginine and proline metabolism (FDR: 6.3629×10^{-43}); glyoxylate and dicarboxylate metabolism (FDR: 1.8621×10^{-38}); glutathione metabolism (FDR: 2.8086×10^{-38}); aminoacyl-tRNA biosynthesis (FDR: 1.1784×10^{-37}); pyruvate metabolism (FDR: 5.0802×10^{-34}) and citrate cycle (FDR: 1.0634×10^{-32}). (**B**) MSEA of differential plasma metabolites in NET patients. X-axis represents the fold enrichment of each metabolite set and the colour of the bars indicates the raw p-value. Thirty-four metabolite sets were significantly enriched (FDR < 0.05). Aspartate metabolism (Q = 28.809), arginine and proline metabolism (Q = 38.272), urea cycle (Q = 40.438), glycolysis (Q = 31.619) and glucose and alanine cycle (Q = 31.619) were the most significantly enriched metabolite sets.

3. Discussion

Our study provides, for the first time, a comprehensive metabolomic profile of NETs, assessed by multiplatform untargeted metabolomic profiling of plasma samples in a large cohort of patients with advanced disease, and identifies a distinctive metabolic signature of potential clinical use. The integration of metabolic data acquired by GC, CE and LC coupled to MS identified 155 differential compounds between NETs and non-cancer patients. ROC and OPLS-DA analysis revealed 49 specific metabolites of diagnostic potential, 48 of which significantly contributed to the model after adjustment for other potential confounding variables such as gender, age, glycemia, creatinine levels and concomitant drug therapy. The unsupervised heatmap cluster plot of these 48 metabolites clearly identified two distinct clusters, one encompassing all NET patients and the other one all non-cancer individuals. Although biochemical assessment of several peptides is currently used in the clinic for the diagnosis and follow-up of NET patients, the use of general tumor markers such as chromogranin A (CgA) or neuron-specific enolase (NSE) is not recommended for screening nor are they sufficiently reliable as sole diagnostic procedures as they may be increased in several other oncological and non-oncological conditions [26]. In this context, the high diagnostic accuracy of the identified metabolites in our study may provide very valuable new tools to improve the specific detection of NETs.

Differential metabolites identified were related with classical cancer pathways (apoptosis, cell cycle) and NET signaling (tryptophan metabolism, angiogenesis or the mTOR pathway). In addition, we identified 32 novel enriched metabolic pathways in NETs related with the TCA cycle, and with arginine, pyruvate or glutathione metabolism, which have distinct implications in oncogenesis. To date, only two small studies have partially explored the metabolomic profile of NET patients, but none of them using plasma samples as main source of metabolic analysis. Kinross et al. conducted a prospective pilot study that analyzed urine samples of 28 gastroenteropancreatic NETs by 1H-NMR spectroscopic profiling. Distinct metabolomic phenotypes were identified by primary tumor site (small bowel versus pancreatic NEN) and function, and they also observed that variations in hippurate metabolism strongly contributed to the class description. This study had, however, important limitations such as the limited sample size, the substantial age gap between control and tumor populations, and the lack of control of other potential confounding variables such as gender, renal function or concomitant drug therapy [22]. More recently, Imperiale et al. assessed the metabolomic fingerprint of 46 small intestinal NET primary tumors and 18 liver NET metastases by 1H-NMR spectroscopy, as compared to 30 normal small intestine and liver samples, and results suggested alterations in crucial metabolic pathways such as the tricarboxylic acid cycle (TCA cycle). Our study also shows an increase in the TCA cycle activity, reflected by the high availability of isocitrate/citrate compounds in the plasma of NET patients (+64%). Moreover, high levels of glucose (+174%), glutamine (+69%) and fatty acids, which fuel the TCA cycle further support the hypothesis of TCA upregulation in NETs. Studies conducted in other hormone-dependent tumors, such as prostate cancer, emphasize the relevance of altered intermediary metabolism in malignant transformation. More specifically, the metabolic transformation of citrate-producing normal cells to citrate-oxidizing malignant cells has been implicated in oncogenesis. Citrate oxidation in the TCA cycle to produce ATP has important implications on cellular bioenergetics, cell growth, apoptosis, lipogenesis and angiogenesis [27]. The TCA cycle is a convergence point in the cellular respiration machinery, strictly regulated to fulfill cell bioenergetics, biosynthetic and redox balance requirements. Although several tumors types are characterized by a marked deregulation of TCA enzymes [28], its involvement in cancer metabolism remains incompletely understood.

The presence of high levels of isocitrate/citrate in plasma of NET patients could also derive from exported citrate from the mitochondrial pool to be used for lipogenesis [29]. Our data show a characteristic lipidome in NET patients, mainly represented by the enrichment of glycerophospholipids, fatty acids and sterols. More specifically, oxidized lysoglycerophospholipids (oxLPCs) were found with increased abundances in NETs indicating a strong oxidative stress in these tumors, as well as high levels of oxidized derivatives of arachidonic acid (HETE), and a major decrease in carnitine levels [25]. L-carnitine is an essential metabolite, critical for the bidirectional transport of long-chain fatty acyl and the acyl coenzyme A between the cytosol and the mitochondria, which has been considered a bottleneck in the metabolism control of cancer cells [30,31]. Recent reports suggest that the carnitine system is essential for the metabolic adaptation of cancer cells, which obtain energy from beta-oxidation of lipids. Thus, low levels of carnitine in plasma of NET patients may be related to the active carnitine system in tumor mitochondria and the upregulation of beta-oxidation pathways.

Arachidonic acid (AA) is a polyunsaturated fatty acid, which is subsequently metabolized through three different enzymatic pathways (cyclooxygenase (COX), lipoxygenase (LOX) and cytochrome (CYP) P450) leading to a wide variety of lipid mediators (HETE) involved in multiple physiological and pathophysiological processes [32]. In addition to high levels of AA in plasma of NET patients, we have also found an important increase of eicosanoids derivatives, intimately related to inflammatory responses [32].

Tumor progression is also dependent on cholesterol metabolism as proliferating cells increase cholesterol uptake. Cancer cells adapt the high requirements of intracellular cholesterol through different mechanisms including the endogenous production of cholesterol and fatty acids, a reduction of their efflux through transporters or an increase in the uptake of low-density lipid particles [33]. Plasma of NET patients show high content of sterols, mainly bile acids, and cholesterol derivatives such as vitamin D, biliverdin and bilirubin, suggesting the dependency of NETs on lipid and sterol metabolism.

Vascularization and angiogenesis have a particular relevance in NET development and progression [34], and several differential metabolites identified in our study may contribute to the angiogenesis switch. For example, arginine was found to be upregulated in NETs and is the main source of nitric oxide. NO exhibits both anti- and pro-tumoral effects, and is deeply involved in the regulation of angiogenesis, apoptosis, cell cycle, invasion and metastasis [35]. Similarly, lysophosphatidic acid, HETEs or biliverdin induce angiogenesis by upregulation of VEGFA, VEGFC, IL-1β and IL-8 [36–38]. Overall, these findings further support the relevant role that angiogenesis plays in the pathogenesis of NETs.

The mTOR pathway is also critical in NETs [39]. In fact, an mTOR inhibitor, everolimus, has demonstrated antiproliferative activity in these neoplasms and is approved for the treatment of advanced gastrointestinal, pancreatic and lung NETs [40,41]. In our study, we detected very high plasma levels of arginine (+243%) and glutamine (+69%) in NET patients, which are, together with leucine, stimulators of mTOR via the regulation complex. Moreover, an increased abundance of phosphatidylcholine (+51% PC(32:0), +39% PC(38:2) and −28% PC(38:5)) was also observed, the synthesis of which is promoted by mTORC1. Interestingly, the mTOR pathway has been associated with cancer through its role in the regulation of polyamine dynamics [42]. Polyamine levels are associated with a reduction of apoptosis and an increase of cancer cell proliferation and expression of metastasis-related genes, although the mechanisms underlying these effects have not been well defined [43]. Of note, we detected a 38% increase in the acetylspermidine polyamine, illustrating the relevance of polyamines metabolism inNETs. Recently, Chalishazar et al. observed that MYC-driven small-cell lung cancer (SCLC) preferentially depends on arginine-regulated pathways, including polyamine biosynthesis and mTOR pathway activation [44]. ASS1, which indirectly produces arginine in the urea cycle, is often decreased or even abolished through epigenetic silencing in many cancers, including SCLC. Moreover, ASS1 knock-down results in increased mTOR activity and in arginine auxotrophy. Thus, arginine deprivation could be a promising therapeutic strategy for cancers that depend on arginine for their survival [3]. Finally, we also found abundance differences in hypoxanthine. Low levels of hypoxanthine both in urine and plasma samples are often observed in cancer patients, especially in patients with advanced disease stages [45], as hyperproliferative tissues require increased DNA synthesis. Accordingly, we observed a 34% decrease of hypoxanthine abundance in the plasma of NET patients. The underlying mechanism of hypoxanthine downregulation in NET patients is unclear, but it is plausible that alterations in purine metabolism may occur during tumor progression. Consistent with this hypothesis, an inverse correlation was found in our study between hypoxanthine plasma levels and the tumor proliferative rate or Ki-67 index ($r2 = -0.243$, $p = 0.033$).

One of the strengths of our study is that it was performed in plasma samples of a large and homogeneous population, uniformly and prospectively collected and analyzed. Moreover, a set of metabolites was validated in a target analysis with a different analytical platform in the same cohort. Nevertheless, results shall be further investigated in an independent NET patient cohort and metabolomic profiling of patients with exocrine tumors of similar tissue origin (lung and gastrointestinal carcinomas) would be very helpful to validate the specific metabolomic profile of NETs and to confirm the diagnostic potential of the metabolic signature identified. Moreover, complementary -omic approaches, such as exome, transcriptome or methylome of these patients, are needed to further understand the underlying mechanisms in NET development and progression. In particular, the metabolomic profile could be combined with complementary analytical approaches in plasma such as cell-free nucleic acids profiling that might be particularly useful for early diagnostics and patient stratification for personalized clinical management. Plasma -omic profiling has the additional advantage of providing a dynamic characterization of disease biology, which could be eventually utilized, beyond accompanying diagnostics, for targeted prevention or screening, individualized treatment strategies, therapeutic monitoring and prediction of patient's outcome.

4. Material and Methods

4.1. Study Population

The study population included patients with advanced, well-differentiated NETs of lung or gastrointestinal origin. Main clinical and pathological features of the study population are summarized in Table S1. Blood samples were obtained for metabolomic analysis from 77 NET patients and 68 non-cancer individuals as the control group. The distribution of gender, age and body mass index (BMI) was similar in the NET and non-cancer cohorts (Table S11). Peripheral blood was collected in sodium EDTA tubes according to standard procedures and fractionated at 3000 rpm for 5 min. Plasma layer was recovered in sterile cryotubes, frozen and stored until use at $-80\ °C$. The study protocol was approved by the institutional ethics committee and all patients provided informed consent prior to study entry.

4.2. Multiplatform Metabolic Fingerprinting

A multiplatform non-targeted metabolomics approach was performed to provide a wide coverage of the metabolome under study. Samples were analyzed according to standard protocols through different separation techniques coupled to mass spectrometry: capillary electrophoresis 7100 coupled to a MS with time-of-flight analyzer, TOF-MS 6224 (Agilent Technologies, Santa Clara, CA, USA) (CE−MS), HPLC system 1290 Infinity II coupled with 6545 QTOF MS detector (Agilent Technologies, Santa Clara, CA, USA) (LC−MS) and GC system 7890A coupled to a mass spectrometer 5975C (Agilent Technologies, Santa Clara, CA, USA) (GC−MS) [46–48] (see File S1).

4.2.1. Data Processing

The raw data obtained by CE−MS were processed with MassHunter Profinder software version B.08.00, applying the Molecular Feature Extraction (MFE) and Find by Ion (FbI) function by Recursive Feature Extraction (RFE). For LC−MS, the raw data were reprocessed by the MFE with MassHunter Qualitative (B.06.00, Agilent Software, Santa Clara, USA) and DA Reprocessor Offline Utilities B.05.00 (Agilent) and Mass Profiler Professional software (B.14.9 Agilent Software, Santa Clara, USA) to find coeluting adducts and aligned and filtered the data. Raw data files from GC−MS analysis were converted to the appropriate format for quantitative analysis through MassHunter Workstation GC−MS Translator (B.04.01) and deconvolution was carried out through Agilent MassHunter Unknowns Analysis Tool 7.0. For more specific details, see File S1.

4.2.2. Statistical Analysis

After correction (see File S1), the data underwent a Quality Assurance procedure, data normality for every platform was assessed by Kolmogorov−Smirnov and Shapiro−Wilk tests and Levene's test was used to test for variance ratio. To determine the statistical significance of each metabolite separately, differences between non-cancer individuals and NET cases were evaluated by applying Student's t-test ($p \leq 0.05$) using MATLAB (R2015a, MathWorks, Natick, MA, USA). Benjamini−Hochberg multiple post-correction method was applied to all p-values to control the false positive rate at level $\alpha = 0.05$.

Multivariate analysis (MVA) was performed in SIMCA 15.0 (Sartorius Stedim Biotech) OPLS-DA model built was used to assess the S-plot and, for variable selection, volcano plots of variable importance in projection (VIP) score and p(corr) [49]. For more specific details, see Supplementary Material and Methods.

4.2.3. Annotation and Compound Identification

An initial annotation of features from LC−MS and CE−MS based on the m/z of the compounds showing significant differences in class separation was performed by CEU Mass Mediator tool [50] (see File S1).

To confirm the annotation of the compounds, LC−MS/MS analysis was carried out by data independent analysis (DIA) and the identification of each metabolite was achieved by manual MS/MS spectra interpretation. Some CE−MS annotations could also be confirmed through in-source fragmentation obtained at high fragmentor voltage (200 V) [51] (see File S1).

4.2.4. Targeted Analysis

Standards used and the corresponding sources are included in Table S12. Two calibration curves were prepared according to the solubility of the standards. An aqueous mixture containing arginine, 1-methyladenosine, biliverdin and bilirubin was diluted to 6 different concentration levels (ca. 1 ng/mL to 1 µg/mL) and another mixture containing 5-hydroxyindoleacetic acid, 3-hydroxydodecanedioic acid, linoleoylcarnitine, oleoylcarnitine, sphingosine-1-phosphate, 3-hydroxydodecanoic acid, HETE, ursodeoxycholic acid and ursodeoxycholic acid 3-sulfate was diluted in MeOH/EtOH (1:1). Plasma samples were prepared using the same protocol used for untargeted HPLC/MS analysis [47]. Targeted analysis was performed on an Agilent 1290 Infinity UHPLC (Agilent Technologies, Waldbronn, Germany) system coupled with an Agilent 6460 Triple Quadrupole Mass Spectrometer with an electrospray ionization (ESI) source (HPLC QqQ MS/MS). In the final method, chromatographic separation of compounds was achieved with a Zorbax C8 Eclipse Plus column (Agilent Technologies, 2.1 × 150 mm, 1.8 µm) thermostated at 55 °C. For individual analytes, MS-related parameters were tuned by the Agilent MassHunter Optimizer (software version B.07.00, Agilent Software, Santa Clara, USA) using authentic standards for reference. MassHunter Optimizer automatically optimized the data acquisition parameters for MRM (multiple-reaction monitoring) mode. System control and initial chromatogram review were performed with Agilent MassHunter Qualitative software (version B.08.00, Agilent Software, Santa Clara, USA). Data reprocessing were carried out using Agilent MassHunter QQQ Quantitative software program (version B.09.00, Agilent Software, Santa Clara, USA). Metabolites were quantified according to the response factor of the respective calibration curve.

4.3. Metabolite Classification and Pathway Analysis

The discriminant metabolites summarized in Table 1 were classified by biochemical classes and by their relationship with specific molecular pathways (apoptosis, cell cycle, angiogenesis, mTOR pathway, Warburg effect, oxidative stress, tryptophan metabolism, collagen metabolism, carnitine metabolism, methionine cycle, arachidonic acid metabolism, urea cycle, polyamines, and heme metabolism). In order to refine the identification of aberrant molecular pathways in NET patients we analysed our data by Metabolite Pathway Analysis (MPA) and Metabolite Set Enrichment Analysis (MSEA) using MetaboAnalyst 4.0 platform (http://www.metaboanalyst.ca/, accessed on 8 December 2020) [52]. The databases of reference employed were KEGG homo sapiens (Oct 2019) and SMPD [53,54].

4.4. Clinical and Molecular Data Analysis

In order to evaluate the diagnostic potential of metabolites, Receiver Operating Characteristic (ROC) curves and Area Under the Curve (AUCs) were assessed, and sensitivity and specificity values were calculated (according to Youden Index) [55]. Associations with relevant clinical features (age, gender, BMI, glycemia and creatinine plasma levels) and common concomitant medications selected by their putative influence in metabolomics (Table S13) were assessed in metabolites with AUC > 0.85, using Fisher's exact test, chi-squared test or Pearson correlation, as appropriate ($p < 0.05$ were considered significant). Next, logistic regression models were built for each metabolite adjusting for age, gender, glycemia, creatinine plasma levels and significant patient medication selected from association analysis. Metabolites with AUC > 0.85 were considered as potential biomarkers. Additionally, differential metabolites from OPLS-DA models with VIP >1.0 and $|p(corr)| > 0.5$ were also considered as potential biomarkers.

4.5. Heatmap and Hierarchical Clustering

Heatmaps were conducted with the log10 value of each metabolite levels in plasma samples. Unsupervised hierarchical clustering was performed for metabolites and patients using Pearson correlation and average as linkage method. Both were conducted using the Morpheus Software (Broad Institute; https://software.broadinstitute.org/morpheus, accessed on 18 December 2020).

5. Conclusions

In conclusion, untargeted plasma metabolomic profiling of NET patients, that integrated metabolic data acquired by GC, CE and LC in both polarity modes coupled to MS, has identified a distinct metabolic signature of potential clinical use. Indeed, our study has identified and validated a reduced set of metabolites of high diagnostic accuracy that may improve the specific detection of NETs. Differential metabolites were related with classical cancer pathways (apoptosis, cell cycle) and NET signaling (tryptophan metabolism, angiogenesis, mTOR). In addition, MPA/MSEA analysis of these metabolites has revealed new enriched metabolic pathways in NETs, related with the TCA cycle and with arginine, pyruvate or glutathione metabolism, which have distinct implications in oncogenesis and may open innovative avenues of clinical research, including the identification of potential novel targets for therapy. This is to our knowledge the most comprehensive metabolic profiling study performed to date in NETs and provides very valuable information to develop useful biomarkers for the management of these patients in clinical practice.

Supplementary Materials: The following are available online at https://www.mdpi.com/article/10.3390/cancers13112634/s1, Figure S1: PCA-X score plots, Figure S2: PCA-X unsupervised models, Figure S3: NETs show a specific metabolomic profile, Figure S4: Pathway analysis of 48 metabolites with diagnostic potential, Figure S5: Pathway classification of differential plasma metabolites in NET patients, Table S1: Clinical, biochemical and pathological features of NET population, Table S2: Metabolites with biomarker potential in NET patients, Table S3: Drugs as potential confounding variables, Table S4: Logistic regression analysis of metabolites with biomarker potential in the plasma of NET patients, Table S5: Metabolites candidates for targeted validation, Table S6: Validation of potential diagnostic biomarkers of NETs, Table S7: Relevant metabolic pathways related to the identified differential plasma metabolites in NET patients by Metabolite Pathway Analysis (MPA), Table S8: Relevant metabolic pathways related to the identified differential plasma metabolites in NET patients by Metabolite Set Enrichment Analysis (MSEA), Table S9: Relevant metabolic pathways related to the diagnostic biomarker metabolites in NET patients by Metabolite Pathway Analysis (MPA), Table S10: Relevant metabolic pathways related to the diagnostic biomarker metabolites in NET patients by Metabolite Set Enrichment Analysis (MSEA), Table S11: Distribution of age, gender and body mass index in NET patients and non-cancer individuals, Table S12: Standards and the corresponding sources used for the targeted analysis, Table S13: Drug intake of NET and non-cancer patients, File S1: Supplementary Materials and Methods.

Author Contributions: B.S., A.L.-L., C.B. and R.G.-C. designed the study. A.L.-L. performed and C.B. and A.L.-G. oversaw the laboratory analysis. B.S., A.L.-L., A.L.-P., C.C.-P. and A.L.-G. ana-lyzed the data. A.L.S., B.G.-C., M.C.R.-M., P.E.-O., J.S., B.R.-C. and R.R., contributed to interpret-ing and discussing the results and to writing the manuscript. All authors have read and agreed to the published version of the manuscript.

Funding: This work was partially funded by Pfizer, Project G1808 from the Spanish National Taskforce on Neuroendocrine Tumors (GETNE), Ministry of Science, Innovation and Universities of Spain (MICINN) and FEDER funding (Ref. RTI2018-095166-B-I00) and Autonomous Community of Madrid (NOVELREN-CM. Ref: B2017/BMD3751). B.S. is funded by AECC (POSTDO46SOLD, Spain). A.L.-L. thanks CEU-International Doctoral School (CEINDO) for her fellowship. A.L.-P. is funded by CAM (PEJD-2019-PRE/BMD-17058, Progama de Empleo Juvenil (YEI), co-funded by European Union (ERDF/ESF, "Investing in your future"). C.C.-P. was partially funded by CAM (PEJD-2016-PRE/BMD-2666). B.R.-C. was partially funded by CAM (PEJD-2017-PRE/BMD-4981). ALS is funded by Instituto de Salud Carlos III (Contrato Rio Hortega). M.R.-M. is funded by AECC (CLSEN19003RIES).

Institutional Review Board Statement: The study was conducted according to the guidelines of the Declaration of Helsinki, and the study protocol was approved by the Ethics Committee of Hospital Universitario 12 de Octubre (Madrid, Spain) (protocol code 19/159, date of approval 25 June 2019).

Informed Consent Statement: Written informed consent was provided by patients prior to study participation.

Data Availability Statement: The data presented in this study are available from the corresponding author upon reasonable request.

Acknowledgments: We would like to thank all patients for the plasma samples provided for this study.

Conflicts of Interest: The authors declare no conflict of interest.

References

1. Cairns, R.A.; Harris, I.S.; Mak, T.W. Regulation of Cancer Cell Metabolism. *Nat. Rev. Cancer* **2011**, *11*, 85–95. [CrossRef] [PubMed]
2. Suhre, K.; Shin, S.-Y.; Petersen, A.-K.; Mohney, R.P.; Meredith, D.; Wägele, B.; Altmaier, E.; CARDIoGRAM; Deloukas, P.; Erdmann, J.; et al. Human Metabolic Individuality in Biomedical and Pharmaceutical Research. *Nature* **2011**, *477*, 54–60. [CrossRef] [PubMed]
3. Sullivan, L.B.; Gui, D.Y.; Vander Heiden, M.G. Altered Metabolite Levels in Cancer: Implications for Tumour Biology and Cancer Therapy. *Nat. Rev. Cancer* **2016**, *16*, 680–693. [CrossRef] [PubMed]
4. Newgard, C.B. Metabolomics and Metabolic Diseases: Where Do We Stand? *Cell Metab.* **2017**, *25*, 43–56. [CrossRef]
5. Oakman, C.; Tenori, L.; Claudino, W.M.; Cappadona, S.; Nepi, S.; Battaglia, A.; Bernini, P.; Zafarana, E.; Saccenti, E.; Fornier, M.; et al. Identification of a Serum-Detectable Metabolomic Fingerprint Potentially Correlated with the Presence of Micrometastatic Disease in Early Breast Cancer Patients at Varying Risks of Disease Relapse by Traditional Prognostic Methods. *Ann. Oncol. Off. J. Eur. Soc. Med. Oncol.* **2011**, *22*, 1295–1301. [CrossRef]
6. Martín-Blázquez, A.; Díaz, C.; González-Flores, E.; Franco-Rivas, D.; Jiménez-Luna, C.; Melguizo, C.; Prados, J.; Genilloud, O.; Vicente, F.; Caba, O.; et al. Untargeted LC-HRMS-Based Metabolomics to Identify Novel Biomarkers of Metastatic Colorectal Cancer. *Sci. Rep.* **2019**, *9*, 20198. [CrossRef]
7. Bertini, I.; Cacciatore, S.; Jensen, B.V.; Schou, J.V.; Johansen, J.S.; Kruhøffer, M.; Luchinat, C.; Nielsen, D.L.; Turano, P. Metabolomic NMR Fingerprinting to Identify and Predict Survival of Patients with Metastatic Colorectal Cancer. *Cancer Res.* **2012**, *72*, 356–364. [CrossRef]
8. Deja, S.; Porebska, I.; Kowal, A.; Zabek, A.; Barg, W.; Pawelczyk, K.; Stanimirova, I.; Daszykowski, M.; Korzeniewska, A.; Jankowska, R.; et al. Metabolomics Provide New Insights on Lung Cancer Staging and Discrimination from Chronic Obstructive Pulmonary Disease. *J. Pharm. Biomed. Anal.* **2014**, *100*, 369–380. [CrossRef]
9. González-Flores, E.; Serrano, R.; Sevilla, I.; Viúdez, A.; Barriuso, J.; Benavent, M.; Capdevila, J.; Jimenez-Fonseca, P.; López, C.; Garcia-Carbonero, R. SEOM Clinical Guidelines for the Diagnosis and Treatment of Gastroenteropancreatic and Bronchial Neuroendocrine Neoplasms (NENs) (2018). *Clin. Transl. Oncol.* **2019**, *21*, 55–63. [CrossRef]
10. Hofland, J.; Kaltsas, G.; de Herder, W.W. Advances in the Diagnosis and Management of Well-Differentiated Neuroendocrine Neoplasms. *Endocr. Rev.* **2020**, *41*, 371–403. [CrossRef]
11. Dasari, A.; Shen, C.; Halperin, D.; Zhao, B.; Zhou, S.; Xu, Y.; Shih, T.; Yao, J.C. Trends in the Incidence, Prevalence, and Survival Outcomes in Patients With Neuroendocrine Tumors in the United States. *JAMA Oncol.* **2017**, *3*, 1335–1342. [CrossRef]
12. Nuñez-Valdovinos, B.; Carmona-Bayonas, A.; Jimenez-Fonseca, P.; Capdevila, J.; Castaño-Pascual, Á.; Benavent, M.; Barrio, J.J.P.; Teule, A.; Alonso, V.; Custodio, A.; et al. Neuroendocrine Tumor Heterogeneity Adds Uncertainty to the World Health Organization 2010 Classification: Real-World Data from the Spanish Tumor Registry (R-GETNE). *Oncologist* **2018**, *23*, 422–432. [CrossRef]
13. Stevenson, M.; Lines, K.E.; Thakker, R.V. Molecular Genetic Studies of Pancreatic Neuroendocrine Tumors: New Therapeutic Approaches. *Endocrinol. Metab. Clin. N. Am.* **2018**, *47*, 525–548. [CrossRef]
14. Brandi, M.L.; Gagel, R.F.; Angeli, A.; Bilezikian, J.P.; Beck-Peccoz, P.; Bordi, C.; Conte-Devolx, B.; Falchetti, A.; Gheri, R.G.; Libroia, A.; et al. Guidelines for Diagnosis and Therapy of MEN Type 1 and Type 2. *J. Clin. Endocrinol. Metab.* **2001**, *86*, 5658–5671. [CrossRef]
15. Thakker, R.V. Multiple Endocrine Neoplasia Type 1 (MEN1) and Type 4 (MEN4). *Mol. Cell. Endocrinol.* **2014**, *386*, 2–15. [CrossRef]
16. Rednam, S.P.; Erez, A.; Druker, H.; Janeway, K.A.; Kamihara, J.; Kohlmann, W.K.; Nathanson, K.L.; States, L.J.; Tomlinson, G.E.; Villani, A.; et al. Von Hippel-Lindau and Hereditary Pheochromocytoma/Paraganglioma Syndromes: Clinical Features, Genetics, and Surveillance Recommendations in Childhood. *Clin. Cancer Res. Off. J. Am. Assoc. Cancer Res.* **2017**, *23*, e68–e75. [CrossRef]
17. Jett, K.; Friedman, J.M. Clinical and Genetic Aspects of Neurofibromatosis 1. *Genet. Med. Off. J. Am. Coll. Med. Genet.* **2010**, *12*, 1–11. [CrossRef]
18. Larson, A.M.; Hedgire, S.S.; Deshpande, V.; Stemmer-Rachamimov, A.O.; Harisinghani, M.G.; Ferrone, C.R.; Shah, U.; Thiele, E.A. Pancreatic Neuroendocrine Tumors in Patients with Tuberous Sclerosis Complex. *Clin. Genet.* **2012**, *82*, 558–563. [CrossRef]

19. Scarpa, A.; Chang, D.K.; Nones, K.; Corbo, V.; Patch, A.-M.; Bailey, P.; Lawlor, R.T.; Johns, A.L.; Miller, D.K.; Mafficini, A.; et al. Whole-Genome Landscape of Pancreatic Neuroendocrine Tumours. *Nature* **2017**, *543*, 65–71. [CrossRef]
20. Jiao, Y.; Shi, C.; Edil, B.H.; de Wilde, R.F.; Klimstra, D.S.; Maitra, A.; Schulick, R.D.; Tang, L.H.; Wolfgang, C.L.; Choti, M.A.; et al. DAXX/ATRX, MEN1, and MTOR Pathway Genes Are Frequently Altered in Pancreatic Neuroendocrine Tumors. *Science* **2011**, *331*, 1199–1203. [CrossRef]
21. Karpathakis, A.; Dibra, H.; Pipinikas, C.; Feber, A.; Morris, T.; Francis, J.; Oukrif, D.; Mandair, D.; Pericleous, M.; Mohmaduvesh, M.; et al. Prognostic Impact of Novel Molecular Subtypes of Small Intestinal Neuroendocrine Tumor. *Clin. Cancer Res.* **2016**, *22*, 250–258. [CrossRef]
22. Kinross, J.M.; Drymousis, P.; Jiménez, B.; Frilling, A. Metabonomic Profiling: A Novel Approach in Neuroendocrine Neoplasias. *Surgery* **2013**, *154*, 1185–1192; discussion 1192–1193. [CrossRef]
23. Imperiale, A.; Poncet, G.; Addeo, P.; Ruhland, E.; Roche, C.; Battini, S.; Cicek, A.E.; Chenard, M.P.; Hervieu, V.; Goichot, B.; et al. Metabolomics of Small Intestine Neuroendocrine Tumors and Related Hepatic Metastases. *Metabolites* **2019**, *9*, 300. [CrossRef]
24. Fahy, E.; Subramaniam, S.; Murphy, R.C.; Nishijima, M.; Raetz, C.R.H.; Shimizu, T.; Spener, F.; van Meer, G.; Wakelam, M.J.O.; Dennis, E.A. Update of the LIPID MAPS Comprehensive Classification System for Lipids. *J. Lipid Res.* **2009**, *50*, S9–S14. [CrossRef]
25. López-López, Á.; Godzien, J.; Soldevilla, B.; Gradillas, A.; López-Gonzálvez, Á.; Lens-Pardo, A.; La Salvia, A.; Del Carmen Riesco-Martínez, M.; García-Carbonero, R.; Barbas, C. Oxidized Lipids in the Metabolic Profiling of Neuroendocrine Tumors - Analytical Challenges and Biological Implications. *J. Chromatogr. A* **2020**, *1625*, 461233. [CrossRef]
26. Sansone, A.; Lauretta, R.; Vottari, S.; Chiefari, A.; Barnabei, A.; Romanelli, F.; Appetecchia, M. Specific and Non-Specific Biomarkers in Neuroendocrine Gastroenteropancreatic Tumors. *Cancers* **2019**, *11*, 1113. [CrossRef]
27. Costello, L.C.; Franklin, R.B. The Intermediary Metabolism of the Prostate: A Key to Understanding the Pathogenesis and Progression of Prostate Malignancy. *Oncology* **2000**, *59*, 269–282. [CrossRef]
28. Anderson, N.M.; Mucka, P.; Kern, J.G.; Feng, H. The Emerging Role and Targetability of the TCA Cycle in Cancer Metabolism. *Protein Cell* **2018**, *9*, 216–237. [CrossRef]
29. Jiang, L.; Boufersaoui, A.; Yang, C.; Ko, B.; Rakheja, D.; Guevara, G.; Hu, Z.; DeBerardinis, R.J. Quantitative Metabolic Flux Analysis Reveals an Unconventional Pathway of Fatty Acid Synthesis in Cancer Cells Deficient for the Mitochondrial Citrate Transport Protein. *Metab. Eng.* **2017**, *43*, 198–207. [CrossRef]
30. Reuter, S.E.; Evans, A.M. Carnitine and Acylcarnitines. *Clin. Pharmacokinet.* **2012**, *51*, 553–572. [CrossRef]
31. Gao, B.; Lue, H.-W.; Podolak, J.; Fan, S.; Zhang, Y.; Serawat, A.; Alumkal, J.J.; Fiehn, O.; Thomas, G.V. Multi-Omics Analyses Detail Metabolic Reprogramming in Lipids, Carnitines, and Use of Glycolytic Intermediates between Prostate Small Cell Neuroendocrine Carcinoma and Prostate Adenocarcinoma. *Metabolites* **2019**, *9*, 82. [CrossRef] [PubMed]
32. Powell, W.S.; Rokach, J. Biosynthesis, Biological Effects, and Receptors of Hydroxyeicosatetraenoic Acids (HETEs) and Oxoeicosatetraenoic Acids (Oxo-ETEs) Derived from Arachidonic Acid. *Biochim. Biophys. Acta* **2015**, *1851*, 340–355. [CrossRef] [PubMed]
33. Huang, B.; Song, B.-L.; Xu, C. Cholesterol Metabolism in Cancer: Mechanisms and Therapeutic Opportunities. *Nat. Metab.* **2020**, *2*, 132–141. [CrossRef] [PubMed]
34. La Rosa, S.; Uccella, S.; Finzi, G.; Albarello, L.; Sessa, F.; Capella, C. Localization of Vascular Endothelial Growth Factor and Its Receptors in Digestive Endocrine Tumors: Correlation with Microvessel Density and Clinicopathologic Features. *Hum. Pathol.* **2003**, *34*, 18–27. [CrossRef]
35. Keshet, R.; Erez, A. Arginine and the Metabolic Regulation of Nitric Oxide Synthesis in Cancer. *Dis. Model. Mech.* **2018**, *11*. [CrossRef]
36. Xu, Y. Targeting Lysophosphatidic Acid in Cancer: The Issues in Moving from Bench to Bedside. *Cancers* **2019**, *11*, 1523. [CrossRef]
37. Panagiotopoulos, A.A.; Kalyvianaki, K.; Castanas, E.; Kampa, M. Eicosanoids in Prostate Cancer. *Cancer Metastasis Rev.* **2018**, *37*, 237–243. [CrossRef]
38. Chau, L.-Y. Heme Oxygenase-1: Emerging Target of Cancer Therapy. *J. Biomed. Sci.* **2015**, *22*, 22. [CrossRef]
39. Lamberti, G.; Brighi, N.; Maggio, I.; Manuzzi, L.; Peterle, C.; Ambrosini, V.; Ricci, C.; Casadei, R.; Campana, D. The Role of MTOR in Neuroendocrine Tumors: Future Cornerstone of a Winning Strategy? *Int. J. Mol. Sci.* **2018**, *19*, 747. [CrossRef]
40. Yao, J.C.; Shah, M.H.; Ito, T.; Bohas, C.L.; Wolin, E.M.; Van Cutsem, E.; Hobday, T.J.; Okusaka, T.; Capdevila, J.; de Vries, E.G.E.; et al. Everolimus for Advanced Pancreatic Neuroendocrine Tumors. *N. Engl. J. Med.* **2011**, *364*, 514–523. [CrossRef]
41. Yao, J.C.; Fazio, N.; Singh, S.; Buzzoni, R.; Carnaghi, C.; Wolin, E.; Tomasek, J.; Raderer, M.; Lahner, H.; Voi, M.; et al. Everolimus for the Treatment of Advanced, Non-Functional Neuroendocrine Tumours of the Lung or Gastrointestinal Tract (RADIANT-4): A Randomised, Placebo-Controlled, Phase 3 Study. *Lancet Lond. Engl.* **2016**, *387*, 968–977. [CrossRef]
42. Zabala-Letona, A.; Arruabarrena-Aristorena, A.; Martín-Martín, N.; Fernandez-Ruiz, S.; Sutherland, J.D.; Clasquin, M.; Tomas-Cortazar, J.; Jimenez, J.; Torres, I.; Quang, P.; et al. MTORC1-Dependent AMD1 Regulation Sustains Polyamine Metabolism in Prostate Cancer. *Nature* **2017**, *547*, 109–113. [CrossRef]
43. Gerner, E.W.; Meyskens, F.L. Polyamines and Cancer: Old Molecules, New Understanding. *Nat. Rev. Cancer* **2004**, *4*, 781–792. [CrossRef]
44. Chalishazar, M.D.; Wait, S.J.; Huang, F.; Ireland, A.S.; Mukhopadhyay, A.; Lee, Y.; Schuman, S.S.; Guthrie, M.R.; Berrett, K.C.; Vahrenkamp, J.M.; et al. MYC-Driven Small-Cell Lung Cancer Is Metabolically Distinct and Vulnerable to Arginine Depletion. *Clin. Cancer Res. Off. J. Am. Assoc. Cancer Res.* **2019**, *25*, 5107–5121. [CrossRef]

45. Kim, K.; Yeo, S.-G.; Yoo, B.C. Identification of Hypoxanthine and Phosphoenolpyruvic Acid as Serum Markers of Chemoradiotherapy Response in Locally Advanced Rectal Cancer. *Cancer Res. Treat. Off. J. Korean Cancer Assoc.* **2015**, *47*, 78–89. [CrossRef]
46. Naz, S.; Garcia, A.; Rusak, M.; Barbas, C. Method Development and Validation for Rat Serum Fingerprinting with CE-MS: Application to Ventilator-Induced-Lung-Injury Study. *Anal. Bioanal. Chem.* **2013**, *405*, 4849–4858. [CrossRef]
47. Godzien, J.; Kalaska, B.; Adamska-Patruno, E.; Siroka, J.; Ciborowski, M.; Kretowski, A.; Barbas, C. Oxidized Glycerophosphatidylcholines in Diabetes through Non-Targeted Metabolomics: Their Annotation and Biological Meaning. *J. Chromatogr. B Analyt. Technol. Biomed. Life. Sci.* **2019**, *1120*, 62–70. [CrossRef]
48. Dudzik, D.; Zorawski, M.; Skotnicki, M.; Zarzycki, W.; Kozlowska, G.; Bibik-Malinowska, K.; Vallejo, M.; García, A.; Barbas, C.; Ramos, M.P. Metabolic Fingerprint of Gestational Diabetes Mellitus. *J. Proteomics* **2014**, *103*, 57–71. [CrossRef]
49. Wheelock, Å.M.; Wheelock, C.E. Trials and Tribulations of 'omics Data Analysis: Assessing Quality of SIMCA-Based Multivariate Models Using Examples from Pulmonary Medicine. *Mol. Biosyst.* **2013**, *9*, 2589–2596. [CrossRef]
50. Gil de la Fuente, A.; Godzien, J.; Fernández López, M.; Rupérez, F.J.; Barbas, C.; Otero, A. Knowledge-Based Metabolite Annotation Tool: CEU Mass Mediator. *J. Pharm. Biomed. Anal.* **2018**, *154*, 138–149. [CrossRef]
51. Mamani-Huanca, M.; Gradillas, A.; Gil de la Fuente, A.; López-Gonzálvez, Á.; Barbas, C. Unveiling the Fragmentation Mechanisms of Modified Amino Acids as the Key for Their Targeted Identification. *Anal. Chem.* **2020**, *92*, 4848–4857. [CrossRef] [PubMed]
52. Chong, J.; Wishart, D.S.; Xia, J. Using MetaboAnalyst 4.0 for Comprehensive and Integrative Metabolomics Data Analysis. *Curr. Protoc. Bioinforma.* **2019**, *68*, e86. [CrossRef] [PubMed]
53. Kanehisa, M.; Goto, S.; Sato, Y.; Kawashima, M.; Furumichi, M.; Tanabe, M. Data, Information, Knowledge and Principle: Back to Metabolism in KEGG. *Nucleic Acids Res.* **2014**, *42*, D199–D205. [CrossRef] [PubMed]
54. Jewison, T.; Su, Y.; Disfany, F.M.; Liang, Y.; Knox, C.; Maciejewski, A.; Poelzer, J.; Huynh, J.; Zhou, Y.; Arndt, D.; et al. SMPDB 2.0: Big Improvements to the Small Molecule Pathway Database. *Nucleic Acids Res.* **2014**, *42*, D478–D484. [CrossRef]
55. Habibzadeh, F.; Habibzadeh, F.; Habibzadeh, P.; Yadollahie, M. On Determining the Most Appropriate Test Cut-off Value: The Case of Tests with Continuous Results. *Biochem. Med.* **2016**, *26*, 297–307. [CrossRef]

Article

A Novel Two-Lipid Signature Is a Strong and Independent Prognostic Factor in Ovarian Cancer

Liina Salminen [1], Elena Ioana Braicu [2], Mitja Lääperi [3], Antti Jylhä [3], Sinikka Oksa [4], Sakari Hietanen [1], Jalid Sehouli [2], Hagen Kulbe [2], Andreas du Bois [5], Sven Mahner [6], Philipp Harter [5], Olli Carpén [7,8], Kaisa Huhtinen [7], Johanna Hynninen [1] and Mika Hilvo [3,*]

1. Department of Obstetrics and Gynecology, Turku University Hospital and University of Turku, 20521 Turku, Finland; litusa@utu.fi (L.S.); sakari.hietanen@utu.fi (S.H.); mijohy@utu.fi (J.H.)
2. Department of Gynecology, Charité—Universitätsmedizin Berlin, Corporate Member of Freie Universität Berlin, Humboldt Universität zu Berlin, and Berlin Institute of Health, Campus Virchow Klinikum, 13353 Berlin, Germany; elena.braicu@charite.de (E.I.B.); jalid.sehouli@charite.de (J.S.); hagen.kulbe@charite.de (H.K.)
3. Zora Biosciences Oy, 02150 Espoo, Finland; mitja@laaperi.com (M.L.); antti.jylha@thl.fi (A.J.)
4. Satasairaala Central Hospital, Department of Obstetrics and Gynecology, 28500 Pori, Finland; sinikka.oksa@satasairaala.fi
5. Department of Gynecology and Gynecologic Oncology, Kliniken Essen-Mitte, 45136 Essen, Germany; prof.dubois@googlemail.com (A.d.B.); p.harter@gmx.de (P.H.)
6. Department of Obstetrics and Gynecology, University Hospital, LMU Munich, 80337 Munich, Germany; sven.mahner@med.uni-muenchen.de
7. Institute of Biomedicine and FICAN West Cancer Centre Cancer Research Unit, University of Turku, 20521 Turku, Finland; olli.carpen@helsinki.fi (O.C.); kaisa.huhtinen@utu.fi (K.H.)
8. Department of Pathology and Research Program in Systems Oncology, University of Helsinki and Helsinki University Hospital, 00014 Helsinki, Finland
* Correspondence: mika.hilvo@zora.fi; Tel.: +358-50-5347782

Simple Summary: Most ovarian cancer patients initially show a response to primary treatments, but the development of refractory disease is a major problem. Currently, there are no blood-based prognostic biomarkers, and the prognosis of a patient is determined by the International Federation of Gynecology and Obstetrics (FIGO) stage and residual disease after cytoreductive surgery. In this study, we developed and validated a novel test based on the ratio of two circulatory lipids that enables the prognostic stratification of ovarian cancer patients at the time of diagnosis, prior to any oncological treatments. The translational relevance of this test is to find those patients with poor prognosis early on, and to identify patients that are at high risk of recurrence despite complete cytoreduction. Thus, the test enables the early direction of novel targeted therapies to those ovarian cancer patients at greatest risk of recurrence and death.

Abstract: Epithelial ovarian cancer (EOC) generally responds well to oncological treatments, but the eventual development of a refractory disease is a major clinical problem. Presently, there are no prognostic blood-based biomarkers for the stratification of EOC patients at the time of diagnosis. We set out to assess and validate the prognostic utility of a novel two-lipid signature, as the lipidome is known to be markedly aberrant in EOC patients. The study consisted of 499 women with histologically confirmed EOC that were prospectively recruited at the university hospitals in Turku (Finland) and Charité (Berlin, Germany). Lipidomic screening by tandem liquid chromatography–mass spectrometry (LC-MS/MS) was performed for all baseline serum samples of these patients, and additionally for 20 patients of the Turku cohort at various timepoints. A two-lipid signature, based on the ratio of the ceramide Cer(d18:1/18:0) and phosphatidylcholine PC(O-38:4), showed consistent prognostic performance in all investigated study cohorts. In the Turku cohort, the unadjusted hazard ratios (HRs) per standard deviation (SD) (95% confidence interval) were 1.79 (1.40, 2.29) for overall and 1.40 (1.14, 1.71) for progression-free survival. In a Charité cohort incorporating only stage III completely resected patients, the corresponding HRs were 1.59 (1.08, 2.35) and 1.53 (1.02, 2.30). In linear-mixed models predicting progression of the disease, the two-lipid signature showed higher performance (beta per SD increase 1.99 (1.38, 2.97)) than cancer antigen 125 (CA-125, 1.78 (1.13, 2.87)). The two-lipid signature was able to identify EOC patients with an especially poor prognosis at the time of diagnosis, and also showed promise for the detection of disease relapse.

Keywords: ovarian cancer; lipidomics; lipid; prognosis; ceramide; phospholipid; plasmalogen; biomarker; patient stratification; outcome; personalized medicine

1. Introduction

Lipids play a fundamental role in the function of normal cells. They enable chemical energy storage, cellular signaling, cell–cell interactions in tissues, and adequate function of cell membranes, subsequently regulating cell survival, proliferation, and death [1]. These mechanisms are also tightly associated with and modified in oncogenic processes, particularly cell transformation, tumor progression, and metastasis [1,2]. As an emerging hallmark of cancer [3], metabolic and lipidomic dysregulation has attracted scientific interest, and there is increasing evidence on the utility of lipidomics in the discovery of circulating cancer biomarkers as well as disease mechanism exploration in tumors [4].

Comprehensive circulatory metabolic and lipidomic alterations have been described in epithelial ovarian cancer (EOC) [5], which is a malignancy with a generally unfavorable outcome. With a five-year survival expectancy below 50%, EOC still has the highest mortality among gynecological cancers, despite the recent advances in oncological treatments [6]. Although most patients initially show response to primary treatment, the frequent development of refractory disease is a major problem. Cancer antigen 125 (CA-125) is a well-validated biomarker used in the diagnostics and treatment monitoring of EOC; however, it has remained of little utility in the prognostic evaluation of EOC patients in the clinical setting [7]. As circulatory prognostic biomarkers are lacking, the prognostic stratification of patients is currently based on the International Federation of Gynecology and Obstetrics (FIGO) stage and residual disease after cytoreductive surgery [8]. In addition, patients with mutations in the homologous recombination repair (HR) genes have been shown to possess a better prognosis than those without mutations [9]. As of now, targeted therapies, i.e., poly (ADP-ribose) polymerase (PARP) inhibitors and anti-vascular endothelial growth factor (VEGF) monoclonal antibodies, have become part of the standard treatment regimen of EOC patients and consequently, patient selection and the timely administration of targeted treatments have emerged as new challenges in clinical care [10,11].

The investigation of circulatory lipidomic changes/aberrations may enable the prognostic evaluation of cancer patients. Recently, a distinct plasma three-lipid signature (ceramide, sphingomyelin, and phosphatidylcholine) was associated with the poor overall survival of patients with castration-resistant prostatic cancer [12]. Equivalently, two ceramide species and 14 phospholipids quantified from the plasma of patients with liver cancer were associated with increased mortality in another contemporary study [13]. Studies evaluating the feasibility of lipidomics in the prognostic stratification of EOC patients are scarce but promising. Specifically, unsaturated phospholipids and ceramide species have been suggested to play important and complex roles in EOC patient outcomes [14,15]. Phospholipids have been shown to augment ovarian cancer invasion and metastasis, i.e., by activating proteolytic enzymes, while ceramides are known to form more complex sphingolipids, which are bioactive lipids suspected to boost the survival of cancerous cells and facilitate tumor progression [16,17]. In the current study, we set out to evaluate the prognostic utility of a novel two-lipid signature test in patients with EOC. The test builds on the ratio of a circulatory ceramide (d18:1/18:0) (Cer(d18:1/18:0)) and a plasmalogen (PC(O-38:4)), of which the former has previously been detected in increasing and the latter in decreasing concentrations in the sera of EOC patients [15]. In addition, the ability of the two-lipid signature test to detect early disease recurrence was evaluated with longitudinal lipid measurements.

2. Materials and Methods

2.1. Patient Cohorts and Samples

Global lipidomic analysis was performed for serum samples obtained in 3 independent ovarian cancer study cohorts, 1 from the Turku University Hospital (Turku, Finland) and 2 from the Charité University Hospital (Berlin, Germany) (Table 1). In the Turku cohort, patients with histologically confirmed invasive EOC were prospectively recruited at the University Hospital of Turku, Turku, Finland, in 2009–2019 (ClinicalTrials.gov identifier: NCT01276574). In addition, 114 patients with histologically confirmed benign gynecological diseases (benign tumors, inflammatory processes, and endometriosis) were included in the study. A team of gynecologic oncologists evaluated patients diagnosed with ovarian cancer, and if the tumor was considered primarily unresectable, patients received neoadjuvant chemotherapy with subsequent interval debulking surgery [18]. Finally, a set of 20 high-grade serous carcinoma (HGSOC) patients were selected from the Turku cohort for the longitudinal lipidomic analyses (Table S1). For these patients, samples were collected before each cycle of chemotherapy and during follow up until first relapse.

Table 1. Clinical characteristics of the study cohorts.

Characteristic	Subgroup	Turku	Charité 1	Charité 2	Charité 3
Malignant (N)		197	51	104	147
Age (years)		66 (59–72)	61 (56–68)	65 (57–70)	59 (50–67)
Histology	Serous	156	48	95	147
	Mucinous	13		1	
	Endometroid	16	1	2	
	Clear-cell	12		1	
	Other/unknown		2	5	
Stage	I	31			2
	II	12		1	5
	III	102	51	67	99
	IV	50		24	31
	NA	2		12	10
Follow-up time (years)	Death	2.6 (1.5–3.9)	3.6 (1.9–6.0)	1.6 (0.9–2.3)	3.2 (1.7–4.3)
	Progression	1.3 (0.8–2.2)	2.0 (1.0–3.6)	1.2 (0.8–1.8)	1.5 (0.8–2.8)
Benign (N)		114			98
Age (years)		55 (45–68)			41 (31–55)

The cohorts from Charité were prospectively included in the Tumor Bank Ovarian Cancer (www.toc-network.de, accessed on 6 April 2021). The patients in the first Charité cohort (Charité 1) also participated the LION (Lymphadenectomy in Ovarian Neoplasms) clinical trial [19], and consisted exclusively of stage III, completely cytoreduced patients (Table 1). The second cohort from Charité (Charité 2) (Table 1) was a patient cohort where the prognostic serum samples were obtained either before or after neoadjuvant chemotherapy, whereas all the samples for the prognostic analyses from Turku and Charité 1 cohorts were obtained before any oncological treatments. For additional validation, we used baseline serum lipidomics data from a third, previously published [15,20] Charité cohort, Charité 3 (Table 1).

For all cohorts, the FIGO stage was determined according to the FIGO 2014 guidelines. The operating team carefully assessed the amount of residual disease after cytoreductive surgery, if present. The primary chemotherapy regimen consisted of carboplatin and taxane. The response to primary treatment and progression were defined according to the Response Evaluation Criteria in Solid Tumors (RECIST) guidelines [21]. Second-line medical treatment was chosen individually for each patient based on the timing of the relapse (platinum sensitive vs. resistant) and included chemotherapy indicated for the treatment of relapsed EOC [11]. The progression-free survival (PFS) was calculated from the time of diagnosis to disease relapse.

2.2. Lipidomic and Conventional Biomarker Analyses

A global lipidomic screening method was used to analyze the samples. In brief, 10 µL of samples were used for the extraction of the lipids using a modified Folch extraction [22]. The analysis was performed on a hybrid triple quadrupole/linear ion trap mass spectrometer (QTRAP 5500, AB Sciex, Concords, Canada) equipped with ultra-high-performance liquid chromatography (UHPLC) (Nexera-X2, Shimadzu, Kyoto, Japan). Chromatographic separation was performed on an Acquity BEH C18, 2.1 × 50 mm id. 1.7 µm column (Waters Corporation, Milford, MA, USA). The data were collected using a scheduled multiple reaction monitoring (sMRM™) algorithm [23]. The lipidomic data were processed using Analyst and MultiQuant 3.0 software (AB Sciex), and the area or height ratios of the analytes and their corresponding internal standard peaks were normalized with the IS amount and the sample volume. The details of the chromatography and mass spectrometry conditions have been described previously [15].

For patients recruited at the Turku University Hospital, the serum CA-125 (U/mL) concentrations were determined from serum samples with the electrochemiluminescence method on the Cobas e 601 instrument or a Modular E170 automatic analyzer (Roche Diagnostics GmbH, Mannheim, Germany). For Charité patients, CA-125 was measured using Elecsys CA 125 II assay (Roche Diagnostics GmbH, Mannheim, Germany). The serum human epididymis protein 4 (HE4) (pmol/L) concentrations were determined with the enzyme immunoassay method according to the instructions of the manufacturer (Fujirebio Diagnostics Inc., Malvern, PA, USA).

2.3. Statistical Analyses

Baseline characteristics of the cohorts were described using medians (interquartile range, IQR) for continuous variables. Two-group comparisons were performed by calculating the mean relative difference between the groups, and the p-values were determined by parametric t-tests on log-transformed concentrations. The selection procedure for the prognostic lipid ratio has been described in detail in Figure S1. Uni- and multivariate Cox proportional hazard regression models were used to determine hazard ratios (HRs) and 95% confidence intervals for the associations of lipids and clinical measurements with overall and progression-free survival of the patients. The effects were expressed per increase in standard deviation of the biomarkers. The changes over time were investigated using linear mixed models with random intercepts, using the lme4 (version 1.1–23) package. R version 4.0.2 was used for all statistical analyses.

3. Results

3.1. Selection of a Prognostic Two-Lipid Signature

We carried out a global lipidomic analysis of pretreatment serum samples from 197 ovarian cancer patients and 114 benign controls (Turku study, Table 1). The results were in line with previous findings showing that ovarian cancer profoundly affects the lipidome, and that the lipid alterations are already observable in early-stage patients (Table S2). Furthermore, the results ratified that a large number of the lipids is associated with the overall and progression-free survival of the patients (Table S2), and from these a single prognostic lipid ratio was constructed. The selection was performed by taking from lipids showing association with ovarian cancer and survival, a lipid ratio that showed highest C-statistic for overall survival (Figure S1, Table S2). The lipids selected for this two-lipid signature were PC(O-38:4) and Cer(d18:1/18:0), and the Cer(d18:1/18:0)/PC(O-38:4) ratio showed higher C-statistics for both overall and progression-free survival than these lipids individually (Table S2).

3.2. Prognostic Value of the Two-Lipid Signature Test

Cox proportional hazards models for overall and progression-free survival were constructed in all study cohorts to evaluate the prognostic performance of the Cer(d18:1/18:0) / PC(O-38:4) lipid signature when measured prior to surgery or any other oncological treatments. For overall survival, the point estimates for HRs per standard deviation ranged from 1.40 to 2.12 and the C-statistic from 0.592 to 0.707 in the full study cohorts (Turku, Charité 1, Charité 2, Charité 3), as well as in the subcohorts of patients without macroscopic residual disease after surgery (R0) (Table 2). Adjusted HR point estimates varied from 1.10 to 2.16, and the C-statistic from 0.626 to 0.735. Importantly, pretreatment CA-125 value was not indicative of overall survival in any of the study cohorts (Table 2).

For progression, the HR point estimates were between 1.22 and 1.53 in the cohorts, and the C-statistic from 0.524 to 0.644 (Table 2). A C-statistic of 0.615 was recorded for the Charité 1 cohort which included only stage III R0 resected patients (Table 2). Again, CA-125 values were more modest, ranging from 0.418 to 0.592 (Table 2). In the Turku study we had HE4 data available for the majority of the patients, and this clinically established biomarker also showed worse performance than the two-lipid signature, except for the R0 population (Table S3).

Table 2. Hazard ratios of the two-lipid signature and cancer antigen 125 (CA-125) for overall and progression-free survival.

Endpoint	Study	(Sub)Group	Cer(d18:1/18:0)/PC(O-38:4)					CA-125				
			Ev+	Ev−	HR (95% CI) [a]	C-Statistic [a]	HR (95% CI) [b]	C-Statistic [b]	Ev+	Ev−	HR (95% CI) [a]	C-Statistic [a]	HR (95% CI) [b]	C-Statistic [b]
Death	Turku	All	90	93	1.79 (1.40, 2.29)	0.707	1.72 (1.32, 2.25)	0.717	85	92	1.02 (0.79, 1.32)	0.486	0.79 (0.60, 1.06)	0.655
	Turku	No residual tumor	21	58	2.12 (1.26, 3.55)	0.648	2.16 (1.20, 3.86)	0.735	21	58	0.86 (0.52, 1.44)	0.532	0.60 (0.35, 1.04)	0.703
	Charité 1	Stage III, no residual tumor	33	17	1.59 (1.08, 2.35)	0.592	1.67 (1.12, 2.47)	0.626	32	16	1.21 (0.82, 1.78)	0.534	1.38 (0.91, 2.08)	0.628
	Charité 2	All	28	76	1.95 (1.31, 2.88)	0.694	1.73 (1.11, 2.71)	0.706	25	74	1.39 (0.94, 2.05)	0.421	1.26 (0.81, 1.98)	0.580
	Charité 3	All	77	66	1.40 (1.12, 1.74)	0.630	1.10 (0.88, 1.38)	0.722	76	64	1.18 (0.92, 1.50)	0.544	1.02 (0.78, 1.32)	0.706
	Charité 3	No residual tumor	42	46	1.40 (1.03, 1.91)	0.621	1.23 (0.91, 1.68)	0.672	42	44	1.21 (0.87, 1.68)	0.550	1.17 (0.84, 1.64)	0.633
Progression	Turku	All	122	61	1.40 (1.14, 1.71)	0.644	1.28 (1.04, 1.57)	0.700	118	59	1.35 (1.08, 1.69)	0.585	1.01 (0.76, 1.34)	0.667
	Turku	No residual tumor	34	45	1.28 (0.88, 1.87)	0.524	1.13 (0.71, 1.78)	0.747	34	45	1.29 (0.87, 1.92)	0.418	0.83 (0.50, 1.37)	0.740
	Charité 1	Stage III, no residual tumor	34	16	1.53 (1.02, 2.30)	0.615	1.55 (1.03, 2.32)	0.629	34	14	1.39 (0.95, 2.05)	0.592	1.51 (1.00, 2.28)	0.603
	Charité 2	All	49	55	1.27 (0.90, 1.80)	0.589	1.31 (0.90, 1.90)	0.625	46	53	1.22 (0.88, 1.67)	0.560	1.45 (1.01, 2.09)	0.592
	Charité 3	All	84	59	1.22 (0.99, 1.51)	0.563	1.09 (0.87, 1.37)	0.633	82	58	1.15 (0.91, 1.44)	0.541	1.04 (0.82, 1.32)	0.580
	Charité 3	No residual tumor	58	30	1.32 (1.00, 1.73)	0.567	1.22 (0.92, 1.62)	0.632	57	29	1.09 (0.84, 1.42)	0.476	1.02 (0.78, 1.34)	0.544

Hazard ratios (HRs) are expressed per increase in standard deviation. Ev+, event; Ev−, no event; CI, confidence interval. [a] Unadjusted models, [b] adjusted with age in all cohorts and additionally by stage in the Turku and Charité 2 studies and additionally by success of tumor removal in the Charité 3 cohort.

3.3. Risk Tables for Ovarian Cancer Patients

To illustrate the clinical relevance of the two-lipid signature, patients were split by quartiles of this ratio. The two lowest quartiles were combined to place focus on the highest quartiles. Event rates in these quartiles were calculated at 1, 3, and 5 years in the Turku as well as Charité 1 cohorts. It was apparent that within one year of the diagnosis, the lipid signature already showed consistent risk prediction for progression-free survival, and the difference between the lowest and highest quartiles remained until the five-year follow-up (Figure 1). For death, the separation of the risk in the Charité 1 study became apparent only after three years, whereas it was already evident in the Turku cohort during the first year (Figure 1). For comparison, CA-125 was categorized similarly and in general the results were less consistent than for the lipid ratio (Figure 1).

Cer(d18:1/18:0) / PC(O-38:4)		Quartile	1 year			3 years			5 years		
			Event	No event	%	Event	No event	%	Event	No event	%
Turku	Death	Q1-Q2	1	48	2%	7	29	19%	11	10	52%
		Q3	4	54	7%	20	26	43%	28	9	76%
		Q4	17	54	24%	43	17	72%	47	8	85%
	Progression	Q1-Q2	9	39	19%	22	16	58%	22	7	76%
		Q3	14	43	25%	38	9	81%	40	4	91%
		Q4	38	31	55%	59	5	92%	62	2	97%
Charité 1	Death	Q1-Q2	2	23	8%	6	19	24%	11	11	50%
		Q3	0	13	0%	5	8	38%	7	5	58%
		Q4	0	12	0%	7	5	58%	8	3	73%
	Progression	Q1-Q2	4	20	17%	11	11	50%	14	5	74%
		Q3	3	10	23%	6	7	46%	10	2	83%
		Q4	7	6	54%	11	1	92%	11	0	100%

CA-125		Quartile	1 year			3 years			5 years		
			Event	No event	%	Event	No event	%	Event	No event	%
Turku	Death	Q1-Q2	3	33	8%	9	16	36%	11	6	65%
		Q3	12	54	18%	29	20	59%	33	7	83%
		Q4	6	64	9%	29	33	47%	37	13	74%
	Progression	Q1-Q2	9	25	26%	13	10	57%	13	5	72%
		Q3	23	42	35%	44	9	83%	46	3	94%
		Q4	27	43	39%	58	10	85%	61	4	94%
Charité 1	Death	Q1-Q2	1	23	4%	6	18	25%	10	10	50%
		Q3	0	11	0%	6	5	55%	8	3	73%
		Q4	0	13	0%	5	8	38%	7	5	58%
	Progression	Q1-Q2	3	21	13%	10	12	45%	13	5	72%
		Q3	7	5	58%	10	1	91%	11	0	100%
		Q4	4	9	31%	8	5	62%	11	1	92%

Figure 1. Survival in different quartiles (Q1–Q4) of the lipid ratio (**upper panel**) or CA-125 (**lower panel**).

Based on the Turku cohort data, a Cox regression model incorporating lipid ratio quartiles and the clinically important risk factors, i.e., FIGO stage and residual tumor after surgery, was constructed to predict the risk of progression after one year and the risk of death within five years of the diagnosis and biomarker measurements. As expected, for stage I–II tumors the risk was generally much lower than for stage III–IV tumors (Figure 2). However, for this population, after five years the risk of death was almost three-fold higher in the highest quartile of the lipid signature when compared to the lowest quartiles. Importantly, the stage III–IV R0 resected patients belonging to the highest biomarker quartile had a higher risk for both worse progression-free and overall survival than patients with a suboptimal surgery result but belonging to the low biomarker quartiles (Figure 2). When only HGSOC patients were included, the results were similar except for stage I–II patients that had higher overall risk (Figure S2).

Stage	Residual Tumor	PROGRESSION (1 YEAR)			DEATH (5 YEARS)		
III–IV	> 10 mm	23	33	44	52	73	90
III–IV	0–10 mm	22	30	41	41	60	81
III–IV	0 mm	20	28	39	1	48	69
I–II	All	1	2	3	6	9	17
	Lipid ratio quartile	Q1–Q2	Q3	Q4	Q1–Q2	Q3	Q4

Figure 2. Risk (%) of progression in 1 year or death in 5 years, based on the lipid ratio quartile, stage, and success of tumor removal in surgery in the Turku cohort.

3.4. Results From Longitudinal Analyses

For 20 patients of the Turku cohort, we analyzed serum samples taken at the time of diagnosis and in different phases during the course of the disease. The results from the linear mixed models analyzing the interaction of biomarkers and possible progression of the disease revealed that the two-lipid signature was more consistently elevated at the time of progression (beta per SD increase 1.99 (95% confidence interval (CI) 1.38, 2.97)) than CA-125 (beta per SD increase 1.78 (95% CI 1.13, 2.87)). Indeed, when the results were plotted for all the patients, it was evident that for two cases (subjects 1 and 10) the lipid signature was clearly elevated at the time of progression, whereas CA-125 showed no or minor elevation (Figure 3 and Figure S3). Importantly, there were no cases where CA-125 had shown elevation during progression but the lipid signature did not (Figure S3).

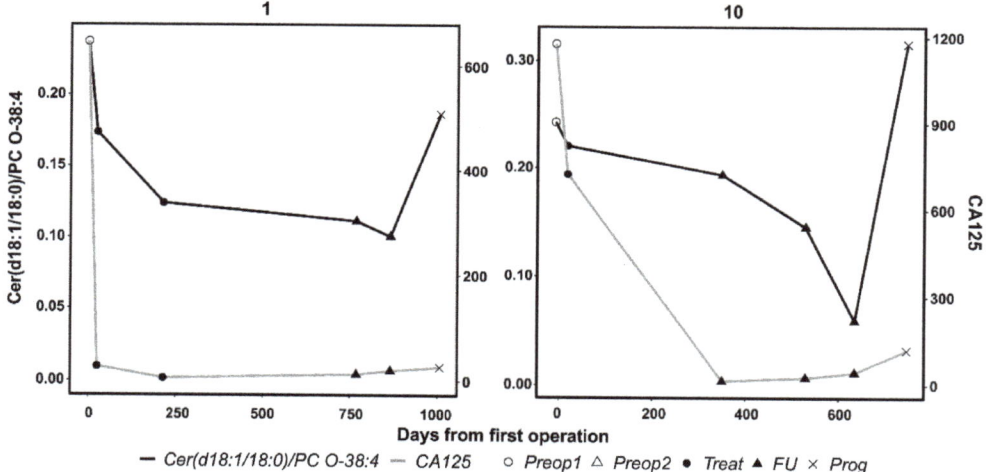

Figure 3. Longitudinal lipidomic and CA-125 analyses of two exemplified ovarian cancer patients. At progression, the two-lipid signature was strongly elevated whereas CA-125 showed none or minor elevation. Preop1, before first surgery; Preop2, before second surgery; Treat, on-treatment; FU, follow-up; Prog, progression.

4. Discussion

As of now, prognostic tools for evaluating the outcome of EOC patients are limited to clinical and histopathological variables, while no biomarkers are in routine use. In the current study, we introduced for the first time a two-lipid signature in the prognostic stratification of EOC patients. For cardiovascular risk prediction, it has been shown that distinct lipid ratios outperform single lipid concentrations in the prognostic performance [24,25]. For this reason and to construct as simple a test as possible, we decided to select a single lipid ratio for an ovarian cancer prognostic test. Above all, the lipid test identified especially poor-outcome patients at the time of diagnosis, before any oncological treatments. Unsurprisingly, CA-125 was not significantly associated with the survival of EOC patients.

Patient selection and the most advantageous implementation of precision medicine is of increasing importance in the regular treatment of EOC patients. In ovarian cancer patients with HGSOC histology, a homologous recombination deficiency (HRD) indicates better prognosis and response to platinum and PARP inhibition therapies [26]. Regarding antiangiogenic agents, there is currently no clear consensus on which patients should receive it and more importantly, which patients should not, although bevacizumab is generally seen as a part of the standard treatment of EOC patients [11]. The strength of the current two-lipid signature test is the capacity of finding the poor prognostic patients early: notably, the test functioned in the Charité 1 cohort with completely surgically debulked stage III patients. When identified early at diagnosis, the poor prognosis patients can be offered comprehensive genetic testing and targeted therapies within clinical trials. Altogether, a better prognostic evaluation of individual EOC patients could aid clinicians in directing these treatments more effectively and also improve patient counselling as the modern surgical and oncological treatments are utterly demanding.

In all patients with EOC, major effort is directed at reaching optimal cytoreduction (0 mm residual disease) as the presence of macroscopic residual disease has been associated with an especially poor prognosis [27]. Indeed, complex ultra-radical surgery including peritonectomy and multiorgan resections have become conventional as the prognostic benefit is generally considered to outweigh the increase in perioperative complications and morbidity [28]. Intriguingly, in the present study, optimally cytoreduced patients in the highest lipid test quartile had an inferior survival outcome compared to patients in the lower lipid test quartiles, regardless of the presence of macroscopic residual disease. These findings suggest that the tumor biology and/or composition of the non-macroscopic residual tumor may have an equally important role in prognosis as the surgery outcome. Of note, in addition to PC(O-38:4), decreased concentrations of a large number of other plasmalogen lipids showed prognostic value. We have previously shown that downregulation of the peroxisome-associated *ABCD1* (ATP Binding Cassette Subfamily D Member 1) gene as well as altered serum lipids and metabolites related to peroxisomal disorders are associated with poor survival in ovarian cancer patients [15,20]. Since the biosynthesis of plasmalogens occurs in peroxisomes [29], it is possible that peroxisomal dysfunction in ovarian cancer cells might explain the downregulation of plasmalogens. In our previous study it was revealed that the alterations in serum ceramides, i.e., their increase or decrease, is dependent on the fatty acyl side chain, and Cer(d18:1/18:0) showed the strongest elevation due to ovarian cancer [15]. Intriguingly, our present data showed that this same ceramide lipid is also the most prognostic one, implying that Cer(d18:1/18:0) is the most important ceramide both diagnostically and prognostically. Thus, it appears that instead of global ceramide upregulation, the alterations are lipid-specific. The possible peroxisomal dysfunction as well as the biological role of Cer(d18:1/18:0) in ovarian cancer warrant mechanistic studies. Further studies on the treatment response are required, as rendering a patient a non-responder for operative treatment a priori is unjustified without robust evidence.

The potential of the two-lipid signature test to detect disease recurrence was estimated in a proof-of-concept manner; however, the exploratory results of this study suggest that aberrations in the concentration of circulatory lipids are already present early-on in tumor development. Indeed, the two-lipid signature test might improve the follow up and early detection of disease recurrence, although it is unclear whether the treatment of early, asymptomatic recurrence is beneficial in the era of novel targeted therapies. It is tempting to hypothesize that the two-lipid signature test might similarly improve the detection of early stage EOC.

The strength of this study is the robust study configuration, as the results were tested in one cohort and validated in three additional, separate cohorts. The LION trial implemented a strict study protocol, which emphasizes the independent prognostic value of the two-lipid signature test. A limitation of the study is that the HR status was not available from our study sets, and the correlation of the lipid test and HRD remains to be assessed in future studies. In addition, the exploration of multiomic profiling might be another feasible method to increase the prognostic potential of the lipidomic analyses [30]. Another limitation is the low number of patients included in the longitudinal analyses; however, the longitudinal analyses were carried out in a proof-of-concept manner and the results need to be further validated in larger patient cohorts. In addition, the lipidomic method utilized is only semi-quantitative, and for clinical use a fully quantitative and analytically validated method has to be developed. However, for one component of the lipid signature, i.e., Cer (d18:1/18:0), a clinically validated method already exists [31], and the addition of another lipid component to the method is feasible. A quantitative method is also needed to confirm the calibration of the risk models between different study cohorts.

5. Conclusions

EOC continues to present a challenge for clinicians and scientists alike. Novel, robust, and affordable biomarkers are needed to improve the detection and monitoring of the disease, and also for the optimal allocation of precision drugs and complex surgical treatment. The analysis of circulatory lipids presents a non-invasive method, which differentiates patients with an especially aggressive disease from those with a more favorable disease outcome. Thus, the lipid signature may serve as a novel tool for treatment stratification, which will be an important topic for future studies.

Supplementary Materials: The following are available online at https://www.mdpi.com/article/10.3390/cancers13081764/s1, Figure S1: Selection of the prognostic lipid ratio, Figure S2: Risk (%) of progression in 1 year or death in 5 years, Figure S3: Longitudinal analysis of the two-lipid signature and CA-125 in 20 ovarian cancer patients, Table S1: Characteristics of the Turku longitudinal patient cohort, Table S2: Global serum lipidome, CA125 and HE4., Table S3: Prognostic performance of HE4 in the Turku cohort.

Author Contributions: Research concept and design: L.S., E.I.B., J.S., A.d.B., S.M., P.H., O.C., K.H., J.H., M.H. Data acquisition: L.S., E.I.B., A.J., S.O., S.H., H.K., K.H., J.H. Statistical analyses: M.H., M.L. Drafting of the manuscript: L.S., M.H. Critical revision of the manuscript for key intellectual content: E.I.B., M.L., A.J., S.O., S.H., J.S., H.K., A.d.B., S.M., P.H., O.C., K.H., J.H. All authors approved the version to be published and agree to be accountable for all aspects of the work in ensuring that questions related to the accuracy or integrity of any part of the work are appropriately investigated and resolved. All authors have read and agreed to the published version of the manuscript.

Funding: L.S. and J.H. were financially supported by the Clinical Research (VTR) Fund of the Turku University Hospital, K.H. by the Cancer Society of South-West Finland, and O.C. by Finnish Cancer Society, Sigrid Juselius Foundation, and Finska Läkaresälskapet.

Institutional Review Board Statement: The study was conducted according to the guidelines of the Declaration of Helsinki, and was approved by the Institutional Ethics Committee of the HOSPITAL DISTRICT OF SOUTHWEST FINLAND (protocol code 145/1801/2015, 20 October 2015) for the Turku cohort and of the ETHIK COMMISION OF THE CHARITÉ UNIVERSITÄTSMEDIZIN (protocol code 207/2003, January 2004) for the Charité cohorts.

Informed Consent Statement: Informed consent was obtained from all subjects involved in the study.

Data Availability Statement: The data presented in this study are available for academic research on request from the corresponding author. The data are not publicly available due to the European General Data Protection Regulation (GDPR).

Conflicts of Interest: M.H. is employed by Zora Biosciences Oy, which holds patent disclosures for diagnostic and prognostic tests of ovarian cancer using lipids. S.M. reports research support, advisory board, honoraria, and travel expenses from AbbVie, AstraZeneca, Clovis, Eisai, GlaxoSmithKline, Medac, MSD, Novartis, Olympus, PharmaMar, Pfizer, Roche, Sensor Kinesis, Teva, and Tesaro. A.D. reports honoraria for lectures and/or advisory boards from Roche, Clovis, GSK/Tesaro, Astra Zeneca, Biocad, Genmab, and Amgen.

References

1. Perrotti, F.; Rosa, C.; Cicalini, I.; Sacchetta, P.; Del Boccio, P.; Genovesi, D.; Pieragostino, D. Advances in Lipidomics for Cancer Biomarkers Discovery. *Int. J. Mol. Sci.* **2016**, *17*, 1992. [CrossRef]
2. Zhao, G.; Cardenas, H.; Matei, D. Ovarian Cancer—Why Lipids Matter. *Cancers* **2019**, *11*, 1870. [CrossRef] [PubMed]
3. Hanahan, D.; Weinberg, R.A. Hallmarks of cancer: The next generation. *Cell* **2011**, *144*, 646–674. [CrossRef]
4. Yan, F.; Zhao, H.; Zeng, Y. Lipidomics: A promising cancer biomarker. *Clin. Transl. Med.* **2018**, 21–23. [CrossRef] [PubMed]
5. Ke, C.; Hou, Y.; Zhang, H.; Fan, L.; Ge, T.; Guo, B.; Zhang, F.; Yang, K.; Wang, J.; Lou, G.; et al. Large-scale profiling of metabolic dysregulation in ovarian cancer. *Int. J. Cancer* **2014**. [CrossRef] [PubMed]
6. Torre, L.A.; Trabert, B.; DeSantis, C.E.; Miller, K.D.; Samimi, G.; Runowicz, C.D.; Gaudet, M.M.; Jemal, A.; Siegel, R.L. Ovarian cancer statistics, 2018. *CA Cancer J. Clin.* **2018**, *68*, 284–296. [CrossRef] [PubMed]
7. Karam, A.K.; Karlan, B.Y. Ovarian cancer: The duplicity of CA125 measurement. *Nat. Rev. Clin. Oncol.* **2010**, *7*, 335–339. [CrossRef]
8. Ledermann, J.A.; Raja, F.A.; Fotopoulou, C.; Gonzalez-Martin, A.; Colombo, N.; Sessa, C. Newly diagnosed and relapsed epithelial ovarian carcinoma: ESMO Clinical Practice Guidelines for diagnosis, treatment and follow-up †. *Ann. Oncol.* **2013**, *24*, 24–32. [CrossRef]
9. Norquist, B.M.; Brady, M.F.; Harrell, M.I.; Walsh, T.; Lee, M.K.; Gulsuner, S.; Bernards, S.S.; Casadei, S.; Burger, R.A.; Tewari, K.S.; et al. Mutations in homologous recombination genes and outcomes in ovarian carcinoma patients in GOG 218: An NRG oncology/Gynecologic oncology group study. *Clin. Cancer Res.* **2018**, *24*, 777–783. [CrossRef]
10. Monk, B.J.; Pujade-Lauraine, E.; Burger, R.A. Integrating bevacizumab into the management of epithelial ovarian cancer: The controversy of front-line versus recurrent disease. *Ann. Oncol.* **2013**, *24*, x53–x58. [CrossRef]
11. Colombo, N.; Sessa, C.; Du Bois, A.; Ledermann, J.; McCluggage, W.G.; McNeish, I.; Morice, P.; Pignata, S.; Ray-Coquard, I.; Vergote, I.; et al. ESMO-ESGO consensus conference recommendations on ovarian cancer: Pathology and molecular biology, early and advanced stages, borderline tumours and recurrent disease. *Int. J. Gynecol. Cancer* **2019**, *29*, 728–760. [CrossRef]
12. Lin, H.M.; Mahon, K.L.; Weir, J.M.; Mundra, P.A.; Spielman, C.; Briscoe, K.; Gurney, H.; Mallesara, G.; Marx, G.; Stockler, M.R.; et al. A distinct plasma lipid signature associated with poor prognosis in castration-resistant prostate cancer. *Int. J. Cancer* **2017**, *141*, 2112–2120. [CrossRef]
13. Cotte, A.K.; Cottet, V.; Aires, V.; Mouillot, T.; Rizk, M.; Vinault, S.; Binquet, C.; De Barros, J.P.P.; Hillon, P.; Delmas, D. Phospholipid profiles and hepatocellular carcinoma risk and prognosis in cirrhotic patients. *Oncotarget* **2019**, *10*, 2161–2172. [CrossRef]
14. Bachmayr-Heyda, A.; Aust, S.; Auer, K.; Meier, S.M.; Schmetterer, K.G.; Dekan, S.; Gerner, C.; Pils, D. Integrative Systemic and Local Metabolomics with Impact on Survival in High-Grade Serous Ovarian Cancer. *Clin. Cancer Res.* **2017**, *23*, 2081–2092. [CrossRef] [PubMed]
15. Braicu, E.I.; Darb-Esfahani, S.; Schmitt, W.D.; Koistinen, K.M.; Heiskanen, L.; Pöhö, P.; Budczies, J.; Kuhberg, M.; Dietel, M.; Frezza, C.; et al. High-grade serous carcinoma patients exhibit profound alterations in lipid metabolism. *Oncotarget* **2017**, *8*, 102912–102922. [CrossRef] [PubMed]
16. Shida, D.; Takabe, K.; Kapitonov, D.; Milstien, S.; Spiegel, S. Targeting SphK1 as a New Strategy against Cancer. *Curr. Drug Targets* **2008**, *9*, 662–673. [CrossRef] [PubMed]
17. Pyragius, C.E.; Fuller, M.; Ricciardelli, C.; Oehler, M.K. Aberrant lipid metabolism: An emerging diagnostic and therapeutic target in ovarian cancer. *Int. J. Mol. Sci.* **2013**, *14*, 7742–7756. [CrossRef] [PubMed]
18. Salminen, L.; Nadeem, N.; Jain, S.; Grènman, S.; Carpén, O.; Hietanen, S.; Oksa, S.; Lamminmäki, U.; Pettersson, K.; Gidwani, K.; et al. A longitudinal analysis of CA125 glycoforms in the monitoring and follow up of high grade serous ovarian cancer. *Gynecol. Oncol.* **2020**, *156*, 689–694. [CrossRef]
19. Harter, P.; Sehouli, J.; Lorusso, D.; Reuss, A.; Vergote, I.; Marth, C.; Kim, J.-W.; Raspagliesi, F.; Lampe, B.; Aletti, G.; et al. A Randomized Trial of Lymphadenectomy in Patients with Advanced Ovarian Neoplasms. *N. Engl. J. Med.* **2019**, *380*, 822–832. [CrossRef]
20. Hilvo, M.; De Santiago, I.; Gopalacharyulu, P.; Schmitt, W.D.; Budczies, J.; Kuhberg, M.; Dietel, M.; Aittokallio, T.; Markowetz, F.; Denkert, C.; et al. Accumulated metabolites of hydroxybutyric acid serve as diagnostic and prognostic biomarkers of ovarian high-grade serous carcinomas. *Cancer Res.* **2016**, *76*. [CrossRef] [PubMed]
21. John, G.; Rustin, S.; Vergote, I.; Eisenhauer, E.; Pujade-lauraine, E.; Quinn, M.; Thigpen, T.; Bois, A.; Kristensen, G. Definitions for Response and Progression in Ovarian Cancer Clinical Trials Incorporating RECIST 1.1 and CA 125 Agreed by the Gynecological Cancer Intergroup (GCIG). *Int. J. Gynecol. Cancer* **2011**, *21*, 419–423. [CrossRef]
22. Folch, J.; Lees, M.; Sloane Stanley, G.H. A simple method for the isolation and purification of total lipides from animal tissues. *J. Biol. Chem.* **1957**, *226*, 497–509. [CrossRef]
23. Weir, J.M.; Wong, G.; Barlow, C.K.; Greeve, M.A.; Kowalczyk, A.; Almasy, L.; Comuzzie, A.G.; Mahaney, M.C.; Jowett, J.B.M.; Shaw, J.; et al. Plasma lipid profiling in a large population-based cohort. *J. Lipid Res.* **2013**, *54*, 2898–2908. [CrossRef] [PubMed]
24. Laaksonen, R.; Ekroos, K.; Sysi-Aho, M.; Hilvo, M.; Vihervaara, T.; Kauhanen, D.; Suoniemi, M.; Hurme, R.; März, W.; Scharnagl, H.; et al. Plasma ceramides predict cardiovascular death in patients with stable coronary artery disease and acute coronary syndromes beyond LDL-cholesterol. *Eur. Heart J.* **2016**, *37*, 1967–1976. [CrossRef]

25. Hilvo, M.; Meikle, P.J.; Pedersen, E.R.; Tell, G.S.; Dhar, I.; Brenner, H.; Schöttker, B.; Lääperi, M.; Kauhanen, D.; Koistinen, K.M.; et al. Development and validation of a ceramide- and phospholipid-based cardiovascular risk estimation score for coronary artery disease patients. *Eur. Heart J.* **2020**, *41*, 371–380. [CrossRef] [PubMed]
26. Takaya, H.; Nakai, H.; Takamatsu, S.; Mandai, M.; Matsumura, N. Homologous recombination deficiency status-based classification of high-grade serous ovarian carcinoma. *Sci. Rep.* **2020**, *10*, 1–8. [CrossRef]
27. Du Bois, A.; Reuss, A.; Pujade-Lauraine, E.; Harter, P.; Ray-Coquard, I.; Pfisterer, J. Role of surgical outcome as prognostic factor in advanced epithelial ovarian cancer: A combined exploratory analysis of 3 prospectively randomized phase 3 multicenter trials: By the arbeitsgemeinschaft gynaekologische onkologie studiengruppe ovarialkarzin. *Cancer* **2009**, *115*, 1234–1244. [CrossRef] [PubMed]
28. Aletti, G.D.; Dowdy, S.C.; Gostout, B.S.; Jones, M.B.; Stanhope, C.R.; Wilson, T.O.; Podratz, K.C.; Cliby, W.A. Aggressive Surgical Effort and Improved Survival in Advanced-Stage Ovarian Cancer. *Obstet. Gynecol.* **2006**, *107*, 77–85. [CrossRef] [PubMed]
29. Messias, M.C.F.; Mecatti, G.C.; Priolli, D.G.; De Oliveira Carvalho, P. Plasmalogen lipids: Functional mechanism and their involvement in gastrointestinal cancer. *Lipids Health Dis.* **2018**, *17*, 1–12. [CrossRef]
30. Clifford, C.; Vitkin, N.; Nersesian, S.; Reid-Schachter, G.; Francis, J.A.; Koti, M. Multi-omics in high-grade serous ovarian cancer: Biomarkers from genome to the immunome. *Am. J. Reprod. Immunol.* **2018**, *80*, 1–10. [CrossRef]
31. Kauhanen, D.; Sysi-Aho, M.; Koistinen, K.M.; Laaksonen, R.; Sinisalo, J.; Ekroos, K. Development and validation of a high-throughput LC-MS/MS assay for routine measurement of molecular ceramides. *Anal. Bioanal. Chem.* **2016**, *408*, 3475–3483. [CrossRef] [PubMed]

Article

Immune Monitoring in Melanoma and Urothelial Cancer Patients Treated with Anti-PD-1 Immunotherapy and SBRT Discloses Tumor Specific Immune Signatures

Annabel Meireson [1,2], Simon J. Tavernier [3,4], Sofie Van Gassen [5,6], Nora Sundahl [1,7], Annelies Demeyer [1,2], Mathieu Spaas [1,7], Vibeke Kruse [1,8], Liesbeth Ferdinande [9], Jo Van Dorpe [1,9], Benjamin Hennart [10,11], Delphine Allorge [10,11], Filomeen Haerynck [3], Karel Decaestecker [1,12], Sylvie Rottey [1,8], Yvan Saeys [1,5,6], Piet Ost [1,7] and Lieve Brochez [1,2,*]

1. Cancer Research Institute Ghent (CRIG), Ghent University, 9000 Ghent, Belgium; annabel.meireson@ugent.be (A.M.); nora.sundahl@ugent.be (N.S.); annelies.demeyer@ugent.be (A.D.); mathieu.spaas@ugent.be (M.S.); vibeke.kruse@ugent.be (V.K.); jo.vandorpe@ugent.be (J.V.D.); karel.decaestecker@ugent.be (K.D.); sylvie.rottey@ugent.be (S.R.); yvan.saeys@ugent.be (Y.S.); piet.ost@ugent.be (P.O.)
2. Dermatology Research Unit, Ghent University Hospital, 9000 Ghent, Belgium
3. Centre for Primary Immunodeficiency Ghent, Primary Immune Deficiency Research Lab, Department of Internal Medicine and Pediatrics, Jeffrey Modell Diagnosis and Research Centre, Ghent University Hospital, 9000 Ghent, Belgium; simon.tavernier@irc.vib-ugent.be (S.J.T.); filomeen.haerynck@ugent.be (F.H.)
4. VIB Center for Inflammation Research, Unit of Molecular Signal Transduction in Inflammation, 9000 Ghent, Belgium
5. VIB Center for Inflammation Research, Unit of Data Mining and Modeling for Biomedicine, 9000 Ghent, Belgium; sofie.vangassen@ugent.be
6. Department of Applied Mathematics, Computer Science and Statistics, Ghent University, 9000 Ghent, Belgium
7. Department of Radiation Oncology and Experimental Cancer Research, Ghent University Hospital, 9000 Ghent, Belgium
8. Department of Medical Oncology, Ghent University Hospital, 9000 Ghent, Belgium
9. Department of Pathology, Ghent University Hospital, 9000 Ghent, Belgium; liesbeth.ferdinande@ugent.be
10. Unité Fonctionnelle de Toxicologie, CHU Lille, F-59000 Lille, France; benjamin.hennart@CHRU-LILLE.FR (B.H.); delphine.allorge@CHRU-LILLE.FR (D.A.)
11. ULR 4483-IMPact de l'Environnement Chimique sur la Santé Humaine (IMPECS), Université de Lille, F-59000 Lille, France
12. Department of Urology, Ghent University Hospital, 9000 Ghent, Belgium
* Correspondence: lieve.brochez@ugent.be

Simple Summary: Currently available biomarkers for response to checkpoint inhibitors are incomplete and predominantly focus on tumor tissue analysis e.g., tumor mutational burden, programmed cell death-ligand 1 (PD-L1) expression. Biomarkers in peripheral blood would allow a more dynamic monitoring and could offer a way for sequential adaptation of treatment strategy. We conducted an in-depth analysis of baseline and on-treatment systemic immune features in a cohort of stage III/IV melanoma and stage IV urothelial cancer (UC) patients treated with anti-programmed cell death-1 (anti-PD-1) therapy combined with stereotactic body radiotherapy (SBRT) in a similar regimen/schedule. Baseline immunity was clearly different between these two cohorts, indicating a less active immune landscape in UC patients. This study also detected signatures of proliferation in the CD8+ T-cell compartment pre-treatment and early after anti-PD-1 initiation that were positively correlated with clinical outcome in both tumor types. In addition our data support the biological relevance of PD-1/PD-L1 expression on circulating immune cell subsets, especially in melanoma.

Abstract: (1) Background: Blockade of the PD-1/PD-L1 pathway has revolutionized the oncology field in the last decade. However, the proportion of patients experiencing a durable response is still limited. In the current study, we performed an extensive immune monitoring in patients with stage III/IV melanoma and stage IV UC who received anti-PD-1 immunotherapy with SBRT. (2) Methods: In total 145 blood samples from 38 patients, collected at fixed time points before and during treatment,

were phenotyped via high-parameter flow cytometry, luminex assay and UPLC-MS/MS. (3) Results: Baseline systemic immunity in melanoma and UC patients was different with a more prominent myeloid compartment and a higher neutrophil to lymphocyte ratio in UC. Proliferation (Ki67$^+$) of CD8$^+$ T-cells and of the PD-1$^+$/PD-L1$^+$ CD8$^+$ subset at baseline correlated with progression free survival in melanoma. In contrast a higher frequency of PD-1/PD-L1 expressing non-proliferating (Ki67$^-$) CD8$^+$ and CD4$^+$ T-cells before treatment was associated with worse outcome in melanoma. In UC, the expansion of Ki67$^+$ CD8$^+$ T-cells and of the PD-L1$^+$ subset relative to tumor burden correlated with clinical outcome. (4) Conclusion: This study reveals a clearly different immune landscape in melanoma and UC at baseline, which may impact immunotherapy response. Signatures of proliferation in the CD8$^+$ T-cell compartment prior to and early after anti-PD-1 initiation were positively correlated with clinical outcome in both cohorts. PD-1/PD-L1 expression on circulating immune cell subsets seems of clinical relevance in the melanoma cohort.

Keywords: immunotherapy; anti-PD-1; melanoma; urothelial cancer; immune monitoring; blood biomarkers

1. Introduction

New insights in immuno-oncology and the subsequently developed immunotherapies have caused a major breakthrough in the oncology field in the last decade, creating the hope of curing (metastatic) cancer. Despite the encouraging results, the proportion of patients experiencing a durable response is still limited. In 2018 about 43% of cancer patients in the United States were eligible for checkpoint inhibitor therapy compared to 1.5% in 2011, while the estimated percentage of response only modestly increased from 0.14% to 12.4% in the same time period [1]. Combination strategies are currently being tested in different cancer types in an attempt to improve response rates [2,3], but the combination of cytotoxic lymphocyte antigen 4 (CTLA-4) blockade and programmed cell death receptor 1 (PD-1) blockade is well recognized to inevitably elicit higher toxicity and also implies a higher cost. Both from the patient's and healthcare budget's perspective there is a need for new translational insights that could help to optimize current immunotherapies.

Up to date predictive biomarkers have mainly been identified in tumor tissue. The immunohistochemical expression of PD-L1 is currently one of the most widely used biomarkers and, high expression has been correlated with response to PD-1/PD-L1 immunotherapy [4,5]. However, a systematic evaluation of 45 Food and Drug Administration (FDA) approved trials involving 15 tumor types demonstrated that PD-L1 expression was predictive in only 28.9% of cases [6]. High tumor mutational burden is also associated with better response [7,8] and this finding led to FDA approval for checkpoint inhibition in patients with microsatellite instability-high or mismatch repair-deficient solid tumors, irrespective of cancer type [9,10]. Patients who respond to anti-PD-1 therapy exhibit a tumor micro-environment that is enriched for interferon γ (IFNγ) and tumor infiltrating lymphocytes (TILs), the so called 'hot' tumors [11–13].

Blood-based biomarkers have been far less reported and have not yet entered clinical practice, although they could have the benefit of a dynamic monitoring during the treatment course with the possibility to adapt immunotherapeutic strategies.

In the current study, we performed immune monitoring in patients with inoperable stage III/IV melanoma and patients with stage IV UC who received anti-PD-1 immunotherapy combined with SBRT. The immune landscape before and during treatment was compared between tumor types and the relation to clinical outcome was investigated.

2. Materials and Methods

2.1. Patient Samples

The biospecimens evaluated in this study were obtained from patients with melanoma or UC who participated in two separate clinical trials (Figure 1a,b). A phase 2 trial included

20 patients with unresectable stage III or stage IV metastatic melanoma who were treated in the first line with anti-PD-1 (nivolumab) and SBRT (NCT02821182) [14]. The samples from metastatic UC patients were collected during a randomized phase I trial with SBRT administered either prior to the first anti-PD-1 cycle (arm A: SBRT prior to any treatment with pembrolizumab, $n = 9$), or during anti-PD-1 treatment (arm B: SBRT prior to the third pembrolizumab cycle, $n = 9$) [15]. Both trials were approved by the Ethics Committee of Ghent University Hospital and are registered on Clinicaltrials.gov (resp. NCT02821182 and NCT02826564). At fixed time points through treatment, peripheral blood samples (EDTA and serum tubes) were collected from melanoma ($n = 85$) and UC patients ($n = 60$) respectively. Peripheral blood mononuclear cells (PBMCs) were isolated via Lymphoprep centrifugation and stored in liquid nitrogen using standard methods.

Tumor burden was assessed using CT/MRI or PET-CT scan of the chest, abdomen and pelvis at baseline, after the fourth cycle of anti-PD-1 and after every fifth cycle (melanoma) or third cycle (UC) thereafter until the end of treatment. Tumor burden was defined as the sum of the longest diameters for a maximum of five target lesions and up to two lesions per organ. For lymph nodes the shortest axis was measured. Clinical responses were determined based on Response Evaluation Criteria in Solid Tumors (RECIST) 1.1 criteria. Disease control was achieved in 12 melanoma patients (complete response, CR ($n = 3$); partial response; PR ($n = 6$) and stable disease, SD ($n = 3$)) while 8 patients showed progressive disease (non-responder). In UC, no objective responders were observed in arm A, while of 4 patients in arm B achieved a complete or partial response (CR: $n = 1$; PR: $n = 3$).

2.2. Flow Cytometry

Cryopreserved PBMCs were thawed and washed in RPMI 1640 medium supplemented with Glutamax (2.05 mM), 10% FCS and penicillin (100 U/mL)-streptomycin (100 µg/mL). Cells were stained with monoclonal antibodies labeled with fluorochromes. A complete list of the used antibodies can be found in Table S3. In a first step, 2.5×10^6 cells were stained with FcR blocking reagent for blocking of unspecific binding of antibodies (130-059-901, Miltenyi, Madrid, Spain) and a mixture of Fixable Viability dye eFluor 506 (65-0866-14, eBioscience, San Diego, CA, USA) and antibodies against surface markers in PBS and BD Horizon Brilliant Stain Buffer (563794, BD Biosciences, San Jose, CA, USA), incubated for 30 min at 4 °C and washed. In a second step, cells were fixed and permeabilized with Foxp3 Transcription Factor Staining Buffer Set (00-5523-00, eBioscience, San Diego, CA, USA), and subsequently stained intracellularly for 30 min at RT. Labeled cell suspensions were acquired on a BD FACSymphony flowcytometer (BD Biosciences, San Jose, CA, USA) and data was analyzed with FlowJo 10.6.2 software (Ashland, OR, USA). Gating strategies are depicted in Figure S1.

The frequency of neutrophils and lymphocytes in white blood cells was determined for all of the samples using automated blood cell counting equipment (Sysmex XE-5000, Norderstedt, Germany) during routine lab evaluations.

2.3. High Dimensional Data Analysis of Flow Cytometry Data

2.3.1. t-SNE

Live CD8$^+$ T cells were gated in FlowJo v10.6.2 and exported as separate fcs files for melanoma and UC. Populations before and during treatment were randomly downsampled and subsequently concatenated into 1 file (total events melanoma: 1.234.633 events; total events urothelial cancer arm A: 689.057 events; total events urothelial cancer arm B: 979.821 events). Next, concatenated samples were analyzed via t-distributed stochastic neighborhood embedding (t-SNE) in FlowJo v10.6.2. Opt-SNE was applied as learning configuration, with perplexity set to 30 and iterations to 1000. The colors in the heatmap represent the measured means intensity value of Ki67 in a given cluster.

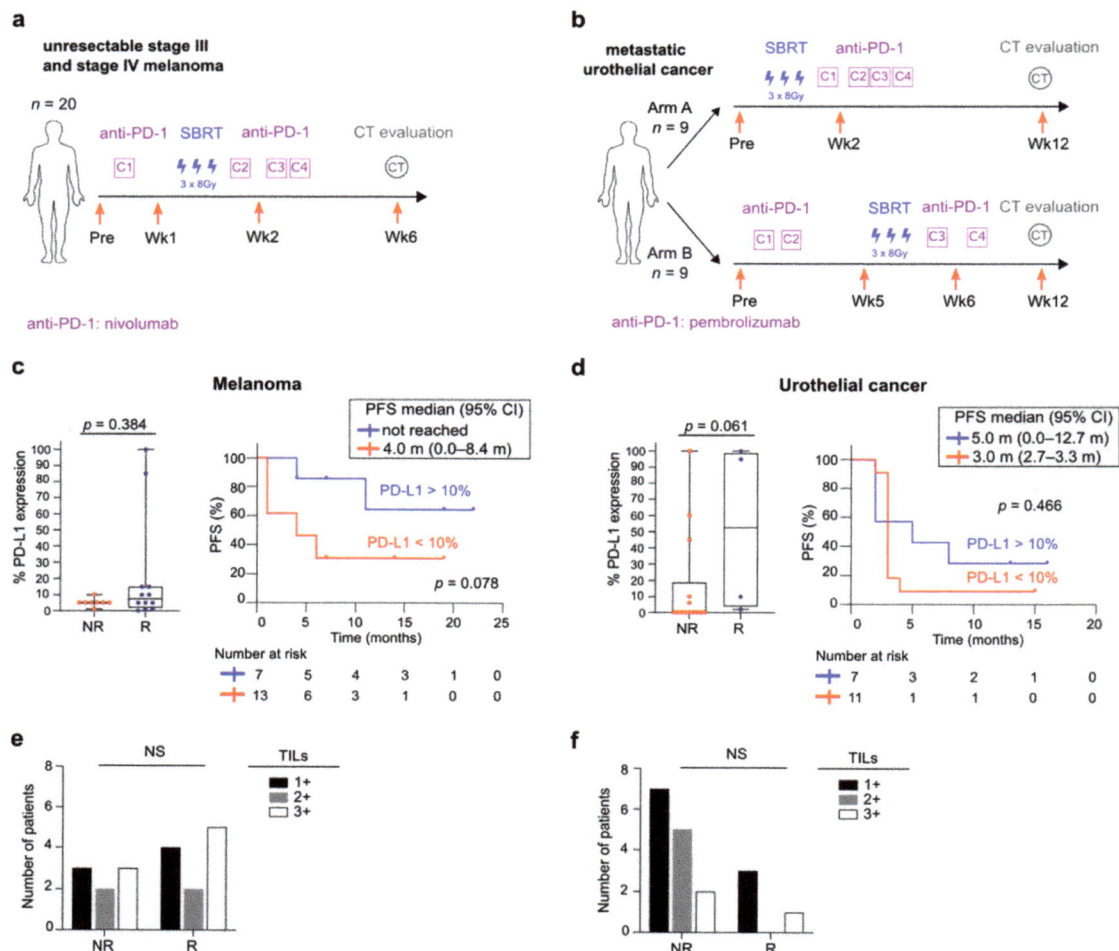

Figure 1. Overview of the clinical trial treatment strategy and PD-L1 and TIL quantification. (**a**) Schematic of design of phase 2 clinical trial in unresectable stage III and stage IV melanoma receiving a combination of anti-PD-1 and SBRT. (**b**) Schematic of design of randomized phase 1 trial combining anti-PD-1 with either sequential (Arm A) or concomitant SBRT (Arm B) in metastatic UC. Red arrows indicate time of blood collection. (**c**) Boxplots with tumoral PD-L1 expression in non-responders and responders (left) and Kaplan–Meier estimate of PFS stratified according to tumoral PD-L1 expression (right) in melanoma and (**d**) in UC. Whiskers of boxplots extend to the minimum and maximum data point, with the horizontal line indicating the median. p value calculated using two-sided Mann–Whitney U test (left) and log-rank test (right). (**e**) TIL quantification in non-responders and responders in melanoma and (**f**) in UC. TILs were evaluated semi quantitatively: 1+, sporadic TILs; 2+, moderate number of TILs; 3+, abundant occurrence of TILs. p value calculated using Fisher's exact test. Pre, pre-treatment; Wk, week; Gy, gray; SBRT, stereotactic body radiotherapy; CT, computed tomography; NR, non-responder; R, responder; PFS, progression free survival; NS, not significant; TIL, tumor infiltrating lymphocytes.

2.3.2. FlowSOM

The melanoma and UC datasets were analyzed separately, following the same pipeline. The fcs files were first cleaned by manual gating in FlowJo, after which the data was imported in R. An aggregate was generated with approximately 3 million cells, with an equal number of cells subsampled at random without repetition from each sample (melanoma: 85 samples with 35,295 cells each, urothelial cancer arm B: 36 samples with 83,334 cells each). This aggregate was then used to train a FlowSOM model with a 15 by 15 grid (225 clusters) and 30 metaclusters. Thirteen markers were taken along for the clustering: CD3, CD4, CD8, CD25, CD19, CD56, HLA-DR, CD123, CD33, CD11b, CD14, CD16 and FoxP3.

Once the model was built, all samples were fully mapped onto the model, resulting in a cluster and metacluster assignment for each cell. From this mapping, the cluster and metacluster abundances per sample were extracted. Additionally, for 6 markers (CTLA-4, Ki67, IDO, PD-1, PD-L1 and HLA-DR), a positivity threshold was determined by manual gating. We used these thresholds to determine the abundance of each possible subpopulation in each (meta-) cluster. A subpopulation was defined by being either positive, negative or neutral (both positive and negative cells included) for each of the markers, resulting in 729 potential combinations per (meta-) cluster. As many of these combinations would not occur in reality, these subpopulations were then filtered, only keeping those where at least 5 samples had at least 30 cells. This resulted in a total of 76,039 features describing the immune profile of melanoma samples and 70,648 features for the urothelial cancer samples.

2.4. Cytokine Measurement

Magnetic luminex assay (R&D systems, Minneapolis, MN, USA) was performed on cryopreserved serum samples according to manufacturer's instructions using a customized panel, including CXCL9, CXCL10, MICA, MICB, ULBP-1, ULBP-2/5/6, ULBP-3, ULBP-4 and s100B. Serum concentrations were measured on a Bio-Plex 200 Array Reader (Bio-Rad, Hercules, CA, USA).

2.5. UPLC-MS/MS

Tryptophan (Trp) and its metabolite kynurenine (Kyn) were quantified according to previously published methods [16,17], with slight modifications. Cryopreserved serum samples (50 µL) were extracted using 50 µL acetonitrile containing Trp-D5 (50 µM, CDN Isotopes, Pointe-Claire, QC, Canada) as an internal standard. The samples were centrifuged (8 min, 11,800 rpm, 4 °C) and the supernatants (50 µL) were added to deionized water (600 µL). Fifteen µL of this mixture was injected in an ultra-high-performance liquid chromatography system coupled to tandem mass spectrometry detector (UPLC-MS/MS, Acquity TQ-S Detector, Waters, Milford, MA, USA) equipped with a HSS C18 column. Ions of each analyzed compound were detected in a positive ion mode using multiple reaction monitoring.

2.6. Scoring of PD-L1 and Tumor Infiltrating Cells

Formalin-fixed, paraffin-embedded (FFPE) tumor samples were collected at time of surgical resection before start of systemic treatment in melanoma and UC patients. 4 µm-thick FFPE tissue sections were subjected to heat-induced antigen retrieval and incubated with primary monoclonal antibodies against PD-L1: clone SP263 (Ventana Medical Systems Inc., Tucson, AZ, USA) for melanoma samples and clone 22C3 (Agilent Technologies, Santa Clara, CA, USA) for UC samples. Samples were visualized with 3,3′-diaminobenzidine (DAB) chromogen and hematoxylin counterstain and cover-slipped for review. Scoring of PD-L1 was conducted by 2 pathologists blinded to patient characteristics. In melanoma sections, the percentage of tumor cells with membranous PD-L1 staining was scored (0–100%). In UC sections, the percentage of tumor cells and any tumor infiltrating mononuclear inflammatory cells with membranous PD-L1 staining was scored (0–100%).

The abundance of intraepithelial TILs was determined on H&E stained sections. This morphological assessment of TILs within tumor nests was evaluated semi quantitatively: 1+, sporadic TILs; 2+, moderate number of TILs; 3+, abundant occurrence of TILs. For dichotomization, the TILs score was categorized into 'low' (1+ or 2+) and 'high' (3+). TILs were assessed on 19 melanoma patients as the only available specimens for the 20th patient was a cytological sample.

2.7. Statistics

To compare longitudinal immunologic effects, p-values for each measured immune feature were calculated using a Wilcoxon matched-pairs signed-ranks test. Associations between immune features and treatment response were identified by Mann-Whitney U tests comparing the frequencies of phenotypes between responders and non-responders. Progression free survival (PFS) was defined as the time from inclusion to disease progression, death or the last follow-up, whichever occurred first. PFS curves were estimated using the Kaplan-Meier method by dichotomizing immune phenotypes of interest through their median value. Survival curves between patients with high (above the median) and low (below the median) frequencies of the immune feature of interest were compared using a Log-Rank test. Cox regression models have been used to perform univariate analysis. Correlations between continuous variables were determined by Spearman's r coefficient. A chi square test was employed to test for association between two categorical variables. Fold change in proliferation was calculated by dividing the frequency of Ki67$^+$ T-cells in on-treatment samples to the frequency of Ki67$^+$ T-cells at pre-treatment. Statistical analyses were performed using IBM SPSS v26 and all tests were performed two-sided; $p < 0.05$ was considered to be statistically significant. Graphs were plotted with Graphpad Prism (GraphPad software Inc., San Diego, CA, USA). For FlowSOM analysis, a Wilcoxon Rank Sum test was executed in R to compare responders and non-responders, after which the features were ranked by p-value.

3. Results

3.1. Overview of Patient Cohorts

Blood samples (n = 145) of 20 melanoma patients and 18 UC patients treated with anti-PD-1 therapy combined with SBRT were included (NCT02821182 and NCT02826564). The design of the clinical trials and time points of blood sample collection is schematically presented in Figure 1a,b. The clinical results of these trials have been reported elsewhere [14,15]. Detailed patient characteristics are described in Tables S1 and S2.

The median age in the melanoma cohort was 60.5 years (34.0–80.0 years) and 68.0 years (50.0–84.0 years) in the UC cohort (Mann-Whitney U test, $p = 0.055$). Age was not correlated with clinical outcome in the melanoma cohort (Mann-Whitney U test, $p = 0.473$). In the UC cohort, median age was higher in responders (75.5 years, (71.0–84.0 years)) compared to non-responders (61.0 years (50.0–79.0 years), Mann-Whitney U test, $p = 0.018$). The median tumor burden was lower in melanoma patients compared to UC patients (23.5 mm (10.0–100.0 mm) versus 45.8 mm (12.10–106.90 mm), Mann-Whitney U test, $p = 0.033$). Median baseline tumor burden was not different in responders versus non-responders in the melanoma cohort. Responders in the UC cohort tended to have lower median baseline tumor burden compared to non-responders (arm B: 29.45 mm (12.10–46.90 mm) versus 60.50 mm (44.70–75.00 mm), Mann-Whitney U test, $p = 0.032$, for arm A+B $p = 0.101$). Prior systemic treatment had been administered in 2/20 melanoma patients (anti-CTLA-4 and BRAF-targeted therapy). In the UC cohort, 13/18 patients had been treated with one or more platinum-based chemotherapies prior to enrollment in the study.

Tumoral PD-L1 expression was not significantly related to response or PFS in melanoma or UC patients (Figure 1c,d). No difference in baseline TILs was found between responders and non-responders (Figure 1e,f).

3.2. Differences in Baseline Immunity between Melanoma and UC Cohort

Significant differences in the baseline immune landscape were observed between the melanoma and UC cohort (Figure 2a,b).

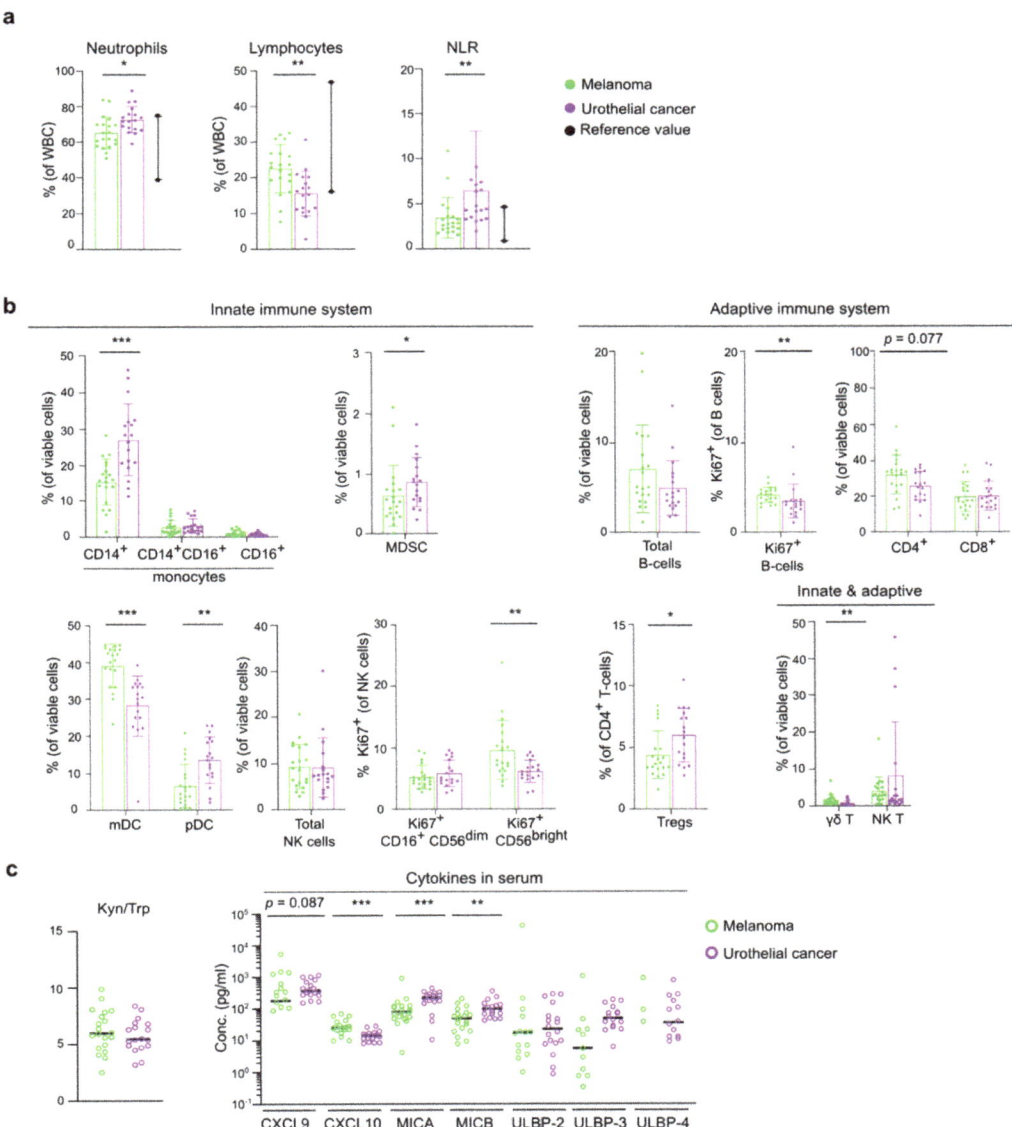

Figure 2. Baseline systemic immunity differs between melanoma and urothelial cancer. (**a**) Frequency of neutrophils, lymphocytes and neutrophil-to-lymphocyte ratio in melanoma and UC. Reference values are depicted in black. (**b**) Frequency of immune cell populations of innate and adaptive immune system. Error bar denotes ± SD. (**c**) (left) Ratio of serum concentrations of kynurenine (Kyn) on tryptophan (Trp), presented values are Kyn/Trp x 100. (right) Serum concentrations of T-cell activating chemokines CXCL9 and CXCL10 and concentrations of ligands for NK cell activing receptor NKG2D: MICA, MICB, ULBP-2, ULBP-3 and ULBP-4. Concentrations out of the range of detection could not be depicted. p value calculated using two-sided Mann-Whitney U test. * $p < 0.05$, ** $p < 0.01$, *** $p < 0.001$. WBC, white blood cells; MDSC, myeloid-derived suppressor cells; mDC, myeloid dendritic cells; pDC, plasmacytoid dendritic cells; Tregs, regulatory T-cells; ND, not detectable.

The neutrophil-to-lymphocyte ratio (NLR) was higher in UC compared to melanoma. Melanoma patients had a clearly higher lymphocyte frequency and more γδ T-cells and proliferating (Ki67 expressing) B-cells compared to UC patients. There was no significant difference in the frequency of CD4$^+$ and CD8$^+$ T-cells between the two cohorts, except for the frequency of regulatory T-cells (Tregs, as defined by CD25$^+$ Foxp3$^+$ CD4$^+$ cells [18]) which was lower in melanoma patients.

In the UC cohort, the frequency of neutrophils, classical CD14$^+$ monocytes, plasmacytoid dendritic cells (pDCs) and myeloid-derived suppressor cells (MDSCs) was higher compared to the melanoma cohort. Notably, the frequency of CD16$^+$ monocytes correlated negatively with tumor burden in UC patients (Spearman's CC: -0.627, $p = 0.005$). No significant differences were observed in the total NK cell population but the percentage of Ki67$^+$ CD56^{bright} NK was lower in the UC cohort.

The baseline concentration of IFNγ-inducible chemokine CXCL10 was higher in melanoma, whereas higher serum concentrations of MICA and MICB-both soluble NKG2D ligands-were detected in UC (Figure 2c). No differences in baseline kynurenine to tryptophan ratio (Kyn/Trp) were observed.

Altogether these data indicate a more favorable baseline immune landscape in the melanoma cohort compared to UC patients.

3.3. Early Systemic Immune Changes after Anti-PD-1 Treatment Initiation

To study early dynamic changes in systemic immunity upon anti-PD-1 initiation, blood samples after 1 cycle in the melanoma cohort (collected at week 1) and after 2 cycles in the UC cohort arm B (collected at week 5) were examined.

While a significant increase in the Ki67 expressing CD8$^+$ T-cell population was observed, the increases in Ki67$^+$ CD8$^+$ T-cell subsets co-expressing the checkpoint molecules PD-L1 or CTLA-4 were even more pronounced (Figure 3a,b). In both tumor types, Ki67$^+$ CD8$^+$ T-cells seemed to peak after 1 or 2 cycles of anti-PD-1 therapy (Figure 3c,d). Interestingly, in the melanoma cohort the percentage of Ki67$^+$ CD8$^+$ T-cells, Ki67$^+$ PD-L1$^+$ CD8$^+$ T-cells and especially Ki67$^+$ PD-1$^+$ CD8$^+$ T-cells at baseline and for the former two also at week 1 were positively correlated with PFS (Figure 3e). In UC, PFS correlated with the increase of Ki67$^+$ CD8$^+$ T-cells to tumor burden and with the increase of Ki67$^+$ PD-L1$^+$ CD8$^+$ T-cells to tumor burden (Figure 3f). In melanoma the increase of Ki67$^+$(PD-L1$^+$) CD8$^+$ T-cells to tumor burden did not correlate with PFS.

In a subset of 7 melanoma patients with an additional blood sample collected during anti-PD-1 treatment at a median time interval of 6 months (range: 3–16 months) after start of treatment. Ki67$^+$ CD8$^+$ T-cells co-expressing PD-L1 or CTLA-4 had returned to baseline levels.

Global high-dimensional mapping of flow cytometry data via the t-SNE algorithm provided more insights into this proliferating subset of CD8$^+$ T-lymphocytes. t-SNE analysis revealed a highly Ki67-positive CD8$^+$ T-cell cluster, already present before treatment in melanoma and UC (Figure 3g,h). Compared to the total CD8$^+$ T-cell population this Ki67$^+$ CD8$^+$ T-cell cluster demonstrated enriched expression of the T-cell activation marker HLA-DR and the immune checkpoint molecule IDO1 (Figure 3i,j). A variable expression for CTLA-4, PD-1 and its ligand PD-L1 was present in this cluster, with cells either expressing or lacking these markers. To assess the dynamics of this cluster during therapy, we manually gated on this cluster in the individual t-SNE map of each patient on each time point. Independent of response to immunotherapy, the frequency of Ki67$^+$ CD8$^+$ T-cells increased at week 1 and this was maintained at week 6 in melanoma patients (Figure 3k). In UC patients, the increase in Ki67 expressing CD8$^+$ T-cells tended to be higher in responders (Figure 3l).

In addition, a significant increase in serum CXCL10 and Kyn/Trp was observed after 1 cycle of anti-PD-1 in the melanoma cohort (Figure 3m). The magnitude of these increases was not significantly different between responders and non-responders. The increases in CXCL10 and Kyn/Trp were not significant in the UC cohort (Figure 3n).

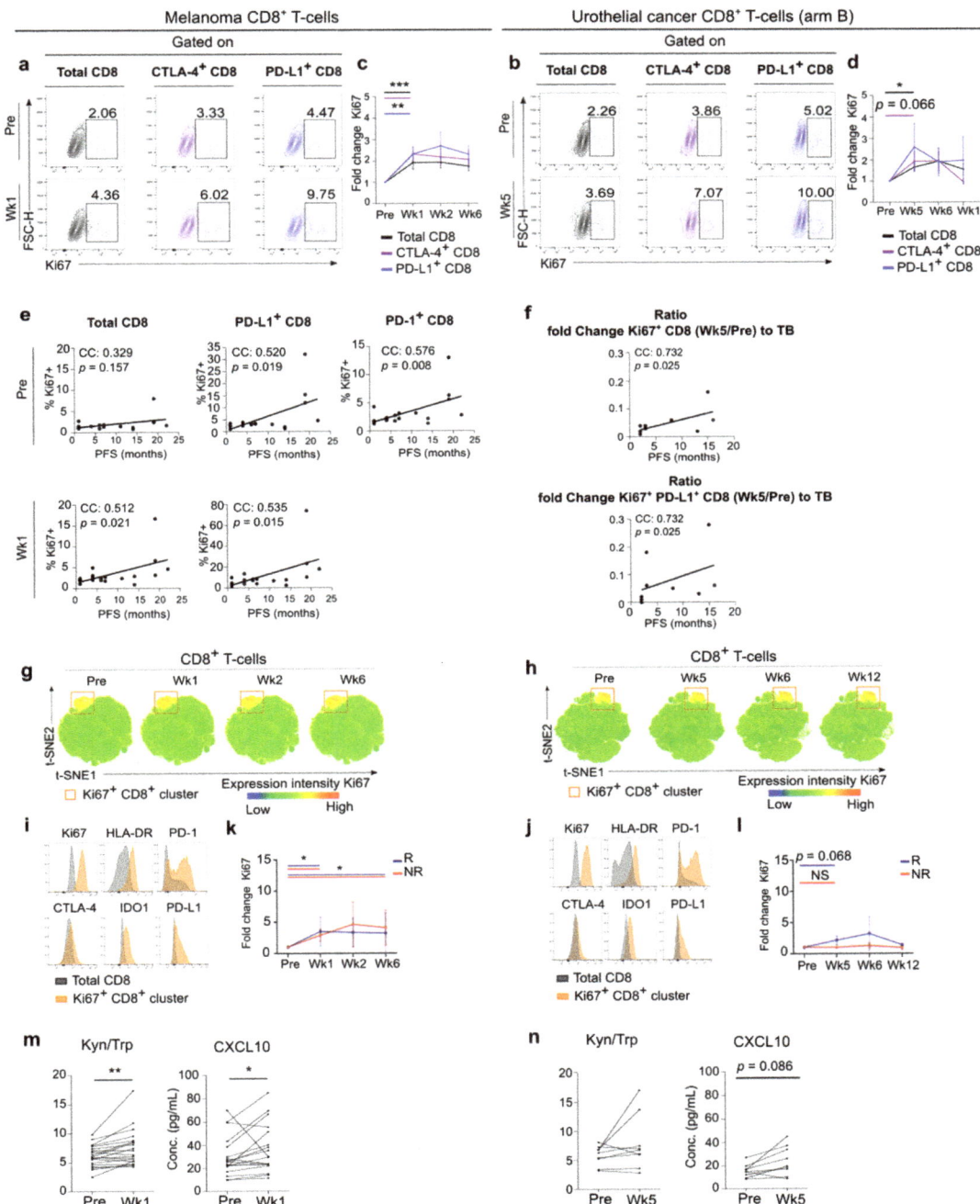

Figure 3. Early upregulation of proliferating CD8+ T-cells in response to anti-PD-1. (**a**) Contour plots representing Ki67 expression in CD8+ T-cell subsets at pre-treatment (Pre) and after 1 cycle of anti-PD-1 (Week 1) in 12 independent melanoma patients. (**b**) Contour plots representing Ki67 expression in CD8+ T-cell subsets at pre-treatment and after 2 cycles of anti-

PD-1 (Week 5) in 9 independent UC patients (arm B). (c) Lineplot with fold induction of Ki67 expression in CD8$^+$ T-cell subsets at indicated times in melanoma (n = 20) and (d) in UC (arm B, n = 9). Data shown are relative to pre-treatment samples. Error bar denotes mean ± SEM. p value calculated using two-sided Wilcoxon matched-pairs test. * $p < 0.05$, ** $p < 0.01$. (e) Spearman correlation of PFS to Ki67 expression in the total CD8$^+$ T-cell population (left), PD-L1$^+$ CD8$^+$ T-cells (middle) and PD-1$^+$ CD8$^+$ T-cells (right) at pre-treatment (up) and after 1 cycle of anti-PD-1 (Week 1, down) in melanoma. (f) Spearman correlation of PFS to the ratio of the fold change of Ki67 increase on CD8$^+$ T-cells (week 5 on pre-treatment) to tumor burden (up) and to the ratio of the fold change of Ki67 increase on PD-L1$^+$ CD8$^+$ T-cells (week 5 on pre-treatment) to tumor burden (down) in UC. (g) t-distributed stochastic neighbor embedding (t-SNE) map of CD8$^+$ T-cells overlaid with the expression level of Ki67 as a heat map in melanoma and (h) in UC. (i) Phenotypic description of the Ki67$^+$ cluster in the CD8$^+$ T-cell t-SNE map of melanoma and (j) of UC. Histograms depict expression profile of functional markers in the Ki67$^+$ CD8$^+$ cluster (orange) compared to total CD8$^+$ T-cell population (grey). (k) Lineplot with fold induction of Ki67$^+$ cells in CD8$^+$ T-cell t-SNE map of non-responders (NR) and responders (R) to anti-PD-1 at indicated times in melanoma and (l) in UC. Data shown are relative to pre-treatment samples. Error bar denotes ± SEM. p value calculated using two-sided Wilcoxon matched-pairs test. * $p < 0.05$, ** $p < 0.01$. (m) Lineplots with ratio of concentrations of serum Kyn and Trp (×100) and concentration of CXCL10 at indicated times in melanoma and (n) in UC. p value calculated using two-sided Wilcoxon matched-pairs test. ** $p < 0.01$. Wk, week; NS, not significant, TB, tumor burden.

To conclude, via a manual gating approach and an unsupervised clustering approach we report marked invigoration of CD8$^+$ T-cell subsets that have enriched expression of the activation marker HLA-DR and variably express immune checkpoint molecules. upon anti-PD-1 treatment initiation These proliferating CD8$^+$ T-cell populations peaked after 1 to 2 cycles of anti-PD-1 in both melanoma and UC patients. Altogether these data point to the possible clinical significance of baseline Ki67$^+$ CD8$^+$ T-cells and mainly the PD-1 expressing subset in melanoma. In UC the early increase of Ki67$^+$ CD8$^+$ T-cells and of the PD-L1 expressing subset relative to tumor burden seems to be crucial for PFS.

3.4. Systemic Immune Changes after SBRT

To explore the impact of SBRT, the dynamics of immune cell frequencies before and after SBRT were investigated (resp. blood samples collected at week 1 and 2 in melanoma and week 5 and 6 in UC arm B). As described above, proliferation of the T-cell compartment peaked at the first on-treatment blood sample, which was collected before SBRT administration. No additional increases in (Ki67 expressing) T-cell subsets were detected after SBRT in melanoma nor in UC. In melanoma, modest increases in the serum concentration of CXCL10 and Kyn/Trp were observed, while the frequency of B-cells decreased (Figure S2a), but these changes were not significantly different to the observed trend before SBRT. In the UC cohort, these changes could not be confirmed (Figure S2b).

3.5. FlowSOM Analysis to Discover Immune Signatures Correlating with Clinical Outcome

In order to detect discrete differences in the systemic immune response between responders and non-responders, we applied the algorithm FlowSOM to the flow cytometry dataset. FlowSOM, a powerful clustering-based technique to explore cellular heterogeneity, generates a Minimum Spanning Tree with each node existing of a group of phenotypically related cells [19]. The 85 fcs files of melanoma patients were concatenated into one single FlowSOM tree for all individuals (Figure 4a). The frequency of cells assigned to a specific metacluster or cluster were compared between responders and non-responders before and during treatment. No differences between responders and non-responders in the percentage of cells assigned to a specific metacluster or cluster were noticed. We further explored differences between melanoma responders and non-responders by investigating the extent of (co-) expression of 6 functional markers (CTLA-4, Ki67, IDO, PD-1, PD-L1 and HLA-DR) in the FlowSOM clusters. Features distinguishing responders and non-responders were predominantly found in the T-cell compartment.

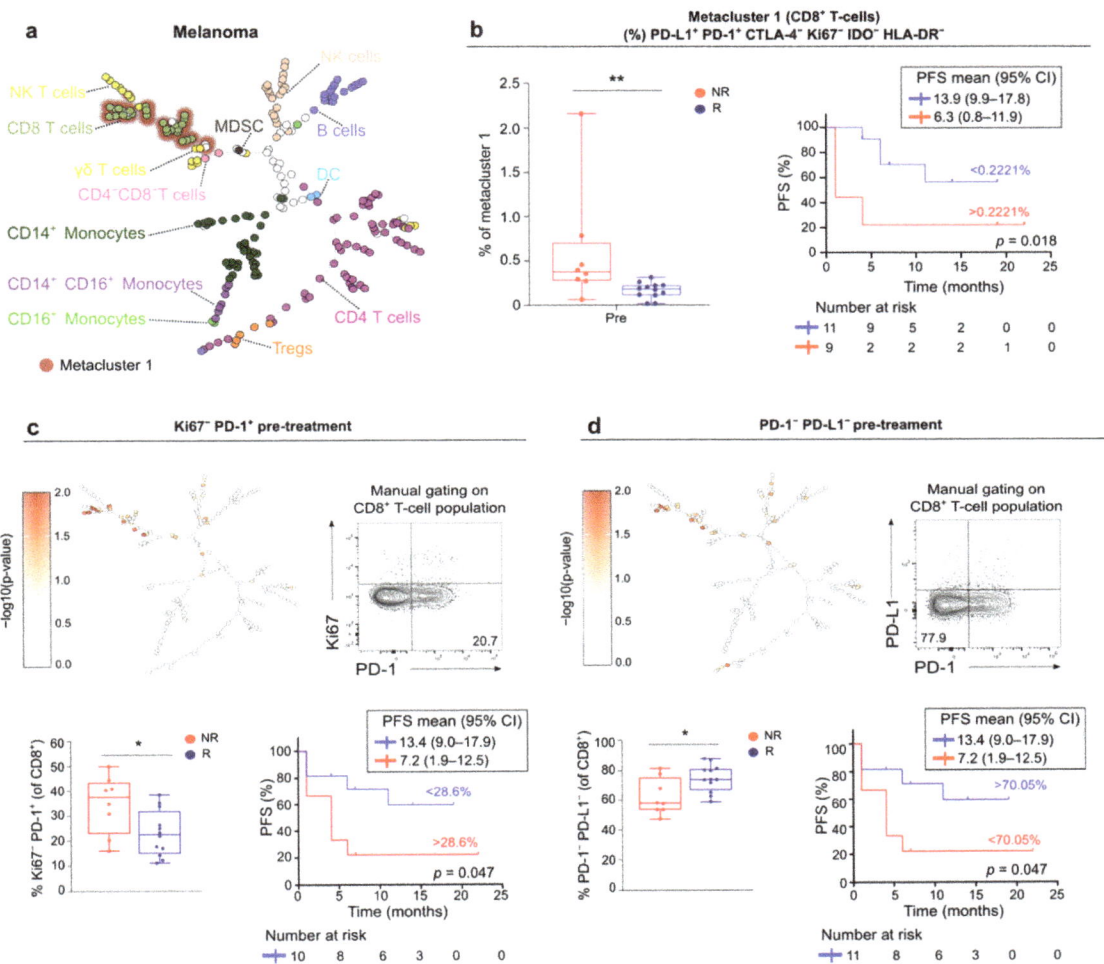

Figure 4. Pre-treatment expression of PD-1 and PD-L1 in non-proliferating CD8$^+$ T-cells correlates with non-response to anti-PD-1 in melanoma. (**a**) FlowSOM tree of concatenated flow cytometry data of PBMCs from 20 melanoma patients (85 samples). (**b**) (left) Boxplot of the pre-treatment expression of indicated signature in metacluster 1 (CD8$^+$ T-cells) in non-responders (NR) and responders (R). p value calculated using two-sided Mann-Whitney U test. ** $p < 0.01$. (right) Kaplan-Maier estimate of PFS stratified according to low (<0.2221%) or high (>0.2221%) expression of PD-L1$^+$ PD-1$^+$ CTLA-4$^-$ Ki67$^-$ IDO$^-$ HLA-DR$^-$ in metacluster 1. p value calculated using log-rank test. (**c**,**d**) (top left) Melanoma FlowSOM tree depicting differences in expression of the indicated signature in clusters between non-responders and responders at pre-treatment. $-\log10(p$ values) are plotted on FlowSOM tree showing the significantly over- or underrepresented clusters in non-responders versus responders. (top right) Contour plot ($n = 10$) representing manual gating strategy on total CD8$^+$ T-cell population to confirm FlowSOM signature. (below left) Boxplots with expression of manually gated signature in non-responders (NR) and responders (R), p value in boxplots calculated using two-sided Mann-Whitney U test. * $0.01 < p < 0.05$. (below right) Kaplan-Maier estimate of PFS stratified according to low or high expression of indicated signature. p value calculated using log-rank test. (**b**–**d**). Whiskers of boxplots extend to the minimum and maximum data point, with the horizontal line indicating the median.

We first focused on features in (meta-) clusters corresponding to CD8$^+$ T-cells. Flow-SOM assigned CD8$^+$ T-cells to 1 single metacluster (metacluster 1). Within this metacluster, the baseline expression of PD-L1$^+$ PD-1$^+$ CTLA-4$^-$ Ki67$^-$ IDO$^-$ HLA-DR$^-$ was higher in non-responding patients (Figure 4b). The frequency of CD8$^+$ T-cells with this phenotype was associated with worse PFS (Figure 4b, Log-Rank analysis, p = 0.018). Further, multiple CD8$^+$ T-cell clusters showed differential expression of Ki67$^-$ PD-1$^+$ and PD-1$^-$ PD-L1$^-$ between responders and non-responders (Figure 4c,d). We manually gated on Ki67$^-$ PD-1$^+$ CD8$^+$ T-cells, which confirmed higher frequencies in non-responders. In contrast, manual gating on PD-1$^-$ PD-L1$^-$ showed decreased expression in non-responders compared to responders. These signatures were inversely correlated with each other (Spearman's CC: -0.965, $p < 0.001$) and were both associated with PFS (Figure 4c,d).

In the CD4$^+$ T-cell compartment, 51 signatures were detected to be differentially expressed pre-treatment between responders and non-responders in melanoma (cluster 204, cluster 205, cluster 206 and cluster 220). Notably, all signatures involved PD-L1 expression and were highly interrelated (Figure 5a,b). The majority of signatures distinguishing responders from non-responders were expressed in cluster 205, which is a HLA-DR positive CD25$^-$ FoxP3$^-$ CD4$^+$ T-cell population (Figure 5c). Non-responders had an increased expression of PD-L1 in this cluster compared to responders (p = 0.0041, Figure 5d). PD-L1$^+$ CD4$^+$ cells in cluster 205 of non-responders were negative for expression of CTLA-4 or Ki67 and were HLA-DR^{dim} (Figure 5e). This phenotype of CD4$^+$ T-cells could be confirmed via a manual gating approach. Non-responders did not only have a higher frequency of this subset of CD4$^+$ T-cells at baseline but also during treatment (Figure 5f).

Since FlowSOM assigned Tregs to the same metacluster as other CD4$^+$ T-cell populations in melanoma, Treg clusters were analyzed separately by defining them as one metacluster. Co-expression patterns in this Treg metacluster (including cluster 207, cluster 208, cluster 221, cluster 222 and cluster 223) were investigated. Non-responders were found to have less Tregs with phenotype HLA-DR$^+$ PD-L1$^-$ IDO$^-$ during treatment (Figure S3a). This was confirmed via a manual gating approach (Figure S3b).

A similar strategy was applied to the UC cohort, concatenating 35 fcs files of the 9 arm B patients into one single FlowSOM tree (Figure 6a). Other than in the melanoma cohort, analysis in the UC cohort predominantly revealed alterations in (meta-) clusters corresponding to monocytes. Cluster 215 containing non-classical CD14$^-$ CD16$^+$ monocytes was overrepresented in responders before and during treatment (Figure 6b). The frequency of cells in metacluster 28, which includes cluster 215, was different between responders and non-responders at week 5 and week 12 (Figure 6c). Cluster 202, cluster 216, cluster 217 and cluster 218 contain CD14$^+$ CD16$^+$ monocytes and were overrepresented in responders at week 12 (metacluster 23, Figure 6d). Baseline metacluster 23 and metacluster 28 were both inversely correlated with baseline tumor burden (resp. Spearman's CC: -0.817, $p = 0.007$ and Spearman's CC: -0.833, $p = 0.005$), and also inversely correlated with the serum Kyn/Trp ratio (resp. Spearman's CC: -0.817, $p = 0.007$ and Spearman's CC: -0.900, $p = 0.001$). Furthermore, 3 additional clusters with classical CD14$^+$ CD16$^-$ monocytes were overrepresented in responders (cluster 208 before treatment $p = 0.0159$, cluster 210 and cluster 224 at week 12, both $p = 0.0159$), although not reflected on metacluster level (Figure 6b). In addition enhanced expression of proliferation marker Ki67 in cluster 123 (corresponding to CD56^{bright} NK cells) at week 12 was found to be associated with lower Kyn/Trp ratio and better response (Figure 6e,f).

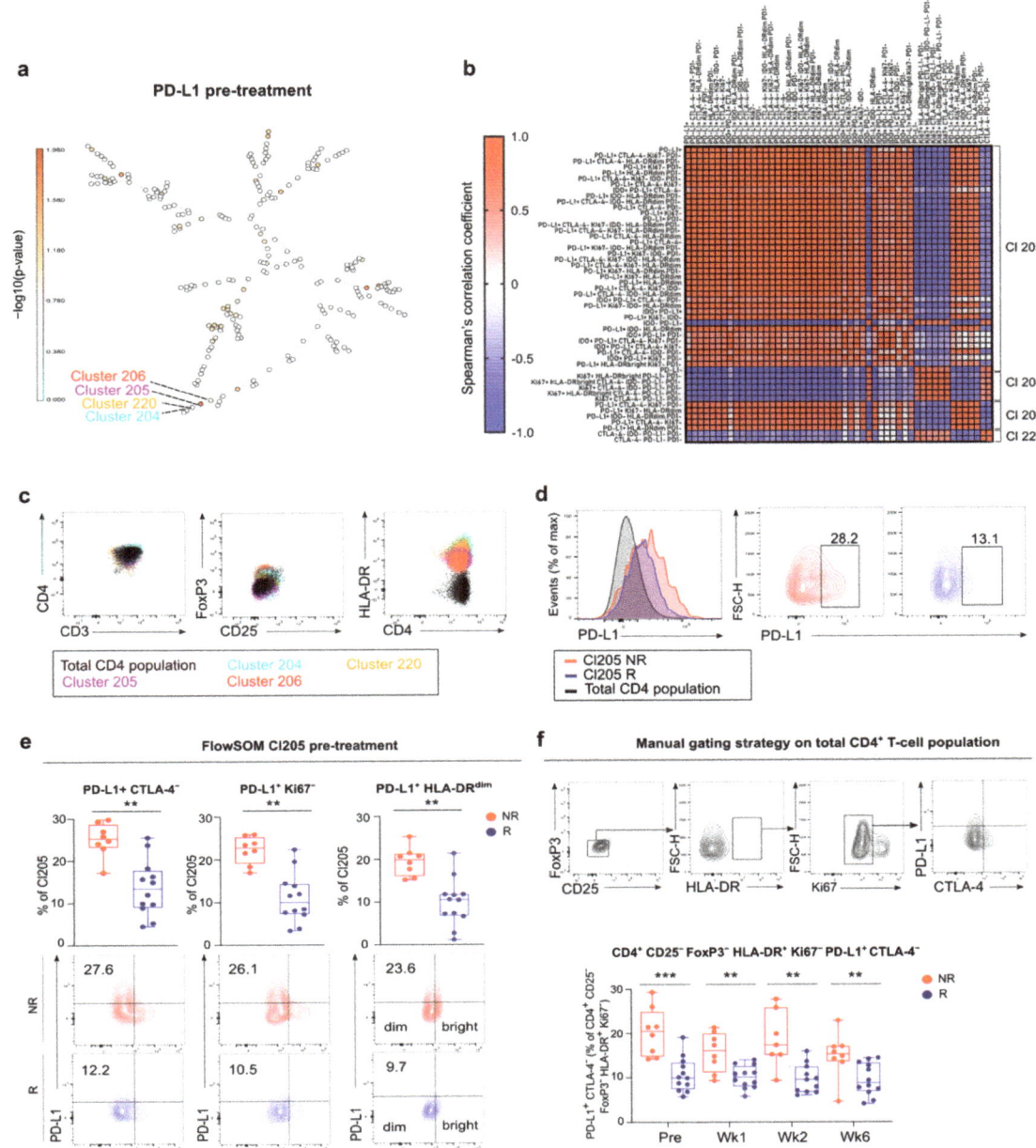

Figure 5. Pre-treatment expression of PD-L1 in non-proliferating CD4$^+$ T-cells correlate with non-response to anti-PD-1 in melanoma patients. (**a**) Melanoma FlowSOM tree representing differences in PD-L1 expression in clusters between non-responders and responders at pre-treatment. −log10(p-values) are plotted on tree showing the significantly over- or underrepresented clusters in non-responders versus responders. (**b**) Correlation matrix of pre-treatment signatures (co-) expressing PD-L1 in FlowSOM clusters corresponding to CD4$^+$ T-cells. Colored boxes represent Spearman's correlation with

a significance of $p < 0.05$. Red to blue represents correlation coefficients ranging from 1 to -1, respectively. (**c**) Representative flow plots of 10 independent melanoma patients with the phenotype of indicated clusters. (**d**) Histogram and contour plots with PD-L1 expression in cluster 205 of non-responders (NR, $n = 5$) versus cluster 205 of responders (R, $n = 5$) versus the total CD4$^+$ T-cell population ($n = 10$). (**e**) (top) Boxplots with the frequency of expression of PD-L1 combined with CTLA-4, Ki67 or HLA-DR in cluster 205 in non-responders (NR) and responders (R). (bottom) Contour plots with indicated signatures in cluster 205 in non-responders (NR, $n = 5$) and responders (R, $n = 5$). (**f**) (top) Contour plots representing manual gating strategy of PD-L1$^+$ CTLA-4$^-$ Ki67$^-$ HLA-DR$^+$ CD4$^+$ T-cells. (bottom) Boxplots with frequency of manually gated signature in CD4$^+$ T-cell population at indicated times. (**e**,**f**). Whiskers of boxplots extend to the minimum and maximum data point, with the horizontal line indicating the median. p value calculated using two-sided Mann-Whitney U test. ** $p < 0.01$, *** $p < 0.001$. Wk, week.

Altogether, the results obtained by FlowSOM analysis highlight distinct signatures in melanoma and UC that correlate with clinical outcome. In melanoma, these signatures were predominantly found in the lymphoid compartment and mainly involved different baseline expression patterns of PD-1 and/or PD-L1: the expression of PD-L1/PD-1 in non-proliferating (Ki67$^-$) CD8$^+$ and CD4$^+$ T cells was associated with worse clinical outcome. In the UC cohort signatures with a higher frequency of (non-) classical monocytes were found to be correlated with response, but also had a strong inverse correlation with tumor burden.

3.6. Link between Blood and Tumor Micro-Environment

We explored possible associations between the systemic immune landscape and the TILs score/PD-L1 expression in the tumor. In melanoma, patients with a high TILs score (score 3 versus score 1 and 2) had a significantly lower frequency of circulating PD-L1$^+$ CD4$^+$ T-cells and PD-L1$^+$ PD-1$^+$ CD4$^+$ T-cells (Figure S4a). This could not be confirmed in the UC cohort (Figure S4b). No correlation between PD-L1 staining in the tumor micro-environment and systemic immune features could be observed for both cohorts.

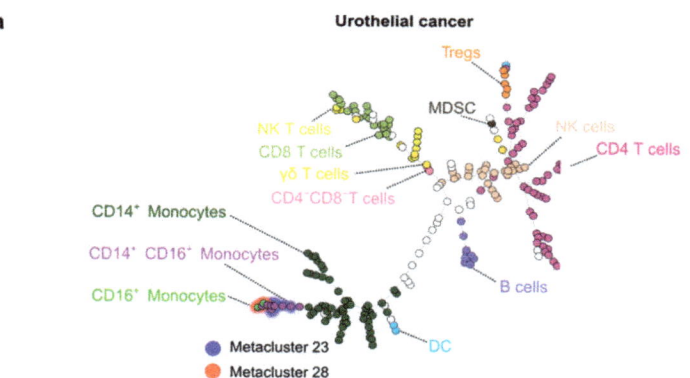

Figure 6. *Cont.*

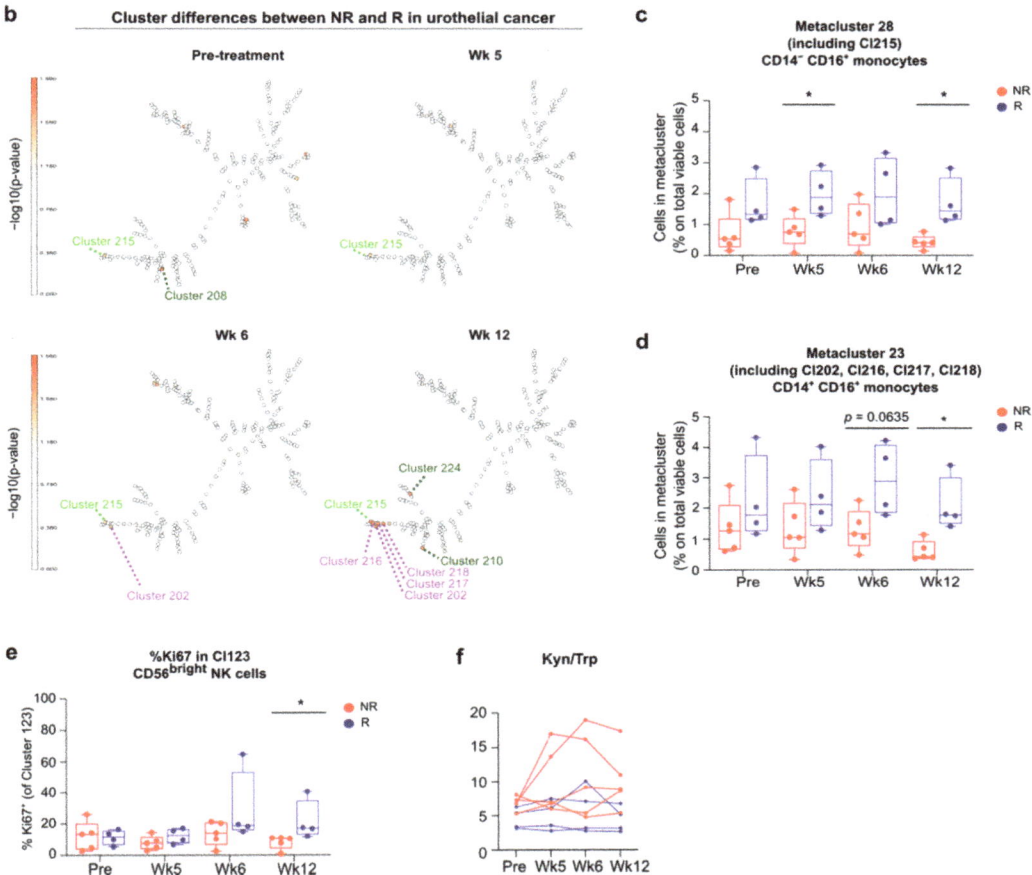

Figure 6. Increased frequency of monocytes associates to response in urothelial cancer. (**a**) FlowSOM tree of concatenated flow cytometry data of PBMCs from 9 UC patients (arm B, 36 samples). (**b**) UC FlowSOM trees depicting differences in the percentage of cells assigned to clusters between non-responders and responders at pre-treatment, week 5, week 6 and week 12 of anti-PD-1 treatment. $-\log10(p$ values) are plotted on trees showing the significantly over- or underrepresented clusters in non-responders versus responders. Colors of cluster numbers correspond with immune cell populations in a. (**c**) Boxplots with percentages of metacluster 28 corresponding to CD14$^-$ CD16$^+$ monocytes in non-responders (NR) and responders (R) at indicated times. (**d**) Boxplots with percentages of metacluster 23 corresponding to CD14$^+$ CD16$^+$ monocytes in non-responders (NR) and responders (R) at indicated times. (**e**) Boxplots with percentages of Ki67 expression in cluster 123 corresponding to CD56^{bright} NK cells in non-responders (NR) and responders (R) at indicated times. (**f**) Lineplots with the ratio of concentrations of serum Kyn and Trp ($\times 100$) in non-responders (NR) and responders (R) at indicated times. (**c**–**e**) Whiskers of boxplots extend to the minimum and maximum data point, with the horizontal line indicating the median. p value calculated using two-sided Mann-Whitney U test. * $0.01 < p < 0.05$.

4. Discussion

In this study we conducted an in-depth analysis of baseline and on-treatment systemic immune features in a cohort of melanoma and UC patients treated with anti-PD-1 therapy combined with SBRT in a similar design.

Baseline immunity (before start of treatment) was clearly different between these two cohorts, supporting a less active immune landscape in UC compared to melanoma. NLR was significantly higher in the UC cohort. Variations in baseline NLR have been reported between tumor types and increased pre-treatment NLR has been linked to worse outcome in patients treated with immunotherapy [20]. The NLR is considered as a marker reflecting the balance between inflammation state (pro-tumoral) and adaptive immune surveillance and response (anti-tumoral). UC patients also had higher frequencies of classical monocytes and immunosuppressive MDSCs. In the melanoma cohort, cells of the lymphoid lineage were higher as reflected by higher frequencies of lymphocytes in total, γδ T-cells and proliferating B-cells. In line with this, higher serum concentrations of CXCL10, an IFNγ-inducible chemokine involved in T-cell recruitment to the tumor [21,22], were measured in melanoma compared to UC. In contrast, serum concentrations of soluble MICA and MICB were higher in UC patients. MICA and MICB are ligands for the activating receptor NKG2D and their soluble form has been implicated in the perturbation of effector immune cell function and the stimulation of MDSCs [23].

The observation of a distinct baseline systemic immunity in the 2 cohorts may play a prominent role in the different response rates to immunotherapy. The objective response rate (ORR) of anti-PD-1 monotherapy reported in inoperable stage III/IV melanoma is around 42–45%, while ORR reported in chemotherapy-refractory metastatic urothelial cancer is considerably lower (15–28.6%) [24–27]. Currently, our understanding of intrinsic factors such as tumor type and burden, patient age and sex, and extrinsic factors such as prior systemic treatments that shape the immune system is far from complete. Tumor mutational burden has been linked to response to immunotherapy and varies across tumor types, with melanoma constituting the largest neoantigen repertoire [8,28]. Both patients' age and sex were evidenced to impact the driver mutations that arise during tumorigenesis, with younger and female patients accumulating driver mutations that are less readily presented by MHC molecules [29]. In contrast, in a meta-analysis including more than 10,000 patients treated with immunotherapy for several types of advanced cancers, a higher relative reduction of the risk of death was observed in male compared to female patients [30]. Since a higher tumor mutational burden has been reported in male patients [31,32] and this is a predictor of benefit from immune checkpoint inhibitors [33,34], this could be a possible explanation for improved overall survival rates in male patients. Aging has been reported to accompany certain immune changes such as a decrease in the number and functionality of naïve $CD8^+$ T-cells [35,36] and reduced phagocytic function and HLA-II expression of DCs [37], indicating elder individuals have an impaired T-cell response to cross-presented antigens (immunosenescence). Nevertheless, a large multicentric study reported that older melanoma patients had better response to anti-PD-1 treatment compared to younger patients [38]. In our study median age in the UC cohort was higher in responders compared to non-responders (75.5 versus 61.0 years). Age was not correlated with NLR in the melanoma nor the UC cohort, which is consistent with other reports [39,40]. The depicted reference values of neutrophils and lymphocytes (Figure 2a) further support baseline differences per tumor type independent of sex and age.

Importantly, the majority of UC patients received prior chemotherapy and one third even received two or three treatment lines before trial inclusion, which may have altered the immune landscape considerably. The impact of these immunological alterations on immunotherapy response is unclear. A number of recent studies hypothesize that chemotherapy may sensitize tumors for immunotherapy whereas others postulate that chemotherapy negatively impacts myelopoiesis, induces inflammation and increased expression of immunosuppressive molecules such as IDO and PD-L1 [41–45].

The observations in this study demonstrate important differences in baseline immunity between and within tumor types and these may be important determinants for immunotherapy response. Better insights into the various intrinsic and extrinsic factors that shape this baseline immunity may be relevant in order to gain further insights how to optimize immunotherapy response across various cancer types.

Pathological response predictive for clinical outcome to immunotherapy has been reported early after initiation of anti-PD-1 in melanoma [34,46] and the accumulation of $CD8^+$ T-cells expressing inhibitory receptors (exhausted T-cells, T_{ex}) was detected in the peripheral blood within 3 weeks after immunotherapy initiation [46–48]. In the current study, we observed increased proliferation of CD8$^+$ T-cells in the blood as early as 7 days after anti-PD-1 treatment initiation in melanoma patients. Similar increases in Ki67$^+$ CD8$^+$ T-cells were detected after one or two treatment cycles in UC patients. Proliferating CD8$^+$ T-cells were positive for the activation marker HLA-DR and for IDO and had variable expression of checkpoint molecules such as PD-1, PD-L1 and CTLA-4.These findings are in line with previous data in NSCLC and melanoma, where anti-PD-1 was reported to revitalize an already existing T-cell response consisting of primed (tumor-specific) CD8$^+$ T-cells that had become exhausted due to chronic antigen stimulation [46–48]. It has been hypothesized by Huang et al. that reinvigoration of T_{ex} occurs in the peripheral blood prior to migrating into the tumor as supported by a single peak of PD-1-blockade-induced immune reinvigoration despite on-going treatment [46,47]. In line with this, proliferating CD8$^+$ T-cells in the current study peaked early in the PBMC compartment and declined upon further anti-PD-1 administration.

No clear immune boost effect could be observed after SBRT in these 2 small patient cohorts except from a moderate increase in CXCL10 in the melanoma cohort.

In melanoma, proliferation of the total CD8$^+$ T-cell population, PD-L1$^+$ CD8$^+$ T-cells and PD-1$^+$ CD8$^+$ T-cells at baseline were correlated with prolonged PFS. The former two populations were also correlated with PFS after one cycle of anti-PD-1 (PD-1 expression was not measurable beyond baseline presumably due to anti-PD-1 treatment preventing the in vitro added PD-1 antibodies from binding their epitopes). In contrast, FlowSOM analysis supports a negative impact of baseline PD-1/PD-L1 expression in non-proliferating (Ki67$^-$) T-helper (CD25$^-$ Foxp3$^-$ CD4$^+$) and cytotoxic T-cells (CD8$^+$). A negative prognostic effect of PD-L1 expressing CD8$^+$ T-cells in melanoma has been reported in the context of anti-CTLA-4 immunotherapy and also in early stage melanoma without systemic treatment [49,50]. FlowSOM analysis also revealed PD-1/PD-L1 co-expression on circulating CD8$^+$ T-cells. This has been described before, and PD-1 and PD-L1 were shown to bind in cis with high affinity in in vitro lentivirally transduced cell cultures, including Jurkat Cells, evidencing this interaction can also occur on T-cells in vivo [51]. These in cis PD-1/PD-L1 interactions on CD8$^+$ T-cells might reflect functional inactivation, which would explain the enhanced co-expression of PD-1 and PD-L1 on CD8$^+$ T-cells in non-responders observed in this study. In addition PD-L1/PD-1 co-expressing CD4$^+$ T-cells in blood tend to be related to a lower TILs score at the level of tumor micro-environment in our melanoma cohort. In UC patients, the expansion of proliferating (Ki67$^+$) CD8$^+$ T-cells and its PD-L1$^+$ subset relative to tumor burden was correlated with longer PFS.

These data support that the size of the proliferating cytotoxic T-cell compartment and its expansion is closely involved in the immunotherapy response. As UC patients have lower baseline lymphocyte counts compared to melanoma, the actual magnitude of the expansion might be important for response initiation. In addition, in arm B of the UC cohort tumor burden was significantly lower in responders versus non-responders, which may explain why the ratio is of importance in the UC cohort. Huang et al. have reported that the magnitude of the reinvigoration of T_{ex} as a ratio to pre-treatment tumor burden was correlated with clinical outcome in immunotherapy in melanoma [47]. The fact that tumor burden in arm B of the UC cohort was significantly lower in responders compared to non-responders, may be a reason why this ratio was related to response only in the UC cohort in this study. Our data are also supported by data from the neo-adjuvant setting where a single injection of pembrolizumab in resectable stage III or IV melanoma patients resulted in the expansion of Ki67$^+$ PD-1$^+$ CTLA-4$^+$ CD8$^+$ T-cells in the peripheral blood of patients 7 days post injection. This Ki67$^+$ CD8$^+$ T-cell population was demonstrated to be present in the blood before start of the treatment and supports the reinvigorating properties of anti-PD-1 therapy on a preexisting immune response [52]. In the study of Huang et al. the CD8$^+$ T-cell population responding to anti-PD-1 treatment was characterized as CD45^{lo} CD27^{hi}, containing cells with high expression of CTLA-4, 2B4 and PD-1. Moreover this population was Eomes^{hi} and T-bet^{lo}, which is consistent with an exhausted T-cell phenotype [47]. Although the proliferating CD8$^+$ T-cells in our study had higher expression of the activation marker HLA-DR compared to the non-proliferating CD8$^+$ T-cells, they also had higher IDO expression and variable expression of PD-1 and PD-L1. The expression of these immune checkpoint molecules has been shown to be a possible physiological negative feedback mechanism upon immune stimulation [45,53]. These data may explain conflicting results on the prognostic value of checkpoint molecules expressed on immune cells.

These data also underline the relevance of analyzing PD-1/PD-L1 expression on circulating T-cell subsets. Whereas PD-1 is predominantly expressed on lymphocytes, its ligand PD-L1 has been detected on a variety of cells in the tumor microenvironment including conventional DCs, macrophages, MDSCs, and extracellular vesicles [54–57]. Blockade of PD-L1 signaling on immune cells (especially DCs and macrophages) was demonstrated to be critical for an optimal anti-tumor immune response, as opposed to/in addition to cancer-cell intrinsic PD-L1 expression [55,56]. This may explain the inconsistent observations on the role of tumor PD-L1 expression in predicting response to PD-1 blockade, and why its absence does not preclude response [58]. Although PD-L1 expression in tumor tissue has been related to response to PD-1 blockade, a systematic evaluation of 45 FDA-approved trials involving 15 tumor types demonstrated that PD-L1 expression was predictive in only 28.9% of cases [6]. PD-L1 expression on circulating T-cells is less studied. Pre-treatment PD-L1 expression on peripheral CD8$^+$ and CD4$^+$ T-cells was associated with worse outcome in melanoma patients receiving CTLA-4 blockade [49]. We previously reported that the frequency of circulating PD-L1$^+$ CD8$^+$ T-cells in early-stage melanoma was an independent prognostic marker. High frequencies of PD-L1$^+$ CD8$^+$ T-cells were associated with other immune suppressive features including increased Kyn/Trp ratio (implying increased IDO1 activity) and increased MDSCs and Tregs [50]. Together with the observation in the current study that the level of PD-L1 on circulating CD4$^+$ and CD8$^+$ T-cells is of importance for the outcome of anti-PD-1 treatment, these findings suggest that PD-L1 expression in the lymphocyte compartment might be an important blood biomarker in cancer patients receiving PD-1 blockade.

FlowSom analysis in the UC cohort revealed higher frequencies of monocytes in responding UC patients. High frequencies of non-classical CD14$^-$ CD16$^+$ monocytes and intermediate CD14$^+$ CD16$^+$ monocytes were closely correlated with lower tumor burden at baseline. The percentage of proliferating CD56^{bright} NK cells was also found to be increased in responding UC patients at week 12. Intratumoral CD56^{bright} NK cells have been previously reported to be associated with improved survival outcomes in localized stage bladder cancer [59]. At week 12 responding UC patients also had lower levels of Kyn/Trp, suggesting decreased activity of IDO1, an enzyme that is implicated in acquired immune tolerance [57,60].

The immunotherapy field in oncology is rapidly changing with superior long-term results of combination immunotherapy in melanoma and renal cell carcinoma [61] and very promising results in melanoma in the neoadjuvant setting that seem to be extendable to other tumor types [62,63]. Moreover the number of clinical trials with new immune targets is increasing e.g., TIM-3, LAG-3, GITR, TIGIT. Immune monitoring of peripheral blood is attractive for dynamic monitoring of the immune system, which ideally could lead to a strategy of treatment adaptation in order to optimize response. In the current study blood signatures before and during treatment with anti-PD-1 therapy combined with SBRT were investigated. Whether the observed signatures related to clinical outcome are applicable in daily practice and can be extrapolated to other immunotherapy regimens such as the combination of anti-PD-1 with anti-CTLA4 needs to be further investigated. Distinct cellular mechanisms of anti-PD-1 or anti-CTLA-4 monotherapy compared to combination therapy have been detected in the peripheral blood [64,65] and anti-CTLA4 monotherapy has been shown to induce some immune landscape changes in blood that are considered negative for response on subsequent anti-PD-1 treatment [52]. These immune monitoring data can provide relevant insights in how to optimize immunotherapy strategy.

5. Conclusions

Despite the limitations of small sample sizes, use of cryopreserved samples and multiple testing in the FlowSom analysis, this study clearly reveals a different baseline immune landscape in melanoma and UC which may be of importance for immunotherapy response. The intrinsic (host and/or tumor related) and extrinsic factors (e.g., prior treatments) that shape this immune landscape are currently incompletely understood. Better insights in these determinants may be important to gain new insights for optimizing immunotherapy outcome. This study also reports signatures of proliferation in the $CD8^+$ T-cell compartment prior to and early after anti-PD-1 initiation that were positively correlated with clinical outcome. Moreover our data support the clinical relevance of PD-1/PD-L1 expression on circulating immune cell subsets in melanoma.

Supplementary Materials: The following are available online at https://www.mdpi.com/article/10.3390/cancers13112630/s1, Figure S1: Gating strategies of immune cell populations analyzed in this study, Figure S2: Systemic immune changes after SBRT, Figure S3: $HLA-DR^+$ $PD-L1^-$ IDO^- expressing regulatory T-cells are upregulated in responding melanoma patients, Figure S4: Abundance of TILs is linked with blood PD-L1 and PD-1 expression on $CD4^+$ T-cells, Table S1: Patient characteristics of melanoma patients, Table S2: Patient characteristics of urothelial cancer patients, Table S3: List of monoclonal antibodies for flow cytometry.

Author Contributions: A.M., S.J.T., N.S., P.O. and L.B. designed the study. N.S., M.S., V.K., K.D., S.R., P.O. and L.B. provided patient samples and clinical information. A.M., S.J.T. and B.H. conducted experiments. L.F. and J.V.D. evaluated the immunohistochemical staining. S.V.G. and Y.S. performed FlowSOM analysis. A.M. and S.V.G. performed statistical evaluations and created the figures in the manuscript. S.J.T., A.D., D.A., F.H. provided scientific advice. A.M., A.D. and L.B. wrote the paper with input from all of the authors. All authors have read and agreed to the published version of the manuscript.

Funding: This work was supported by the Innovation and Clinical Research Foundation of Ghent University Hospital and Kom op tegen Kanker (Stand up to Cancer, grant number 12294), the Flemish cancer society. S.J.T. is a beneficiary of a postdoctoral BOF grant and is a postdoctoral fellow with the Fund for Scientific Research Flanders. S.V.G. is an ISAC Marylou Ingram Scholar and supported by an FWO postdoctoral research grant (Research Foundation—Flanders).

Institutional Review Board Statement: The study was conducted according to the guidelines of the Declaration of Helsinki, and approved by the Institutional Review Board (or Ethics Committee) of University Hospital Ghent (melanoma cohort: protocol code 2016/0540 and 24/05/2016; UC cohort: protocol code 2016/0661 and 07/06/2016).

Informed Consent Statement: Informed consent was obtained from all subjects involved in the study.

Data Availability Statement: Data is contained within the article or Supplementary Material.

Acknowledgments: The authors thank the patients and their family for their participation in this study and Els Van Maelsaeke for her excellent technical assistance. We would like to thank the VIB Flow Core for training, support and access to the instrument park.

Conflicts of Interest: N.S. reports travel grants from Merck Sharpe & Dohme, Astellas, Bayer and Bristol-Myers Squibb. V.K. provided consultation, attended advisory boards, and/or provided lectures for Roche, Bristol-Myers Squibb, Merck Sharp & Dohme, Novartis, Amgen and Sanofi. S.R. received a research grant from MSD and ROCHE. P.O. received a research grant from Ferring, Merck, Varian, Bayer; consultancy fees from Ferring, Bayer, Janssen, Sandoz, Sanofi. The remaining authors declare no competing interests.

References

1. Haslam, A.; Prasad, V. Estimation of the Percentage of US Patients with Cancer Who Are Eligible for and Respond to Checkpoint Inhibitor Immunotherapy Drugs. *JAMA Netw. Open* **2019**, *2*, e192535. [CrossRef]
2. Larkin, J.; Chiarion-Sileni, V.; Gonzalez, R.; Grob, J.J.; Cowey, C.L.; Lao, C.D.; Schadendorf, D.; Dummer, R.; Smylie, M.; Rutkowski, P.; et al. Combined Nivolumab and Ipilimumab or Monotherapy in Untreated Melanoma. *N. Engl. J. Med.* **2015**, *373*, 23–34. [CrossRef]

3. Sharma, P.; Siefker-Radtke, A.; de Braud, F.; Basso, U.; Calvo, E.; Bono, P.; Morse, M.A.; Ascierto, P.A.; Lopez-Martin, J.; Brossart, P.; et al. Nivolumab Alone and with Ipilimumab in Previously Treated Metastatic Urothelial Carcinoma: CheckMate 032 Nivolumab 1 mg/kg Plus Ipilimumab 3 mg/kg Expansion Cohort Results. *J. Clin. Oncol.* **2019**, *37*, 1608–1616. [CrossRef]
4. Rosenberg, J.E.; Hoffman-Censits, J.; Powles, T.; van der Heijden, M.S.; Balar, A.V.; Necchi, A.; Dawson, N.; O'Donnell, P.H.; Balmanoukian, A.; Loriot, Y.; et al. Atezolizumab in patients with locally advanced and metastatic urothelial carcinoma who have progressed following treatment with platinum-based chemotherapy: A single-arm, multicentre, phase 2 trial. *Lancet* **2016**, *387*, 1909–1920. [CrossRef]
5. Taube, J.M.; Klein, A.; Brahmer, J.R.; Xu, H.; Pan, X.; Kim, J.H.; Chen, L.; Pardoll, D.M.; Topalian, S.L.; Anders, R.A. Association of PD-1, PD-1 ligands, and other features of the tumor immune microenvironment with response to anti-PD-1 therapy. *Clin. Cancer Res.* **2014**, *20*, 5064–5074. [CrossRef] [PubMed]
6. Davis, A.A.; Patel, V.G. The role of PD-L1 expression as a predictive biomarker: An analysis of all US Food and Drug Administration (FDA) approvals of immune checkpoint inhibitors. *J. Immunother. Cancer* **2019**, *7*, 1–8. [CrossRef] [PubMed]
7. Yarchoan, M.; Hopkins, A.; Jaffee, E.M. Tumor Mutational Burden and Response Rate to PD-1 Inhibition. *N. Engl. J. Med.* **2017**, *377*, 2500–2501. [CrossRef] [PubMed]
8. Schumacher, T.N.; Schreiber, R.D. Neoantigens in cancer immunotherapy. *Science* **2015**, *348*, 69–74. [CrossRef] [PubMed]
9. Lemery, S.; Keegan, P.; Pazdur, R. First FDA Approval Agnostic of Cancer Site—When a Biomarker Defines the Indication. *N. Engl. J. Med.* **2017**, *377*, 1409–1412. [CrossRef]
10. Prasad, V.; Kaestner, V.; Mailankody, S. Cancer Drugs Approved Based on Biomarkers and Not Tumor Type-FDA Approval of Pembrolizumab for Mismatch Repair-Deficient Solid Cancers. *JAMA Oncol.* **2018**, *4*, 157–158. [CrossRef]
11. Tumeh, P.C.; Harview, C.L.; Yearley, J.H.; Shintaku, I.P.; Taylor, E.J.M.; Robert, L.; Chmielowski, B.; Spasic, M.; Henry, G.; Ciobanu, V.; et al. PD-1 blockade induces responses by inhibiting adaptive immune resistance. *Nature* **2014**, *515*, 568–571. [CrossRef]
12. Ayers, M.; Lunceford, J.; Nebozhyn, M.; Murphy, E.; Loboda, A.; Kaufman, D.R.; Albright, A.; Cheng, J.D.; Kang, S.P.; Shankaran, V.; et al. IFN-γ–related mRNA profile predicts clinical response to PD-1 blockade. *J. Clin. Investig.* **2017**, *127*, 2930–2940. [CrossRef] [PubMed]
13. Gide, T.N.; Quek, C.; Menzies, A.M.; Tasker, A.T.; Shang, P.; Holst, J.; Madore, J.; Lim, S.Y.; Velickovic, R.; Wongchenko, M.; et al. Distinct Immune Cell Populations Define Response to Anti-PD-1 Monotherapy and Anti-PD-1/Anti-CTLA-4 Combined Therapy. *Cancer Cell* **2019**, *35*, 238–255. [CrossRef] [PubMed]
14. Sundahl, N.; Seremet, T.; Van Dorpe, J.; Neyns, B.; Ferdinande, L.; Meireson, A.; Brochez, L.; Kruse, V.; Ost, P. Phase 2 Trial of Nivolumab Combined with Stereotactic Body Radiation Therapy in Patients with Metastatic or Locally Advanced Inoperable Melanoma. *Int. J. Radiat. Oncol.* **2019**, *104*, 828–835. [CrossRef]
15. Sundahl, N.; Vandekerkhove, G.; Decaestecker, K.; Meireson, A.; De Visschere, P.; Fonteyne, V.; De Maeseneer, D.; Reynders, D.; Goetghebeur, E.; Van Dorpe, J.; et al. Randomized Phase 1 Trial of Pembrolizumab with Sequential Versus Concomitant Stereotactic Body Radiotherapy in Metastatic Urothelial Carcinoma. *Eur. Urol.* **2019**, *75*, 707–711. [CrossRef] [PubMed]
16. Yamada, K.; Miyazaki, T.; Shibata, T.; Hara, N.; Tsuchiya, M. Simultaneous measurement of tryptophan and related compounds by liquid chromatography/electrospray ionization tandem mass spectrometry. *J. Chromatogr. B* **2008**, *867*, 57–61. [CrossRef]
17. Schefold, J.C.; Zeden, J.-P.; Fotopoulou, C.; von Haehling, S.; Pschowski, R.; Hasper, D.; Volk, H.-D.; Schuett, C.; Reinke, P. Increased indoleamine 2,3-dioxygenase (IDO) activity and elevated serum levels of tryptophan catabolites in patients with chronic kidney disease: A possible link between chronic inflammation and uraemic symptoms. *Nephrol. Dial. Transplant.* **2009**, *24*, 1901–1908. [CrossRef]
18. Rodríguez-Perea, A.L.; Arcia, E.D.; Rueda, C.M.; Velilla, P.A. Phenotypical characterization of regulatory T cells in humans and rodents. *Clin. Exp. Immunol.* **2016**, *185*, 281–291. [CrossRef]
19. Van Gassen, S.; Callebaut, B.; Van Helden, M.J.; Lambrecht, B.N.; Demeester, P.; Dhaene, T.; Saeys, Y. FlowSOM: Using self-organizing maps for visualization and interpretation of cytometry data. *Cytom. Part A* **2015**, *87*, 636–645. [CrossRef]
20. Jiang, T.; Qiao, M.; Zhao, C.; Li, X.; Gao, G.; Su, C.; Ren, S.; Zhou, C. Pretreatment neutrophil-to-lymphocyte ratio is associated with outcome of advanced-stage cancer patients treated with immunotherapy: A meta-analysis. *Cancer Immunol. Immunother.* **2018**, *67*, 713–727. [CrossRef]
21. Harlin, H.; Meng, Y.; Peterson, A.C.; Zha, Y.; Tretiakova, M.; Slingluff, C.; McKee, M.; Gajewski, T.F. Chemokine expression in melanoma metastases associated with CD8+ T-cell recruitment. *Cancer Res.* **2009**, *69*, 3077–3085. [CrossRef]
22. Huang, B.; Han, W.; Sheng, Z.-F.; Shen, G.-L. Identification of immune-related biomarkers associated with tumorigenesis and prognosis in cutaneous melanoma patients. *Cancer Cell Int.* **2020**, *20*, 1–15. [CrossRef]
23. Dhar, P.; Wu, J.D. NKG2D and its ligands in cancer. *Curr. Opin. Immunol.* **2018**, *51*, 55–61. [CrossRef] [PubMed]
24. Sharma, P.; Callahan, M.K.; Bono, P.; Kim, J.; Spiliopoulou, P.; Calvo, E.; Pillai, R.N.; Ott, P.A.; de Braud, F.; Morse, M.; et al. Nivolumab monotherapy in recurrent metastatic urothelial carcinoma (CheckMate 032): A multicentre, open-label, two-stage, multi-arm, phase 1/2 trial. *Lancet Oncol.* **2016**, *17*, 1590–1598. [CrossRef]
25. Sharma, P.; Retz, M.; Siefker-Radtke, A.; Baron, A.; Necchi, A.; Bedke, J.; Plimack, E.R.; Vaena, D.; Grimm, M.-O.; Bracarda, S.; et al. Nivolumab in metastatic urothelial carcinoma after platinum therapy (CheckMate 275): A multicentre, single-arm, phase 2 trial. *Lancet Oncol.* **2017**, *18*, 312–322. [CrossRef]

26. Bellmunt, J.; de Wit, R.; Vaughn, D.J.; Fradet, Y.; Lee, J.-L.; Fong, L.; Vogelzang, N.J.; Climent, M.A.; Petrylak, D.P.; Choueiri, T.K.; et al. Pembrolizumab as Second-Line Therapy for Advanced Urothelial Carcinoma. *N. Engl. J. Med.* **2017**, *376*, 1015–1026. [CrossRef]
27. Vuky, J.; Balar, A.V.; Castellano, D.; O'Donnell, P.H.; Grivas, P.; Bellmunt, J.; Powles, T.; Bajorin, D.; Hahn, N.M.; Savage, M.J.; et al. Long-Term Outcomes in KEYNOTE-052: Phase II Study Investigating First-Line Pembrolizumab in Cisplatin-Ineligible Patients With Locally Advanced or Metastatic Urothelial Cancer. *J. Clin. Oncol.* **2020**, *38*, 2658–2666. [CrossRef] [PubMed]
28. Alexandrov, L.B.; Nik-Zainal, S.; Wedge, D.C.; Aparicio, S.A.J.R.; Behjati, S.; Biankin, A.V.; Bignell, G.R.; Bolli, N.; Borg, A.; Børresen-Dale, A.-L.; et al. Signatures of mutational processes in human cancer. *Nature* **2013**, *500*, 415–421. [CrossRef]
29. Castro, A.; Pyke, R.M.; Zhang, X.; Thompson, W.K.; Day, C.-P.; Alexandrov, L.B.; Zanetti, M.; Carter, H. Strength of immune selection in tumors varies with sex and age. *Nat. Commun.* **2020**, *11*, 1–9. [CrossRef]
30. Conforti, F.; Pala, L.; Bagnardi, V.; Pas, T.D.; Martinetti, M.; Viale, G.; Gelber, R.D.; Goldhirsch, A. Cancer immunotherapy efficacy and patients' sex: A systematic review and meta-analysis. *Lancet Oncol.* **2018**, *19*, 737–746. [CrossRef]
31. Gupta, S.; Artomov, M.; Goggins, W.; Daly, M.; Tsao, H. Gender Disparity and Mutation Burden in Metastatic Melanoma. *JNCI J. Natl. Cancer Inst.* **2015**, *107*. [CrossRef] [PubMed]
32. Xiao, D.; Pan, H.; Li, F.; Wu, K.; Zhang, X.; He, J. Analysis of ultra-deep targeted sequencing reveals mutation burden is associated with gender and clinical outcome in lung adenocarcinoma. *Oncotarget* **2016**, *7*, 22857–22864. [CrossRef]
33. Allen, E.M.V.; Miao, D.; Schilling, B.; Shukla, S.A.; Blank, C.; Zimmer, L.; Sucker, A.; Hillen, U.; Foppen, M.H.G.; Goldinger, S.M.; et al. Genomic correlates of response to CTLA-4 blockade in metastatic melanoma. *Science* **2015**, *350*, 207–211. [CrossRef]
34. Rozeman, E.A.; Hoefsmit, E.P.; Reijers, I.L.M.; Saw, R.P.M.; Versluis, J.M.; Krijgsman, O.; Dimitriadis, P.; Sikorska, K.; van de Wiel, B.A.; Eriksson, H.; et al. Survival and biomarker analyses from the OpACIN-neo and OpACIN neoadjuvant immunotherapy trials in stage III melanoma. *Nat. Med.* **2021**, *27*, 256–263. [CrossRef] [PubMed]
35. Renkema, K.R.; Li, G.; Wu, A.; Smithey, M.J.; Nikolich-Žugich, J. Two separate defects affecting true naive or virtual memory T cell precursors combine to reduce naive T cell responses with aging. *J. Immunol. 1950* **2014**, *192*, 151–159. [CrossRef] [PubMed]
36. Briceño, O.; Lissina, A.; Wanke, K.; Afonso, G.; von Braun, A.; Ragon, K.; Miquel, T.; Gostick, E.; Papagno, L.; Stiasny, K.; et al. Reduced naïve CD8+ T-cell priming efficacy in elderly adults. *Aging Cell* **2016**, *15*, 14–21. [CrossRef]
37. Agrawal, A.; Agrawal, S.; Gupta, S. Dendritic cells in human aging. *Exp. Gerontol.* **2007**, *42*, 421–426. [CrossRef] [PubMed]
38. Kugel, C.H.; Douglass, S.M.; Webster, M.R.; Kaur, A.; Liu, Q.; Yin, X.; Weiss, S.A.; Darvishian, F.; Al-Rohil, R.N.; Ndoye, A.; et al. Age Correlates with Response to Anti-PD1, Reflecting Age-Related Differences in Intratumoral Effector and Regulatory T-Cell Populations. *Clin. Cancer Res.* **2018**, *24*, 5347–5356. [CrossRef] [PubMed]
39. Bartlett, E.K.; Flynn, J.R.; Panageas, K.S.; Ferraro, R.A.; Sta. Cruz, J.M.; Postow, M.A.; Coit, D.G.; Ariyan, C.E. High neutrophil-to-lymphocyte ratio (NLR) is associated with treatment failure and death in patients who have melanoma treated with PD-1 inhibitor monotherapy. *Cancer* **2020**, *126*, 76–85. [CrossRef] [PubMed]
40. Matsukane, R.; Watanabe, H.; Minami, H.; Hata, K.; Suetsugu, K.; Tsuji, T.; Masuda, S.; Okamoto, I.; Nakagawa, T.; Ito, T.; et al. Continuous monitoring of neutrophils to lymphocytes ratio for estimating the onset, severity, and subsequent prognosis of immune related adverse events. *Sci. Rep.* **2021**, *11*, 1–11. [CrossRef] [PubMed]
41. Cornen, S.; Vivier, E. Chemotherapy and tumor immunity. *Science* **2018**, *362*, 1355–1356. [CrossRef]
42. Ruscetti, M.; Leibold, J.; Bott, M.J.; Fennell, M.; Kulick, A.; Salgado, N.R.; Chen, C.-C.; Ho, Y.-J.; Sanchez-Rivera, F.J.; Feucht, J.; et al. NK cell-mediated cytotoxicity contributes to tumor control by a cytostatic drug combination. *Science* **2018**, *362*, 1416–1422. [CrossRef]
43. Zhang, P.; Su, D.-M.; Liang, M.; Fu, J. Chemopreventive agents induce programmed death-1-ligand 1 (PD-L1) surface expression in breast cancer cells and promote PD-L1-mediated T cell apoptosis. *Mol. Immunol.* **2008**, *45*, 1470–1476. [CrossRef]
44. Qin, X.; Liu, C.; Zhou, Y.; Wang, G. Cisplatin induces programmed death-1-ligand 1(PD-L1) over-expression in hepatoma H22 cells via Erk /MAPK signaling pathway. *Cell. Mol. Biol.* **2010**, *56*, 1366–1372.
45. Brochez, L.; Chevolet, I.; Kruse, V. The rationale of indoleamine 2,3-dioxygenase inhibition for cancer therapy. *Eur. J. Cancer* **2017**, *76*, 167–182. [CrossRef]
46. Huang, A.C.; Orlowski, R.J.; Xu, X.; Mick, R.; George, S.M.; Yan, P.K.; Manne, S.; Kraya, A.A.; Wubbenhorst, B.; Dorfman, L.; et al. A single dose of neoadjuvant PD-1 blockade predicts clinical outcomes in resectable melanoma. *Nat. Med.* **2019**, *25*, 454–461. [CrossRef] [PubMed]
47. Huang, A.C.; Postow, M.A.; Orlowski, R.J.; Mick, R.; Bengsch, B.; Manne, S.; Xu, W.; Harmon, S.; Giles, J.R.; Wenz, B.; et al. T-cell invigoration to tumour burden ratio associated with anti-PD-1 response. *Nature* **2017**, *545*, 60–65. [CrossRef] [PubMed]
48. Kamphorst, A.O.; Pillai, R.N.; Yang, S.; Nasti, T.H.; Akondy, R.S.; Wieland, A.; Sica, G.L.; Yu, K.; Koenig, L.; Patel, N.T.; et al. Proliferation of PD-1+ CD8 T cells in peripheral blood after PD-1-targeted therapy in lung cancer patients. *Proc. Natl. Acad. Sci. USA* **2017**, *114*, 4993–4998. [CrossRef] [PubMed]
49. Jacquelot, N.; Roberti, M.P.; Enot, D.P.; Rusakiewicz, S.; Ternès, N.; Jegou, S.; Woods, D.M.; Sodré, A.L.; Hansen, M.; Meirow, Y.; et al. Predictors of responses to immune checkpoint blockade in advanced melanoma. *Nat. Commun.* **2017**, *8*, 1–13. [CrossRef]
50. Brochez, L.; Meireson, A.; Chevolet, I.; Sundahl, N.; Ost, P.; Kruse, V. Challenging PD-L1 expressing cytotoxic T cells as a predictor for response to immunotherapy in melanoma. *Nat. Commun.* **2018**, *9*, 1–3. [CrossRef] [PubMed]
51. Zhao, Y.; Harrison, D.L.; Song, Y.; Ji, J.; Huang, J.; Hui, E. Antigen-Presenting Cell-Intrinsic PD-1 Neutralizes PD-L1 in cis to Attenuate PD-1 Signaling in T Cells. *Cell Rep.* **2018**, *24*, 379–390. [CrossRef] [PubMed]

52. Krishnamoorthy, M.; Lenehan, J.G.; Maleki Vareki, S. Neoadjuvant Immunotherapy for High-Risk, Resectable Malignancies: Scientific Rationale and Clinical Challenges. *J. Natl. Cancer Inst.* **2021**. [CrossRef] [PubMed]
53. Kinter, A.L.; Godbout, E.J.; McNally, J.P.; Sereti, I.; Roby, G.A.; O'Shea, M.A.; Fauci, A.S. The common gamma-chain cytokines IL-2, IL-7, IL-15, and IL-21 induce the expression of programmed death-1 and its ligands. *J. Immunol.* **2008**, *181*, 6738–6746. [CrossRef]
54. Dammeijer, F.; van Gulijk, M.; Mulder, E.E.; Lukkes, M.; Klaase, L.; van den Bosch, T.; van Nimwegen, M.; Lau, S.P.; Latupeirissa, K.; Schetters, S.; et al. The PD-1/PD-L1-Checkpoint Restrains T cell Immunity in Tumor-Draining Lymph Nodes. *Cancer Cell* **2020**, *38*, 685–700. [CrossRef] [PubMed]
55. Peng, Q.; Qiu, X.; Zhang, Z.; Zhang, S.; Zhang, Y.; Liang, Y.; Guo, J.; Peng, H.; Chen, M.; Fu, Y.-X.; et al. PD-L1 on dendritic cells attenuates T cell activation and regulates response to immune checkpoint blockade. *Nat. Commun.* **2020**, *11*, 1–8. [CrossRef]
56. Lin, H.; Wei, S.; Hurt, E.M.; Green, M.D.; Zhao, L.; Vatan, L.; Szeliga, W.; Herbst, R.; Harms, P.W.; Fecher, L.A.; et al. Host expression of PD-L1 determines efficacy of PD-L1 pathway blockade-mediated tumor regression. *J. Clin. Investig.* **2018**, *128*, 805–815. [CrossRef]
57. Chen, G.; Huang, A.C.; Zhang, W.; Zhang, G.; Wu, M.; Xu, W.; Yu, Z.; Yang, J.; Wang, B.; Sun, H.; et al. Exosomal PD-L1 contributes to immunosuppression and is associated with anti-PD-1 response. *Nature* **2018**, *560*, 382–386. [CrossRef]
58. Sunshine, J.; Taube, J.M. PD-1/PD-L1 inhibitors. *Curr. Opin. Pharmacol.* **2015**, *23*, 32–38. [CrossRef]
59. Mukherjee, N.; Ji, N.; Hurez, V.; Curiel, T.J.; Montgomery, M.O.; Braun, A.J.; Nicolas, M.; Aguilera, M.; Kaushik, D.; Liu, Q.; et al. Intratumoral CD56bright natural killer cells are associated with improved survival in bladder cancer. *Oncotarget* **2018**, *9*, 36492–36502. [CrossRef]
60. Meireson, A.; Devos, M.; Brochez, L. IDO Expression in Cancer: Different Compartment, Different Functionality? *Front. Immunol.* **2020**, *11*, 2340. [CrossRef]
61. Hayashi, H.; Nakagawa, K. Combination therapy with PD-1 or PD-L1 inhibitors for cancer. *Int. J. Clin. Oncol.* **2020**, *25*, 818–830. [CrossRef] [PubMed]
62. Rozeman, E.A.; Menzies, A.M.; van Akkooi, A.C.J.; Adhikari, C.; Bierman, C.; van de Wiel, B.A.; Scolyer, R.A.; Krijgsman, O.; Sikorska, K.; Eriksson, H.; et al. Identification of the optimal combination dosing schedule of neoadjuvant ipilimumab plus nivolumab in macroscopic stage III melanoma (OpACIN-neo): A multicentre, phase 2, randomised, controlled trial. *Lancet Oncol.* **2019**, *20*, 948–960. [CrossRef]
63. van Dijk, N.; Gil-Jimenez, A.; Silina, K.; Hendricksen, K.; Smit, L.A.; de Feijter, J.M.; van Montfoort, M.L.; van Rooijen, C.; Peters, D.; Broeks, A.; et al. Preoperative ipilimumab plus nivolumab in locoregionally advanced urothelial cancer: The NABUCCO trial. *Nat. Med.* **2020**, *26*, 1839–1844. [CrossRef] [PubMed]
64. Wei, S.C.; Anang, N.-A.A.S.; Sharma, R.; Andrews, M.C.; Reuben, A.; Levine, J.H.; Cogdill, A.P.; Mancuso, J.J.; Wargo, J.A.; Pe'er, D.; et al. Combination anti–CTLA-4 plus anti–PD-1 checkpoint blockade utilizes cellular mechanisms partially distinct from monotherapies. *Proc. Natl. Acad. Sci. USA* **2019**, *116*, 22699–22709. [CrossRef] [PubMed]
65. Woods, D.M.; Laino, A.S.; Winters, A.; Alexandre, J.; Freeman, D.; Rao, V.; Adavani, S.S.; Weber, J.S.; Chattopadhyay, P.K. Nivolumab and ipilimumab are associated with distinct immune landscape changes and response-associated immunophenotypes. *JCI Insight* **2020**, *5*, e137066.

Article

Immune Monitoring during Therapy Reveals Activitory and Regulatory Immune Responses in High-Risk Neuroblastoma

Celina L. Szanto [1,2,†], Annelisa M. Cornel [1,2,†], Sara M. Tamminga [2], Eveline M. Delemarre [2], Coco C. H. de Koning [1,2], Denise A. M. H. van den Beemt [1,2], Ester Dunnebach [1,2], Michelle L. Tas [1], Miranda P. Dierselhuis [1], Lieve G. A. M. Tytgat [1], Max M. van Noesel [1,3], Kathelijne C. J. M. Kraal [1], Jaap-Jan Boelens [4], Alwin D. R. Huitema [1,5,6] and Stefan Nierkens [1,2,*]

1. Princess Máxima Center for Pediatric Oncology, Utrecht University, 3584 CS Utrecht, The Netherlands; C.L.Szanto-2@prinsesmaximacentrum.nl (C.L.S.); a.m.cornel@umcutrecht.nl (A.M.C.); C.C.H.deKoning@umcutrecht.nl (C.C.H.d.K.); D.A.M.vandenBeemt@umcutrecht.nl (D.A.M.H.v.d.B.); E.Dunnebach-2@umcutrecht.nl (E.D.); M.Tas@prinsesmaximacentrum.nl (M.L.T.); M.P.Dierselhuis@prinsesmaximacentrum.nl (M.P.D.); G.A.M.Tytgat@prinsesmaximacentrum.nl (L.G.A.M.T.); M.M.vanNoesel@prinsesmaximacentrum.nl (M.M.v.N.); K.C.J.Kraal@prinsesmaximacentrum.nl (K.C.J.M.K.); A.D.R.Huitema-2@umcutrecht.nl (A.D.R.H.)
2. Center for Translational Immunology, University Medical Center Utrecht, Utrecht University, 3584 CX Utrecht, The Netherlands; s.m.tamminga@amsterdamumc.nl (S.M.T.); E.M.Delemarre@umcutrecht.nl (E.M.D.)
3. Division Imaging and Cancer, University Medical Center Utrecht, Utrecht University, 3584 CX Utrecht, The Netherlands
4. Stem Cell Transplantation and Cellular Therapies Program, Department of Pediatrics, Memorial Sloan Kettering Cancer Center, New York, NY 10065, USA; boelensj@mskcc.org
5. Department of Pharmacy and Pharmacology, Netherlands Cancer Institute, 1066 CX Amsterdam, The Netherlands
6. Department of Clinical Pharmacy, University Medical Center Utrecht, Utrecht University, 3584 CX Utrecht, The Netherlands
* Correspondence: S.Nierkens-2@prinsesmaximacentrum.nl
† These authors contributed equally to this work.

Simple Summary: Neuroblastoma is a type of childhood cancer accounting for approximately 15% of childhood cancer deaths. Despite intensive treatment, including immunotherapy, prognosis of high-risk neuroblastoma is poor. Increasing amounts of research show that the fighting capacity of the immune system is very important for the outcome of neuroblastoma patients. Therefore, we investigated the fighting capacity of immune cells in blood at diagnosis and during the different phases of therapy. In this study, we observed both processes that stimulate and processes that decrease fighting capacity of immune cells in neuroblastoma patients during therapy. Despite this, we show that overall fighting capacity of the immune system of neuroblastoma patients is impaired at diagnosis as well as during therapy. In addition, we observed a lot of variation between patients, which might explain differences in therapy efficacy between patients. This study provides insight for improvement of therapy timing as well as new therapy strategies enhancing immune cell fighting capacity.

Abstract: Despite intensive treatment, including consolidation immunotherapy (IT), prognosis of high-risk neuroblastoma (HR-NBL) is poor. Immune status of patients over the course of treatment, and thus immunological features potentially explaining therapy efficacy, are largely unknown. In this study, the dynamics of immune cell subsets and their function were explored in 25 HR-NBL patients at diagnosis, during induction chemotherapy, before high-dose chemotherapy, and during IT. The dynamics of immune cells varied largely between patients. IL-2- and GM-CSF-containing IT cycles resulted in significant expansion of effector cells (NK-cells in IL-2 cycles, neutrophils and monocytes in GM-CSF cycles). Nonetheless, the cytotoxic phenotype of NK-cells was majorly disturbed at the start of IT, and both IL-2 and GM-CSF IT cycles induced preferential expansion of suppressive regulatory T-cells. Interestingly, proliferative capacity of purified patient T-cells was impaired at diagnosis as well as during therapy. This study indicates the presence of both immune-enhancing as

well as regulatory responses in HR-NBL patients during (immuno)therapy. Especially the double-edged effects observed in IL-2-containing IT cycles are interesting, as this potentially explains the absence of clinical benefit of IL-2 addition to IT cycles. This suggests that there is a need to combine anti-GD2 with more specific immune-enhancing strategies to improve IT outcome in HR-NBL.

Keywords: neuroblastoma; immune monitoring; anti-GD2; IL-2; GM-CSF; ASCT; immunotherapy; dinutuximab

1. Introduction

Neuroblastoma (NBL) is the most common extracranial solid tumor in children, accounting for approximately 15% of all pediatric oncology deaths [1]. Patients are stratified as low, intermediate or high risk (HR), depending on various factors (e.g., age, tumor stage, and several genetic components, such as MYCN amplification) [2]. HR-NBL patients are treated with multimodal therapy consisting of chemotherapy, high-dose chemotherapy followed by autologous stem cell transplantation (ASCT), resection of the tumor, local radiation, and maintenance immunotherapy (IT) consisting of the anti-GD2 monoclonal antibody, often combined with the cytokines IL-2 and GM-CSF, and isotretinoin acid [3–5]. Despite intensive treatment, 5 year event-free survival (EFS) is <50% [6,7].

Dinutuximab, the monoclonal antibody used in NBL IT, targets GD2 on the surface of NBL cells and signals antibody-dependent cell-mediated cytotoxicity (ADCC) and complement-dependent cytotoxicity (CDC) [3]. The rationale to alternately add GM-CSF and IL-2 to the IT cycles was to increase expansion and functional activity of natural killer (NK) cells, lymphocytes, monocytes/macrophages, and neutrophils. This was mainly supported by in vitro data indicating superior cytotoxic effects when combining dinutuximab with these cytokines [8,9]. Even though IT increased 2 year EFS and overall survival (OS) [3], relapses are still observed in the majority of patients.

The dose, timing, and chosen immunotherapeutic compound combinations are currently highly empirical and do not take patients' immune status into account. Fast immune reconstitution during chemotherapy and higher absolute lymphocyte and monocyte counts have been associated with improved overall outcome in multiple cancers [10–12]. Nassin et al. showed that most patients with HR-NBL do not have full immune reconstitution at the start of IT (based on total white blood cell count (WBC), hemoglobin, and platelet, absolute neutrophil, lymphocyte and monocyte counts) and that immune recovery may correlate with disease-related outcomes [13]. Relatively fast NK-cell recovery early after ASCT was an important rationale for timing of IT early after transplantation [14]. Nonetheless, more detailed evaluation of NK-cell subsets showed that most cells are immature, cytokine-releasing (CD56bright, CD16+/−) rather than cytotoxic (CD56dim, CD16+). This may suggest suboptimal timing of dinutuximab IT early after transplantation, as cytotoxic NK-cells are mainly responsible for anti-GD2-dependent ADCC [13]. Nonetheless, to date, the potential effect of the IT regimen on shifting to the mature NK-cell phenotype has not been addressed.

Another important observation came from a phase III clinical trial where no additive effect of IL-2 administration on outcome of high-risk NBL patients was observed [4]. It is hypothesized that this may be the result of masking of the positive effects of IL-2 (e.g., on NK-cell expansion and functionality) by preferential regulatory T-cell (Treg) expansion [4,13], an effect known when administering (low dose) IL-2 to patients with autoimmune diseases [15]. Nevertheless, studies addressing this observation during NBL IT are lacking.

It may be hypothesized that post-ASCT immune reconstitution occurs with disparate kinetics in different patients, which may affect treatment efficacy of immune-targeting therapy. Comprehensive understanding of the status of the immune system in these patients may be instrumental for further development of immunotherapeutic interventions after ASCT. However, no studies have monitored the immune status in NBL patients during chemotherapy and IT and included functional analysis. Therefore, we monitored the immune status in NBL patients during chemo- and immunotherapy. In addition, the effect of IL-2 and GM-CSF on leukocyte and lymphocyte subpopulations and their (effector) cell functions during IT were studied.

2. Materials and Methods

2.1. Patients and Treatment

HR-NBL patients diagnosed between January 2015 and January 2018 treated in the Princess Máxima Center for Pediatric Oncology (Utrecht, The Netherlands) or Uniklinik Köln (Cologne, Germany) were included in this study. Patients were treated following the same treatment protocol based on N5/N6 chemotherapy (Dutch NBL2009 trial [16] and NB2013-HR pilot GPOH/DCOG trial; N5 = cisplatin, etoposide, vindesine, N6 = vincristine, dacarbacin, ifosfamide, doxorubicin). Staging was performed according to the International NBL Staging System (INSS) [17]. MYCN and ALK amplification status was determined with FISH, SNP-array was used for the determination of CNVs in 1p and 17q. The study was approved by the Medical Ethical Committees (Academic Medical Center, Amsterdam, the Netherlands; NL50762.018.14 and the University of Cologne, German trial 2013-004481-34). Written informed consent was obtained from the parents or guardians before enrollment in accordance with the Declaration of Helsinki.

2.2. Sample Collection

Peripheral blood samples (EDTA) were transported to the laboratory at room temperature (RT), and a Trucount cell subset enumeration tube was analyzed using flow cytometry within 24 h after blood withdrawal. Plasma was isolated after centrifugation and stored at -80 °C until analysis. Peripheral blood mononuclear cells (PBMCs) were isolated using Ficoll density gradient centrifugation, frozen in fetal calf serum (Bodinco, Alkmaar, The Netherlands) containing 10% dimethyl sulphoxide (Sigma-Aldrich, St. Louis, MO, USA), and stored in liquid nitrogen in the UMCU biobank until use in experiments. Frozen control donor PBMCs, taken from healthy adult volunteers, served as control group.

In Utrecht, peripheral blood samples were taken at diagnosis (1 sample from 7 patients), after each N5/N6 cycle (1–3 samples from 18 patients), before the high-dose (HD) chemotherapy regimen (1 sample from 7 patients), at start of IT (1 sample from 7 patients) and after 3 and 6 cycles of IT (1–2 samples from 8 patients) as depicted in Figure S1. In Cologne, peripheral blood samples were taken at start of IT and every 2 weeks during IT cycle 1–5. Samples were shipped at RT to the laboratory in Utrecht and processed within 24 h as described above.

2.3. Treg and NK-Cell Phenotyping

PBMCs were thawed and stained with either Treg or NK-cell discriminating antibodies. The Treg panel was comprised of the following extracellular antibodies: CD3-AF700, CD4-eFluor780, CD8-PE-Cy7, CD25-PE, CD127-BV421, CD45RO-BV711 (Biolegend, Biolegends, Koblenz; Germany). For intracellular staining, cells were permeabilized after extracellular staining, using the eBioscience kit (Thermo Fisher Scientific, Darmstadt, Germany) and stained for FOXP3 expression. The NK-cell panel comprised of CD3-AF700, CD19-eFluor780, CD56-PE-Cy7, CD16-BV510, CD45RO-BV711, TCRVα24-PE, TCRVβ11-FITC (Biolegend). All samples were measured within 24 h after staining on a BD LSR Fortessa (BD Biosciences, Heidelberg, Germany). All flow cytometry data were analyzed with FlowJo software version 10.6.0 (Tree Star, Ashland, OR, USA). Output CSV documents were further analyzed using RStudio (version 1.2.1335).

2.4. Proliferation Assay

To assess proliferation of T-cells, PBMCs were thawed, labelled with Celltrace Violet (CTV) (ThermoFisher Scientific)) and cultured in a round-bottom 96-well plate for 3 days at 37 °C and 5% CO_2. 25,000 PBMCs were cultured in duplicates in the presence of anti-CD3 (0.5 μg/mL, 16-0037-81; ThermoFisher Scientific), or without stimuli. On day 3, supernatants were collected (pooled from duplos) and stored (as described in Section 2.6). Proliferation of PBMCs was analyzed using flow cytometry.

2.5. Suppression Assay

Patient and healthy-donor (HD) CD4+CD25highCD127low Tregs were sorted using BD FACSAria™. Tregs were added to CTV-labelled effector cells at an effector-to-target ratio (E:T) of 2:1 in a crossover manner: (1) Tregs patient + effector cells patient; (2) Tregs patient + effector cells HD; (3) Tregs HD + effector cells patient; (4) Tregs HD + effector cells HD. Then, 96-well plates were coated with anti-CD3 (16-0037-81; ThermoFisher) to provide a proliferation stimulus. At day 3, the proliferation of effector cells was analyzed with flow cytometry.

2.6. Protein Profiling

Supernatant from the proliferation assays was collected after 3 days of culture, and stored at −80 °C until cytokine measurement. Interferon-γ (IFN-γ), tumor necrosis factor α (TNF-α), soluble IL-2R, IL-2, IL-10, IL-13, and IL-17 were measured using multiplex immunoassays (Luminex Technology, Austin, TX, USA). The multiplex immunoassay was performed as described previously by the MultiPlex Core Facility (MPCF) of the UMCU [18]. Out-of-range (OOR</OOR>) and extrapolated values were systematically replaced using the following procedure. The LLOQ (lower limit of quantification) and ULOQ (upper limit of quantification) were retrieved for the measured analytes of the experiment. The LLOQ and ULOQ values were retrieved per analyte by the MPCF. The lowest measurement was compared with LLOQ for each marker, to retrieve the lowest values for all measured markers. The same was performed for the highest value. OOR< data were replaced by the lowest value divided by 2. OOR> data were replaced by highest value times 2. The same procedure was performed for extrapolated data. For some markers, there are no LLOQ and ULOQ obtained yet. In that case, the lowest and highest measurements within the experiment were used for the replacement of OOR and extrapolated data.

Plasma samples were analyzed using the Proseek Multiplex Immuno-oncology immunoassay panel (Olink Biosciences, Uppsala, Sweden). Proseek is a high-throughput multiplex immunoassay based on proximity extension assay (PEA) technology that enables the analysis of 92 immuno-oncology-related biomarkers simultaneously. In short, PEA technology makes use of antibody pairs linked with matching DNA-oligonucleotides per protein of interest. These oligonucleotides hybridize when brought into proximity after binding the protein and are extended by DNA polymerase, thereby forming PCR targets. These targets are quantified by real-time PCR. Obtained results are expressed in normalized protein expression (NPX) values, which are in a log2 scale.

2.7. Statistics

Statistical analysis of absolute cell numbers and Treg expansion during IT was performed using the Mann–Whitney U test, comparing differences between groups before and after administration of IL-2 and GM-CSF. Hierarchical clustering analyses, presented as heatmaps, were based on Ward's method and pairwise correlation distance. Heatmaps were generated using the heatmap.2 function from the gplots package [19]. To identify significant differences between protein levels before and after IL-2 and GM-CSF IT cycles, the Wilcoxon signed rank test was performed with correction for multiple testing according to Benjamini and Hochberg [20] for IL-2 cycles and the Mann–Whitney test with correction for multiple testing [20] for GM-CSF cycles. RStudio Project Software (version 1.2.1335) [21] was used for statistical analyses. Adjusted p-values of < 0.05 were considered statistically significant.

3. Results

3.1. Patient Characteristics

Twenty-five patients were included in this study (Table 1) with a median age at diagnosis of 3.9 years (range 0.3–10.8). A slight majority (56%, $n = 14$) had at least a partial response after induction chemotherapy. These patients continued therapy following the HR treatment protocol. Nonresponders (44%, $n = 11$) received additional chemotherapy (2–4 N8 cycles (etoposide, topotecan, cyclophosphamide)), and 14% ($n = 4$) received ^{131}I-metaiodobenzylguanidine (^{131}I-MIBG) therapy. Twenty out of 25 patients received HD chemotherapy followed by ASCT, seventy percent ($n = 14/20$) of patients received HD busulfan and melphalan (Bu-Mel) and 30% ($n = 7/20$) received HD carboplatin, etoposide, and melphalan. Following ASCT, 80% ($n = 6/20$) received dinutuximab IT in combination with cytokines. The four patients who did not receive IT had progressive disease. The mean time from ASCT to start IT was 137 days (range 108–193 days). The median time of follow-up for surviving patients was 2.14 years (range 0.65–3.67). The median event-free survival (EFS) was 1.65 years (range 0.11–3.67).

Table 1. Patient characteristics and time of sampling.

Patient Characteristics	Total ($n = 25$)
Gender	
male	14 (56%)
female	11 (44%)
Median age at diagnosis, year, (range)	3.9 (0.3–10.8)
Stage 3 disease	1 (4%)
Stage 4 disease	24 (96%)
Genetics	
MYCN	
Neg	18 (72%)
Gain	2 (8%)
Amp	5 (20%)
1p	
normal	14 (56%)
partial loss	9 (36%)
loss	1 (4%)
gain	1 (4%)
17q	
normal	1 (4%)
partial gain	10 (40%)
gain	11 (44%)
unknown	3 (12%)
ALK mutation	
Yes	5 (20%)
no	16 (64%)
gain	1 (4%)
unknown	3 (12%)
CR or PR after induction chemotherapy ($3\times$ N5/N6)	14 (56%)
HD + ASCT	20 (80%)
Conditioning Regimen	
Busulfan/melphalan	14/20 (70%)
Carboplatin/etoposide/melphalan	6/20 (30%)
CD34+ cell dose $\times 10^6$/kg, (range)	2.47 (0.59–21.73)
Immunotherapy	16 (64%)
Time to immunotherapy, d, (range)	137 (108–193)
Event: progression or relapse	7 (31%)
Event: Refractory Disease	3 (14%)
Event: Toxicity	1 (5%)
Alive at last FU	14 (56%)
Median EFS, year (range)	1.65 (0.11–3.67)
Median follow-up OS, year, (range)	2.14 (0.65–3.67)

Abbreviations: CR, complete response; PR, partial response; HD, high-dose; ASCT, autologous stem cell transplantation; FU, follow-up; EFS, event-free survival; OS, overall survival.

3.2. Immune Profiles at Diagnosis, during Induction Chemotherapy, and before High-Dose Chemotherapy Show Broad Variation between Patients

In the period before ASCT, large variations were observed between patients and between treatment cycles within individual patients in absolute leukocyte, lymphocyte, monocyte, neutrophil, eosinophil and specific lymphocyte subsets (B-cells, NK-cells, and T-cells) (Figure 1). Absolute neutrophil counts fluctuated most, peaking after the first N5/N6 chemotherapy cycle. B-cells decreased after the first round of N5/N6 chemotherapy and remained low during chemotherapy. Absolute lymphocyte counts remained similar between patients, while NK-cells and T-cells showed a large variation between patients. No correlation was found between absolute lymphocyte counts and occurrence of an event or MYCN status.

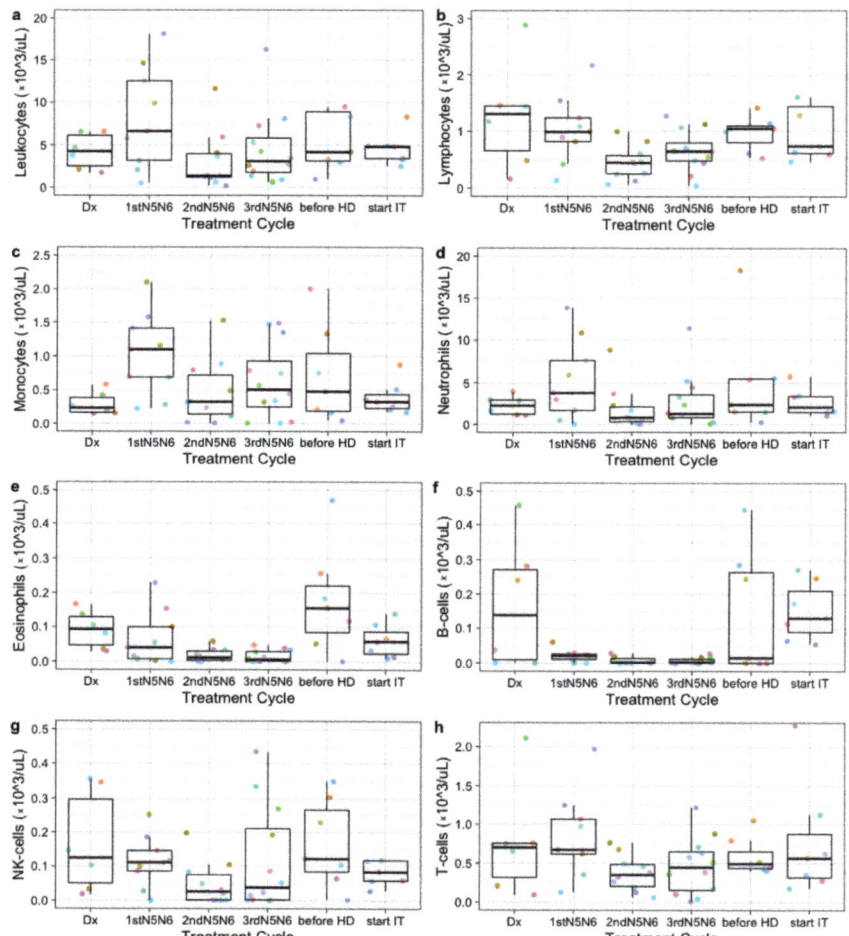

Figure 1. Immune profiles at diagnosis, during induction chemotherapy, and before high-dose conditioning. Each colored dot indicates absolute counts from one patient ($\times 10^3$/uL). Absolute leukocyte (**a**), lymphocyte (**b**), monocyte (**c**), neutrophil (**d**), eosinophil (**e**), B cell (**f**), NK cell (**g**), and T cell (**h**) numbers are shown at diagnosis (Dx), after the 1st, 2nd, and 3rd round of N5/N6 induction chemotherapy, before high-dose chemotherapy (before HD), and at start of immunotherapy (start IT) from 6, 9, 10, 12, 7, and 4 patients respectively.

3.3. Immune Profiles during Immunotherapy Show Effect of IL-2 and GM-CSF on Leukocyte and Lymphocyte Subsets

To determine whether the in vitro effects of IL-2 and GM-CSF are also observed in vivo, immune profiles were generated during IT. In concordance with the rationale, total lymphocyte counts increased significantly after IL-2-containing IT cycles ($p = 0.01$), due to an increase of NK-cells ($p < 0.01$) (Figure 2 and Figure S2). IL-2 had no effect on total CD3+ T-cells ($p = 0.67$), CD19+ B cells ($p = 0.70$), and monocytes ($p = 0.57$). Neutrophils decreased significantly after IL-2 administration ($p = 0.01$), while eosinophils showed a trend towards increased numbers in peripheral blood after IL-2 ($p = 0.19$).

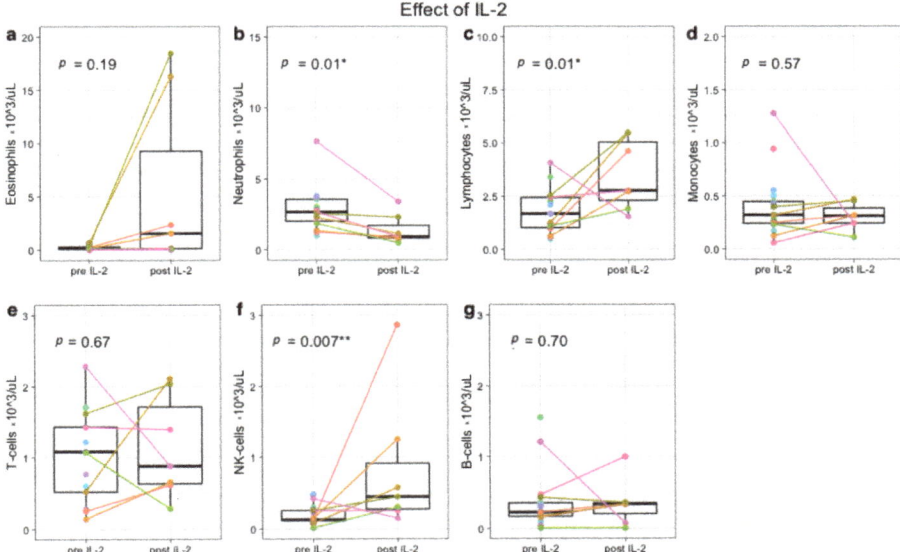

Figure 2. Immune profiles before and after IL-2-containing immunotherapy cycles. Each colored dot indicates absolute counts from one patient ($\times 10^3$ cells/uL). From 5 patients, samples were paired before IL-2 (day 1 IT cycle 2 or 4) and after IL-2 (day 15 IT cycle 2 or 4). In total, 7 paired samples are depicted (colored lines), because two patients were monitored in both IL-2 cycles. Nine single measurements from 9 other patients were included, resulting in a total of 14 patients (11 in study, 3 leftover material during IT). Absolute eosinophil (**a**), neutrophil (**b**), lymphocyte (**c**), monocyte (**d**), T-cell (**e**), NK-cell (**f**), and B-cell numbers (**g**) are shown. * $p < 0.05$, ** $p < 0.001$.

GM-CSF-containing IT cycles increased total lymphocytes ($p = 0.05$) and monocytes ($p = 0.03$), and a trend towards increased neutrophils ($p = 0.07$). GM-CSF had no effect on total CD3+ T-cell ($p = 0.28$), NK-cells ($p = 0.12$), and CD19+ B cells ($p = 0.19$) (Figure 3 and Figure S3). In addition, administration of GM-CSF resulted in a notable increase of eosinophils ($p < 0.001$).

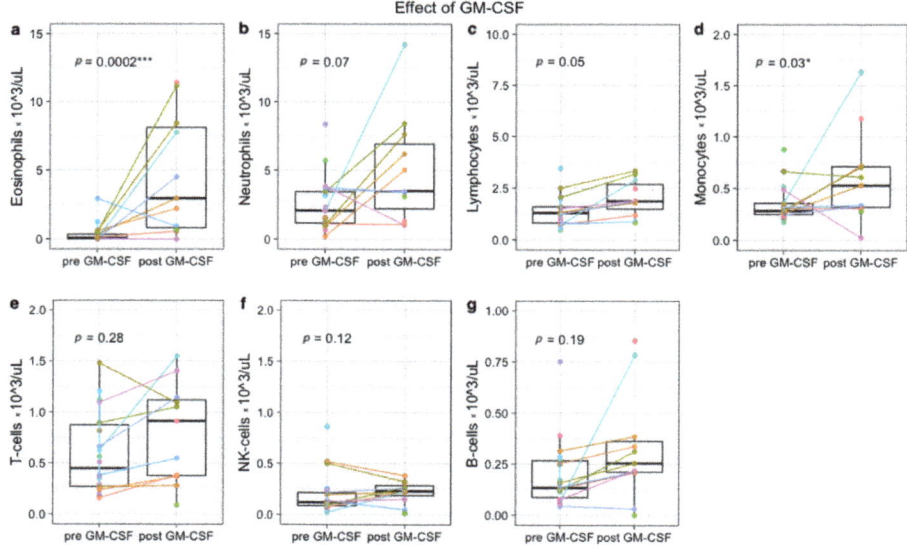

Figure 3. Immune profiles before and after GM-CSF-containing immunotherapy cycle. Each colored dot indicates absolute counts from one patient ($\times 10^3$ cells/uL). From 5 patients, samples were paired before GM-CSF (day 1 IT cycle 1, 3 or 5) and after GM-CSF (day 15 IT cycle 1, 3 or 5). In total, 9 paired samples are depicted (colored lines), because two patients were monitored during all 3 GM-CSF cycles. Twelve single measurements from 12 other patients were included, resulting in a total of 17 patients (11 in study, 6 left over material during IT). Absolute eosinophil (**a**), neutrophil (**b**), lymphocyte (**c**), monocyte (**d**), T-cell (**e**), NK-cell (**f**), and B-cell numbers (**g**) are shown. * $p < 0.05$, *** $p < 0.0001$.

3.4. Plasma Protein Profiling Further Supports IL-2 and GM-CSF Mediated Immune Engagement during Immunotherapy

Olink protein analysis was subsequently performed in plasma samples of 6 patients to determine protein profiles along the IT course. Protein profiling showed distinct patterns between pre- and post-IL-2 and pre- and post-GM-CSF-containing IT cycles. Unsupervised clustering resulted in complete separation of protein profiles pre- and post-IL-2-containing IT cycles (Figure S4A) and partial separation of protein profiles pre- and post-GM-CSF-containing IT cycles (Figure S4B).

Even though the sample sizes are too small to observe statistically significant differences upon IL-2-containing IT, increases can be observed in many NK-cell activation-associated markers, including GZMA/B/H, KIR3DL1, and NCR1 (all $p = 0.18$), IFN-γ ($p = 0.34$), CASP-8 ($p = 0.17$), and KLRD1 ($p = 0.32$) (Figure 4). Upon GM-CSF-containing IT cycles, significant increases in several neutrophil-, monocyte-, and eosinophil-associated factors, including CCL23 ($p = 0.046$), CCL17 ($p = 0.015$), CXCL11 ($p = 0.037$), and MCP-4 ($p = 0.015$) are observed (Figure 5).

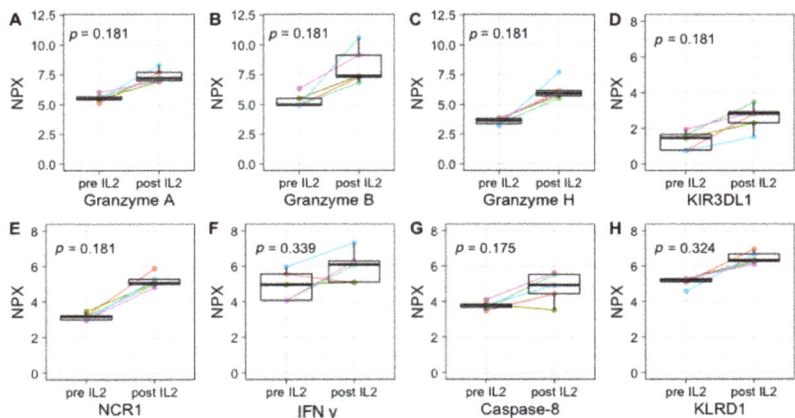

Figure 4. Upregulation of NK-cell activation-associated protein markers upon IL-2-containing immunotherapy cycles. Plasma protein concentration of GZMA/B/H ($p = 0.181$) (**A–C**), and KLRD1 ($p = 0.324$) (**D**), NCR1 ($p = 0.181$) (**E**), IFN-y ($p = 0.339$) (**F**), CASP-8 ($p = 0.175$) (**G**), and KLRD1 ($p = 0.324$) (**H**) pre- and post-IL-2-containing IT cycles. Protein expression is shown as normalized protein expression (NPX). In total, 5 paired samples are shown, as two patients were monitored during both IT cycles.

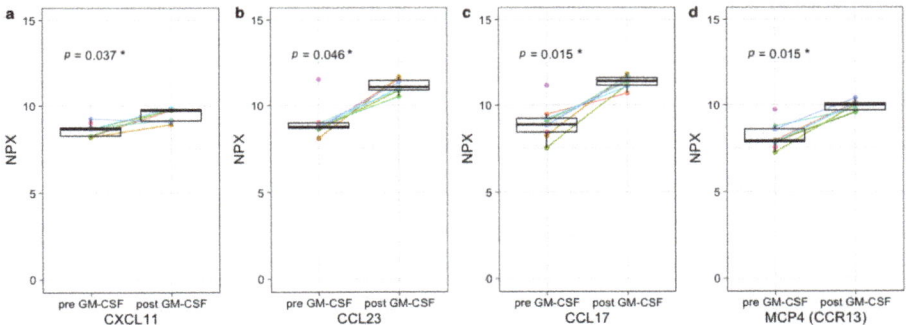

Figure 5. Upregulation of neutrophil-, monocyte-, and eosinophil-associated factors upon GM-CSF-containing immunotherapy cycles. Plasma protein concentration of CXCL11 ($p = 0.037$) (**a**), CCL23 ($p = 0.046$) (**b**), CCL17 ($p = 0.015$) (**c**), and MCP-4 (**d**) ($p = 0.015$) pre- and post-GM-CSF-containing IT cycles. Protein expression is shown as normalized protein expression (NPX). In total, 7 paired samples are shown, as two patients were monitored during all three IT cycles. Two single measurements from patients pre-GM-CSF were included, resulting in a total of 9 patients pre- and 7 post-GM-CSF. * $p < 0.05$.

3.5. NK-Cell Phenotype Varies Widely between Patients and Is Suboptimal for Efficient Dinutuximab-Mediated Cytotoxicity

As mentioned, the timing of IT in the NBL treatment protocol is established based on the observation of relatively fast NK-cell recovery early after ASCT [14]. Fast NK-cell recovery was observed based on absolute cell numbers (Figure 1g). However, even though variation is large, the balance between absolute numbers of mature, cytotoxic NK-cells ($CD56^{dim}CD16^{+}$) known to be mainly responsible for anti-GD2-dependent ADCC [13] and immature, cytokine-releasing NK-cells ($CD56^{bright}CD16^{-}$) was majorly disturbed at diagnosis and during all phases of the treatment protocol [22] (Figure 6a).

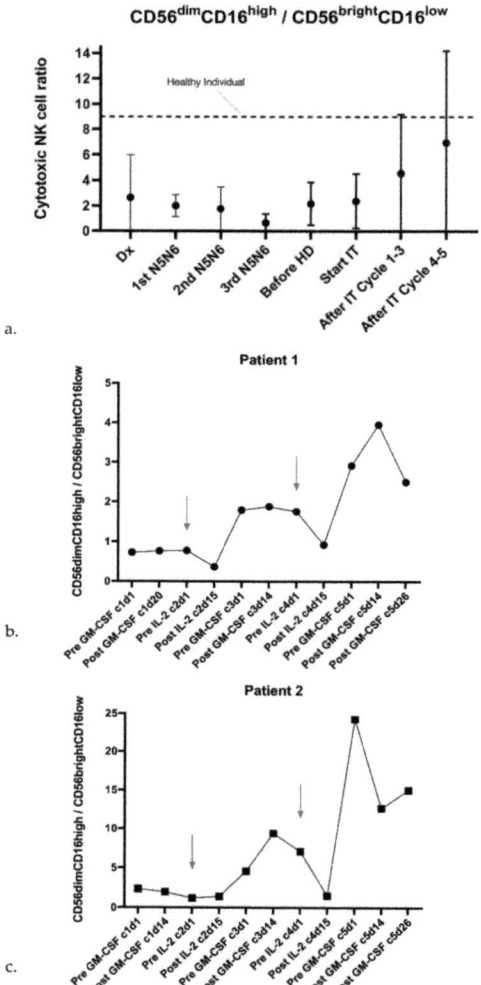

Figure 6. The cytotoxic $CD56^{dim}CD16^+/CD56^{bright}CD16^-$ NK-cell ratio during HR-NBL therapy. (**a**) The ratio of absolute CD56dimCD16+ and CD56brightCD16- Trucount cell numbers is highly variable between patients and is decreased at diagnosis and during therapy of HR-NBL patients. Dx: $n = 7$, 1st N5/N6: $n = 11$, 2nd N5/N6: $n = 10$; 3rd N5/N6: $n = 11$, before HD: $n = 7$, start IT: $n = 7$, After IT Cycle 1–3: $n = 10$, After IT cycle 4–5: $n = 8$. The dotted line reflects the reference value of the cytotoxic NK-cell ratio of healthy individuals [22]. (**b**,**c**) In-depth monitoring of the fraction of $CD56^{dim}CD16^+$ and $CD56^{bright}CD16^-$ in two patients during the IT course shows an increase in cytotoxic ($CD56^{dim}CD16^+$) NK-cell phenotype after IL-2-containing IT cycles. In patient 1, the ratio remains below the normal cytotoxic NK-cell ratio of 9, whereas the ratio of patient 1 reaches normal values after the first IL-2-containing IT cycle and is increased after the second IL-2-containing IT cycle. Red arrows indicate start of IL-2-containing therapy cycles.

As plasma levels of NK-cell activation-associated markers increased upon IL-2-containing IT cycles, the NK-cell phenotype of two patients was subsequently assessed along the IT course. In both patients, we observed a major shift towards the mature, cytotoxic $CD56^{dim}CD16^+$ phenotype after both IL-2-containing IT cycles (Figure 6b,c). The $CD56^{dim}CD16^+/CD56^{bright}CD16^-$

ratio of patient 1 remained lower than the ratio of 9–9.5 in healthy controls [23], whereas the ratio of patient 2 reached a normal (IL-2 cycle 1) or superior (IL-2 cycle 2) NK-cell ratio.

3.6. Preferential Treg Expansion and Impaired T-Cell Proliferation during Therapy

Even though no significant changes were observed in absolute CD3+ T-cell levels after IL-2- or GM-CSF-containing IT cycles, it is suggested that cytokine therapy can shift the phenotype of CD3+ T-cells. To explore this effect during IT, extensive phenotyping of the CD3+ T-cell fraction was performed. Administration of IL-2 in this study massively increased the frequency of circulating CD4+CD25^{high}CD127^{dim} FOXP3+ Tregs (Figure 7a,b). In addition, GM-CSF also increased the frequency of Tregs, although to a lower extent than IL-2 (Figure 7b). These data were supported by an increased trend in plasma levels of IL-10 (GM-CSF: $p = 0.144$, IL-2: $p = 0.339$) (Figure 7c).

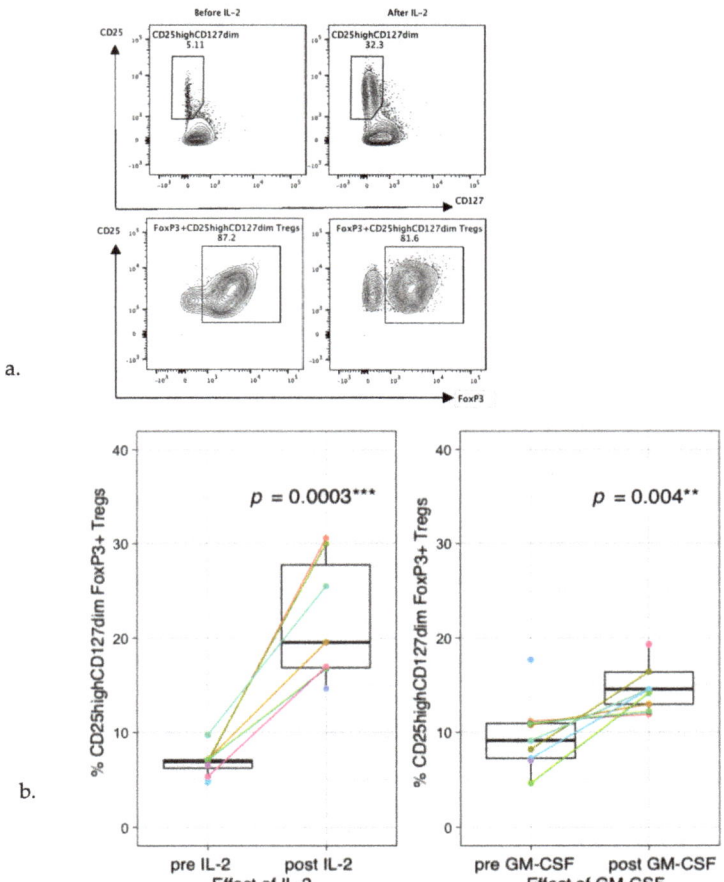

Figure 7. *Cont.*

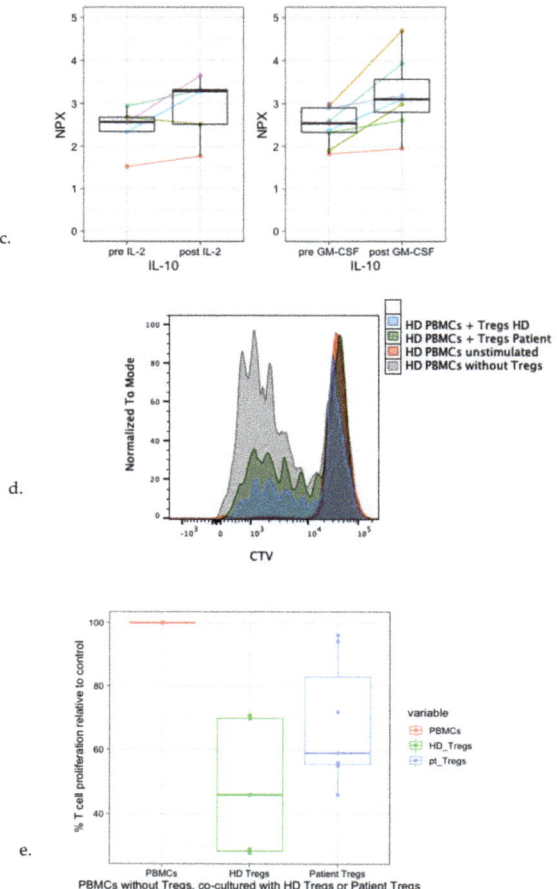

Figure 7. Regulatory T-cell profiles and their suppressive capacity during immunotherapy. (**a**) Example of gating of CD25^{high}CD127^{dim} cells within the CD3+CD4+ T-cell population (upper panels) and gating of FoxP3 within the CD25^{high}CD127^{dim} cell population before and after IL-2 administration (lower panels). (**b**) Percentages of Tregs (within CD3+CD4+ T-cell population) increase 4–5-fold after IL-2 administration (left) and increase 1–2-fold after GM-CSF administration (right). (**c**) Plasma IL-10 levels pre- and post-IL-2 ($p = 0.339$) (left) and GM-CSF ($p = 0.144$) (right). Protein expression is shown as normalized protein expression (NPX). IL-2: In total, 5 paired samples are shown, as two patients were monitored during both IT cycles. GM-CSF: In total, 7 paired samples are shown, as two patients were monitored during all three IT cycles. Two single measurements from patients pre-GM-CSF were included, resulting in a total of 9 patients pre- and 7 post-GM-CSF. (**d**) CTV staining of PBMCs of a healthy donor co-cultured without Tregs (grey), with patient Tregs (green), or healthy-donor Tregs (blue), or unstimulated (red) at an effector-to-target ratio of 2:1. (**e**) Relative percentages of proliferation of HD CD3+ T-cells co-cultured with patient Tregs (blue) or HD Tregs (green) compared to proliferation without Tregs (red). CD3+ T-cell proliferation was measured in patient 1 (during cycle 2 and 4), patient 2 (during cycles 1, 2 and 5) and patient 3 (during cycle 1 and 2). HD = healthy donor, PT = patient. ** $p < 0.001$, *** $p < 0.0001$.

To subsequently determine whether patient Tregs are functional, a Treg crossover suppression assay was performed in which patient Tregs from different IT time points were co-cultured with healthy-donor PBMCs. Healthy-donor PBMC proliferation was decreased upon co-culture with patient Tregs, indicating their suppressive capacity, even though suppressive capacity seems to be decreased when compared with healthy-donor Tregs (Figure 7d,e). In 2 of the 7 measurements (patient 1 cycle 2 and patient 3 cycle 1), no T-cell suppression was noticed.

To assess functionality of the CD3+ T-cell fraction in terms of proliferative capacity during IT, PBMCs were stimulated for three days with anti-CD3. Interestingly, anti-CD3-mediated T-cell proliferation was impaired in the majority of patients at different IT time points (Figure 8a). This was supported by decreased levels of secreted cytokines in stimulated patient PBMCs as compared to healthy-donor PBMCs (Figure 8b). Possible interference of CD25+CD127low Tregs or low-density eosinophils on T-cell proliferation was ruled out by performing additional T-cell proliferation assays without these cell populations.

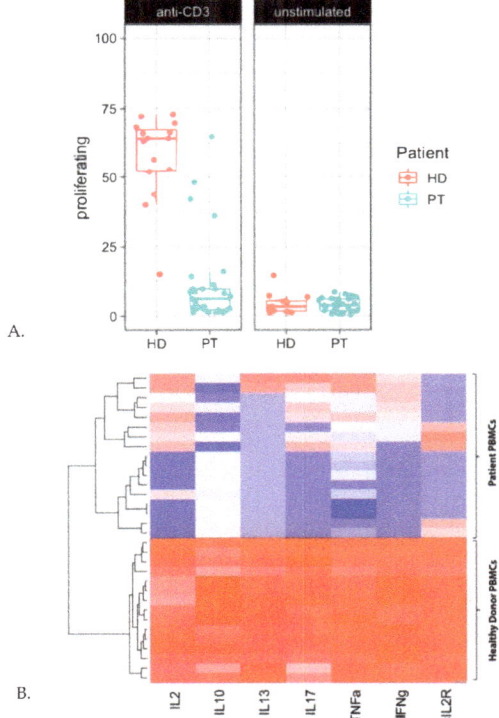

Figure 8. T-cell proliferation is impaired at diagnosis as well as during therapy in HR-NBL. (**A**) PBMCs of healthy donors (HD) (red) and patients (PT) (blue) were stimulated with anti-CD3 (0.5 µg/mL). T-cell proliferation of each individual sample is shown (duplos were pooled); PBMCs HD ($n = 8$), PBMCs patients ($n = 12$). (**B**) Supernatants (HD $n = 15$, patients $n = 17$) were analyzed using Luminex-based multiplex immunoassays. The heatmap shows the log concentration of IL-2, IL-10, IL-13, IL-7, TNF-α, IFN-γ and soluble IL-2R, with low levels indicated in blue and high levels indicated in red.

4. Discussion

Absolute lymphocyte counts, relative presence of subsets, and their phenotypical characteristics are rarely monitored in NBL patients and not used as prognostic criteria or treatment guidance, largely due to a lack of knowledge on clinical significance. In the present study, we show that immune profiles of HR-NBL patients are already disturbed (reduced levels of CD3+, CD56+, and CD19+ lymphocyte subsets) at diagnosis when compared to age-matched controls [24]. This is in line with Tamura et al. [25], who also reported that lower levels of immune cells at diagnosis may predict poor prognosis in patients with NBL. As HR-NBL often disseminates to the bone marrow, it is hypothesized that the decreased immune cell levels are most likely caused by tumor replacement and/or by tumor-related suppressive factors present in the bone marrow niche [25,26]. This is supported by studies observing lower leukocyte [26] or monocyte and lymphocyte [25] levels in patients with bone marrow metastases.

Moreover, we confirm data from Chung et al. [26] showing that the decrease in total leukocytes and lymphocytes in children with HR-NBL is even more pronounced after chemotherapy. We however observed a large interpatient variability between chemotherapy cycles; while B cells are completely depressed during all stages of N5/N6 chemotherapy, the numbers of monocytes, NK and T lymphocytes differed enormously. Whether these variations correlate to clinical outcome will be subject of follow-up studies with larger cohorts.

The effect of chemotherapeutic agents on the immune compartment should be kept in mind when combining IT with re-induction chemotherapy in relapsed/refractory patients. The effect of chemotherapy on IT efficacy is paradoxal, as levels of effector cells are often affected. On the other hand, targeting of immunosuppressive immune subsets and increased immunogenicity of tumor cells are described as processes to enhance IT efficacy [27–29]. Timing and chemotherapeutic compound selection are key to maximize the effect of IT in refractory/relapsed patients.

When subsequently looking into the functionality of T-cells at diagnosis and during the therapy regimen, we noticed hampered proliferation and cytokine secretion upon anti-CD3-mediated T-cell stimulation. In line with this, impaired PHA mitogenesis at diagnosis and during NBL therapy has been observed in several studies [30,31]. Helson et al. [31] and Pelizzo et al. [32] showed hampered PHA-mediated T-cell mitogenesis when cultures were supplemented with serum of NBL-patients, or mesenchymal stromal cells (MSCs) from HR-NBL patients, respectively. This indicates the presence of both local and systemic immune modulation by the NBL tumor. Several factors have been described that are able to modulate T-cell functionality, including TGF-β, Indoleamine-pyrrole 2,3-dioxygenase (IDO), and arginase [33,34]. The depletion of arginine by arginase [33] leads to T-cell cycle arrest, impaired proliferation, and reduced activation [35,36]. Although impaired T-cell proliferation is already noticed at diagnosis, it should be noted that immune function may be further inhibited by intensive treatment. In-depth phenotyping, proteomics, and pathway-analysis of T-cells during HR-NBL treatment is necessary to unravel mechanisms responsible for T-cell dysfunctionality as a first step to develop strategies to counteract this effect.

The effect of the IT regimen on NK-cell phenotype is largely unknown. Even though variation between patients is considerable, our data indicate that the cytotoxic NK-cell ratio increased during IT. We observed a delayed increase of the cytotoxic ratio in two patients upon IL-2-containing IT cycles. However, the NK-cell phenotype ratio of the majority of patients is still decreased at the end of IT, which suggests suboptimal IT timing. The observed differential effect of GM-CSF- and IL-2-containing IT cycles on the cytotoxic NK-cell ratio indicates that this is an effect induced by IL-2 rather than dinutuximab itself.

To our knowledge, this is the first study to show beneficial effects of GM-CSF and IL-2 addition to IT cycles in HR-NBL patients on both NK-cells (increased cytotoxic NK-cell ratio and plasma levels of NK-cell-associated factors (e.g., granzymes, KLRD1, NCR1, IFN-γ, CASP-8, KLRD1)), as well as on myeloid cells (based on plasma levels of neutrophil/monocyte-associated factors (e.g., CXCL11, CCL17, CCL23, and MCP4)). Nonetheless, Ladenstein and colleagues [4] recently concluded from a phase III clinical trial that there is no additive effect of IL-2 administration on outcome of HR-NBL patients. We noticed a strong increase of CD127dimCD25highFOXP3+ Tregs after IL-2, and to a lesser extent, also GM-CSF administration. This increase has been described before [37]; however, in many cases without confirming FOXP3 positivity, this may be expected based on results from autoimmune patients [15] where (low dose) IL-2 is administered to induce Tregs. Previously, preclinical data showed that Tregs inhibit anti-NBL immune responses before and after ASCT [38–40]. Using functional suppression assays in a crossover format, we showed that these Tregs also maintain their suppressive capacity at multiple time points during IT. Together, these data suggest that the beneficial effects of IL-2 may be masked by preferential Treg expansion.

The observation of increased NK-cell cytotoxicity during IL-2-containing IT cycles in our opinion substantiates the need to replace IL-2 during dinutuximab IT with other non-Treg engaging (immuno)therapeutic compounds/strategies to maximize IT efficacy. First of all, the start of IT can be delayed to allow further recovery of the NK-cell fraction. However, the observation that the NK-cell phenotype is already disturbed at diagnosis, together with the risk of the tumor to expand before the start of IT, are arguments against postponement of IT. A second strategy would be to combine dinutuximab with soluble factors more specifically activating NK-cells, for example, Lirilumab, an anti-KIR antibody currently tested in the ESMART trial from the ITCC (ClinicalTrials.gov Identifier: NCT02813135). In addition, NKTR-214, a CD122-biased cytokine agonist designed to preferentially activate and expand effector CD8+ T- and NK-cells over Tregs via the heterodimeric IL-2 receptor pathway (IL-2R-$\beta\gamma$) [41], is an interesting candidate to replace IL-2 [42]. Combining dinutuximab with IL-15 is also of interest, as this cytokine is known to specifically expand and mature NK-cells, without affecting Treg expansion [43,44]. The delayed effect of IL-2 on the cytotoxic NK-cell ratio observed in this study may substantiate an approach in which NK-cell engaging therapy is provided prior to dinutuximab-based IT. A third strategy would be to combine IT with an adoptive NK-cell therapy at the start of IT to maximize effector cell function, either via an autologous (ClinicalTrials.gov Identifiers: NCT02573896, NCT04211675) or allogeneic (haploidentical) [45] strategy (ClinicalTrials.gov Identifier: NCT03242603). The advantage of using allogeneic cells is the potential to select a mismatched donor to maximize anti-tumor effect. On the other hand, the risk of graft rejection and mismatch-related adverse events in allogeneic settings is a clear disadvantage compared to the use of an autologous, ex vivo-expanded, cell product.

Immune monitoring of HR-NBL patients comes with some limitations. The availability of patient samples was limited by dropout of patients from the study after relapse/progression of disease, transfer to other trials, failure of blood sampling, and logistical issues. In this study, immune status was monitored in peripheral blood only, which provides markers that would be easily translatable to monitoring protocols in the clinic. Nevertheless, information on tumor-infiltrating lymphocytes (TILs), and monitoring lymphocytes in tissues, would help to elucidate the mechanisms of (resistance to) therapy, and indicate whether markers at the tumor site are systemically reflected in the blood. Multinational collaborations in NBL cohorts are needed to allow for a larger sample size to confirm the findings from this study and relate them to clinical parameters and outcome.

5. Conclusions

(Functional) immune monitoring in HR-NBL patients revealed the presence of both immune-enhancing and immune regulatory effects during the therapy course. The immune-enhancing effects observed upon IL-2-containing IT cycles, despite simultaneous Treg expansion, clearly demonstrate the potential of combining dinutuximab with other NK-cell engaging strategies. In addition, the observed systemic T-cell dysfunction at diagnosis as well as during HR-NBL therapy highlights another mechanism, besides lack of MHC-I expression and immune checkpoint expression, that should be unraveled to generate long-term anti-NBL immune responses and immunological memory needed to prevent relapse.

Supplementary Materials: The following are available online at https://www.mdpi.com/article/10.3390/cancers13092096/s1, Figure S1: Schematic overview of sampling time points during the HR-NBL treatment course; Figure S2: Percentages of cell types based on trucount data before and after IL-2-containing immunotherapy cycles; Figure S3: Percentages of cell types based on trucount data before and after GM-CSF-containing immunotherapy cycle; Figure S4: Clustering of immune-oncology-related plasma protein concentrations of patients pre- and post-IL2- and GM-CSF-containing immunotherapy cycles.

Author Contributions: S.N., J.-J.B. and C.L.S. designed the study, and A.M.C. and C.L.S. wrote the manuscript. K.C.J.M.K., M.M.v.N., L.G.A.M.T., M.P.D., M.L.T. selected patients for the study and provided critical comments. C.L.S., S.M.T., C.C.H.d.K., D.A.M.H.v.d.B., E.D. performed experiments. C.L.S. and A.M.C. analyzed the data with support from E.M.D. and A.D.R.H. provided critical comments, and all authors read and approved the manuscript. All authors have read and agreed to the published version of the manuscript.

Funding: This work was supported by the Villa Joep Foundation (IWOV-Actief.51391.180034).

Institutional Review Board Statement: The study was conducted according to the guidelines of the Declaration of Helsinki, and approved by the Ethics Committee of the Academic Medical Center Amsterdam (NL50762.018.14, approval data 11 November 2015) and the University of Cologne, Germany (2013-004481-34, approval date 2 May 2015).

Informed Consent Statement: Informed consent was obtained from all subjects involved in the study.

Data Availability Statement: The data presented in this study are available upon request from the corresponding author. The data are not publicly available due to privacy restrictions.

Acknowledgments: The authors thank Frank Berthold and Barbara Hero from Uniklinik Köln to provide blood samples for this study.

Conflicts of Interest: The authors declare no conflict of interest.

References

1. Howlader, N.; Noone, A.M.; Krapcho, M.; Miller, D.; Brest, A.; Yu, M.; Ruhl, J.; Tatalovich, Z.; Mariotto, A.; Lewis, D.R.; et al. (Eds.) *SEER Cancer Statistics Review, 1975–2016*; National Cancer Institute: Bethesda, MD, USA, 2016.
2. Cohn, S.L.; Pearson, A.D.J.; London, W.B.; Monclair, T.; Ambros, P.F.; Brodeur, G.M.; Faldum, A.; Hero, B.; Iehara, T.; Machin, D.; et al. The International Neuroblastoma Risk Group (INRG) classification system: An INRG task force report. *J. Clin. Oncol.* **2009**, *27*, 289–297. [CrossRef] [PubMed]
3. Yu, A.L.; Gilman, A.L.; Ozkaynak, M.F.; London, W.B.; Kreissman, S.G.; Chen, H.X.; Smith, M.; Anderson, B.; Villablanca, J.G.; Matthay, K.K.; et al. Anti-GD2 Antibody with GM-CSF, Interleukin-2, and Isotretinoin for Neuroblastoma. *N. Engl. J. Med.* **2010**, *363*, 1324–1334. [CrossRef] [PubMed]
4. Ladenstein, R.; Pötschger, U.; Valteau-Couanet, D.; Luksch, R.; Castel, V.; Yaniv, I.; Laureys, G.; Brock, P.; Michon, J.M.; Owens, C.; et al. Interleukin 2 with anti-GD2 antibody ch14.18/CHO (dinutuximab beta) in patients with high-risk neuroblastoma (HR-NBL1/SIOPEN): A multicentre, randomised, phase 3 trial. *Lancet Oncol.* **2018**, *19*, 1617–1629. [CrossRef]
5. Yu, A.L.; Gilman, A.L.; Ozkaynak, M.F.; Naranjo, A.; Diccianni, M.B.; Gan, J.; Hank, J.A.; Batova, A.; London, W.B.; Tenney, S.C.; et al. Long-term follow-up of a Phase III Study of ch14.18 (Dinutuximab) + Cytokine Immunotherapy in Children with High-risk Neuroblastoma: COG Study ANBL0032. *Clin. Cancer Res.* **2021**, *18*. [CrossRef]
6. Park, J.R.; Bagatell, R.; London, W.B.; Maris, J.M.; Cohn, S.L.; Mattay, K.M.; Hogarty, M. Children's Oncology Group's 2013 blueprint for research: Neuroblastoma. *Pediatr. Blood Cancer* **2013**, *60*, 985–993. [CrossRef]
7. Pinto, N.R.; Applebaum, M.A.; Volchenboum, S.L.; Matthay, K.K.; London, W.B.; Ambros, P.F.; Nakagawara, A.; Berthold, F.; Schleiermacher, G.; Park, J.R.; et al. Advances in risk classification and treatment strategies for neuroblastoma. *J. Clin. Oncol.* **2015**, *33*, 3008–3017. [CrossRef]
8. Masucci, G.; Ragnhammar, P.; Wersäll, P.; Mellstedt, H. Granulocyte-monocyte colony-stimulating-factor augments the interleukin-2-induced cytotoxic activity of human lymphocytes in the absence and presence of mouse or chimeric monoclonal antibodies (mAb 17-1A). *Cancer Immunol. Immunother.* **1990**, *31*, 231–235. [CrossRef]
9. Hank, J.A.; Surfus, J.; Sondel, P.M.; Robinson, R.R.; Mueller, B.M.; Reisfeld, R.A.; Cheung, N.K. Augmentation of Antibody Dependent Cell Mediated Cytotoxicity following in Vivo Therapy with Recombinant Interleukin 2. *Cancer Res.* **1990**, *50*, 5234–5239.
10. Thoma, M.D.; Huneke, T.J.; DeCook, L.J.; Johnson, N.D.; Wiegand, R.A.; Litzow, M.R.; Hogan, W.J.; Porrata, L.F.; Holtan, S.G. Peripheral blood lymphocyte and monocyte recovery and survival in acute leukemia postmyeloablative allogeneic hematopoietic stem cell transplant. *Biol. Blood Marrow Transplant.* **2012**, *18*, 600–607. [CrossRef]
11. Galvez-Silva, J.; Maher, O.M.; Park, M.; Liu, D.; Hernandez, F.; Tewari, P.; Nieto, Y. Prognostic Analysis of Absolute Lymphocyte and Monocyte Counts after Autologous Stem Cell Transplantation in Children, Adolescents, and Young Adults with Refractory or Relapsed Hodgkin Lymphoma. *Biol. Blood Marrow Transplant.* **2017**, *23*, 1276–1281. [CrossRef]
12. Kim, H.T.; Armand, P.; Frederick, D.; Andler, E.; Cutler, C.; Koreth, J.; Alyea, E.P.; Antin, J.H.; Soiffer, R.J.; Ritz, J.; et al. Absolute lymphocyte count recovery after allogeneic hematopoietic stem cell transplantation predicts clinical outcome. *Biol. Blood Marrow Transplant.* **2015**, *21*, 873–880. [CrossRef]
13. Nassin, M.L.; Nicolaou, E.; Gurbuxani, S.; Cohn, S.L.; Cunningham, J.M.; LaBelle, J.L. Immune Reconstitution Following Autologous Stem Cell Transplantation in Patients with High-Risk Neuroblastoma at the Time of Immunotherapy. *Biol. Blood Marrow Transplant.* **2018**, *24*, 452–459. [CrossRef]

14. Scheid, C.; Pettengell, R.; Ghielmini, M.; Radford, J.A.; Morgenstern, G.R.; Stern, P.L.; Crowther, D. Time-course of the recovery of cellular immune function after high-dose chemotherapy and peripheral blood progenitor cell transplantation for high-grade non-Hodgkin's lymphoma. *Bone Marrow Transplant.* **1995**, *15*, 901–906.
15. Ye, C.; Brand, D.; Zheng, S.G. Targeting IL-2: An unexpected effect in treating immunological diseases. *Signal Transduct. Target. Ther.* **2018**, *3*, 1–10. [CrossRef]
16. Dutch Childhood Oncology Group (DCOG). *DCOG NBL 2009 Treatment Protocol for Risk Adapted Treatment of Children with Neuroblastoma Admendment 1*; 2012.
17. Brodeur, G.M.; Pritchard, J.; Berthold, F.; Carlsen, N.L.; Castel, V.; Castelberry, R.P.; De Bernardi, B.; Evans, A.E.; Favrot, M.; Hedborg, F. Revisions of the international criteria for neuroblastoma diagnosis, staging and response to treatment. *Prog. Clin. Biol. Res.* **1994**, *385*, 363–369. [CrossRef]
18. De Jager, W.; Prakken, B.J.; Bijlsma, J.W.J.; Kuis, W.; Rijkers, G.T. Improved multiplex immunoassay performance in human plasma and synovial fluid following removal of interfering heterophilic antibodies. *J. Immunol. Methods* **2005**, *300*, 124–135. [CrossRef]
19. Warnes, G.J.; Bolker, B.; Bonebakker, L.; Gentleman, G.; Liaw, W.H.A.; Lumley, T.; Maechler, M.; Magnusson, A.; Moeller, S.; Schwartz, M.; et al. gplots: Various R Programming Tools for Plotting Data. *R Package Version* **2019**, *2*, 1.
20. Url, S.; Society, R.S.; Society, R.S. Controlling the False Discovery Rate: A Practical and Powerful Approach to Multiple Testing. *J. R. Stat. Soc. Ser. B* **1995**, *57*, 289–300.
21. R Core Team. *R: A Language and Environment for Statistical Computing*; R Foundation for Statistical Computing: Vienna, Austria, 2018.
22. Angelo, L.S.; Banerjee, P.P.; Monaco-Shawver, L.; Rosen, J.B.; Makedonas, G.; Forbes, L.R.; Mace, E.M.; Orange, J.S. Practical NK cell phenotyping and variability in healthy adults. *Immunol. Res.* **2015**, *62*, 341–356. [CrossRef]
23. Cooper, M.A.; Fehniger, T.A.; Caligiuri, M.A. The biology of human natural killer-cell subsets. *Trends Immunol.* **2001**, *22*, 633–640. [CrossRef]
24. Tosato, F.; Bucciol, G.; Pantano, G.; Putti, M.C.; Sanzari, M.C.; Basso, G.; Plebani, M. Lymphocytes subsets reference values in childhood. *Cytom. Part A* **2015**, *87*, 81–85. [CrossRef]
25. Tamura, A.; Inoue, S.; Mori, T.; Noguchi, J.; Nakamura, S.; Saito, A.; Kozaki, A.; Ishida, T.; Sadaoka, K.; Hasegawa, D.; et al. Low Multiplication Value of Absolute Monocyte Count and Absolute Lymphocyte Count at Diagnosis May Predict Poor Prognosis in Neuroblastoma. *Front. Oncol.* **2020**, *10*, 1–9. [CrossRef]
26. Chung, H.S.; Higgins, G.R.; Siegel, S.E.; Seeger, R.C. Abnormalities of the immune system in children with neuroblastoma related to the neoplasm and chemotherapy. *J. Pediatr.* **1977**, *90*, 548–554. [CrossRef]
27. Cornel, A.M.; Mimpen, I.L.; Nierkens, S. MHC class I downregulation in cancer: Underlying mechanisms and potential targets for cancer immunotherapy. *Cancers* **2020**, *12*, 1760. [CrossRef]
28. Domingos-pereira, S.; Galliverti, G.; Hanahan, D.; Nardelli-haefliger, D. Intravaginal CpG as tri-therapy towards efficient regression of genital HPV16 tumors. *J. Immunol. Ther. Cancer* **2019**, *1*, 1–7.
29. Chitadze, G.; Lettau, M.; Luecke, S.; Wang, T.; Janssen, O.; Fürst, D.; Mytilineos, J.; Wesch, D.; Oberg, H.H.; Held-Feindt, J.; et al. NKG2D- and T-cell receptor-dependent lysis of malignant glioma cell lines by human γδ T cells: Modulation by temozolomide and A disintegrin and metalloproteases 10 and 17 inhibitors. *Oncoimmunology* **2016**, *5*, 1–13. [CrossRef]
30. Rosanda, C.; De Bernardi, B.; Pasino, M.; Bisogni, M.C.; Maggio, A.; Haupt, R.; Tonini, G.P.; Ponzoni, M. Immune Evaluation of 50 Children with Neuroblastoma at Onset. *Med. Pediatr. Oncol.* **1982**. [CrossRef]
31. Helson, L.; Shou, L.; Tauber, J. Lymphocyte transformation in children with neuroblastoma. *J. Natl. Cancer Inst.* **1976**, *57*, 721–722. [CrossRef]
32. Pelizzo, G.; Veschi, V.; Mantelli, M.; Croce, S.; Di Benedetto, V.; D'Angelo, P.; Maltese, A.; Catenacci, L.; Apuzzo, T.; Scavo, E.; et al. Microenvironment in neuroblastoma: Isolation and characterization of tumor-derived mesenchymal stromal cells. *BMC Cancer* **2018**, *18*, 1176. [CrossRef]
33. Mussai, F.; Egan, S.; Hunter, S.; Webber, H.; Fisher, J.; Wheat, R.; McConville, C.; Sbirkov, Y.; Wheeler, K.; Bendle, G.; et al. Neuroblastoma arginase activity creates an immunosuppressive microenvironment that impairs autologous and engineered immunity. *Cancer Res.* **2015**, *75*, 3043–3053. [CrossRef]
34. Wang, Q.; Ding, G.; Xu, X. Immunomodulatory functions of mesenchymal stem cells and possible mechanisms. *Histol. Histopathol.* **2016**, *31*, 949–959. [CrossRef] [PubMed]
35. Rodriguez, P.C.; Quiceno, D.G.; Ochoa, A.C. L-arginine availability regulates T-lymphocyte cell-cycle progression. *Blood* **2007**, *109*, 1568–1573. [CrossRef] [PubMed]
36. Zea, A.H.; Rodriguez, P.C.; Culotta, K.S.; Hernandez, C.P.; DeSalvo, J.; Ochoa, J.B.; Park, H.J.; Zabaleta, J.; Ochoa, A.C. L-Arginine modulates CD3ζ expression and T cell function in activated human T lymphocytes. *Cell. Immunol.* **2004**, *232*, 21–31. [CrossRef] [PubMed]
37. Troschke-Meurer, S.; Siebert, N.; Marx, M.; Zumpe, M.; Ehlert, K.; Mutschlechner, O.; Loibner, H.; Ladenstein, R.; Lode, H.N. Low CD4$^+$/CD25$^+$/CD127$^-$ regulatory T cell- and high INF-γ levels are associated with improved survival of neuroblastoma patients treated with long-term infusion of ch14.18/CHO combined with interleukin-2. *Oncoimmunology* **2019**, *8*. [CrossRef] [PubMed]
38. Jing, W.; Gershan, J.A.; Johnson, B.D. Depletion of CD4 T cells enhances immunotherapy for neuroblastoma after syngeneic HSCT but compromises development of antitumor immune memory. *Blood* **2009**, *113*, 4449–4457. [CrossRef]

39. Jing, W.; Yan, X.; Hallett, W.H.D.; Gershan, J.A.; Johnson, B.D. Depletion of CD25+ T cells from hematopoietic stem cell grafts increases posttransplantation vaccine-induced immunity to neuroblastoma. *Blood* **2011**. [CrossRef]
40. Johnson, B.D.; Jing, W.; Orentas, R.J. CD25+ regulatory T cell inhibition enhances vaccine-induced immunity to neuroblastoma. *J. Immunother.* **2007**, *30*, 203–214. [CrossRef]
41. Charych, D.H.; Hoch, U.; Langowski, J.L.; Lee, S.R.; Addepalli, M.K.; Kirk, P.B.; Sheng, D.; Liu, X.; Sims, P.W.; VanderVeen, L.A.; et al. NKTR-214, an Engineered Cytokine with Biased IL2 Receptor Binding, Increased Tumor Exposure, and Marked Efficacy in Mouse Tumor Models. *Clin. Cancer Res.* **2016**, *22*, 680–690. [CrossRef]
42. Diab, A.; Tannir, N.M.; Bentebibel, S.E.; Hwu, P.; Papadimitrakopoulou, V.; Haymaker, C.; Kluger, H.M.; Gettinger, S.N.; Sznol, M.; Tykodi, S.S.; et al. Bempegaldesleukin (NKTR-214) plus Nivolumab in Patients with Advanced Solid Tumors: Phase I Dose-Escalation Study of Safety, Efficacy, and Immune Activation (PIVOT-02). *Cancer Discov.* **2020**, *10*, 1158–1173. [CrossRef]
43. Waldmann, T.A. The biology of interleukin-2 and interleukin-15: Implications for cancer therapy and vaccine design. *Nat. Rev. Immunol.* **2006**, *6*, 595–601. [CrossRef]
44. Heinze, A.; Grebe, B.; Bremm, M.; Huenecke, S.; Munir, T.A.; Graafen, L.; Frueh, J.T.; Merker, M.; Rettinger, E.; Soerensen, J.; et al. The Synergistic Use of IL-15 and IL-21 for the Generation of NK Cells From CD3/CD19-Depleted Grafts Improves Their ex vivo Expansion and Cytotoxic Potential against Neuroblastoma: Perspective for Optimized Immunotherapy Post Haploidentical Stem Cell Trans. *Front. Immunol.* **2019**, *10*, 1–20. [CrossRef]
45. Modak, S.; Le Luduec, J.B.; Cheung, I.Y.; Goldman, D.A.; Ostrovnaya, I.; Doubrovina, E.; Basu, E.; Kushner, B.H.; Kramer, K.; Roberts, S.S.; et al. Adoptive immunotherapy with haploidentical natural killer cells and Anti-GD2 monoclonal antibody m3F8 for resistant neuroblastoma: Results of a phase I study. *Oncoimmunology* **2018**, *7*, 1–10. [CrossRef]

Article

Plasma Nucleosomes in Primary Breast Cancer

Michal Mego [1,2,*], Katarina Kalavska [2], Marian Karaba [3], Gabriel Minarik [4], Juraj Benca [3,5], Tatiana Sedlackova [4], Paulina Gronesova [6], Dana Cholujova [6], Daniel Pindak [3,7], Jozef Mardiak [1] and Peter Celec [4]

1. 2nd Department of Oncology, Faculty of Medicine, Comenius University and National Cancer Institute, 83310 Bratislava, Slovakia; jmardiak@gmail.com
2. Translational Research Unit, Faculty of Medicine, Comenius University and National Cancer Institute, 83310 Bratislava, Slovakia; katarina.kalavska@nou.sk
3. Department of Oncosurgery, National Cancer Institute, 83310 Bratislava, Slovakia; marian.karaba@nou.sk (M.K.); juraj.benca@nou.sk (J.B.); daniel.pindak@nou.sk (D.P.)
4. Institute of Molecular Biomedicine, Faculty of Medicine, Comenius University, 81372 Bratislava, Slovakia; gabriel.minarik@gmail.com (G.M.); tatiana.sedlackova@gmail.com (T.S.); petercelec@gmail.com (P.C.)
5. Department of Medicine, St. Elizabeth University, 81102 Bratislava, Slovakia
6. Biomedical Center, Slovak Academy of Sciences, 84505 Bratislava, Slovakia; paulina.gronesova@gmail.com (P.G.); dana.cholujova@savba.sk (D.C.)
7. Department of Oncosurgery, Slovak Medical University, 83101 Bratislava, Slovakia
* Correspondence: misomego@gmail.com or michal.mego@nou.sk; Tel.: +421-2-59378366; Fax: +421-2-54774943

Received: 7 August 2020; Accepted: 8 September 2020; Published: 10 September 2020

Simple Summary: Nucleosomes composed of DNA and histone proteins enter the extracellular space and end eventually in the circulation when cells die. In blood plasma, they could represent a nonspecific marker of cell death, potentially useful for noninvasive monitoring of cancer. The aim of this study was to analyze circulating nucleosomes in relation to patient/tumor characteristics and prognosis in nonmetastatic breast cancer. This study included 92 patients with breast cancer treated with surgery. Plasma nucleosomes were detected in samples taken in the morning on the day of surgery. Circulating nucleosomes were positively associated with the systemic inflammation but not with other patient/tumor characteristics. Patients with lower nucleosomes had lower risk of disease recurrence compared to patients with higher nucleosomes. Our data suggest that plasma nucleosomes in nonmetastatic breast cancer are associated with systemic inflammation and might have a prognostic value. The underlying mechanisms require further studies.

Abstract: When cells die, nucleosomes composed of DNA and histone proteins enter the extracellular space and end eventually in the circulation. In plasma, they might serve as a nonspecific marker of cell death, potentially useful for noninvasive monitoring of tumor dynamics. The aim of this study was to analyze circulating nucleosomes in relation to patient/tumor characteristics and prognosis in primary breast cancer. This study included 92 patients with breast cancer treated with surgery for whom plasma isolated was available in the biobank. Plasma nucleosomes were detected in samples taken in the morning on the day of surgery using Cell Death Detection ELISA kit with anti-histone and anti-DNA antibodies. Circulating nucleosomes were positively associated with the systemic inflammatory index (SII), but not with other patient/tumor characteristics. Patients with high SII in comparison to low SII had higher circulating nucleosomes (by 59%, $p = 0.02$). Nucleosomes correlated with plasma plasminogen activator inhibitor-1, IL-15, IL-16, IL-18, and hepatocyte growth factor. Patients with lower nucleosomes had significantly better disease-free survival (HR = 0.46, $p = 0.05$). In a multivariate analysis, nucleosomes, hormone receptor status, HER2 status, lymph node involvement, and tumor grade were independent predictors of disease-free survival. Our data suggest that plasma nucleosomes in primary breast cancer are associated with systemic inflammation and might have a prognostic value. The underlying mechanisms require further studies.

Keywords: primary breast cancer; circulating nucleosomes; circulating tumor cells; plasminogen activator inhibitor-1; cytokines

1. Introduction

Breast cancer is the most common diagnosed cancer and the leading cause of cancer death among women in developed countries [1]. Despite advances in cancer prevention, diagnoses, and treatment, still approximately 5% of patients are diagnosed with metastatic disease, and 20–30% of initially primary breast cancer develops metastasis subsequently, during the course of the disease.

Extracellular DNA (ecDNA), also called cell-free DNA, is present in blood plasma in various forms [2]. EcDNA in the circulation of cancer patients contains tumor DNA from the primary tumor, metastasis, or circulating tumor cells, as well as healthy host cells mostly of hematopoietic origin [3–5]. Plasma ecDNA is partially free unbound DNA and, so, sensitive to rapid cleavage, but it also can be protected as ecDNA hidden in apoptotic bodies and/or bound to proteins such as histones in the form of nucleosomes [5].

Nucleosomes are composed of DNA wound around histone proteins and represent the basic structural unit of chromatin in the nucleus [6]. After cell death, membranes and nuclei disintegrate and cell-free nucleosomes can get into the circulation. Plasma nucleosomes might serve as a nonspecific biomarker of cell death [7]. This might be of interest in patients not only with autoimmune diseases, but also with sepsis or cancer [8–10]. The prognostic value of the concentration of circulating nucleosomes was shown in several types of cancer including lung, pancreatic, or colorectal cancer [11–15]. For example, in pancreatic cancer, high nucleosome levels during treatment, but not pretherapeutic levels, correlate with time to progression [16]. Similarly, in non-small cell lung cancer, high baseline nucleosome level and/or during chemotherapy was associated with poor response to treatment and these data suggested that circulating nucleosomes are a valuable tool for early prediction of chemotherapy efficacy in cancer patients [17]. However, when it comes to primary breast cancer, data in the published literature are limited.

In this study, we aimed to analyze circulating nucleosomes in relation to patients/tumor characteristics and prognosis in primary breast cancer.

2. Methods

2.1. Study Patients

This study included 92 primary breast cancer patients (stage I–III) treated with surgery from March to November 2012, for whom plasma isolated in the morning on the day of surgery was available in the biobank. This study represents a substudy of a translational trial that aimed to evaluate prognostic value of circulating tumor cells in primary breast cancer [18]. Study eligibility criteria and study details were described previously [18]. The study was approved by the Institutional Review Board (IRB) of the National Cancer Institute of Slovakia (TRUSK002, 20.6.2011). Each participant provided signed informed consent before study enrollment.

2.2. Detection of Circulating Tumor Cells (CTCs) in Peripheral Blood

CTCs were detected in peripheral blood by a quantitative real-time polymerase chain reaction (qRT-PCR)-based assay of peripheral blood as described previously [18–20].

2.3. Plasma Isolation

Venous peripheral blood samples were collected in EDTA-treated tubes in the morning on the day of surgery and centrifuged at $1000\times g$ for 10 min at room temperature within 2 h of venipuncture and processed as described previously [21].

2.4. Quantification of Circulating Nucleosomes

The commercially available Cell Death Detection kit (Roche, Basel, Switzerland) was used for the measurement of nucleosomes. Briefly, 20 mL of plasma was mixed with biotin-labeled anti-histone and peroxidase-conjugated anti-DNA antibodies. After incubation and washing, the substrate for the peroxidase enzyme was added. Absorbance was measured at 405 nm in arbitrary units after stopping the reaction. Interassay and intra-assay coefficients of variation were below 10% and 5%, respectively.

2.5. Measurement of DD, TF, uPA, and PAI-1 in Plasma

Plasma tissue factor (TF), d-dimer (DD), urokinase plasminogen activator (uPA), and plasminogen activator inhibitor-1 (PAI-1) were analyzed using enzyme-linked immunosorbent assays (ELISA) as described previously [21].

2.6. Plasma Cytokines and Angiogenic Factors Analysis

Plasma samples were analyzed for 51 plasma cytokines and angiogenic factors: TGF-β1, TGF-β2, TGF-β3, IFN-α2, IL-1α, IL-2Rα, IL-3, IL-12p40, IL-16, IL-18, CTACK, Gro-α, HGF, LIF, MCP-3, M-CSF, MIF, MIG, β-NGF, SCF, SCGF-β, SDF-1α, TNF-β, TRAIL, IL-1β, Il-1RA, IL-2, IL-4, IL-5, IL-6, IL-7, IL-8, IL-9, IL-10, IL-12, IL-13, IL-15, IL-17, Eotaxin, FGF basic, G-CSF, GM-CSF, IFN-γ, IP-10, MCP-1, MIP-1α, MIP-1β, PDGF bb, RANTES, TNF-α, VEGF using predesigned panels as described previously and were available for subset of patients (Bio-Plex Pro TGF-β assay, Bio-Plex Pro Human Cytokine 21- and 27-plex immunoassays; Bio-Rad Laboratories, Hercules, CA, USA) [22]. The large panel of cytokines was analyzed as data were available from the previous study [22].

2.7. Complete Blood Count and Inflammation-Based Scores

Complete blood count (CBC) and CBC-derived inflammation-based scores were calculated as described previously [23,24]. For CBC-derived inflammation-based scores, identical cut-off values as published previously for metastatic breast cancer patients were used [23,24]. Data for calculation of NLR, PLR, MLR, SII were available for 54, 52, 48, and 52 patients, respectively.

2.8. Statistical Analysis

The characteristics of patients is summarized using mean (range) for continuous variables and frequency (percentage) for categorical variables. The median follow-up period was calculated as the median observation time among all patients and among those who were still alive at the time of their last follow-up. Disease-free survival (DFS) was calculated from the date of blood sampling to the date of disease recurrence (locoregional or distant), secondary cancer, death, or last follow-up. DFS was estimated using the Kaplan–Meier product limit method and compared between groups by log-rank test. For survival analysis, circulating nucleosomes were dichotomized to "low" or "high" (nucleosome level below vs. above mean, respectively). Univariate analyses with Chi squared or Fisher's exact test were performed to find associations between prognostic factors.

A multivariate Cox proportional hazards model for DFS was used to assess differences in outcome on the basis of the nucleosomes status (above mean vs. below mean), hormone receptor status (positive for either vs. negative for both), HER-2 status (positive or negative), axillary lymph node involvement (N0 vs. N+), grade (grade 3 vs. grade 1 and 2). Stepwise regression techniques were used to build multivariate models using a significance level of 0.10 to remain in the model. All p values presented are two-sided, and associations were considered significant if the p value was less than or equal to 0.05. Statistical analyses were performed using NCSS 11 Statistical Software (2016, NCSS, LLC., Kaysville, UT, USA, ncss.com/software/ncss).

3. Results

3.1. Patients' Characteristics

The study population consisted of 92 primary breast cancer patients with a median age of 60 years (range: 25–83 years). The patient characteristics are shown in Table 1. There were 79 (85.9%) patients with estrogen receptor-positive (ER) and/or progesterone receptor-positive (PR) tumors, and 16 (17.4%) patients with HER2/neu-positive tumors.

Table 1. Patients' characteristics.

Variable	N	%
All Patients	92	100.0
T-stage		
T1	58	63.0
>T1	34	37.0
Histology		
IDC	76	82.6
other	16	17.4
Grade		
low and intermediate	49	53.3
high grade	41	44.6
unknown	2	2.2
Lymph nodes		
N0	57	62.0
N+	34	37.0
unknown	1	1.1
LVI		
present	69	75.0
absent	23	25.0
Hormone receptor status (cut-off 1%)		
negative for both	13	14.1
positive for either	79	85.9
Estrogen receptor-positive (cut-off 1%)		
negative	16	17.4
positive	76	82.6
Progesterone receptor-positive (cut-off 1%)		
negative	25	27.2
positive	67	72.8
HER2 status		
positive	16	17.4
negative	76	82.6
P53 status		
negative	59	64.1
positive	32	34.8
unknown	1	1.1
BCL-2		
negative	27	29.3
positive	65	70.7
unknown		
Ki67 status (cut-off 14%)		
<14%	48	52.2
>14%	44	47.8

Table 1. *Cont.*

Variable	N	%
Molecular subtype		
Luminal A	43	46.7
Luminal B	36	39.1
HER2+	1	1.1
Triple-negative (TN)	12	13.0
CTC EP		
negative	75	81.5
positive	17	18.5
CTC EMT		
negative	76	82.6
positive	16	17.4
CTC ANY		
negative	62	67.4
positive	30	32.6

Abbreviations: CTC EP, circulating tumor cells with epithelial phenotype; CTC EMT, circulating tumor cells with epithelial–mesenchymal transition phenotype; CTC ANY, circulating tumor cells irrespective of phenotype; LVI, lymphovascular invasion.

3.2. Association between Nucleosomes and Patient/Tumor Characteristics

The characteristics of patients and the associations with circulating nucleosomes are shown in Table 2. The concentration of circulating nucleosomes was not associated with any patient/tumor characteristics except the systemic inflammatory index (SII), where patients with high SII had significantly higher levels of circulating nucleosomes compared to patients with low SII (0.17 vs. 0.27, $p = 0.02$). There was also a trend for higher level of circulating nucleosomes in patients with high neutrophil/lymphocyte ratio ($p = 0.07$). There was no association between molecular subtype and plasma nucleosomes, even if molecular subtypes of breast cancer were further segregated by tumor grade. We also analyzed association of chronic medication/comorbidities (Appendix A, Table A1) and circulating nucleosomes, but we found no association.

Table 2. Association between nucleosomes and patient/tumor characteristics.

Variable	N	Mean	SEM	Median	*p*-Value
All	92	0.18	0.02	0.13	NA
T-stage					
T1	58	0.20	0.02	0.14	0.30
>T1	34	0.15	0.03	0.13	
Histology					
invasive ductal carcinoma	76	0.19	0.02	0.13	0.61
other	16	0.15	0.04	0.13	
Grade					
low and intermediate	49	0.20	0.03	0.14	0.91
high grade	41	0.16	0.03	0.13	
unknown	2				
Lymph nodes					
N0	57	0.18	0.02	0.12	0.10
N+	34	0.19	0.03	0.17	
unknown	1				
Lymphovascular invasion					
absent	69	0.18	0.02	0.13	0.20
present	23	0.19	0.04	0.16	

Table 2. Cont.

Variable	N	Mean	SEM	Median	p-Value
Hormone receptor status (cut-off 1%)					
negative for both	13	0.12	0.05	0.10	0.17
positive for either	79	0.19	0.02	0.14	
HER2 status					
negative	76	0.19	0.02	0.13	0.91
positive	16	0.17	0.04	0.15	
P53 status					
negative	59	0.19	0.02	0.14	0.51
positive	32	0.17	0.03	0.13	
unknown	1				
BCL-2					
negative	27	0.15	0.03	0.13	0.52
positive	65	0.20	0.02	0.13	
Ki67 status (cut-off 14%)					
<14%	48	0.20	0.03	0.15	0.38
>14%	44	0.16	0.03	0.13	
unknown					
Molecular subtype					
Luminal A	43	0.21	0.03	0.15	0.22
Luminal B	36	0.17	0.03	0.14	
HER2+	1	0.25	0.17	0.25	
Triple-negative (TN)	12	0.10	0.05	0.09	
CTC EP					
negative	75	0.18	0.02	0.13	0.44
positive	17	0.17	0.04	0.14	
CTC EMT					
negative	76	0.19	0.02	0.13	0.78
positive	16	0.16	0.04	0.13	
CTC ANY					
negative	62	0.19	0.02	0.13	0.19
positive	30	0.18	0.03	0.15	
NLR (neutrophil/lymphocyte ratio) *					
<3	43	0.18	0.03	0.12	0.07
>3	11	0.26	0.06	0.17	
PLR (platelet/lymphocyte ratio) *					
<210	43	0.19	0.03	0.12	0.71
>210	9	0.21	0.07	0.15	
MLR (monocyte/lymphocyte ratio) *					
<0.34	40	0.20	0.03	0.12	0.60
>0.34	8	0.15	0.07	0.14	
SII (systemic inflammatory index) *					
<836	40	0.17	0.03	0.10	**0.02**
>836	12	0.27	0.06	0.17	

Abbreviations: CTC EP, circulating tumor cells with epithelial phenotype; CTC EMT, circulating tumor cells with epithelial–mesenchymal transition phenotype; CTC ANY, circulating tumor cells irrespective of phenotype. * Data for calculation of NLR, PLR, MLR, SII were available for 54, 52, 48, and 52 patients, respectively; NA, not applicable. p-Values < 0.05 are written in Bold.

3.3. Association between Nucleosomes and Plasma Cytokines

Patients with nucleosomes above mean in peripheral blood had significantly elevated plasma IL-16 ($p = 0.005$), IL-18 ($p = 0.0004$), and hepatocyte growth factor ($p = 0.043$), as compared to patients with nucleosomes below mean, while there was an inverse correlation between nucleosomes and IL-15

(p = 0.036). There was also a trend for higher IFN-α2 (p = 0.055) and RANTES (p = 0.053) in patients with higher nucleosome level (Table 3).

Table 3. Association between nucleosomes and plasma cytokines.

Variable	N	Mean	SEM	Median	p-Value
IFN_a2 (ng/mL)					
nucleosomes low	57	101.6	3.2	102.2	0.055
nucleosomes high	26	114.5	4.7	114.7	
IL_16 (ng/mL)					
nucleosomes low	58	349.2	19.8	330.9	**0.005**
nucleosomes high	27	446.7	29.0	419.5	
IL_18 (ng/mL)					
nucleosomes low	57	60.4	12.9	33.8	**0.0004**
nucleosomes high	26	120.0	19.1	69.7	
HGF (ng/mL)					
nucleosomes low	58	760.8	184.7	222.7	**0.043**
nucleosomes high	27	1312.2	270.8	438.6	
M_CSF (ng/mL)					
nucleosomes low	58	10.7	1.5	6.9	0.066
nucleosomes high	27	14.8	2.2	12.4	
IL_15 (ng/mL)					
nucleosomes low	39	22.0	2.1	16.7	**0.036**
nucleosomes high	22	13.6	2.9	12.8	
RANTES (ng/mL)					
nucleosomes low	58	8890.2	816.2	7644.8	0.053
nucleosomes high	27	7185.2	1196.2	4352.3	

Abbreviations: SEM, standard error of the mean. p-Values < 0.05 are written in Bold.

3.4. Nucleosomes and Coagulation

There was no association between circulating nucleosomes and DD, TF, and/or uPA, while patients with nucleosomes above mean had significantly elevated levels of plasma PAI-1 (Table 4).

Table 4. Association between nucleosomes and coagulation.

Variable	N	Mean	SEM	Median	p-Value
Tissue factor (pg/mL)					
nucleosomes low	61	66.2	2.2	60.4	0.464
nucleosomes high	31	62.1	3.0	60.0	
D-dimer (ng/mL)					
nucleosomes low	61	412.3	53.6	312.2	0.394
nucleosomes high	31	552.3	75.2	401.7	
uPA (ng/mL) *					
nucleosomes low	59	4.8	0.5	3.8	0.925
nucleosomes high	31	4.6	0.6	3.7	
PAI_1 (pg/mL) *					
nucleosomes low	59	285.5	21.6	269.2	**0.042**
nucleosomes high	31	387.4	29.8	305.2	

Abbreviations: SEM, standard error of the mean; uPA, urokinase plasminogen activator; PAI-1, plasminogen activator inhibitor-1. * uPA and PAI-1 were not determined in two patients. p-Values < 0.05 are written in Bold.

3.5. Prognostic Value of Nucleosomes on Disease-Free Survival in Primary Breast Cancer

At a median follow-up time of 55.0 months (range = 4.9–76.7 months), 23 patients (25.0%) had experienced a DFS event, and 15 patients (16.3%) had died. Herein, we present DFS analysis due to the immaturity of overall survival data. Patients with lower than mean nucleosomes had significantly better disease-free survival (HR = 0.46, 95% CI 0.19–1.12, p = 0.05) (Figure 1). The prognostic value of circulating nucleosomes was most pronounced in lymph node-positive disease with high proliferation rate and in patients with detectable circulating tumor cells with epithelial-to-mesenchymal transition, but negative for epithelial circulating tumor cells (Table 5). In a multivariate analysis, nucleosomes, hormone receptor status, HER2 status, lymph node involvement, and tumor grade were independent predictors of disease-free survival (Table 6).

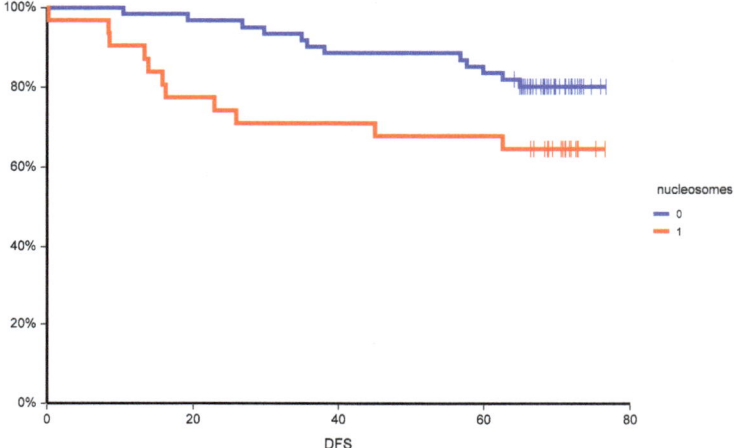

Figure 1. Kaplan–Meier estimates of probabilities of disease-free survival according to plasma nucleosome status in primary breast cancer patients (n = 92). HR = 0.46. 95% CI 0.19–1.12, p = 0.05, 0—nucleosomes below mean, 1—nucleosomes above mean.

Table 5. Prognostic value of nucleosomes on disease-free survival in primary breast cancer (nucleosomes dichotomized below vs. above mean).

Variable	HR	95% CI Low	95% CI High	p-Value
All	0.46	0.19	1.12	**0.05**
T-stage				
T1	0.29	0.08	1.02	**0.04**
>T1	0.56	0.14	2.19	0.33
Histology				
IDC	0.35	0.14	0.91	**0.01**
other	0	0	0	0.33
Grade				
low and intermediate	0.31	0.07	1.34	0.09
high grade	0.48	0.15	1.55	0.15
Lymph nodes				
N0	0.86	0.15	4.92	0.86
N+	0.36	0.13	1.04	**0.03**

Table 5. Cont.

Variable	HR	95% CI Low	95% CI High	p-Value
Lymphovascular invasion				
absent	0.46	0.14	1.53	0.15
present	0.54	0.15	1.97	0.31
Hormone receptor status (cut-off 1%)				
negative for both	0.36	0.04	3.33	0.21
positive for either	0.41	0.15	1.15	0.06
HER2 status				
negative	0.55	0.19	1.6	0.23
positive	0.3	0.06	1.55	0.09
P53 status				
negative	0.48	0.16	1.44	0.13
positive	0.39	0.08	1.86	0.20
BCL-2				
negative	0.25	0.05	1.17	**0.02**
positive	0.59	0.19	1.85	0.33
Ki67 status (cut-off 14%)				
<14%	0.73	0.11	4.71	0.72
>14%	0.35	0.12	1	**0.02**
CTC EP				
negative	0.31	0.12	0.83	**0.01**
positive	0	0	0	0.18
CTC EMT				
negative	0.68	0.23	2.03	0.46
positive	0.17	0.03	0.9	**0.01**
CTC ANY				
negative	0.45	0.13	1.52	0.14
positive	0.5	0.14	1.87	0.27

p-Values < 0.05 are written in Bold.

Table 6. Multivariate analysis of factors associated with disease-free survival.

Variable	HR	95% CI Low	95% CI High	p-Value
Nucleosomes above mean vs. below mean	2.67	1.12	6.36	**0.0268**
Hormone receptor status (cut-off 1%) positive for either vs. negative for both	0.30	0.11	0.80	**0.0164**
HER2 status amplified vs. nonamplified	3.06	1.21	7.79	**0.0187**
Lymph nodes positive vs. negative	6.56	2.50	17.21	**0.0001**
Grade grade 3 vs. grade 1 and 2	2.85	1.14	7.08	**0.0246**

p-Values < 0.05 are written in Bold.

Circulating nucleosomes added prognostic value also to prognostic value of CTC_EMT, where double-positive patients (positive for both CTC_EMT and high-circulating nucleosomes) had worse prognosis compared to all other groups of patients (Figure 2).

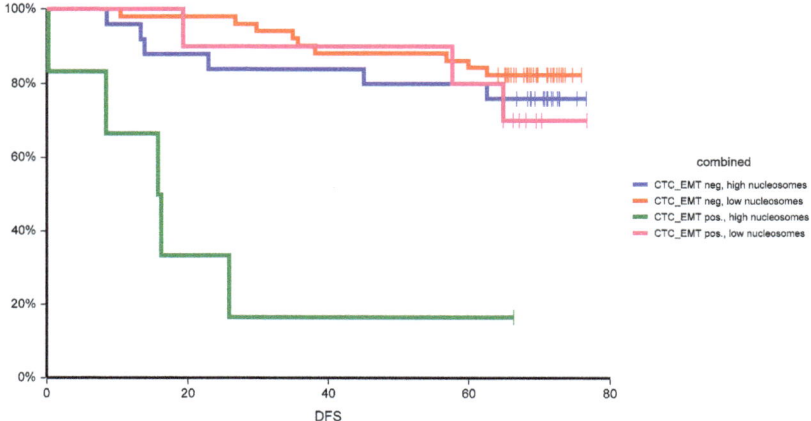

Figure 2. Kaplan–Meier estimates of probabilities of disease-free survival according to plasma nucleosome status and CTC_EMT in primary breast cancer patients (n = 92). Patients positive for CTC_EMT and high level of circulating nucleosomes had significantly worse survival compared to all other groups (p = 0.0000003).

4. Discussion

In this translational study, circulating nucleosomes showed neither an association with basic patient/tumor characteristics nor a correlation to CTCs. The origin of circulating nucleosomes is unclear and likely complex [25]. While there is no correlation between CTCs and SII and/or neutrophil/lymphocyte ratio [23,24], this study showed for the first time an association between plasma nucleosomes and SII. Patients with high SII had significantly higher level of nucleosomes. Similarly, there was a trend of higher nucleosomes in patients with high neutrophil/lymphocyte ratio, however, the neutrophil/lymphocyte ratio is part of the SII.

Tumor-induced systemic changes in immune cells contribute to cancer progression and metastasis. Various forms of ecDNA including extracellular nucleosomes and naked ecDNA differ in their cytotoxic and proinflammatory effects [26]. For example, histones in the nucleosomes induce proinflammatory signaling via toll-like receptors (TLR2/4), with subsequent production of TNF-α, IL-6, IL-10, and myeloperoxidase, but they exhibit TLR-independent cytotoxicity as well [26–28]. On the other hand, the ecDNA as part of the nucleosomes is recognized by the TLR9 [29]. In our study, nucleosomes were associated with several proinflammatory cytokines, suggesting the association of circulating nucleosomes with systemic inflammation. Histones in the nucleosomes could induce formation of neutrophil extracellular traps (NETs), which contain nucleosomes and stimulate further NETs production in a positive feedback loop [27]. On the other hand, nucleosomes could induce different inflammatory pathways, as they, in contrast to histones, seem not to be cytotoxic to the endothelium [28]. The analyzed nucleosomes could be from tumor cells, but also from the released NETs. This would explain the observed association between circulating nucleosomes and systemic inflammation in primary breast cancer patients. NETs contain nuclear DNA and proteins that possess antibacterial characteristics crucial for fighting pathogens [30,31]. The same NETs, however, also induce intravascular coagulation [32] and their overproduction can lead to autoimmune diseases [33]. While circulating ecDNA correlates with activation of coagulation [34], we for the first time describe this association for circulating nucleosomes. Further research is needed to uncover if nucleosomes directly activate PAI-1, or if high PAI-1 is a marker of coagulation activation in more aggressive disease that leads to release of more nucleosomes.

Data on the prognostic value of plasma nucleosomes in breast cancer is limited. In a small study, nucleosomes were elevated in locally confined and metastatic breast cancer in comparison to healthy individuals. During neoadjuvant chemotherapy, patients with no change of a local disease had significantly higher pretherapeutic concentrations of nucleosomes than patients in remission [14]. In another study, plasma nucleosomes were higher in primary breast cancer patients when compared to healthy controls, and similarly to our study, there was no association between nucleosomes and patient/tumor characteristics [15]. Circulating nucleosomes were, however, not able to discriminate between benign and malignant breast lesions [35]. Their concentration was found to be associated with lymph node-positive breast cancer and the presence of distant metastases [35].

In our study, we observed an inferior outcome of primary breast cancer patients with high plasma nucleosomes. This is in contrast to a previous study, where elevated plasma nucleosomes were associated with a better prognosis in both node-negative and node-positive early breast cancer [15]. However, the nucleosome detection method as well as the cut-off value to discriminate "low" and "high" plasma nucleosomes was different compared to our trial and therefore, these differences in results could be due to these factors. In our trial, the prognostic value of nucleosomes was consistent in various subgroups, however, it was most pronounced in poor prognostic subgroups such as lymph node-positive disease with high proliferation rate and in patients with detectable circulating tumor cells with epithelial-to-mesenchymal transition. The prognostic value of circulating nucleosomes was independent from established prognostic markers and was confirmed in a multivariate analysis. Moreover, when we combined two circulating biomarkers, circulating tumor cells, and circulating nucleosomes, we were able to uncover a subgroup of patients with extremely poor prognosis with two-year DFS of only 33.3%.

Our study has some limitations. The major one is small sample size, especially for associations between inflammatory indexes and nucleosomes. This is associated with decreased statistical power of analyses and increased confidence intervals of results. Other limitations represent the data availability for analysis of association between circulating nucleosomes and various clinic–pathological parameters, which further decreases statistical robustness and could have an impact on study results. Circulating plasma nucleosomes increase in non-neoplastic disease processes including inflammation, autoimmune diseases, sepsis, and stroke. When we analyzed association between chronic medication/comorbidities and circulating nucleosomes, no association was found, however, none of our patients received anti-inflammatory drugs and/or had inflammatory disease that could affect study results. Another limitation is lack of follow-up analysis on patient samples collected postsurgery to examine whether the presurgery baseline levels of circulating plasma nucleosomes were altered postsurgery and whether this alteration in circulating nucleosome levels is correlated with decrease in systemic inflammatory index.

5. Conclusions

In conclusion, in this translational study, we have shown for the first time that circulating nucleosomes are associated with systemic inflammation and activation of coagulation in primary breast cancer. More importantly, we proved their prognostic value. While it is clear that the underlying mechanisms of nucleosome release, their origin, and their fate require further studies, we suggest that the quantification of plasma nucleosomes could be added to the established prognostic markers in breast cancer. Future trials should focus on validation of these results to establish prognostic utility of plasma circulation nucleosomes in addition to established prognostic factors.

Author Contributions: Conceptualization, P.C., and M.M.; Data curation, G.M., T.S., and K.K.; Formal analysis, G.M., T.S., P.G., D.C., and P.C.; Funding acquisition, P.C., G.M., and M.M.; Investigation, M.K., G.M., J.B., and M.M.; Methodology, G.M., T.S., and P.C.; Project administration, P.C. and M.M.; Resources, J.M., K.K., M.K., J.B., and M.M.; Validation, P.C. and M.M.; Visualization, K.K., D.P., and J.M.; Writing—Original draft, P.C., and M.M.; Writing—Review & editing, all authors. All authors have read and agreed to the published version of the manuscript.

Funding: This research was funded by the Slovak Research and Development Agency (APVV), grant number APVV-16-0010, APVV-16-0178, by ERA-NET EuroNanoMed II INNOCENT and by Scientific Grant Agency (VEGA), contracts No. 1/0724/11, 1/0044/15, 1/0271/17, and 2/0052/18.

Acknowledgments: We would like to acknowledge Denisa Manasova for her excellent technical help. We are grateful to all patients for their participation in the study.

Conflicts of Interest: The authors declare no conflict of interest.

Appendix A

Table A1. Drug history in the last 6 months.

Chronic Medication	N	%
NSAID	0	0.0
Corticosteroids	0	0.0
L-thyroxin	8	8.7
ACEi	11	12.0
Sartans	14	15.2
Betablockers	27	29.3
Statins	15	16.3
Metformin	4	4.3
Insulin	2	2.2
LMWH	7	7.6
Warfarin	0	0.0

Abbreviations: ACEi, angiotensin-converting enzyme inhibitors; LMWH, low-molecular-weight heparin.

References

1. Bray, F.; Ferlay, J.; Soerjomataram, I.; Siegel, R.L.; Torre, L.A.; Jemal, A. Global cancer statistics 2018: GLOBOCAN estimates of incidence and mortality worldwide for 36 cancers in 185 countries. *CA Cancer J. Clin.* **2018**, *68*, 394–424. [CrossRef]
2. McAnena, P.; Brown, J.A.; Kerin, M.J. Circulating Nucleosomes and Nucleosome Modifications as Biomarkers in Cancer. *Cancers* **2017**, *9*, 5. [CrossRef] [PubMed]
3. Lui, Y.Y.; Chik, K.W.; Chiu, R.W.; Ho, C.Y.; Lam, C.W.; Lo, Y.M. Predominant hematopoietic origin of cell-free DNA in plasma and serum after sex-mismatched bone marrow transplantation. *Clin. Chem.* **2002**, *48*, 421–427. [CrossRef] [PubMed]
4. Sun, K.; Jiang, P.; Chan, K.C.; Wong, J.; Cheng, Y.K.; Liang, R.H.; Chan, W.K.; Ma, E.S.; Chan, S.L.; Cheng, S.H.; et al. Plasma DNA tissue mapping by genome-wide methylation sequencing for noninvasive prenatal, cancer, and transplantation assessments. *Proc. Natl. Acad. Sci. USA* **2015**, *112*, E5503–E5512. [CrossRef] [PubMed]
5. Kustanovich, A.; Schwartz, R.; Peretz, T.; Grinshpun, A. Life and death of circulating cell-free DNA. *Cancer Biol.* **2019**, *20*, 1057–1067. [CrossRef] [PubMed]
6. Khorasanizadeh, S. The nucleosome: From genomic organization to genomic regulation. *Cell* **2004**, *116*, 259–272. [CrossRef]
7. Jiang, N.; Reich, C.F., 3rd; Monestier, M.; Pisetsky, D.S. The expression of plasma nucleosomes in mice undergoing in vivo apoptosis. *Clin. Immunol.* **2003**, *106*, 139–147. [CrossRef]
8. Holdenrieder, S.; Nagel, D.; Schalhorn, A.; Heinemann, V.; Wilkowski, R.; von Pawel, J.; Raith, H.; Feldmann, K.; Kremer, A.E.; Muller, S.; et al. Clinical relevance of circulating nucleosomes in cancer. *Ann. N.Y. Acad. Sci.* **2008**, *1137*, 180–189. [CrossRef]
9. Zeerleder, S.; Zwart, B.; Wuillemin, W.A.; Aarden, L.A.; Groeneveld, A.B.; Caliezi, C.; van Nieuwenhuijze, A.E.; van Mierlo, G.J.; Eerenberg, A.J.; Lammle, B.; et al. Elevated nucleosome levels in systemic inflammation and sepsis. *Crit. Care Med.* **2003**, *31*, 1947–1951. [CrossRef]
10. Pisetsky, D.S. The complex role of DNA, histones and HMGB1 in the pathogenesis of SLE. *Autoimmunity* **2014**, *47*, 487–493. [CrossRef]

11. Bauden, M.; Pamart, D.; Ansari, D.; Herzog, M.; Eccleston, M.; Micallef, J.; Andersson, B.; Andersson, R. Circulating nucleosomes as epigenetic biomarkers in pancreatic cancer. *Clin. Epigenet.* **2015**, *7*, 106. [CrossRef] [PubMed]
12. Rahier, J.F.; Druez, A.; Faugeras, L.; Martinet, J.P.; Gehenot, M.; Josseaux, E.; Herzog, M.; Micallef, J.; George, F.; Delos, M.; et al. Circulating nucleosomes as new blood-based biomarkers for detection of colorectal cancer. *Clin. Epigenet.* **2017**, *9*, 53. [CrossRef] [PubMed]
13. Wang, K.; Shan, S.; Wang, S.; Gu, X.; Zhou, X.; Ren, T. HMGB1-containing nucleosome mediates chemotherapy-induced metastasis of human lung cancer. *Biochem. Biophys. Res. Commun.* **2018**, *500*, 758–764. [CrossRef] [PubMed]
14. Stoetzer, O.J.; Fersching, D.M.; Salat, C.; Steinkohl, O.; Gabka, C.J.; Hamann, U.; Braun, M.; Feller, A.M.; Heinemann, V.; Siegele, B.; et al. Prediction of response to neoadjuvant chemotherapy in breast cancer patients by circulating apoptotic biomarkers nucleosomes, DNAse, cytokeratin-18 fragments and survivin. *Cancer Lett.* **2013**, *336*, 140–148. [CrossRef] [PubMed]
15. Kuroi, K.; Tanaka, C.; Toi, M. Plasma Nucleosome Levels in Node-Negative Breast Cancer Patients. *Breast Cancer* **1999**, *6*, 361–364. [CrossRef] [PubMed]
16. Wittwer, C.; Boeck, S.; Heinemann, V.; Haas, M.; Stieber, P.; Nagel, D.; Holdenrieder, S. Circulating nucleosomes and immunogenic cell death markers HMGB1, sRAGE and DNAse in patients with advanced pancreatic cancer undergoing chemotherapy. *Int. J. Cancer* **2013**, *133*, 2619–2630. [CrossRef] [PubMed]
17. Holdenrieder, S.; Stieber, P.; von Pawel, J.; Raith, H.; Nagel, D.; Feldmann, K.; Seidel, D. Circulating nucleosomes predict the response to chemotherapy in patients with advanced non-small cell lung cancer. *Clin. Cancer Res.* **2004**, *10*, 5981–5987. [CrossRef] [PubMed]
18. Mego, M.; Karaba, M.; Minarik, G.; Benca, J.; Silvia, J.; Sedlackova, T.; Manasova, D.; Kalavska, K.; Pindak, D.; Cristofanilli, M.; et al. Circulating Tumor Cells With Epithelial-to-mesenchymal Transition Phenotypes Associated With Inferior Outcomes in Primary Breast Cancer. *Anticancer Res.* **2019**, *39*, 1829–1837. [CrossRef]
19. Mego, M.; Cierna, Z.; Janega, P.; Karaba, M.; Minarik, G.; Benca, J.; Sedlackova, T.; Sieberova, G.; Gronesova, P.; Manasova, D.; et al. Relationship between circulating tumor cells and epithelial to mesenchymal transition in early breast cancer. *BMC Cancer* **2015**, *15*, 533. [CrossRef]
20. Mego, M.; Mani, S.A.; Lee, B.N.; Li, C.; Evans, K.W.; Cohen, E.N.; Gao, H.; Jackson, S.A.; Giordano, A.; Hortobagyi, G.N.; et al. Expression of epithelial-mesenchymal transition-inducing transcription factors in primary breast cancer: The effect of neoadjuvant therapy. *Int. J. Cancer* **2012**, *130*, 808–816. [CrossRef]
21. Mego, M.; Karaba, M.; Minarik, G.; Benca, J.; Sedlackova, T.; Tothova, L.; Vlkova, B.; Cierna, Z.; Janega, P.; Luha, J.; et al. Relationship between circulating tumor cells, blood coagulation, and urokinase-plasminogen-activator system in early breast cancer patients. *Breast J.* **2015**, *21*, 155–160. [CrossRef]
22. Mego, M.; Cholujova, D.; Minarik, G.; Sedlackova, T.; Gronesova, P.; Karaba, M.; Benca, J.; Cingelova, S.; Cierna, Z.; Manasova, D.; et al. CXCR4-SDF-1 interaction potentially mediates trafficking of circulating tumor cells in primary breast cancer. *BMC Cancer* **2016**, *16*, 127. [CrossRef] [PubMed]
23. De Giorgi, U.; Mego, M.; Scarpi, E.; Giordano, A.; Giuliano, M.; Valero, V.; Alvarez, R.H.; Ueno, N.T.; Cristofanilli, M.; Reuben, J.M. Association between circulating tumor cells and peripheral blood monocytes in metastatic breast cancer. *Adv. Med. Oncol.* **2019**, *11*, 1758835919866065. [CrossRef] [PubMed]
24. Miklikova, S.; Minarik, G.; Sedlackova, T.; Plava, J.; Cihova, M.; Jurisova, S.; Kalavska, K.; Karaba, M.; Benca, J.; Smolkova, B.; et al. Inflammation-Based Scores Increase the Prognostic Value of Circulating Tumor Cells in Primary Breast Cancer. *Cancers* **2020**, *12*, 1134. [CrossRef] [PubMed]
25. Thierry, A.R.; El Messaoudi, S.; Gahan, P.B.; Anker, P.; Stroun, M. Origins, structures, and functions of circulating DNA in oncology. *Cancer Metastasis Rev.* **2016**, *35*, 347–376. [CrossRef] [PubMed]
26. Chaudhary, S.; Mittra, I. Cell-free chromatin: A newly described mediator of systemic inflammation. *J. Biosci.* **2019**, *44*, 32. [CrossRef] [PubMed]
27. Abrams, S.T.; Zhang, N.; Manson, J.; Liu, T.; Dart, C.; Baluwa, F.; Wang, S.S.; Brohi, K.; Kipar, A.; Yu, W.; et al. Circulating histones are mediators of trauma-associated lung injury. *Am. J. Respir. Crit. Care Med.* **2013**, *187*, 160–169. [CrossRef]
28. Xu, J.; Zhang, X.; Pelayo, R.; Monestier, M.; Ammollo, C.T.; Semeraro, F.; Taylor, F.B.; Esmon, N.L.; Lupu, F.; Esmon, C.T. Extracellular histones are major mediators of death in sepsis. *Nat. Med.* **2009**, *15*, 1318–1321. [CrossRef]

29. Kang, T.H.; Mao, C.P.; Kim, Y.S.; Kim, T.W.; Yang, A.; Lam, B.; Tseng, S.H.; Farmer, E.; Park, Y.M.; Hung, C.F. TLR9 acts as a sensor for tumor-released DNA to modulate anti-tumor immunity after chemotherapy. *J. Immunother. Cancer* **2019**, *7*, 260. [CrossRef]
30. Yang, H.; Biermann, M.H.; Brauner, J.M.; Liu, Y.; Zhao, Y.; Herrmann, M. New Insights into Neutrophil Extracellular Traps: Mechanisms of Formation and Role in Inflammation. *Front. Immunol.* **2016**, *7*, 302. [CrossRef]
31. Brinkmann, V.; Reichard, U.; Goosmann, C.; Fauler, B.; Uhlemann, Y.; Weiss, D.S.; Weinrauch, Y.; Zychlinsky, A. Neutrophil extracellular traps kill bacteria. *Science* **2004**, *303*, 1532–1535. [CrossRef] [PubMed]
32. McDonald, B.; Davis, R.P.; Kim, S.J.; Tse, M.; Esmon, C.T.; Kolaczkowska, E.; Jenne, C.N. Platelets and neutrophil extracellular traps collaborate to promote intravascular coagulation during sepsis in mice. *Blood* **2017**, *129*, 1357–1367. [CrossRef] [PubMed]
33. Kaplan, M.J.; Radic, M. Neutrophil extracellular traps: Double-edged swords of innate immunity. *J. Immunol.* **2012**, *189*, 2689–2695. [CrossRef]
34. Von Meijenfeldt, F.A.; Burlage, L.C.; Bos, S.; Adelmeijer, J.; Porte, R.J.; Lisman, T. Elevated Plasma Levels of Cell-Free DNA During Liver Transplantation Are Associated With Activation of Coagulation. *Liver Transpl.* **2018**, *24*, 1716–1725. [CrossRef] [PubMed]
35. Roth, C.; Pantel, K.; Muller, V.; Rack, B.; Kasimir-Bauer, S.; Janni, W.; Schwarzenbach, H. Apoptosis-related deregulation of proteolytic activities and high serum levels of circulating nucleosomes and DNA in blood correlate with breast cancer progression. *BMC Cancer* **2011**, *11*, 4. [CrossRef] [PubMed]

 © 2020 by the authors. Licensee MDPI, Basel, Switzerland. This article is an open access article distributed under the terms and conditions of the Creative Commons Attribution (CC BY) license (http://creativecommons.org/licenses/by/4.0/).

Article

Tracking the Progression of Triple Negative Mammary Tumors over Time by Chemometric Analysis of Urinary Volatile Organic Compounds

Mark Woollam [1,2], Luqi Wang [2,3], Paul Grocki [1,2], Shengzhi Liu [2,3], Amanda P. Siegel [1,2], Maitri Kalra [4], John V. Goodpaster [1], Hiroki Yokota [2,3,5] and Mangilal Agarwal [1,2,6,*]

1. Department of Chemistry and Chemical Biology, Indiana University-Purdue University, Indianapolis, IN 46202, USA; mwoollam@iu.edu (M.W.); pgrocki@iu.edu (P.G.); apsiegel@iupui.edu (A.P.S.); jvgoodpa@iupui.edu (J.V.G.)
2. Integrated Nanosystems Development Institute, Indiana University-Purdue University, Indianapolis, IN 46202, USA; luqiwang@ccmu.edu.cn (L.W.); liu441@iupui.edu (S.L.); hyokota@iupui.edu (H.Y.)
3. Department of Biomedical Engineering, Indiana University-Purdue University, Indianapolis, IN 46202, USA
4. Hematology and Oncology, Ball Memorial Hospital, Indiana University Health, Muncie, IN 47303, USA; mkalra@IUHealth.org
5. Biomechanics and Biomaterials Research Center, Indiana University-Purdue University, Indianapolis, IN 46202, USA
6. Department of Mechanical & Energy Engineering, Indiana University-Purdue University, Indianapolis, IN 46202, USA
* Correspondence: agarwal@iupui.edu

Citation: Woollam, M.; Wang, L.; Grocki, P.; Liu, S.; Siegel, A.P.; Kalra, M.; Goodpaster, J.V.; Yokota, H.; Agarwal, M. Tracking the Progression of Triple Negative Mammary Tumors over Time by Chemometric Analysis of Urinary Volatile Organic Compounds. *Cancers* 2021, 13, 1462. https://doi.org/10.3390/cancers13061462

Academic Editor: Fabrizio Bianchi

Received: 3 February 2021
Accepted: 17 March 2021
Published: 23 March 2021

Publisher's Note: MDPI stays neutral with regard to jurisdictional claims in published maps and institutional affiliations.

Copyright: © 2021 by the authors. Licensee MDPI, Basel, Switzerland. This article is an open access article distributed under the terms and conditions of the Creative Commons Attribution (CC BY) license (https://creativecommons.org/licenses/by/4.0/).

Simple Summary: Volatile organic compounds (VOCs) in urine have been shown to be potential biomarkers for breast cancer. However, how urinary VOCs change upon the course of tumor progression has never been studied. The aim of our study was to identify changes in VOC profiles corresponding to mammary tumor (triple negative cells) presence and progression in mice models of induced breast cancer. Urine samples were collected from mice prior to tumor injection and from days 2–19 after. VOC models constructed by linear discriminant analysis had high ability to distinguish tumor-bearing mice from control and determine the week of urine collection after tumor injection. Principal component regression analysis demonstrated that VOCs could predict the number of days since tumor injection. VOCs identified from these analyses correspond to metabolic pathways dysregulated by breast cancer and previous biomarker investigations. It is anticipated that these findings can be translated into human research for early detection of breast cancer recurrence.

Abstract: Previous studies have shown that volatile organic compounds (VOCs) are potential biomarkers of breast cancer. An unanswered question is how urinary VOCs change over time as tumors progress. To explore this, BALB/c mice were injected with 4T1.2 triple negative murine tumor cells in the tibia. This typically causes tumor progression and osteolysis in 1–2 weeks. Samples were collected prior to tumor injection and from days 2–19. Samples were analyzed by headspace solid phase microextraction coupled to gas chromatography–mass spectrometry. Univariate analysis identified VOCs that were biomarkers for breast cancer; some of these varied significantly over time and others did not. Principal component analysis was used to distinguish Cancer (all Weeks) from Control and Cancer Week 1 from Cancer Week 3 with over 90% accuracy. Forward feature selection and linear discriminant analysis identified a unique panel that could identify tumor presence with 94% accuracy and distinguish progression (Cancer Week 1 from Cancer Week 3) with 97% accuracy. Principal component regression analysis also demonstrated that a VOC panel could predict number of days since tumor injection ($R^2 = 0.71$ and adjusted $R^2 = 0.63$). VOC biomarkers identified by these analyses were associated with metabolic pathways relevant to breast cancer.

Keywords: volatile organic compounds (VOCs); gas chromatography (GC); mass spectrometry (MS); headspace solid phase microextraction (HS-SPME); breast cancer biomarkers; principal component analysis (PCA); linear discriminant analysis (LDA); principal component regression (PCR)

1. Introduction

Breast cancer is estimated to comprise 30% of total diagnosed cancer cases for women in 2021: over 280,000 patients will be diagnosed with and over 40,000 patients will die from breast cancer [1]. Accurate and efficient screening/diagnostics is crucial, as the earlier breast cancer is detected, the more efficacious the treatment [2]. Biopsies are used for diagnostic confirmation and pathological grading. Breast cancer staging is based on tumor size and number of lymph nodes affected, but aggressiveness depends on its dynamic rate of change. Breast cancer imaging techniques are used for monitoring tumor progression and morphological responses to treatment [3]. These tools are expensive, unreliable, cause harm through exposure to radiation [4] and can only detect morphological changes six to eight weeks after treatment [5]. There is a growing need for a less invasive and more accurate tool for early detection at the time of diagnosis or for cancer recurrence. An accurate and noninvasive assay to diagnose and monitor tumor progression could aid in patient decision making after diagnosis and possibly during treatment.

Previous studies have demonstrated canine's ability to detect the presence of prostate [6], lung [7], breast [8,9] and ovarian [10] cancer in biosamples with high accuracy [11,12]. Canines noninvasively detect volatile metabolites generated by the disease condition, allowing them to accurately detect cancer by scent. Based on canine results, groups have used headspace solid phase microextraction (HS-SPME) or other extraction techniques coupled with gas chromatography-mass spectrometry (GC-MS) to conduct untargeted analyses of volatile organic compounds (VOCs). Cancer dysregulates metabolic pathways to enable tumor growth [13]. The biological rationale for exploiting VOCs is they are by- or end products of these dysregulated pathways [14]. Furthermore, VOCs can be noninvasively sampled and detected in human biofluids including sweat, saliva, blood, breath and urine. Groups have previously identified VOC biomarkers for lung [15], prostate [16–19] and breast cancer [20], even classifying unique cancers from each other [21,22].

There has been an interest in using VOCs and other types of molecular biomarkers [23–25] in urine or breath for breast cancer diagnostics. One group analyzing alveolar breath by GC-MS found ten VOCs, that could distinguish breast cancer in patients with sensitivity = 75.3% and specificity = 84.8% [20]. Another group also detected a unique biosignature of six VOCs for breast cancer in human urine via unsupervised multivariate statistical analysis [26]. That group subsequently implemented a central composite design to optimize method parameters and used them to identify ten additional breast cancer volatile biomarkers that had accuracy of >90% [27]. The present study analyzes urinary VOCs in mice with breast cancer using GC-MS not simply to identify VOCs of breast cancer, but to learn how VOCs change in different conditions associated with breast cancer. For example, we previously analyzed changing patterns in VOCs caused by tumor location [28] or effect of treatments [29–31]. Interestingly, our murine research has previously reported almost half the VOCs tentatively identified by a different research group analyzing human breath for VOCs of breast cancer [20] and for the sixteen VOCs reported in the *Silva* papers, which incubated under different conditions, about half if including isomers and other highly similar molecules. Additionally, urinary VOC biomarkers for breast cancer and tumor progression may be useful in conjunction with circulating biomarkers of breast cancer progression. For an example, *Ibrahim* et al. found the receptor activator of nuclear factor-κB ligand (RANKL) to be a potential circulating biomarker for bone metastasis [32]. Correlating the urinary biomarkers from this study to RANKL and other circulating biomarkers which may be identified would help validate the candidates identified in this study. Additionally, circulating tumor and urinary VOCs can be coupled, which could also potentially improve classification accuracies and acceptance of alternative assays for breast cancer diagnosis, prognosis and monitoring the efficacy of treatment.

All of our previous studies analyzed samples from mice collected at the same time point (three weeks after tumor injection); none analyzed VOCs at intermediate points of time. There are many murine models of cancer progression, but in the current study, BALB/c mice were injected with 4T1.2 cells into the tibia. Tumors injected in this way progress and typically produce osteolytic lesions in 1–2 weeks [33]. A previous study by some of the authors, using such a model, found decreases in bone stiffness which would be indicative of osteolytic lesions after one week and before two weeks [34]. Detailed and additional biological information regarding this cohort of mice has been previously published [30]. For the current study, urine samples were collected from day 2–day 19 and grouped by number of weeks after injection. Analyses included comparisons between samples by week collected and regression analysis of samples by day collected to observe trends. Analysis of VOCs over time may help identify differences in VOC patterns to determine which biomarkers are better predictors of late-stage cancers and which are equally effective at identifying early-stage cancers.

2. Materials and Methods

2.1. Materials and Instrumentation

Female BALB/c mice (6 weeks old) were purchased from Harlan Laboratories, Indianapolis, IN, USA. 4T1.2 tumor cells were acquired from Dr. R. Anderson at the Peter MacCallum Cancer Institute (Melbourne, Victoria, Australia). Glass Pasteur pipettes were used for urine collection and purchased from Thermo Scientific (Waltham, MA, USA). 10 mL headspace vials were purchased from Thermo Scientific. Guanidine Hydrochloride (GHCl) (pH = 8.5) was purchased from Sigma Aldrich (St. Louis, MO, USA) and was used as a major urinary protein (MUP) denaturing agent. Two-centimeter polydimethylsiloxane/carboxen/divinylbenzene (PDMS/CAR/DVB) SPME fibers (Supelco; Bellefonte, PA, USA) were employed to concentrate and extract VOCs. A 7890A GC system coupled to an Agilent (Santa Clara, CA, USA) 7200 Accurate-Mass Quadrupole time-of-flight (QTOF) MS system with a PAL autosampling system (CTC Analytics; Raleigh, NC, USA) was used to analyze VOCs. An Agilent Ultra Inert HP-5ms, GC column of 30 m in length, 250 μm internal diameter and 0.25 μm film thickness was utilized. MATLAB (R2020a; Natick, MA, USA) and Origin (Northampton, MA, USA) were used in generating figures for chemometric analyses.

2.2. Tumor Injection and Urine Collection

A total of 20 mice were injected with 4T1.2 cells (triple negative mammary tumors) into the right tibia [30]. Prior to tumor injection, urine was collected from the mice to serve as the Control group. Urine was collected from the 20 mice the day following tumor injection and over the course of three weeks. It is important to note that not all mice provided urine samples at each time point. All experimental procedures followed the Guiding Principles in the Care and Use of Animals supported by the American Physiological Society and were approved by the Indiana University Animal Care and Use Committee (protocol code: SC292R; date of approval: 30 May 2019). Mice were kept in cages at ambient temperature and fed the same diet (mouse-chow ad libitum). Mice were transferred to a cage where the floor was covered in parafilm during urine collection. Urine was collected over dry ice using glass Pasteur pipettes into glass centrifuge tubes and centrifuged at 3000 RPM. A total of 50 μL was transferred to a 10 mL headspace vial and stored in a −80 °C freezer.

2.3. HS-SPME and GC-MS QTOF Analysis

Urinary VOCs were detected through headspace analysis utilizing a SPME fiber and GC-MS QTOF. The SPME fiber was conditioned before the first sample each day and between each run. As the mice gave limited amounts of urine, only one aliquot was analyzed per mouse. GHCl was added in a 1:1 volumetric ratio one hour prior to GC analysis to denature the MUPs in mouse urine that bind VOCs [35]. Next, the sample was agitated at 250 rpm and heated to 60 °C for 30 min. Then, the SPME fiber was inserted into the vial for an additional 30 min (same agitation rate and temperature). The fiber was injected into the GC inlet at 250 °C for two minutes to thermally desorb the VOCs. The chromatographic protocol involved maintaining the oven temperature at 40 °C for two minutes followed by a ramp to 100 °C at a rate of 8 °C/min, a 15 °C/min ramp to 120 °C, an 8 °C/min to 180 °C, a 15 °C/min to 200 °C and finally an 8 °C/min ramp to 260 °C. An external reference standard was run each day to verify instrument reproducibility.

2.4. Data Treatment and Chemometric Analyses

Deconvolution and spectral alignment of chromatographic peaks based on similarities in mass-to-charge ratio (m/z) and retention time were performed in MassHunter Profinder. Features identified as silanes/siloxanes (products of SPME degradation) were removed. VOCs that did not appear in at least 50% of either Control or Cancer Weeks 1–3 samples were also excluded. To normalize the data, MS Total Useful Signal (MSTUS) was calculated and applied to remove unwanted non-biological intraclass variation [36]. Finally, MSTUS values were autoscaled (z-scored) to obtain a matrix with similar signal range. Univariate statistical analysis was implemented (two-tailed Student's T-test) on the Control group against tumor-bearing mice with urine collected on days 2, 5 and 6 (Week 1), urine collected on days 8, 9, 12 and 13 (Week 2) and finally, urine collected on days 16, 17 and 19 (Week 3). These analyses were implemented to identify VOCs differentially expressed (p-value < 0.05) between Cancer Weeks 1–3 and Control as well as between Cancer Week 1 and Cancer Week 3. p-values were adjusted utilizing the Benjamini–Hochberg procedure [37] to account for false discovery rates (FDR). Hierarchical heatmaps were generated for VOCs statistically significant by p-value < 0.05 for Control against Weeks 1–3 and Week 1 against Week 3 to visualize changes in VOC concentration induced by cancer injection and progression.

Principal component analysis (PCA) was performed using all VOCs with p-value < 0.05 between the Control group and all Cancer groups. PCA was also implemented on a smaller group of VOCs with the lowest p-value (for Cancer/Control and Cancer Week 1/Week 3) to separate Control, Cancer Week 1 and Cancer Week 3. The matrix of VOCs was then subject to supervised linear discriminant analysis coupled with forward feature selection (iterative LDA, iLDA) [38] to build predictive classification models. PCA and iLDA were performed independently of each other in parallel. iLDA was used to develop VOC panels separating Control vs. Cancer, Cancer Week 1 vs. Cancer Week 3 and lastly, Cancer Week 1 vs. Cancer Week 3 vs. Control. Leave one out cross validation (LOOCV) and fivefold cross validation (partitioned 1000 times, median value utilized) were performed to determine if the models were overfit [39]. Receiver operator characteristic (ROC) curves for each model were built to visualize classification accuracies. If the area under the curve (AUC) of the ROC differed more than 0.10 between the training and cross validation data sets, the model was deemed overfit.

2.5. Regression Analyses

To further investigate the ability of individual VOCs to monitor mammary tumor progression, linear regression analysis was undertaken for VOCs with p-value < 0.05 (Cancer Week 1 vs. Cancer Week 3). Here, Cancer samples were analyzed by day of urine collection after tumor injection. Principal component regression (PCR) analysis was also implemented to identify if a panel of VOCs can track mammary tumor progression by days after injection. PCR proceeds by running PCA on the table of the explanatory (input) variables. Then, an Ordinary Least Squares regression is completed on a group of principal components selected by the user. Finally, PCR computes the parameters of the model that correspond to the explanatory (input) variables. The number of principal components utilized was varied and tested to ensure the production of a stable model. The first iteration of analysis was followed by dimension reduction to only include VOCs that significantly contributed toward the PCR model (p-value < 0.05). Analysis of variance (ANOVA) was implemented to determine if there was a significant linear correlation between the independent and the dependent variables. Determination coefficients (R^2 value), regression coefficients and standard errors were used to assess the degree of correlation.

2.6. VOC Identification and Metabolic Pathway Analysis

After data screening and analysis, volatiles were identified using MassHunter Profinder and MassHunter Unknowns Analysis with the NIST17 library. The data set produced through Profinder contained average retention times, retention time span and the mass spectra of the VOCs. Utilizing these quantifiers, VOCs in Profinder were found in Unknowns Analysis. Features in Unknowns Analysis were assigned a match factor from the NIST17 library; VOCs identified with a match factor greater than 70 and an appropriate experimental non-polar retention index (NPRI) value were deemed tentatively identified. Experimental NPRI was determined using an instrument-specific calibration curve [28,29]. The Human Metabolome Database [40] and Kyoto Encyclopedia of Genes and Genomes Pathways [41] were used to aid in interpreting the relevance of VOCs in the context of cancer metabolism.

3. Results

3.1. Urine Collection, Spectral Alignment and Data Normalization

An illustration of the experimental procedure that was implemented to identify VOC biomarkers of mammary tumor progression in mouse urine can be visualized in Figure 1. A total of 65 urine samples were collected, aliquoted and analyzed from four different sample classes over the course of three weeks (Control (20), Cancer Week 1 (12), Cancer Week 2 (15) and Cancer Week 3 (18)). Spectral alignment of sample chromatograms generated a matrix of 250 VOCs which were subject to chemometric analyses after removing silanes/siloxanes and volatiles not detected in at least half of either Control or Cancer samples classes.

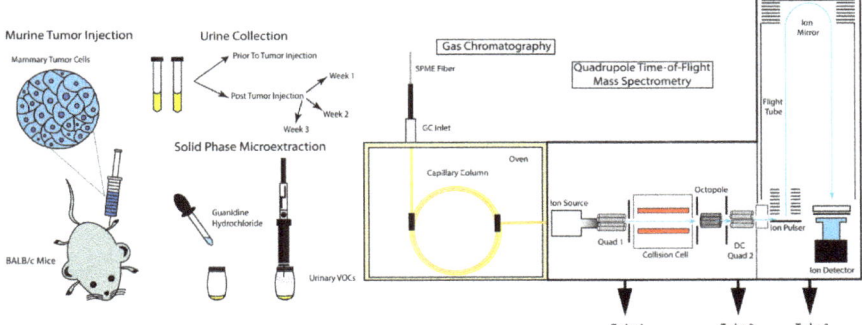

Figure 1. Illustration of murine tumor injection, mouse urine sample collection, sample treatment and analysis via solid phase microextraction coupled to gas chromatography-quadrupole time-of-flight mass spectrometry to identify volatile organic compound (VOC) biomarkers for breast cancer and tumor progression.

3.2. Univariate Statistical Analysis

After normalization, Student's T-test was performed between Control and All Cancer classes and identified 44 out of 250 VOCs with p-value < 0.05. Additionally, the T-test found 37 VOCs with p-value < 0.05 when applied between the Cancer Week 1 and Week 3 sample classes. These VOCs are identified and listed in Table S1 with their corresponding name, retention time and p-values. After adjusting p-values for FDR, 18 VOCs with p-value < 0.05 were found between Control and Cancer (italicized in Table S1) and 3 VOCs were identified with p-value < 0.05 between Week 1 and Week 3 (underlined in Table S1). Of the 44 VOCs identified with p-value < 0.05 between Control and Weeks 1–3, 37 features were downregulated and 7 upregulated. With regards to the 37 significant VOCs between Cancer Week 1 and Week 3, 24 were downregulated and 13 upregulated. Downregulated features were more significant than upregulated ones for both comparisons. Hierarchical heatmaps were generated using statistically significant VOCs between each comparison and are shown in Figure 2a,b. The heatmap corresponding to Control vs. Weeks 1–3, shows low intraclass variation and high interclass variation (Figure 2a). Figure 2b shows many VOCs are downregulated in Week 3, indicating the concentration of these VOCs decreases as cancer progresses. The heatmap displays high intraclass variation for VOCs with $p < 0.05$ detected in Week 2, indicating some mice progressed faster than others, or possibly some mice had undergone tumor-induced osteolysis and some had not yet [30], but no imaging or invasive studies were undertaken during intermediate time points, so this is not confirmed. In Figure 2, abbreviations for VOCs that show high statistical significance and are utilized for further analyses are indicated.

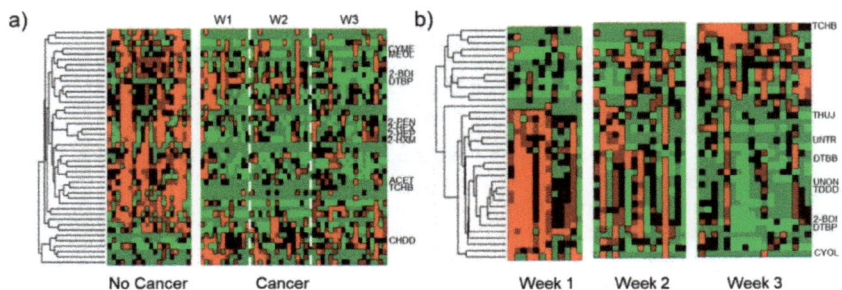

Figure 2. Hierarchical heatmaps for the VOCs identified with p-value < 0.05 in the (**a**) Cancer Weeks 1–3 vs. Control and (**b**) Cancer Week 1 vs. Cancer Week 3 comparisons. These plots show an abundant number of VOCs differentially expressed due to the presence of cancer and tumor progression. Full names of VOCs used for further analyses (but here abbreviated) are enumerated in the text and all VOCs shown in the heatmap and associated p values are listed in Table S1.

3.3. Multivariate Classification Analyses

PCA was utilized to visualize global patterns in the data. PCA using all 44 VOCs with p-value < 0.05 between Cancer Weeks 1–3 and Control can be observed in Figure 3a. Along the first two principal components, all Cancer samples were separated from Control samples with 98% sensitivity and 95% specificity. A smaller panel of 10 VOCs with low p-values for both tumor presence and progression were selected using an ad hoc approach and PCA was run using this smaller set (Figure 3b). For this analysis, Cancer Week 2 was excluded as it is intermediary and the goal was to observe significant differences at the two endpoints. Cancer Week 1 and Week 3 samples were separated from Control samples with sensitivity = 97% and specificity = 90%. Principal component 1 demonstrates sample separation between Cancer and Control samples and Principal component 2 strongly contributed toward sample separation between Cancer Week 1 and Cancer Week 3. VOCs used in Figure 3b are labeled in Table S1 with an asterisk (*).

Even though PCA produced a reasonable separation, this required a relatively large number of VOCs. To build a predictive classification model and decrease the number of VOCs used, iLDA was used to distinguish Cancer Weeks 1–3 from Control samples. Knowledge-based feature selection was implemented by limiting the analysis to ketones, aromatics and terpenes as these functional groups have been previously reported by our team to be potential biomarkers [28,29]. A panel of five VOCs (cymene (CYME), acetone (ACET), 2-heptanone (2-HEP), 2,5-cyclohexadiene-1,4-dione, 2,6-bis(1,1-dimethylethyl)- (CHDD) and 2-hexanone, 5-methyl (2-HXM)) could classify tumor presence (Cancer Weeks 1–3 from Control) with an AUC equal to 0.99 in the training set (sensitivity = 98% and specificity = 95%) (the First LDA Model). The one-dimensional LDA plot can be observed in Figure 4a and it is clear the first linear discriminant accounted for the significant differences between the two sample classes. Data perturbation techniques were implemented to test the robustness of the classification models [39,42]. LOOCV (AUC = 0.97) and fivefold cross validation (AUC = 0.98) showed values similar to the training data, demonstrating the model was not overfit. The respective ROC curves can be seen in Figure 4b and the two-dimensional LDA plot is illustrated in Figure 4c.

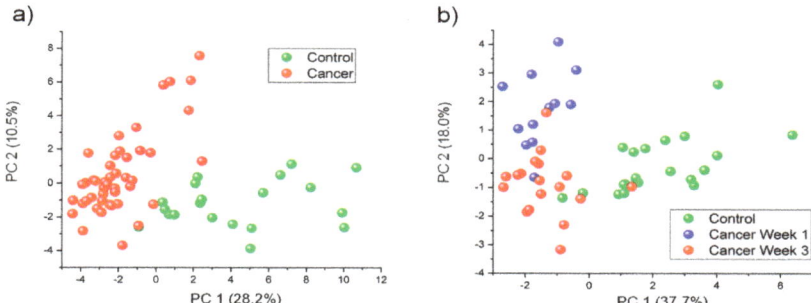

Figure 3. (a) PCA using all VOCs with $p < 0.05$ between all Cancer samples and Control samples (44 VOCs). This panel can distinguish Cancer Weeks 1–3 from Control with high accuracy. (b) PCA using 10 VOCs separating Cancer Week 1, Cancer Week 3 and Control samples. This smaller panel of VOCs can separate these classes with good accuracy.

iLDA was also implemented on a subset of ketones, terpenes and aromatics to model changes in murine VOCs between Cancer Week 1 and Cancer Week 3. iLDA identified a different biosignature of five VOCs (p-Cymen-8-ol (CYOL), 1,3,5-Undecatriene (UNTR), 8,8,9-Trimethyl-deca-3,5-diene-2,7-dione (TDDD), 2,4-Di-tert-butylphenol (DTBP) and 2-Butanone, 3,3-dimethyl- (2-BDI)) that classified Cancer Week 1 from Cancer Week 3 with 100% accuracy in the training data set (AUC = 1.0) (The Second LDA Model). Use of the first linear discriminant led to a perfect separation and one-dimensional LDA box/whisker plots can be observed in Figure 4d. LOOCV (AUC = 0.94) and fivefold cross validation (AUC = 0.97) were implemented and showed the model was not overfit (ROC in Figure 4e). Cancer Week 2 samples were tested using this model and the two-dimensional LDA plot can be seen in Figure 4f, which shows that some of the samples clustered in between Cancer Week 1 and Week 3, some clustered with Cancer Week 1 and some clustered with Cancer Week 3. This mirrors the results from the hierarchical heatmap in Figure 2b. Week 2 samples were not included in the statistical analysis comparing Weeks 1 and 3, shown in Figure 4d,e.

Next, the team undertook iLDA to identify a single panel capable of distinguishing all three sample classes of interest (Control, Cancer Week 1 and Cancer Week 3). iLDA was undertaken and applied on all VOCs that were differentially expressed for all sample comparisons (including VOCs only significant in Week 3 of Cancer). LDA utilizing five compounds (damascenone, 1,3,5-trichlorobenzene (TCHB), linalool, DTBP and 2-hexanone (2-HEX)) led to accurate classification of all three sample classes (the Third, LDA Model). Linalool is a linear monoterpenoid and damascenone an isoprenoid lipid that are not listed in Table S1 because although statistically significant for Control/Week 3 classification, they were not univariately significant for either All Cancer/Control or the Cancer Week 1/Week 3 comparison. Using this third LDA model, Cancer samples were distinguished from Control samples with an AUC equal to 0.98, sensitivity = 100% and specificity = 95% (LOOCV AUC = 0.97 and fivefold cross validation = 0.95). Alternatively, Cancer Week 1 was classified from Week 3 with AUC = 0.99, sensitivity = 100% and specificity = 92% (LOOCV AUC = 0.97 and fivefold cross validation AUC = 0.97). The ROC curves can be seen in Figure 4g,h, while the two-dimensional LDA plot can be observed in Figure 4i. Urine samples collected from the Week 2 cohort were tested using this classification model and the LDA plot is shown in Figure S1. Week 2 samples are 100% distinguished from Control samples and again cluster in between Cancer Week 1 and Cancer Week 3. However, many of the mouse urine samples were classified as either Cancer Week 1 or Cancer

Week 3, showing that some of the mice in Cancer Week 2 had tumors that may have progressed faster.

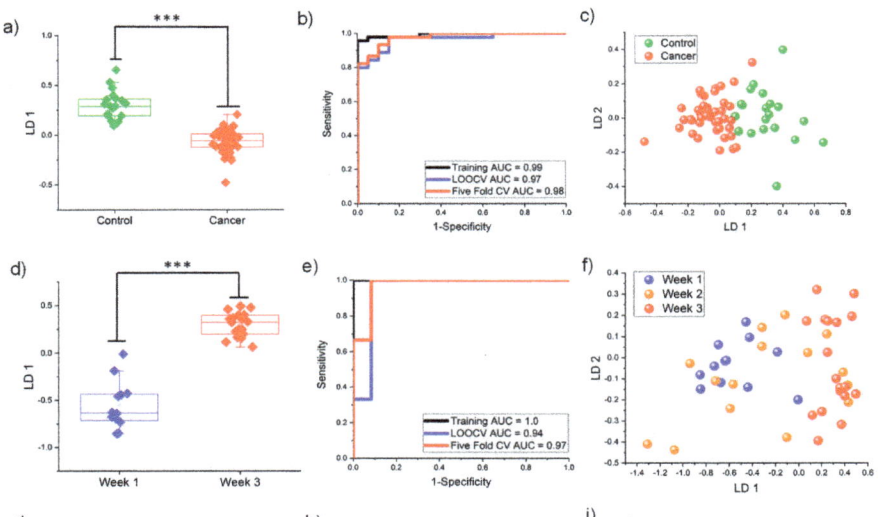

Figure 4. (**a**–**c**) First LDA Model distinguishes all Cancer (Weeks 1–3) from Control with high accuracy using five VOCs showing results in (**a**) one dimension, (**b**) the receiver operator characteristic (ROC), and (**c**) First LDA model in two dimensions. (**d**–**f**) Second LDA Model distinguishes Cancer Week 1 from Cancer Week 3 with high accuracy showing results using five VOCs in (**d**) one dimension, (**e**) the ROC for Week 1 vs. Week 3, and (**f**) Second LDA model in two dimensions. Week 2 samples shown in (**f**) not included in (**d**) LD1 or (**e**) ROC analysis. (**g**–**i**) Third LDA Model separates Cancer Week 1, Cancer Week 3 and Control with high accuracy. The ROC curves are shown for (**g**) Cancer Weeks 1 and 3 vs. Control and (**h**) the Cancer Week 1 vs. Week 3. (**i**) The third LDA model in two dimensions. *** $p < 0.001$.

3.4. Linear and Principal Component Regression Analysis

Linear regression analysis was undertaken on individual VOCs (37 with p-value by Student's t-test < 0.05 between Cancer Week 1 and Week 3) to look for significant trends with respect to the day on which urine was collected after tumor injection. Determination coefficients (R^2 value), regression coefficients and standard errors for all 37 individual VOCs can be observed in Supplementary Table S2. ANOVA determined that 23 out of the 37 VOCs had statistically significant trends by linear regression (p-value < 0.05). Of the 23 VOCs, only 5 VOCs had a positive regression coefficient while the other 18 features had negative regression coefficients. Even though statistically significant correlations were observed, none of the VOCs had an adjusted R^2 greater than 0.35. Next, PCR was implemented on the same 37 VOCs. After employing PCR on all 37 principal components, a relatively good fit was obtained ($R^2 = 0.94$), but the adjusted R^2 value was much lower (0.61), indicating too many variables were being utilized. A scatter plot showing the R^2 and adjusted R^2 value as a function of the number of principal components utilized for this model can be seen in Figure 5a, which indicates that analyzing more than 19 principal components will result in an overfit model. PCR using the first 19 components resulted in a linear correlation with $R^2 = 0.82$, adjusted $R^2 = 0.68$ and root mean square error (RMSE) = 3.3 (Figure 5b). The 95% confidence interval for the linear regression model is additionally illustrated in Figure 5b. The standardized coefficients for all 37 VOCs can be observed in Table S2. To increase the stability of the model, PCR was implemented using only the 19 VOCs (principal component loadings) statistically significantly contributing toward the first iteration of regression analysis (Figure 5b). Again, when plotting the R^2 and adjusted R^2 value as a function of the number of principal components (Figure 5c), utilizing more than 10 principal components results in an overfit model. The first 10 principal components resulted in a linear correlation with $R^2 = 0.71$, adjusted $R^2 = 0.63$ and RMSE = 3.6 (Figure 5d). The 95% confidence interval for this PCR model is can also be observed in Figure 5d. The standardized coefficients for all 19 VOCs for this model can also be observed in Table S2.

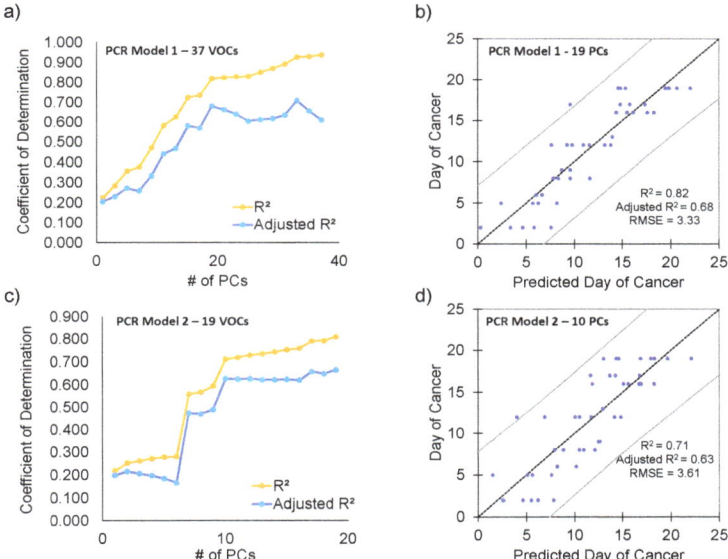

Figure 5. Principal component regression (PCR) analysis using models starting with 37 and 19 VOCs, respectively, identify leaner models in an iterative fashion. (**a**) Coefficient of determination plotted against the number of principal components utilized for the principal component regression (PCR) analysis using 37 VOCs with *p*-value < 0.05 (Cancer Week 1 vs. Cancer Week 3). (**b**) PCR analysis using the first 19 principal components with calculated R^2 equal to 0.82 and adjusted R^2 equal to 0.68. (**c**) Coefficient of determination plotted against the number of principal components utilized for the PCR analysis using 19 VOCs significantly contributing toward the linear correlation. (**d**) PCR analysis using the first 10 principal components results in a more stable model which could predict the number of days after tumor injection with R^2 equal to 0.71 and adjusted R^2 equal to 0.63.

4. Discussion

The heatmaps (Figure 2) are consistent with previous studies in murine models of breast cancer [28,29]; more VOCs are downregulated by cancer. However, previous studies found roughly twice as many volatile terpenes and terpenoids among the cancer biomarkers as ketones [28]. In the current study, ketones were observed to be more consistently dysregulated by mammary tumors, regardless of when the urine was collected (Week 1, Week 2 or Week 3). PCA and iLDA had the ability to classify all Cancer samples with over 95% classification accuracy (Figures 3a and 4a–c) using different panels of VOCs. Multivariate analyses could even distinguish the three sample classes of interest (Cancer

Week 1, Cancer Week 3 and Control) with over 90% accuracy (Figures 3b and 4g–i) based on different VOCs. These models show the ability of VOCs to classify any type of Cancer and distinguish progression by week after tumor injection with outstanding accuracy.

PCR of all 37 VOCs with p-value < 0.05 and utilizing the first 19 principal components resulted in a stable linear model (Figure 5a,b). Upon limiting principal component regression to 19 VOCs, a model utilizing the first ten principal components resulted in a model with greater stability (Figure 5c,d). Both of the principal component regression models presented have a higher degree of linear correlation relative to any individual VOC (Table S2). The 19 VOCs identified from principal component regression as useful for tracking tumor progression by day correspond to the results from multivariate classification of Cancer by week. For example, five of the ten VOCs (DTBP, 2-BDI, CYOL, THUJ and TCHB) that contribute toward the separation of the Control, Cancer Week 1 and Week 3 in PCA (Figure 3b) were identified by PCR. On the other hand, four of the five VOCs identified by LDA (CYOL, TDDD, 2-BDI and DTBP) to separate Cancer Week 1 from Cancer Week 3 (Figure 4d–f) were found to be significant by PCR. Two of the VOCs identified by LDA to classify Control, Cancer Week 1 and Week 3 (TCHB and DTBP, shown in Figure 4g–i) contributed toward PCR.

Ketones were the most frequent functional group detected as significantly dysregulated as shown in Table S1. This is consistent with our previous analysis which showed ketones were depleted in tumor-bearing mice [28]. Ketones have been previously reported by *Silva* et al. to be potential markers for breast cancer as they are products of lipid peroxidation [43]. Ketones and other carbonyls have been reported to be markers for prostate cancer [16,18], lung cancer [44,45] and diabetes [46]. Another study by the authors showed two ketones reported in this study, 2-HEP and 2-PEN, were enriched in urine samples collected from mice receiving an antitumoral treatment [47] (treatment was bone loading, a simulated form of exercise). Furthermore, in vitro analyses showed upon treatment with 2-HEP and 2-PEN, hypothalamic neuronal cells had reduced tumor cell viability accompanied with elevated levels of aralkylamine N-acetyltransferase (AANAT) and tyrosine hydrogenase (TH). AANAT and TH are rate-limiting enzymes that produce melatonin and dopamine, which have been shown to have a role in tumor suppression [47]. These studies show the potential antitumor capability of ketones, which is intriguing as they were downregulated by mammary tumors in this study.

Volatile terpenes/terpenoids (VTs) were found to be depleted in Cancer samples (specifically in Week 3). VTs were previously identified to be potential markers of mammary tumors in mice [28,29] and are synthesized in vivo by the mevalonate (MVA) pathway. The MVA pathway has not only been shown to be dysregulated by cancer, but also to play a significant role in tumor growth and transformation [48,49]. The change in VT profiles may be due to osteolytic lesions likely forming by Week 3. Down-regulation of the MVA pathway through the use of a class of HMG-CoA reductase inhibitors known as statins is known to slow osteolysis in breast cancer models [30,50,51]. Cholesterol, an end product downstream of the MVA pathway, has also been reported to play a role in tumor growth [52]. The current authors previously have shown a correlation between the dysregulation of urinary VTs and the upregulation of cholesterol in mice [30]. This is important, as both VTs and cholesterol are products of the MVA pathway. Lastly, there is a link between VTs as markers of cancer and potential treatments, because they have demonstrated inhibitory effects against cancer [53–55].

2,4-Di-tert-butylphenol (DTBP) and 3,5-di-tert-Butyl-4-hydroxybenzaldehyde (DTBB), which are classified as phenolic antioxidizing agents [40], are two aromatic VOCs identified in the study. DTBB does not show any significant differences between Control samples and Cancer Week 1 but becomes increasingly depleted by Week 3. On the other hand, DTBP is significantly enriched in Cancer Week 1 relative to Control samples but becomes progressively depleted by Cancer Week 3. These aromatic VOCs are potentially of interest because natural phenolic antioxidant agents are secondary metabolites and their ability to serve as an anticancer therapy has been previously analyzed. Studies have shown that several groups of phenolic antioxidants inhibit the growth and proliferation of tumor cells, but the mechanism of action has not been entirely elucidated [56]. Given the previously demonstrated antitumor capability, it is unsurprising to see their depletion induced by tumor progression and/or the formation of osteolytic lesions.

Taken together, these results show that VOCs can not only classify tumor-bearing mice, but also accurately track progression. Limitations of this study include the relatively small number of samples analyzed. Additionally, the metabolic variation of all mice was relatively controlled: BALB/c mice were the same age, given the same tumors (triple negative 4T1.2 cells), kept in the same environment and fed the same diet. Women have varied breast cancer tumors, larger variability in age, exercise, diet and other lifestyle factors. However, it is evident most of the VOCs in the murine study do not represent the tumors themselves, but the metabolic response to tumors; a human metabolic response to tumors would be expected to also present significant similarities despite differences in tumor type and the other variations noted above. In the future, it may be fruitful to combine VOC analysis with other predictors of tumor progression such as the identified circulating biomarker for bone metastasis [32] and others which have not yet been identified.

5. Conclusions

It is important, when translating results from murine models to human studies, to recollect that murine samples may represent late stages in cancer tumor progression. This study found tumor injection to the tibia led to many VOCs being dysregulated as early as the first week after injection. Some VOCs remained relatively constant over the course of the study, while others were insignificant early in the study but were dysregulated later in the study. This is hypothesized to be because of cancer progression and/or the formation of osteolytic lesions induced by mammary tumor injection/progression. It is hoped these findings can be translated into human research for early detection of breast cancer recurrence as 20–30% of patients with early breast cancer will experience relapse with distant metastatic disease and bone metastases being the most common presentation at the time of recurrence [2]. Further human research can be focused on finding if the urinary VOCs can be detected before the radiographic appearance of lesions on the bone scans or PET scans, or exploring if there is any correlation between the levels of VOCs and the extent of tumor burden.

Supplementary Materials: The following are available online at https://www.mdpi.com/2072-6694/13/6/1462/s1, Table S1: List of VOCs with p-value < 0.05 between Cancer Weeks 1–3 vs. Control (tumor presence) or Cancer Week 1 vs. Cancer Week 3 (tumor progression) with corresponding abbreviations, retention times and p-values for both comparisons (NS = no statistical significance, upward facing arrows show upregulation in all Cancer samples or in Cancer Week 3 while downward facing arrows signify downregulation, * denotes VOCs used for PCA in Figure 3b, underline denotes p-value < 0.05 between Cancer Week 1 and Week 3 after FDR adjustment, italics denotes p-value < 0.05 between Cancer and No Cancer after FDR adjustment). Table S2. Regression analysis results of individual 37 VOCs identified as significantly different (Cancer Week 1 and Week 3), standardized regression coefficients for the same 37 VOCs analyzed by PCR in Figure 5b and standardized regression coefficients for the 19 VOCs analyzed by PCR in Figure 5d. Figure S1: 2D LDA plot with Cancer Week 2 samples tested utilizing the LDA model initially built to discriminate Control, Cancer Week 1 and Cancer Week 3.

Author Contributions: Conceptualization: M.A., H.Y. and A.P.S.; Data curation M.W., P.G., J.V.G. and A.P.S.; Formal analysis M.W., L.W., S.L., P.G., M.W., J.V.G. and A.P.S.; Funding acquisition: M.A. and H.Y.; Investigation M.K., M.A. and H.Y.; Methodology: L.W., M.W., S.L., P.G., M.K., J.V.G. and A.P.S.; Project administration: M.A., H.Y. and A.P.S.; Resources: M.A., M.K. and H.Y.; Supervision: M.A., H.Y., M.K., J.V.G. and A.P.S.; Validation: M.W., P.G. and A.P.S.; Visualization: J.V.G., M.W. and P.G.; Roles/Writing—original draft: M.W., P.G. and A.P.S.; Writing—review & editing: M.W., P.G., S.L., A.P.S., M.K., J.V.G., H.Y. and M.A. All authors have read and agreed to the published version of the manuscript.

Funding: This research was funded by U.S. National Institutes of Health (NIH)/National Institute of Arthritis and Musculoskeletal and Skin Diseases (grant # R01AR052144) and the National Science Foundation (grant # 1852105 and 1502310).

Institutional Review Board Statement: All experimental procedures followed the Guiding Principles in the Care and Use of Animals supported by the American Physiological Society and were approved by the Indiana University Animal Care and Use Committee (protocol code: SC292R; date of approval: 30 May 2019). No humans were utilized in this study.

Informed Consent Statement: Not applicable.

Data Availability Statement: The data presented in this study, are available on request from the corresponding author.

Acknowledgments: The authors would like to thank Solveig Naumann for helping with optimization of procedure and Paula Angarita-Rivera for assisting with analyzing urine samples via GC–MS QTOF. The authors would finally, like to acknowledge the National Science Foundation (grant # 1502310) and Agilent Technologies.

Conflicts of Interest: The authors declare no conflict of interest.

References

1. Siegel, R.L.; Miller, K.D.; Fuchs, H.E.; Jemal, A. Cancer Statistics, 2021. *CA A Cancer J. Clin.* **2021**, *71*, 7–33. [CrossRef]
2. Kennecke, H.; Yerushalmi, R.; Woods, R.; Cheang, M.C.U.; Voduc, D.; Speers, C.H.; Nielsen, T.O.; Gelmon, K. Metastatic Behavior of Breast Cancer Subtypes. *JCO* **2010**, *28*, 3271–3277. [CrossRef]
3. Diagnostic & Monitoring Tests | Cancer Testing | UNM Cancer Center. Available online: http://cancer.unm.edu/cancer/cancer-info/testing-overview/diagnostics-monitoring-tests/ (accessed on 11 May 2019).
4. Radhakrishna, S.; Agarwal, S.; Parikh, P.M.; Kaur, K.; Panwar, S.; Sharma, S.; Dey, A.; Saxena, K.K.; Chandra, M.; Sud, S. Role of Magnetic Resonance Imaging in Breast Cancer Management. *South Asian J. Cancer* **2018**, *7*, 69–71. [CrossRef] [PubMed]
5. Thoeny, H.C.; Ross, B.D. Predicting and Monitoring Cancer Treatment Response with Diffusion-Weighted MRI. *J. Magn. Reson. Imaging* **2010**, *32*, 2–16. [CrossRef] [PubMed]
6. Taverna, G.; Tidu, L.; Grizzi, F.; Torri, V.; Mandressi, A.; Sardella, P.; La Torre, G.; Cocciolone, G.; Seveso, M.; Giusti, G.; et al. Olfactory System of Highly Trained Dogs Detects Prostate Cancer in Urine Samples. *J. Urol.* **2015**, *193*, 1382–1387. [CrossRef]
7. Hackner, K.; Errhalt, P.; Mueller, M.R.; Speiser, M.; Marzluf, B.A.; Schulheim, A.; Schenk, P.; Bilek, J.; Doll, T. Canine Scent Detection for the Diagnosis of Lung Cancer in a Screening-like Situation. *J. Breath Res.* **2016**, *10*, 046003. [CrossRef]
8. McCulloch, M.; Jezierski, T.; Broffman, M.; Hubbard, A.; Turner, K.; Janecki, T. Diagnostic Accuracy of Canine Scent Detection in Early- and Late-Stage Lung and Breast Cancers. *Integr. Cancer* **2006**, *5*, 30–39. [CrossRef] [PubMed]
9. Gordon, R.T.; Schatz, C.B.; Myers, L.J.; Kosty, M.; Gonczy, C.; Kroener, J.; Tran, M.; Kurtzhals, P.; Heath, S.; Koziol, J.A.; et al. The Use of Canines in the Detection of Human Cancers. *J. Altern. Complement Med.* **2008**, *14*, 61–67. [CrossRef]
10. Human Ovarian Carcinomas Detected by Specific Odor—György Horvath, Gunvor Af Klinteberg Järverud, Sven Järverud, István Horváth. 2008. Available online: https://journals.sagepub.com/doi/abs/10.1177/1534735408319058 (accessed on 8 June 2020).
11. Jezierski, T.; Walczak, M.; Ligor, T.; Rudnicka, J.; Buszewski, B. Study of the Art: Canine Olfaction Used for Cancer Detection on the Basis of Breath Odour. Perspectives and Limitations. *J. Breath Res.* **2015**, *9*, 027001. [CrossRef]
12. Jenkins, E.K.; DeChant, M.T.; Perry, E.B. When the Nose Doesn't Know: Canine Olfactory Function Associated With Health, Management, and Potential Links to Microbiota. *Front. Vet. Sci.* **2018**, *5*. [CrossRef]
13. Hanahan, D.; Weinberg, R.A. Hallmarks of Cancer: The next Generation. *Cell* **2011**, *144*, 646–674. [CrossRef] [PubMed]
14. Janfaza, S.; Khorsand, B.; Nikkhah, M.; Zahiri, J. Digging Deeper into Volatile Organic Compounds Associated with Cancer. *Biol. Methods Protoc.* **2019**, *4*. [CrossRef]
15. Saalberg, Y.; Wolff, M. VOC Breath Biomarkers in Lung Cancer. *Clin. Chim. Acta* **2016**, *459*, 5–9. [CrossRef]
16. Khalid, T.; Aggio, R.; White, P.; De Lacy Costello, B.; Persad, R.; Al-Kateb, H.; Jones, P.; Probert, C.S.; Ratcliffe, N. Urinary Volatile Organic Compounds for the Detection of Prostate Cancer. *PLoS ONE* **2015**, *10*, e0143283. [CrossRef] [PubMed]
17. Gao, Q.; Su, X.; Annabi, M.H.; Schreiter, B.R.; Prince, T.; Ackerman, A.; Morgas, S.; Mata, V.; Williams, H.; Lee, W.-Y. Application of Urinary Volatile Organic Compounds (VOCs) for the Diagnosis of Prostate Cancer. *Clin. Genitourin. Cancer* **2019**, *17*, 183–190. [CrossRef] [PubMed]
18. Lima, A.R.; Pinto, J.; Azevedo, A.I.; Barros-Silva, D.; Jerónimo, C.; Henrique, R.; de Lourdes Bastos, M.; Guedes de Pinho, P.; Carvalho, M. Identification of a Biomarker Panel for Improvement of Prostate Cancer Diagnosis by Volatile Metabolic Profiling of Urine. *Br. J. Cancer* **2019**, *121*, 857–868. [CrossRef]
19. Lima, A.R.; Araújo, A.M.; Pinto, J.; Jerónimo, C.; Henrique, R.; de Bastos, M.L.; Carvalho, M.; de Pinho, P.G. Discrimination between the Human Prostate Normal and Cancer Cell Exometabolome by GC-MS. *Sci. Rep.* **2018**, *8*, 1–12. [CrossRef]
20. Phillips, M.; Cataneo, R.N.; Saunders, C.; Hope, P.; Schmitt, P.; Wai, J. Volatile Biomarkers in the Breath of Women with Breast Cancer. *J. Breath Res.* **2010**, *4*, 026003. [CrossRef]

21. Peng, G.; Hakim, M.; Broza, Y.Y.; Billan, S.; Abdah-Bortnyak, R.; Kuten, A.; Tisch, U.; Haick, H. Detection of Lung, Breast, Colorectal, and Prostate Cancers from Exhaled Breath Using a Single Array of Nanosensors. *Br. J. Cancer* **2010**, *103*, 542–551. [CrossRef]
22. Lima, A.R.; Pinto, J.; Carvalho-Maia, C.; Jerónimo, C.; Henrique, R.; de Bastos, M.L.; Carvalho, M.; Guedes de Pinho, P. A Panel of Urinary Volatile Biomarkers for Differential Diagnosis of Prostate Cancer from Other Urological Cancers. *Cancers* **2020**, *12*, 2017. [CrossRef]
23. Beretov, J.; Wasinger, V.C.; Millar, E.K.A.; Schwartz, P.; Graham, P.H.; Li, Y. Proteomic Analysis of Urine to Identify Breast Cancer Biomarker Candidates Using a Label-Free LC-MS/MS Approach. *PLoS ONE* **2015**, *10*. [CrossRef]
24. Cala, M.; Aldana, J.; Sánchez, J.; Guio, J.; Meesters, R.J.W. Urinary Metabolite and Lipid Alterations in Colombian Hispanic Women with Breast Cancer: A Pilot Study. *J. Pharm. Biomed. Anal.* **2018**, *152*, 234–241. [CrossRef] [PubMed]
25. Hirschfeld, M.; Rücker, G.; Weiß, D.; Berner, K.; Ritter, A.; Jäger, M.; Erbes, T. Urinary Exosomal MicroRNAs as Potential Non-Invasive Biomarkers in Breast Cancer Detection. *Mol. Diagn.* **2020**, *24*, 215–232. [CrossRef] [PubMed]
26. Silva, C.L.; Passos, M.; Câmara, J.S. Solid Phase Microextraction, Mass Spectrometry and Metabolomic Approaches for Detection of Potential Urinary Cancer Biomarkers—A Powerful Strategy for Breast Cancer Diagnosis. *Talanta* **2012**, *89*, 360–368. [CrossRef] [PubMed]
27. Silva, C.L.; Perestrelo, R.; Silva, P.; Tomás, H.; Câmara, J.S. Implementing a Central Composite Design for the Optimization of Solid Phase Microextraction to Establish the Urinary Volatomic Expression: A First Approach for Breast Cancer. *Metabolomics* **2019**, *15*, 64. [CrossRef] [PubMed]
28. Woollam, M.; Teli, M.; Angarita-Rivera, P.; Liu, S.; Siegel, A.P.; Yokota, H.; Agarwal, M. Detection of Volatile Organic Compounds (VOCs) in Urine via Gas Chromatography-Mass Spectrometry QTOF to Differentiate Between Localized and Metastatic Models of Breast Cancer. *Sci. Rep.* **2019**, *9*, 2526. [CrossRef] [PubMed]
29. Woollam, M.; Teli, M.; Liu, S.; Daneshkhah, A.; Siegel, A.P.; Yokota, H.; Agarwal, M. Urinary Volatile Terpenes Analyzed by Gas Chromatography–Mass Spectrometry to Monitor Breast Cancer Treatment Efficacy in Mice. *J. Proteome Res.* **2020**, *19*, 1913–1922. [CrossRef]
30. Wang, L.; Wang, Y.; Chen, A.; Teli, M.; Kondo, R.; Jalali, A.; Fan, Y.; Liu, S.; Zhao, X.; Siegel, A.; et al. Pitavastatin Slows Tumor Progression and Alters Urine-Derived Volatile Organic Compounds through the Mevalonate Pathway. *FASEB J.* **2019**, *33*, 13710–13721. [CrossRef]
31. Fan, Y.; Jalali, A.; Chen, A.; Zhao, X.; Liu, S.; Teli, M.; Guo, Y.; Li, F.; Li, J.; Siegel, A.; et al. Skeletal Loading Regulates Breast Cancer-Associated Osteolysis in a Loading Intensity-Dependent Fashion. *Bone Res.* **2020**, *8*, 9. [CrossRef]
32. Ibrahim, T.; Ricci, M.; Scarpi, E.; Bongiovanni, A.; Ricci, R.; Riva, N.; Liverani, C.; De Vita, A.; La Manna, F.; Oboldi, D.; et al. RANKL: A Promising Circulating Marker for Bone Metastasis Response. *Oncol. Lett.* **2016**, *12*, 2970–2975. [CrossRef]
33. Tulotta, C.; Groenewoud, A.; Snaar-Jagalska, B.E.; Ottewell, P. Animal Models of Breast Cancer Bone Metastasis. *Bone Res. Protoc.* **2019**, 309–330. [CrossRef]
34. Liu, S.; Fan, Y.; Chen, A.; Jalali, A.; Minami, K.; Ogawa, K.; Nakshatri, H.; Li, B.-Y.; Yokota, H. Osteocyte-Driven Downregulation of Snail Restrains Effects of Drd2 Inhibitors on Mammary Tumor Cells. *Cancer Res.* **2018**. [CrossRef] [PubMed]
35. Kwak, J.; Grigsby, C.C.; Rizki, M.M.; Preti, G.; Köksal, M.; Josue, J.; Yamazaki, K.; Beauchamp, G.K. Differential Binding between Volatile Ligands and Major Urinary Proteins Due to Genetic Variation in Mice. *Physiol. Behav.* **2012**, *107*, 112–120. [CrossRef] [PubMed]
36. Li, B.; Tang, J.; Yang, Q.; Li, S.; Cui, X.; Li, Y.; Chen, Y.; Xue, W.; Li, X.; Zhu, F. NOREVA: Normalization and Evaluation of MS-Based Metabolomics Data. *Nucleic Acids Res.* **2017**, *45*, W162–W170. [CrossRef] [PubMed]
37. Benjamini, Y.; Hochberg, Y. Controlling the False Discovery Rate: A Practical and Powerful Approach to Multiple Testing. *J. R. Stat. Soc. Ser. B* **1995**, *57*, 289–300. [CrossRef]
38. Siegel, A.P.; Daneshkhah, A.; Hardin, D.S.; Shrestha, S.; Varahramyan, K.; Agarwal, M. Analyzing Breath Samples of Hypoglycemic Events in Type 1 Diabetes Patients: Towards Developing an Alternative to Diabetes Alert Dogs. *J. Breath Res.* **2017**, *11*, 026007. [CrossRef] [PubMed]
39. Babyak, M.A. What You See May Not Be What You Get: A Brief, Nontechnical Introduction to Overfitting in Regression-Type Models. *Psychosom. Med.* **2004**, *66*, 411–421. [PubMed]
40. Wishart, D.S.; Tzur, D.; Knox, C.; Eisner, R.; Guo, A.C.; Young, N.; Cheng, D.; Jewell, K.; Arndt, D.; Sawhney, S.; et al. HMDB: The Human Metabolome Database. *Nucleic Acids Res.* **2007**, *35*, D521–D526. [CrossRef]
41. Qiu, Y.-Q. KEGG Pathway Database. In *Encyclopedia of Systems Biology*; Dubitzky, W., Wolkenhauer, O., Cho, K.-H., Yokota, H., Eds.; Springer: New York, NY, USA, 2013; pp. 1068–1069. ISBN 978-1-4419-9863-7.
42. Xia, J.; Broadhurst, D.I.; Wilson, M.; Wishart, D.S. Translational Biomarker Discovery in Clinical Metabolomics: An Introductory Tutorial. *Metabolomics* **2013**, *9*, 280–299. [CrossRef]
43. Silva, C.L.; Perestrelo, R.; Silva, P.; Tomás, H.; Câmara, J.S. Volatile Metabolomic Signature of Human Breast Cancer Cell Lines. *Sci. Rep.* **2017**, *7*. [CrossRef]
44. Li, M.; Yang, D.; Brock, G.; Knipp, R.J.; Bousamra, M.; Nantz, M.H.; Fu, X.-A. Breath Carbonyl Compounds as Biomarkers of Lung Cancer. *Lung Cancer* **2015**, *90*, 92–97. [CrossRef]
45. Santos, P.M.; del Nogal Sánchez, M.; Pozas, Á.P.C.; Pavón, J.L.P.; Cordero, B.M. Determination of Ketones and Ethyl Acetate—a Preliminary Study for the Discrimination of Patients with Lung Cancer. *Anal. Bioanal. Chem.* **2017**, *409*, 5689–5696. [CrossRef]

46. Minh, T.D.C.; Blake, D.R.; Galassetti, P.R. The Clinical Potential of Exhaled Breath Analysis For Diabetes Mellitus. *Diabetes Res. Clin. Pr.* **2012**, *97*, 195–205. [CrossRef]
47. Wu, D.; Fan, Y.; Liu, S.; Woollam, M.D.; Sun, X.; Murao, E.; Zha, R.; Prakash, R.; Park, C.; Siegel, A.P.; et al. Loading-Induced Antitumor Capability of Murine and Human Urine. *FASEB J.* **2020**. [CrossRef]
48. Clendening, J.W.; Pandyra, A.; Boutros, P.C.; Ghamrasni, S.E.; Khosravi, F.; Trentin, G.A.; Martirosyan, A.; Hakem, A.; Hakem, R.; Jurisica, I.; et al. Dysregulation of the Mevalonate Pathway Promotes Transformation. *PNAS* **2010**, *107*, 15051–15056. [CrossRef]
49. Pampalakis, G.; Obasuyi, O.; Papadodima, O.; Chatziioannou, A.; Zoumpourlis, V.; Sotiropoulou, G. The KLK5 Protease Suppresses Breast Cancer by Repressing the Mevalonate Pathway. *Oncotarget* **2013**, *5*, 2390–2403. [CrossRef]
50. Ahn, K.S.; Sethi, G.; Chaturvedi, M.M.; Aggarwal, B.B. Simvastatin, 3-Hydroxy-3-Methylglutaryl Coenzyme A Reductase Inhibitor, Suppresses Osteoclastogenesis Induced by Receptor Activator of Nuclear Factor-KB Ligand through Modulation of NF-KB Pathway. *Int. J. Cancer* **2008**, *123*, 1733–1740. [CrossRef]
51. Mandal, C.C.; Ghosh-Choudhury, N.; Yoneda, T.; Choudhury, G.G.; Ghosh-Choudhury, N. Simvastatin Prevents Skeletal Metastasis of Breast Cancer by an Antagonistic Interplay between P53 and CD44. *J. Biol. Chem.* **2011**, *286*, 11314–11327. [CrossRef] [PubMed]
52. Llaverias, G.; Danilo, C.; Mercier, I.; Daumer, K.; Capozza, F.; Williams, T.M.; Sotgia, F.; Lisanti, M.P.; Frank, P.G. Role of Cholesterol in the Development and Progression of Breast Cancer. *Am. J. Pathol.* **2011**, *178*, 402–412. [CrossRef] [PubMed]
53. Gould, M.N. Prevention and Therapy of Mammary Cancer by Monoterpenes. *J. Cell. Biochem.* **1995**, *59*, 139–144. [CrossRef] [PubMed]
54. Paduch, R.; Kandefer-Szerszeń, M.; Trytek, M.; Fiedurek, J. Terpenes: Substances Useful in Human Healthcare. *Arch. Immunol. Ther. Exp.* **2007**, *55*, 315. [CrossRef] [PubMed]
55. Neighbors, J.D. The Mevalonate Pathway and Terpenes: A Diversity of Chemopreventatives. *Curr. Pharm. Rep.* **2018**, *4*, 157–169. [CrossRef]
56. Jafari, S.; Saeidnia, S.; Abdollahi, M. Role of Natural Phenolic Compounds in Cancer Chemoprevention via Regulation of the Cell Cycle. *Curr. Pharm. Biotechnol.* **2014**, *15*, 409–421. [CrossRef] [PubMed]

MDPI
St. Alban-Anlage 66
4052 Basel
Switzerland
Tel. +41 61 683 77 34
Fax +41 61 302 89 18
www.mdpi.com

Cancers Editorial Office
E-mail: cancers@mdpi.com
www.mdpi.com/journal/cancers

www.ingramcontent.com/pod-product-compliance
Lightning Source LLC
LaVergne TN
LVHW070127100526
838202LV00016B/2243